CREEP PROBLEMS IN STRUCTURAL MEMBERS

NORTH-HOLLAND SERIES IN

APPLIED MATHEMATICS AND MECHANICS

EDITORS:

H. A. LAUWERIER

Institute of Applied Mathematics
University of Amsterdam

W. T. KOITER

Laboratory of Applied Mechanics
Technical University, Delft

VOLUME 7

NORTH-HOLLAND PUBLISHING COMPANY – AMSTERDAM • LONDON

CREEP PROBLEMS IN STRUCTURAL MEMBERS

BY

Yu. N. RABOTNOV

Faculty of Mathematics and Mechanics,
University of Moscow

Translated from the Russian by

TRANSCRIPTA SERVICE LTD, LONDON

English translation edited by

F. A. LECKIE

Department of Engineering, University of Leicester

1969

NORTH-HOLLAND PUBLISHING COMPANY – AMSTERDAM • LONDON

Originally entitled:
POLZUCHEST ELEMENTOV KONSTRUKTSII
(Nauka, Moscow, 1966)

PUBLISHERS:
NORTH-HOLLAND PUBLISHING COMPANY – AMSTERDAM
NORTH-HOLLAND PUBLISHING COMPANY LTD. – LONDON

SOLE DISTRIBUTORS FOR THE WESTERN HEMISPHERE:
WILEY INTERSCIENCE DIVISION
JOHN WILEY & SONS, INC. – NEW YORK

Library of Congress Catalog Card Number 73–79718

SBN 7204 2357 0

PRINTED IN THE NETHERLANDS

Preface

The theory of creep is a section of the mechanics of a deformed solid which has been developed in recent years and takes its place alongside such sections of mechanics as the theories of elasticity and plasticity. We shall apply the term 'creep' to the totality of effects which can be explained on the assumption that the relationship between stress and strain is time dependent, either explicitly or in terms of certain operators. Creep properties are observed in various kinds of materials including metals, plastics, ores, concrete, natural and artificial stone and ice. Obviously the physical mechanisms of creep are quite different in these various materials. Indeed the similarity between the superficial aspects of creep in different materials is only apparent and the phenomenon demands different methods of phenomenological description in different materials.

Mechanics is concerned with the phenomenological description of the creep process and the fundamental mathematical relationships are formulated on the basis of macro-experimental results. At the present level of development of the science physical concepts are of indirect rather than direct use in mechanics. It is, therefore, natural to treat creep mechanics separately as a subject which operates with the usual concepts of the mechanics of solids and to introduce a certain selection of governing parameters in a formal way. These parameters describe what may be termed the structural state of the material, though we are not always able to relate some particular parameter to actual peculiarities of structure which are recorded by physical methods.

Creep theory is concerned with a much more complicated set of phenomena than is the theory of elasticity or even the theory of plasticity and has not reached the same level of logical perfection as have the theories of elasticity and of ideal plasticity. The situation in regard to creep theory may be compared with that of the theory of plasticity of a work-hardening substance; in both fields the tendency to describe the totality of properties of a real body by a single system of equations does not lead to a sufficiently definite result. In both plasticity and creep it is possible to construct various simplified equations which can describe a limited range of effects. Creep theory is concerned with a great variety of materials, often of unstable properties and in order to solve practical problems it has been necessary to include in the theory approximate schemes which encompass the general outlines of the effects.

There is not as yet a unified theory of creep that is applicable to all materials and the author considers that such a theory is inherently impossible. This book is primarily concerned with creep in metals at high temperatures. The corresponding theory was originally developed in connection with turbine manufacture but it has subsequently been applied to nuclear power engineering, chemical engineering, aviation and jet propulsion. In selecting the title of the book the author wished to emphasise that it would not present a physical theory of creep. Considerable attention is paid to the analysis of experimental results because the object of the analysis is to assess whether the material tested can be applied in some particular design working under specific conditions. The same experimental data could be used to draw conclusions concerning the causes of greater or less resistance to creep or, in the long run, to develop new and improved materials. This aspect is not considered at all in the present work. Neither was it possible to give full details of practical methods of calculating creep and long term strength of machine components under actual service conditions, which could appropriately form part of quite a different kind of book.

Two chapters are only indirectly related to the main content of the book; one of these contains a brief outline of the theory of plasticity and the second the fundamentals of the theory of hereditary elasticity or viscoelasticity. Both these theories have certain concepts and methods in common with the theory of creep in metals.

The author is grateful for valuable discussions with his students and colleagues about many of the points of view here presented. He is particularly grateful to G.V.Ivanov, N.I.Malinin, S.T.Mileyko, V.S.Namestnikov, Yu.V. Nemirovskiy and S.A.Shesterikov for reading parts of the manuscript and for valuable comments.

Contents

Introduction

A characteristic feature of recent developments in the mechanics of deformed solids is the tendency to extend its physical basis and to improve the phenomenological schemes which underlie the theory. As hitherto, the task of mechanics is to study the motion of real bodies by the methods of mechanics of a continuous medium, by the procedure of creating models which reproduce the fundamental and most important properties of real objects.

In the last ten years there have been extensive developments in the theory of plasticity, which deals with the irreversible deformation of solids. In most commonly used metals, at moderate temperatures, to a first approximation, irreversible deformations depend only on the magnitude of the forces acting and on the order of their application but not on their duration or rate of application. This statement is not entirely accurate for there are certain time-dependent effects even at normal temperatures, but they are insignificant and are neglected in the theory of plasticity. However, the sequence in which various actions are performed upon a body is not a matter of indifference in the theory of plasticity and, therefore, the concepts of "before" and "after" have to be introduced so that time does figure in the theory, though not quantitatively and, therefore, any monotonic function of time may be considered instead of time itself.

At present the limits of applicability of the theory of plasticity are clearly outlined so that we need not consider them in more detail here. Subsequently we shall return to the question of plasticity because the theories which form the main subject of our exposition were developed in close relationship with the theory of plasticity, both subjects having many fundamental concepts in common. The point is that in many materials under service conditions the relationship between applied forces and displacements is essentially time dependent. This dependence is evident for example in that deformation at constant load is not constant but increases with time until it becomes excessive if the load is applied long enough. On the other hand if a product undergoes enforced deformation and is then held by clamps which maintain the deformation unaltered, the clamp reactions diminish with time so that, for example, a shrink fit can give way. We shall include all effects of this kind under the general term of "creep" even though the term is often given a narrower meaning.

Creep in metals at high temperatures is probably of the greatest practical interest. For example, the behaviour of carbon steel at normal temperatures and moderate stress is well described by the theory of elasticity, and at higher stresses by the theory of plasticity but it behaves quite differently at a temperature of about 450°C. At this temperature the steel ceases to obey Hooke's law even at low stress and its stress-strain curve is essentially dependent on the rate of extension, and therefore, it is simply not possible to speak of a stress-strain curve which forms the basis of the theory of plasticity. When a constant load is applied the specimen continues to deform, and it is said to creep. Creep is observed at any stress, the rate of creep is often highly dependent on the magnitude of the stress but there is no definite creep limit, that is a stress below which there is no creep. Products that have to operate at high temperatures are made of special steels and alloys which have better creep resistance, but for any alloy there is a temperature above which creep becomes considerable and inevitably leads to failure after a certain time, depending on the applied stress. Austentic steels with high contents of alloying elements such as chromium and nickel can operate reliably at temperatures of the order of 600°C. Alloys based on nickel or cobalt are used at higher temperatures. Further increase in heat resistance of technical alloys can be achieved by using metals of higher melting point such as molybdenum, tungsten and niobium.

Practical strength calculations on components of steam and later gas turbines instigated the search for a theoretical description of the creep process in order to calculate the strain after a certain time and the conditions of failure. Unlike ordinary strength calculations which have the object of determining the absolute strength of a component, the object of creep calculations is to ensure that the component will not fail to perform its structural function within a certain period of time. Thus any product which is designed for creep conditions is intended to have a certain creep life.

The following procedure was used when it first became necessary in engineering practice to allow for creep. Specimens were tested to determine the nominal creep limit at a particular temperature, that is the maximum stress at which the creep deformation in a time equal to the creep life of the product does not exceed a certain predetermined nominal value, such as 1%. It can happen that when the stress has been applied for a certain time the specimen fails even though the deformation is less than the permitted value. It is then necessary to determine the long term ultimate strength, that is the maximum stress which does not cause the specimen to fail within a given time. Then the permissible stress is determined, using one of the above ultimate strengths divided by a safety factor and the component is designed by the normal method for this value of permissible stress. This method of calculation is very approximate and the safety factor is excessively high. The following requirements should be applied to structures which operate at high temperature: 1) the total deformation should not exceed a certain value, defined in accordance with the structural function of the component; 2) failure should not occur during the service life. Thus the objects of the theory are to develop methods of determining the deformation in bodies subject to creep and of determining the time to failure.

High molecular compounds provide another important field of application of creep theory. The extensive use that is made of plastics in modern industry has given rise to greater interest in their mechanical properties, which are highly time dependent. So long as plastics were used only for lightly loaded components which were not specially responsible for the strength of the structure as a whole, the study of creep in these materials formed part of the general investigation of their physical-mechanical properties and it was not specially necessary to predict their life under service conditions. However, even at normal temperatures many organic materials display appreciable creep which is immediately obvious to the observer of mechanical properties. Accordingly, this problem has for some time attracted the attention of numerous investigators but their conclusions are mainly qualitative and provide insufficient basis for a mechanical theory. The position with organic and polymeric materials is further complicated because their characteristics are unstable and highly dependent on the method of manufacture, on the time elapsed between manufacture and testing, on humidity and on other factors.

Creep effects are observed in certain structural materials such as concrete. Creep of concrete is highly dependent on its age, in the course of time it shrinks and its volume changes. However, a theory of creep in concrete which allows for these effects has been developed. Similar effects are observed in soils, in structures of ice and in other examples which we shall not consider here.

Creep theory has developed in different parallel directions and it is still difficult to consider it as a unified whole. During the twenties the term rheology appeared, it is derived from the Greek word rheo, to flow. The name was applied to the science which describes all the flow effects that are observed in solids. It was based on studies by numerous physical chemists on the flow of highly viscous and fine particulate systems. It was found that Newton's law of viscous flow does not apply to these materials, viscous forces are not linearly dependent on the applied stress and substances which were considered to be liquids were found to have certain properties of solids, such as elasticity and ultimate shear strength. In attempting to describe substances of this kind the basic idea was to unite the properties of elasticity and viscosity, to describe them both by the same equations which, in particular cases, become the equations of the theory of elasticity or of the hydrodynamics of viscous liquids. The simplest models of this kind were proposed long ago by Maxwell, Kelvin, Voigt and others. It was later found that these simple models do not describe the real properties of materials but only provide a qualitative illustration of the observed facts. More complex models were developed in which the elastic and viscous components were combined in various ways. With these models it was found possible to describe satisfactorily the behaviour of materials with a linear stress-strain relationship. For a certain range of materials, provided the stress is not too high, it can now be stated that linear theory gives a satisfactory first approximation, so that the formal apparatus of linear theory, which has been developed in sufficient breadth and detail, finds practical application. Nevertheless this field of application remains fairly narrow. In most plastics it is necessary to take account of non-linear effects and frequently they are governing. The development of non-linear theories has only recently commenced.

The behaviour of metals at high temperatures is essentially different from that of plastics for which the usual rheological schemes for metals are clearly unsuitable. So-called delayed elasticity occupies a considerable place in linear theories. A part of the deformation, which is not viscous flow deformation, is reversible, that is it disappears when the load has been removed for a certain time, which may be quite long. In metals the reversible part of the deformation is relatively small and can often be neglected. In materials that display creep the relationship between deformation and stress is markedly non-linear. For example, if the relationship between rate and stress is approximated by a power function, the value of the power exponent can range from 3 to 12. In such cases no linear approximation can be suitable. Therefore, the theories of creep in metals and of rheology have developed quite independently of one another. Development of the theory of creep in metals was more influenced by the theory of plasticity, particularly as the fundamental results in creep theory were first determined by scientists who were also working on plasticity theory.

1

The Fundamental Concepts of the Mechanics of a Solid Body

The usual techniques of the mechanics of a continuous medium are used to describe the motion of a medium in a state of creep, that is tensors of stress, deformation and rate-of-deformation are introduced. In general the displacements of points in a body can be significant in relation to its initial dimensions, and then the deformations should be described by a tensor of finite deformations and the boundary conditions should also be given on the deformed surface of the body. The theory of finite deformations of an elastic body has been relatively little developed, the fundamental equations of this theory have only recently been derived and considerable difficulties can arise in the solution of particular problems. Special difficulties are encountered in attempting to construct the theory of creep at finite deformations. Therefore, we shall consider here two types of problem:

a) Deformations are small enough for their squares to be neglected. Displacements are small and boundary conditions are given on the surface of the undeformed body.

b) Deformation is unlimited, but the mechanical behaviour of the body is described by local characteristics, for instance, by the relationships between stress and rate of deformation. The formal techniques are similar in the two cases.

This chapter includes certain fundamental information concerning tensors in a three-dimensional Euclidian space and concerning the tensors that are encountered in the mechanics of a solid, namely the tensors of stress, deformation and rate-of-deformation. The presentation makes no claim to completeness or strictness, the chapter is rather an assembly of definitions, terms and formulae to which reference will be made in the subsequent presentation. Certain additional information about general theory is given in later chapters where the need for it arises.

1. Tensors in three-dimensional Euclidian space

Let us select an arbitrary system of Cartesian coordinates in three-dimensional space denoting them, for uniformity, x_1, x_2, x_3. To be specific, we select a right handed system of coordinates as shown in fig. 1. Looking along the positive direction of the x_3 axis the shortest transition from axis x_1 to axis x_2 is by positive, counter-clockwise rotation. Unit vectors on the coordinate axes are denoted $\mathbf{e}_1$, $\mathbf{e}_2$, $\mathbf{e}_3$, respectively.

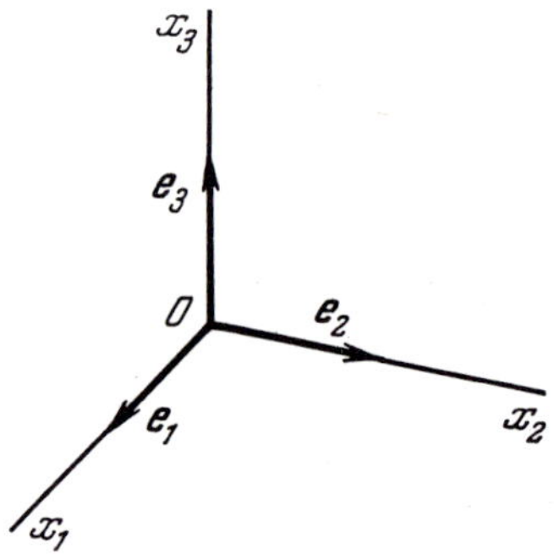

Fig. 1.

An arbitrary vector **a** with components a_1, a_2, a_3 can be represented as follows:

$$\mathbf{a} = \sum_{i=1}^{3} a_i \mathbf{e}_i \, .$$

For brevity we shall omit the summation sign in denoting that summation is carried out from $i = 1$ to $i = 3$. Thus the short notation of the above formula becomes

$$\mathbf{a} = a_i \mathbf{e}_i \,. \tag{1.1}$$

Now let us suppose that we wish to convert from the original coordinate system to a new one with axes x'_1, x'_2 and x'_3 the corresponding unit vectors being $\mathbf{e}'_1$, $\mathbf{e}'_2$ and $\mathbf{e}'_3$. The components of the vector **a** relative to the new axes are a'_1, a'_2 and a'_3. In the new axes the vector **a** can be represented by an expression analogous to (1.1), we may therefore write

$$\mathbf{a} = a_i \mathbf{e}_i = a'_k \mathbf{e}'_k \,.$$

Multiply both sides of this equation by $\mathbf{e}'_i$. As $\mathbf{e}'_k \mathbf{e}'_i = 0$ where $i \neq k$ and $\mathbf{e}'_i \mathbf{e}'_i = 1$ we obtain

$$a'_i = a_k (\mathbf{e}'_i \mathbf{e}_k) \,.$$

Introduce the notation

$$\mathbf{e}'_i \mathbf{e}_k = \beta_{ik} \,.$$

When the coordinate axes are rotated the conversion formulae for the vector components are of the following form

$$a'_i = \beta_{ik} a_k \,. \tag{1.2}$$

In precisely the same way we can express vector components in the original axes in terms of components in the new axes. We then obtain

$$a_i = \beta_{ki} a'_k \,. \tag{1.3}$$

Unit vectors of the coordinate axes in the new and original systems are expressed in terms of one another in a precisely similar way. Thus the vector $\mathbf{e}'_i$ can always be represented by a linear combination of vectors $\mathbf{e}_1$, $\mathbf{e}_2$ and $\mathbf{e}_3$ consequently

$$\mathbf{e}'_i = \alpha_{is} \mathbf{e}_s \,.$$

To determine the coefficients α_{is} multiply the above equation by $\mathbf{e}_k$ to obtain

$$\alpha_{ik} = \mathbf{e}'_i \mathbf{e}_k \,.$$

Consequently, $\alpha_{ik} = \beta_{ik}$ and the unit vectors are converted by the same formulae (1.2) and (1.3) in which a_i and a_i' are replaced by $\mathbf{e}_i$ and $\mathbf{e}_i'$, respectively.

For convenience we give separately the conversion formulae for unit vectors on the coordinate axes.

$$\mathbf{e}_i' = \beta_{ik}\mathbf{e}_k\,, \tag{1.4}$$

$$\mathbf{e}_i = \beta_{ki}\mathbf{e}_k'\,. \tag{1.5}$$

Formulae (1.2) and (1.3) may be taken as defining a vector. Instead of considering the geometrical interpretation of a vector as a length with a direction, we shall understand by the term vector the totality of three quantities a_1, a_2 and a_3 which are related to the given system of coordinates and which are converted to another system by formula (1.2).

This formal definition, which is not associated with intuitive concepts, is convenient in that it provides a natural introduction for the more general concept of a tensor, for which there is no simple geometrical analogy. We shall use the term second-rank tensor for a totality of quantities A_{ij} which have the property that if $\mathbf{a}$ is an arbitrary vector in their turn the quantities $A_{ij}a_j$ or $A_{ji}a_j$ are components of the vector.

From this definition we can readily derive conversion formulae for tensor components when the coordinate axes are rotated. Let the tensor components relative to the new axes be A_{ij}'. By definition $A_{ij}'a_j'$ are components relative to the new axes of the vector of which the components in the old axes were $A_{ij}a_j$. Consequently, these quantities are related by the vector component conversion formulae (1.2)

$$A_{ij}'a_j' = \beta_{ik}A_{ks}a_s\,.$$

But in its turn

$$a_s = \beta_{js}a_j'\,.$$

Thus we have obtained

$$A_{ij}'a_j' = \beta_{ik}\beta_{js}A_{ks}a_j'\,.$$

This equation is valid for arbitrary values of a_j', therefore it follows that

$$A_{ij}' = \beta_{ik}\beta_{js}A_{ks}\,. \tag{1.6}$$

It can readily be shown that the inverse conversion is performed by the formulae

$$A_{ij} = \beta_{ki}\beta_{sj}A'_{ks} \,. \tag{1.7}$$

We could have immediately defined a second rank tensor as a totality of nine quantities A_{ij} which are associated with the given system of coordinates and which change according to formulae (1.6) and (1.7) when the coordinates are converted. Symmetric tensors, that is those for which $A_{ij} = A_{ji}$, are often encountered in practical applications.

Certain consequences follow from the definition of a second rank tensor.

1) Given two arbitrary vectors **a** and **b** the expression

$$A_{ij}a_ib_j$$

is invariant, i.e., it is a quantity which does not alter when the coordinates are altered. Indeed it is a scalar product of a vector with components $A_{ij}b_j$ and a vector **a** or a vector with components $A_{ij}a_j$ and a vector **b**. Conversely, given the totality of quantities A_{ij} such that $A_{ij}a_ib_j$ is invariant whatever the vectors **a** and **b** then the quantities A_{ij} form a second rank tensor.

2) Given two arbitrary vectors **a** and **b** the quantities

$$C_{ij} = a_ib_j$$

form a second rank tensor.

Indeed, let us take a third arbitrary vector **r**. The expression $C_{ij}r_j$ is equal to $a_ib_jr_j = a_i(\mathbf{br})$. This is a component of a vector which differs from **a** by a scalar vector product (**br**); thus by definition the quantities C_{ij} are components of a tensor. A tensor formed in this way is termed multiplicative, the operation of generating a multiplicative tensor from two vectors is termed their dyadic product.

3) The totality of Kronecker symbols forms a tensor. Indeed

$$\delta_{ij}a_j = a_i \,.$$

This tensor is termed a unit tensor. Note that

$$\delta_{ij} = \mathbf{e}_i\mathbf{e}_j = \mathbf{e}'_i\mathbf{e}'_j \,.$$

The components of a unit tensor do not alter when the coordinates are altered.

4) Given two tensors A_{ij} and B_{ij}, quantities of the type $A_{is}B_{sj}$ are components of a second rank tensor. Indeed consider the expression

$$A_{is}B_{sj}a_j ,$$

where a_j are components of an arbitrary vector. In accordance with the definition of a tensor B_{ij} the product $B_{sj}a_j$ is a component of a vector which we shall denote b_s, consequently

$$A_{is}B_{sj}a_j = A_{is}b_s .$$

But this is a vector component, in accordance with the definition of the tensor A_{ij} and this demonstrates the tensor nature of the quantities $A_{is}B_{sj}$. We may proceed further on this path, adding for example a third tensor C_{ij} to give expressions of the form

$$A_{is}B_{sk}C_{kj} ,$$

which in their turn are components of a second rank tensor and so on.

If A_{ij} is a second rank tensor then A_{ji} is also a second rank tensor. Obviously, the sum or difference of the components of two tensors is a tensor. Therefore, the quantities $\frac{1}{2}(A_{ij}+A_{ji})$ and $\frac{1}{2}(A_{ij}-A_{ji})$ are components of a tensor. The first of them is symmetric and its components do not alter on changing the sequence of the subscripts. The second is antisymmetric and its components change sign when the subscripts are reversed. Thus a second rank tensor can always be considered as the sum of a symmetric and an antisymmetric tensor

$$A_{ij} = \tfrac{1}{2}(A_{ij}+A_{ji}) + \tfrac{1}{2}(A_{ij}-A_{ji}) .$$

In the same way as we have defined a second rank tensor we can introduce the concept of a tensor of arbitrary rank n. This is the totality of the quantities

$$A_{i_1 i_2 \ldots i_n} ,$$

associated with a given system of coordinates and having properties such that the quantities

$$A_{i_1 i_2 \ldots i_n} a_{i_k} \qquad (1 \leqslant k \leqslant n) ,$$

where a_{i_k} are components of an arbitrary vector, form a tensor of rank $n-1$. In precisely the same way we can proceed from a tensor of rank $n-1$ to a tensor of rank $n-2$ so that we finally obtain a first rank tensor, that is a vector. When the coordinates are changed the conversion formulae for an n rank tensor can be derived by a procedure analogous to that used for a second rank tensor. These formulae are of the form

$$A'_{i_1 i_2 \ldots i_n} = \beta_{i_1 j_1} \beta_{i_2 j_2} \ldots \beta_{i_n j_n} A_{j_1 j_2 \ldots j_n} \,. \tag{1.8}$$

Just as in the case of a second rank tensor, in making the inverse conversion it is necessary to change the sequence of the suffixes of $\beta_{i_s j_s}$.

It follows from the definition of an n rank tensor that, given arbitrary vectors $\mathbf{a}^{(1)}, \mathbf{a}^{(2)}, \ldots, \mathbf{a}^{(n)}$, the expression $A_{i_1 i_2 \ldots i_n} a_{i_1}^{(1)} a_{i_2}^{(2)} \ldots a_{i_n}^{(n)}$ is invariant. Since this expression is invariant the quantities $A_{i_1 i_2 \ldots i_n}$ form a tensor.

Given two tensors, the tensor $A_{i_1 i_2 \ldots i_n}$ of rank n and the tensor $B_{j_1 j_2 \ldots j_m}$ of rank m the totality of the quantities

$$A_{i_1 i_2 \ldots i_n} B_{j_1 j_2 \ldots j_m}$$

forms a tensor of rank $n+m$ which is termed the product of the tensors $A_{i_1 i_2 \ldots i_n}$ and $B_{j_1 j_2 \ldots j_m}$. Indeed select an arbitrary vector $\mathbf{a}^{(s)}$ and form the expressions

$$A_{i_1 i_2 \ldots i_n} B_{j_1 j_2 \ldots j_m} a_{i_s}^{(s)} \qquad (s \leqslant n) \,,$$

which depend on $n+m-1$ subscripts. To demonstrate the tensor nature of these expressions take a new arbitrary vector and form the expressions

$$A_{i_1 i_2 \ldots i_n} B_{j_1 j_2 \ldots j_m} a_{i_s}^{(s)} a_{i_k}^{(k)} \qquad (k \neq s, k \leqslant n, s \leqslant n)$$

or

$$A_{i_1 i_2 \ldots i_n} B_{j_1 j_2 \ldots j_m} a_{i_s}^{(s)} b_{j_k}^{(k)} \qquad (s \leqslant n, k \leqslant m) \,.$$

These quantities now depend on $n+m-2$ subscripts. Repeating the operation $n+m$ times we finally obtain the following expression

$$(A_{i_1 i_2 \ldots i_n} a_{i_1}^{(1)} \ldots a_{i_n}^{(n)})(B_{j_1 j_2 \ldots j_m} b_{j_1}^{(1)} \ldots b_{j_m}^{(m)}) \,.$$

Each of the quantities within brackets is a scalar invariant, which provides the necessary demonstration.

The operation of multiplying tensors leads to tensors of higher rank and the way in which lower rank tensors are obtained from a given tensor or totality to them will be considered in the next section.

2. Contraction of tensors. Invariants

We shall show that the sum of the components of a second rank tensor with identical subscripts

$$A_{ii} = A_{ij}\delta_{ij}$$

is invariant. Indeed by the conversion formula (1.6)

$$A'_{ii} = \beta_{ik}\beta_{il}A_{il} \, .$$

But by formula (1.5)

$$\mathbf{e}_k = \beta_{ik}\mathbf{e}'_i \, , \qquad \mathbf{e}_l = \beta_{il}\mathbf{e}'_i \, .$$

Multiplying these expressions together we obtain

$$\mathbf{e}_k\,\mathbf{e}_l = \beta_{ik}\,\beta_{il} = \delta_{kl} \, .$$

Therefore

$$A'_{ii} = \delta_{kl}A_{kl} = A_{ii} \, .$$

The operation of obtaining the scalar invariant of a tensor in this way is termed contraction.

Given an arbitrary tensor of rank n we can derive from it a tensor of rank $n-2$ by making any two subscripts equal to one another and summating from one to three by this subscript. Indeed, let us convert the tensor to new axes by formulae (1.8) putting $i_n = i_{n-1}$ and summing by this subscript; we obtain

$$A'_{i_1 i_2 \ldots i_{n-2} i_n i_n} = \beta_{i_1 j_1}\beta_{i_2 j_2} \cdots \beta_{i_n j_{n-1}}\beta_{i_n j_n}A_{j_1 j_2 \ldots j_{n-1} j_n}$$
$$= \beta_{i_1 j_1} \cdots \beta_{i_{n-2} j_{n-2}}A_{j_1 j_2 \ldots j_{n-2} j_n j_n} \, ,$$

as

$$\beta_{i_n j_{n-1}} \beta_{i_n j_n} = \delta_{j_{n-1} j_n} .$$

Hence it follows that the quantities

$$A_{i_1 i_2 \ldots i_{n-2} i_n i_n}$$

form a tensor of rank $n-2$.

If n is a whole number, subsequent contraction of the tensor ultimately leads to a second rank tensor and the next contraction to an invariant. If we perform this operation for various pairs of subscripts we will, in general, obtain different invariants which are termed linear invariants. If we multiply the tensor by itself and contract the derived multiplicative vector of rank $2n$ we ultimately obtain a system of quadratic invariants, and so on. The number of invariants of a tensor constructed in this way may be indefinitely great but the number of independent invariants is finite. For example, if we consider the vector $\mathbf{a}_i$ as a second rank tensor the quadratic invariant of the vector is a linear invariant of the multiplicative tensor $a_i a_j$, this is the square of the length of the vector $a_i a_i$. Generating a multiplicative tensor of indefinitely high even order $a_i a_j \ldots a_s$ and contracting by arbitrary pairs of subscripts, by each contraction we obtain a scalar factor $a_i a_i$; therefore, the square of the length of the vector is the sole independent invariant, all the other invariants being powers of it.

We shall not consider here the construction of a complete system of independent invariants of an arbitrary tensor but will consider only the important practical case of a symmetrical second rank tensor for which there are three independent invariants. To explain this consider the problem of referring a symmetrical tensor to the principal axes. We shall seek a direction given by unit vector $\mathbf{v}$ for which

$$A_{ij} \nu_j = A \nu_i .$$

Here A is scalar. Write this system of equations to define the unknowns ν_j and A in the following way

$$(A_{ij} - \delta_{ij} A) \nu_j = 0 . \tag{2.1}$$

Equating to zero the determinant of this system of equations which is linear relative to the unknowns ν_j we obtain a cubic equation for A. Denote the three roots of this equation by $A^{(1)}$, $A^{(2)}$ and $A^{(3)}$. In (2.1) put $A = A^{(k)}$ then we can determine the direction cosines $\nu_j^{(k)}$. The quantities $A^{(k)}$ are termed the principal values of the tensor, the directions $\nu^{(k)}$ the principal directions.

Let us first suppose that the values of $A^{(k)}$ are real and different. The first of these statements will be demonstrated, the second is not essential. It follows from (2.1) that

$$\nu_i^{(k)} = \frac{1}{A^{(k)}} A_{ij} \nu_j^{(k)} .$$

Multiplying both sides by $\nu_i^{(l)}$ $(l \neq k)$ and summing by the index i we obtain

$$\nu_i^{(k)} \nu_i^{(l)} = \frac{1}{A^{(k)}} A_{ij} \nu_j^{(k)} \nu_i^{(l)} . \tag{2.2}$$

On the other hand if we start from the equation

$$\nu_i^{(l)} = \frac{1}{A^{(l)}} A_{ij} \nu_i^{(l)} ,$$

and multiply both sides by $\nu_i^{(k)}$ we obtain

$$\nu_i^{(l)} \nu_i^{(k)} = \frac{1}{A^{(l)}} A_{ij} \nu_j^{(l)} \nu_i^{(k)} . \tag{2.3}$$

But $A_{ij} \nu_j^{(k)} \nu_i^{(l)} = A_{ij} \nu_j^{(l)} \nu_i^{(k)}$ because the tensor A_{ij} is symmetrical. Therefore, eqs. (2.2) and (2.3) are simultaneously possible only when $A_{ij} \nu_j^{(l)} \nu_i^{(k)} = 0$ because we have assumed that $A^{(k)} \neq A^{(l)}$. Hence it follows that

$$\nu_i^{(k)} \nu_i^{(l)} = 0 . \tag{2.4}$$

Therefore, if the principal values are different the principal directions of the tensor are mutually orthogonal.

We shall now show that the principal values of the tensor are always real. Assume the opposite, for instance that $A^{(3)}$ is real and $A^{(1)}$ and $A^{(2)}$ are complex quantities. Being roots of an equation with real coefficients $A^{(1)}$ and $A^{(2)}$ are conjugate complex quantities. Substituting $A^{(1)}$ and $A^{(2)}$ into (2.1) we also obtain conjugate complex values for $\nu_i^{(1)}$ and $\nu_i^{(2)}$

$$\nu_i^{(1)} = n_i + \sqrt{-1} m_i ,$$

$$\nu_i^{(2)} = n_i - \sqrt{-1} m_i .$$

Substituting into the condition of orthogonality (2.4) we obtain

$$\nu_i^{(1)}\nu_i^{(2)} = \sum (n_i^2 + m_i^2) = 0\,.$$

But this equation is impossible and consequently the principal values cannot be complex.

As we have already mentioned, the case of equal roots presents no difficulties. Suppose for example that $A^{(1)} = A^{(2)}$. This means that when $A = A^{(1)}$ the system (2.1) contains only a single independent equation, its solution remains indeterminate and after applying the normalisation condition $\nu_i\nu_i = 1$ we obtain a totality of vectors $\mathbf{v}^{(1)}$ which depend on a single parameter. These vectors are orthogonal to the vector $\mathbf{v}^{(3)}$ and consequently, lie in a plane perpendicular to $\mathbf{v}^{(3)}$. Any two mutually orthogonal vectors $\mathbf{v}^{(1)}$ and $\mathbf{v}^{(2)}$ that lie in this plane can be taken as vectors that correspond to the principal direction. If finally, $A^{(1)} = A^{(2)} = A^{(3)}$ any orthogonal axes can be taken as principal. This is possible only if $A_{ij} = A\delta_{ij}$. A tensor of this kind is termed spherical.

The principal values of a tensor are invariants and a symmetrical second rank tensor can have no other independent invariants, but the principal system of independent invariants can be selected in various ways. If the tensor is referred to axes which are oriented in the principal directions its matrix will be

$$\begin{vmatrix} A^{(1)} & 0 & 0 \\ 0 & A^{(2)} & 0 \\ 0 & 0 & A^{(3)} \end{vmatrix}.$$

The first invariant of the tensor A_{ii} is equal to the sum of the principal values $A^{(1)} + A^{(2)} + A^{(3)}$, the second invariant $A_{ij}A_{ij}$ is the sum of the squares of the principal values and the third invariant $A_{ij}A_{js}A_{si}$ is the sum of the cubes. Let us denote the invariants by the same letters as the tensor components using Roman numerals in the subscript which denote the order of the invariant. Then,

$$A_{\mathrm{I}} = A_{ii} = A^{(1)} + A^{(2)} + A^{(3)}\,,$$

$$A_{\mathrm{II}} = A_{ij}A_{ij} = (A^{(1)})^2 + (A^{(2)})^2 + (A^{(3)})^2\,,$$

$$A_{\mathrm{III}} = A_{ij}A_{js}A_{si} = (A^{(1)})^3 + (A^{(2)})^3 + (A^{(3)})^3\,.$$

Symmetrical invariants of different orders can be introduced by other methods. Thus if to the invariant A_{II} we add the square of the first invariant with any constant factor we obtain another quadratic invariant. To the cubic invariant we may add any linear combination of A_{I}^3 and $A_{\mathrm{I}}A_{\mathrm{II}}$.

Another natural method of constructing a fundamental system of invariants is as follows. By expanding the determinant (2.1) we obtain a cubic equation for A. The coefficient of A to the third power is -1, the coefficients of successive powers are known symmetrical functions of the roots of the equation, that is invariants of first, second and third power respectively.

Write these equations as follows

$$A^3 - I_{\mathrm{I}}A^2 + I_{\mathrm{II}}A - I_{\mathrm{III}} = 0\,,$$

by the usual formulae of the theory of symmetrical functions we have

$$I_{\mathrm{I}} = A_{\mathrm{I}}\,,$$

$$I_{\mathrm{II}} = \tfrac{1}{2}(A_{\mathrm{I}}^2 - A_{\mathrm{II}})\,, \tag{2.5}$$

$$I_{\mathrm{III}} = \tfrac{1}{6}(A_{\mathrm{I}}^3 - 3A_{\mathrm{I}}A_{\mathrm{II}} + 2A_{\mathrm{III}})\,.$$

3. Mutual basis. Covariant and contravariant components of a tensor

The advantage of tensor notation, particularly in the mechanics of continuous media, becomes evident when we convert from Cartesian coordinates to arbitrary coordinates which are, in general, not orthogonal. Let us assign the directions of the coordinate axes to three non-coplanar vectors $\mathbf{e}_1$, $\mathbf{e}_2$, $\mathbf{e}_3$ which in general are not necessarily unit vectors. In addition to the fundamental basis $\mathbf{e}_i$ we introduce the mutual basis $\mathbf{e}^i$, that is a set of three vectors which are associated with the fundamental basis vectors, by the following relationships

$$\mathbf{e}_i\mathbf{e}^j = \delta_i^j \tag{3.1}$$

(we shall denote the Kronecker symbols δ_i^j). For this it suffices to take

$$\mathbf{e}^1 = \frac{\mathbf{e}_2 \times \mathbf{e}_3}{g}\,, \qquad \mathbf{e}^2 = \frac{\mathbf{e}_3 \times \mathbf{e}_1}{g}\,, \qquad \mathbf{e}^3 = \frac{\mathbf{e}_1 \times \mathbf{e}_2}{g}\,,$$

$$g = \mathbf{e}_1(\mathbf{e}_2 \times \mathbf{e}_3)\,.$$

The arbitrary vector **a** can now be represented in two ways, either

$$\mathbf{a} = a^i \mathbf{e}_i \,, \tag{3.2}$$

or

$$\mathbf{a} = a_i \mathbf{e}^i \,. \tag{3.3}$$

The quantities a_i are termed covariant components of the vector, the quantities a^i contravariant components. The origin of these terms is clear from the following. Let us convert the fundamental basis by the formula

$$\mathbf{e}'_i = \alpha_i^j \mathbf{e}_j \,. \tag{3.4}$$

Here α_i^j is an arbitrary matrix with determinant different from zero. The inverse conversion is

$$\mathbf{e}_i = \beta_i^j \mathbf{e}'_j \,. \tag{3.5}$$

The matrix β_i^j is inverse with respect to α_i^j, its elements are algebraic complements to the corresponding elements of the matrices of the primary conversion divided by the determinant of this matrix. We shall now show how the mutual basis vector is then converted. Put

$$\mathbf{e}^{s'} = \gamma_k^s \mathbf{e}^k \,.$$

Multiply both sides of this equation by $\mathbf{e}_i$. In view of the condition (3.1) only one term remains in the right-hand side and we obtain

$$\gamma_i^s = \mathbf{e}^{s'} \mathbf{e}_i \,.$$

Now use the conversion formula (3.5)

$$\gamma_i^s = \beta_i^j \mathbf{e}^{s'} \mathbf{e}'_j = \beta_i^j \delta_j^s = \beta_i^s \,.$$

Thus the mutual basis vectors are converted by the formula

$$\mathbf{e}^{i'} = \beta_j^i \mathbf{e}^j \,, \tag{3.6}$$

and, as can easily be verified

$$\mathbf{e}^i = \alpha^i_j \mathbf{e}^{j'} . \tag{3.6'}$$

Thus in converting the fundamental bases of vectors the mutual bases are obtained by the inverse conversion formulae.

Let us now consider how the covariant and contravariant components of a vector are converted. It follows from the definition of a vector that

$$\mathbf{a} = a^k \mathbf{e}_k = a^{i'} \mathbf{e}'_i .$$

Using expression (3.5) we rewrite this identity as follows:

$$a^{i'} \mathbf{e}'_i = a^k \beta^s_k \mathbf{e}'_s .$$

Multiplying both sides by $\mathbf{e}^{i'}$, we obtain

$$a^{i'} = \beta^i_k a^k . \tag{3.7}$$

Thus the contravariant components of the vector are converted by the formula of the inverse conversion of the basis vectors, and this is why they are so named. Similarly, it can be shown that the covariant components are converted by the direct conversion formulae of the basis, namely

$$a'_i = \alpha^k_i a_k . \tag{3.8}$$

Subscripts are commonly used to denote covariant quantities and superscripts for contravariant. Writing the vector in terms of one or other component we always summate by recurrent indices, one of which is a superscript and the other a subscript. In what follows we shall continue to apply this rule when we encounter more complicated expressions which contain several superscripts and subscripts. The convenience that results from simultaneous use of fundamental and mutual bases can be explained by the following example. Suppose that we are given two vectors **a** and **b**,

$$\mathbf{a} = a_i \mathbf{e}^i = a^i \mathbf{e}_i ,$$

$$\mathbf{b} = b_i \mathbf{e}^i = b^i \mathbf{e}_i .$$

The scalar product of these vectors may be constructed in various ways. If we assign both vectors covariant components we obtain

$$\mathbf{ab} = a_i b_j \mathbf{e}^i \mathbf{e}^j .$$

The products $\mathbf{e}^i\mathbf{e}^j$ are not zero and unity respectively, as in the case of orthogonal Cartesian coordinates. A similar result is obtained if both vectors are assigned contravariant components. But if one vector is assigned covariant components and the other contravariant, in view of expression (3.1) the scalar product has the simplest form

$$\mathbf{ab} = a_i b^i = a^i b_i \, .$$

Taking the conversion formulae (3.7) and (3.8) as defining a vector we can introduce the concept of a tensor of arbitrary rank. This is the totality of quantities $A'^{\cdots ab\cdots}_{ij\cdots}$ which, when the basis is changed, are converted by the formula

$$A'^{\cdots ab\cdots}_{ij\cdots} = \alpha_i^s \alpha_j^k \ldots \beta_l^a \beta_m^b \ldots A^{\cdots lm\cdots}_{sk\cdots} \, . \tag{3.9}$$

The tensor $A^{\cdots ab\cdots}_{ij\cdots}$ that is introduced in this way is mixed, it is covariant with respect to the subscripts $i, j, \ldots$ and contravariant with respect to the superscripts $a, b, \ldots$. Tensors may also be either purely covariant or contravariant. Defining a vector as a certain geometrical image or form we arrive at the concept of covariant and contravariant components of a vector, that is of different notations of one and the same quantity. Because of its formal nature, the definition of a tensor that we have given does not yet contain a principle by which to establish the equivalence of different notations of one and the same tensor, that is, which would permit conversion from covariant indices to contravariant and vice versa. To make this possible we proceed in the same way as in section 1. We define the contravariant second rank tensor A^{ij} as the totality of quantities which have the properties that if a_j are covariant components of an arbitrary vector $\mathbf{a}$ then the quantities

$$b^i = A^{ij} a_j \tag{3.10}$$

are contravariant components of the vector $\mathbf{b}$. This definition is equivalent to that given above. Indeed on conversion to another basis

$$A'^{ij} a'_j = \beta_k^i \beta_l^j A^{kl} \alpha_j^s a_s = \delta_l^s A^{kl} a_s \beta_k^i = A^{ks} a_s \beta_k^i = b^k \beta_k^i = b^{i'} \, .$$

A covariant second rank tensor is similarly defined

$$b_i = A_{ij} a^j \tag{3.11}$$

and a mixed tensor

$$b_i = A_{i.}^{\cdot j} a_j \, , \; b^i = A^{i.}_{\cdot j} a^j \, . \tag{3.12}$$

We shall say that A^{ij}, A_{ji}, $A^{i}_{\cdot j}$ represent contravariant, covariant and mixed components of one and the same tensor if the quantities b_i and b^i in formulae (3.10), (3.11), (3.12) represent covariant and contravariant components of one and the same vector **b**, whatever the vector **a** with components a_i or a^i. Hence we can derive formulae for conversion from covariant to contravariant tensor components and vice versa. Use for example formula (3.10) and the first of the formulae (3.12) and represent the vector **b** in the following way:

$$\mathbf{b} = A^{ij} a_j \mathbf{e}_i = A_{s.}^{\cdot j} a_j \mathbf{e}^s \, .$$

Multiplying both sides of this equation by $\mathbf{e}_k$, we obtain

$$A^{ij} a_j (\mathbf{e}_i \mathbf{e}_k) = A_{s.}^{\cdot j} a_j \delta_k^s = A_{k}^{\cdot j} a_j \, .$$

Introduce the notation

$$\mathbf{e}_i \mathbf{e}_k = g_{ik} = g_{ki}$$

and equate factors with identical values of a_j. Thus we obtain the following formula for converting from contravariant to mixed components

$$A_{i.}^{\cdot j} = A^{kj} g_{ik} \, . \tag{3.13}$$

In precisely the same way we convert from mixed components to covariant

$$A_{ij} = A_{i.}^{\cdot k} g_{jk}$$

and from contravariant to covariant

$$A_{ij} = A^{ks} g_{ik} g_{js} \, .$$

Quantities g_{ij} define a so-called metric tensor. The tensor nature of these quantities is easily seen from their definition if we use the basis conversion formula (3.4). The operation of replacing the contravariant index by the covariant, or "lowering the index", is performed by means of the metric tensor.

In this sense formula (3.13) is common, and the rule for lowering indices is that we multiply the basic tensor by the metric and contract by a pair of indices so that the rank of the tensor is maintained but one of the indices is lowered. For the inverse conversion from covariant to contravariant quantities we need contravariant components of the metric tensor g^{ij}. Solving eq. (3.13) relative to A^{kj} we can write

$$A^{ij} = A_{k}^{\cdot j} g^{ik} . \tag{3.14}$$

Here g^{ik} represents elements of a matrix which is inverse with respect to g_{ik}. It is evident that

$$g^{ik} = \mathbf{e}^i \mathbf{e}^k .$$

The operation of raising the index follows the same rule as that of lowering it, only using contravariant components of the metric tensor. Note that the vector components are converted in a completely similar way, namely

$$a_i = g_{ik} a^k , \qquad a^i = g^{ik} a_k .$$

The relationship between the fundamental and mutual basis is established as follows:

$$\mathbf{e}_i = g_{ik} \mathbf{e}^k , \qquad \mathbf{e}^i = g^{ik} \mathbf{e}_k .$$

Finally note that the operation of lowering or raising the index can be applied to the metric tensor itself, namely

$$g_i^j = g_{ik} g^{jk} = \mathbf{e}_i \mathbf{e}^j = \delta_i^j .$$

Thus mixed components of a metric tensor form a unit tensor.

The tensor invariants are constructed by contraction of a metric tensor and can be represented in various equivalent forms. For instance, for a second rank symmetrical tensor

$$A_{\mathrm{I}} = A_{ij} g^{ij} = A^{ij} g_{ij} = A_i^j g_j^i ,$$

$$A_{\mathrm{II}} = A_{ik} A_{js} g^{ij} g^{ks} = A_i^k A_j^s g_s^i g_k^j = A^{ik} A^{js} g_{is} g_{kj}$$

$$= A^{ij} A_{ij} = A_i^j A_j^i .$$

4. Differential operations

The most natural selection of reference system in three-dimensional space is as follows. Assume that the radius vector **r** of a point is given as a function of three curvilinear coordinates of the point, the parameters x^1, x^2, x^3. The derivatives of the radius vector **r** with respect to the coordinates define three directions at each point which can always be taken as the directions of the axes of the local system of coordinates. Take as basis vectors derivatives of the vector **r** with respect to the variables x^1, x^2, x^3 respectively. Denote these derivatives by an index that corresponds to the coordinate with respect to which differentiation is performed, putting it after a comma

$$\mathbf{e}_i = \frac{\partial \mathbf{r}}{\partial x^i} = \mathbf{r}_{,i}\,. \tag{4.1}$$

In this case basis conversion denotes conversion from one system of curvilinear coordinates x^i to another $x^{i'}$, and here the Jacobian conversion is assumed different from zero. Denote by $\mathbf{r}'$ the radius vector of the point expressed as a function of the new coordinates $x^{i'}$. The basis vectors for the new system of coordinates are expressed as follows:

$$\mathbf{e}'_i = \frac{\partial \mathbf{r}'}{\partial x^{i'}} = \mathbf{r}'_{,i}\,.$$

Bearing in mind that the new coordinates are known functions of the old apply the rule for differentiation of a complex function

$$\mathbf{e}'_i = \mathbf{r}'_{,i} = \mathbf{r}_{,j}\frac{\partial x^j}{\partial x^{i'}} = \frac{\partial x^j}{\partial x^{i'}}\mathbf{e}_j\,.$$

Therefore, the direct conversion coefficients α_i^j which figure in formula (3.4) are derivatives of the old coordinates with respect to the corresponding new ones

$$\alpha_i^j = \frac{\partial x^j}{\partial x^{i'}}\,.$$

Correspondingly, the inverse conversion coefficients are

$$\beta_i^j = \frac{\partial x^{j'}}{\partial x^i}\,.$$

Inverse conversion is possible because the Jacobian differs from zero.

As we have seen the operation of differentiation with respect to a coordinate leads to the generation of covariant quantities: putting an index after the comma to denote differentiation with respect to the coordinate we may proceed as with normal covariant indices. Now write an expression for the differential radius vector **dr**

$$\mathbf{dr} = dx^i \mathbf{r}_{,i} = dx^i \mathbf{e}_i . \tag{4.2}$$

The indices of the coordinates, and consequently coordinates of the differentials have been written as superscripts. Formula (4.2) shows that dx^i are indeed contravariant components of the vector **dr**. This is easily confirmed by a direct check. Indeed

$$dx^{i'} = \frac{\partial x^{i'}}{\partial x^j} dx^j = \beta^{i'}_j dx^j .$$

Squaring eq. (4.2) we obtain on the left-hand side the square of linear element, that is an invariant quantity which defines the local metric, namely

$$ds^2 = g_{ij} dx^i dx^j .$$

This justifies the use of the term metric tensor for the totality of the quantities g_{ij}

$$g_{ij} = \mathbf{e}_i \mathbf{e}_j = \mathbf{r}_{,i} \mathbf{r}_{,j} .$$

We can now construct covariant components of the vector **dr**, namely

$$dx_i = g_{ij} dx^j .$$

These quantities are not total differentials, consequently there are no functions x_i which may be taken as coordinates of a point in space and increments of which are covariant components of **dr**; transition from contravariant quantities to covariant has meaning only for an infinitesimally small neighbourhood of the point.

Now consider the arbitrary vector **a** which is a function of the coordinates of the points in space, that is a vector field. For a vector at any point we shall give its contravariant components with respect to the local basis

$$\mathbf{a} = a^i \mathbf{e}_i .$$

Determine the differential of the vector **a**, that is its increment on transition to an infinitesimally close point **r** + d**r**. In differentiating we should take into account not only that the quantities a^i are functions of the coordinates but that the basis vectors also vary from point to point and should be considered as variables. Therefore,

$$\mathrm{d}\mathbf{a} = \left(\frac{\partial a^i}{\partial x^j}\mathbf{e}_i + a^k \frac{\partial \mathbf{e}_k}{\partial x^j}\right) \mathrm{d}x^j .$$

But $\partial \mathbf{e}_k / \partial x^j$ is a vector which is the second derivative of the vector **r** with respect to x^k and x^j

$$\frac{\partial \mathbf{e}_k}{\partial x^j} = \mathbf{r}_{,kj} .$$

Any vector in space can, in its turn, be represented as a linear combination of basis vectors, therefore we may take

$$\frac{\partial \mathbf{e}_k}{\partial x^j} = \Gamma^i_{jk} \mathbf{e}_i .$$

Now the expression for the differential of the vector **a** may be written as follows:

$$\mathrm{d}\mathbf{a} = \left(\frac{\partial a^i}{\partial x^j} + \Gamma^i_{jk} a^k\right) \mathbf{e}_i \mathrm{d}x^j .$$

The expressions in brackets are termed covariant differential components of the vector **a** and are denoted by subscripts after the comma

$$a^i_{,j} = \frac{\partial a^i}{\partial x^j} + \Gamma^i_{jk} a^k . \tag{4.3}$$

The expression for the differential d**a** is now written as follows

$$\mathrm{d}\mathbf{a} = a^i_{,j} \mathbf{e}_i \mathrm{d}x^j . \tag{4.4}$$

The differential of the vector **a** would have exactly the same form if the basis were constant but then in formula (4.4) the covariant derivatives would be replaced by ordinary differential coefficients. The differential d**r** may be arbitrary, formula (4.4) shows that the totality of the covariant derivatives of the covariant vector components form a mixed second rank tensor. If the basis vectors are constant, i.e., the examination leads to a Cartesian system of coordinates, the covariant derivatives coincide with the ordinary ones. It remains to determine the quantities Γ^i_{jk} which are termed Christoffel symbols. It should be noted that although these quantities are characterised by three indices they do not form a third rank tensor, that is when the coordinates are altered they are not converted by the conversion formula for tensor components. By the definition of these quantities

$$\mathbf{r}_{,kj} = \Gamma^i_{kj}\mathbf{e}_i \,.$$

Multiplying both sides of this equation by $\mathbf{e}_s$ we obtain

$$\Gamma^i_{kj} g_{is} = \mathbf{r}_{,s}\mathbf{r}_{,kj} \,.$$

We shall consider, formally, that the operation of lowering the index by means of a metric tensor also applies to the Christoffel symbols. Therefore, the sum in the left-hand side may be denoted Γ_{kjs}. The right-hand side is expressed in terms of derivatives of the metric tensor components. For this we write the following identities

$$\frac{\partial g_{kj}}{\partial x^s} = \frac{\partial}{\partial x^s}(\mathbf{r}_{,k}\mathbf{r}_{,j}) = \mathbf{r}_{,ks}\mathbf{r}_{,j} + \mathbf{r}_{,k}\mathbf{r}_{,js} \,,$$

$$\frac{\partial g_{ks}}{\partial x^j} = \frac{\partial}{\partial x^j}(\mathbf{r}_{,k}\mathbf{r}_{,s}) = \mathbf{r}_{,kj}\mathbf{r}_{,s} + \mathbf{r}_{,k}\mathbf{r}_{,js} \,,$$

$$\frac{\partial g_{sj}}{\partial x^k} = \frac{\partial}{\partial x^k}(\mathbf{r}_{,s}\mathbf{r}_{,j}) = \mathbf{r}_{,sk}\mathbf{r}_{,j} + \mathbf{r}_{,s}\mathbf{r}_{,jk} \,.$$

It is now easily shown that

$$\Gamma_{kjs} = \Gamma^i_{kj} g_{is} = \frac{1}{2}\left(\frac{\partial g_{ks}}{\partial x^j} + \frac{\partial g_{sj}}{\partial x^k} - \frac{\partial g_{kj}}{\partial x^s}\right) \,. \tag{4.5}$$

Bearing in mind that g^{ij} are elements of a matrix, inverse with respect to g_{ij}, then by the rule for raising indices we obtain

$$\Gamma^i_{kj} = \Gamma_{kjs} g^{is} \,. \tag{4.6}$$

By definition the Christoffel symbols Γ^i_{jk} are contravariant components of the vector $\partial \mathbf{e}^k/\partial x^j$. Converting to covariant components we may write

$$\frac{\partial \mathbf{e}_k}{\partial x^j} = \Gamma_{jki}\mathbf{e}^i .$$

In a similar way we may derive formulae for the derivatives of mutual basis vectors.

Write the identity

$$\mathbf{e}_s \mathbf{e}^k = g^k_s .$$

The quantities g^k_s are either zero or unity and, therefore, in differentiating with respect to x^j we obtain

$$\frac{\partial \mathbf{e}_s}{\partial x^j}\mathbf{e}^k + \mathbf{e}_s \frac{\partial \mathbf{e}^k}{\partial x^j} = 0 .$$

Expressing $\partial \mathbf{e}_s/\partial x^j$ in terms of the mutual basis vector we obtain

$$\Gamma_{jsi} g^{ik} + \mathbf{e}_s \frac{\partial \mathbf{e}^k}{\partial x^j} = 0 .$$

The quantity $\mathbf{e}_s(\partial \mathbf{e}^k/\partial x^j)$ is a covariant component of the vector $\partial \mathbf{e}^k/\partial x^j$; consequently

$$\frac{\partial \mathbf{e}^k}{\partial x^j} = -\Gamma^k_{ji}\mathbf{e}^i .$$

We can now define the operation of covariant differentiation of covariant vector components. Indeed given that $\mathbf{a} = a_i \mathbf{e}^i$ then

$$\mathrm{d}\mathbf{a} = \left(\frac{\partial a_i}{\partial x^j}\mathbf{e}^i + a_i \frac{\partial \mathbf{e}^i}{\partial x^j}\right)\mathrm{d}x$$

or

$$\mathrm{d}\mathbf{a} = a_{i,j}\mathbf{e}^i \,\mathrm{d}x^j . \tag{4.7}$$

Here

$$a_{i,j} = \frac{\partial a_i}{\partial x^j} - \Gamma^k_{ij} a_k . \tag{4.8}$$

Now consider the covariants of the second rank tensor A_{ij} which is a function of the point and an arbitrary variable vector **a**. By definition expressions of the form $A_{ij}a^j$ are covariant components of a new vector **b**:

$$b_i = A_{ij} a^j .$$

Calculate the covariant derivatives of the quantities b_i by formula (4.8)

$$b_{i,s} = \frac{\partial}{\partial x^s}(A_{ij}a^j) - \Gamma^k_{is}(A_{kj}a^j) = \frac{\partial A_{ij}}{\partial x^s} a^j + A_{ij}\frac{\partial a^j}{\partial x^s} - \Gamma^k_{is} A_{kj} a^j .$$

Replace the normal differential coefficient $\partial a^j/\partial x^s$ by the covariant derivatives by formula (4.3) namely

$$\frac{\partial a^j}{\partial x^s} = a^j_{,s} - \Gamma^j_{sk} a^k .$$

Grouping the terms and altering the places of the upper indices k and j we obtain

$$b_{i,s} = \left(\frac{\partial A_{ij}}{\partial x^s} - \Gamma^k_{is} A_{kj} - \Gamma^k_{sj} A_{ik}\right) a^j + A_{ij} a^j_{,s} .$$

The expression in brackets is termed the covariant derivative of the components of the second rank tensor A_{ij} and is denoted

$$A_{ij,s} = \frac{\partial A_{ij}}{\partial x^s} - \Gamma^k_{is} A_{kj} - \Gamma^k_{sj} A_{ik} . \tag{4.9}$$

Here the normal rule for differentiating a product applies to covariant differentiation of vector components, namely

$$b_{i,s} = A_{ij,s} a^j + A_{ij} a^j_{,s} . \tag{4.10}$$

If the tensor has either mixed or contravariant components the corresponding terms which depend on the Christoffel symbols appear with inverse sign, namely

$$A^i_{j,s} = \frac{\partial A^i_j}{\partial x^s} + \Gamma^i_{ks} A^k_j - \Gamma^k_{js} A^i_k\,,$$

$$A^{ij}_{,s} = \frac{\partial A^{ij}}{\partial x^s} + \Gamma^i_{ks} A^{kj} + \Gamma^j_{ks} A^{ik}\,. \qquad (4.11)$$

Covariant derivatives of tensors of any rank are defined in a similar way.

Note that covariant derivatives of the components of a metric tensor are zero. Indeed by formula (4.9)

$$g_{ij,s} = \frac{\partial g^{ij}}{\partial x^s} - \Gamma^k_{is} g_{kj} - \Gamma^k_{sj} g_{ik} = \frac{\partial g_{ij}}{\partial x^s} - \Gamma_{isj} - \Gamma_{sji}\,.$$

But it will be directly seen from formula (4.5) that the right-hand side is identically zero, therefore

$$g_{ij,s} = 0\,.$$

Similarly

$$g^{ij}_{,s} = 0\,.$$

Therefore, in a certain sense the metric tensor is constant. Using this property it is easy to show that a common rule applies in raising and lowering the indices in expressions for the covariant derivatives of a vector or tensor of arbitrary rank. For instance

$$a^i_{,j} = a_{k,j} g^{ki}\,, \qquad a_{i,j} = a^k_{,j} g_{ki}\,.$$

5. Theory of deformation in a continuous medium

By way of application of the general theory of tensors presented above, let us consider the problem of a stressed state in a continuous medium. A continuous medium is a hypothetical medium which is studied in mechanics, replacing real substances such as solids, liquids or gases in mechanical models. Continuous media are introduced so as to make it possible to consider infinitesimally small elements of volume and consequently, to apply infinitesimal calculus to mechanics. There have been numerous attempts to provide a stricter foundation for the mechanics of a continuous medium on the basis of discrete concepts, but the introduction of micro-objects, namely material points with certain systems of bonds, is of itself an abstraction which is only a very nominal representation of the real structure of materials.

Let there be x^i coordinate points of a body in its initial state, in which no external force is applied. A precise definition of initial state is only required for the formulation of physical laws which describe the behaviour of the body under load, but any state could be considered as initial in constructing a general theory of stress and deformation. Let the radius vector of a point M in the initial state be $\mathbf{r}(x^i)$. Let us consider that when deformation occurs the coordinate lines of the system x^i remain materially related to the body, so that in one of the successive states, which we shall term deformed, the point M of the material is displaced in space to the point M', its radius vector becomes $\mathbf{r}'$ but the curvilinear coordinates of x^i remain the same. The difference $\mathbf{r}'-\mathbf{r}=\mathbf{u}$ is termed the displacement of the point. The square of the linear element in the undeformed state is determined as follows:

$$\mathrm{d}s^2 = \mathrm{d}\mathbf{r}^2 = g_{ij}\mathrm{d}x^i\mathrm{d}x^j .$$

After deformation the square of the linear element is

$$\mathrm{d}s'^2 = \mathrm{d}\mathbf{r}'^2 = g'_{ij}\mathrm{d}x^i\mathrm{d}x^j .$$

Denote by e the relative elongation of the linear element due to deformation

$$e = \frac{\mathrm{d}s' - \mathrm{d}s}{\mathrm{d}s} .$$

Knowing the quantities g_{ij} and g'_{ij} we can calculate the relative elongation of any infinitesimally small segment originating from the point M and, therefore, the quantity g'_{ij} may be taken as characteristic of local deformation. The deformation tensor e_{ij} is the term applied to half the difference between the metric tensor in the deformed and in the initial states.

$$e_{ij} = \tfrac{1}{2}(g'_{ij} - g_{ij}) . \tag{5.1}$$

The relative elongation, or strain of the element $\mathrm{d}s$ is found by the deformation tensor using the equation

$$(1+e)^2 - 1 = 2\,\frac{e_{ij}\,\mathrm{d}x^i\,\mathrm{d}x^j}{\mathrm{d}s^2}\,. \tag{5.2}$$

For finite deformations the definition of deformation tensor that we have given is not unique, other authors have developed other points of view. In what follows we shall have no further need for a general theory of finite deformations and therefore, we proceed directly to the case of small deformations. The physical limitation that the deformation is small signifies that the relative elongation of any linear element is small compared with unity, therefore, we may neglect the quantity e^2 in the left-hand side of formula (5.2) and obtain

$$e = \frac{e_{ij}\,\mathrm{d}x^i\,\mathrm{d}x^j}{\mathrm{d}s^2}\,.$$

Now suppose that we know the displacement vector $\mathbf{u}$, given the components relative to the local basis

$$\mathbf{u} = u^i \mathbf{e}_i = u_i \mathbf{e}^i\,.$$

Then $\mathbf{dr}' = \mathbf{dr} + \mathbf{du}$. Here

$$\mathbf{dr} = \mathbf{e}_i \mathrm{d}x^i\,,$$

$$\mathbf{du} = u_{i,j}\mathbf{e}^i \mathrm{d}x^j\,.$$

Now calculate the square of the linear element in the deformed state

$$\mathrm{d}s'^2 = \mathbf{dr}'^2 = (\mathbf{dr} + \mathbf{du})^2 = \mathbf{dr}^2 + 2\mathbf{dr}\,\mathbf{du} + \mathbf{du}^2\,.$$

Introducing the values $d\mathbf{r}$ and $d\mathbf{u}$ we obtain

$$\mathrm{d}s'^2 = \mathrm{d}s^2 + 2u_{i,j}\mathrm{d}x^i\mathrm{d}x^j + u_{k,i}u_{s,j}g^{ks}\mathrm{d}x^i\mathrm{d}x^j\,,$$

or

$$2e_{ij}\mathrm{d}x^i\mathrm{d}x^j = (2u_{i,j} + u_{k,i}u_{s,j}g^{ks})\mathrm{d}x^i\mathrm{d}x^j\,. \tag{5.3}$$

The tensor e_{ij} is symmetrical, whereas the expressions in brackets in the right-hand side represent the components of an asymmetrical tensor in view of the first term, since $u_{i,j} \neq u_{j,i}$. However, the quadratic form in the right-hand side of the equation remains unaltered if we replace the tensor by its symmetrical part. Equating coefficients of the two identically equal quadratic forms we obtain the following expression for the component of a finite deformation tensor,

$$e_{ij} = \tfrac{1}{2}(u_{i,j} + u_{j,i} + u_{k,i}u_{s,j}g^{ks}) . \tag{5.4}$$

The antisymmetrical part of the tensor, which figures in the right-hand side of eq. (5.3), is termed the rotation tensor,

$$\omega_{ij} = \tfrac{1}{2}(u_{i,j} - u_{j,i}) . \tag{5.5}$$

The case of small displacements is of particular interest in practical applications. We shall term a displacement small when the quantities $u_{i,j}$ are small and their derivatives are negligible. Then the expressions for the components of the deformation tensor simplify and we obtain

$$e_{ij} = \tfrac{1}{2}(u_{i,j} + u_{j,i}) . \tag{5.6}$$

Formula (5.5) for a rotation tensor is the same for both small and finite deformations.

6. Theory of stress

The stress vector on an area which passes through a given point, and which is given by a unit vector $\mathbf{v}$ normal to this area, is defined as the limit of the ratio of the force acting on the area to the area when the area converges to a point. In selecting a system of coordinates we thereby define, at any point, three planes each of which is defined by the basis vectors $\mathbf{e}_i$. The normals to these areas are vectors of the mutual basis. We shall denote the areas by the same indices as apply to the normals to them. Thus the area which passes through the basis vectors $\mathbf{e}_2$ and $\mathbf{e}_3$ is denoted by the index 1 because the normal to this area is the vector $\mathbf{e}^1$ of the mutual basis. Denote by $\mathbf{S}^i$ the stress vector on the area with number i. Now consider a plane with normal $\mathbf{v}$ and consider the tetrahedron shown in fig. 2. Denote the area of the faces contained between the basis vectors by $d\omega_i$, the area of the fourth face is denoted $d\omega$, let the vector stress on this face be $\mathbf{S}$. The sum of the forces acting on the tetrahedron should be zero, therefore

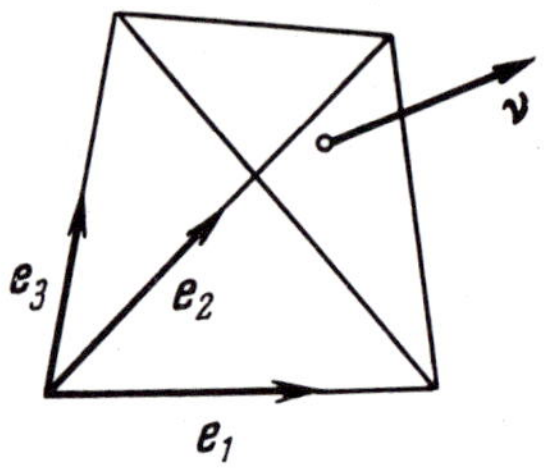

Fig. 2.

$$\mathbf{S}\mathrm{d}\omega - \mathbf{S}^i \mathrm{d}\underset{i}{\omega} = 0\,. \tag{6.1}$$

The vectorised area $d\omega$ is $\mathbf{v}\,\mathrm{d}\omega$, similarly the vectorised area $\mathrm{d}\underset{i}{\omega}$ is the product of $\mathrm{d}\omega_i$ and unit vector in the direction $\mathbf{e}^i$, that is the product with a vector $\mathbf{e}^i/|\mathbf{e}^i|$ but $\mathbf{e}^i\mathbf{e}^i = g^{ii}$, consequently the modulus of $\mathbf{e}^i$ is $\sqrt{g^{ii}}$ and

$$\mathrm{d}\underset{i}{\boldsymbol{\omega}} = \frac{1}{\sqrt{g^{ii}}}\,\mathbf{e}^i\,\mathrm{d}\underset{i}{\omega}\,.$$

In this formula we do not sum with respect to the index i. We know that the sum of the vectors of areas which form a closed polyhedron is zero, therefore

$$\mathbf{v}\mathrm{d}\omega = \frac{\mathrm{d}\underset{i}{\omega}}{\sqrt{g^{ii}}}\,\mathbf{e}^i\,.$$

Here the right-hand side contains a sum with respect to the index i. From the relationships given above it follows that the quantities $1/\sqrt{g^{ii}}\,(\partial\omega^i/\partial\omega)$ are covariant components of unit vector on the normal $\mathbf{v}$; therefore we may write

$$\mathrm{d}\underset{i}{\omega} = \sqrt{g^{ii}}\,\nu_i\,\mathrm{d}\omega\,.$$

Writing this into the condition of equilibrium (6.1) and cancelling $\mathrm{d}\omega$ we obtain

$$\mathbf{S} = \sqrt{g^{ii}}\,\mathbf{S}^i \nu_i\,.$$

Each of the vectors $\mathbf{S}^i$ can be given three components with respect to the covariant basis, therefore we take

$$\sqrt{g^{ii}}\mathbf{S}^i = \sigma^{ij}\mathbf{e}_j \, .$$

The previous formula is rewritten as

$$\mathbf{S} = \sigma^{ij}\nu_i \mathbf{e}_j \, . \tag{6.2}$$

Formula (6.2) defines a stress vector on any area, given the normal $\mathbf{v}$; it follows from the structure of this formula that the quantities σ^{ij} form a contravariant tensor termed the stress tensor. Now suppose that certain surface forces $\mathbf{T}$ and volumetric forces $\mathbf{F}$ act on the body. Consider the equilibrium of an arbitrary volume V bounded by a surface Σ. Formula (6.2) gives the force per unit area of the surface Σ and, therefore, the condition that the principal vector is zero is

$$\int_V \mathbf{F} \mathrm{d}V + \int_\Sigma \sigma^{ij}\mathbf{e}_j \nu_i \mathrm{d}\Sigma = 0 \, .$$

Assume for the time being that $\sigma^{ij}\mathbf{e}_j = \mathbf{t}^i$ and convert the second integral by Green's formula

$$\int_\Sigma \mathbf{t}^i \nu_i \mathrm{d}\Sigma = \int \operatorname{div} \mathbf{t}^i \mathrm{d}V .$$

But

$$\operatorname{div} \mathbf{t}^i = \mathbf{t}^i_{,l} = (\sigma^{ij}\mathbf{e}_j)_{,i} = \sigma^{ij}_{,i}\mathbf{e}_j \, .$$

Therefore, the condition that the principal vector is zero reduces to

$$\int_V (\sigma^{ij}_{,i}\mathbf{e}_j + \mathbf{F}) \mathrm{d}V = 0 \, .$$

Since this condition is fulfilled for any volume V, the subintegral expression should be zero and we obtain

$$\sigma^{ij}_{,j} + F^i = 0 \, . \tag{6.3}$$

Now determine the principal moment of the system of forces that acts on the volume V and equate it to zero

$$\int_V \mathbf{F} \times \mathbf{r} \mathrm{d}V + \int_\Sigma \sigma^{ij} \mathbf{e}_j \nu_i \times \mathbf{r} \mathrm{d}\Sigma = 0 .$$

Writing a second integral by means of vectors $\mathbf{t}^i$ and applying Green's formula we arrive at the following condition of equilibrium of torques

$$\int_V \{(\mathbf{F} + \operatorname{div} \mathbf{t}^i) \times \mathbf{r} + \mathbf{t}^i \times \mathbf{r}_{,i}\} \, \mathrm{d}V = 0 .$$

But the expression in the first subintegral brackets becomes zero because of (6.3) and the following condition remains

$$\mathbf{t}^i \times \mathbf{e}_i = \sigma^{ij} \mathbf{e}_j \times \mathbf{e}_i = 0 .$$

It follows from this that $\sigma^{ij} = \sigma^{ji}$ and the stress tensor is symmetrical.

7. Cartesian coordinates. Determination of displacements

The presentation of the theory of stress and deformation in a continuous medium in its most general tensor form for an arbitrary system of curvilinear non-orthogonal coordinates is useful in the theory of finite deformations and also in certain special sections of mechanics of a solid body, for instance the theory of envelopes. The general formulae that are derived will be used later to formulate the equations of equilibrium in curvilinear coordinates and to calculate deformation components. However, it is not always advisable to make a consecutive analysis in the most general form, because there is no point in introducing unnecessary mathematical complications in presenting physical theories and, therefore, we shall mainly use rectangular, Cartesian coordinates. In these coordinates the fundamental and mutual bases coincide, the difference between the covariant and contravariant components of vectors and tensors disappears. Therefore, in what follows we shall use only subscripts and summate with respect to subscripts as necessary.

Indices after the comma no longer denote covariants but ordinary differentiation, which amounts to the same. Many formulae that are derived for Cartesian coordinates remain valid in the general case, but this is not always true and care must be taken in making such extensions. For instance, following (6.3) and the notes concerning indices, the equation of equilibrium of an element in Cartesian coordinates can be written

$$\sigma_{ij,j} + F_i = 0 . \tag{7.1}$$

However, if eq. (7.1) is derived directly in Cartesian coordinates it is still not evident how to define the components of a tensor so as to obtain eq. (6.3) for contravariant components.

In what follows we shall denote Cartesian coordinates of points in space, in the same way as curvilinear coordinates, in terms of x_i. These are Cartesian coordinates of material points in the medium in the undeformed state and at the same time coordinates of points in space relative to a fixed system of coordinates. However, in developing the theory of deformations we decided to consider that the coordinates of a material point remain unaltered during deformation and, therefore, the coordinate lines of the system, which represents the system of mutually orthogonal straight lines, are curved and extended during deformation. Hence it follows that the coordinates x_i are also curvilinear coordinates of points on the body after deformation. This double role of Cartesian coordinates does not give rise to difficulty inasmuch as we consider the deformations to be small. Indeed when we introduced a local basis in section 5 and determined covariant and contravariant components of the corresponding quantities, we proceeded from the metrics of the undeformed body given by the tensors g_{ij}. With similar justification we could select a fundamental basis, associated with the deformed system of coordinates, that is with the metric tensor g'_{ij}. Both approaches are possible in the theory of finite deformations and different methods are used in different variants of the theory. In formulating the equations of equilibrium we should, of course, always deal with the metric of the deformed medium.

If deformations are small and squares of derivatives with respect to displacement are neglected, it makes insignificant difference which metric is used and, therefore, in fact we are never concerned with the curvilinear system of coordinates into which the initial Cartesian system is converted by deformation. All the quantities relate to the initial basis, given by a set of three orthogonal unit vectors of constant direction.

The components of the deformation tensor which define the distortions of the metric after deformation are now given by the formula

$$e_{ij} = \tfrac{1}{2}(u_{i,j} + u_{j,i}) \,,$$

and components of rotation by

$$\omega_{ij} = \tfrac{1}{2}(u_{i,j} - u_{j,i}) \,.$$

Here u_i are the usual components of the displacement vector on the coordinate axes, the indices after the comma denote normal differentiation with respect to the coordinate.

Now let the components of the deformation tensor be given functions of the coordinates x_i, the question arises whether displacements can be determined. Note firstly that six components of deformation are expressed in terms of three displacements and, therefore, they cannot be independent. Geometrically, this signifies that arbitrary distortion of the metric can have the result that this is no longer a metric of Euclidian space. Obviously, if we set the problem of determining the displacement for a given deformation we incidentally determine the conditions which the quantities e_{ij} must satisfy.

Assume for the moment that besides e_{ij} we also know the rotation tensor ω_{ij}. Then

$$u_{i,j} = e_{ij} + \omega_{ij}$$

and the displacements are found by simple integration

$$u_i = \int (e_{ij} + \omega_{ij})\,\mathrm{d}x_j\,.$$

We need to exclude the quantity ω_{ij} from this integral. Consider first the notation used in this formula. Suppose that at the point M_0 we know the displacement u_{i0} and also the components of rotation ω_{ij0}. Let the coordinates of the point M_0 be x_{i0}. We require to determine the displacement at the point M_1 with coordinates x_{i1}. Then

$$u_i = u_{i0} + \int_{M_0}^{M_1} e_{ij}\,\mathrm{d}x_j + \int_{M_0}^{M_1} \omega_{ij}\,\mathrm{d}x_j\,. \tag{7.2}$$

We integrate along an arbitrary curve which joins the points M_0 and M_1, x_i are the instantaneous coordinates of points on this curve. The coordinates x_{i1} of the point M_1 are considered constant, consequently

$$\mathrm{d}x_j = -\,\mathrm{d}(x_{j1} - x_j)\,.$$

Introducing this expression into the last integral of eq. (7.2) we obtain

$$u_i = u_{i0} + \int_{M_0}^{M_1} e_{ij}\,\mathrm{d}x_j - \int_{M_0}^{M_1} \omega_{ij}\,\mathrm{d}(x_{j1} - x_j)\,.$$

Applying the formula for integration by parts we obtain

$$u_i = u_{i0} + \int_{M_0}^{M_1} e_{ij} \mathrm{d}x_j - \omega_{ij}(x_{j1} - x_j)\Big|_{M_0}^{M_1} + \int_{M_0}^{M_1} (x_{j1} - x_j)\,\omega_{ij,k}\,\mathrm{d}x_k \, .$$

The quantities $\omega_{ij,k}$ are expressed in terms of derivatives with respect to deformation components so that, as can be readily shown, the following identity exists

$$\omega_{ij,k} \equiv e_{ik,j} - e_{kj,i} \, .$$

Therefore

$$u_i = u_{i0} + \omega_{ij0}(x_{j1} - x_{j0})$$

$$+ \int_{M_0}^{M_1} [e_{ik} + (x_{j1} - x_j)(e_{ik,j} - e_{kj,i})] \, \mathrm{d}x_k \, . \qquad (7.3)$$

In order for the integral in formula (7.3) to be independent of the path of integration, the subintegral expression should be a total differential. This condition leads to the identity

$$e_{ik,js} + e_{kj,is} = e_{is,jk} + e_{sj,ik} \, . \qquad (7.4)$$

Eqs. (7.4) are St. Venant's conditions of deformation compatibility, of these equations six are different. If the compatibility conditions are fulfilled, formulae (7.3) define the displacements, to the accuracy of the displacements of the solid body. The constants u_{i0} define progressive displacements and ω_{ij0} represent rigid rotation.

8. Certain further properties of the stress tensor

The general theory of referring a symmetrical second rank tensor to the principal axes, in application to the stress tensor, can be used to find three mutually orthogonal areas subject only to tensile or compressive stresses σ_1, σ_2 and σ_3. We shall number the principal stresses in such a way that

$$\sigma_1 \geqslant \sigma_2 \geqslant \sigma_3 \, ,$$

although we shall be obliged to depart from this rule in certain cases. These quantities are of fundamental importance in formulating various physical relationships and, therefore, it will often be convenient to refer the tensor to the principal axes and to deal with its three principal values. Numbering the principal axes in the same way as the principal values we denote by $\mathbf{e}_1$, $\mathbf{e}_2$ and $\mathbf{e}_3$ unit vectors on the principal axes. Given a certain area with a normal $\boldsymbol{\nu} = \nu_i \mathbf{e}_i$, by formula (6.2) the stress on this area is

$$\mathbf{S} = \sigma_1 \nu_1 \mathbf{e}_1 + \sigma_2 \nu_2 \mathbf{e}_2 + \sigma_3 \nu_3 \mathbf{e}_3 .$$

The normal stress on the area is

$$\sigma_\nu = \mathbf{S}\boldsymbol{\nu} = \sigma_1 \nu_1^2 + \sigma_2 \nu_2^2 + \sigma_3 \nu_3^2 . \tag{8.1}$$

It follows from formula (8.1) that σ_1 is the maximum and σ_3 the minimum value of the normal stress on any area that passes through the given point.

The absolute value of the tangential stress on the area with normal $\boldsymbol{\nu}$ is

$$\tau = \sqrt{\mathbf{S}^2 - \sigma_\nu^2} ,$$

or

$$\tau^2 = \sigma_1^2 (\nu_1^2 - \nu_1^4) + \sigma_2^2 (\nu_2^2 - \nu_2^4) + \sigma_3^2 (\nu_3^2 - \nu_3^4)$$
$$- 2\sigma_1\sigma_2\nu_1^2\nu_2^2 - 2\sigma_2\sigma_3\nu_2^2\nu_3^2 - 2\sigma_3\sigma_1\nu_3^2\nu_1^2 . \tag{8.2}$$

Now let us determine the maximum tangential stress. We shall seek the maximum of τ^2 as function of the direction cosines of the normal $\boldsymbol{\nu}$ which are related as follows

$$\nu_1^2 + \nu_2^2 + \nu_3^2 = 1 .$$

Applying the method of indeterminate Lagrange factors we obtain the following equations

$$\sigma_1^2 (2\nu_1 - 4\nu_1^3) - 4\sigma_1\sigma_2\nu_1\nu_2^2 - 4\sigma_1\sigma_3\nu_1\nu_3^2 + 2\lambda\nu_1 = 0$$
$$\cdots\cdots\cdots\cdots\cdots\cdots\cdots\cdots\cdots\cdots$$

or, bearing in mind (8.1),

$$[\sigma_1(\sigma_1 - 2\sigma_\nu) + \lambda]\,\nu_1 = 0\,,$$

$$[\sigma_2(\sigma_2 - 2\sigma_\nu) + \lambda]\,\nu_2 = 0\,,$$

$$[\sigma_3(\sigma_3 - 2\sigma_\nu) + \lambda]\,\nu_3 = 0\,.$$

The case, for instance, when $\nu_1 = \nu_2 = 0$ is without interest, we obtain the principal stress for which $\tau^2 = 0$. Assume that $\nu_1 \neq 0$ and $\nu_2 \neq 0$. Excluding λ from the first two equations we obtain

$$\sigma_1^2 - \sigma_2^2 - 2\sigma_\nu(\sigma_1 - \sigma_2) = 0\,.$$

If $\sigma_1 \neq \sigma_2$ then

$$\sigma_\nu = \frac{\sigma_1 + \sigma_2}{2}\,,$$

and consequently, $\nu_1^2 = \nu_2^2 = \frac{1}{2}$, $\nu_3 = 0$. Substituting into the formula for τ we find

$$\tau_{12} = \pm\frac{\sigma_1 - \sigma_2}{2}$$

for an area that is inclined at an angle of 45° to the coordinate planes 13 and 23. Similarly

$$\tau_{23} = \pm\frac{\sigma_2 - \sigma_3}{2}\,,$$

$$\tau_{31} = \pm\frac{\sigma_3 - \sigma_1}{2}\,.$$

The maximum tangential stress $\tau_{max} = (\sigma_1 - \sigma_3)/2$. The principal tangential stresses found in this way do not alter if we superpose an omnidirectional tension or compression on the stressed state of the body. Each of the principal stresses is then altered by the same amount, but the differences between them remain unaltered. Experimental results and theoretical considerations indicate that the course of irreversible deformation in metals is not altered when a hydrostatic component is superposed on the fundamental stressed state; therefore, the principal tangential stresses represent an important physical characteristic of the stressed state. It should be noted that only two of them are independent because of the obvious identity

$$\tau_{12} = \tau_{23} = \tau_{31} = 0 .$$

Here, in the expressions for τ_{ij}, we must everywhere use either subscripts or superscripts.

Following the general rule given in section 2 we can determine the system of stress tensor invariants, namely $S_{\mathrm{I}} = \sigma_{ii}$, $S_{\mathrm{II}} = \sigma_{ij}\sigma_{ij}$, $S_{\mathrm{III}} = \sigma_{ij}\sigma_{jk}\sigma_{ki}$.

The idea of isolating the hydrostatic component from the stressed state, as being unimportant in formulating the physical relationships of the theories of plasticity and creep, leads to the concept of the stress tensor deviator which is defined as follows:

$$\bar{\sigma}_{ij} = \sigma_{ij} - \sigma\delta_{ij} .$$

Here $\sigma = \frac{1}{3} S_{\mathrm{I}} = \frac{1}{3} \sigma_{ii}$. The stress tensor deviator remains unaltered on the application of hydrostatic stress. Indeed, let us put

$$\sigma'_{ij} = \sigma_{ij} + \delta_{ij} p .$$

Then

$$\sigma' = \tfrac{1}{3} \sigma'_{ii} = \sigma + p .$$

Consequently,

$$\bar{\sigma}'_{ij} = \sigma_{ij} + \delta_{ij} p - \delta_{ij} (\sigma + p) = \bar{\sigma}_{ij} .$$

The first invariant of the deviator is zero, the two other invariants are of fundamental importance for the theories of plasticity and creep, and, therefore, we shall consider them in more detail. The second invariant of the deviator is defined as follows

$$\bar{S}_{\mathrm{II}} = (\sigma_{ij} - \delta_{ij}\sigma)(\sigma_{ij} - \delta_{ij}\sigma) = \sigma_{ij}\sigma_{ij} - 2\sigma\sigma_{ij}\delta_{ij} + \delta_{ij}\delta_{ij}\sigma^2 .$$

Whence

$$\bar{S}_{\mathrm{II}} = S_{\mathrm{II}} - \tfrac{1}{3} S_{\mathrm{I}}^2 = S_{\mathrm{II}} - 3\sigma^2 . \tag{8.3}$$

If we express $\bar{S}_{II}$ in terms of the principal stresses we obtain

$$\bar{S}_{II} = \tfrac{1}{3}\,[(\sigma_1 - \sigma_2)^2 + (\sigma_2 - \sigma_3)^2 + (\sigma_3 - \sigma_1)^2] \,. \tag{8.4}$$

A simple and vivid interpretation of the quantity $\bar{S}_{II}$ was given by Roche and Eichinger. Let us calculate the stress of an area which is equally inclined to the three principal axes, that is one for which $\nu_1 = \nu_2 = \nu_3 = 1/\sqrt{3}$. We shall term this area an octahedral area and the stresses octahedral stresses. From formula (8.1) on an octahedral area

$$\sigma_\nu = \tfrac{1}{3}\,(\sigma_1 + \sigma_2 + \sigma_3)\,,$$

by formula (8.2)

$$\tau_0^2 = \tfrac{1}{9}\,[(\sigma_1 - \sigma_2)^2 + (\sigma_2 - \sigma_3)^2 + (\sigma_3 - \sigma_1)^2] \,. \tag{8.5}$$

We shall term the quantity τ_0 the octahedral tangential stress. Therefore

$$\bar{S}_{II} = 3\tau_0^2 \,. \tag{8.6}$$

The quantities σ and τ_0 are normal and tangential stresses on the octahedral area, which are naturally related to the two first invariants of the stress tensor; the normal octahedral stress is equal to one third of the first invariant, the tangential octahedral stress differs only by a constant factor from the square root of the second invariant of the deviator. This selection of the basic system of invariants is convenient in many problems of the theory of plasticity. Very often, instead of the octahedral tangential stress, we shall use the stress intensity σ_0, which differs from τ_0 only by a factor. This factor is selected in such a way that under uniaxial tensile stress σ the quantity σ_0 equals the stress σ. For this we must have

$$\tau_0 = \frac{\sqrt{2}}{3}\,\sigma_0\,,$$

and consequently

$$\sigma_0^2 = \tfrac{3}{2}\,\bar{S}_{II} = \tfrac{1}{2}\,[(\sigma_1 - \sigma_2)^2 + (\sigma_2 - \sigma_3)^2 + (\sigma_3 - \sigma_1)^2] \,. \tag{8.7}$$

Now let us establish a relationship between the three invariants of the stress tensor deviator and the actual tensor invariants. By definition

$$\bar{S}_{\mathrm{III}} = (\sigma_{ik} - \sigma\delta_{ik})(\sigma_{kj} - \sigma\delta_{kj})(\sigma_{ji} - \sigma\delta_{ji}) =$$

$$= \sigma_{ik}(\sigma_{kj} - \sigma\delta_{kj})(\sigma_{ji} - \sigma\delta_{ji}) - \sigma\delta_{ik}(\sigma_{kj} - \sigma\delta_{kj})(\sigma_{ji} - \sigma\delta_{ji}) .$$

Now multiply the first two brackets together and note that the factor δ_{ik} before the second product of the brackets can be omitted by substituting i for k in the bracket. We obtain

$$\bar{S}_{\mathrm{III}} = \sigma_{ik}[\sigma_{kj}\sigma_{ji} - \sigma(\sigma_{kj}\delta_{ji} + \sigma_{ji}\delta_{kj}) + \sigma^2\delta_{kj}\delta_{ji}]$$

$$- \sigma(\sigma_{ij} - \sigma\delta_{ij})(\sigma_{ji} - \sigma\delta_{ji}) = S_{\mathrm{III}} - 2\sigma S_{\mathrm{II}} + 3\sigma^3 - \sigma\bar{S}_{\mathrm{II}} .$$

Bearing in mind (8.3) we finally obtain

$$\bar{S}_{\mathrm{III}} = S_{\mathrm{III}} - S_{\mathrm{I}}S_{\mathrm{II}} - \tfrac{2}{9}S_{\mathrm{I}}^3 . \tag{8.8}$$

Instead of considering the third invariant of the stress tensor, we can consider another quantity which together with σ and σ_0 fully determines the stressed state to an accuracy within the rotation of the principal axes in space and consequently, together with σ and σ_0 it forms a complete system of invariants of the stress tensor. The analysis given below introduces this quantity in a natural way.

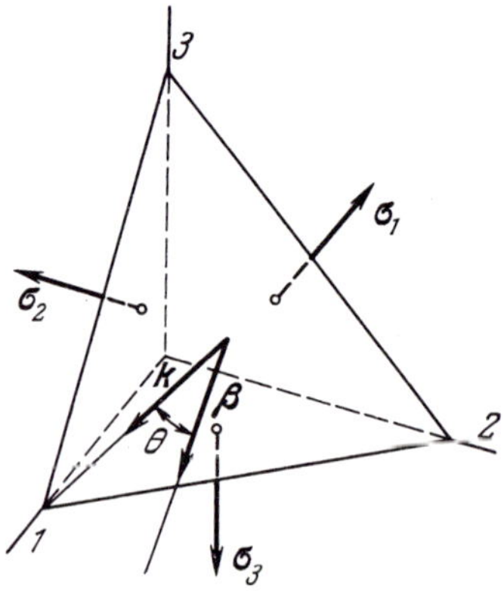

Fig. 3.

It is evident that the stress tensor deviator has the same principal axes as the tensor itself; the principal components of the stress tensor deviator are

$$\bar{\sigma}_1 = \sigma_1 - \sigma , \qquad \bar{\sigma}_2 = \sigma_2 - \sigma , \qquad \bar{\sigma}_3 = \sigma_3 - \sigma .$$

Apply to the body a hydrostatic stress of intensity $(-\sigma)$, then the principal areas are subjected to stresses $\bar{\sigma}_1$, $\bar{\sigma}_2$ and $\bar{\sigma}_3$ and the normal stress on the octahedral area disappears leaving only the octahedral tangential stress τ_0. Consider the equilibrium of a tetrahedron contained between the principal planes and the octahedral plane (fig. 3). If the area of the face in the octahedral plane is ω the area of each of the other three faces is $\omega/\sqrt{3}$. Formulating the equation of equilibrium of the tetrahedron we obtain

$$\bar{\sigma}_i = \sqrt{3}\,\tau_0 \beta_i .$$

Here $\boldsymbol{\beta}$ is a unit vector in the octahedral plane which governs the direction of the octahedral tangential stress. The direction of the vector $\boldsymbol{\beta}$ may be given in terms of the angle between this vector and some fixed direction in the octahedral plane. For example, let us draw a line from the centre of the octahedral area to the point of intersection with the principal axis 1; the direction of this line is $\mathbf{k}$. Put $\boldsymbol{\beta}\mathbf{k} = \cos\theta$. It is evident from geometrical considerations that $k_1 = 2/\sqrt{6}$, $k_2 = k_3 = -1/\sqrt{6}$. The components of the vector $\boldsymbol{\beta}$ are found from the equations

$$\beta_1^2 + \beta_2^2 + \beta_3^2 = 1 ,$$

$$\beta_1 + \beta_2 + \beta_3 = 0 ,$$

$$2\beta_1 - \beta_2 - \beta_3 = \sqrt{6} \cos\theta .$$

The first equation is the condition of normalisation which expresses the fact that the vector $\boldsymbol{\beta}$ is in the octahedral plane and, consequently, is orthogonal to the vector of the normal which forms equal angles with the three principal axes. The third equation is obtained from the condition $\beta_k = \cos\theta$.

Solving these equations relative to β_1 we find

$$\beta_1 = \sqrt{\tfrac{2}{3}} \cos\theta .$$

From considerations of symmetry it follows that

$$\beta_2 = \sqrt{\tfrac{2}{3}} \cos\left(\theta - \frac{2\pi}{3}\right),$$

$$\beta_3 = \sqrt{\tfrac{2}{3}} \cos\left(\theta - \frac{4\pi}{3}\right).$$

Now we can write the following formulae for the principal stresses

$$\begin{aligned} \sigma_1 &= \sigma + \tfrac{2}{3}\sigma_0 \cos\theta, \\ \sigma_2 &= \sigma + \tfrac{2}{3}\sigma_0 \cos\left(\theta - \frac{2\pi}{3}\right), \\ \sigma_3 &= \sigma + \tfrac{2}{3}\sigma_0 \cos\left(\theta - \frac{4\pi}{3}\right). \end{aligned} \tag{8.9}$$

The angle θ is termed the angle of similarity of the stress tensor deviators. If two tensors have the same principal axes and their angles θ are equal, their deviators differ only by a common factor. Now let us calculate the third invariant of the deviator. In the principal axes

$$\bar{S}_{\mathrm{III}} = (\sigma_1 - \sigma)^3 + (\sigma_2 - \sigma)^3 + (\sigma_3 - \sigma)^3 .$$

By means of formulae (8.9) we obtain

$$\bar{S}_{\mathrm{III}} = \tfrac{2}{9}\sigma_0^3 \cos 3\theta . \tag{8.10}$$

We note a further interpretation of the second invariant deviator, due to V. V. Novozhilov [89]. Let us calculate the root mean square value of the tangential stress on the surface of a sphere. For a sphere $\nu_i = x_i/r$ we calculate $\int \tau^2 \mathrm{d}\omega$, where τ is given by formula (8.2); we encounter two types of integrals

$$I_1 = \int \left(\frac{x^2}{r^2} - \frac{x^4}{r^4}\right) \mathrm{d}\omega \quad \text{and} \quad I_2 = 2\int \frac{x^2 y^2}{r^4}\,\mathrm{d}\omega .$$

We can readily convert I_1 to the form

$$I_1 = \int \frac{x^2y^2 + x^2z^2}{r^4}\,\mathrm{d}\omega = I_2 .$$

because for spheres $\int \frac{x^2y^2}{r^4}\,\mathrm{d}\omega = \int \frac{x^2z^2}{r^4}\,\mathrm{d}\omega$. Hence it follows that

$$\int \tau^2 \mathrm{d}\omega = \frac{I_2}{2}\{(\sigma_1-\sigma_2)^2+(\sigma_2-\sigma_3)^2+(\sigma_3-\sigma_1)^2\}\,.$$

Therefore, the octahedral tangential stress differs from the root mean square tangential stress only by a constant factor.

9. Rate-of-deformation tensor. Invariants of deformation and rate of deformation tensors

In formulating the fundamental physical relationships which characterise particular real bodies we must consider the rate-of-deformation tensor as well as the deformation tensor. Let the radius vector of a material point be $\mathbf{r}(x^i, t)$, the initial undeformed state corresponds to the instant $t = 0$, $\mathbf{r}(x^i, 0) = \mathbf{r}_0$. As in section 5 the curvilinear coordinates of a material point M remain unaltered. The components of the metric tensor g_{ij} are now functions of the coordinates and time and we shall denote by g_{ij0} the metric tensor in the initial undeformed state. In the initial state the element $\mathrm{d}s_0$ is of length $\mathrm{d}s$ at time t, and length $\mathrm{d}s + \delta\mathrm{d}s$ at time $t + \delta t$; $\delta\mathrm{d}s/\delta t$ is the rate of elongation of the segment which, at the given instant, has a length $\mathrm{d}s$, the rate of deformation of this segment is

$$\epsilon = \frac{\delta\mathrm{d}s}{\delta t\,\mathrm{d}s}\,.$$

But $\mathrm{d}s = \mathrm{d}s_0\,(1+e)$, $\delta\mathrm{d}s/\mathrm{d}t = \dot{e}\mathrm{d}s_0$, therefore

$$\epsilon = \frac{\dot{e}}{1+e}\,. \tag{9.1}$$

Differentiating formula (5.2) we obtain

$$\dot{e}(1+e) = \frac{\dot{e}_{ij}\mathrm{d}x^i\mathrm{d}x^j}{\mathrm{d}s^2}\,.$$

Whence

$$\epsilon = \frac{\dot{e}_{ij}\mathrm{d}x^i\mathrm{d}x^j}{\mathrm{d}s^2}\frac{1}{(1+e)^2}. \tag{9.2}$$

We shall denote the quantities $\dot{e}_{ij}$ by ϵ_{ij} and term them components of the rate-of-deformation tensor. The vector $\mathbf{r}$ can be represented in the form

$$\mathbf{r}(x^i, t) = \mathbf{r}_0(x^i) + \mathbf{u}(x^i, t)\,.$$

The rate vector $\mathbf{v} = \dot{\mathbf{u}}$. Differentiating expression (5.4) we obtain the following expression for the components of the rate-of-deformation tensor

$$\epsilon_{ij} = \tfrac{1}{2}\,[v_{i,j} + v_{j,i} + (v_{k,i}u_{s,j} + u_{k,i}v_{s,j})g^{ks}]\,. \tag{9.3}$$

In the case of small displacements

$$\epsilon_{ij} = \tfrac{1}{2}\,(v_{i,j} + v_{j,i})\,, \tag{9.4}$$

that is the components of the rate-of-deformation tensor are expressed in terms of the rate vector in just the same way as components of the deformation tensor are expressed in terms of the displacement vector. If the derivatives of the displacements are small then $e \ll 1$ and in the numerator of formula (9.2) we may ignore the quantity e as compared with unity. Then

$$\epsilon = \frac{\dot{e}_{ij}\mathrm{d}x^i\mathrm{d}x^j}{\mathrm{d}s^2} = \frac{\epsilon_{ij}\mathrm{d}x^i\mathrm{d}x^j}{\mathrm{d}s^2}, \tag{9.5}$$

and the analogy between the rate and deformation tensors becomes complete. However, it still does not follow from the assumption $e \ll 1$ that the derivatives of displacement are small and that terms of the form $v_{k,i}u_{s,j}$ in formula (9.3) may be neglected.

A system of invariants of deformation and rate-of-deformation tensors is constructed in just the same way as the system of stress tensor invariants, all the considerations of section 8 apply equally to these two tensors. We shall denote by E_{I}, E_{II} and E_{III} the invariants of the deformation tensor, and by H_{I}, H_{II} and H_{III} the invariants of the rate-of-deformation tensor. In the case of small deformations we introduce the quantities

$$e = \tfrac{1}{3}\,E_{\mathrm{I}} \quad \text{and} \quad \epsilon = \tfrac{1}{3}\,H_{\mathrm{I}}\,, \tag{9.6}$$

and here E_{I} and H_{I} represent the relative change in volume and rate of relative change in volume respectively. No confusion should arise from the circumstance that the notations e and ϵ were previously used to denote the relative elongation and rate of deformations of the element $\mathrm{d}s$ are now used to denote invariants of the corresponding tensors.

By definition of the tensor ϵ_{ij} it follows that

$$\epsilon = \dot{e} .$$

In exactly the same way as the stress tensor deviator was determined we introduce deviators of the deformation and rate-of-deformation tensors

$$\bar{e}_{ij} = e_{ij} - e\delta_{ij} , \quad \bar{\epsilon}_{ij} = \epsilon_{ij} - \epsilon\delta_{ij} .$$

The second invariants of the deviators are determined by formulae completely analogous to formula (8.4),

$$\begin{aligned} \bar{E}_{\mathrm{II}} &= \tfrac{1}{3} \left[(e_1 - e_2)^2 + (e_2 - e_3)^2 + (e_3 - e_1)^2\right] , \\ \bar{H}_{\mathrm{II}} &= \tfrac{1}{3} \left[(\epsilon_1 - \epsilon_2)^2 + (\epsilon_2 - \epsilon_3)^2 + (\epsilon_3 - \epsilon_1)^2\right] . \end{aligned} \tag{9.7}$$

However, in place of the quantity τ_0 which is determined by formula (8.6) we introduce an octahedral shear γ_0 and octahedral rate-of-shear η_0 by means of the expressions

$$\bar{E}_{\mathrm{II}} = \tfrac{3}{4} \gamma_0^2 , \quad \bar{H}_{\mathrm{II}} = \tfrac{3}{4} \eta_0^2 . \tag{9.8}$$

The difference between formulae (9.8) and (8.6) arises because components of the deformation tensor with unlike indices represent not shears, but halves of shears. Following up the idea according to which the quantity τ_0 was interpreted as a tangential stress on the octahedral plane we can formulate the problem of determining the maximum shear in a system of axes, one of which is normal to the octahedral plane. This maximum shear is indeed the octahedral shear γ_0, the direction of the corresponding axis which lies in the octahedral plane is termed the direction of octahedral shear. In stressed state theory all the considerations were of a completely descriptive character, but the results do not depend on this descriptive representation, they express certain properties of symmetrical second rank tensors, irrespective of their physical nature. Therefore, we shall now proceed in a purely formal way. In addition to the stress intensity we introduce the concepts of intensity of deformation and of rate-of-deformation. These are denoted e_0 and ϵ_0 and differ from γ_0 and η_0 by a numerical factor, which is selected in such a way that under uniaxial tension a rod of incompressible material e_0 acquires an elongation e, ϵ_0 is the rate of deformation of the elongation ϵ. In uniaxial tension $e_1 = e_2 = e_3 = -\frac{1}{2}e$; the condition that has been set will be fulfilled if

$$e_0^2 = \tfrac{2}{3}\bar{E}_{\mathrm{II}}, \qquad \epsilon_0^2 = \tfrac{2}{3}\bar{H}_{\mathrm{II}}. \tag{9.9}$$

Comparing (9.9) with (9.8) we find

$$e_0 = \frac{\gamma_0}{\sqrt{2}}, \qquad \epsilon_0 = \frac{\eta_0}{\sqrt{2}}.$$

The direction of the octahedral rate of shear forms an angle φ in the octahedral plane with the direction of the projection of axis number 1 on the octahedral plane. This angle plays the same part as the angle θ in the theory of the stressed state and is expressed in terms of the invariants H_{II} and H_{III}. By analogy with (8.9) we obtain the following formulae

$$\begin{aligned} \epsilon_1 &= \epsilon + \epsilon_0 \cos\varphi, \\ \epsilon_2 &= \epsilon + \epsilon_0 \cos\left(\varphi - \frac{2\pi}{3}\right), \\ \epsilon_3 &= \epsilon + \epsilon_0 \cos\left(\varphi - \frac{4\pi}{3}\right). \end{aligned} \tag{9.10}$$

Similar expressions are obtained for the principal components of the deformation tensor, e_1, e_2 and e_3,

$$\begin{aligned} e_1 &= e + e_0 \cos\psi, \\ e_2 &= e + e_0 \cos\left(\psi - \frac{2\pi}{3}\right), \\ e_3 &= e + e_0 \cos\left(\psi - \frac{4\pi}{3}\right). \end{aligned} \tag{9.11}$$

Note that e_0 and ϵ_0 are uniform first power functions of the components of the corresponding tensors, but are not linear functions. Therefore, it does not follow from the definition $\epsilon_{ij} = \dot{e}_{ij}$ that $e_0 = \epsilon_0$. This equation is fulfilled only in the case of so-called proportional strain when $\bar{e}_{ij} = \bar{a}_{ij}\tau(t)$ where $\bar{a}_{ij}$ is a constant tensor. Here $\bar{\epsilon}_{ij} = \bar{a}_{ij}\dot{\tau}(t)$ and consequently, at every instant the ratios of the components of the rate-of-deformation deviator coincide with the ratios of the components of the deformation tensor deviator.

$$\epsilon_{11} : \epsilon_{22} : \dots : \epsilon_{31} = e_{11} : e_{22} : \dots : e_{31} \, .$$

So far we have considered a case in which the displacements of points in the body are small, so that we could neglect changes in metrics during deformation. However, the formal theory of rate-of-deformation tensor is fully applicable to the case of indefinitely great deformations of the medium provided that we refrain from using a system of coordinates that is associated with the body and which is deformed together with it. Fixing a system of coordinates x_i in space we shall consider the field of the velocity vector $\mathbf{v}(x_i)$, related, not to a particular material point, but to a fixed point in space, through which the material passes. Thus, we arrive at a definition of the rate-of-deformation tensor (9.4) which is now in no way associated with deformations as such. This is the usual point of view for hydrodynamics.

10. An elastic body

Consider a body loaded by surface forces **T** and bulk forces **F**. The displacements of its points are u_i. Suppose that the external forces acquire increments δT_i and δF_i so that displacements of the points alter by amounts δu_i. If δT_i and δF_i are infinitesimally small, the work performed by external forces on additional deformation is expressed as

$$\delta A = \int_V F_i \delta u_i \mathrm{d}V + \int_\Sigma T_i \delta u_i \mathrm{d}\Sigma \, .$$

But the superficial forces are expressed in terms of values of the components of the stress tensor at points on the surface

$$T_i = \sigma_{ij} \nu_j \, .$$

Introducing T_i into the expression for δA and converting the surface integral into a bulk integral by Green's formula we obtain

$$\delta A = \int_V [F_i \delta u_i + (\sigma_{ij} \delta u_i)_{,j}] \, dV$$

$$= \int_V [(F_j + \sigma_{ij,j}) \delta u_i + \sigma_{ij} \delta u_{i,j}] \, dV .$$

Because of the equilibrium equations the term in round brackets ( ) becomes zero and we obtain

$$\delta A = \int_V \sigma_{ij} \delta u_{i,j} \mathrm{d}V . \tag{10.1}$$

The quantities

$$\delta e_{ij} = \tfrac{1}{2} (\delta u_{i,j} + \delta u_{j,i})$$

represent the increments of the deformation components caused by change in the stressed state on altering the external forces. Because of symmetry of the tensor σ_{ij} we can rewrite formula (10.1) as follows:

$$\delta A = \int_V \sigma_{ij} \delta e_{ij} \mathrm{d}V .$$

Now put

$$\sigma_{ij} \delta e_{ij} = \delta U .$$

We shall now consider two states of the body: an initial state in which it is subject to forces T_{i0} and F_{i0}, the stresses being σ_{ij0} and the deformations e_{ij0}, and a final state corresponding to the forces T_{i1} and F_{i1}, stresses σ_{ij1} and deformation e_{ij1}. Transition from the initial to the final state can follow various paths depending on the way in which we alter the external forces. The deformation of each element can be conveniently represented in the following way. Consider a nine-dimensional space referred to an orthogonal Cartesian system of coordinates. We lay out on the coordinate axes the values of the deformation components e_{ij} so that the deformation tensor e_{ij} is in accordance with a nine-dimensional space vector. This representation of a tensor by means of a vector is of limited value, it is not invariant because the same deformation tensor related to other axes will correspond to another vector which is, however, of the same length because the square of the vector length is the second invariant of the tensor. However, the length of the vector is its only invariant, whereas the tensor has three invariants, and in this sense the representation of a tensor in terms of a multidimensional vector restricts the possibilities. Considering the vector e_{ij} as the radius vector of a point in nine-dimensional space we represent the deformed state of the element by a point N in nine-dimensional space; the initial state corresponds to the point N_0, the final state to the point N_1, the process of deformation of the element is represented by movement from N_0 to N_1 along a certain curve. The work performed by external forces applied to unit volume is then expressed as

$$U = \int_{N_0}^{N_1} \sigma_{ij} \mathrm{d}e_{ij}, \tag{10.2}$$

and the total work as

$$\tilde{U} = \int_V U \mathrm{d}V. \tag{10.3}$$

We shall say that a body is elastic if the quantity U is independent of the deformation path. To ensure this the subintegral expression in formula (10.2) should be a total differential, that is we should have

$$\sigma_{ij} = \frac{\partial U}{\partial e_{ij}}. \tag{10.4}$$

The function $U(e_{ij})$ is termed the elastic potential.

Simultaneously with spatial deformation we may consider a nine-dimensional space of stresses, the coordinate points of which are equal to the components of the stress tensor. The deformation path N_0N_1 corresponds to the loading path M_0M_1, the point N of the deformation path with coordinates e_{ij} corresponds to the point M of the loading path with coordinates σ_{ij}. By analogy with (10.2) consider the curvilinear integral

$$\Phi = \int_{M_0}^{M_1} e_{ij} \mathrm{d}\sigma_{ij}. \tag{10.5}$$

It is found that if there is an elastic potential $U(e_{ij})$, then there is also a function Φ, which is independent of the loading path. Indeed

$$e_{ij}\mathrm{d}\sigma_{ij} = \mathrm{d}(e_{ij}\sigma_{ij}) - \sigma_{ij}\mathrm{d}e_{ij} \,.$$

Consequently

$$\Phi = e_{ij}\sigma_{ij} - \int \sigma_{ij}\mathrm{d}e_{ij} \,.$$

As there is a unique point correspondence between the loading and deformation paths the integral is taken along the path N_0N and is consequently equal to U. Therefore

$$\Phi = e_{ij}\sigma_{ij} - U \,. \tag{10.6}$$

Hence it follows that

$$e_{ij} = \frac{\partial \Phi}{\partial \sigma_{ij}} \,. \tag{10.7}$$

Of course, in using formula (10.7) Φ must be expressed as a function of σ_{ij}, which is always possible because from (10.4) we can find e_{ij} as function of σ_{ij} and substitute these expressions into $U(e_{ij})$ and $\sigma_{ij}e_{ij}$. The function $\Phi(\sigma_{ij})$ is termed the additional work. This nomenclature can be explained by reference to the example of a non-linear body subject to uniaxial stress. If the curve of the dependence of stress on deformation is as shown in fig. 4, U is represented by the hatched area and Φ by the area above the line that is required to complete the rectangle formed by the abscissus and ordinate of the extreme point.

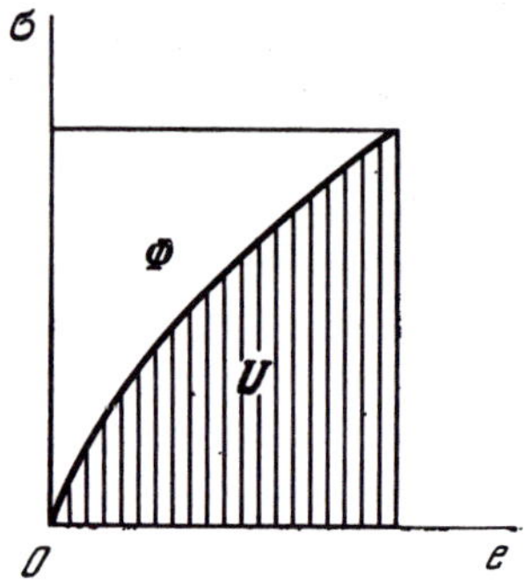

Fig. 4.

The simplest assumption concerning the form of the function U is that it is a quadratic function of the deformation components.

Put

$$U = \frac{1}{2} E_{ijkl} e_{ij} e_{kl} . \tag{10.8}$$

The quantities E_{ijkl} form a fourth rank tensor, symmetrical relative to the indices ij and kl, i.e. it is such that $E_{ijkl} = E_{jikl}, E_{ijkl} = E_{ijlk}$. We shall term this tensor the modulus of elasticity. From formula (10.4)

$$\sigma_{ij} = E_{ijkl} e_{kl} . \tag{10.9}$$

Here it is borne in mind that on differentiating (10.8) identical terms appear twice. From formula (10.4) it follows that

$$\frac{\partial \sigma_{ij}}{\partial e_{rs}} = \frac{\partial \sigma_{rs}}{\partial e_{ij}} .$$

Applying this relationship to (10.9) we obtain a still further condition of symmetry for the tensor E_{ijkl}, namely

$$E_{ijkl} = E_{klij} .$$

It can readily be shown that the number of different moduli of elasticity is 21. Indeed six components of the stress are expressed in a linear manner in terms of six components of deformation and the total number of coefficients is 36. But because of symmetry of the matrix only 21 of them are different.

In particular cases the presence of so-called elastic symmetry can be used to reduce the number of moduli of elasticity but this can be done only for vectors with one particular selection of coordinate axes.

Solving (10.9) relative to e_{ij} we obtain another form of the law of linear elasticity

$$e_{ij} = c_{ijkl} \sigma_{kl} .$$

The quantity U can now be represented as follows:

$$U = \frac{1}{2} \sigma_{ij} e_{ij} = \frac{1}{2} c_{ijkl} \sigma_{ij} \sigma_{kl} .$$

For a linearly elastic body the additional work Φ is equal to the elastic potential U

$$\Phi = U .$$

11. An isotropic elastic body

An isotropic body is one in which the properties are independent of the orientation of the element under consideration. The equations relating stresses to deformations in an isotropic body should be of the same form whatever the selection of coordinate axes and consequently, the elastic potential should depend on the invariants of the deformation tensor. Let us put

$$U = U[E_{\mathrm{I}}, E_{\mathrm{II}}, E_{\mathrm{III}}] .$$

Here

$$E_{\mathrm{I}} = e_{ij}\delta_{ij} ,$$

$$E_{\mathrm{II}} = e_{ij}e_{ij} ,$$

$$E_{\mathrm{III}} = e_{ij}e_{jk}e_{ki} .$$

Hence, from formulae (10.4) it follows that everywhere

$$\sigma_{ij} = \frac{\partial U}{\partial E_{\mathrm{I}}} \delta_{ij} + 2\frac{\partial U}{\partial E_{\mathrm{II}}} e_{ij} + 3\frac{\partial U}{\partial E_{\mathrm{III}}} e_{ik}e_{kj} . \tag{11.1}$$

If the body is linearly-elastic the function U should be a quadratic function of the strain components in the isotropic body and is necessarily of the following form:

$$U = \tfrac{1}{2}\lambda E_{\mathrm{I}}^2 + \mu E_{\mathrm{II}} .$$

Expression (11.1) assumes the following form:

$$\sigma_{ij} = \lambda E_{\mathrm{I}}\delta_{ij} + 2\mu e_{ij} . \tag{11.2}$$

The constants λ and μ have the dimensions of stress and are termed Lamé constants. Forming a first invariant from both parts of (11.2) we obtain

$$\sigma = \tfrac{1}{3} S_{\mathrm{I}} = (\lambda + \tfrac{2}{3}\mu) E_{\mathrm{I}}$$

or $\sigma = 3Ke$.

The quantity

$$K = \lambda + \tfrac{2}{3}\mu$$

is termed the bulk modulus of elasticity.

Eq. (11.2) solved relative to e_{ij} is often written in the form

$$e_{ij} = \frac{1+\nu}{E}\left[\sigma_{ij} - \delta_{ij}\frac{\nu}{1+\nu}\sigma\right]. \tag{11.3}$$

The modulus of elasticity E and Poisson's ratio ν are related to the Lamé constants as follows

$$E = \frac{\mu(3\lambda + 2\mu)}{\lambda + \mu}, \qquad \nu = \frac{\lambda}{2(\lambda + \mu)},$$

or

$$\lambda = \frac{E\nu}{(1+\nu)(1-2\nu)}, \qquad \mu = \frac{E}{2(1+\nu)}.$$

Note also that

$$K = \frac{E}{3(1-2\nu)}.$$

Returning to the case of non-linear elastic material note that in formula (11.1) the non-linearity figures in two ways, on the one hand $\partial U/\partial E_{\mathrm{I}}$, $\partial U/\partial E_{\mathrm{II}}$, $\partial U/\partial E_{\mathrm{III}}$ are functions of invariants and on the other hand the tensor σ_{ij} is a linear combination of three tensors δ_{ij}, e_{ij} and $e_{ik}e_{kj}$.

In describing the behaviour of real materials it is often possible to consider U as a function of only two invariants of the deformation tensor, namely, E_{I} and E_{II}. Here the components of the stress and deformation tensors are related by linear tensor expressions

$$\sigma_{ij} = \frac{\partial U}{\partial E_{\mathrm{I}}}\delta_{ij} + 2\frac{\partial U}{\partial E_{\mathrm{II}}}e_{ij}, \tag{11.4}$$

similar to the relationships of the linear theory of elasticity but with the difference that the moduli of elasticity are no longer constant but are functions of the deformation tensor invariants.

There is reason to consider the second invariant of the deformation tensor deviator instead of the second invariant of the deformation tensor, namely

$$\bar{E}_{\mathrm{II}} = E_{\mathrm{II}} - \tfrac{1}{3} E_{\mathrm{I}}^2 .$$

Instead of (11.4) we then obtain the following expressions

$$\sigma_{ij} = \left(\frac{\partial U}{\partial E_{\mathrm{I}}} - \frac{\partial U}{\partial \bar{E}_{\mathrm{II}}} \tfrac{2}{3} E_{\mathrm{I}} \right) \delta_{ij} + 2 \frac{\partial U}{\partial \bar{E}_{\mathrm{II}}} e_{ij} .$$

Hence it follows that

$$\sigma_{ij} = \frac{\partial U}{\partial E_{\mathrm{I}}} \delta_{ij} + 2 \frac{\partial U}{\partial \bar{E}_{\mathrm{II}}} \bar{e}_{ij} . \tag{11.5}$$

Let us form the first invariant from both parts of these expressions, we obtain

$$\sigma = \frac{\partial U}{\partial E_{\mathrm{I}}} . \tag{11.6}$$

Eq. (11.6) expresses the law of elasticity for bulk deformation. The relationship between σ_{ij} and e_{ij} can now be written

$$\bar{\sigma}_{ij} = 2 \frac{\partial U}{\partial \bar{E}_{\mathrm{II}}} \bar{e}_{ij} ; \tag{11.7}$$

the expressions that are obtained may be termed the law of elasticity for shear deformation. Squaring eq. (11.7) and contracting the left- and right-hand sides with respect to i and j we obtain

$$\bar{S}_{\mathrm{II}} = 4 \left(\frac{\partial U}{\partial \bar{E}_{\mathrm{II}}} \right)^2 \bar{E}_{\mathrm{II}} ,$$

or

$$\tau_0 = \frac{\partial U}{\partial \bar{E}_{\mathrm{II}}} \gamma_0 . \tag{11.8}$$

Eq. (11.7) can be rewritten

$$\bar{\sigma}_{ij} = \frac{2\tau_0}{\gamma_0}\bar{e}_{ij} = \frac{2}{3}\frac{\sigma_0}{e_0}\bar{e}_{ij}\,. \tag{11.9}$$

Here in view of (11.8), $\sigma_0 = f(e, e_0)$. Using (11.6) we can express e_0 in terms of σ_0 and σ, so that $e_0 = h(\sigma, \sigma_0)$, then from (11.9) the deformation deviator is expressed in terms of the stress deviator by formulae

$$\bar{e}_{ij} = \frac{3}{2}\frac{e_0}{\sigma_0}\bar{\sigma}_{ij}\,.$$

The next step in simplifying the non-linear law of elasticity is that the bulk deformation is assumed independent of the shear and consequently in eq. (11.6) $\partial U/\partial E_{\text{I}}$ is independent of $\bar{E}_{\text{II}}$. Then the elastic potential may be represented as the sum of two functions, of E_{I} and of E_{II}:

$$U = U_1(E_{\text{I}}) + U_2(\bar{E}_{\text{II}})\,.$$

In eq. (11.9) $\sigma_0 = f(e_0)$; therefore, the elastic properties of the material are fully determined by giving two functions: the dependence of the stress intensity on the shear intensity and the dependence of the hydrostatic stress (the octahedral normal stress) on the relative change in volume.

To determine the functions $\sigma_0 = f(e_0)$ tests must be made under conditions of pure deviator stress, for example, pure shear.

The case of incompressible material requires special consideration. For this $E_{\text{I}} = 0$, consequently, U can be a function of only the invariants $E_{\text{II}} = \bar{E}_{\text{II}}$ and $E_{\text{III}} = \bar{E}_{\text{III}}$. If onto this stressed state we superpose a hydrostatic stress the amount of work remains unaltered and, therefore, the stress tensor is determined to the accuracy of the hydrostatic component. Formally, in (11.1) we must take $\partial U/\partial E_{\text{I}} = \sigma$ and we obtain

$$\bar{\sigma}_{ij} = 2\frac{\partial U}{\partial \bar{E}_{\text{II}}}e_{ij} + 3\frac{\partial U}{\partial \bar{E}_{\text{III}}}e_{ik}e_{kj}\,. \tag{11.10}$$

For it to be possible to use formula (11.9) we must know either σ_0 as function of e_0 or τ_0 as function of γ_0, it does not matter which. If the material is incompressible, then in pure tension the relative elongation $e = e_0$ and the tensile stress σ is σ_0 and the dependence of σ_0 on e_0 is given directly by the tension curve. If the material is compressible the relationship between σ_0

and e_0 does not coincide with that between σ and e in pure tension and the tension curve must be recalculated to determine the required relationships. In pure shear there is no change in volume and the τ–γ diagram differs from the σ_0–e_0 diagram only in scale. In pure shear the τ_0–γ_0 diagram coincides exactly with the τ–γ diagram. Indeed if for example σ_{12} is equal to τ, then $e_{12} = \frac{1}{2}\gamma$ and it follows from (11.9) that

$$\frac{\tau}{\gamma} = \frac{\tau_0}{\gamma_0} .$$

From this standpoint use of the quantities τ_0 and γ_0 as fundamental characteristics has certain advantages.

In the linear theory of elasticity

$$\frac{\tau_0}{\gamma_0} = \mu = \text{const.}$$

The notation G is often used instead of μ for the shear modulus. In the linear case expression (11.9) may be written as

$$\bar{\sigma}_{ij} = 2G\bar{e}_{ij} . \tag{11.11}$$

Note that in the linear theory of elasticity the angles of similarity of the stress and deformation tensor deviators are equal,

$$\theta = \psi .$$

This follows from comparison of formulae (8.9), (9.11) and (11.11).

2

Theory of Plasticity

Theory of plasticity is the name given to the mechanical theory, the object of which is to describe the behaviour under load of materials which undergo irreversible deformation with the condition that the relationship between deformation and strain is independent of time. This is true, to a first approximation, for metals in general engineering use at normal temperatures. Therefore, the theory of plasticity is concerned with describing the behaviour of steel, brass, aluminium alloys and certain other metals and alloys at normal and subnormal temperatures. At higher temperatures plastic deformation is often accompanied by creep. Strictly speaking, in studying creep it is often convenient to separate plastic deformation from creep deformation.

Creep theory developed after plasticity theory and borrowed many fundamental principles from it. Therefore, before presenting material that relates directly to creep it seems advisable to give a brief review of the present state of the theory of plasticity applied to work-hardening materials. The object of this synopsis is to present the ideas and methods which underlie the formulation of the equations of the theory of plasticity and not the application of these equations in simplified form to the solution of particular problems. In particular, we shall not consider the question of ideal plasticity at all.

The elementary physical facts about plasticity and also the fundamental information about plasticity in a state of complex stress are assumed to be known to the reader.

12. The deformation theory of plasticity

The deformation theory of plasticity due to Nadai [300] and Hencky [227] is an extension to a plastic body of the stress/deformation relationship which is established in the non-linear theory of elasticity. The fundamental assumptions that are usually made in deriving the equations are:

1. that volumetric deformation observes the law of linear elasticity; and that the plastic potential U is independent of a third invariant. These assumptions are not strictly necessary but they underlie a common variant of deformation theory termed the Il'yushin [43] theory of small elastic-plastic deformations. The law of linear elasticity for volumetric deformation is

$$\sigma = 3Ke \,. \tag{12.1}$$

The relationship between the stress and deformation deviators assumes the form of (11.9), namely

$$\bar{\sigma}_{ij} = 2\frac{\tau_0}{\gamma_0}\bar{e}_{ij} = 2G_s\bar{e}_{ij} \,. \tag{12.2}$$

Here

$$G_s = G_s(\gamma_0) \quad \text{or} \quad G_s = G_s(\tau_0) \,. \tag{12.3}$$

The quantity G_s may be termed the plastic shear modulus; for an elastic body formula (12.2) remains valid and here $G = \mu$.

For further development of the theory it is necessary to determine the limits of applicability of expressions (12.1) - (12.3) and to establish the conditions under which plastic deformation takes place and what occurs during unloading. The simplest way is to study an example. With the limitations which have been assumed in giving the complete characteristics of plasticity it is necessary to know the bulk modulus K and function (12.3). It is simplest to determine the form of this function from tests in pure shear. Suppose, for example, that $\sigma_{12} = \tau \neq 0$, all the remaining components of the stress tensor being zero. The first invariant σ is zero, the stress tensor coincides with its deviator. From formula (12.1) $e = 0$, consequently, the deformation tensor coincides with its deviator and it follows from eq. (12.2) that the only component of the deformation tensor that differs from zero is $e_{12} = \frac{1}{2}\gamma$. Calculating the octahedral stress and the octahedral shear we find

$$\tau_0 = \sqrt{\tfrac{2}{3}}\,\tau \,, \qquad \gamma_0 = \sqrt{\tfrac{2}{3}}\,\gamma \,.$$

By formula (12.3) the relationship between stress and shear is

$$\tau = \sqrt{\tfrac{3}{2}}\, f\left(\sqrt{\tfrac{2}{3}}\,\gamma\right).$$

Conversely, given the experimental curve of the relationship between τ and γ, by simple change of scale we can convert it to a curve of the relationship between τ_0 and γ_0. A pure shear diagram may be obtained, for example, by torsion tests on thin-walled tubes.

The usual method of studying the mechanical properties of material is to make tension tests on a prismatic specimen. In order to determine the function $f(\gamma_0)$ from the tension curve it is necessary to know the bulk modulus K and to recalculate on the basis of plasticity equations. For an incompressible material in tension we take

$$\sigma_{11} = \sigma_1\,, \qquad \sigma_{22} = \sigma_{33} = \sigma_{12} = \sigma_{23} = \sigma_{31} = 0\,, \qquad \sigma = \tfrac{1}{3}\sigma_1\,,$$

$$e_{11} = e_1\,, \qquad e_{22} = e_{33} = -\tfrac{1}{2} e_1\,, \qquad e_{12} = e_{23} = e_{31} = 0\,.$$

Consequently

$$\tau_0 = \frac{\sqrt{2}}{3}\sigma_1\,, \qquad \gamma_0 = \frac{1}{\sqrt{2}} e_1\,.$$

In this case the relationship between τ_0 and γ_0 is given by the tension curve with appropriate scale conversion. As we have seen, formula (12.2) can also be represented in the form

$$\bar{\sigma}_{ij} = \frac{2}{3}\frac{\sigma_0}{e_0}\bar{e}_{ij}\,. \tag{12.4}$$

The quantities σ_0 and e_0 were defined in such a way that in pure tension $\sigma_0 = \sigma_1$, $e_0 = e_1$ if the material is incompressible. Therefore, eq. (12.4) is convenient and we take it that in it σ_0 are functions of e_0 and then the form of these functions gives the tension curve directly. If compressibility of the material must be allowed for the tension curve must be reconstructed to obtain the dependence of σ_0 on e_0. Methods of performing this reconstruction are available in the literature of the subject.

The fundamental experimental fact that is observed with uniaxial loading is as follows. So long as we move along the deformation curve from the origin of coordinates in the direction shown by the arrow (fig. 5), i.e. so long

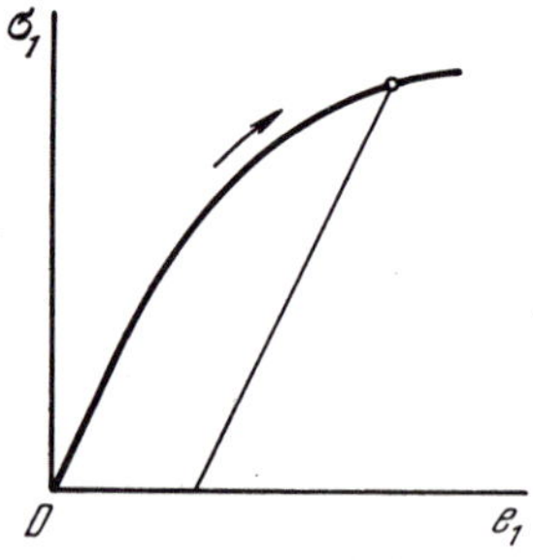

Fig. 5.

as σ_1 and e_1 increase, the relationship between σ_1 and e_1 is given by the plastic deformation diagram. The relationship between stress and deformation is single-valued in the sense that deformation may be either fast or slow, only moving in the reverse direction, or the unloading of the specimen is forbidden. When the stress is relieved the material returns to the elastic state; on unloading the relationship between stress and deformation is a straight line which passes through the point on the deformation curve at which unloading commences. The law of elasticity during unloading is not an accurate physical law, and experimental deviations from it are observed. But any theory of plasticity is concerned with a hypothetical ideal medium and not with real bodies, whose behaviour is always described by some approximation. Following Il'yushin [43] let us now suppose that in the case of a complex stress state the relationship between octahedral stresses and deformations is the same as in the uniaxial case, plastic deformation occurs when σ_0 and consequently e_0 increase; when the stress diminishes, unloading occurs and the elasticity law applies. Therefore, if eqs. (12.1) - (12.3) are valid for plastic deformation when $d\sigma_0 > 0$, on unloading when $d\sigma_0 < 0$ the stress deformation relationship is given by the expressions

$$\bar{\sigma}_{ij} - \bar{\sigma}'_{ij} = 2G(\bar{e}_{ij} - \bar{e}'_{ij}) ,$$

$$\sigma - \sigma' = 3K(e - e') , \quad \text{or} \quad \sigma = 3Ke .$$

Here the primes denote stresses and deformations at the instant of starting to unload.

The deformation theory of plasticity, formulated in this way, cannot be considered as a pure physical theory which is applicable to all possible cases of unloading. One of several objections to it is as follows. Suppose, that by continuous plastic deformation, i.e. by a process of deformation such that at any instant $\mathrm{d}e_0 > 0$, we have achieved a condition that is characterised by the quantities σ_{ij} and e_{ij}. Now perform an infinitely small neutral deformation, namely, a deformation for which $e_0 = \text{const}$ and $\mathrm{d}e_0 = 0$. The strain increments are $\mathrm{d}e_{ij}$ and they are related by the linear expression

$$\bar{e}_{ij}\mathrm{d}\bar{e}_{ij} = 0 .$$

The increment in the stress deviator is obtained by differentiating formula (12.2)

$$\mathrm{d}\bar{\sigma}_{ij} = 2G_s(e_0)\mathrm{d}\bar{e}_{ij} .$$

Since the case $\mathrm{d}e_0 > 0$ is that of plastic deformation and $\mathrm{d}e_0 < 0$ that of unloading, neutral strain when $\mathrm{d}e_0 = 0$ may also be considered as a limiting case of unloading, and for unloading

$$\mathrm{d}\bar{\sigma}_{ij} = 2G\,\mathrm{d}\bar{e}_{ij} .$$

Here G is the modulus of elasticity, whereas in the previous formula G_s is the plastic shear modulus and dependent on e_0. Obviously in reality the transition from plastic deformation to unloading must be continuous and so the result that we have obtained is unacceptable.

In what follows it will be convenient to use geometrical terminology. We shall represent the deformation tensor by a vector $\mathbf{e}$ in nine-dimensional space. The components of the vector relative to some orthogonal bases in this space are equal to the components of the deformation tensor. Any loading process may be represented as a motion of the end of the deformation vector along a certain curve termed the deformation path. Note that because of symmetry of the deformation tensor we can consider six-dimensional space instead of nine-dimensional. Moreover, by assuming that the laws of plasticity are expressed by relationships between deviators we may consider that the space of stress deviators can be determined. However, we shall not proceed further with this point.

We can construct a surface $e_0 = \text{const}$ through each point M in the deformation space which corresponds to a particular value of e_0. We shall term this the surface of deformation. If loading is continued from the point M three cases are possible:

a) the increment of the vector $\mathrm{d}\mathbf{e}$ forms an acute angle with the direction of the normal $\mathbf{n}$ to the surface of deformation at the point M, consequently, $\mathrm{d}e_0 > 0$ and additional plastic strain occurs;

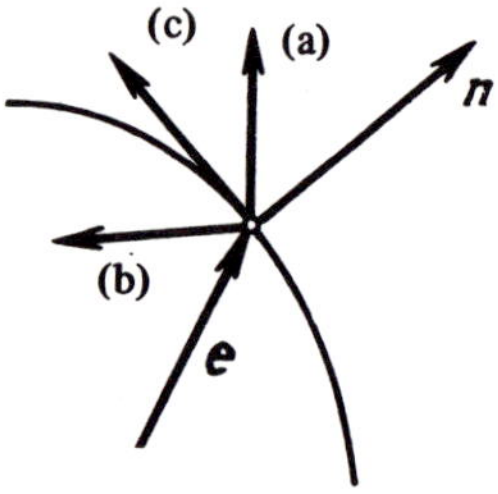

Fig. 6.

b) the angle between **de** and **n** is greater than $\pi/2$ and unloading occurs;

c) the vector **de** lies in the tangential plane to the surface of deformation and deformation is neutral.

All three cases are shown schematically in fig. 6. Let us at the same time consider nine-dimensional stress space: a vector $\boldsymbol{\sigma}$ in this space is termed a stress vector and its components are components of the stress tensor. Suppose that deformation in the stress space corresponds to the loading path and the surface of deformation corresponds to the loading surface. From the standpoint of the strain theory which we have formulated it does not matter whether we consider the space of deformation or of stress, neutral deformation corresponds to neutral loading.

Numerous experimental results are available for the case of neutral loading. The simplest experimental scheme is as follows. A thin-walled tubular specimen is extended in the plastic region by a stress σ, with constant tensile stress a small torque is applied to it setting up a tangential stress $d\tau$. In combined tension and torsion the condition of constancy of octahedral stress is written

$$\sigma^2 + 3\tau^2 = \text{const} .$$

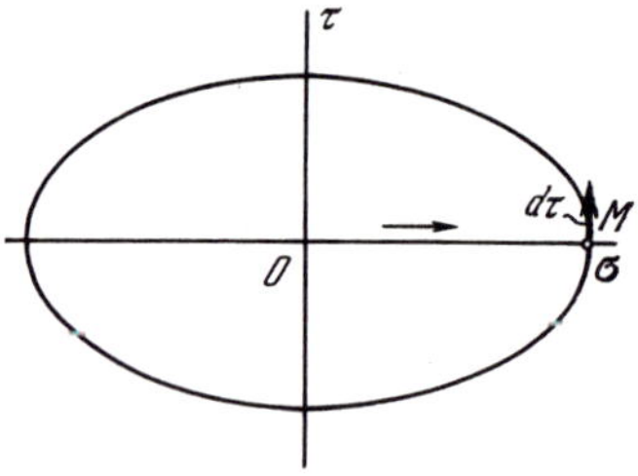

Fig. 7.

Since the stressed state is characterised by two quantities σ and τ the loading space is a plane and the loading surfaces are similar ellipses (fig. 7). In this case the loading path in tension is represented by the segment OM on the abscissus; applying a tangential stress $\mathrm{d}\tau$ at constant σ we move through an infinitesimally small segment of the arc of the ellipse, i.e. we perform neutral loading. The angle of twisting and the tangential stress are then related by the expression

$$\mathrm{d}\tau = G^* \mathrm{d}\gamma \, .$$

According to deformation theory the quantity G^* should be equal to $G_s(\gamma_0)$ where γ_0 is the octahedral shear for point M. If we consider that neutral loading is accompanied only by elastic strain, then $G^* = G$. Tests carried out with low plastic tensile strain have shown that the second assumption is justified and it may be considered, to a sufficient degree of accuracy, that $G^* = G$. However, tests made with considerable initial deformation indicate that $G^* < G$, although it is still considerably greater than is predicted by deformation theory. Fig. 8 shows an experimental curve of the relationship between $\Delta\gamma$ and $\Delta\tau$ and also straight lines corresponding to the cases $G^* = G_s$ and $G^* = G$ (thin solid lines $G^* = G_s$, dashed lines $G^* = G$).

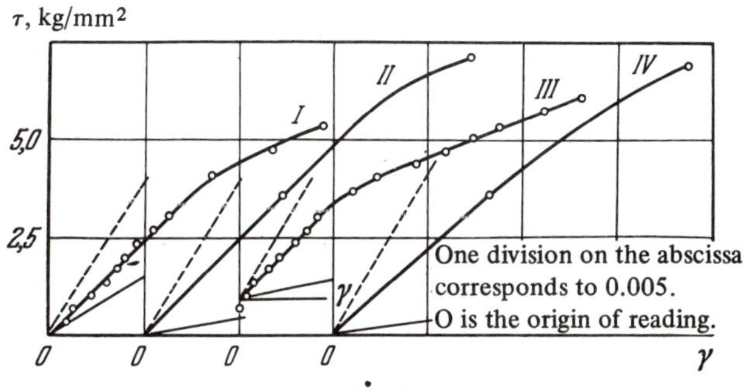

Fig. 8.

13. Deformation theory with proportional loading

In analysing experimental results for plastic deformation under conditions of complex stressing, Il'yushin [43] noticed that the deformation theory equations give sufficiently good results if during loading all the components of the stress deviator vary in proportion to one and the same parameter α

which is termed the loading parameter. This means that we may take

$$\bar{\sigma}_{ij} = \bar{\sigma}_{ij}^{0} \alpha \, .$$

Here α is a parameter which varies steadily from zero to unity, σ_{ij}^{0} is the stress in the final state. Hence it obviously follows that $\tau_0 = \alpha \tau_0^0$. Here the deformation deviator also alters in proportion to a single parameter β. Indeed according to (12.2)

$$\bar{e}_{ij} = \frac{1}{2G_s(\alpha \tau_0^0)} \alpha \bar{\sigma}_{ij}^{0} \, .$$

We put

$$\bar{e}_{ij}^{0} = \frac{1}{2G_s(\tau_0^0)} \bar{\sigma}_{ij}^{0} \, .$$

Then

$$\bar{e}_{ij} = \beta e_{ij}^{0} \, ,$$

where

$$\beta = \alpha \, \frac{G_s(\tau_0)}{G_s(\alpha \tau_0)} \, .$$

Note that during proportional change of the stress tensor component, the components of the deformation tensor (and not of the deformation deviator) do not alter proportionately.

Proportional loading is comparatively easy to achieve experimentally, and, therefore, the laws of the deformation theory of plasticity can be checked under these conditions. For carbon and low alloy steels in the tempered and normalised states, curves of the dependence of τ_0 on γ_0 coincide for different kinds of stressed state, but some systematic deviations are observed in other materials. For example, the tension and compression curves of aluminium and magnesium alloys do not coincide. We must, therefore, consider that octahedral shear depends not only on octahedral stress but also on the first or third invariant of the stress tensor (see Yagn [158] and Prager [333]).

It is not yet clear to what extent relationships of the deformation theory of plasticity are valid when the loading is not proportional. Experimental results [31] indicate that the theory provides a good description of the process of proportional loading and that it also gives satisfactory results for loading which is nearly proportional. The very term "loading which is nearly proportional" is very imprecise and one way of making the concept more definite will be considered in the next section, but for the present we shall consider the extent to which conditions of proportional loading can be maintained during deformation of a real body by external forces. Putting $\mathbf{T} = \mathbf{T}^0\alpha$, $\mathbf{F} = \mathbf{F}^0\alpha$ we satisfy the equations of equilibrium and the boundary conditions, if we put $\sigma_{ij} = \sigma_{ij}^0\alpha$, where α is a constant for all points in the body. The bulk deformation is related to the hydrostatic pressure by Hooke's law, consequently

$$e = \frac{1}{3K}\sigma^0\alpha = e^0\alpha .$$

On the other hand, $\bar{e}_{ij} = \beta\bar{e}_{ij}^0$ and $\beta = \alpha[G_s(\tau_0^0)/G_s(\alpha\tau_0^0)]$, therefore

$$e_{ij} = \delta_{ij}e^0\alpha + e_{ij}^0\beta = e_{ij}^0\beta + e^0(\beta - \alpha)\delta_{ij} .$$

Introducing these expressions into the equations for combined deformation we find that, generally speaking, if e_{ij}^0 satisfy the equations of compatibility then e_{ij} do not satisfy them. The self-evident conditions for which the equations of compatibility are satisfied during deformation are as follows. Firstly, β is independent of the coordinates and secondly either $\alpha = \beta$, or $e^0 = 0$. The first condition is fulfilled when $G_s(\tau_0)$ is a power function of τ_0. If $\gamma_0 = a\tau_0^k$, then $G_s = \tau_0^{1-k}/a$, consequently, $\beta = \alpha^k$. When $k = 1$ the second condition is fulfilled for $e^0 \neq 0$ but this is a trivial case of a linearly elastic body which is of no interest to us. If $k \neq 1$ necessarily $e^0 = 0$, i.e. the material is incompressible. Therefore, sufficient conditions for the achievement of proportional loading during proportional change in the external applied forces are:

1. there should be a power relationship between octahedral stress and octahedral shear;
2. the material should be incompressible (Il'yushin's [43] theorem).

The first of these conditions is extremely restrictive as it requires a power relationship with the same exponent throughout the range of change of τ_0 and γ_0 and does not admit a piecewise-power approximation to the true relationship. In particular the formulation of the problem of proportional loading is without meaning for an ideal plastic body, for which τ_0 remains constant.

14. Drucker's postulate of work-hardening

We have shown that the deformation theory of plasticity does not satisfy the condition of continuity of the relationships between stress and deformation on transition from loading to unloading. In analysing this and other theories of plasticity we may encounter similar circumstances in which certain consequences of the theory appear to be contradictory or physically inacceptable. We therefore require simple criteria which must, as a minimum requirement, be satisfied by any theory of plasticity. Drucker [194-199] has attempted to formulate such a criterion in the form of the following work-hardening postulate. Consider a stressed state σ_{ij}^* represented in a nine-dimensional space by a point M^* which is the end of the stress vector $\boldsymbol{\sigma}^*$. Through the point M^* there passes a closed loading surface S^*, i.e. a surface which separates the region of elastic state from that of plastic. This means that motion of the end of the vector $\boldsymbol{\sigma}^*$ outside the volume that is bounded by the surface S^* corresponds to additional plastic deformation. Let us also consider a second stressed state σ_{ij} to which corresponds a point M with radius vector $\boldsymbol{\sigma}$. On transition from point M^* to point M the additional stress is $\boldsymbol{\sigma}-\boldsymbol{\sigma}^*$. Now suppose that we have left point M^* and returned to it through a closed path, which may lie partially outside the surface S^*. The stress increment $\mathrm{d}\boldsymbol{\sigma}$ always corresponds to a certain deformation increment $\mathbf{de}$. Drucker's postulate asserts that the work of additional stresses on a closed loading path is not negative,

$$\oint (\sigma_{ij}-\sigma_{ij}^*)\mathrm{d}e_{ij} \geqslant 0 . \tag{14.1}$$

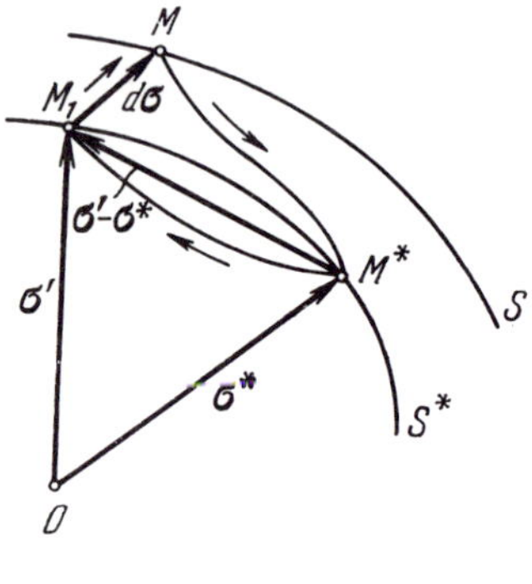

Fig. 9.

The simplest consequence of Drucker's postulate is as follows. Consider a closed path shown in fig. 9. The point M^* which corresponds to the stressed state $\boldsymbol{\sigma}^*$ lies inside the surface S^* or, in the extreme case, it lies on this surface as shown in the sketch. We may proceed from the point M^* to the point M_1, which lies on the surface S^*, by any curve that joins these two points and which lies entirely within the surface S^*. The stress $\boldsymbol{\sigma}'$ is then altered by the increment $\mathrm{d}\boldsymbol{\sigma}$ and passes outside the surface S^*, i.e., it is associated with plastic deformation so that we reach a new point M which is infinitesimally close to the loading path. Through this point there passes a new flow surface S, consequently, return from M to M^* by any path that lies within S means unloading. On the segments M^*M_1 and MM^* the deformations are elastic, on the segment M_1M they include elastic and plastic components.

$$\mathrm{d}\mathbf{e} = \mathrm{d}\mathbf{e}^{\mathrm{e}} + \mathrm{d}\mathbf{e}^{\mathrm{p}} .$$

The work of the additional stress on elastic displacements is zero for a closed loading path. Indeed

$$\oint (\boldsymbol{\sigma} - \boldsymbol{\sigma}^*)\mathrm{d}\mathbf{e}^{\mathrm{e}} = \oint \boldsymbol{\sigma}\,\mathrm{d}\mathbf{e}^{\mathrm{e}} - \boldsymbol{\sigma}^* \oint \mathrm{d}\mathbf{e}^{\mathrm{e}} .$$

In this expression the first integral is the total work, which in accordance with the definition of elasticity is zero for a closed path, the second integral is also zero because the elastic deformations are single-valued. Therefore, irreversible work is performed only during plastic deformation and it follows from (14.1) that

$$(\sigma'_{ij} - \sigma^*_{ij})\mathrm{d}e^{\mathrm{p}}_{ij} \geqslant 0 . \tag{14.2}$$

Now let $\sigma'_{ij} = \sigma^*_{ij}$, we leave the point M^* for an infinitesimally close point M, all the deformations being plastic. We shall term this path an active loading path. We then return from point M to M^* by an arbitrary path located entirely within S. Once again irreversible work is performed only on the plastic component of deformation, consequently

$$\mathrm{d}\sigma_{ij}\mathrm{d}e^{\mathrm{p}}_{ij} \geqslant 0 . \tag{14.3}$$

The condition (14.2) signifies the vector $\boldsymbol{\sigma} - \boldsymbol{\sigma}^*$, where $\boldsymbol{\sigma}^*$ is the radius vector of any point within S^*, forms an acute angle with the direction of the vector $\mathrm{d}\mathbf{e}^{\mathrm{p}}$. If the loading surface is smooth and we can construct a unique tangential

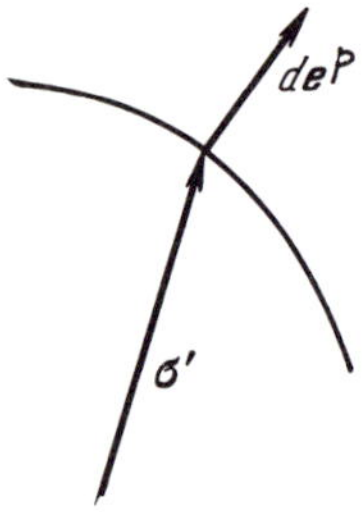

Fig. 10.

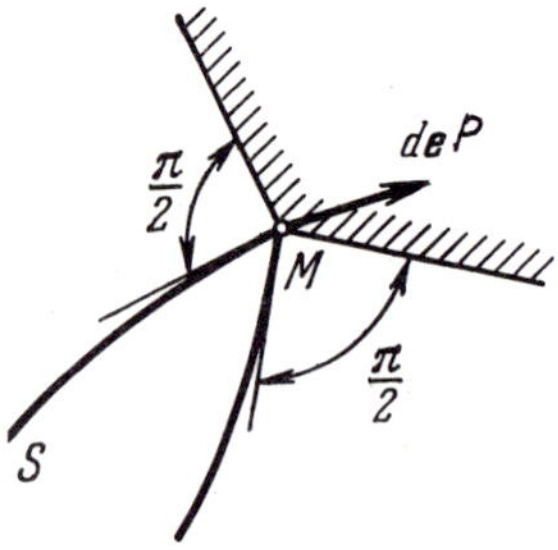

Fig. 11.

hyperplane at the point M_1, then the vector $\mathbf{de}^p$ must necessarily be in the direction of the normal to the surface S^* (fig. 10), whatever the magnitude and direction of $\mathbf{d\sigma}$. The surface lies entirely on one side of the tangential hyperplane and is consequently convex.

The assumption that the surface S is smooth at the point M is not essential and in some variants of the theory of plasticity it is taken that the point M is at a corner. This means that the tangential surfaces form a cone with its apex at the point M (fig. 11). It follows from Drucker's postulate (condition 14.2) that the vector must lie within the cone formed by the normals to the surface S in the neighbourhood of the point M. In this case no other limitations on the direction of the vector $\mathbf{de}^p$ follow from Drucker's postulate. As Drucker himself has repeatedly pointed out, his postulate in no way follows from the laws of thermodynamics and is not a thermodynamic postulate. Drucker uses the term "quasi-thermodynamic".

The inequality (14.1) signifies that the work of any additional action on the displacement that it produces is either positive or zero for a closed cycle, whereas it follows from the first law of thermodynamics that the work of the total system of forces acting on a body is not negative. Consequently, additional actions cannot be extracted from the energy system. This property characterises a certain stability of the material, as is evident from the simple example of the tension diagram. If $d\sigma de > 0$ it means that the increment of e corresponds to the increment of σ of the same sign and the material is work-hardened. But if $d\sigma de < 0$ the stress diminishes with increase in elongation, the material is unstable in the sense that a rod subjected to a constant tensile force cannot be in equilibrium. Therefore, fulfillment of Drucker's postulate may be considered as defining a class of stable material.

15. Possible boundaries of applicability of the deformation theory of plasticity

Following Budiansky [176] let us find for which loading paths the equations of the deformation theory of plasticity satisfy Drucker's postulate. For simplicity we shall consider that the material is incompressible and then the fundamental equations of the Nadai-Hencky theory are

$$e_{ij} = \frac{1}{2G_s}\,\bar{\sigma}_{ij}\,.$$

The elastic deformation components are expressed in a similar way, only we must use the elasticity modulus G in place of the plastic shear modulus G_s

$$e^{\mathrm{e}}_{ij} = \frac{1}{2G}\,\bar{\sigma}_{ij}\,.$$

Subtracting the elastic parts from the total deformation we obtain expressions for the plastic deformation

$$e^{\mathrm{p}}_{ij} = \tfrac{1}{2}\left(\frac{1}{G_s} - \frac{1}{G}\right)\bar{\sigma}_{ij}\,. \tag{15.1}$$

Now calculate the increment in the plastic deformation with given stress increments of σ_{ij}

$$\mathrm{d}e^{\mathrm{p}}_{ij} = \tfrac{1}{2}\left(\frac{1}{G_s} - \frac{1}{G}\right)\mathrm{d}\bar{\sigma}_{ij} + \bar{\sigma}_{ij}\,\tfrac{1}{2}\,\mathrm{d}\left(\frac{1}{G_s}\right).$$

The quantity G_s is a function of τ_0 and

$$\gamma_0 = \frac{\tau_0}{G_s}\,.$$

Differentiating this expression we obtain

$$\mathrm{d}\gamma_0 = \tau_0\,\mathrm{d}\left(\frac{1}{G_s}\right) + \frac{1}{G_s}\,\mathrm{d}\tau_0\,.$$

Whence

$$\mathrm{d}\left(\frac{1}{G_s}\right) = \frac{\mathrm{d}\tau_0}{\tau_0}\left(\frac{\mathrm{d}\gamma_0}{\mathrm{d}\tau_0} - \frac{1}{G_s}\right).$$

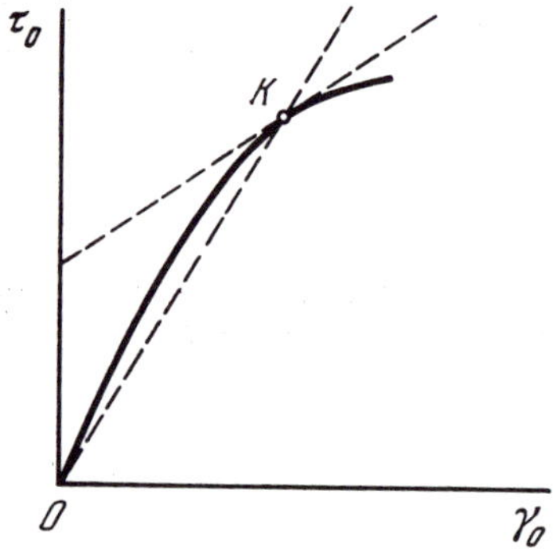

Fig. 12.

Turning to the diagram of the relationship between τ_0 and γ_0 (fig. 12) we notice that $\mathrm{d}\tau_0/\mathrm{d}\gamma_0$ is the deformation curve at the point $K(\gamma_0, \tau_0)$. This quantity is termed the tangential shear modulus

$$G_t = \frac{\mathrm{d}\tau_0}{\mathrm{d}\gamma_0}\,.$$

The quantity G_s, which is the value of the tangent of a line passing from the origin of coordinates to the point K, is often termed the secant shear modulus. Now the expression for $\mathrm{d}e_{ij}^{\mathrm{p}}$ can be rewritten

$$\mathrm{d}e_{ij}^{\mathrm{p}} = \tfrac{1}{2}\left(\frac{1}{G_s} - \frac{1}{G}\right)\mathrm{d}\bar{\sigma}_{ij} + \tfrac{1}{2}\left(\frac{1}{G_t} - \frac{1}{G_s}\right)\frac{\bar{\sigma}_{ij}}{\tau_0}\,\mathrm{d}\tau_0\,. \tag{15.2}$$

Subsequently, in place of τ_0 it will be convenient to use another quantity s, which is the square-root of the second invariant of the stress tensor deviator

$$s^2 = \bar{S}_{\mathrm{II}} = \tfrac{1}{3}\,\tau_0^2\,.$$

Instead of the space represented by the stress tensor σ_{ij} we shall consider one represented by the stress deviator $\bar{\sigma}_{ij}$. For simplicity, we shall again make the space nine-dimensional, even though the deviator has only five independent components so that we might consider only a five-dimensional space, as certain authors do. As before we denote by $\boldsymbol{\sigma}$ the radius vector of a point M in the space but now the only components of the vector $\boldsymbol{\sigma}$ are σ_{ij} and not $\bar{\sigma}_{ij}$. The quantity s has a simple geometrical meaning, it is the length of the vector that represents the stress deviator. In expression (15.2) s may be substituted for τ_0 and we obtain

$$\mathrm{d}e^{\mathrm{p}}_{ij} = \frac{1}{2}\left(\frac{1}{G_s} - \frac{1}{G}\right)\mathrm{d}\bar{\sigma}_{ij} + \frac{1}{2}\left(\frac{1}{G_t} - \frac{1}{G_s}\right)\frac{\bar{\sigma}_{ij}}{s}\,\mathrm{d}s\,. \tag{15.2'}$$

Let us check whether condition (14.3) of Drucker's postulate is fulfilled. Since the material is assumed incompressible, on a displacement $\mathrm{d}e^{\mathrm{p}}_{ij}$ only the deviator part of the stress tensor performs work, consequently we have

$$\mathrm{d}\bar{\sigma}_{ij}\mathrm{d}e^{\mathrm{p}}_{ij} \geqslant 0\,.$$

Introducing expression (15.2′) we obtain

$$\mathrm{d}\bar{\sigma}_{ij}\mathrm{d}e^{\mathrm{p}}_{ij} = \frac{1}{2}\left(\frac{1}{G_s} - \frac{1}{G}\right)\mathrm{d}\bar{\sigma}_{ij}\mathrm{d}\bar{\sigma}_{ij} + \frac{1}{2}\left(\frac{1}{G_t} - \frac{1}{G_s}\right)\frac{\bar{\sigma}_{ij}\mathrm{d}\bar{\sigma}_{ij}}{s}\,\mathrm{d}s\,.$$

But $\bar{\sigma}_{ij}\mathrm{d}\bar{\sigma}_{ij} = s\,\mathrm{d}s$, therefore

$$\mathrm{d}\bar{\sigma}_{ij}\mathrm{d}e^{\mathrm{p}}_{ij} = \frac{1}{2}\left(\frac{1}{G_s} - \frac{1}{G}\right)\mathrm{d}\bar{\sigma}_{ij}\mathrm{d}\bar{\sigma}_{ij} + \frac{1}{2}\left(\frac{1}{G_t} - \frac{1}{G_s}\right)\mathrm{d}s^2\,.$$

This expression is positive when $G > G_s > G_t$ which is always fulfilled if the material work-hardens steadily and there are no inflection points on the deformation curve. Therefore, the limitations that follow from Drucker's postulate can only relate to the shape of the loading surface which separates the region of active plastic deformation from that of unloading. In the first place it is obvious that the loading surface cannot be smooth, for on a smooth loading surface the direction of the plastic strain increment vector depends only on the stress vector $\boldsymbol{\sigma}$ and not on $\mathrm{d}\boldsymbol{\sigma}$. The second term in formula (15.2′) satisfies this condition, indeed the quantity s alters with any change in σ_{ij}, and the corresponding deformation component $\mathrm{d}e^{\mathrm{p}}_{ij}$ alters in proportion to $\bar{\sigma}_{ij}$. However, the first term depends directly on $\mathrm{d}\bar{\sigma}_{ij}$, and not indirectly through the quantity s, and therefore the direction of the vector $\mathrm{d}\boldsymbol{e}^{\mathrm{p}}$ depends on $\mathrm{d}\boldsymbol{\sigma}$, therefore, the point M is a corner point on the loading surface. Accordingly, let us suppose that the tangents to the loading surface at the point M form a cone. Let the angle between any generator of the cone and the radius vector of the point M be β (fig. 13). Select the stress increments in such a way that the angle between the vectors $\mathrm{d}\boldsymbol{\sigma}$ and $\boldsymbol{\sigma}$ is α, then the angle between the plastic deformation increment and the vector $\boldsymbol{\sigma}$ is δ.

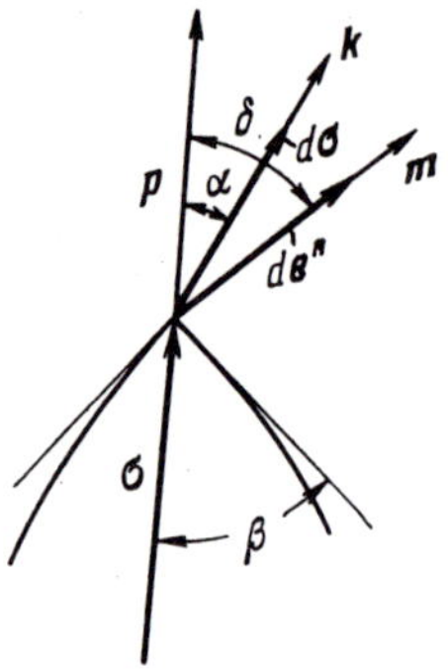

Fig. 13.

The condition that the plastic deformation increment vector lies inside the cone of normals to the loading surface reduces to fulfillment of the following inequality

$$\beta + \delta \leqslant \frac{\pi}{2} . \tag{15.3}$$

Now put

$$\boldsymbol{\sigma} = s\mathbf{p} , \quad \mathrm{d}\boldsymbol{\sigma} = \mathbf{k}\mathrm{d}q , \quad \mathrm{d}\mathbf{e}^{\mathrm{p}} = \mathbf{m}\mathrm{d}r . \tag{15.4}$$

Unit vectors **p**, **k** and **m** are shown in fig. 13. For the moment suppose that

$$\tfrac{1}{2}\left(\frac{1}{G_s} - \frac{1}{G}\right) = A , \quad \tfrac{1}{2}\left(\frac{1}{G_t} - \frac{1}{G_s}\right) = B ,$$

write (15.2′) in the form of a vector equation

$$\mathrm{d}\mathbf{e}^{\mathrm{p}} = A\,\mathrm{d}\boldsymbol{\sigma} + B\,\frac{\boldsymbol{\sigma}}{s}\,\mathrm{d}s .$$

Here introduce expressions (15.4) noticing first that $s\mathrm{d}s = \boldsymbol{\sigma}\mathrm{d}\boldsymbol{\sigma} = s\mathrm{d}q \cos\alpha$. We obtain

$$\mathbf{m}\mathrm{d}r = (\mathbf{k}A + \mathbf{p}B\cos\alpha)\,\mathrm{d}q . \tag{15.5}$$

Squaring both sides of eq. (15.5) we find that

$$\frac{\mathrm{d}q}{\mathrm{d}r} = \frac{1}{[A^2 + (B^2 + 2AB)\cos^2\alpha]^{1/2}} .$$

Now multiply (15.5) by unit vector **p**. We obtain

$$\cos\delta \, \mathrm{d}r = (A + B)\cos\alpha \, \mathrm{d}q \, .$$

Now we may write an expression for cos δ, namely

$$\cos\delta = \frac{(A + B)\cos\alpha}{[A^2 + (B^2 + 2AB)\cos^2\alpha]^{1/2}} \, .$$

Put

$$N = \frac{A + B}{A} = \frac{\dfrac{1}{G_t} - \dfrac{1}{G}}{\dfrac{1}{G_s} - \dfrac{1}{G}} \, .$$

The previous formula is rewritten as follows

$$\cos\delta = \frac{N\cos\alpha}{[1 + (N^2 - 1)\cos^2\alpha]^{1/2}} \, . \qquad (15.6)$$

Inequality (15.3) may be replaced by the inequality

$$\cos\delta \geqslant \sin\beta$$

or

$$\frac{N^2\cos^2\alpha}{1 + (N^2 - 1)\cos^2\alpha} \geqslant \sin^2\beta \, .$$

Rewrite the latter inequality as

$$\frac{N^2}{N^2 + \tan^2\alpha} \geqslant \frac{\tan^2\beta}{1 + \tan^2\beta} \, .$$

Hence it follows that

$$\tan\alpha \leqslant \frac{N}{\tan\beta} \, . \qquad (15.7)$$

No conclusions can be drawn from inequality (15.7) concerning the shape of the loading surface in the neighbourhood of the point M, it only limits the selection of loading path for which the relationships of the deformation theory are valid. The angle β, which we shall term the angle of plasticity, at the point M can depend on the stress σ_{ij} and also on the history of loading. Inequality (15.7) shows that the deformation theory equations do not contradict Drucker's postulate so long as the additional load vector $\mathbf{d\sigma}$ is in the direction of the region that is bounded by the cone of which the generator forms an angle α with the radius vector of the point M. If the direction of the stress deviator increment vector $\mathbf{d\sigma}$ is within the loading surface, that is within the cone that is formed by tangents to the loading surface, then elastic unloading occurs. If the additional load vector passes out into the space between the two cones the relationship between stress and deformation increments should be established on the basis of certain additional considerations. Now let us suppose that, at least in the neighbourhood of the point M, the loading surface depends only on the coordinates of this point, i.e. on the vector $\boldsymbol{\sigma}$ and does not depend on the loading path. Then the plasticity cone will be a circular cone with its centre line passing through the origin of coordinates, just as in the case of proportional loading. This reasoning is not entirely strict. Indeed the formulation of the deformation theory of plasticity is such that all components which represent the stress vector are fully equivalent to one another and the equations contain no elements that could introduce asymmetry. However, the loading surface should be determined independently and no limitations can be applied to its form in advance. Therefore, the assumption that the plasticity cone is a cone of rotation is not necessary, although it is quite natural. Our problem is to determine the possible boundaries of validity of the Nadai-Hencky theory. The true limits can only be established on the basis of experimental results, but more definite conclusions can only be obtained by adopting certain hypotheses and therefore we shall introduce the simplest, most vivid hypotheses. Assume that the dependence of the loading surface on the coordinates of the point M is continuous and consider two situations which can arise on transition from point M of the stress deviator space to an infinitesimally neighbouring point M'. The angle between the plasticity cones at the points M and M' is infinitesimally small, the difference between the plasticity angles is also infinitesimally small and therefore, we consider that the generators of the two cones are parallel. In the first case the loading path that corresponds to transition from M to M' lies inside the loading surface of the point M', the return path from M' to M represents unloading. In the second case the segment MM' lies outside the plasticity cone of the point M' and, therefore, the path from M' to M is accompanied by plastic deformation, which contradicts the definition of unloading. Therefore, if we consider that the surface of plasticity does not depend on the history of loading the possible paths of loading should be limited by the condition that $\alpha \leqslant \beta$. Suppose that in (15.7) $\alpha = \beta$, we find a value β_0 for which the angle α is maximum, namely

$$\tan \beta_0 = N^{\frac{1}{2}} . \tag{15.8}$$

Now suppose that we have solved some problem in the theory of plasticity by means of the deformation theory equations. This means that for each point in the body we have constructed the unloading path and consequently know the angle α between the radius vector of the point in the stress space (deviator) and the tangent to the loading path. On the other hand, at each point we know the quantity N. If $\alpha < \beta_0$, and β_0 is given by formula (15.8), application of deformation theory to the problem may be considered justified in the sense that the requirements of Drucker's postulate are not infringed.

16. Two-dimensional model of work-hardening body

The foregoing considerations concerning the boundaries of applicability of the equations of the deformation theory of plasticity and concerning the need in this case to assume the existence of an angular point on the loading surface have a graphic physical basis. Real polycrystalline material is a random conglomeration of crystalline grains, plastic slip develops only along certain unfavourably located slip systems and is accompanied by plastic deformation. When the nature of the stressed state alters, i.e. there is deviation from proportional loading, some new elements enter the plastic state and others that were plastically deformed are unloaded and return to the elastic state. Later on, we shall become familiar with certain theories of plasticity which allow for the structure of the material, but for the present we return to consideration of a simple model which can be used to explain the general properties of elastic-plastic systems. Apparently, work-hardening plastic bodies also possess this property. In this connection, as Hencky [227] has shown, the properties of a work-hardening material during extension can be modelled by a rod system consisting of elastic-plastic elements of a material not subject to work-hardening.

As a model of a work-hardening body we select a circular thin-walled tube of elastic-plastic material that is not subject to work-hardening. The tube is bent by two moments M_x and M_y acting in perpendicular planes.

Denote the radius of the tube R, the wall thickness δ, the modulus of

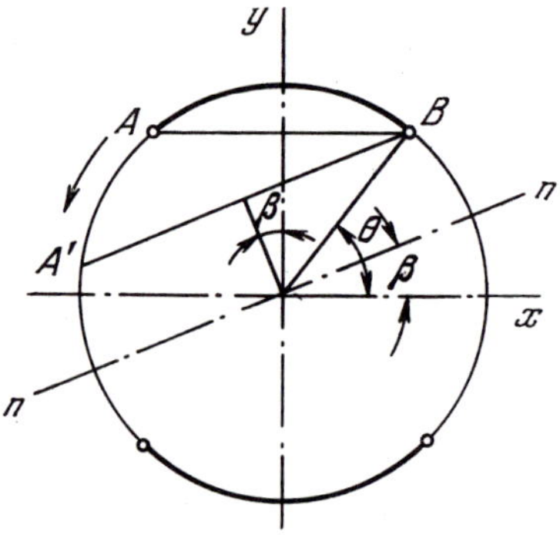

Fig. 14.

elasticity of the material E, the yield point σ_T. We shall consider loading to be proportional when M_x and M_y vary proportionately. The neutral axis then retains its original position and it evidently suffices to consider bending of the tube by the moment $M = \sqrt{M_x^2 + M_y^2}$ applied in a constant plane. Without loss of generality we may consider that the moment M_x acts alone in the plane yOz and the x axis is the neutral axis. Provided that the stress nowhere exceeds the yield point the curvature of the bent axis depends on the torque in the following way:

$$k = \frac{M}{\pi R^3 \delta E} \, .$$

Plastic deformations mainly occur in sections that are most remote from the neutral axis, that is on the y axis, and when the moment increases symmetrical arc-shaped plastic zones are formed, as shown in zone AB of fig. 14. Denote by θ the polar angle of the boundary of the plastic zone. The deformation at points A and B is σ_T/E, therefore

$$kR \sin \theta = \frac{\sigma_T}{E} \, .$$

If we define the position of points on the circle by a polar angle ψ then in the elastic zone, when $\psi < \theta$

$$\sigma = \sigma_T \frac{\sin \psi}{\sin \theta} \, . \tag{16.1}$$

Now calculate the bending moment in the section by the formula

$$M = 4\int_0^{\frac{\pi}{2}} \delta\sigma R^2 \sin\psi \, d\psi$$

and here consider that in the plastic region $\sigma = \sigma_T$ and in the elastic region the stress is given by eq. (16.1). By integration we obtain

$$M = R^2\delta\sigma_T \frac{2\theta + \sin 2\theta}{\sin\theta} .$$

Replace the moments and changes of curvature by the corresponding dimensionless quantities

$$Q = \frac{M}{R^2\delta\sigma_T} , \qquad q = \frac{ER}{\sigma_T} k = \frac{1}{\sin\theta} .$$

The relationship between Q and q is then

$$\begin{aligned} &Q = \pi q && \text{in the elastic region when } q < 1 , \\ &Q = (2\theta + \sin 2\theta)q && \text{in the elastic-plastic region when } q > 1 . \end{aligned} \tag{16.2}$$

Fig. 15 shows a diagram of the relationship between Q and q, it is very similar to the plasticity diagram for a work-hardening material. The quantity Q is analogous to stress, or better to the deviator part of the stress, whereas q is the corresponding deformation. The modulus of elasticity G is equal to π, the plastic modulus G_s is $2\theta + \sin 2\theta$. This model corresponds to a body with

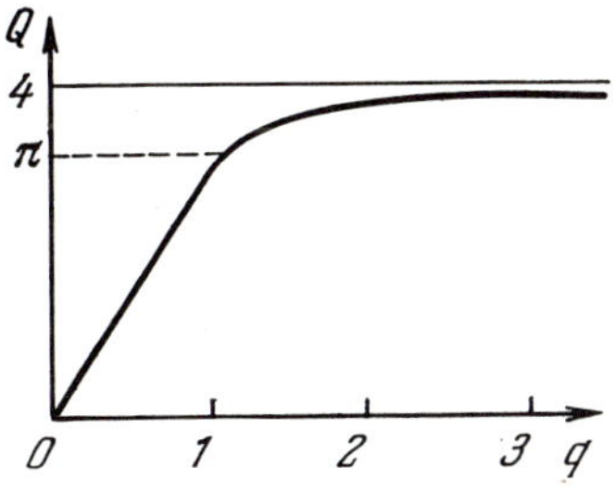

Fig. 15.

initial isotropy. Indeed, relationships (16.2) are valid whatever the plane in which the bending moment M acts, when the moment increases bending always occurs in the same plane. We can now easily return to the general case of bending about an arbitrary axis and proportionally increasing torques M_x and M_y act on the tube. The corresponding curvatures are k_x and k_y, with proportional loading we always have $k_x : k_y = M_x : M_y$. Converting to the corresponding dimensionless quantities

$$Q_x = G_s q_x\,, \qquad Q_y = G_s q_y\,. \tag{16.3}$$

Here G_s depends on the quantity $q = \sqrt{q_x^2 + q_y^2}$ or, which is equivalent, on $Q = \sqrt{Q_x^2 + Q_y^2}$ if $q < 1$ and $Q < \pi$ respectively, and in formula (16.3) we must take $G_s = G = \pi$. Expressions (16.3) are analogous to the equations of the Nadai-Hencky deformation theory of plasticity, Q^2 and q^2 correspond to the quadratic invariants of the stress and deformation tensors, Q and q are analogues of octahedral shear and octahedral tangential stress. The stress deviator space corresponds to a plane. If we lay out Q_x and Q_y on the coordinate axes the condition of reaching a plastic state

$$Q = \sqrt{Q_x^2 + Q_y^2} = \pi$$

defines a circle in the plane which corresponds to the initial surface of flow (fig. 16). Any path of proportional loading corresponds to a line issuing from the origin of the coordinates.

Eqs. (16.3) were derived for the case of proportional loading, but they are also valid for certain other types of loading. Assume that proportional loading has been applied up to a certain value $Q > \pi$. For simplicity assume that

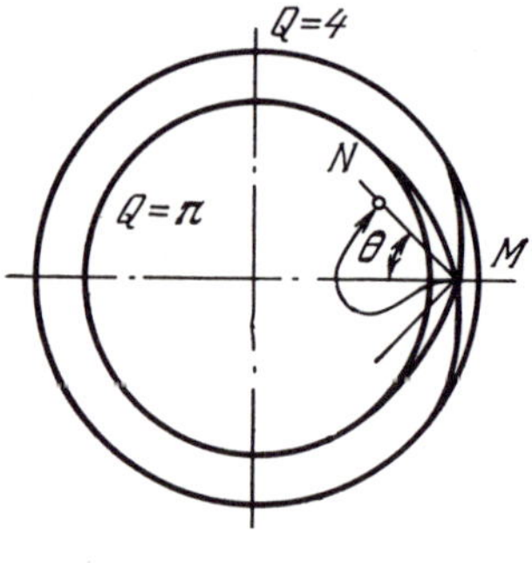

Fig. 16.

loading took place along the Q_x axis to the point M. The boundary of the plastic region in the tube is here given by the angle θ. Now load additionally, so altering not only Q_x but also Q_y, which was zero in the first stage of loading. The neutral axis of curvature is now no longer the x axis but a certain straight line nn' (fig. 14). Depending on the relationship between Q_x and Q_y, that is depending on the loading path, the plastic region AB (and a symmetrical region on the other side) may be increased continuously and partial unloading may occur. In the first case eqs. (16.3) of the deformation theory remain valid. Consider the extreme case when the point A moves in the direction of the arrow to the position A' and the point B remains unaltered. If the plastic zone becomes $A'B$ the neutral axis nn' should be parallel to the chord $A'B$, denote the angle between nn' and the x axis by β. The bending moment relative to this axis is found from formulae (16.2) by substituting into them $\theta - \beta$ in place of θ. Thus we obtain

$$Q_x = Q(\theta - \beta) \cos \beta\,, \qquad Q_y = Q(\theta - \beta) \sin \beta\,. \tag{16.4}$$

Eqs. (16.4) are parametric equations of the limiting loading path issuing from the point M for which the expressions of the deformation theory (16.3) are still valid. This limiting path is illustrated diagrammatically in fig. 16. The symmetrical curve corresponds to motion of the boundary B of the plastic region.

In the neighbourhood of the point M

$$\Delta Q_x = \frac{\mathrm{d}Q}{\mathrm{d}\theta} \cos \beta\,(-\Delta\beta) - Q \sin \beta\,\Delta\beta\,,$$

$$\Delta Q_y = \frac{\mathrm{d}Q}{\mathrm{d}\theta} \sin \beta\,(-\Delta\beta) + Q \cos \beta\,\Delta\beta\,,$$

and since β is small we have

$$\Delta Q_x = -\frac{\mathrm{d}Q}{\mathrm{d}\theta}\,\Delta\beta\,, \qquad \Delta Q_y = Q\,\Delta\beta\,.$$

Note that

$$\frac{\mathrm{d}Q}{\mathrm{d}\theta} = \frac{\mathrm{d}Q}{\mathrm{d}q}\frac{\mathrm{d}q}{\mathrm{d}\theta} = \frac{\mathrm{d}Q}{\mathrm{d}q}\,q \,\mathrm{ctan}\,\theta\,.$$

The quantity $\mathrm{d}Q/\mathrm{d}q = 2\theta - \sin 2\theta$ represents the tangential modulus G_t on the deformation curve (fig. 15). We may now write

$$\Delta Q_x = -G_t q \operatorname{ctan} \theta \, \Delta\beta \,, \qquad \Delta Q_y = G_s q \, \Delta\beta \,.$$

The slope of the limiting paths that pass through the point M for which the equations of the deformation theory remain valid is defined as

$$\frac{\Delta Q_x}{\Delta Q_y} = \pm \frac{G_s}{G_t} \tan \theta \,.$$

The corresponding angle α is $\pi/2$ when $\theta = 0$ and $\theta = \pi/2$, its minimum value of $76°28'$ occurs when $\theta = 57°30'$.

Now consider the conditions of unloading after proportional loading. Evidently here too proportional moment reduction is not the only possible method of elastic unloading, it is only necessary that rotation of the section that results from unloading should take place about an axis which does not intersect the plastic region. Let us term this the axis of unloading. Assume that loading was proportional along the x axis to a dimensionless torque value of Q; unloading occurs as a result of additional rotation about an axis that forms an angle φ with the x axis. Denote the change in dimensionless bending moments that results from unloading by Q'_x and Q'_y so that during unloading

$$Q_x = Q - Q'_x \,, \qquad Q_y = -Q'_y \,.$$

During unloading the material is deformed elastically and, therefore, Q'_x and Q'_y are components of torque directed along the unloading axis, consequently, we may put

$$Q'_x = -\pi\omega \cos\varphi \,, \qquad Q'_y = -\pi\omega \sin\varphi \,.$$

The change of stress that results from unloading at a point with polar angle ψ is then given by

$$\sigma' = -\omega\sigma_{\mathrm{T}} \sin(\psi - \varphi) \,.$$

If we fix the value of φ and alter ω unloading occurs along a straight line that passes through the point M. The equation of this line is

$$Q_x = Q - \pi\omega \cos\varphi \,, \qquad Q_y = -\pi\omega \sin\varphi \,. \tag{16.5}$$

The stress is then the difference between the stress at the instant of unloading and the stress σ', namely

$$\sigma = \sigma_{\mathrm{T}} [1 - \omega \sin (\psi - \varphi)] \qquad (0 \leqslant \alpha \leqslant \pi - \theta),$$
$$\sigma = \sigma_{\mathrm{T}} \left[\frac{\sin \psi}{\sin \theta} - \omega \sin (\psi - \varphi) \right] \qquad (0 \leqslant \varphi \leqslant \theta,\ \pi - \theta \leqslant \varphi \leqslant \pi). \tag{16.6}$$

In order for unloading to be elastic throughout the section the unloading axis must not intersect the plastic region, that is $\varphi \leqslant \theta$. The limiting unloading line corresponds to $\varphi = \theta$ and forms an angle θ with the Q_x axis at the point M. The result is shown diagrammatically in fig. 17. Region I lies within the loading contour, which in our model replaces the loading surface of nine-dimensional space: the point M is a point on the loading contour which corresponds to the conical point of the hypersurface of loading. The angle 2α corresponds to the cone of possible directions of increment of the stress deviator, for which the equations of the deformation theory remain valid (region II). The straight lines passing through the point M at an angle β to the axis form the boundary of the loading contour. In order to find the entire contour, we consider rectilinear unloading paths (16.5) when $\varphi < \theta$. Elastic unloading continues until the stress at any point again reaches the yield point. Assume that this occurs at a point in the initial plastic region. From the first of formulae (16.6) the greatest negative stress occurs when $\psi - \varphi = \pi/2$, this stress is $\sigma_{\mathrm{T}}(1-\omega)$. Equating this to $-\sigma_{\mathrm{T}}$ we obtain $\omega = 2$. The boundary of the region of elastic deformation is determined by formulae (16.5) taking $\omega = 2$ and considering φ a variable,

$$Q_x = Q - 2\pi \cos \varphi, \qquad Q_y = -2\pi \sin \varphi.$$

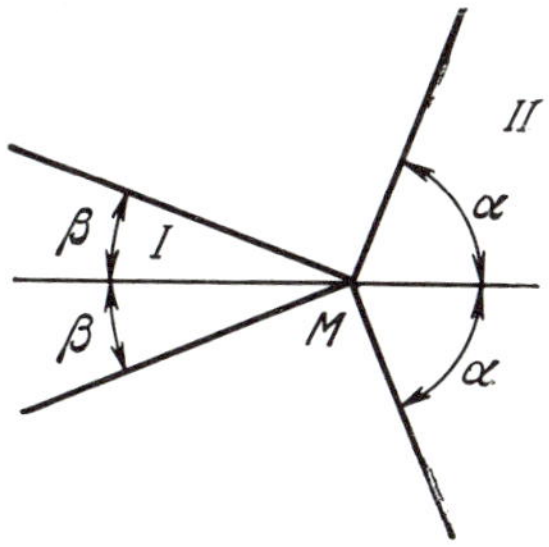

Fig. 17.

This is the equation of a circle of radius 2π with centre at the point $M(Q, 0)$. Now consider the possibility of determining plastic deformation during unloading in this region, which remained elastic during the first unloading. From the second of formulae (16.6) the condition of reaching the extremal value of stress is

$$\frac{\cos\psi}{\sin\theta} - \omega\cos(\psi - \varphi) = 0\,.$$

Equating this stress to the yield point with minus sign we obtain

$$\frac{\sin\psi}{\sin\theta} - \omega\sin(\psi - \theta) = -1\,.$$

Square and add the above equations, eliminate the angle ψ and obtain a relationship between φ and θ, namely

$$\frac{1}{\sin^2\theta} + \omega^2 - 2\omega\frac{\cos\varphi}{\sin\theta} = 1\,. \tag{16.7}$$

When this condition is fulfilled eqs. (16.5) define a circle of radius π with centre at the point with coordinates $Q_x = Q - \pi/\sin\theta$, $Q_y = 0$. Indeed

$$Q_x - Q + \frac{\pi}{\sin\theta} = -\pi\omega\cos\varphi + \frac{\pi}{\sin\theta}\,,$$

$$Q_y = -\pi\omega\sin\varphi\,.$$

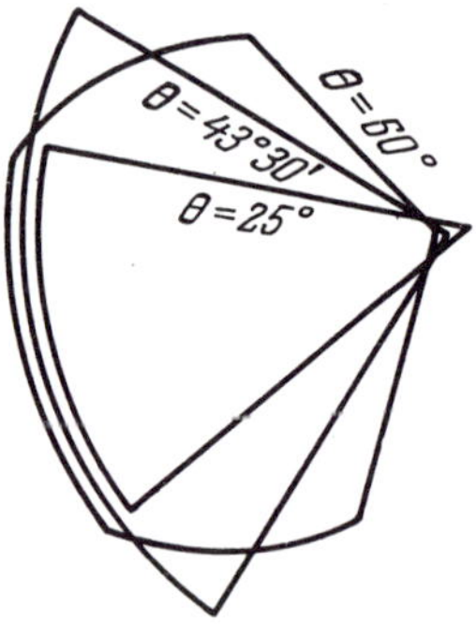

Fig. 18.

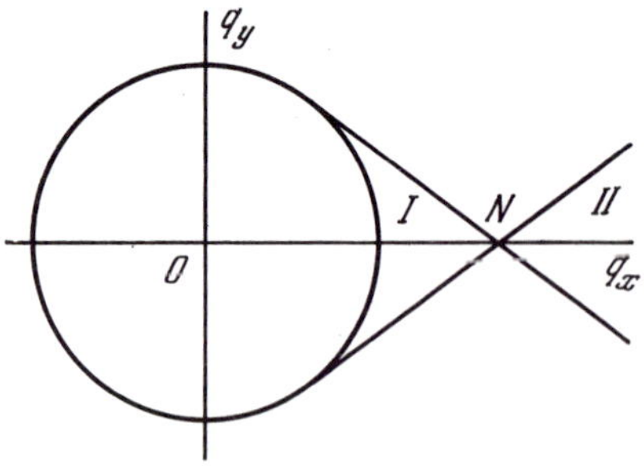

Fig. 19.

Squaring and adding, bearing in mind (16.7) we obtain

$$\left(Q_x - Q + \frac{\pi}{\sin\theta}\right)^2 + Q_y^2 = \pi^2 .$$

Therefore, the loading contour is bounded by two straight lines and arcs of a single circle (the first) when $\theta \leqslant 43°30'$ and two circles when $\theta > 43°30'$. These contours are constructed for several values of θ in fig. 18.

Now consider the representation of the deformation plane (fig. 19). The initial boundary of plasticity is represented by a circle $q_x^2 + q_y^2 = 1$, the point M corresponds to the point N, with abscissus $q = 1/\sin\theta$. With elastic loading $\Delta q_y/\Delta q_x = \Delta Q_y/\Delta Q_x$, consequently, the limiting straight lines of elastic unloading form an angle θ with the abscissus and are tangential to the initial circle. Find the relationship between the increments of the quantities q_x, q_y and Q_x, Q_y in the region where the equations of the deformation theory are valid. From (16.3) $q_x = (1/G_s)Q_x$. We shall consider that G_s is given as function of $Q = \sqrt{Q_x^2 + Q_y^2}$. Then

$$\Delta q_x = \frac{\Delta Q_x}{G_s} + Q_x \frac{\mathrm{d}}{\mathrm{d}Q}\left(\frac{1}{G_s}\right) \Delta Q ;$$

differentiating the quantity $q = Q/G_s$ with respect to Q we obtain

$$\frac{\mathrm{d}q}{\mathrm{d}Q} = \frac{1}{G_s} + Q \frac{\mathrm{d}}{\mathrm{d}Q}\left(\frac{1}{G_s}\right),$$

but $\mathrm{d}q/\mathrm{d}Q = 1/G_t$, therefore

$$\frac{\mathrm{d}}{\mathrm{d}Q}\left(\frac{1}{G_s}\right) = \frac{1}{Q}\left(\frac{1}{G_t} - \frac{1}{G_s}\right).$$

Introducing this into the expression for Δq_x we obtain

$$\Delta q_x = \frac{1}{G_s}\Delta Q_x + \left(\frac{1}{G_t} - \frac{1}{G_s}\right)\frac{Q_x}{Q^2}(Q_x \Delta Q_x + Q_y \Delta Q_y) .$$

Similarly

$$\Delta q_y = \frac{1}{G_s}\Delta Q_y + \left(\frac{1}{G_t} - \frac{1}{G_s}\right)\frac{Q_y}{Q^2}(Q_x \Delta Q_x + Q_y \Delta Q_y) .$$

In the neighbourhood of the point M we must take $Q_x = Q$ and $Q_y = 0$, therefore

$$\Delta q_x = \frac{\Delta Q_x}{G_t}, \qquad \Delta q_y = \frac{\Delta Q_y}{G_s}.$$

The ratio of the deformation increments is

$$\frac{\Delta q_x}{\Delta q_y} = \frac{G_t}{G_s} \frac{\Delta Q_y}{\Delta Q_x}. \tag{16.8}$$

For limiting directions of the deformation theory when $\Delta Q_x : \Delta Q_y = G_s/G_t \tan\theta$, we obtain

$$\frac{\Delta q_y}{\Delta q_x} = \tan\theta .$$

Thus the equations of the deformation theory remain valid when the stress increments lie within the external angle formed by the tangent to the initial flow curve (region II in fig. 19). Certain other conclusions which follow from consideration of this model will be considered in section 20.

17. **Flow theory for a smooth loading surface. Isotropic work-hardening**

Quite recently, as a result of Budiansky's investigations [176] it has been found possible to extend the limits of applicability of the deformation theory, as we have illustrated with a simple model. The obvious contradictions to which this theory leads have led to the development of theories based on a fundamentally different approach. The starting point of the deformation theory of plasticity is the non-linear theory of elasticity, it is a generalisation of this theory to the case of non-elastic material. In flow theories, on the other hand, a relationship is established between deformation increments and the stressed state. If the stressed state is considered to vary as a function of some steadily increasing parameter we may treat this paramcter as time and introduce the rate-of-deformation into consideration. In the theory of plasticity, unlike the theory of creep, the selection of time scale is immaterial.

In introducing the concept of surface of loading we did not consider how this surface alters during plastic deformation. The surface that passes through the point M may depend not only on the coordinates of the point, i.e. on the stress, but also on the deformation, possibly in an extremely complicated way. This relationship may be allowed for by choice of the loading path that leads to the point M.

Suppose that the equation of the surface of loading that passes through the point M can be written

$$f(\sigma_{ij}) - k^2 = 0\,.$$

The functions $f(\sigma_{ij})$ and the constant k^2 may depend on the deformation and also in any way, however complicated, on the loading and deformation paths. If the stresses acquire increments $\mathrm{d}\sigma_{ij}$ such that

$$\mathrm{d}f = \frac{\partial f}{\partial \sigma_{ij}}\,\mathrm{d}\sigma_{ij} > 0\,,$$

a plastic deformation $\mathrm{d}e^{\mathrm{p}}_{ij}$ results. In section 14 it was shown that the vector of plastic deformation increment is in the direction of the normal to the surface of loading, and, therefore, we may write

$$\dot{e}^{\mathrm{p}}_{ij} = \lambda\,\frac{\partial f}{\partial \sigma_{ij}}\,.$$

Here, as was discussed above, we have converted from increments to rates. In order to define the form of the factor λ we assume a linear relationship between $\mathrm{d}e^{\mathrm{p}}_{ij}$ and $\mathrm{d}\sigma_{ij}$. This means that we assume that the principle of linear superposition is valid for small increments of stress and plastic deformation. Drucker [196] showed that this hypothesis is not a necessary condition, but follows from the postulate of plasticity, but we shall not give his demonstration. In accordance with the proposed hypothesis we can resolve the increment of the vector $\boldsymbol{\sigma}$ into two components: $\mathbf{d}\boldsymbol{\sigma}'$ in a tangential direction to the surface of loading and $\mathbf{d}\boldsymbol{\sigma}''$ in the direction of the normal. The first component corresponds to neutral loading which does not cause plastic deformation. Consequently, the increment of plastic deformation can only depend on the normal component of the vector $\mathbf{d}\boldsymbol{\sigma}$, which is proportional to the quantity $(\partial f/\partial \sigma_{kl})\,\mathrm{d}\sigma_{kl}$. Therefore, the factor λ must also be proportional to this quantity. As a result we obtain

$$\dot{e}^{\mathrm{p}}_{ij} = H\,\frac{\partial f}{\partial \sigma_{ij}}\,\frac{\partial f}{\partial \sigma_{kl}}\,\dot{\sigma}_{kl}\,. \tag{17.1}$$

The quantity H in formula (17.1) may be a function of stress and deformation, it also depends on the paths of loading and deformation, just like the function f. Note that in formula (17.1) any function of f may be substituted for the loading function f, inasmuch as we have free choice of the function H.

The simplest possible assumption is that the functions f and H depend only on the second invariant of the stress deviator. In view of what we have said we may take it that

$$f = \bar{S}_{\mathrm{II}} = S_{\mathrm{II}} - \tfrac{1}{3} S_{\mathrm{I}}^2 .$$

Then

$$\frac{\partial f}{\partial \sigma_{ij}} = 2(\sigma_{ij} - \sigma\delta_{ij}) = 2\bar{\sigma}_{ij} .$$

The expression $(\partial f/\partial \sigma_{kl})\, \dot{\sigma}_{kl}$ is the derivative of f with respect to time, that is with respect to $\bar{S}_{\mathrm{II}}$, or, converting to octahedral stress

$$\dot{f} = \dot{\bar{S}}_{\mathrm{II}} = 6\tau_0 \dot{\tau}_0 .$$

Introducing this into (17.1) we find

$$\dot{\bar{e}}_{ij}^{\mathrm{p}} = \bar{\sigma}_{ij} h(\tau_0) \dot{\tau}_0 . \tag{17.2}$$

Equations of this kind were proposed by Prager [334]. In the case of proportional loading we may consider the quantity τ_0 either as the parameter of loading or as time in the sense explained above. The condition of proportional loading is that $\bar{\sigma}_{ij}/\tau_0 = \text{const}$, therefore, eq. (17.2) may be integrated, namely

$$\bar{e}_{ij}^{\mathrm{p}} = \bar{\sigma}_{ij} \frac{1}{\tau_0} \int_0^{\tau_0} \tau_0 h(\tau_0)\, \mathrm{d}\tau_0 .$$

Comparing these formulae with eqs. (15.1) of the deformation theory we find that they coincide and

$$\frac{1}{\tau_0} \int_0^{\tau_0} \tau_0 h(\tau_0)\, \mathrm{d}\tau_0 = \frac{1}{2}\left(\frac{1}{G_s} - \frac{1}{G}\right) .$$

Thus, from eq. (17.2) there follows the existence of a single diagram of the relationship between τ_0 and γ_0 which is valid for proportional loading. Multiplying by τ_0 and differentiating, after self-evident modifications we obtain

$$h(\tau_0) = \frac{1}{2\tau_0}\left(\frac{1}{G_t} - \frac{1}{G}\right). \tag{17.3}$$

Therefore, the function $h(\tau_0)$ is easily found if we know the stress plastic strain relationship obtained on a uniaxially deformed specimen of the given material. Adding the elastic deformation and using (17.3) we may write the flow theory equation as

$$\dot{\bar{e}}_{ij} = \frac{1}{2G}\dot{\bar{\sigma}}_{ij} + \frac{1}{3K}\dot{\sigma}\delta_{ij} + \frac{\bar{\sigma}_{ij}}{2\tau_0}\left(\frac{1}{G_t} - \frac{1}{G}\right). \tag{17.4}$$

In deriving expressions (17.1) we considered that the functions f retain their form during plastic loading, whereas the quantity k alters in such a way that the surface of loading always passes through the point of loading, and alters whilst maintaining similarity. As we saw in section 15 in the case of an incompressible material of which the plastic behaviour is independent of the first invariant of the stress tensor, instead of considering stress space we may consider a space of the stress deviator; in this space the surface $\bar{S}_{II} = k^2$ is a sphere with centre at the origin of coordinates and of continually increasing radius. A plastic body whose properties are described in this way is termed a body with isotropic work-hardening. If an octahedral stress τ_0' is achieved during deformation followed by unloading, then on repeated loading through any path the plastic state is reached when the octahedral stress becomes equal to τ_0'. In particular, extension of the specimen raises the yield point in compression. This consequence of eq. (17.1) contravens the experimentally ob-

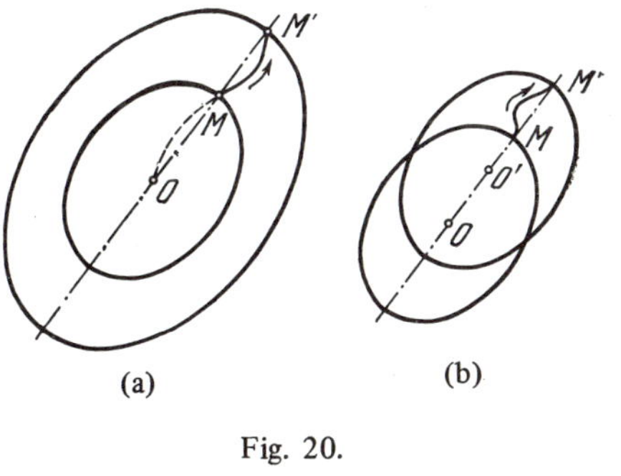

Fig. 20.

served Bauschinger effect, namely, that when the yield point is raised in tension it is reduced in compression. According to the simplest idealised scheme the reduction in the yield point in compression is the same as the increase in the yield point in tension. In order to formulate equations of a theory of the type of flow which allows for the ideal Bauschinger effect, we must assume that the surface of loading is displaced parallel to itself following the point of loading. Successive loading surfaces are shown schematically in fig. 20 for cases of (a) isotropic work-hardening and (b) the ideal Bauschinger effect, the initial surface of loading being the same in both cases.

18. Theory of flow with translational work-hardening

Translational work-hardening, which corresponds to the ideal Bauschinger effect, is defined as work-hardening of the material in which the surface of loading is progressively displaced whilst retaining its shape and size. If the equation of the initial surface of loading is $f(\sigma_{ij}) = k^2$, during parallel transfer the equation of the next surface of loading is

$$f(\sigma_{ij} - s_{ij}) = k^2 . \tag{18.1}$$

Here s_{ij} is a tensor of which the components in the stress space are the coordinates of the centre of the surface of loading. Obviously, the tensor s_{ij} should be associated with plastic deformation. The simplest assumption (due to Ishlinskiy [46]) is that we take into consideration

$$s_{ij} = ce^{\mathrm{p}}_{ij} \tag{18.2}$$

and as function $f(\sigma_{ij})$ we take the second invariant of the deviator. In the uniaxial stressed state eq. (18.1) directly establishes a linear relationship between the stress and plastic deformation; therefore, this is the equation of the theory of plasticity with linear work-hardening.

Independently of and simultaneously with Ishlinskiy, Prager [335] proposed an analogous hypothesis, termed the hypothesis of kinetic work-hardening, because it can be illustrated with a simple kinematic model. For clarity consider a two-dimensional case in which the surface of loading corresponds to the contour of loading. Let us consider that we have a frame, of the shape of the contour of loading, which can be freely displaced in the stress plane, but only in translation as there are guides that prevent rotation. In the plane there is a pin that moves along the loading path. If there is no friction between pin and frame, as the pin moves in an arbitrary direction at an acute angle to the direction of the external normal to the contour of the frame the latter moves in the direction of the normal; therefore, the displacement of the centre of the frame is in the same direction as the increment of plastic deformation.

It remains to suppose that this displacement is proportional to the increment of plastic deformation, that is, to assume that expression (18.2) applies The considerations that have been developed for a plane model must now be transferred to nine-dimensional space.

Note that in considering particular problems of the theory of plasticity we may consider a subspace with a smaller number of dimensions rather than considering the complete stress space. For example, a plane stressed state is characterised by three stresses, consequently, it suffices to consider a three-dimensional space of stress, and if we are concerned with principal stresses, then with a two-dimensional space for which Prager's kinematic model that we have described would appear to be directly applicable. However, in a plane stressed state deformation is not plane and the use of a two-dimensional kinematic model can be misleading, as was shown by Budiansky (see discussion on Prager's article [335]). More detailed investigation of the shape of the law of kinematic work-hardening in particular subspaces of the stress space is contained in the article by Shield and Ziegler [351].

A further variant of the hypothesis of translational work-hardening was proposed by Ziegler [385], according to which the tensor s_{ij} is defined in the following way:

$$\dot{s}_{ij} = \dot{\mu}(\sigma_{ij} - s_{ij}) .$$

When $\sigma_{ij} = 0$, s_{ij} is correspondingly also zero, therefore, the kinematic model is retained in any subspace of the stress space. A more general standpoint was developed by Kadashevich and Novozhilov [48]. These authors suppose that the tensors s_{ij} and e^{p}_{ij} are related by formulae of the same type as in the strain theory of plasticity, namely

$$e^{\mathrm{p}}_{ij} = \frac{1}{2G^*} s_{ij} . \tag{18.3}$$

Here G^* is a function of the invariants of the tensor s_{ij}. In particular examples the authors considered, as in the Nadai-Hencky theory of plasticity, that G^* depends only on the second invariant of the deviator of the tensor s_{ij}.

The essential non-linearity of expressions (18.3) permitted the authors to give a much better description of the behaviour of real material; theoretical predictions were in very good agreement with certain experimental results.

19. Singular surfaces of loading

In considering the limits of applicability of the theory of plasticity, we showed that successive surfaces of loading may be singular; the occurrence of conical points on the loading surface at the point of loading was illustrated by a model. Certain new theories of plasticity which we consider below lead to singularities of this kind. Here, we shall consider another kind of singularity. The initial loading surface may already consist of parts of several smooth surfaces which form edges where they intersect. On the edges the directions of the normals to the surface of loading are indeterminate, therefore, all that can be said about the direction of increment of the vector of plastic strain is that it lies in a plane normal to the edge and is included between the normals to the surfaces which intersect to form the edge. During the process of plastic deformation that is accompanied by work-hardening, the stress surface either extends maintaining a similar shape or is displaced parallel to itself. Here unlike the case considered above, new singular features do not occur, the smooth parts of the surface of loading remain smooth. The simplest and most natural example of this singular surface is obtained if we generalise the St. Venant-Tresca theory of plasticity to the case of work-hardening material. In this theory the criterion of achieving the plastic state is the maximum value of the tangential stress. We shall consider that the maximum tangential stress also defines the secondary yield point, that is the condition of transition to the plastic state on repeated loading after the occurrence of plastic deformation and unloading. The equation of the surface of loading is

$$\tau_{\max}^2 = k^2 \, .$$

The constant k is always selected so that the surface passes through the point of loading. In the general case the expression for the maximum tangential stress in terms of the components of the stress tensor is extremely complicated and of no practical value: the St. Venant type of theory shows to advantage only when the principal directions are known in advance and we may consider a three-dimensional space of the principal stresses in place of the nine-dimensional stress space. If $\sigma_1 > \sigma_2 > \sigma_3$ then $\tau_{\max} = (\sigma_1 - \sigma_3)/2$ and the condition that $\tau_{\max}$ is constant is written as

$$\sigma_1 - \sigma_3 = \pm\, 2k\,. \tag{19.1}$$

This is the equation of pairs of planes parallel to the axis 2. In the space of principal stresses we cannot relate the axes to the quantities σ_1, σ_2 and σ_3 without infringing the inequality $\sigma_1 \geqslant \sigma_2 \geqslant \sigma_3$; therefore, we denote the principal stresses by σ_ξ, σ_η and σ_ζ; at different points in the space of the principal stresses the relationships between them are different and accordingly, some stresses or other should be named σ_1, σ_2 and σ_3. In place of the single condition (19.1) we obtain six conditions

$$\begin{aligned} \sigma_\xi - \sigma_\eta &= \pm\, 2k\,, \\ \sigma_\eta - \sigma_\zeta &= \pm\, 2k\,, \\ \sigma_\zeta - \sigma_\xi &= \pm\, 2k\,. \end{aligned} \tag{19.2}$$

The six planes which satisfy eqs. (19.2) form a hexahedral prism, which is the surface of loading. Suppose that we are on the edges of the prism, corresponding to the first of the conditions (19.2) when $f = \sigma_\xi - \sigma_\eta$. The derivatives of the functions f are $\partial f/\partial \sigma_\xi = 1$, $\partial f/\partial \sigma_\eta = -1$, $\partial f/\partial \sigma_\zeta = 0$. By formula (17.1) we obtain

$$\begin{aligned} \dot{e}^{\mathrm{p}}_\xi &= H(\dot{\sigma}_\xi - \dot{\sigma}_\eta)\,, \\ \dot{e}^{\mathrm{p}}_\eta &= -H(\dot{\sigma}_\xi - \dot{\sigma}_\eta)\,, \\ \dot{e}^{\mathrm{p}}_\zeta &= 0\,. \end{aligned} \tag{19.3}$$

It is natural to suppose that the function H in its turn depends on the maximum tangential stress, that is on the quantity $\sigma_\xi - \sigma_\eta$; then the expressions for the rate of plastic deformation can be integrated and we obtain

$$\begin{aligned} e^{\mathrm{p}}_\xi &= h(\sigma_\xi - \sigma_\eta)\,, \\ e^{\mathrm{p}}_\eta &= -h(\sigma_\xi - \sigma_\eta)\,, \\ e^{\mathrm{p}}_\zeta &= 0\,. \end{aligned} \tag{19.4}$$

Thus although we took as a basis the expressions of the flow theory, we have finally derived formulae for the relationship between plastic deformations and stresses, as in the deformation theory of plasticity. The mechanical meaning of formula (19.4) is both simple and obvious; plastic deformation is pure shear in the plane $\xi O \eta$, with active plastic deformation the magnitude of this shear depends uniquely on the corresponding tangential stress, this relationship is maintained so long as the inequality $\sigma_\xi > \sigma_\zeta > \sigma_\eta$ is observed. As soon as this inequality is infringed another pair of the three stresses σ_ξ, σ_η and σ_ζ must be taken as maximum and minimum stresses respectively and formulae (19.3) are then also valid for other indices. These may again be integrated. Suppose for example that we have arrived on the face of the surface of loading $\sigma_\xi - \sigma_\zeta = \pm 2k$. Then integrating expressions which are analogous with expressions (19.3) we obtain

$$e_\xi^{\mathrm{p}} = h(\sigma_\xi - \sigma_\zeta) + e_\xi^{\mathrm{p}\prime} ,$$

$$e_\eta^{\mathrm{p}} = e_\eta^{\mathrm{p}\prime} ,$$

$$e_\zeta^{\mathrm{p}} = -h(\sigma_\xi - \sigma_\zeta) .$$

Here the primes denote plastic deformations which have accumulated during the interval when the point of loading was on the face $\sigma_\xi - \sigma_\eta = \pm 2k$, and which are calculated by formulae (19.4).

The case in which the point of loading remains on the edge of the loading surface requires special consideration. If for example, $\sigma_\xi = \sigma_\eta > \sigma_\zeta$, two conditions are fulfilled simultaneously, namely, $\sigma_\xi - \sigma_\zeta = \pm 2k$ and $\sigma_\eta - \sigma_\zeta = \pm 2k$. Fig. 21 shows a section of a prism in the neighbourhood of the edge of an octahedral plane, the normals to the prism surfaces lie in this place.

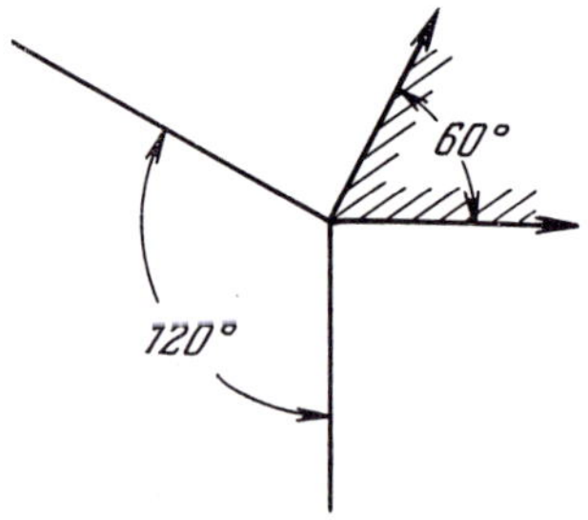

Fig. 21.

At the edge between faces the normals to the faces form an angle which contains all possible directions of the increments of plastic deformation. A vector that is within this angle may be represented as a linear combination of vectors of the extreme normals with positive coefficients. The rate of plastic deformation of which the vector is normal to the first plane $\sigma_\xi - \sigma_\zeta = +2k$ is

$$\dot{e}^p_\xi = H_1(\dot{\sigma}_\xi - \dot{\sigma}_\zeta)\,, \quad \dot{e}^p_\eta = 0\,, \quad \dot{e}^p_\zeta = -H_1(\dot{\sigma}_\xi - \dot{\sigma}_\zeta)\,.$$

In exactly the same way the rate that is normal to the second face $\sigma_\eta - \sigma_\zeta = +2k$

$$\dot{e}^p_\xi = 0\,, \quad \dot{e}^p_\eta = H_2(\dot{\sigma}_\eta - \dot{\sigma}_\zeta)\,, \quad \dot{e}_\zeta = -H_2(\dot{\sigma}_\eta - \dot{\sigma}_\zeta)\,.$$

Adding these rates we obtain

$$\dot{e}^p_\xi = H_1(\dot{\sigma}_\xi - \dot{\sigma}_\zeta)\,,$$

$$\dot{e}^p_\eta = H_2(\dot{\sigma}_\xi - \dot{\sigma}_\zeta)\,,$$

$$\dot{e}^p_\zeta = -(H_1 + H_2)(\dot{\sigma}_\xi - \dot{\sigma}_\zeta)\,.$$

Here σ_ξ is written in place of σ_η. The quantity $(H_1 + H_2)$ should be considered as a function of $\sigma_\xi - \sigma_\zeta$. Indeed this case in which $\sigma_\xi = \sigma_\eta$ is, in effect, uniaxial compression. As we have considered that plastic deformation is independent of the hydrostatic part of the stress tensor let us apply to the body a hydrostatic compression of intensity $-\sigma_\xi$. Then the stresses on the ξ and η axes become zero and the compressive stress $-(\sigma_\xi - \upsilon_\zeta)$ remains in the ζ direction.

The deformation in the direction of the ζ axis is uniquely dependent on this stress, namely, $e^p_\zeta = -h(\sigma_\xi - \sigma_\zeta)$. But this expression should be obtained by integrating the last of the flow equations given above. The final result is written as

$$e_\zeta = -h(\sigma_\xi - \sigma_\zeta)\,, \quad e_\xi = \lambda h(\sigma_\xi - \sigma_\zeta)\,,$$
$$e_\eta = (1 - \lambda)h(\sigma_\xi - \sigma_\zeta)\,. \tag{19.5}$$

The parameter λ remains indeterminate in expressions (19.5). Its significance is that the condition $\sigma_\xi = \sigma_\eta$ limits the selection of possible stressed states;

under these conditions there must be a certain free deformation in order for the equations of compatibility of deformations to be fulfilled. On the other hand, certain consequences of the conclusion that the deformation is indeterminate at the singular point on the loading surface are contrary to experimental results. For example, in simple tension or compression transverse deformations may be of any value provided only that the volume remains constant. This physically inacceptable result is an unavoidable consequence of the very extensive schematisation of the true state of affairs. In actuality we may probably speak of considerable instability of transverse deformation in tension or compression. This means that on applying small transverse stresses in some transverse direction we reach a condition in which all the transverse plastic deformation occurs in this direction. Test results of this kind do not exist and we consider it more correct to consider the theory of flow with a piecewise-linear loading surface as an approximation to a more realistic physical theory based on a smooth loading surface. As usual, we indicate the extreme consequences of the approximation that has been adopted, which are clearly contradictory to experience and common sense, but very often the results of calculation by this piecewise-linear theory are sufficiently satisfactory and the error that they introduce is amply justified by their great simplicity.

Here we shall not consider in detail the model of translational work-hardening with a singular surface of loading. The simple diagram shown in fig. 22 (due to Hodge [22]) illustrates the difference between kinematic and isotropic work-hardening in the two-dimensional case. The number 1 denotes the original surface of loading, the straight line *OPM* represents the path of loading. With isotropic work-hardening the hexagon that is formed by the

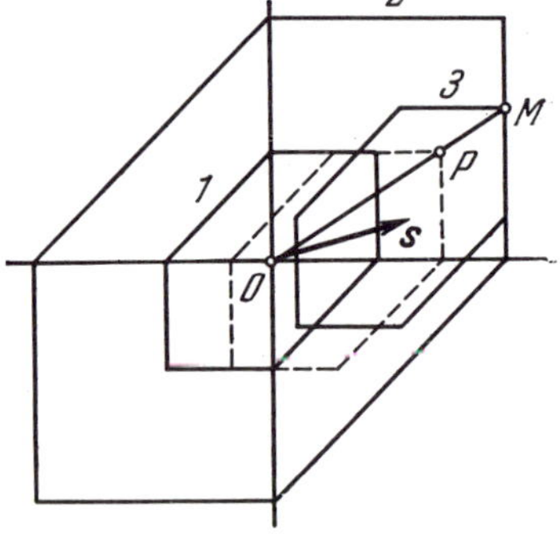

Fig. 22.

surface of loading extends uniformly and during proportional loading the position of the point M remains unchanged in relation to the apex of the hexagon, only the scale changes. The vector of plastic deformation increment is always in the direction of the normal to the side of the hexagon which contains the point M; therefore, the vector of total plastic deformation is always in the direction of the abscissa. If work-hardening is linear the plastic deformation is proportional to the change in linear dimensions of the surface of loading. In the case of translational work-hardening the origin of the surface of loading first moves to the right until the point of loading reaches the point P. The corresponding position of the surface of loading is shown dotted. Over the segment OP of the path of loading the theories of isotropic work-hardening and of translational work-hardening both predict the same magnitude and direction of the plastic deformation vector. Over the segment PM of the path of loading, the loading surface moves in the same direction so that the apex of the hexagon always remains on the path of loading. The number 3 denotes the position of the surface of loading at the point M. If work-hardening is considered linear the radius vector s of the centre of the hexagon represents the plastic deformation to some scale. We shall not derive the corresponding formula for plastic deformation but will merely note that in calculating deformation, for example, in the case of a plane stressed state we cannot limit ourselves to a two-dimensional model, but must consider motion of the surface of loading in a nine-dimensional space, as has already been mentioned in section 18. This problem is considered in detail in the article by Khuan-Ke-chzhi [144].

The use of a St. Venant prism as the surface of loading is not the only example of application of piecewise-linear function of loading to the theory of plasticity. In solving particular problems certain of the stresses can usually be made zero and it then suffices to consider a subspace of the stress space. Further, it is sometimes possible to predict the change of stress, that is to isolate the region in the stress space in which the paths of loading are located. A smooth surface can always be more or less accurately approximated by a polygon. The general theory of plastic flow (developed by Hodge [232]) is based on this arbitrary linear approximation of the surface of loading.

It should be noted that in the case of proportional loading the hypothesis of translational work-hardening does not lead to equations of the deformation theory. This qualification is necessary in connection with the widely held opinion that the deformation theory is universally valid for proportional loading.

20. Slip theory

The theories of plasticity that we have considered so far are based on hypotheses of a formal nature, which are selected without taking the actual structure of the metal into account. This approach has advantages and disadvantages. On the one hand, the general laws of plasticity formulated for an arbitrary body irrespective of its physical nature, permit a single theory to cover the plasticity of metal, the limiting equilibrium of soils, brittle fracture of concrete structures, and so on. On the other hand when we come to particularise these general laws to describe some real medium we are obliged to use macro-experimental results and to describe the effects without explaining them. In 1949, Batdorf and Budiansky [164] proposed a new theory of plasticity which they termed the slip theory. The basic idea of this theory is as follows. A real metal is a conglomeration of random crystalline grains. Plastic deformation of a single crystal results from shear in a particular direction in a particular crystallographic plane. Let $\mathbf{n}$ be the normal to the slip plane, $\boldsymbol{\beta}$ the direction of slip in this plane. The combination of the two unit vectors $\mathbf{n}$ and $\boldsymbol{\beta}$ defines what is termed the slip system. If the tangential stress $\tau_{n\beta}$ in the plane of $\mathbf{n}$ and direction of $\boldsymbol{\beta}$ exceeds the yield point the crystal is deformed plastically in pure shear $\gamma_{n\beta}$. The quantity $\gamma_{n\beta}$ is a fully determinate function of $\tau_{n\beta}$. In real crystals there are always several possible slip systems which correspond to close-packed planes and directions in which neighbouring atoms are closest together. Thus, crystals with a face-centred cubic lattice have 12 slip systems. Slip usually occurs in several systems simultaneously, and shear in one slip system tends to strengthen the other systems. Thus the mechanism of plastic deformation accompanied by work-hardening is fairly complicated even in a single crystal. In the simplest scheme, proposed by Batdorf and Budiansky, each grain is supposed to have only one slip system. If a certain stressed state is set up in the body, when the stress in certain grains becomes sufficiently great the tangential stress in the slip system is greater than the ultimate stress and the grain undergoes plastic shear deformation. The true picture is fairly complicated, deformation is only possible provided neighbouring grains do not hinder motion. The macroscopic effect of plastic deformation of the body as a whole is observed when chains of plastically deformed grains are formed. In the initial stages of deformation it may be supposed that certain units remain in the elastic state and are displaced and rotate as a whole. Hence the great heterogeneity of plastic deformation, particularly in materials with only slight work-hardening where deformation is accompanied by localisation of shears into certain regions [37].

In the slip theory these difficulties are overcome by introducing certain

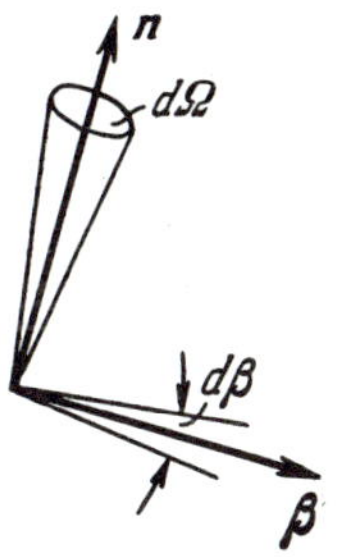

Fig. 23.

simple hypotheses. Let us arbitrarily select two mutually perpendicular directions **n** and **β**. If the number of crystal grains in the volume under consideration is large there is always a certain number of grains for which the normal to the plane of possible slip (which is postulated to be unique), lies within a cone with axis **n** and solid angle $\mathrm{d}\Omega$ (fig. 23). Such grains will be said to have a slip plane **n**. If the material is initially isotropic the grain distribution is random and there are no preferential directions relative to the volume of grains with slip plane **n** in the total volume of the body which is subject to uniform stress proportional to $\mathrm{d}\Omega$. If the number of grains with slip plane **n** is large, in some of them the slip direction lies within the angle $\mathrm{d}\beta$ with bisector **β** and their relative volume is proportional to $\mathrm{d}\beta$. Therefore, the volume of grains which have a slip system **nβ** is proportional to $\mathrm{d}\Omega\,\mathrm{d}\beta$. A tangential stress $\tau_{n\beta}$ acts in the slip system **nβ** and the corresponding grains undergo pure shear deformation $\gamma^{\mathrm{p}}_{n\beta} = F(\tau_{n\beta})$. We now introduce the hypothesis that the deformation of grains with slip system **nβ** causes the same total deformation of the body, which is proportional to the relative volume of the corresponding grains, namely

$$\mathrm{d}\gamma^{\mathrm{p}}_{n\beta} = F(\tau_{n\beta})\,\mathrm{d}\Omega\,\mathrm{d}\beta\ .$$

Accordingly, deformation of the body as a whole is a result of the superposition of an infinite number of pure shear deformations for all possible slip systems **nβ**. In order to calculate this deformation let us convert to components of the deformation tensor relative to fixed axes x_1, x_2, x_3, using the conversion formulae of second rank tensor components (1.7). Taking the directions **n** and **β** as directions 1 and 2 in the new system of coordinates we must take all values of e_{ij} equal to zero except $e_{12} = \frac{1}{2}\gamma_{n\beta}$. Then

$$\mathrm{d}e^{\mathrm{p}}_{ij} = (\beta_{1i}\beta_{2j} + \beta_{2i}\beta_{1j})\,\mathrm{d}e_{12}\ .$$

But $\beta_{1i} = n_i$, $\beta_{2i} = \beta_i$, consequently

$$de^{p}_{ij} = \tfrac{1}{2}\,(n_i\beta_j + n_j\beta_i)F(\tau_{n\beta})\,d\Omega\,d\beta\,.$$

The deformation formulae are

$$e^{p}_{ij} = \tfrac{1}{2}\int_{\Omega}\int_{-\frac{\pi}{2}}^{+\frac{\pi}{2}} (n_i\beta_j + n_j\beta_i)F(\tau_{n\beta})\,d\Omega\,d\beta\,. \qquad (20.1)$$

Here the tangential stress is calculated by the formula

$$\tau_{n\beta} = \sigma_{ij}n_i\beta_j\,. \qquad (20.2)$$

For practical calculations of deformation the directions **n** and **β** are given in terms of Euler angles through which integration is performed. The fundamental difficulty here is that $F(\tau)$ differs from zero when $\tau > \tau_T$, consequently the integrals in formula (20.1) do not extend over the entire surface of a sphere of unit radius but only over a certain region of it. Batdorf and Budiansky's first article gave a solution of the problem of determining the function F from a given tensile-test curve. Considering that deformation occurs only along axis 1, by formula (20.2) we obtain $\tau_{n\beta} = \sigma_{11}n_1\beta_1$ and as plastic deformation occurs when $\tau_{n\beta} > \tau_T$ we should have

$$n_1\beta_1 > \frac{\tau_T}{\sigma_{11}}\,.$$

This condition defines the region of integration. The authors presupposed that the function $F(\tau)$ is of the form

$$F(\tau) = \sum_{1}^{N} a_n\left(\frac{\tau}{\tau_T} - 1\right)^n\,.$$

Then the dependence of the plastic elongation on normal stress is

$$e^{p}_{11} = \sum_{1}^{N} a_n g_n\left(\frac{\sigma_{11}}{2\tau_T}\right)\,.$$

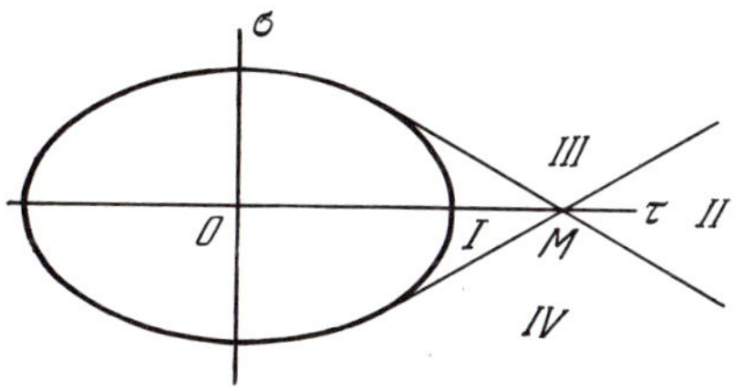

Fig. 24.

The functions g_n were calculated by a very laborious numerical integration for the case $N = 5$.

According to the slip theory plastic deformation starts when the yield point is reached in certain of the slip systems; therefore, the initial surface of loading corresponds to the condition of maximum Tresca tangential stress. Indeed if $\tau_{max} = \tau_T$ there is always a group of crystal grains for which this stress is a tangential stress in the slip system. For successive surfaces of loading the point of loading is conical. Investigations based on eqs. (20.1) are extremely complicated; Cicala [184] completely analysed the simple case of a thin-walled tube which is first stretched and then twisted. We shall give only the final results, without reproducing the calculations which are fairly difficult. As there are two stresses in the test the results may be represented in planes with coordinates σ and τ (fig. 24). The initial surface of loading is an ellipse, the equation of which is

$$\sigma^2 + 4\tau^2 = \text{const} .$$

The loading path that is considered passes from point O along the abscissus to point M, and then along a line issuing from point M. The tangents to the initial ellipse from point M divide the σ–τ plane into four regions. If the stress increments are such that the loading point lies within region I, elastic unloading occurs. If there are plastic deformation increments when the stress increases, corresponding to region II, the deformations are determined by the formulae of the deformation theory of plasticity; finally, for loading paths that lie in regions III and IV the relationships between stresses and deformations are given by formulae (20.1), in a fairly complicated form. The author obtained the final result for orthogonal additional loading, for which the loading path from point M is orthogonal to OM; this means that there is additional torsion with constant tensile force. For this case

$$\frac{\Delta\tau}{\Delta\gamma} = G^* = \frac{G}{1 + \frac{3}{2} G \left(\frac{1}{E_s} - \frac{1}{E_t} \right)} . \tag{20.3}$$

Comparisons with experimental results which have been made by numerous authors show that formula (20.3) gives low results for G^* [125, 165].

The qualitative theory developed by Ciçala is very similar to that which we found for the model considered in section 16. The arrangement of the regions shown in fig. 24 is entirely analogous to the arrangement shown in fig. 19; although of course fig. 19 relates to deformation planes; in the stress planes the sides of the angles I and II are not extensions of one another and are not tangential to the initial curve. This similarity between the qualitative results is not surprising. Both the Batdorf-Budiansky theory and the model that we have considered, represent the body as a collection of elastic-plastic elements to which the simplest of properties are allotted. In the slip theory these elements are idealised crystal grains with one unique slip system. During active plastic deformation, the tangential stress and shear are related by a unique functional relationship and, therefore, the behaviour of the body as a whole is described by certain equations of the same type as those of the deformation theory, until all the elements are subject to deformation. When the stress is increased plastic deformation can occur in new elements but unloading should not then occur. This position corresponds to additional loading within the angle II. On additional loading within the regions III and IV, some of the elements are additionally loaded, and plastic deformations may occur in new elements, but certain of the plastically deformed elements are unloaded and return to the elastic state. This is the reason for the complexity of the analysis for regions III and IV.

As the authors themselves have recognised [165] the Batdorf-Budiansky slip theory as a whole has not been confirmed experimentally. For though certain qualitative effects are predicted the quantitative results are not entirely satisfactory. However, the basic idea of this theory has been developed in various forms by a number of authors. It should be noted that as early as 1938, Taylor [370] attempted to relate the properties of a polycrystalline aggregate to the properties of its component grains. Here the fundamental idea was that arbitrary deformation that is not accompanied by change in volume can be represented as the result of shears in five slip systems, and since the number of slip systems in the crystal grain is more than five it can be deformed in various ways. Further development of Taylor's idea is contained in the work of Bishop and Hill [171] and variants of the slip theory developed by Lin [285, 286] are also relevant.

21. Model of a plane body

It is almost impossible to make practical use of the slip theory eqs. (20.1) because of their complexity. As yet practical application of the theory to calculations has been limited to problems of stability, for which it has been possible to show that the relationship between stress increments is such that unloading does not occur and the slip theory equations reduce to the deformation theory equations. At the same time, the hypotheses that underlie the slip theory are not so reliable as to make it worth while to seek accurate solutions. To simplify the fundamental relationships of the theory and to obtain more graphic qualitative results, we shall, following Klyushnikov [59], consider a certain model two-dimensional medium, for which $e_{33} = 0$, $\sigma_{33} = \frac{1}{2}(\sigma_{11} + \sigma_{22})$. Then $\bar{\sigma}_{11} = -\bar{\sigma}_{22} = \frac{1}{2}(\sigma_{11} - \sigma_{22})$, $e^{\mathrm{p}}_{11} + e^{\mathrm{p}}_{22} = 0$ because of the condition of incompressibility, and therefore we may put $e^{\mathrm{p}}_{11} = -e^{\mathrm{p}}_{22} = \frac{1}{2}(e^{\mathrm{p}}_{11} - e^{\mathrm{p}}_{22})$. We shall consider that this material can be deformed only by shears in planes normal to the planes $x_1 x_2$.

The following tangential stress occurs over a certain area, which is at an angle ω to the x_1 axis

$$\tau = \tfrac{1}{2}(\sigma_{11} - \sigma_{22}) \sin 2\omega + \sigma_{12} \cos 2\omega \,. \tag{21.1}$$

In elements for which τ in the direction of possible slip exceeds the yield point τ_{T}, plastic shear occurs (which is proportional to $\mathrm{d}\omega$) and forms part of the total deformation of the volume. Therefore,

$$\mathrm{d}\gamma = F(\tau)\,\mathrm{d}\omega \,.$$

The corresponding components of plastic deformation on the axes are

$$\mathrm{d}e^{\mathrm{p}}_{11} = -\mathrm{d}e^{\mathrm{p}}_{22} = \tfrac{1}{2}(\mathrm{d}e^{\mathrm{p}}_{11} - \mathrm{d}e^{\mathrm{p}}_{22}) = \tfrac{1}{2}\,\mathrm{d}\gamma \sin 2\omega = \tfrac{1}{2} F(\tau) \sin 2\omega\,\mathrm{d}\omega$$

$$\mathrm{d}e^{\mathrm{p}}_{12} = \tfrac{1}{2}\,\mathrm{d}\gamma \cos 2\omega = \tfrac{1}{2} F(\tau) \cos 2\omega\,\mathrm{d}\omega \,.$$

As coordinates in the two-dimensional stress space take the quantities $Q_1 = \frac{1}{2}(\sigma_{11} - \sigma_{22})$ and $Q_2 = \sigma_{12}$ corresponding to the deformations $q_1 = \frac{1}{2}(e^{\mathrm{p}}_{11} - e^{\mathrm{p}}_{22})$ and $q_2 = e^{\mathrm{p}}_{12}$. The initial surface of loading is represented in these coordinates by a circle $Q_1^2 + Q_2^2 = \text{const.}$

Plastic deformation found by the formulae

$$q_1 = \tfrac{1}{2} \int F(\tau) \sin 2\omega \, d\omega ,$$
$$q_2 = \tfrac{1}{2} \int F(\tau) \cos 2\omega \, d\omega . \qquad (21.2)$$

Here

$$\tau = Q_1 \sin 2\omega + Q_2 \cos 2\omega . \qquad (21.3)$$

If additional loading is applied from a certain stressed state then

$$\delta q_1 = \tfrac{1}{2} \int F'(\tau) \delta\tau \sin 2\omega \, d\omega ,$$
$$\delta q_2 = \tfrac{1}{2} \int F'(\tau) \delta\tau \cos 2\omega \, d\omega . \qquad (21.4)$$

Here it should be borne in mind that the limits of integration can also alter on additional loading. Apply expressions (21.4) to the case of proportional loading. If β is the angle between the loading line and the Q_1 axis we may put

$$Q_1 = Q \cos\beta , \quad Q_2 = Q \sin\beta ,$$

and consequently

$$\tau = Q \sin(2\omega + \beta) .$$

Instead of resolving the vector $\mathbf{q}$ into the base vectors $\mathbf{i}_1, \mathbf{i}_2$ of the initial coordinate system let us introduce the base vectors $\mathbf{e}_1, \mathbf{e}_2$ which are asso-

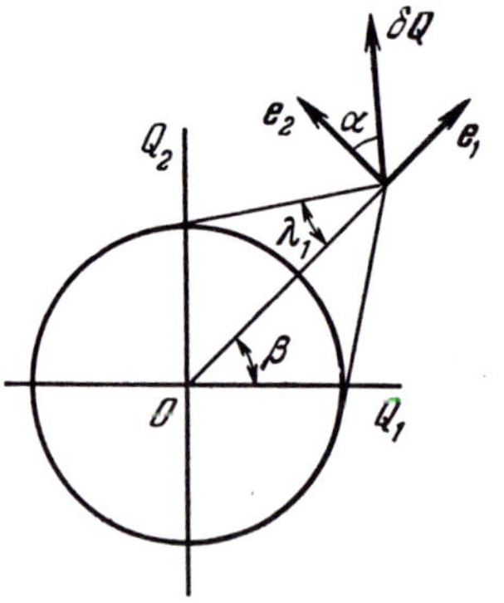

Fig. 25.

ciated with the proportional loading path (fig. 25). In vector form eq. (21.2) is

$$\mathbf{q} = \tfrac{1}{2} \int F(\tau)[\mathbf{i}_1 \sin 2\omega + \mathbf{i}_2 \cos 2\omega]\, d\omega\,.$$

But

$$\mathbf{i}_1 = \mathbf{e}_1 \cos\beta - \mathbf{e}_2 \sin\beta\,, \qquad \mathbf{i}_2 = \mathbf{e}_1 \sin\beta + \mathbf{e}_2 \cos\beta\,.$$

Therefore

$$\mathbf{i}_1 \sin 2\omega + \mathbf{i}_2 \cos 2\omega = \mathbf{e}_1 \sin(2\omega + \beta) + \mathbf{e}_2 \cos(2\omega + \beta)\,.$$

Now put

$$2\omega + \beta = \frac{\pi}{2} + \lambda\,.$$

We obtain

$$\mathbf{q} = \tfrac{1}{4} \int F(\tau)(\mathbf{e}_1 \cos\lambda - \mathbf{e}_2 \sin\lambda)\, d\lambda\,. \tag{21.5}$$

Here $\tau = Q \cos\lambda$ and integration is extended to values of λ for which $\tau > \tau_{\mathrm{T}}$, that is

$$\cos\lambda > \frac{\tau_{\mathrm{T}}}{Q}\,.$$

If we take $\cos\lambda_1 = \tau_{\mathrm{T}}/Q$, the region of integration becomes

$$-\lambda_1 \leqslant \lambda \leqslant \lambda_1\,.$$

Now consider the case of additional loading in an arbitrary direction after proportional unloading. Let the increment of the vector $\mathbf{Q}$ be at an angle α to $\mathbf{e}_2$ and let its magnitude be ϵ. Then

$$\Delta Q_1 = \epsilon \sin(\alpha - \beta)\,, \qquad \Delta Q_2 = \epsilon \cos(\alpha - \beta)\,.$$

Consequently

$$\Delta\tau = \epsilon \cos(\alpha - \beta - 2\omega) = \epsilon \sin(\alpha - \lambda)\,.$$

The formula for the increment of the plastic deformation vector is

$$\Delta \mathbf{q} = \tfrac{1}{4}\, \epsilon \int F'(\tau) \sin(\alpha - \lambda)[\mathbf{e}_1 \cos\lambda - \mathbf{e}_2 \sin\lambda]\, d\lambda\,. \tag{21.6}$$

In using this formula it should be borne in mind that with small additional loading only those elements which were plastically deformed receive additional plastic deformation if $\Delta\tau > 0$. If $\Delta\tau < 0$ elastic unloading occurs. Here the following cases are possible:

1. $\alpha \leqslant -\lambda_1$; then $\alpha - \lambda$ is always negative, and elastic unloading occurs throughout the region $-\lambda_1 < \lambda < \lambda_1$, i.e. $\Delta\tau < 0$. The straight lines $\alpha = -\lambda_1$ define the angle of the loading curve. It is readily confirmed that these lines are tangential to the initial circle.

2. $\alpha \geqslant \lambda_1$; then throughout the region the difference $\alpha - \lambda$ is positive and additional loading occurs everywhere. Expressions (21.4) are obtained from expressions (21.2) in which integration is performed by variation.

3. $-\lambda_1 < \alpha < \lambda_1$. In this case in calculating expressions (21.6) integration is performed between the limits $-\alpha \leqslant \lambda \leqslant \alpha$.

Note that the considerations which arise in analysis of the Klyushnikov model, which reproduce in very simplified form the considerations that arise in analysis of the complete Batdorf-Budiansky model, are fully analogous with those which we applied to investigate bending of an elastic-plastic tube in section 16. There we were concerned only with the condition of unloading and the condition of validity of the deformation theory equations; however, it is not difficult to show what occurs when proportional loading is followed by an arbitrary small additional loading. Returning to fig. 17 and that part of the text of section 16 which examines elastic unloading, we suppose that on unloading the axis of additional rotation forms an angle $\varphi > \theta$ with the x axis and thus intersects the plastic zone. Additional stresses of elastic unloading $\sigma' = -\omega\sigma_T \sin(\psi - \varphi)$ occur only in the elastic region, that is when $-\theta < \psi < \varphi$ and $\pi - \theta < \psi < \pi + \varphi$. Calculating the moments due to these stresses, the results may be written

$$\begin{aligned} \Delta Q_x &= \Delta q_x \frac{G_t(\varphi) + G_t(\theta)}{2} + \Delta q_y (\sin^2\varphi - \sin^2\theta)\,, \\ \Delta Q_y &= \Delta q_x (\sin^2\varphi - \sin^2\theta) + \Delta q_y \frac{G_s(\varphi) + G_s(\theta)}{2}\,. \end{aligned} \tag{21.7}$$

Here, evidently

$$\frac{\Delta q_y}{\Delta q_x} = \tan\varphi\,.$$

Expressions (21.7) are fairly complicated non-linear relationships between increments of stress and deformation, and it is hardly to be expected that the corresponding expressions for real materials should be simple. Both the slip theory and consideration of a simple model lead to rather pessimistic conclusions concerning future progress of the theory of plasticity. In the first place, the basic idea of the slip theory undoubtedly reflects correctly the fundamental mechanism of plastic deformation in polycrystalline metal; the hypotheses that are introduced to derive the equations undoubtedly greatly simplify the true position but still give rise to complicated calculations. The proposed model of an elastic-plastic body seems to be the simplest example of a system of simple elements such that its mechanical behaviour is governed by the possibility of plastic additional loading and elastic unloading of these elements. In both the slip theory and the simple model, integration of the plasticity equations for any particular loading path is difficult. The equations do not lead to qualitative conclusions concerning the nature of the change in the surface of loading with more or less complicated loading paths.

22. Flow with a singular surface of loading

In the foregoing we have elucidated the limits of applicability of the deformation theory of plasticity, and also in considering the slip theory we encountered surfaces of loading which have a conical point at the point of loading. On the other hand, the flow theory is based on the fact that the vector of the plastic deformation increment is normal to the surface of loading. However, at the conical point the direction of the normals becomes indeterminate and therefore some additional hypotheses must be formulated.

In section 19 we have already encountered the laws of flow which use singular surfaces of loading. However, there we were considering angles or edges on the surfaces of loading which exist in the initial state. Subsequent changes in the surface of loading represent a similar extension or transition of the surface as a whole. In the slip theory we encountered another type of work-hardening in which the conical point follows the loading path in a certain way and the geometry of the surface of loading undergoes essential change as plastic deformation develops. One of the possible schemes for describing work-hardening of this kind was given by Saunders [344], his starting point is the concept of a surface of flow formed of plane faces. The fundamental hypothesis is that if the point of loading is located on one of the planes that form the surface of loading, then as the load is increased this, and only this plane moves parallel to itself in a single direction. The increment

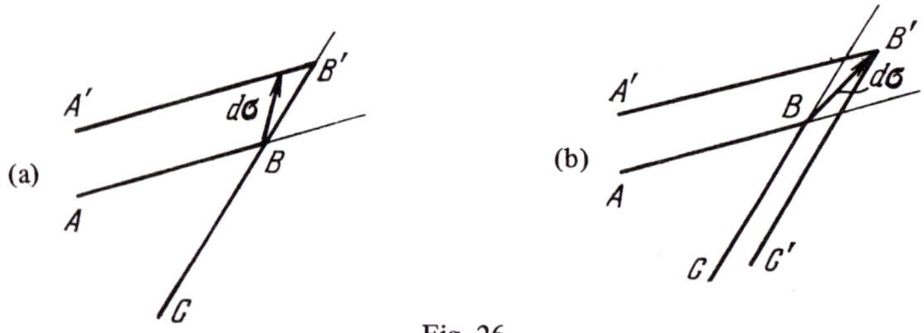

Fig. 26.

of plastic deformation is proportional to the displacement of the corresponding plane. If the point of loading is on an edge, each of the intersecting planes is displaced parallel to itself in a single direction and the increments of plastic deformation are calculated for each plane separately and added. The case when the point of loading is located on an edge is shown schematically in fig. 26. Here *AB* and *BC* are planes of the surface of loading. If the direction of additional loading is outside the angle formed by the planes of loading then only one plane is displaced (fig. 26a). If the direction of the additional loading vector lies within this angle, both planes *AB* and *BC* are displaced (fig. 26b). Now consider the change in a surface of loading that consists of the planes as a whole. Fig. 27 shows the initial surface of loading and the subsequent surface of loading for the given loading path.

If the initial surface of loading is smooth it can always be represented as the envelope of a family of loading planes; we thus obtain an infinite number of linear loading functions. Repeating the considerations for a polyhedral surface of loading we find that in proportional loading each successive surface of loading consists of a cone of tangents to the initial surface from the loading point and part of this surface which is bounded by a cone and a convex volume (fig. 28).

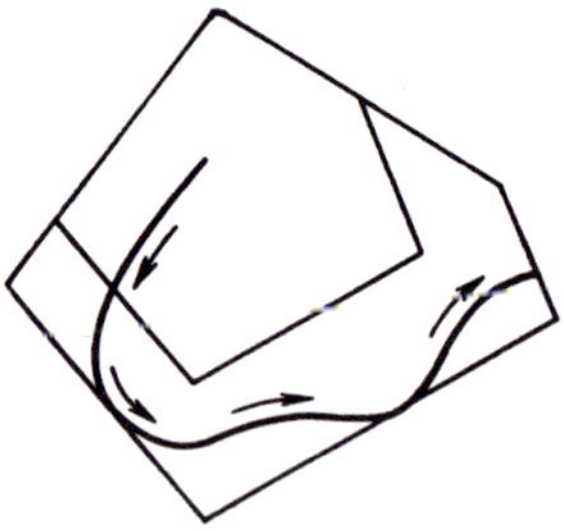

Fig. 27.

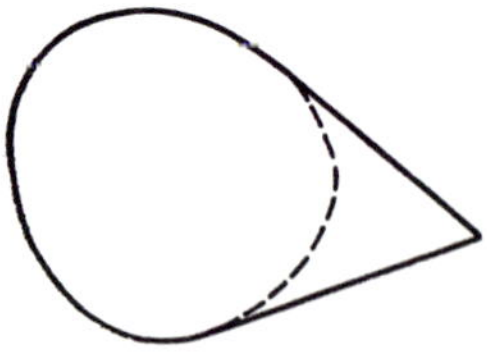

Fig. 28.

It should be noted that Saunders' hypothesis does not, for example, describe the Bauschinger effect. The proposition that plane surfaces of loading are displaced parallel to themselves, and then only when the loading point is on the given surface, is evidently oversimplified. Saunders stipulates the possibility of interaction between loading planes but gives no specific hypothesis to describe this interaction.

The formulae of plastic strain increments on additional loading for a plane loading path according to Saunders' theory were given in the work of Klyushnikov [59]. Klyushnikov [58] proposed a new variant of the flow theory based on the following hypotheses (the formulation relates to a two-dimensional plane of a nine-dimensional stress space):

1. The hypothesis of continuity. If the distance between corresponding points of two loading paths is less than δ then the distance between corresponding points of the deformation path is less than $e(\delta)$ where $e \to 0$ when $\delta \to 0$.

2. During plastic deformation the loading curve alters continuously, and during additional loading points on the plasticity curve in the neighbourhood of the region of additional loading can be displaced only in the direction opposite to the origin of coordinates.

3. The relationship between the vectors $\mathbf{d\sigma}$ and $\mathbf{de}^{\mathrm{p}}$ depends only on local characteristics of the loading curve in the neighbourhood of the loading point.

The fundamental result of Saunders' and Klyushnikov's theories is that, with certain limitations applied to the loading path, the differential relationships of the theory may be integrated to give equations of the deformation type which are applicable to some particular region. As in the case of the slip theory the use of this theory for arbitrary loading paths leads to almost insurmountable difficulties.

3

Linear Viscoelastic Media

The theory of linear viscoelasticity is a natural generalisation of classical trends in solid state mechanics, namely of the theory of elasticity and the hydromechanics of a viscous liquid. The properties of elasticity and viscosity can be combined in various ways to construct numerous hypothetical media which correspond in behaviour, at least qualitatively, to real substances that are intermediate between solids and liquids. Rheological models have been extensively applied in study of the properties of colloidal solutions, suspensions and high-molecular-weight materials, although the physicists and physical chemists who have taken a leading part in the development of theoretical rheology were more interested in qualitative elucidation of the observed effects than in accurate quantitative description. There have been many publications on the development of a formal theory of viscoelasticity and it would appear that certain trends have arisen not so much from practical needs as from the internal logic of development of the science. Lately there has been lively interest in the theory of viscoelasticity, the extensive practical application of polymers has encouraged study of basic methods of calculating structural components made from these materials. The approximations of the linear theory suffice for a certain range of problems and, therefore, existing theory finds direct application. In recent years considerable work has been done on the development of general methods of solving problems of viscoelasticity and on individual more or less difficult problems; the published material is so extensive that a special monograph would be required to provide a reasonably complete presentation.

Linear theories are quite unsuited to the description of creep in metals, which for a long time was the only field of technical interest and so creep theory developed along another path. In effect the first serious application of the theory of linear viscoelasticity was the study of creep in concrete. The application of these ideas to soils, rocks and certain structural materials was somewhat sporadic.

The field of application of the theory of linear viscoelasticity is now much wider, but the limits of its applicability are not yet clearly established. We thought it useful to devote a chapter to this theory, partly to elucidate the different fundamental approaches to the construction of a creep theory and partly to emphasise the fundamental difference between the behaviour of metals and that of polymeric materials and also to give certain information about the creep of polymeric materials as such. For many of these materials the region of linearity is extremely narrow and the theory needs improving to allow for non-linear relationships between stress and deformation. Certain information concerning this kind of generalisation is given in the next chapter.

Only information of the most general kind has been selected for the present chapter, the author has devoted rather more attention to problems on which he has personally worked.

23. The simplest viscoelastic body

The effect of creep, i.e. the increase in deformation of a loaded body with time, has long been known, particularly in organic materials. However, there was no great need to allow for creep in the design of structures and the first observations of creep were purely qualitative. On first acquaintance with creep, for example in the silk threads used in physical instruments, the analogy with viscous flow of liquids is strikingly obvious. In a viscous liquid the stress and rate of strain are, of course, related by Newton's law of viscosity

$$\sigma = 3\eta\dot{\epsilon} .$$

In an elastic material the stress and strain are related by Hooke's law

$$\sigma = Ee .$$

In attempting to describe time-dependent deformation of a solid the first thought is to combine the properties of viscosity and elasticity. This leads to Maxwell's concept of a viscoelastic body for which the relationship between stress and deformation is given by the equation

$$\dot{e} = \frac{\dot{\sigma}}{E} + \frac{\sigma}{3\eta} . \tag{23.1}$$

If the stress acting on the body is constant the rate of deformation is constant. Now suppose that the body is extended by a stress σ_0 after which the deformation e is fixed so that when $t = 0$, then $\sigma = \sigma_0$, when $t > 0$, $\dot{e} = 0$. Taking

$$\frac{3\eta}{E} = \tau ,$$

from (23.1) we obtain

$$\dot{\sigma} + \frac{\sigma}{\tau} = 0 .$$

Integrating this equation with allowance for the initial condition we obtain

$$\sigma = \sigma_0 \exp\left(-\frac{t}{\tau}\right) . \tag{23.2}$$

Formula (23.2) gives the law of stress relaxation with constant deformation. The constant τ is termed the relaxation time.

By combining the properties of elasticity and viscosity in parallel we obtain an equation that differs from (23.1), namely

$$\sigma = Ee + 3\eta\dot{e} = E(e + \tau\dot{e}) . \tag{23.3}$$

A body whose behaviour is described by eq. (23.3) is called a Voigt viscoelastic body. Unlike a Maxwell body, when constant loading is applied a Voigt body does not deform indefinitely, its elongation tends to a value σ_0/E according to the law

$$e = \frac{\sigma_0}{E}\left[1 - \exp\left(-\frac{t}{\tau}\right)\right] . \tag{23.4}$$

When the deformation is maintained constant a Voigt body does not relax, the stress in it remains constant. In this case the time τ is termed the delay time.

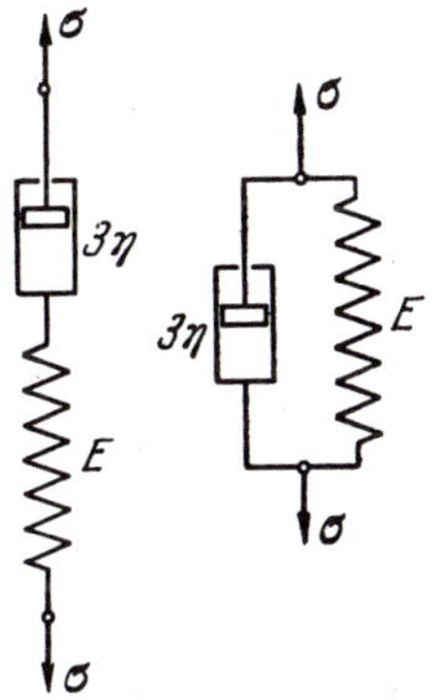

Fig. 29.

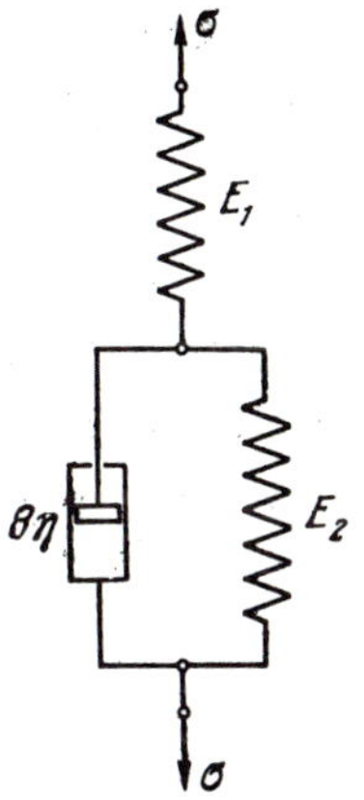

Fig. 30.

To illustrate the behaviour under load of Maxwell and Voigt bodies, consider a model consisting of a viscous resistance (an oil-filled cylinder containing a piston with a small hole) and a spring. As shown in fig. 29 these components may be connected in series or in parallel. If the viscosity of the cylinder and piston is 3η, the spring stiffness E, the applied force σ and the total displacement e, motion of the first model is described by eq. (23.1) and that of the second by eq. (23.3). All kinds of media which display time-dependent mechanical behaviour are commonly modelled in this way and the models are called rheological models. Some authors attach great importance to them and consider that the problem of describing the properties of some particular medium is solved when an adequate rheological model has been found. In reality any model of this kind is only an analogy and not a means of explaining the effect. The structure of materials does not include real components which display ideal elasticity or ideal viscosity; it is sometimes possible to distinguish structural components to which elastic or viscous properties can be attributed but their real relationships and real interactions are very remote from schemes such as those illustrated in fig. 29. It is true that the possibility of constructing a rheological model ensures that the description of the process is reliable and guards against such errors as infringing the law of conservation of energy of the system. Given a rheological model it is always possible to isolate the elastic energy and to determine the positive rate of energy dissipation.

There are probably no materials in nature which follow exactly either eq. (23.1) or (23.3); an ideal Maxwell liquid and an ideal Voigt viscoelastic body represent certain conceivable models with properties which are fairly remote from those of all real bodies. Until recently there have been continued attempts to use Maxwell's equation to describe the properties of soils, certain plastics, rubber-like materials and even metals. The Voigt model has frequently been used to describe energy dissipation in elastic vibration. These attempts do not always agree with experimental results.

One particular combination of spring and viscous resistance gives a scheme whose behaviour, at any rate qualitatively, best reproduces the behaviour of real solids under load. The corresponding rheological model is shown in fig. 30. Denoting the spring stiffnesses by E_1 and E_2 respectively the viscosity of the oil system is 3η. The elongation of the first spring on application of the force σ is

$$e_1 = \frac{\sigma}{E_1} .$$

The elongation of the second spring is related to the force by expression (23.3), namely

$$\sigma = E_2 e_2 + 3\eta \dot{e}_2 .$$

The total elongation is $e = e_1 + e_2$; eliminating e_1 and e_2 we obtain

$$\left(\frac{1}{E_1} + \frac{1}{E_2}\right)\sigma + \frac{3\eta}{E_1 E_2}\dot{\sigma} = e + \frac{3\eta}{E_2}\dot{e} .$$

Putting

$$E_1 = E , \quad \frac{E_1 + E_2}{3\eta} = \lambda , \quad \frac{E_2}{3\eta} = \mu ,$$

we obtain the following relationship between stress and deformation,

$$\dot{\sigma} + \lambda\sigma = E(\dot{e} + \mu e) . \tag{23.5}$$

It is obvious by definition that $\lambda > \mu$; this follows from consideration of the model but it can also be shown directly by applying eq. (23.5) to particular cases of loading. Let the load be applied very rapidly so that $\dot{\sigma}$ is large. Then $\dot{e}$ will also be large, σ and e may be neglected, and we find $\dot{\sigma} = E\dot{e}$. The quantity E is the instantaneous modulus of elasticity. Now suppose that the load acts for a fairly long time, the rate of deformation becomes very slow and the derivatives $\dot{\sigma}$ and $\dot{e}$ in (23.5) are negligibly small. Then from (23.5) we obtain $\sigma = E(\mu/\lambda)e$. Then quantity $E(\mu/\lambda)$ is called the long-term modulus of elasticity, it relates the stress and deformation in the limiting conditions. If a load is suddenly applied and then maintained constant, the original deformation is given by the instantaneous modulus: this deformation increases and tends to a value corresponding to the long-term modulus. It is natural to suppose that meanwhile the external force performs positive work, and for this it is necessary that $\lambda > \mu$.

Eq. (23.5) may be solved relative to either stress or deformation. We shall consider for example, that e is a given function of time. Integrate the resulting equation for $\sigma(t)$. The integral of the homogeneous equation is

$$\sigma = C \exp(-\lambda t) .$$

Applying the method of variation of parameters we find

$$C = E \int_{-\infty}^{t} (\dot{e} + \mu e) \exp(\lambda\tau)\, d\tau =$$

$$= E\left[e(t) \exp \lambda t - (\lambda - \mu) \int_{-\infty}^{t} e(\tau) \exp(\lambda\tau)\, d\tau \right].$$

Then the stress is expressed in terms of the deformation as follows:

$$\sigma = E\left[e - (\lambda - \mu) \int_{-\infty}^{t} e(\tau) \exp[-\lambda(t - \tau)]\, d\tau \right]. \tag{23.6}$$

In an exactly similar way we can express e in terms of σ. The result is

$$e = \frac{1}{E}\left[\sigma + (\lambda - \mu) \int_{-\infty}^{t} \sigma(\tau) \exp[-\mu(t - \tau)]\, d\tau \right]. \tag{23.7}$$

A Volterra type integral eq. (23.6) or (23.7) is fully equivalent to eq. (23.5) with the corresponding initial conditions. The lower limit of integration may be made zero instead of negative infinity when $t < 0$ $e = 0$ and $\sigma = 0$.

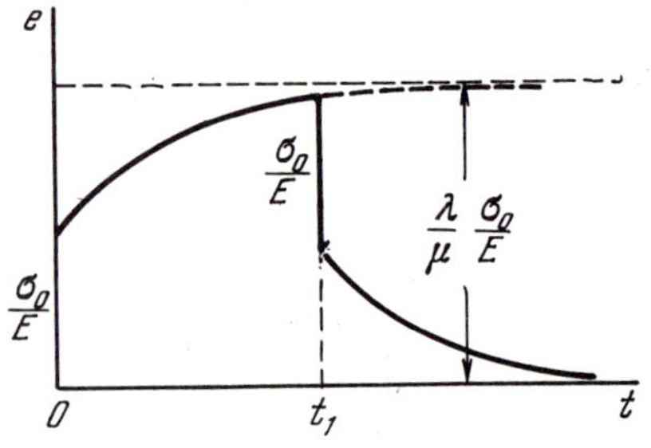

Fig. 31.

Apply the integral eq. (23.7) to the case of a body subject to a constant load applied at the time $t = 0$. Then $\sigma = 0$ when $t < 0$ and $\sigma = \sigma_0 = \text{const}$ when $t > 0$. By formula (23.7)

$$e = \frac{\sigma_0}{E}\left\{1 + \frac{\lambda - \mu}{\mu}[1 - \exp(-\mu t)]\right\}. \tag{23.8}$$

Fig. 31 shows a graph of creep described by eq. (23.8). Now suppose that at the instant t_1 the load is removed so that $\sigma = 0$ when $t > t_1$. By formula (23.7)

$$e = \frac{\sigma_0}{E}\frac{\lambda - \mu}{\mu}\{\exp[-\mu(t - t_1)] - \exp(-\mu t)\}. \tag{23.9}$$

When $t = t_1$ expression (23.9) gives a value of e which differs from that given by formula (23.8) only by the amount of the instantaneous elastic deformation. With unlimited increase in time t the deformation tends to disappear and therefore, all creep deformation is reversible. The body under consideration which is described by eq. (23.5) is called a Kelvin body. Other names for media which are described by eq. (23.5) are encountered in the literature.

If an instantaneous deformation is applied to a Kelvin body the stress that is set up in it is defined by the instantaneous modulus. This stress will relax to a value that corresponds to the deformation with the long-term modulus. Relaxation follows an exponential curve with argument $-\lambda t$ as can readily be shown by integrating eq. (23.6).

Ishlinskiy [44, 45] made a detailed examination of the properties of a Kelvin body and solved various problems. Rzhanitsin [112] also worked on this subject.

24. More general differential deformation laws

The Kelvin model describes the properties of a body that displays after-effects by means of three constants: the instantaneous modulus, the delay time $1/\mu$ and the relaxation time $1/\lambda$. As a rule the behaviour of Kelvin models is in poor quantitative agreement with that of real bodies, creep and relaxation curves are not approximated by exponential functions. The natural way of generalising eq. (23.5) is to take combinations of the higher derivatives of σ and e with respect to time on both left- and right-hand sides. Let us put

$$a_0\sigma + a_1 \frac{\mathrm{d}\sigma}{\mathrm{d}t} + \ldots + a_n \frac{\mathrm{d}^n\sigma}{\mathrm{d}t^n} = b_0 e + b_1 \frac{\mathrm{d}e}{\mathrm{d}t} + \ldots + b_m \frac{\mathrm{d}^m e}{\mathrm{d}t^m}. \tag{24.1}$$

It can be shown that, with certain limitations, a medium described by eq. (24.1) can be modelled by means of springs and viscous resistances, but we shall not develop this point of view.

It is convenient to replace the differential relationships (24.1) by integral relationships as in the preceding paragraph. The following method is used to solve differential equations with constant coefficients. We start from a simple first order differential equation

$$\dot{x} + \beta x = \dot{y}\,. \tag{24.2}$$

Here β is constant, x is the required function of t, y is a given function such that when $t \leqslant 0$, $y = 0$. The solution of this equation, which becomes zero when $t = 0$, is

$$x = y - \beta \int_0^t \exp[-\beta(t-\tau)]\,y(\tau)\,\mathrm{d}\tau = \int_0^t \exp[-\beta(t-\tau)]\,\mathrm{d}y(\tau)\,. \tag{24.3}$$

On the other hand, integrating both sides of eq. (24.2) and allowing for the initial conditions, we obtain

$$x + \beta \int_0^t x(\tau)\,\mathrm{d}\tau = y\,.$$

Introduce the symbolic notation of an integral operator in the following way:

$$\int_0^t x(\tau)\,\mathrm{d}\tau = I^*x \,.$$

Then the preceding equation can be written

$$(1+\beta I^*)x = y \,. \tag{24.4}$$

Formally solve eq. (24.4) relative to x treating the operators as constant factors, we obtain

$$x = \frac{1}{1+\beta I^*}\, y \,. \tag{24.5}$$

The factor by which y is multiplied resembles the expression for a sum of an infinite geometrical progression. If βI^* were a number less than unity the following identity would hold

$$\frac{1}{1+\beta I^*} = 1 - \beta I^* + \beta^2 I^{*2} - \beta^3 I^{*3} + \ldots \tag{24.6}$$

Let us suppose that this identity is valid for operators. By the power n of an operator I^* we understand the operator I^* repeated n times, accordingly

$$I^{*n}y = \int_0^t \int_0^{\tau_n} \ldots \int_0^{\tau_2} y(\tau_1)\,\mathrm{d}\tau_1 \ldots \mathrm{d}\tau_n \,.$$

However, by Cauchy's formula the multiple integral is converted into a normal one and we may write

$$I^{*n}y = \frac{1}{(n-1)!}\int_0^t (t-\tau)^{n-1} y(\tau)\,\mathrm{d}\tau \qquad (n \geqslant 1)\,,$$

$$I^{*0}y = y \,.$$

The formal expression (24.5) which contains the operator I^* is now interpreted as follows

$$x = y - \beta\int_0^t y(\tau)\,\mathrm{d}\tau + \beta^2 \int_0^t y(\tau)(t-\tau)\,\mathrm{d}\tau - \ldots$$

or

$$x = y - \beta \int_0^t \exp\,[-\beta(t-\tau)]\,y(\tau)\,\mathrm{d}\tau\,.$$

Thus by purely formal operations we have again obtained formula (24.3). Unlike a geometrical progression, the series of operators in (24.6) is convergent for any value of β because a series which defines an exponential function is always convergent. For brevity introduce the notation

$$\mathcal{E}^*(\beta)y = \int_0^t \exp\,[\beta(t-\tau)]\,y(\tau)\,\mathrm{d}\tau\,.$$

Expression (24.5) establishes a relationship between the operators I^* and $\mathcal{E}^*(\beta)$, namely

$$\frac{1}{1+\beta I^*} = 1 - \beta\mathcal{E}^*(-\beta)\,. \tag{24.7}$$

Expression (24.7) provides the simplest possible way of solving differential equations with constant coefficients with zero initial conditions provided that the right-hand side of the equation is zero when $t < 0$. In the mechanics of materials this means that the history of the material commences at a certain instant of time and that there is an initial condition before which the material has not been subjected to any influence. Let us slightly alter eqs. (24.2) by taking $\dot{y}$ instead of y in the right-hand side, we then have

$$x = \frac{I^*}{1+\beta I^*}\,y = \frac{1}{\beta}\left[1 - \frac{1}{1+\beta I^*}\right]y\,.$$

In this case by means of formula (24.7) we find

$$x = \mathcal{E}^*(-\beta)y\,.$$

We have thus obtained a formula which is a consequence of (24.7) and which is useful for application

$$\frac{I^*}{1+\beta I^*} = \mathcal{E}^*(-\beta)\,. \tag{24.8}$$

We now return to the problem formulated at the start of this section, namely, to replace differential expression (24.1) by an integral expression. Let p be the greatest of the numbers m and n; multiply both sides of (24.1) by I^{*p}, we obtain

$$\mathcal{A}(I^*)\sigma = \mathcal{B}(I^*)e \,. \tag{24.9}$$

Here $\mathcal{A}$ and $\mathcal{B}$ are polynomials of power p of I^* where

$$\mathcal{A} = a_0 I^{*p} + a_1 I^{*p-1} + \ldots + a_n I^{*p-n} \,,$$

$$\mathcal{B} = b_0 I^{*p} + b_1 I^{*p-1} + \ldots + b_n I^{*p-m} \,.$$

If $n = m$ and the roots of the polynomials $\mathcal{A}$ and $\mathcal{B}$ are real and different, eq. (24.9) can be formally solved with respect to either e or σ. For example, solving it for σ we obtain

$$\sigma = \frac{\mathcal{B}(I^*)}{\mathcal{A}(I^*)}\, e \,.$$

The quotient of two polynomials of power p can be resolved into simple fractions so that

$$\frac{\mathcal{B}(I^*)}{\mathcal{A}(I^*)} = A_0 + \sum_{i=1}^{p} \frac{A_i}{1 + \alpha_i I^*} = A_0 + \sum_{i=1}^{p} A_i [1 - \alpha_i \mathcal{E}^*(-\alpha_i)] \,. \tag{24.10}$$

Consequently, eq. (24.1) is equivalent to the integral equation

$$\sigma = Ae - \int_0^t \Gamma(t-\tau) e(\tau) \mathrm{d}\tau \,. \tag{24.11}$$

Here

$$\Gamma(t-\tau) = \sum_{i=1}^{p} A_i \alpha_i \exp\left[-\alpha_i (t-\tau)\right] \,, \qquad A = \sum_{i=0}^{p} A_i \,.$$

The function $\Gamma(t-\tau)$ will be termed the relaxation function, and a body whose behaviour is described by expression (24.11) is termed a hereditary elastic body, if $m = n$ and the roots of the polynomials $\mathcal{A}$ and $\mathcal{B}$ are real, different and negative.

This last condition means that all values of α_i should be positive. Otherwise with deformation constant the stress increases with time instead of diminishing. With constant e we obtain

$$\sigma = \{A - \sum_{i=1}^{p} A_i [1 - \exp(-\alpha_i t)]\} e \, .$$

When $t = 0$, then $\sigma = Ae$ thus $A = \sum_{i=0}^{\infty} A_i$ is an instantaneous modulus; when $t = \infty$, $\sigma = A_0 e$, consequently A_0 is the long-term modulus. The quantity $1/\alpha_i$ which has the dimensions of time is termed a relaxation time.

Eq. (24.9) can also be solved relative to deformation

$$e = \frac{\mathcal{A}(I^*)}{\mathcal{B}(I^*)} \sigma \, .$$

The quotient is resolved into simple fractions to give

$$\frac{\mathcal{A}(I^*)}{\mathcal{B}(I^*)} = B_0 + \sum_{i=1}^{p} \frac{B_i}{1 + \beta_i I^*} = B_0 + \sum_{i=1}^{p} B_i [1 - \beta_i \mathrm{E}^*(-\beta_i)] \, .$$

The integral relationship between e and σ becomes

$$e = B\sigma + \int_0^t K(t - \tau) \sigma(\tau) \mathrm{d}\tau \, . \qquad (24.12)$$

Here

$$K(t - \tau) = -\sum_{i=1}^{p} B_i \beta_i \exp[-\beta_i (t - \tau)] \, , \qquad B = \sum_{i=0}^{p} B_i \, .$$

We shall call the function $K(t-\tau)$ the creep function, and the relaxation function is its resolvent. Hence at constant stress we obtain the creep law

$$e = \{B - \sum B_i [1 - \exp(-\beta_i t)]\} \sigma \, .$$

If $m \neq n$ things are somewhat different, to explain the essentials it suffices to consider the simple case of a viscoelastic body described by the equation

$$\sigma = E(e + \tau \dot{e}) .$$

Multiplying by I^* we obtain

$$I^*\sigma = E(I^*e + \tau e) . \tag{24.13}$$

Solving this equation for e:

$$e = \frac{I^*}{\tau + I^*} \frac{\sigma}{E} ,$$

or, by formula (24.8)

$$e = \frac{1}{E} \mathcal{E}^* \left(-\frac{1}{\tau} \right) \sigma .$$

If we now solve (24.13) for σ we find

$$\sigma = E \left(e + \frac{\tau}{I^*} e \right) ,$$

that is, we obtain the initial equation because the operator $1/I^*$ is a differential operator. This position always arises when one of the polynomials $\mathcal{A}(I^*)$ or $\mathcal{B}(I^*)$ has zero roots; differential operators remain, together with exponential operators. However, the only case that has a real meaning is that in which a viscous flow is superposed on hereditary elastic strain and consequently $m-n = 1$.

25. The hereditary-elastic body

Expressions of the type of (24.1) are usually taken as the main physical relationships in the rheology of polymeric materials. If the relaxation or creep curves can be approximated by means of a selection of exponential functions then determination of the deformation law, i.e. determination of the coefficients a_k and b_k in expression (24.1) reduces to fulfilling algebraic operations. Indeed suppose that we know the relaxation equation, this means that the quantities A_i and α_i are known. We can then write the right-hand side of expression (24.10); bringing to a common denominator we determine the polynomials $\mathcal{A}$ and $\mathcal{B}$ and consequently also the required coefficients in (24.1). Resolving the reciprocals of the polynomials into partial fractions we find the quantities B_i and β_i which govern the creep law.

However, the integral eq. (24.11) or its inverse eq. (24.12), establishes a more general relationship between stress and deformation than does the differential expression (24.1) if we understand by $K(t-\tau)$ and $\Gamma(t-\tau)$ respectively not the sum of a finite number of exponential functions but some kernel which is an arbitrary function of $t-\tau$ which satisfies only a certain very general condition which we shall now explain. Hereafter we shall write the law of heredity in the following way:

$$e=\frac{1}{E}\left[\sigma+\int_{-\infty}^{t} K(t-\tau)\sigma(\tau)\,\mathrm{d}\tau\right],$$

$$\sigma=E\left[e-\int_{-\infty}^{t} \Gamma(t-\tau)e(\tau)\,\mathrm{d}\tau\right]. \tag{25.1}$$

The form of the law of hereditary elasticity (25.1) was introduced by Volterra [376] on the basis of the following considerations. The total deformation of a body consists of the instantaneous deformation which depends on the stress acting at a given instant and is related to it by Hooke's law and of inherited deformation. If at an instant of time τ a stress $\sigma(\tau)$ is applied which acts for a time $\mathrm{d}\tau$, the material retains a memory of the action of this stress in the form of a certain small deformation $\mathrm{d}e$ which is proportional to the stress $\sigma(\tau)$ and the time $\mathrm{d}\tau$ for which it acted and also depends on the time elapsed from the instant τ when the stress was applied to the present time t, that is on $t-\tau$. To allow for this relationship it is supposed that $\mathrm{d}e$ is proportional to some function $K(t-\tau)$. Therefore

$$\mathrm{d}e=\frac{1}{E}\sigma(\tau)\,\mathrm{d}\tau K(t-\tau)\,.$$

Integrating w.r.t. τ from $-\infty$ to t and adding the instantaneous elastic deformation we arrive at the first of eqs. (25.1).

Note that the function $K(t-\tau)$ should be a diminishing one, the "memory" of the material tending to be erased with time. The dependence of the creep nucleus on the difference between the arguments $t-\tau$ indicates the invariability of the properties of the material with time; the point of origin on the time scale can be varied arbitrarily and the results remain unaltered. The properties of certain materials vary significantly with time, for example, concrete which ages for about two years. In describing the properties of materials that age the kernel should be made a function of the arguments t and τ separately.

The integral relationships (25.1) may be considered as the limiting case of integral relationships considered in the preceding section when the order of the differential relationships between σ and e tends to infinity. In the limit the sum that appears in the definition of $K(t-\tau)$ in formula (24.12) becomes the integral

$$K(t-\tau) = -\int_0^\infty \beta B(\beta) \exp\left[-\beta(t-\tau)\right] d\beta\,.$$

The function $-B(\beta)$ may be termed the creep spectrum. In exactly the same way

$$\Gamma(t-\tau) = \int_0^\infty \alpha A(\alpha) \exp\left[-\alpha(t-\tau)\right] d\alpha\,.$$

The function $A(\alpha)$ is a relaxation spectrum. The formulae described above show that the creep and relaxation functions are Laplace transforms of the corresponding spectra. Hence it follows that the spectra can be determined from formulae of the type

$$\alpha A(\alpha) = \frac{1}{2\pi i} \int_{\gamma - i\infty}^{\gamma + i\infty} \Gamma(x) \exp(\alpha x)\, dx\,.$$

We have shown that an integral expression of type (25.1) follows from the differential expression (24.1) and here the kernels $K(t-\tau)$ and $\Gamma(t-\tau)$ are sums of exponential functions. It can readily be shown that the inverse conclusion is also valid. Assume that $e(t)$ and $\sigma(t)$ are related by the integral expression

$$e = \frac{1}{E}\left[\sigma + \int_0^t K(t-\tau)\sigma(\tau)\, d\tau\right].$$

Here

$$K(t-\tau) = -\sum_{i=1}^{p} B_i\beta_i \exp\left[-\beta_i(t-\tau)\right] .$$

Rewrite the integral expression in symbolic form

$$Ee = \sigma + \sum_{i=1}^{p} B_i\beta_i \mathcal{E}^*(-\beta_i)\sigma . \tag{25.2}$$

Here

$$\mathcal{E}^*(-\beta_i)\sigma = \int_0^t \exp\left[-\beta_i(t-\tau)\right] \sigma(\tau)\,\mathrm{d}\tau .$$

Note that

$$\frac{\mathrm{d}}{\mathrm{d}t}\left[\mathcal{E}^*(-\beta_i)\sigma\right] = \sigma - \beta_i\mathcal{E}^*(-\beta_i)\sigma .$$

Whence it follows that

$$E\frac{\mathrm{d}e}{\mathrm{d}t} = \frac{\mathrm{d}\sigma}{\mathrm{d}t} + \sum B_i\beta_i\sigma - \sum B_i\beta_i^2\,\mathcal{E}^*(-\beta_i)\sigma ,$$

$$E\frac{\mathrm{d}^2e}{\mathrm{d}t^2} = \frac{\mathrm{d}^2\sigma}{\mathrm{d}t^2} + \sum B_i\beta_i\frac{\mathrm{d}\sigma}{\mathrm{d}t} - \sum B_i\beta_i^2\sigma + \sum B_i\beta_i^3\,\mathcal{E}^*(-\beta_i)\sigma ,$$

. . . .

Formulating the derivatives of e with respect to t to order p, eliminate from these expressions and from eq. (25.2) the quantities $\mathcal{E}^*(-\beta_i)\sigma$ we obtain an expression of the type of (24.1) in which $m = n = p$.

26. **Conditions of a closed Volterra cycle**

We obtain still more general stress deformation relationships than those given by formula (25.1) if we consider that the kernels K and Γ depend on the arguments t and τ in an arbitrary way so that for instance

$$e = \frac{1}{E}\left[\sigma + \int_{-\infty}^{t} K(t,\tau)\sigma(\tau)\,\mathrm{d}\tau\right] . \tag{26.1}$$

The properties of bodies that are described by eq. (26.1) are not invariants relative to change in the time-scale origin; ageing processes or else loss of work-hardening occur in these bodies. Concrete is a typical example of a material which ages. There is no reason to suppose that the modulus of instantaneous elasticity is constant in materials that age and, therefore, in the general case $E = E(t)$.

Now suppose that when $t < 0$ $\sigma = 0$, at time $t = 0$ a stress σ is applied which is a periodic function of time t of period ω, such that $\sigma(t + \omega) = \sigma(t)$. We shall show that if $K(t, \tau) = K(t - \tau)$ the quantity e given by eq. (26.1) (with E constant) differs infinitesimally from a periodic function if t is large enough. Indeed by formula (26.1)

$$Ee(t + \omega) = \sigma(t) + \int_0^{t+\omega} K(t + \omega - \tau)\sigma(\tau)\mathrm{d}\tau .$$

Substitute τ' for the variable of integration, putting $\tau = \tau' + \omega$ we obtain

$$Ee(t + \omega) = \sigma(t) + \int_{-\omega}^{t} K(t - \tau')\sigma(\tau')\mathrm{d}\tau'$$

or

$$E[e(t + \omega) - e(t)] = \int_{-\omega}^{0} K(t - \tau)\sigma(\tau)\mathrm{d}\tau$$

(the index of τ is omitted). Applying the mean value theorem and denoting by M the maximum absolute value of the function $\sigma(t)$, which is obviously always limited, we obtain the following inequality:

$$E|e(t + \omega) - e(t)| < M\omega K(t + \omega\theta) \qquad (0 < \theta < 1) .$$

It is natural to consider that the kernel $K(t)$ tends to zero when the argument increases indefinitely; therefore, the difference $e(t + \omega) - e(t)$ tends to zero with increasing t and this demonstrates the statement.

If the periodic stress commenced to act when $t = -\infty$ then the strict periodicity of the function $e(t)$ follows directly from the theorem. We shall now demonstrate the converse theorem. Suppose that both $\sigma(t)$ and $e(t)$ are periodic functions of period ω. At time t we have

$$Ee(t) = \sigma(t) + \int_{-\infty}^{t} K(t, \tau)\sigma(\tau)\mathrm{d}\tau ,$$

at time $t + \omega$ we have

$$Ee(t + \omega) = \sigma(t + \omega) + \int_{-\infty}^{t+\omega} K(t + \omega, \tau)\sigma(\tau)\mathrm{d}\tau ,$$

and since $e(t + \omega) = e(t)$, $\sigma(t + \omega) = \sigma(t)$, therefore

$$\int_{-\infty}^{t} K(t, \tau)\sigma(\tau)\mathrm{d}\tau = \int_{-\infty}^{t+\omega} K(t + \omega, \tau)\sigma(\tau)\mathrm{d}\tau .$$

On the right-hand side alter the variable of integration τ to $\tau + \omega$ and obtain

$$\int_{-\infty}^{t} K(t, \tau)\sigma(\tau)\mathrm{d}\tau = \int_{-\infty}^{t} K(t + \omega, \tau + \omega)\sigma(\tau)\mathrm{d}\tau ,$$

whence it follows that

$$K(t, \tau) = K(t + \omega, \tau + \omega) ,$$

and with arbitrary ω this is possible only when the kernel K is a function of the difference between the arguments $t-\tau$. The closed cycle theorem was demonstrated by Volterra.

27. Singular heredity function

As we have seen the integral expressions of the heredity theory of elasticity are equivalent to linear differential equations with constant coefficients if the nucleus is the sum of exponential functions. However, in working out experimental data it is found that a kernel which contains one or several exponential terms is little suited to describing the properties of real materials. Of course, by introducing the spectra of creep and relaxation we can approximate the experimental curve to any required degree of accuracy. However, the selection of elastic and viscous components does not give a suitable model for bodies with imperfect elasticity and other functions must be selected to describe their behaviour.

Boltzmann [173] first formulated the principle of heredity, and assumed a heredity kernel in the form

$$K(t-\tau)=\frac{c}{t-\tau}. \tag{27.1}$$

When the stress is constant we obtain from this the logarithmic law of change of creep deformation with time which has frequently been observed in various materials. However, a kernel of the form (27.1) has very special features when $\tau = t$ and the integral diverges from it. In order to avoid this we alter the form of the Boltzmann kernel putting

$$K(t-\tau)=\frac{c}{t-\tau+\tau_0}.$$

Here τ_0 is a constant with the dimensions of time. In the case of creep, for example, when the stress is constant we obtain by integration

$$e=\sigma\left[\frac{1}{E}+c\ln\frac{t+\tau_0}{\tau_0}\right].$$

It will be seen that in this formula τ_0 cannot be equated to zero.

At the start of creep tests, usually immediately after applying the load, the creep rate becomes so high that it is difficult to measure and loading must be considered a dynamic process. The Boltzmann kernel gives an infinitely high deformation rate at the instant of application of load. However, the singularity is too strong, so that not only does the rate become infinitely great, but so does the deformation itself. This defect is avoided if we take as creep kernel the function $c/(t-\tau)^{\alpha}$ where $0<\alpha<1$. With constant load the creep curve equation is then written

$$e=\sigma\left[\frac{1}{E}+\frac{c}{1-\alpha}t^{1-\alpha}\right]. \tag{27.2}$$

This power law describes creep of different materials very well, at least in the first part of the curve. Duffing [200] used this law to interpret test results on leather driving belts, Erzhanov [29] successfully used a similar approximation for creep curves of a variety of rocks. For most materials the value of the exponent α was close to 0.7 although Duffing gives it a value of 0.8.

The idea of introducing into the equations of the heredity theory, integral operators with kernels of the Abel type which have weakly defined singularities, can be constructed as a kind of generalisation of the idea of visco-elasticity. In section 23 we commenced to construct a rheological model which combined viscous and elastic elements. Another way of constructing the relationship between σ and e which contains as extreme cases the laws of elasticity and of viscous flow in a fluid, is as follows [18]. Put

$$\sigma(t) = \kappa \frac{\mathrm{d}^{\alpha} e(t)}{\mathrm{d}t^{\alpha}} ; \tag{27.3}$$

when $\alpha = 0$ and $\kappa = E$, this is Hooke's law; when $\alpha = 1$ and $\kappa = 3\eta$ we obtain the law of Newtonian viscosity. If α is any number between 0 and 1 the derivative of fractional order in formula (27.3) must be understood in the sense of Liouville, namely eq. (27.3) is a symbolic way of writing the integral expression

$$\sigma(t) = \frac{\kappa}{\Gamma(1-\alpha)} \int_{-\infty}^{t} \frac{\mathrm{d}e(\tau)}{(t-\tau)^{\alpha}} . \tag{27.4}$$

In rheological models of more general form, elements of type (27.3) should be combined with viscous elements (see for example [132]).

Volterra's integral equations with kernels with weak singularities were used by Bronskiy [9] and Slonimskiy [131] to describe delayed-action in rubber. The simplest Abel type kernel reproduces satisfactorily the behaviour of the material immediately after application of load; however, the rate of deformation is sometimes more quickly damped than would appear from (27.2). The tendency to combine the properties of kernels with weak singularities with those of exponential kernels led these authors to construct a kernel of the form

$$K(t-\tau) = \mathrm{const} \times (t-\tau)^{-\alpha} \exp\left[-(t-\tau)^{1-\alpha}\right] . \tag{27.5}$$

A resolvent of nucleus (27.5) has not yet been found although, as Bronskiy showed, a function of the same type as (27.5) is an approximate expression for the resolvent.

A somewhat different kernel of the same character was proposed by A. Rzhanitsyn [112], namely

$$K(t-\tau) = \text{const} \times (t-\tau)^{-\alpha} \exp\left[-\beta(t-\tau)\right] . \tag{27.6}$$

The resolvent of this nucleus is constructed in the work of ref. [17].

To describe the properties of real materials we shall use other kernels which retain singularities of the Abel type but which can be used more flexibly.

28. Exponential operators of arbitrary order

We shall denote by K^* an integral operator of Volterra type with kernel $K(t-\tau)$ which depends on the difference between the arguments. By multiplication of the operator by a certain function we understand calculation of the integral

$$K^*f = \int_{-\infty}^{t} K(t-\tau) f(\tau)\, \mathrm{d}\tau ,$$

if it exists.

Now consider the two integral operators L^* and M^* with kernels $L(t-\tau)$ and $M(t-\tau)$ respectively. The product of the operators L^* and M^* is a new operator termed N^*, the kernel of which is

$$N(t-\tau) = \int_{\tau}^{t} L(t-\xi) M(\xi-\tau)\, \mathrm{d}\xi . \tag{28.1}$$

It can readily be shown that multiplication of operators is commutative:

$$N^* = L^*M^* = M^*L^* .$$

This means that the kernel of the operator N^* can also be calculated by the formula

$$N(t-\tau) = \int_{\tau}^{t} L(\xi-\tau) M(t-\xi)\, \mathrm{d}\xi . \tag{28.2}$$

Indeed substitution of the variable $\zeta = t + \tau - \xi$ converts one formula to the other.

The square of the operator is defined as the product of the operator with itself, the kernel of the operator K^{*2} is

$$\int_{\tau}^{t} K(t-\xi)K(\xi-\tau)\mathrm{d}\xi \,.$$

This is the first iteration of the nucleus $K(t-\tau)$ that is $K^{(2)}(t-\tau)$. Continuing this process we define the power m of the operator K^* or K^{*m} as an operator whose nucleus is an iterated kernel of order m of the kernel of the initial operator.

Now write the Volterra integral equation of the second kind,

$$u(t) + \int_{-\infty}^{t} K(t-\tau)u(\tau)\mathrm{d}\tau = v(t)\,, \tag{28.3}$$

in the following way:

$$(1 + K^*)u = v\,.$$

Solve this equation formally with respect to u and obtain

$$u = \frac{v}{1 + K^*}\,.$$

Treat the expression $1/(1 + K^*)$ as a common fraction in which K^* is a number less than unity and again write formally

$$\frac{1}{1 + K^*} = 1 - \Gamma^* = 1 - K^* + K^{*2} - K^{*3} + \ldots \tag{28.4}$$

The formal solution of eq. (28.3) is

$$u = (1 - \Gamma^*)v\,, \tag{28.5}$$

where

$$\Gamma^* = K^* - K^{*2} + K^{*3} - \ldots \tag{28.6}$$

However, formula (28.5) is not merely formal; indeed the kernel of the operator Γ^*, defined by expression (28.6), is the resolvent of the nucleus $K(t-\tau)$, as (28.6) is an operator method of writing the well known Neumann series for resolvents of the Volterra equation

$$\Gamma(t-\tau) = K(t-\tau) - K^{(2)}(t-\tau) + \ldots$$

Of course, a Neumann series is convergent for any value of the argument $t-\tau$, and so we shall assume that the formal expansion of (28.6) is convergent for all regions in which the operator K^* is defined.

Now consider a kernel of the type [101]:

$$I_\alpha(t-\tau) = \frac{(t-\tau)^\alpha}{\Gamma(1+\alpha)}\,. \tag{28.7}$$

Here $\Gamma(1+\alpha)$ is the gamma-function. We are interested in values of α in the range $-1<\alpha<0$; here already the first iteration $I_\alpha^{(2)}(t-\tau)$ is non-singular, the Neumann series for the resolvents is convergent for all values of the argument and the general theory remains valid. We shall define an Abel operator as one with kernel (28.7). The multiplication theorem is valid for Abel operators, namely

$$I_\alpha^* I_\beta^* = I_{\alpha+\beta+1}^*\,. \tag{28.8}$$

Indeed by definition, the kernel of the operator $I_\alpha^* I_\beta^*$ is

$$\int_\tau^t I_\alpha(t-\xi) I_\beta(\xi-\tau)\,\mathrm{d}\xi = \frac{1}{\Gamma(1+\alpha)\Gamma(1+\beta)} \int_\tau^t (t-\xi)^\alpha (\xi-\tau)^\beta\,\mathrm{d}\xi\,.$$

The substitution

$$z = \frac{\xi-\tau}{t-\tau}$$

converts this latter integral into the form

$$(t-\tau)^{\alpha+\beta+1} \int_0^1 (1-z)^\alpha z^\beta\,\mathrm{d}z = \frac{\Gamma(1+\alpha)\,\Gamma(1+\beta)}{\Gamma(1+\alpha+\beta)} (t-\tau)^{\alpha+\beta+1}\,,$$

and this demonstrates formula (28.8).

Now determine the operator $\mathcal{E}_\alpha^*(\beta)$ as follows

$$\frac{1}{1-\beta I_\alpha^*} = 1 + \beta\,\mathcal{E}_\alpha^*(\beta)\,. \tag{28.9}$$

In other words the kernel of the operator $\mathcal{E}_\alpha^*(\beta)$ is the resolvent kernel I_α, that is

$$\mathcal{E}_\alpha^*(\beta) = I_\alpha^* + \beta I_\alpha^{*2} + \beta^2 I_\alpha^{*3} + \ldots \tag{28.10}$$

Applying formula (28.8) we find that the kernel of the operator $\mathcal{E}_\alpha^*(\beta)$ is

$$\mathcal{E}_\alpha(\beta, t-\tau) = (t-\tau)^\alpha \sum_0^\infty \frac{\beta^n (t-\tau)^{n(1+\alpha)}}{\Gamma[(n+1)(1+\alpha)]}\,. \tag{28.11}$$

The series (28.11) converges for all values of the argument $t-\tau$ for any value of β, provided that $-1 < \alpha < \infty$.

29. Multiplication theorem for $\mathcal{E}_\alpha$-operators

If in a Volterra equation of the second kind the integral operator is the operator $\mathcal{E}_\alpha^*$, that is the kernel is a function given by expression (28.10), the resolvent of the integral equation will also be a function of $\mathcal{E}_\alpha$. In section 24 we demonstrated that exponential kernels have this property. We shall demonstrate that

$$\mathcal{E}_\alpha^*(x)\,\mathcal{E}_\alpha^*(y) = \frac{\mathcal{E}_\alpha^*(x) - \mathcal{E}_\alpha^*(y)}{x-y}\,. \tag{29.1}$$

Eq. (29.1) expresses the multiplication theorem for operators $\mathcal{E}_\alpha^*$ [101]. Note in the first place that if we multiply both sides of eq. (28.10) by I_α^* we obtain

$$I_\alpha^*\,\mathcal{E}_\alpha^*(\beta) = \frac{1}{\beta}\,\mathcal{E}_\alpha^*(\beta) - \frac{1}{\beta} I_\alpha^*\,.$$

Multiplying again by I_α^* we find

$$I_\alpha^{*2}\,\mathcal{E}_\alpha^*(\beta) = \frac{1}{\beta^2}\,\mathcal{E}_\alpha^*(\beta) - \frac{1}{\beta^2} I_\alpha^* - \frac{1}{\beta^2} I_\alpha^{*2}\,,$$

and finally

$$I_\alpha^{*n}\,\mathcal{E}_\alpha^*(\beta)=\frac{1}{\beta^n}\mathcal{E}_\alpha^*(\beta)-\frac{1}{\beta^n}I_\alpha^{*1}-\frac{1}{\beta^{n-1}}I_\alpha^{*2}-\dots-\frac{1}{\beta}I_\alpha^{*n}\,. \tag{29.2}$$

We shall now proceed to demonstrate the multiplication theorem. For definiteness suppose that $x<y$ and write

$$\begin{aligned}\mathcal{E}_\alpha^*(x)\mathcal{E}_\alpha^*(y)&=(I_\alpha^*+xI_\alpha^{*2}+x^2I_\alpha^{*3}+\dots)\mathcal{E}_\alpha^*(y)\\
&=\frac{1}{y}\,\mathcal{E}_\alpha^*(y)-\frac{1}{y}\,I_\alpha^*\\
&\quad+\frac{x}{y^2}\,\mathcal{E}_\alpha^*(y)-\frac{x}{y^2}\,I_\alpha^*-\frac{x}{y}\,I_\alpha^{*2}\\
&\quad+\frac{x^2}{y^3}\,\mathcal{E}_\alpha^*(y)-\frac{x^2}{y^3}\,I_\alpha^*-\frac{x^2}{y^2}\,I_\alpha^{*2}-\frac{x^2}{y}\,I_\alpha^{*3}\\
&\quad+\dots\dots\dots\dots\dots\end{aligned}$$

Summing the right-hand side in columns we note that here

$$1+\frac{x}{y}+\frac{x^2}{y^2}+\dots=\frac{y}{y-x}\,.$$

We obtain

$$\mathcal{E}_\alpha^*(x)\mathcal{E}_\alpha^*(y)=\frac{1}{y-x}\,[\mathcal{E}_\alpha^*(y)-I_\alpha^*-xI_\alpha^{*2}-x^2I_\alpha^{*3}-\dots]\,.$$

Therefore, expression (29.1) is demonstrated.

The case when $x=y$ is an exception. The operator

$$\mathcal{E}_\alpha^{*2}(x)=\frac{\partial\mathcal{E}_\alpha^*(x)}{\partial x} \tag{29.3}$$

has as its kernel a new transcendental function, the series for which is obtained by differentiating the series (28.11) with respect to parameter β.

The multiplication theorem can be used to solve any integral equation for which the kernel is an $\mathcal{E}_\alpha$-function. Suppose, for example, that

$$(1-\kappa\mathcal{E}_\alpha^*(\beta))u = v\,.$$

Hence

$$u = \frac{1}{1-\kappa\mathcal{E}_\alpha^*(\beta)}\,v\,.$$

It is readily shown that

$$\frac{1}{1-\kappa\mathcal{E}_\alpha^*(\beta)} = 1+\kappa\,\mathcal{E}_\alpha^*(\kappa+\beta)\,. \tag{29.4}$$

Indeed bringing to a common denominator we obtain

$$\frac{\mathcal{E}_\alpha^*(\kappa+\beta)-\mathcal{E}_\alpha^*(\beta)}{\kappa} = \mathcal{E}_\alpha^*(\kappa+\beta)\,\mathcal{E}_\alpha^*(\beta)\,.$$

This equation follows from (29.1). When $\beta = 0$, from (29.4) we obtain (28.9). In some cases the hereditary operator can be approximated by a linear sum of $\mathcal{E}_\alpha$ operators, all of the same order. The problem arises of inverting an operator of this kind of the form

$$1-\sum_{i=1}^{n}\kappa_i\mathcal{E}_\alpha^*(\beta_i)\,.$$

Putting

$$\left[1-\sum_{i=1}^{n}\kappa_i\mathcal{E}_\alpha^*(\beta_i)\right]^{-1} = 1+\sum_{s=1}^{n}A_s\mathcal{E}_\alpha^*(r_s)\,. \tag{29.5}$$

Convert to a common numerator and apply the multiplication theorem giving

$$\sum_s A_s\mathcal{E}_\alpha^*(r_s)-\sum_i\kappa_i\mathcal{E}_\alpha^*(\beta_i)-\sum_i\sum_s\kappa_iA_s\,\frac{\mathcal{E}_\alpha^*(\beta_i)-\mathcal{E}_\alpha^*(r_s)}{\beta_i-r_s} = 0\,.$$

Comparing coefficients of operators with identical argument we obtain two series of equations:

$$1+\sum_s \frac{A_s}{\beta_i - r_s} = 0\,, \qquad 1+\sum_i \frac{\kappa_i}{\beta_i - r_s} = 0\,.$$

The second equation indicates that quantities r_s are roots of the equations

$$1+\sum_{i=1}^{n} \frac{\kappa_i}{\beta_i - r} = 0\,. \tag{29.6}$$

These roots must be real and different. Moreover, after the quantities r_s have been found, the coefficients A_s are found by solving the system of linear equations

$$1+\sum_{s=1}^{n} \frac{A_s}{\beta_i - r_s} = 0\,. \tag{29.7}$$

In concluding this paragraph we give without demonstration two further theorems due to Rozovskiy [121] which are useful in solving creep problems of non-uniform and anisotropic media.

Theorem 1. If $\beta_1 \neq \beta_2 \neq \ldots \neq \beta_n$, then

$$\prod_{k=1}^{n} \mathcal{E}_\alpha^*(\beta_k) = \sum_{k=1}^{n} \left[\prod_{\substack{i=1\\ i\neq k}}^{n} (\beta_k - \beta_i)\right]^{-1} \mathcal{E}_\alpha^*(\beta_k)\,. \tag{29.8}$$

Theorem 2. If $\beta_1 \neq \beta_2 \neq \ldots \neq \beta_n$, then

$$\left[1 - \sum_{k=1}^{m} M_k \prod_{i=1}^{n} \mathcal{E}_\alpha^*(\beta_i)\right]^{-1} = 1 + \sum_{k=1}^{n} a_k\, \mathcal{E}_\alpha^*(r_k)\,. \tag{29.9}$$

Here r_k are roots of the equations

$$1+\sum_{k=1}^{m} \frac{B_k}{\beta_k - r} = 0 \quad \left(B_k = \left[\prod_{\substack{i=1\\ i\neq k}}^{n} (\beta_k - \beta_i)^{-1}\right] \sum_{p=k}^{m} M_p\right),$$

and the coefficients a_k are determined by solving the system of m linear equations

$$1 + \sum_{k=1}^{m} (B_i - r_k)^{-1} a_k = 0\,, \qquad B_i \neq r_k\,.$$

The first theorem is demonstrated by successive application of the fundamental multiplication theory; in accordance with the first theorem the sum of $\mathcal{E}_\alpha$-operators which appears in the second theorem is converted to a linear combination of these operators, to which formula (29.5) can be applied.

In conclusion, we give one more formula derived by Rozovskiy [122] which is a generalisation of formulae (29.3), namely

$$\mathcal{E}_\alpha^{*(n)}(x) = \frac{1}{(n-1)!}\,\frac{\partial^{n-1}\,\mathcal{E}_\alpha^*(x)}{\partial x^{n-1}}\,. \tag{29.10}$$

30. Asymptotic formulae for $\mathcal{E}$ functions

The series (28.11) which defines the function $\mathcal{E}_\alpha(\beta, t-\tau)$ converges rather slowly and it is difficult to use in practical calculations. In practice, it is of some interest that when $\alpha = -\frac{1}{2}$ the $\mathcal{E}$-function is expressed in terms of the tabulated probability integral $\Phi(x)$ by the formula

$$\mathcal{E}_{-\frac{1}{2}}(\beta, t) = \frac{1}{\sqrt{\pi}} + \beta \exp(\beta^2 t)[1 + \Phi(\beta\sqrt{t})]\,. \tag{30.1}$$

There are no tables of $\mathcal{E}$-functions for other values of α. It is, therefore, of interest to derive approximate expressions for the functions $\mathcal{E}(\beta, t)$ and also for functions that result from application of the $\mathcal{E}$-operator to certain simple functions. If the load is constant, the result of solving particular problems is usually obtained in the form of a product of $\mathcal{E}$-operators and a constant. It obviously suffices to calculate the expression

$$\mathcal{E}_\alpha^*(-\beta)\,1 = \int_0^t \mathcal{E}_\alpha(-\beta, t-\tau)\,\mathrm{d}\tau\,. \tag{30.2}$$

In all applications the argument of $\mathcal{E}_\alpha$-operators is negative, therefore, we introduced the argument $-\beta$ into formula (30.2) and may therefore consider that $\beta > 0$. Expression (30.2) reduces to

$$\mathcal{E}^*_\alpha(-\beta)1 = \frac{1}{\beta}\,[1 - E_{1+\alpha}(-\beta t^{1+\alpha})] \,. \tag{30.3}$$

Here $E_\mu(x)$ is a Mittag-Loeffler function which is defined by the following series

$$E_\mu(x) = \sum_{n=0}^{\infty} \frac{(-1)^n x^n}{\Gamma(\mu n + 1)} \,. \tag{30.4}$$

The properties of the Mittag-Loeffler function were studied in the work [382]. The following approximation has been proposed for the function $E_{1+\alpha}(x)$ [118, 119]:

$$E_{1+\alpha}(-x) \approx \exp(-\gamma x) \qquad (\gamma = (1+\alpha)^{1+\alpha}) \,. \tag{30.5}$$

Formula (30.3) can now be replaced by the following approximate formula

$$\mathcal{E}^*_\alpha(-\beta)1 \approx \frac{1}{\beta}\,[1 - \exp(-\gamma\beta t^{1+\alpha})] \,. \tag{30.6}$$

When $t \to \infty$ the second term in brackets tends to zero and expression (30.6) tends to a limiting value of $1/\beta$. Therefore, $\mathcal{E}_\alpha$-operators describe creep that is limited at constant load; unlimited creep occurs only when $\beta = 0$ and the $\mathcal{E}_\alpha$-operator is converted to the I_α-operator. In this sense kernels constructed of $\mathcal{E}_\alpha$-functions behave in just the same way as a Bronskiy-Slonimskiy kernel (section 27).

Annin [3] has made a further study of the asymptotic properties of $\mathcal{E}$-operators on the basis of the works [211, 25, 26] which studied the asymptotic properties of certain more general types of function than the Mittag-Loeffler function, namely of the type

$$E_\rho(x;k) = \sum_0^{\infty} \frac{x^n}{\Gamma(k + n\rho^{-1})} \,.$$

It is readily found that

$$\mathcal{E}_\alpha(\beta, t) = t^\alpha E_{\frac{1}{1+\alpha}}(\beta t^{1+\alpha}; 1+\alpha) .$$

For values of α in the range $-1 < \alpha < 1$ and for negative values of the argument the asymptotic representation of the $\mathcal{E}$-function is as follows:

$$\mathcal{E}_\alpha(-\beta, t) \approx -t^\alpha \sum_{n=2}^{\infty} \frac{(-1)^n (\beta t^{1+\alpha})^{-n}}{\Gamma[(1+\alpha) - n(1+\alpha)]}$$

$$= \frac{1}{\beta^2 t^{2+\alpha}} \frac{1}{\Gamma[-(1+\alpha)]} + \ldots \tag{30.7}$$

An asymptotic representation has also been found for the product (30.2)

$$\mathcal{E}_\alpha^*(-\beta)1 \approx \frac{1}{\beta} - t^{1+\alpha} \sum_{n=2}^{\infty} \frac{(-1)^n (\beta t^{1+\alpha})^{-n}}{\Gamma[\alpha + 2 - n(1+\alpha)]}$$

$$= \frac{1}{\beta}\left[1 - \frac{1}{\Gamma(-\alpha)} \frac{1}{\beta t^{1+\alpha}} + \ldots\right]. \tag{30.8}$$

We shall now give an asymptotic resolution of the product of the derivatives of an $\mathcal{E}$-operator and unity; a product of this kind has to be calculated when squares of operators appear during the course of the calculations

$$\frac{\partial \mathcal{E}_\alpha^*(-\beta)}{\partial \beta} 1 \approx -\frac{t^{1+\alpha}}{\beta} \sum_{n=2}^{\infty} \frac{(-1)^n n (\beta t^{1+\alpha})^{-n}}{\Gamma[\alpha + 2 - n(1+\alpha)]}$$

$$= -\frac{2}{\Gamma(-\alpha)} \frac{1}{\beta^3 t^{1+\alpha}} + \ldots \tag{30.9}$$

31. The general problem of the theory of hereditary elasticity. Volterra's principle

As we have seen the law of hereditary elasticity for tension-compression is

$$\sigma = E\left(e - \int_0^t \Gamma(t-\tau)e(\tau)\,\mathrm{d}\tau\right),$$

or

$$e = \frac{1}{E}\left(\sigma + \int_0^t K(t-\tau)\sigma(\tau)\,\mathrm{d}\tau\right).$$

Using the symbolic notation introduced in section 28 we may briefly write

$$\sigma = E(1 - \Gamma^*)e\,, \qquad e = \frac{1}{E}(1 + K^*)\sigma\,.$$

Further, denoting

$$E(1 - \Gamma^*) = \widetilde{E}\,.$$

Then

$$\sigma = \widetilde{E}e\,. \tag{31.1}$$

Now the law of hereditary elasticity is of exactly the same form as the usual Hooke's law except that the elastic constant is replaced by the elastic operator $\widetilde{E}$. This notation is not purely formal, indeed eq. (31.1) can be solved in terms of e, namely

$$e = \frac{1}{\widetilde{E}}\sigma\,.$$

The operator $1/\widetilde{E}$ is equivalent to $1/E \cdot 1/(1-\Gamma^*)$ or $1/E \cdot (1 + K^*)$ as shown in section 28; this defines the operation of dividing a constant quantity by an elastic operator. Multiplication of two different elastic operators follows immediately from the rules for multiplying integral operators given in section 28; consequently, all algebraic operations on operators have a precise meaning.

In the general case of a spatially stressed state the law of hereditary elasticity is readily derived from the usual Hooke's law if the elastic constants are replaced by elastic operators formed in the same way as the operator $\widetilde{E}$. For an isotropic hereditary elastic body

$$\sigma_{ij} = \widetilde{\lambda}\theta + 2\widetilde{\mu}e_{ij}\,. \tag{31.2}$$

Here $\widetilde{\lambda} = \lambda(1 + \Lambda^*)$, $\widetilde{\mu} = \mu(1 + M^*)$; the operators Λ^* and M^* are two different operators of Volterra type with kernels $\Lambda(t-\tau)$ and $M(t-\tau)$ respectively. The elasticity law can be written in another form by replacing the operators $\widetilde{\lambda}$ and $\widetilde{\mu}$ by other operators, for example $\widetilde{E}$, $\widetilde{\nu}$, $\widetilde{G}$ which correspond respectively to Young's modulus, Poisson's ratio and the shear modulus. The relationships between the different elastic operators are written in the same way as those between the elastic constants. For instance

$$\widetilde{G} = \frac{\widetilde{E}}{2(1+\widetilde{\nu})} . \tag{31.3}$$

We have already put $\widetilde{E} = E(1-\Gamma^*)$, now put $\widetilde{\nu} = \nu(1 + N^*)$. The following operations must be performed to find the value of expression (31.3). First, notice that

$$\frac{1}{2(1+\widetilde{\nu})} = \frac{1}{2(1+\nu)}\,\frac{1}{1+\dfrac{\nu}{1+\nu}N^*} = \frac{1-R^*}{2(1+\nu)} .$$

The kernel of the operator R^* is the resolvent of the kernel of the operator $\nu/(1+\nu)N^*$; therefore, to determine R^* we must know the solution of the corresponding integral equation. When R^* has been found we obtain

$$\widetilde{G} = \frac{E}{2(1+\nu)}\,[1-\Gamma^*-R^*+\Gamma^*R^*] .$$

All the operations are performed according to the rules of section 29, if the operators Γ^* and N^* belong to the class of $\mathcal{E}$-operators. In describing the hereditary properties of real media shear after-effects play an essential role whereas bulk after-effects can generally be considered absent. Suppose that the bulk compression operator is constant, this means that

$$\frac{1-2\widetilde{\nu}}{\widetilde{E}} = \frac{1-2\nu}{E} = \text{const} .$$

Hence we can easily find

$$\widetilde{\nu} = \nu\left(1 + \frac{1-2\nu}{2\nu}\Gamma^*\right) . \tag{31.4}$$

In solving static problems of the hereditary theory of elasticity it should first be noted that the fundamental equations of the theory of elasticity retain their form, and only the elastic constants are replaced by elastic operators. Multiplication by $\tilde{\lambda}$ or by $\tilde{\mu}$ is an operation of integration with respect to time whereas solution of problems of the theory of elasticity is associated with integration with respect to coordinates. These operations are interchangeable, therefore, any static problem of the theory of hereditary elasticity can be solved in exactly the same way as the corresponding problem in the normal theory of elasticity except that in the final result the elastic constants must be replaced by elastic operators.

We shall name this principle Volterra's principle because it was formulated by this author. By Volterra's principle, it is possible to obtain extremely simple solutions for problems of theory of hereditary elasticity for which solutions of the ordinary theory of elasticity are written as products of a rational function of the elastic constants and a function of the coordinates. If the kernel is a combination of exponential functions of ordinary or fractional order, or if they are degenerate, it is not difficult to determine the significance of the corresponding operators. For example, all possible problems of transverse bending of plates are solved by two operators: $\tilde{D} = \frac{h^3}{12} \frac{\tilde{E}}{1-\tilde{\nu}^2}$ and $\tilde{\nu}$; in plane problems of the theory of elasticity for simply-connected regions if the forces on the boundary are given, the stresses are independent of time, and the same as for an elastic body in accordance with the usual Levy theorem. There are, however, more complicated problems which lead to the construction of new functions and they will be considered in section 35.

In formulating the equations of the spatial problem of viscoelasticity we considered only the isotropic case. A more general proposition is that the stresses and deformations are related by linear expressions of the most general type which can be written

$$\sigma_{ij} = \tilde{E}_{ijkl} e_{kl} \,.$$

Here

$$\tilde{E}_{ijkl} = E_{ijkl}(1 - \Gamma^*_{ijkl}) \,.$$

From the definition of instantaneous elasticity and the laws of thermodynamics it follows that $E_{ijkl} = E_{klij}$, however, the requirement $\Gamma^*_{ijkl} = \Gamma^*_{klij}$ and consequently, $\tilde{E}_{ijkl} = \tilde{E}_{klij}$ is not strictly speaking necessary [225]. This requirement follows from a certain general quasi-thermodynamic principle, the Onsager principle (see [19]).

The general theory of linear viscoelasticity is constructed on the same lines as the usual theory of elasticity and for it we must demonstrate integral theorems analogous to Betti's theorem. The theorem of uniqueness is demonstrated with several limitations superimposed on the kernel of operators. We shall not give these results here and simply refer to the literature [225, 175].

32. Application of Laplace transforms to problems of the theory of hereditary elasticity

Considerations given above, which led us to the formulation of Volterra's principle, can be presented in a somewhat different form if Laplace transforms are applied to the equations of the theory of hereditary elasticity. It will be remembered that the Laplace transform of function $f(t)$ is a function $\bar{f}(p)$ defined as

$$\bar{f}(p) = \int_0^\infty e^{-pt} f(t)\,\mathrm{d}t\,.$$

The inversion of the Laplace integral is given by the formula

$$f(t) = \frac{\mathrm{d}}{\mathrm{d}t}\,\frac{1}{2\pi i}\int_{\gamma-i\infty}^{\gamma+i\infty} \bar{f}(p)\,\frac{e^{pt}}{p}\,\mathrm{d}p\,.$$

Here γ is a positive number greater than the real part of any of the poles of the functions $\bar{f}(p)$.

The convolution of two functions $g(t)$ and $h(t)$ is called the function $f(t)$, and is defined by the equation

$$f(t) = \int_0^t g(t-\tau)h(\tau)\,\mathrm{d}\tau = \int_0^t g(t)h(t-\tau)\,\mathrm{d}\tau\,.$$

The following theorem about convolutions is demonstrated in operational calculus (see for example [27])

$$\bar{f}(p) = \bar{g}(p)\bar{h}(p)\,.$$

Here $\bar{f}(p)$, $\bar{g}(p)$ and $\bar{h}(p)$ are the Laplace transforms of the functions $f(t)$, $g(t)$ and $h(t)$ respectively.

Now write the system of equations of the theory of hereditary elasticity

$$\sigma_{ij,j} + F_i = 0 ,$$

$$e_{ij} = \tfrac{1}{2}(u_{i,j} + u_{j,i}) ,$$

$$\sigma_{ij} = \tilde{\lambda}\theta\delta_{ij} + 2\tilde{\mu}e_{ij} .$$

The corresponding boundary conditions are

$$\sigma_{ij}\nu_j = T_i^e \quad \text{on } \Sigma_T ,$$

$$u_i = u_i^e \quad \text{on } \Sigma_u .$$

Here Σ_T and Σ_u denote the parts of the body surface on which the corresponding tractions or displacements are given. The case in which mixed boundary conditions are applied to parts of the boundaries, that is when some stresses and some displacements are known, is considered in exactly the same way. Applying Laplace transforms to the equations and to the boundary conditions we obtain

$$\bar{\sigma}_{ij,j} + \bar{F}_i = 0 , \tag{32.1}$$

$$\bar{e}_{ij} = \tfrac{1}{2}(\bar{u}_{i,j} + \bar{u}_{j,i}) . \tag{32.2}$$

An expression of the type $\tilde{\lambda}\theta$ contains a convolution, indeed

$$\tilde{\lambda}\theta = \lambda\left(\theta + \int_0^t \Lambda(t-\tau)\theta(\tau)\,\mathrm{d}\tau\right) .$$

The Laplace transform of this equation is

$$\lambda(\bar{\theta} + \bar{\Lambda}\bar{\theta}) = \lambda(1 + \bar{\Lambda})\bar{\theta} .$$

Thus the transformed hereditary law assumes the form of the normal Hooke's law

$$\sigma_{ij} = \lambda(1 + \bar{\Lambda})\bar{\theta}\delta_{ij} + \mu(1 + \bar{M})\bar{e}_{ij} . \tag{32.3}$$

The boundary conditions are also correspondingly transformed

$$\bar{\sigma}_{ij}\nu_j = \bar{T}_i^e \quad \text{or} \quad \Sigma_T, \qquad \bar{u}_i = \bar{u}_i^e \quad \text{or} \quad \Sigma_u . \tag{32.4}$$

In making this last transformation an important circumstance is that the actual boundary of the body does not alter during deformation. Moreover, the functions $\bar{T}_i^e(p, x_k)$ and $\bar{u}_i^e(p, x_k)$ are known functions only when the type of boundary conditions always remains unaltered and the boundary between the parts of the surface Σ_T and Σ_u is also unaltered. Eqs. (32.1) - (32.3) are equations of the normal theory of elasticity for transforms of functions σ_{ij} and u_i and they contain the parameter p. By solving the static problem of the theory of elasticity with boundary conditions (32.4) we find the quantities $\bar{\sigma}_{ij}(p, x_k)$ and $\bar{u}_i(p, x_k)$ after which the originals are found by the inversion formula, that is the functions $\sigma_{ij}(t, x_k)$ and $u_i(t, x_k)$. It is obvious that the direct symbolic method described above and that based on Laplace transforms are different ways of writing the same operation.

Difficulties arise when the nature of the boundary conditions alters with time. Suppose, for example, that the boundary between Σ_T and Σ_u is variable so that at a boundary point we know u_i when $t < t_1$ and T_i when $t > t_1$. To calculate $\bar{T}_i(p)$ we need to know the quantity T_i both when $t > t_1$ and when $t < t_1$. Methods of solving this kind of problem are easily found when the change in type of boundary conditions occurs suddenly over a certain particular part of the surface. Suppose for example, that a die is pressed onto the boundary of the half-plane, that the region of contact is Ω and when $t = t_1$ the die is removed. Two problems must then be solved. For definiteness consider axis 1 as lying on the normal to the boundary of the half-plane. When $t < t_1$ the boundary conditions are $T_2^e = T_3^e = 0$ everywhere, outside Ω $T_1^e = 0$ and inside Ω, $u_1 = u_1^e(x_2, x_3)$. Solving the corresponding problem we find T_1^e in the region Ω, that is the pressure under the die. When $t > t_1$ $T_1^e = T_2^e = T_3^e = 0$ everywhere; applying the Laplace transform we confirm that everywhere $\bar{T}_2^e = \bar{T}_3^e = 0$, whereas $\bar{T}_1^e = 0$ only outside the region Ω, and within it $\bar{T}_1^e(p) = \int_0^{t_1} T_1^e(t)e^{-pt}\mathrm{d}t \neq 0$. In practice, we can calculate the quantity $\bar{T}_1^e(p)$ because when $t < t_1$ the pressure under the die is known from the solution of the first problem, and, therefore, no fundamental difficulties arise in this case. The position is more complicated when the type of boundary conditions alters continuously, for instance when the die moves along the boundary. Problems of this kind are considered later.

Laplace transforms can also be used to solve dynamic problems of the theory of elasticity. The Laplace transform of the derivative of order n of a function $f(t)$ is, of course, given by the following formula:

$$\int_0^\infty \frac{\mathrm{d}^n f(t)}{\mathrm{d}t^n} e^{-pt}\,\mathrm{d}t = p^n \bar{f}(p) - p^{n-1} f(0) - p^{n-2} f^1(0) - \dots$$

In problems concerning steady state vibrations, which are of particular interest, we may consider that the stresses and displacements and their derivatives with respect to time are zero at the time $t = 0$. The equations of motion of a continuous medium are of the following form:

$$\sigma_{ij,j} + F_i - \rho \ddot{u}_i = 0 \,.$$

Applying the Laplace transform we obtain

$$\bar{\sigma}_{ij,j} + \bar{F}_i - \rho p^2 \bar{u}_i = 0 \,. \tag{32.5}$$

Together with eqs. (32.2) and (32.3) we again obtain a complete system for stresses and deformations.

Laplace transforms are obviously not applicable to problems in which the hereditary kernel is not a function of the difference between the arguments, $t-\tau$ and it is difficult to use in problems with variable boundary conditions. Therefore hereafter we will mainly use Volterra's direct symbolic method.

33. More general forms of kernels

The formulation of Volterra's principle is not associated with a specific selection of kernels for which closed cycle conditions are valid and kernels of more general form sometimes arise in applications, namely

$$K(t, \tau) = h(\tau)\varphi(t - \tau) \tag{33.1}$$

or

$$K(t, \tau) = g(t)\varphi(t - \tau) \,. \tag{33.2}$$

Kernels of form (33.1) were introduced by Arutyunyan [4] to describe creep in concrete. The properties of concrete are essentially dependent on its age, i.e. on the time elapsed since it was laid. The ageing process, i.e. the change in mechanical characteristics with time, can be considered independent of the deformation process. Now bear in mind considerations which led to the formulation of the hereditary principle for a stable material given in section 25. If a stress $\sigma(\tau)$ existed in the time interval from τ to $\tau + \mathrm{d}\tau$, the deformation inherited from this time interval is expressed as

$$\mathrm{d}e = \frac{1}{E}\sigma(\tau)\varphi(t-\tau)\mathrm{d}\tau \, .$$

Here instead of $K(t-\tau)$ we have written $\varphi(t-\tau)$. This factor characterises the extent of loss of memory of the effect relating to the time τ, the memory weakens with increase in the distance $t-\tau$ along the time axis. Now suppose that the properties of the material itself change with time, Reproducing the considerations given above it is natural to assume that the quantity $\mathrm{d}e$ depends not only on the difference $t-\tau$ but also on the condition of the material at the instant τ which is allowed for by a term dependent on τ. We obtain

$$\mathrm{d}e = \frac{1}{E}h(\tau)\varphi(t-\tau)\sigma(\tau)\mathrm{d}\tau \, .$$

We thus arrive at a hereditary law for kernels of type (33.1). Arutyunyan's theory also allows for change in the modulus of elasticity with time, and, therefore, the type of the kernel is somewhat more complicated. This theory is considered in more detail in section 34.

Kernels of type (33.2) are encountered in solving numerous problems of the theory of hereditary elasticity with boundary conditions given on a variable boundary. Functions $g(t)$ can be extracted from the integral so that here we are considering an equation of the type

$$(1 + g(t)K^*)u = v \, . \tag{33.3}$$

Equations with kernels of type (33.1) or (33.2) can only be solved when the kernel degenerates. Without exception, all the published results relate to the case in which the function $\varphi(t-\tau)$ is the sum of exponential functions

$$\varphi = \sum_{i=1}^{n} A_i \exp\left[\lambda_i(t-\tau)\right] \, .$$

Consider, by way of example, the solution of eq. (33.3) with this assumption. Rewrite this equation as follows:

$$u + \sum_{i=1}^{n} g(t) A_i \exp(\lambda_i t) K_i = v \,. \tag{33.4}$$

Here

$$K_i = \int_0^t \exp(-\lambda_i \tau) u(\tau) \mathrm{d}\tau \,.$$

Differentiating (33.4) with respect to time, we obtain

$$\dot{u} + \sum_{i=1}^{n} A_i (\dot{g} + \lambda_i g) \exp(\lambda_i t) K_i + u g(t) \sum_{i=1}^{n} A_i = \dot{v} \,.$$

Repeating the differentiation n times we derive a system of $n + 1$ equations which contain u and v linearly, the derivatives of these quantities to order n and n quantities K_i. Eliminating the latter we obtain a single differential relationship between u and v of order n with variable coefficients. Given one of the functions u or v we can always obtain for the other a differential equation with variable coefficients of order n. In an exactly similar way we analyse the case when the kernel degenerates to formula (33.1).

By way of example, consider a problem solved by Lee, Radok and Woodword [280]. A thin-walled cylinder of material subject to creep is contained in a thin metal envelope (fig. 32). The outer radius is denoted by b, the thickness of the envelope by h. The internal radius a is considered variable and is given as function of t, and $\mathrm{d}a/\mathrm{d}t > 0$. When $r = a(t)$ a time-dependent pressure $q(t)$ is applied. A problem of this kind was formulated in connection with the design of solid-fuel rockets. Suppose that the tube is under conditions of plane strain and the material is incompressible. Introduce the usual notation σ_r and σ_θ as radial and hoop stresses. If $u(r)$ is the radial displacement then $e_r = \mathrm{d}u/\mathrm{d}r$ and $e_\theta = u/r$ are the radial and hoop strains. The differential equation of equilibrium in polar coordinates is of the form

$$\frac{\mathrm{d}\sigma_r}{\mathrm{d}r} + \frac{\sigma_r - \sigma_\theta}{r} = 0 \,. \tag{33.5}$$

The relationship between stresses and deformations for an incompressible material can be written

$$\sigma_\theta - \sigma_r = 2\widetilde{G}(e_\theta - e_r) \,. \tag{33.6}$$

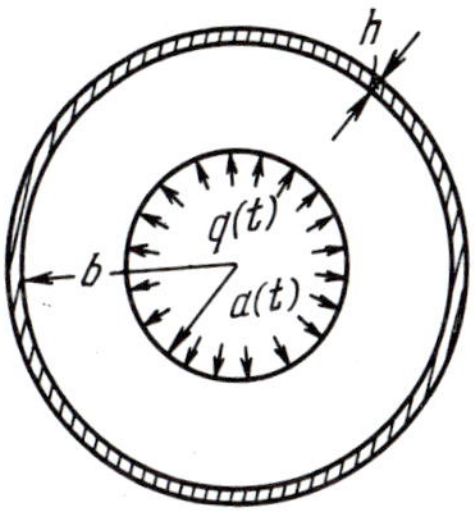

Fig. 32.

On the inner surface when $r = a$, $\sigma_r = -q$ and on the outer surface σ_r is the pressure on the envelope taken with negative sign; the hoop strain of the envelope is

$$e = -\frac{b}{h}\frac{1-\nu^2}{E}\sigma_r(b)\,.$$

Here E and ν are the elastic constants of the material of the envelope. The second boundary conditions are obtained by putting $e = e_\theta(b)$. As the material is incompressible and there is no axial deformation

$$e_r + e_\theta = \frac{\mathrm{d}u}{\mathrm{d}r} + \frac{u}{r} = 0\,.$$

Hence

$$u = \frac{k(t)}{r}, \qquad e_\theta = \frac{k(t)}{r^2}, \qquad e_r = -\frac{k(t)}{r^2}\,.$$

It follows from expression (33.6) that

$$\sigma_\theta - \sigma_r = \frac{4\tilde{G}k}{r^2}\,.$$

Introduce this expression into the conditions of equilibrium (33.5) and integrate this equation, we obtain

$$\sigma_r = -\frac{2\tilde{G}k(t)}{r^2} + k_1(t)\,.$$

Now use the boundary conditions. It follows from the first condition that

$$-q = -\frac{2\widetilde{G}k}{a^2(t)} + k_1(t)\,.$$

Consequently

$$k_1(t) = \frac{2\widetilde{G}k}{a^2(t)} - q\,, \qquad \sigma_r = \left(\frac{1}{a^2} - \frac{1}{r^2}\right) 2\widetilde{G}k - q\,.$$

The second boundary condition leads to the following expression

$$\frac{1}{c}\frac{k(t)}{b^2} = \left(\frac{1}{a^2} - \frac{1}{b^2}\right) 2\widetilde{G}k - q\,. \tag{33.7}$$

Here for brevity we introduce the notation

$$\frac{b}{h}\frac{1-\nu^2}{E} = c\,.$$

Eq. (33.7) is indeed an equation of type (33.2) as the bracket in the right-hand side is a function of t. The authors considered the simple case of a viscoelastic body for which

$$2\widetilde{G} = \alpha\frac{\partial}{\partial t} + \beta\,.$$

Then eq. (33.7) is a simple differential equation for the function $k(t)$, of the following form:

$$\alpha\left(\frac{1}{a^2(t)} - \frac{1}{b^2}\right)\frac{\mathrm{d}k}{\mathrm{d}t} + \left[\beta\left(\frac{1}{a^2(t)} - \frac{1}{b^2}\right) + \frac{1}{b^2c}\right]k = q(t)\,.$$

It is evident that this equation can be integrated by quadrature for any functions $a(t)$ and $q(t)$. The authors made the calculation for the case when $a = \dfrac{a_0}{\sqrt{1-\kappa t}}$ and q constant. In this case the quadratures are performed and the solution is complete. By applying the procedure described in the previous section we can derive a differential equation for the function k for any law of the form

$$\widetilde{G} = \frac{\mathcal{B}(I_0^*)}{\mathcal{A}(I_0^*)}.$$

34. Linear creep of concrete

Let us return to the problem of creep in a medium with time-dependent properties and write the complete equation proposed by Arutyunyan [4] to describe creep of concrete with allowance for ageing.

$$e = \frac{\sigma}{E(t)} - \int_{\tau_1}^{t} \sigma(\tau) \frac{\partial}{\partial \tau} \left[\frac{1}{E(\tau)} + C(t, \tau) \right] \mathrm{d}\tau . \tag{34.1}$$

On the basis of test results the following expression was proposed for the functions $C(t, \tau)$, which is called the measure of creep

$$C(t, \tau) = \varphi(\tau) \{1 - \exp[-\gamma(t - \tau)]\} , \tag{34.2}$$

$$\varphi(\tau) = C_0 + \frac{A_1}{\tau} .$$

Here C_0, A_1 and γ are constants of the material. The lower limit τ_1 of the integral in (34.1) should also be considered as a constant of the material. This is the time commencing from which the behaviour of the concrete is described by the theory of hereditary elasticity.

Formula (34.2) may be made more precise by adding terms containing successive negative integer powers of τ into the expression for $\varphi(\tau)$ and exponential terms in the braces $\{ \}$. We shall not consider here the conditions that led the author to the integral eq. (34.1) and the corresponding expression (34.2) for the kernel. It is important to note that the kernel is degenerate, and therefore, eq. (34.1) can be replaced by a differential equation. If $e(t)$ is the given function the quantity $e(t)E(t) = \sigma_0(t)$ is an instantaneous stress corresponding to the given deformation and eq. (34.1) can be written as

$$\sigma_0 = (1 - K^*)\sigma .$$

Here the kernel of the operator K^* is

$$K(t, \tau) = E(t) \frac{\partial}{\partial t} \{\frac{1}{E(\tau)} + \varphi(\tau)[1 - \exp(-\gamma(t - \tau))]\} . \tag{34.3}$$

After double differentiation of the integral equation with respect to t and certain transformations the integral terms are eliminated and we arrive at the differential equation

$$\frac{d^2\sigma}{dt^2}+\frac{d\sigma}{dt}\left\{\gamma(1+E\varphi)-\frac{E'}{E}\right\}=$$

$$=\frac{d^2\sigma_0}{dt^2}+\frac{d\sigma_0}{dt}\left[\gamma-2\frac{E'}{E}\right]-\sigma_0\left[\frac{E'}{E}\left(\gamma-2\frac{E'}{E}\right)+\frac{E''}{E}\right]. \quad (34.4)$$

With the appropriate initial conditions that follow from the initial integral equation and after certain conversions the solution of this equation is of the form

$$\sigma(t)=\sigma_0(t)-\gamma E\sigma_0(0)\varphi(0)\int_0^t \exp\left[-\eta(\tau)\right]d\tau$$

$$-\gamma E\int_0^t \exp\left[-\eta(\tau)\right]d\tau\int_0^t \sigma_0'(z)\exp\left[\eta(z)\right]\varphi(z)dz\,. \quad (34.5)$$

Here for simplicity of calculation we take $E(t)=E=\text{const}$ and introduce the notation

$$\eta(\tau)=\gamma\int_0^\tau (1+E\varphi(\tau))d\tau\,.$$

If the function $\varphi(\tau)$ is of the form $\varphi(\tau)=C_0+A/\tau$ and the instantaneous stresses are constant, the following formula is obtained for the stress σ:

$$\sigma(t)=\sigma_0\left[1-\gamma E\left(\frac{A}{\tau_1}+C_0\right)\tau_1^p\exp(r\tau_1)\int_{\tau_1}^t\frac{\exp(-r\tau)}{\tau^p}d\tau\right]. \quad (34.6)$$

Here

$$p=\gamma AE\,,\qquad r=\gamma(1+EC_0)\,.$$

According to Arutyunyan's results the value of p lies in the range $0\leqslant p\leqslant 1$ and for most concretes $p\leqslant\frac{1}{2}$. Materials for which p is close to zero are said to age rapidly and in these the kernel is only slightly age-dependent. If p is near to or greater than 0.5 the concrete is slow-ageing. The integral in formula (34.6) is an incomplete gamma function and for brevity we write

$$\Phi(\xi, p) = \int_0^{\xi} e^{-\tau}\tau^{-p}\,\mathrm{d}\tau\,.$$

Using this notation we rewrite formula (34.6) as follows

$$\sigma = \sigma_0 \left\{1 - \gamma E r^{p-1}\left(\frac{A}{\tau_1} + C_0\right) \tau_1^p e^{r\tau_1} \left[\Phi(rt, p) - \Phi(r\tau_1, p)\right]\right\}. \tag{34.6'}$$

Eq. (34.6) describes the relaxation process in a concrete beam in which the elongation is maintained constant.

Eq. (34.1) can evidently be written in the form

$$e = \frac{\sigma}{\widetilde{E}},$$

and in the general case of a spatial problem we may use the notation introduced in section 31.

In considering spatial problems of creep theory an important particular case is that when the operators $\widetilde{\lambda}$ and $\widetilde{\mu}$ differ only by a constant factor. This means that the hereditary elastic properties of the material are given by only one dimensional operator, for instance, the operator $1/\widetilde{E}$ introduced above, while the dimensionless operators become constants. Accordingly Hooke's law may be written

$$\sigma_{ij} = \frac{\widetilde{E}}{E}(\lambda\theta\delta_{ij} + 2\mu e_{ij})\,.$$

Here λ and μ are the usual Lamé constants, and in the general case by e_{ij} we understand expressions of the type

$$e_{ij} = \tfrac{1}{2}(u_{i,j} + u_{j,i}) - e^*_{ij}\,.$$

Here e^*_{ij} is the tensor of deformation associated with, for example, temperature expansion or shrinkage of concrete, and is a bulk change which occurs with time independent of the acting load. Generally speaking this tensor does not satisfy the compatibility equation. The equations of the theory of elasticity in displacements are written

$$\frac{\widetilde{E}}{E}[(\lambda+\mu)\theta_{,i}+\mu\Delta u_i+Q_i]=0\,. \tag{34.7}$$

Here $Q_i=(\lambda+\mu)\theta^*_{,i}+\mu\Delta u_i^*$. The operator factor can be omitted from eq. (34.7) and we obtain the usual equations of the theory of elasticity in displacements. If the conditions of the problem are such that the boundary conditions are given in displacements which are given functions of time, on the boundary, the solution of eq. (34.7) coincides with the solution of the ordinary problem of the theory of elasticity, which may contain time as a parameter if the boundary conditions are time-dependent. Following Arutyunyan we shall describe these displacements as elastic-instantaneous and the corresponding stresses calculated by Hooke's law will be termed elastic-instantaneous stresses σ_{ij}^0. The actual stresses are related to the elastic-instantaneous stresses by the following expression

$$E\sigma_{ij}=\widetilde{E}\sigma_{ij}^0\,,$$

which is identical with expression (34.1). Therefore, formula (34.5) or others following from it provide solutions of relaxation time problems [4].

35. Further applications of Volterra's principle

Now let us return again to kernels which depend on the difference between the arguments and consider more complicated problems in which the solution of the corresponding problem in the theory of elasticity is such that a rational function of the elastic constants does not appear as a factor. One example of this kind, which was considered by Volterra himself [376], is the equilibrium of a sphere of hereditary elastic material with given displacements on the surface. In this case of course [140] the solution of the problem of the normal theory of elasticity may be written as

$$u_i=\varphi_i+(r^2-a^2)\psi_{,i}\,.$$

Here $r^2=x_ix_i$, a is the radius of a sphere, φ_i is a harmonic vector, ψ is a scalar harmonic function associated in a certain way with the vector φ_i. When $r=a$ the second term becomes zero and each of the components of the vector φ_i assumes a value equal to that of the corresponding component of the displacement. From the conditions of the problem these latter are supposed to be known on the surface of a sphere. Introducing φ_i into the Lamé equation, after certain conversions we obtain

$$r\frac{\partial\psi}{\partial r}+\frac{\mu}{\lambda+2\mu}\psi+\frac{\lambda+\mu}{2(\lambda+2\mu)}\omega=0\,. \tag{35.1}$$

Here we take $\omega=\varphi_{i,i}$. We put

$$\frac{\mu}{\lambda+2\mu}=\beta\,,\qquad \frac{\lambda+\mu}{2(\lambda+2\mu)}=A\,.$$

By integration of (35.1) we obtain

$$\psi=-A\int_0^r\left(\frac{\rho}{r}\right)^\beta\omega\frac{\mathrm{d}\rho}{\rho}\,. \tag{35.2}$$

Formula (35.2) must be understood in the following sense. By definition ω is a harmonic function and consequently, it can be represented as a series a typical term of which is the product of a constant and the quantity r^nS_n. Here S_n is a surface spherical function that depends on the coordinates of points on the surface of a unit sphere, for example, on the spherical coordinates. Substituting ω into (35.2) we find that function ψ is represented by a series which differs from the series for ω in that each term is multiplied by $A/(n+\beta)$. Hence it follows that the function ψ is harmonic, as was initially assumed.

Now suppose that the behaviour of the material is described by the equations of the hereditary theory. Then the constants A and β are converted into operators $\widetilde{A}$ and $\widetilde{\beta}=\beta+H^*$. Put $\widetilde{A}\omega=\eta$ and represent $(\rho/r)^{\widetilde{\beta}}$ as follows:

$$\left(\frac{\rho}{r}\right)^{\widetilde{\beta}}=\left(\frac{\rho}{r}\right)^\beta\exp\left(H^*\ln\frac{\rho}{r}\right).$$

Following Volterra introduce the function

$$V(z,t-\tau)=zH(t-\tau)+\frac{z^2}{2!}H^{(2)}(t-\tau)+\frac{z^3}{3!}H^{(3)}(t-\tau)+\ldots$$

Then the kernel of the operator $(\rho/r)^{\widetilde{\beta}}$ is

$$\left(\frac{\rho}{r}\right)^{\beta}\left[1+V\left(\ln\frac{\rho}{r}, t-\tau\right)\right],$$

and in place of (35.2) we obtain the integral equation

$$\psi = -r^{-\beta}\int_0^r \rho^{\beta-1}\left[\eta + \int_0^t \eta(\rho,\tau)V\left(\ln\frac{\rho}{r}, t-\tau\right)\mathrm{d}\tau\right]\mathrm{d}\rho\,. \tag{35.3}$$

In problems like those considered we must introduce the kernel of integral operators which are independent not only of time but also of the coordinates. As a rule, the appropriate operators are obtained by formal resolution into series of certain transcendental functions of operators and coordinates, and as a result the kernels are new transcendental functions. This situation is encountered, for example, in problems of bending of beams on an elastic foundation. A somewhat different method, in which the required solution is constructed directly in series, was proposed by Rozovskiy [120]; we shall not consider it here but refer the reader to the original article. A simpler device for solving certain problems of this type consists in separating the variables. By way of example, consider the problem of transverse bending of beams. The bending equation differs from the differential equation of bending an elastic beam only in that the constant E is replaced by the operator $\widetilde{E} = E(1-\Gamma^*)$, so that we obtain an integral-differential equation

$$(1-\Gamma^*)\frac{\partial^2 v}{\partial z^2} + \alpha^2 v = \frac{M_0(z,t)}{EI}\,. \tag{35.4}$$

Here $M_0(z, t)$ is the bending moment due to the transverse load, $\alpha^2 = P/EI$. For given boundary conditions there exists a succession of fundamental functions v_k each of which satisfies the equation

$$\frac{\mathrm{d}^2 v_k}{\mathrm{d}z^2} + \alpha_k^2 v_k = 0\,. \tag{35.5}$$

Here α_k^2 arc fundamental numbers. We shall seek the solution in the form

$$v = \sum T_k(t)v_k(z)\,.$$

Substitute v into eq. (35.4), bearing in mind (35.5); we obtain

$$\sum [-\alpha_k^2(1-\Gamma^*)T_k + \alpha^2 T_k] v_k = \frac{M_0(z, t)}{EI} .$$

Multiply by v_s and integrate over the length of the beam. Functions v_s are orthogonal and we shall consider them normalised. Introduce the notation

$$\frac{1}{EI}\int_0^t M_0(z, t) v_s(z)\,\mathrm{d}z = m_s(t) .$$

As a result the following integral equation is obtained for T_s:

$$(\alpha^2 - \alpha_s^2 + \alpha_s^2 \Gamma^*) T_s = m_s .$$

Whence

$$T_s = \frac{1}{1 + \dfrac{\alpha_s^2}{\alpha^2 - \alpha_s^2}\Gamma^*} \frac{m_s}{\alpha^2 - \alpha_s^2} . \tag{35.6}$$

36. **Simple dynamic problems**

Problems of free and forced vibrations of elastic systems are also solved by the method of separating the variables. By way of example, consider forced transverse vibrations of a rod. For this purpose, following the general rule in the usual equation for transverse vibration we replace the modulus of elasticity by the corresponding elastic operator. For a rod of constant section F we obtain

$$EI(1-\Gamma^*)\frac{\partial^4 v}{\partial z^4} + \rho F \frac{\partial^2 v}{\partial t^2} = q(z, t) . \tag{36.1}$$

The principal forms of free vibration of an elastic beam are eigenvalues of the eq. (36.2),

$$\frac{\mathrm{d}^4 v_k}{\mathrm{d}z^4} - \lambda_k^4 v_k = 0 \tag{36.2}$$

with corresponding uniform boundary conditions, λ_k^4 are eigenvalues. Putting

$v = \sum T_k(t) v_k(z)$ and introducing into (36.1) using (36.2), we obtain

$$\sum [EI(1 - \Gamma^*)\lambda_k^4 T_k + \rho F \ddot{T}_k] \, v_k = q(z, t) \, .$$

Multiply by $v_s(z)$ and integrate over the length of the beam. Of course, v_k form an orthogonal system of functions which we shall consider to be normalised. Introduce the notation

$$\frac{EI}{\rho F} \lambda_s^4 = \omega_s^2 \, , \; \frac{1}{\rho F} \int_0^l q(z, t) v_s(z) \, \mathrm{d}z = \omega_s^2 q_s(t) \, .$$

In the result the following integral-differential equation is obtained for the functions $T_s(t)$:

$$\ddot{T} + \omega^2 (1 - \Gamma^*) T = \omega^2 q \tag{36.3}$$

(the suffix s is omitted).

The quantity ω that appears in eq. (36.3) is one of the natural frequencies of free vibration of an elastic beam. Eq. (36,3) is encountered not only in the problem of transverse vibration of beams but in all cases where separation of variables is possible, and, therefore, its investigation is of general interest. Matters are particularly simple when the operator Γ^* is the sum of zero-order exponential operators. Put

$$\Gamma = \sum A_s \alpha_s \exp\left[-\alpha_s (t - \tau)\right] \, .$$

Letting $q(t) = \exp(ipt)$, we shall seek a function $T(t)$ in the form $T = a \exp i(pt + \varphi)$. Substituting into (36.3) we obtain

$$a \left\{ -p^2 + \omega^2 \left[1 - \sum \frac{A_s \alpha_s}{a_s + ip} \right] \right\} = \omega^2 \exp(-i\varphi) \, .$$

Separating the real and imaginary parts we find

$$a = \left\{ p^2 \left(\sum \frac{A_s \alpha_s}{\alpha_s^2 + p^2} \right)^2 + \left(\frac{p^2}{\omega^2} - 1 + \sum \frac{A_s \alpha_s^2}{\alpha_s^2 + p^2} \right)^2 \right\}^{-\frac{1}{2}} \tag{36.4}$$

and

$$\tan\varphi = -p\,\frac{\sum \dfrac{A_s\alpha_s}{\alpha_s^2 + p^2}}{\dfrac{p^2}{\omega^2} - 1 + \sum \dfrac{A_s\alpha_s^2}{\alpha_s^2 + p^2}}\,. \tag{36.5}$$

The degenerate case when the operator Γ^* contains terms that represent differentiation operators presents no difficulty. For example, consider a visco-elastic body, i.e. take (see section 24)

$$\Gamma^* = -\alpha\frac{\partial}{\partial t}.$$

Eq. (36.3) reduces to the usual equation of vibration with damping proportional to the first power of velocity.

The same method is applicable to the case when Γ^* is an operator of an arbitrary kernel. Considering as before that $q = \exp(ipt)$, we shall seek a solution in the form $T = a\exp[i(pt + \varphi)]$. Calculate the quantity Γ^*T separately

$$\Gamma^*T = a\exp(i\varphi)\int_{-\infty}^{t} \Gamma(t-\tau)\exp(ip\tau)\,\mathrm{d}\tau\,.$$

Putting $t - \tau = z$, then

$$\int_{-\infty}^{t} \Gamma(t-\tau)\exp(ip\tau)\,\mathrm{d}\tau = \exp(ipt)\int_{0}^{\infty} \Gamma(z)\exp(-ipz)\,\mathrm{d}z$$

$$= \exp(ipt)\left[\int_{0}^{\infty} \Gamma(z)\cos pz\,\mathrm{d}z - i\int_{0}^{\infty} \Gamma(z)\sin pz\,\mathrm{d}z\right].$$

Substituting into (36.3) we obtain

$$a[-p^2 + \omega^2 - \omega^2(\Gamma_1 - i\Gamma_2)] = \omega^2\exp(-i\varphi)\,.$$

Here

$$\Gamma_1 = \int_0^\infty \Gamma(z) \cos pz \, \mathrm{d}z \,, \qquad \Gamma_2 = \int_0^\infty \Gamma(z) \sin pz \, \mathrm{d}z \,.$$

Whence

$$a = \left[\left(-\frac{p^2}{\omega^2} + 1 - \Gamma_1 \right)^2 + \Gamma_2^2 \right]^{-\frac{1}{2}} , \tag{36.6}$$

$$\tan \varphi = \frac{\Gamma_2}{\dfrac{p^2}{\omega^2} - 1 + \Gamma_1} \,. \tag{36.7}$$

37. Complex moduli

The results of the preceding section can be represented more compactly by introducing the concept of complex modulus of elasticity. We start from the usual relationship between stress and deformation of a prismatic rod in tension-compression

$$\sigma = \widetilde{E}e = E(1 - \Gamma^*)e \,.$$

Put $e = e_0 \exp{(ipt)}$ then, as before

$$\sigma = Ee_0 \exp{(ipt)} \left[1 + \int_0^\infty \Gamma(z) \exp{(ipz)} \mathrm{d}z \right] .$$

Introduce the notation

$$E \left[1 + \int_0^\infty \Gamma(z) \exp{(ipz)} \mathrm{d}z \right] = E' + iE''$$

and call the quantity $E' + iE''$ the complex modulus of elasticity. If $e = e_0 \exp{(ipt)}$ then

$$\sigma = [E' + iE'']e \,. \tag{37.1}$$

The meaning of this equation is that σ is also a sinusoidal function of time but shifted so that $\sigma = \sigma_0 \exp{[i(pt - \varphi)]}$ and here

$$\sigma_0 = \sqrt{E'^2 + E''^2} e_0 ,$$

$$\tan \varphi = -\frac{E''}{E'} .$$

The quantities E' and E'' are expressed in the terms of the relaxation kernel as follows:

$$E'(p) = E \left(1 - \int_0^\infty \Gamma(z) \cos pz \, dz\right) ,$$

$$E''(p) = E \int_0^\infty \Gamma(z) \sin pz \, dz . \tag{37.2}$$

Formulae (36.6) and (36.7) can now be rewritten

$$a = \frac{1}{\sqrt{\left(\frac{E'}{E} - \frac{p^2}{\omega^2}\right)^2 + \left(\frac{E''}{E}\right)^2}} ,$$

$$\tan \varphi = - \frac{E''}{E' - E\frac{p^2}{\omega^2}} .$$

If p is sufficiently small compared with ω, i.e. cyclic strain is practically without inertia, then

$$a = \frac{E}{\sqrt{E'^2 + E''^2}} , \qquad \tan \varphi = -\frac{E''}{E'} .$$

The quantities a and φ are determined from experimental results as functions of frequency, therefore, E' and E'' can also be calculated as functions of frequency. Then the relaxation kernel is determined by means of the inverse Fourier transform

$$\Gamma(t) = \frac{2}{\pi} \int_0^\infty \left(\frac{E'(p)}{E} - 1\right) \cos pt \, dp$$

or

$$\Gamma(t) = \frac{2}{\pi} \int_0^\infty \frac{E''(p)}{E} \sin pt \, dp \, .$$

Coincidence, or lack of it between the results obtained by these two formulae is evidence of the suitability or unsuitability of the linear hereditary theory equations to describe vibration of the given material.

38. Creep and relaxation functions

It is often convenient to replace the creep and relaxation kernels by functions

$$\kappa(t) = \int_0^t K(t) \, dt \qquad \text{and} \qquad \gamma(t) = \int_0^t \Gamma(t) \, dt \, , \tag{38.1}$$

which are termed creep and relaxation functions respectively. They can be determined directly from creep tests at constant stress or relaxation tests at constant deformation. Indeed, at constant stress the first of the formulae (25.1) gives

$$e = \frac{\sigma}{E} [1 + \kappa(t)] \, ,$$

and at constant deformation it follows from the second formula that

$$\sigma = Ee \, [1 - \gamma(t)] \, .$$

The integral eqs. (25.1) can now be rewritten in the form

$$\begin{aligned} e &= \frac{1}{E} \left[\sigma + \int_0^t \kappa(t-\tau) \frac{d\sigma(\tau)}{d\tau} \, d\tau \right] , \\ \sigma &= E \left[e - \int_0^t \gamma(t-\tau) \frac{de(\tau)}{d\tau} \, d\tau \right] . \end{aligned} \tag{38.2}$$

In deriving these equations we integrated by parts and used the conditions that when $t = 0$, $\kappa(0) = \gamma(0) = 0$, and also $e(0) = \sigma(0) = 0$. The first condition follows from the definition of the creep relaxation functions and the second follows from the assumption that the material is in its origin condition at zero time.

Now put

$$\kappa^*(t) = \kappa(t) - 1\,,$$

$$\gamma^*(t) = \gamma(t) + 1\,.$$

Then formula (38.2) assumes the form

$$e = \frac{1}{E}\int_0^t \kappa^*(t-\tau)\,\mathrm{d}\sigma(\tau)\,,$$

$$\sigma = E\int_0^t \gamma^*(t-\tau)\,\mathrm{d}e(\tau)\,. \tag{38.3}$$

This compact notation of the hereditary law is often encountered in current literature, the functions κ^* and γ^* are also termed creep and relaxation functions respectively. In order, for example, to apply eq. (38.2) to the case of relaxation at constant deformation we must consider two stages. The first is that of strain increasing from zero to a fixed value e in a short time θ. The value of the integral on the right-hand side that corresponds to this stage is

$$\int_0^\theta \gamma(t-\tau)\,\frac{\mathrm{d}e(\tau)}{\mathrm{d}\tau}\,\mathrm{d}\tau = \gamma(t-\theta')\int_0^\theta \frac{\mathrm{d}e(\tau)}{\mathrm{d}\tau}\,\mathrm{d}\tau\,.$$

Here $|\theta'|<|\theta|$ and disappears together with θ. Proceeding to the limit when $\theta = 0$ and integrating we find that the integral is $\gamma(t)e$. In the second stage $e = \text{const}$, consequently $\mathrm{d}e/\mathrm{d}\tau = 0$ and the corresponding integral becomes zero.

In section 25 the creep and relaxation kernels were represented as Laplace integrals. The corresponding representations of the creep and relaxation functions are

$$\kappa(t) = \int_0^\infty B(\beta) \exp(-\beta t) \mathrm{d}t \,,$$
$$\gamma(t) = -\int_0^\infty A(\alpha) \exp(-\alpha t) \mathrm{d}t \,. \tag{38.4}$$

Since $\kappa(t)$ and $\gamma(t)$ are integrals of the creep and relaxation kernels respectively and one of the kernels is the resolvent of the other, it follows that the spectra of relaxation and creep are related. The integral relationship between $A(\alpha)$ and $-B(\alpha)$ is of fairly complicated form and we do not give it, but refer the reader to Gross' monograph [222].

The complex modulus is expressed in terms of the relaxation function by formulae which follow immediately from (37.2) namely

$$E' + iE'' = E\left[1 - ip \int_0^\infty \gamma(z) \exp(-ipz) \mathrm{d}z\right] , \tag{38.5}$$

whence

$$E' = E\left[1 + p \int_0^\infty \gamma(z) \sin pz \,\mathrm{d}z\right] ,$$

$$E'' = -Ep \int_0^\infty \gamma(z) \cos pz \,\mathrm{d}z \,.$$

Using the second of the formulae (38.4) we can express the complex modulus in terms of the relaxation spectrum. Substituting into (38.5) and altering the order of integration we obtain

$$E' + iE'' = E\left[1 - ip \int_0^\infty \frac{A(\alpha)}{\alpha + ip} \mathrm{d}\alpha\right] . \tag{38.6}$$

Separating the real and imaginary parts we obtain

$$E' = E\left[1 + \int_0^\infty A(\alpha)\,\frac{p^2}{p^2+\alpha^2}\,\mathrm{d}\alpha\right],$$

$$E'' = E\int_0^\infty A(\alpha)\,\frac{p\alpha}{p^2+\mu^2}\,\mathrm{d}\alpha\,. \tag{38.7}$$

In static creep or relaxation tests we determine the creep and relaxation functions directly, in dynamic tests we determine the complex dynamic modulus as function of frequency. The formulae given in this and in the preceding section can be used to calculate the dynamic characteristics of material if its static characteristics are known. Solution of the inverse problem of determining the static characteristics from dynamic test results reduces to inverting the corresponding Fourier integrals.

In working out experimental results the usual tendency is to describe the viscoelastic behaviour of the materials by means of exponential kernels and in selecting the corresponding approximations it is useful to determine the creep and relaxation spectra directly. When the spectrum has been determined, that is, for example, we know the function $-B(\beta)$, the continuous spectrum is replaced by the discrete, this means that the function $B(\beta)$ is defined in the following way. It is zero everywhere except for a finite number of intervals $\beta_k - \eta < \beta < \beta_k + \eta$ and in each of these intervals the function $-B(\beta)$ is $B_k/2\eta$ respectively. Then from formula (38.4) when $\eta \to 0$ we obtain

$$\kappa(t) = \sum B_k \exp(-\beta_k t)\,.$$

Numerous publications deal with the procedure for calculating various viscoelastic characteristics, the fundamental ideas will be found in Gross' monograph mentioned above [222].

39. The moving load problem. Low speeds

Specific problems of the theory of hereditary elasticity arise when a sufficiently extensive body is subject to a load that moves at constant rate and its configuration does not alter with respect to the system of coordinates which moves at the same rate. In the most general formulation the body is a prism of infinite length; the system of coordinates is selected in such a way that the origin moves along the generator at a rate v, the axes x_2 and x_3 lie in the cross-sectional plane of the prism, the axis x_1 is parallel to the generator. In addition to the moving coordinate axes, consider stationary axes which coincide with the moving ones at the time $t = 0$. Denote the coordinates of points relative to the stationary axes by x_i' respectively. The conversion formula from one system of coordinates to the other is

$$x_1 = x_1' - vt\,, \quad x_2 = x_2'\,, \quad x_3 = x_3'\,.$$

Now suppose that external forces, both surface and body, depend only on the coordinates x_i. We seek a steady-state solution, i.e. stress states and displacement distributions that depend on the coordinates x_i and do not explicitly depend on time. Obviously we must suppose that movement commenced sufficiently far away, i.e. the load arrives from infinity. The time dependence of the displacement of the material point is determined by the circumstance that its coordinate x_1' is constant, whereas after a time $\mathrm{d}t$ the coordinate x_1 alters by an amount $\mathrm{d}x_1 = -v\,\mathrm{d}t$. Therefore, the velocity of the material point in the direction x_k is $\partial u_k/\partial t = -v(\partial u_k/\partial x_1)$ and its acceleration is $\partial^2 u_k/\partial t^2 = v^2(\partial^2 u_k/\partial x_1^2)$. We shall consider that the velocity is small enough (compared with the velocity of propagation of elastic waves) that inertia forces resulting from acceleration may be neglected.

The fundamental relationships of the theory of hereditary elasticity

$$\sigma_{ij} = \tilde{\lambda}\theta\delta_{ij} + 2\tilde{\mu}e_{ij}$$

relate to a fixed material point, i.e. to a fixed value of the coordinate x_1'. The quantities e_{ij} are given as functions of the coordinate x_i, their time dependence is determined by the circumstance that $x_1(t) = x_1' - vt$ whereas at the instant of time τ for the same material point we have $x_1(\tau) = \xi = x_1' - v\tau$.

Hence it follows that $t - \tau = -(x - \xi)/v$, $\mathrm{d}\tau = -\mathrm{d}\xi/v$. The operator $\tilde{\lambda}\theta$ assumes the form

$$\tilde{\lambda}\theta = \lambda(1 + \Lambda^*)\theta = \lambda\left[1 + \frac{1}{v}\int_{x_1}^{\infty} \Lambda\left(\frac{\xi - x_1}{v}\right)\theta(\xi, x_2, x_3)\,\mathrm{d}\xi\right]. \qquad (39.1)$$

Denote the new integral operator by Λ^{**}. We shall show that for given conditions the operator of differentiation with respect to the coordinate x_1 and the operator Λ^{**} are commutative. Indeed

$$\frac{\partial}{\partial x_1}(\Lambda^{**}\varphi) = \frac{\partial}{\partial x_1}\left[\frac{1}{v}\int_{x}^{\infty} \Lambda\left(\frac{\xi - x}{v}\right)\varphi(\xi)\,\mathrm{d}\xi\right].$$

After differentiation and integration by parts we obtain

$$\frac{\partial}{\partial x_1}(\Lambda^{**}\varphi) = -\Lambda(\infty)u(\infty) + \Lambda^{**}\frac{\partial u}{\partial x_1}\,. \tag{39.2}$$

In each case it remains to show that $\Lambda(\infty)u(\infty) = 0$. For limited kernels this is self-evident because $u(\infty) = 0$ is a natural condition to impose on displacements in seeking a solution.

Further development of the theory is particularly simple when the operators $\widetilde{\lambda}$ and $\widetilde{\mu}$ are proportional, that is when $\widetilde{\mu} = \mu(1 + \Lambda^*)$. Then the Hooke's law equations assume the form

$$\sigma_{ij} = (1 + \Lambda^{**})(\lambda\theta\delta_{ij} + 2\mu e_{ij})\,. \tag{39.3}$$

We have demonstrated that the operator Λ^{**} is commutative with the operator $\partial/\partial x_k$ so we can introduce the transformed displacements

$$\bar{u}_i = (1 + \Lambda^{**})u_i\,. \tag{39.4}$$

The transformed components of strain are expressed in terms of these by the usual formulae $2\bar{e}_{ij} = \bar{u}_{i,j} + \bar{u}_{j,i}$. The stresses are associated with transformed strains by the usual formulae of Hooke's law. The equations of equilibrium and of compatibility retain their usual form and we arrive at a formulation of the problem of the theory of elasticity. As a result we determine the quantities σ_{ij} and $\bar{u}_i$ after which the displacements can be found by formulae

$$u_i = \frac{\bar{u}_i}{1 + \Lambda^{**}}\,.$$

It is very simple to satisfy the boundary conditions when the moving load is given on the body surface. The boundary conditions are then formulated

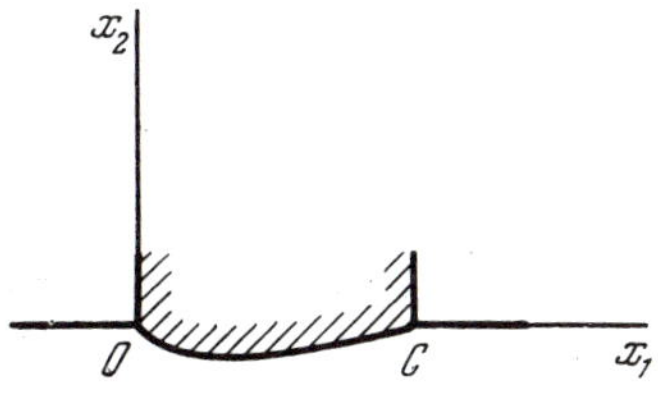

Fig. 33.

just as in the corresponding static problem of theory of elasticity. Mixed problems are much more complicated. Indeed consider for example motion of a rigid die (fig. 33). We are given the displacement u_2 on the segment $0 \leqslant x_1 \leqslant C$, but the quantity u_2 is unknown on segments $x_1 < 0$ and $x_1 > C$, however it is known that $\sigma_{22} = 0$. In order to find $\bar{u}_2$ when $0 \leqslant x_1 \leqslant C$ we need to know the function u_2 not only over this segment but also when $x_1 > C$, as follows from (39.3). Certain possibilities of solving this kind of mixed problem will be considered below.

A certain amount of power is required to maintain a steady-state condition caused by motion of a load at constant rate. The amount of power can be expressed by the formula

$$L = -\int_V \sigma_{ij}\dot{e}_{ij}\mathrm{d}V .$$

But $\dot{e}_{ij} = -v(\partial e_{ij}/\partial x_1)$. Denoting $u_i' = \partial u_i/\partial x_1$, then

$$\frac{\partial e_{ij}}{\partial x_1} = \tfrac{1}{2}(u'_{i,j} + u'_{j,i}) = e'_{ij} .$$

Because of symmetry of the tensor σ_{ij}, we may write

$$L = v \int \sigma_{ij}e'_{ij}\mathrm{d}V = v \int \sigma_{ij}u'_{i,j}\mathrm{d}V . \tag{39.5}$$

Converting the volume integral into one that extends over the surface we obtain

$$L = v \int_\Sigma T_i u_i' \mathrm{d}\Sigma . \tag{39.6}$$

Note that if the body is elastic we have $\sigma_{ij} = E_{ijkl}e_{kl}$ and $\sigma_{ij}e'_{ij} = \partial U/\partial x_1$, where $U = \frac{1}{2} E_{ijkl}e_{ij}e_{kl}$. Consequently by formula (39.5)

$$L = v \int_V \frac{\partial U}{\partial x_1}\mathrm{d}V .$$

Integrating with respect to the volume of an infinite prism with generator parallel to the x_1 axis, and bearing in mind that when $x_1 = \pm\infty$, $e_{ij} = 0$ and consequently $U = 0$, we find that $L = 0$.

The formulation of a solution to the problem of a moving die presents considerable difficulties. In recent years British authors have interested themselves in this problem. In the works [247] and [296] a method of reduction to dual integral equations is used to find solutions, which are in such a form that a numerical result can be obtained for particular conditions, but any kind of qualitative analysis is difficult.

Ivanova [39] attacked the same problem by another method which is based on the use of a complex variable. Difficulties of calculation are encountered in attempting to obtain a numerical result, even in fairly simple exponential nuclei. In the next section we shall consider the idea of this method and its application to a viscoelastic medium of simple properties.

40. **Motion of a die along the boundary of a viscoelastic medium**

The operator Λ^{**} need not be integral. The simplest viscoelastic medium is the degenerate case when $\Lambda^* = \alpha(\partial/\partial t)$. Consequently $\Lambda^{**} = -\alpha\nu(\partial/\partial x_1)$ and the transformed displacements are defined as follows:

$$\bar{u}_i = u_i - \alpha\nu \frac{\partial u_i}{\partial x_1}\,. \tag{40.1}$$

Integrating this differential equation with respect to u_i we obtain the inverse transform

$$u_i = \frac{1}{\alpha\nu} \int_x^\infty \bar{u}_i(\xi) \exp \frac{x_1 - \xi}{\nu}\, d\xi\,. \tag{40.2}$$

Here we allow for the fact that the time $t = -\infty$, when $u_i = 0$ corresponds to the value of the coordinate $x_1 = \infty$.

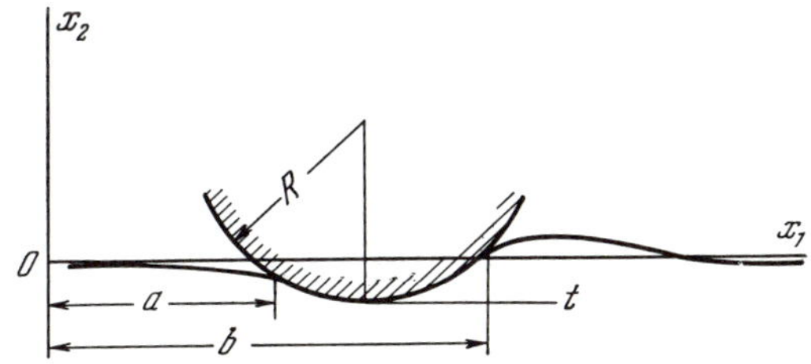

Fig. 34.

As an example of the application of this theory consider the problem of a die moving along the boundary of a viscoelastic half plane. We shall follow the known solution of N. I. Muskhelishvili [77] for the corresponding problem of the theory of elasticity. Suppose that the die is circular and motion is frictionless. Replace the circular arc by a parabola and then the die profile equation is (fig. 34)

$$f(\zeta) = \frac{\zeta^2}{2R} \, .$$

Vertical displacement of points on the boundary of the semiplane on the segment (a, b) follows the die profile, but to determine the stressed state in a viscoelastic medium it is required to solve the auxiliary problem of the theory of elasticity in which the transformed displacements are given. Accordingly we select the transformed die shape

$$\bar{f} = \frac{\zeta^2}{2R} - \alpha\nu \frac{\zeta}{R} \, .$$

If contact takes place over a segment (a, b) the solution of the corresponding plane problem of the theory of elasticity depends on determining the function of a complex variable $\Phi(z)$ which is given by formula [77]:

$$\Phi(z) = \frac{2\mu}{\pi(1+\kappa)\sqrt{(z-a)(b-z)}} \int_a^b \frac{\sqrt{(\zeta-a)(b-\zeta)}}{\zeta - z} f'(\zeta)\,\mathrm{d}\zeta + \frac{D}{\sqrt{(z-a)(b-z)}} \, . \tag{40.3}$$

Here

$$D = \frac{P}{2\pi} \, ,$$

where P is the vertical force acting on the die. The pressure $p(x_1)$ under the die is given by

$$p = \Phi^+(x_1) - \Phi^-(x_1) \, .$$

In our case substitute $f(\zeta)$ in place of $\bar{f}(\zeta)$ and put

$$\bar{f}'(\zeta) = A_1 \zeta + A_0 \, .$$

Therefore, we must calculate the integral

$$\int_a^b \frac{\sqrt{(\zeta-a)(b-\zeta)}}{\zeta-z}(A_1\zeta+A_0)\,d\zeta\,.$$

By the usual method [77] we find

$$\pi i\left\{(A_1 z-A_0)\sqrt{(z-a)(b-z)}\right.$$
$$\left.+i\left[A_1 z^2+\left(A_0-A_1\frac{a+b}{2}\right)z-A_1\frac{(b-a)^2}{8}\right]\right\}.$$

We shall not give a formula for $\Phi(z)$ but give immediately the expression for the pressure

$$p(x_1)=$$
$$=-\frac{4\mu}{1+\kappa}\frac{1}{\sqrt{(x-a)(b-x)}}\left[A_1x_1^2+\left(A_0-A_1\frac{a+b}{2}\right)x_1-A_1\frac{(b-a)^2}{8}+q\right]. \tag{40.4}$$

Here

$$q=\frac{P(1+\kappa)}{4\pi\mu}\,.$$

The constants a and b are found from the finiteness condition of $p(x_1)$ when $x_1=a$ and when $x_1=b$. Omitting obvious calculations we obtain

$$a=-\sqrt{\frac{2q}{A_1}+\frac{2A_0^2}{A_1^2}}-\frac{A_0}{A_1}, \qquad b=\sqrt{\frac{2q}{A_1}+\frac{2A_0^2}{A_1^2}}-\frac{A_0}{A_1}.$$

Returning to the initial notation note that $A_1=1/R$, $A_0=-\alpha v/R$, consequently

$$a=-\sqrt{2qR+2(\alpha v)^2}+\alpha v\,, \qquad b=\sqrt{2qR+2(\alpha v)^2}+\alpha v\,.$$

Now the pressure distribution is given by the expression

$$p(x_1)=-\frac{4\mu}{(1+\kappa)R}\sqrt{(x_1-a)(b-x_1)}\,. \tag{40.5}$$

In conclusion calculate the power required when the die is displaced according to formula (39.6). After certain simple calculations we obtain

$$L = v\frac{a+b}{2}P = \alpha v^2 P\,. \tag{40.6}$$

Formula (40.6) shows that when a roller rolls over a viscoelastic base, it must overcome a frictional rolling torque and the coefficient of rolling friction is αv. Therefore, the result is extremely simple. When a wheel rolls or a die slides over a viscoelastic base the diagram of pressure distribution under the die is the same as in the case of an elastic base. This distribution diagram is displaced in the direction of motion by an amount αv, which represents the lever arm length of the rolling friction couple.

41. Some experimental results on creep in plastics

The results given in this paragraph serve to illustrate the extent to which the theoretical considerations given above can be used to describe the properties of real materials. Numerous authors who have analysed creep test results on various polymeric materials in the vitreous state arrive at the following conclusions.

1. Creep curves are satisfactorily approximated by power functions of time.

2. There is a range of stress within which the deformation after a given time is a linear function of load.

3. Creep deformation is entirely, or almost entirely reversible.

The power dependence of deformation on time was verified from tests lasting up to 100 000 hours [207], it is always incomparably better than the exponential relationship or the logarithmic relationship used by certain authors. However, this indicates that as creep operator we should select the operator $I^*_{-\alpha}$ with nucleus $(t-\tau)^{-\alpha}/\Gamma(1-\alpha)\,(0<\alpha<1)$, then the creep function is

$$\kappa(t) = \frac{t^{1-\alpha}}{\Gamma(2-\alpha)}\,. \tag{41.1}$$

Concerning the linearity of the relationship between e and σ we refer to the work of Findley [208] which describes creep of various plastics by the equation

$$e = a \operatorname{sh} \frac{\sigma}{\sigma_e} + bt^m \operatorname{sh} \frac{\sigma}{\sigma_m} . \tag{41.2}$$

If we take the creep function in the form of (41.1) we find that $\alpha = 1 - m$. Let us find for what values of σ the hyperbolic sine in (41.2) can be replaced by a linear function. The hyperbolic sine x differs from x by 10% when $x \approx 0.77$. Consequently, provided that $\sigma < 0.77\ \sigma_m$, the linear stress dependence can be used in the second term of eq. (41.2) in place of the hyperbolic sine. The following table, taken from the results of [209] for 12 materials gives values of α, σ_m and $\sigma_{\lim} = 0.77\ \sigma_m$, the stress beyond which the error of the linear theory exceeds 10%.

The stresses in the last column are not very small and when products are intended for a long life the designer must select stress in the linear region. Bryzgalin [10] applied the principles of the theory of hereditary elasticity to an anisotropic structural material, glass-plastic AG4S. The material consists of a layer of fibreglass laid in two mutually perpendicular directions and impregnated with resin. The degree of reinforcement is approximately the same in

Material	α	σ_m kg/cm^2	$\sigma_{\lim}$ kg/cm^2
1. Glass textolite with melamine binder	0.981	2460	1900
2. Glass textolite with silicone binder	0.972	548	420
3. Laminated asbestos-plastic	0.733	168	130
4. Glass textolite based on fabric 181 and polyester resin	0.910	980	790
5. As 4, but immersed in water	0.790	914	705
6. Reinforced plastic based on glass and polyester resin	0.810	597	546
7. Glass textolite based on cloth 181 and epoxy resin	0.840	3515	2700
8. As 7, but immersed in water	0.780	773	600
9. Glass textolite based on cloth 1000 and polyester resin	0.900	478	370
10. As 9, but immersed in water	0.810	393	300
11. Glass plastic based on linen cloth and polyester resin	0.800	1547	1190
12. As 11, but immersed in water	0.770	2812	2160

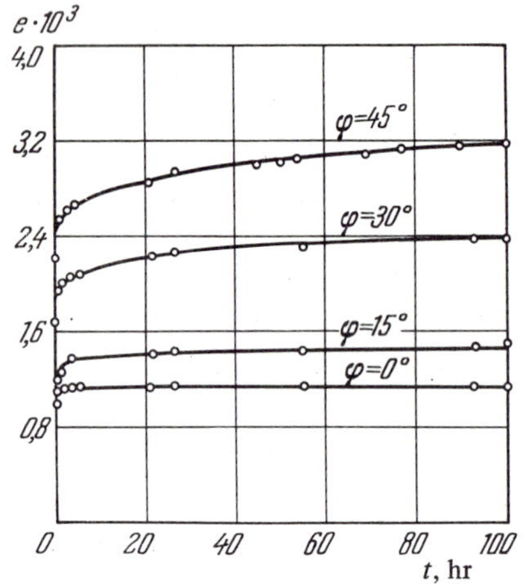

Fig. 35.

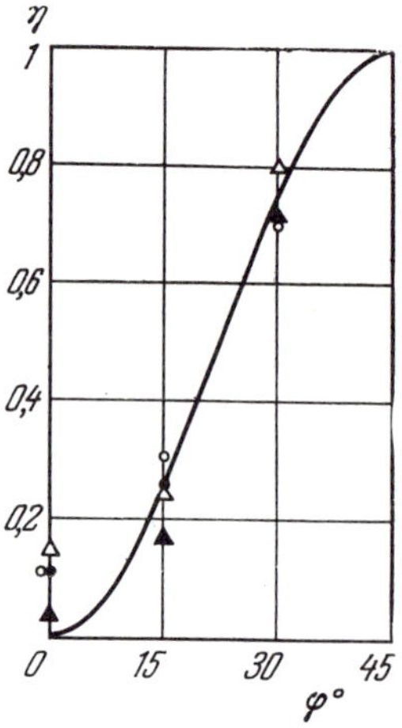

Fig. 36.

both directions. A specimen cut in line with the threads and placed in a testing machine creeps under constant load, but after a few hours the creep practically stops and the creep curve reaches the horizontal asymptote. This is readily understandable, at first the stress becomes distributed between the glass fibre and the resin, but the resin relaxes and gradually the entire load is transferred to the fibreglass which does not itself creep at room temperature. A sample cut from fibreglass plastic at an angle of 45° to the direction of the threads displays intensive undamped creep which is well described by creep functions of the type (41.1). Fig. 35 shows creep tests for specimens cut at various angles φ to the thread direction. The shape of the creep curves suggests the following model of the material. Neglect creep along the fibreglass; if the x and y axes lie along the fibres the material will be considered elastic with respect to the action of stresses σ_x and σ_y. Here it is not only the instantaneous deformations that have elastic characteristics but their limiting values after a fairly long time when the curve has practically reached its asymptote. In relation to shear we shall consider the material to be hereditary elastic. Thus the total system of equations for the plane stressed state is

$$e_x = \frac{1}{E_1}(\sigma_x - \nu_1 \sigma_y), \quad e_y = \frac{1}{E_2}(\sigma_y - \nu_2 \sigma_x), \quad \gamma = \frac{\tau}{\tilde{G}}. \qquad (41.3)$$

Eqs. (41.3) do not necessarily assume the same degree of reinforcement in the x and y directions, otherwise we should have $E_1 = E_2$, $\nu_1 = \nu_2$.

Eqs. (41.3) were checked by comparing the creep deformations for specimens cut at various angles. Fig. 36 shows a theoretical curve of the dependence of creep deformation for a fixed time according to formulae (41.3) and experimental results.

The quantity η is the ratio of the creep deformation after 100 hours for a specimen cut at an angle φ to the direction of the fibres to the creep deformation after the same time with the same stress for a sample cut at an angle of 45°. The scatter of individual test results is particularly high for materials of the glass plastic type and consequently, this agreement between theory and experiment must be considered satisfactory.

4

Creep of Metals. The Fundamental Experimental Facts and Phenomenological Theories

Creep of metals at high temperatures is characterised by two distinctive features. In the first place, most of the creep deformation is irreversible. At relatively low temperatures when the load is removed only part of the total creep deformation is recovered. Secondly, the dependence of creep rate on stress is very non-linear, a small increase in stress can cause a great increase in creep rate. Therefore, the theory of hereditary elasticity considered in Chapter 3 is not applicable to metals and creep theory of metals is founded on quite different bases.

This chapter describes the fundamental experimental facts about the creep with uniaxial stress and also the phenomenological theories which describe this simple case.

42. Fundamental information about creep

Creep effects in metals have aroused considerable interest amongst engineers, firstly because creep is observed in steam turbine components including pipework, discs and blading operating at high temperatures. Discs and blades are stressed by centrifugal forces which cannot be reduced by increasing the thickness; the working stress can only be reduced slightly by designing the disc to have optimum profile. Therefore, there was a direct need to assess the ability of materials to resist creep, and creep tests were widely used for this purpose. Very extensive experimental material in the form of creep curves has been accumulated and is still being accumulated about new alloys.

Standard creep tests are carried out in the following way. A cylindrical specimen is located in a furnace and subjected to a tensile load, usually by a lever device. The furnace temperature is maintained constant to the required degree of accuracy. A strain gauge fitted to the specimen records the elongation at particular instants of time. Many designs of machine are now used for creep testing, they differ mainly in the method of measuring deformation and in the methods of heating and controlling temperature. This is not the place to describe these machines (see for instance [93, 123]), and we shall confine ourselves to making a few remarks about them. In the first place lever type machines ensure that the load is constant, but the stress is not necessarily constant. Indeed during creep the specimen elongates and consequently, its cross-sectional area is reduced. Therefore, with constant applied forces the stress increases during the test. Particular attention was devoted to this point in the early days of creep testing and various devices were proposed to maintain constant stress. However, it had to be assumed that the volume of the specimen remains constant during creep and that the change in cross-sectional area can be determined from the elongation of the specimen. To a first approximation this hypothesis is justified and, therefore, it is not particularly difficult to maintain relatively constant stress, but the influence of the change in cross-sectional area is significant only when the deformation is relatively great. In practical structures only very small creep strains of the order of 1% can be tolerated; modern heat-resistant alloys fracture whilst the elongation is still small and, therefore, creep testing is now restricted to small deformations and there is no point in attempting to correct the load to maintain the stress constant.

The second remark relates to the system of heating. In the usual method the specimen and clamps are placed in a tubular furnace heated by a wire winding or by Silit (silicon carbide) rods. Such a furnace has high thermal inertia, it may take 4 to 6 hours for the specimen to reach a steady temperature and for the test to be started. In actual practice the heating rate may be much faster than this, which may be important in the case of structurally unstable alloys. Therefore, test equipment has been developed in which reflecting mirrors or electron beams are used to heat the specimen, or else current is passed through it. This latter method is the simplest but the properties of some materials are not the same when they are heated by passage of current as when they are heated from the surface.

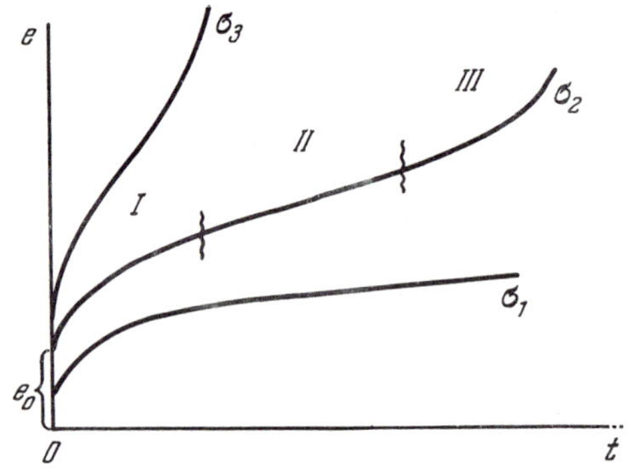

Fig. 37.

Creep test results are plotted as curves of creep against time, called creep curves. Typical creep curves for various stresses $\sigma_1 < \sigma_2 < \sigma_3$ are shown schematically in fig. 37.

The instantaneous deformation that occurs immediately after the load is applied is denoted $e_0(\sigma)$, it can include both elastic and plastic components. There are three distinct parts on the creep curve corresponding to the stress σ_2. Initially, on the segment denoted *I* creep is transient at a diminishing rate. In segment *II* the creep rate reaches a minimum and then the material creeps at constant rate $\dot{e} = \epsilon = \nu(\sigma)$. In segment *III* the creep rate increases rapidly and fracture soon occurs. This subdivision of the creep curve into three zones is somewhat conventional. The linear zone of steady-state creep is not always present and the same fig. 37 shows a creep curve for a higher stress σ_3 in which the second zone forms a region close to the inflection point. At the lower stress σ_1, the creep rate has not yet reached its minimum during the time of test and so the curve corresponds to the first phase of creep.

Some difficulty is encountered in determining the initial deformation e_0 because it depends on the method of applying the load. To make conditions more definite, in the work [30] for example, before creep testing at constant load the load is applied at a constant, strictly fixed rate of stress increase.

43. **Empirical formulae for creep curves**

The fundamental requirement of extrapolating creep curves to longer times faced the earliest investigators of creep and still remains important. In testing specimens of carbon and low-alloy steels, which were used for turbine manufacture in the 1920's and 1930's, the scatter of the experimental results was great, particularly in the initial parts of the creep curve. The only reasonably stable results were those for the dependence of the steady-state creep rate on stress, and so this was the relationship that investigators strove to obtain. Testing programmes were designed to give reasonably reliable results for the straight-line part of the creep curve and for its slope and to give the value of v for a given stress σ.

Note that we are essentially concerned with the possibility of making a straight-line approximation to a certain segment of the curve, the best approximation to select will depend on the length of this segment. Robinson [342] published results of creep tests that lasted for 100 000 hours (12 years) and showed that the creep law altered throughout this time. However, if we are considering a particular range of durations the quantity $v(\sigma)$ can be determined with sufficient accuracy. Different authors have proposed different formulae for $v(\sigma)$, some purely empirical, some based on particular physical considerations. One of the earliest investigators of creep in structural materials was Bailey [161, 162] who proposed the following expression for the function $v(\sigma)$:

$$\epsilon = v(\sigma) = B\sigma^n \ . \tag{43.1}$$

The exponent n is fairly large, for steel it ranges from $n = 3$ to $n = 8$ and sometimes even more. This highly non-linear dependence of creep rate on stress is typical of metals. Obviously no linear theory of the kind described in Chapter 3 can be used. Therefore, creep theory of metals has developed along completely different lines.

It is very simple to determine the constants B and n, by plotting in logarithmic coordinates points that correspond to measured values of v for given values of σ and drawing a straight line through the points.

Let us write expression (43.1) in a somewhat more convenient form, namely

$$\frac{\epsilon}{\epsilon_n} = \left(\frac{\sigma}{\sigma_n}\right)^n \ . \tag{43.2}$$

Here ϵ_n is a constant with the dimensions of rate of deformation, σ_n is a constant with the dimensions of stress. It is evident that

$$B = \epsilon_n \sigma_n^{-n} \ .$$

It may seem unnecessary to introduce two constants ϵ_n and σ_n instead of one, but in solving problems it is often useful to be able to change the scale.

In general let us write the dependence of creep rate on stress in the form

$$\frac{\epsilon}{\epsilon_*} = \hat{v}\left(\frac{\sigma}{\sigma_*}\right).$$

Here ϵ_* and σ_* are appropriately selected constants; the sign $\hat{}$ above a symbol of a function means that the function is dimensionless.

The power relationship (43.1) or (43.2) is often replaced by an exponential relationship

$$\frac{\epsilon}{\epsilon_e} = \exp\left(\frac{\sigma}{\sigma_e}\right). \tag{43.3}$$

This relationship was apparently first given by Ludvik [288], many authors have made extensive use of it and in some respects it is more convenient than a power relationship. The disadvantage of formula (43.3) is that it gives a creep rate different from zero when $\sigma = 0$ and so it cannot give satisfactory results for small values of σ. In order to correct this defect of the exponential formula Nadai [299] proposed to replace the exponential function by a hyperbolic sine. Thus according to Nadai

$$\frac{\epsilon}{\epsilon_e} = 2 \operatorname{sh}\left(\frac{\sigma}{\sigma_e}\right). \tag{43.4}$$

For values of σ which are sufficiently large compared with σ_e formulae (43.4) and (43.3) practically coincide and for small values of σ

$$\frac{\epsilon}{\epsilon_e} \approx 2\frac{\sigma}{\sigma_e}.$$

The hyperbolic sine law was carefully checked by MacVetty [373] and after working out a great deal of experimental material he concluded that formula (43.4) is more accurate than the power formula (43.1).

More recent data on this problem are given in the book [93], the author of which also considers that the exponential creep rate law is more accurate than the power law. Other devices can also be used to correct formula (43.3) for low values of σ. For example, Söderberg [355] proposed the expression

$$\frac{\epsilon}{\epsilon_*} = \exp\left(\frac{\sigma}{\sigma_*}\right) - 1 \; ; \tag{43.5}$$

where $v(\sigma)$ again differs little from a linear function when σ is small.

Our object is to construct a mechanical theory of creep and to apply it to calculations and so we do not attach particular importance to the form in which the function $v(\sigma)$ is given. The selection of a particular expression in a particular problem, over a comparatively narrow stress range, is largely governed by considerations of convenience.

Numerous attempts have been made to formulate equations that describe the creep curve as a whole, some are purely empirical, some follow from particular physical creep theories. It is not part of our task to consider physical creep theories and, therefore, we simply give certain formulae which describe the experimental results reasonably well.

The first systematic study of creep in metals was made by Andrade in 1910 [159]. According to Andrade the length of a specimen varies with time according to the law

$$l = l_0(1 + \beta t^{\frac{1}{3}})e^{kt}. \tag{43.6}$$

Here l_0 is the initial length, β and k are functions of stress and temperature, which were not precisely defined. It should be noted that the author considered finite strains and took special measures to maintain the stress constant. If we consider only stresses for which β and k are sufficiently small the following creep curve equation follows from Andrade's law:

$$e = \beta t^{\frac{1}{3}} + kt \; . \tag{43.7}$$

Here it is assumed that all the deformation is creep deformation.

Therefore, in creep two kinds of flow are superposed on one another: "β-flow" according to Andrade's terminology, which is of diminishing rate and "k-flow" which is of constant rate.

This superposition of two types of flow is also retained in finite deformations. Indeed, if the rate of deformation is related to the instantaneous length of the specimen from eq. (43.6) we obtain

$$\epsilon = \frac{1}{l}\frac{\mathrm{d}l}{\mathrm{d}t} = \frac{\beta}{3}\,\frac{1}{t^{\frac{2}{3}} + \beta t} + k \; .$$

The first term in the right-hand side tends to zero when t increases and only k-flow at constant rate remains.

Andrade and other, mainly British authors, have made numerous investigations to check formula (43.6) and to improve it in application to particular materials and test conditions (see for example [160]).

Eq. (43.7) reproduces actual creep curves fairly satisfactorily, but a more general form of equation is

$$p = S(\sigma)t^m + v(\sigma)t\,. \tag{43.8}$$

Here p is the creep deformation, $p = e - e_0$.

The exponent m is usually close to $\frac{1}{3}$, but for different materials can vary on either side of this value and reach a value of 0.5 or more. For a particular material m is a function of temperature, and it may also depend on the stress. The function $S(\sigma)$ is approximated by the same expressions as $v(\sigma)$.

In many cases we are interested in comparatively short-term creep that forms part of the first segment of the curve. A sufficiently good approximation is obtained for the initial parts of the curve if we retain only the first term in formula (43.8), namely if we take

$$p = S(\sigma)t^m\,. \tag{43.9}$$

Note that formula (43.8) is not claimed to be an extrapolation formula that can predict the amount of creep deformation after a long time from the results of comparatively short term tests.

Amongst other equations which describe both the first and second sections of the creep curve we may mention Oding's equation [93]

$$p = \frac{C\sigma^2 t}{1 + k\sigma t} + v(\sigma)t\,, \tag{43.10}$$

where C and k are constants for a given temperature. According to Oding this equation was checked on 120 creep curves for different grades of steel.

44. Similarity of creep curves

In developing methods of calculating creep and applying them to the study of real objects the fairly considerable scatter of creep test results must be taken into account. A difference of 20% between the deformations found in creep curves taken at the same stress and temperature on specimens of material of a single melt is not considered excessive. To keep the scatter within these limits special measures must be taken to ensure that the initial state of the specimens is strictly controlled. It is, therefore, hardly worth while to attempt very accurate analytical description of all the parts of creep curves and approximating expressions are usually selected on grounds of convenience. Many authors who have been concerned with the development of mechanical creep theories have used the hypothesis of similarity of creep curves. This means that creep deformation can be represented as follows:

$$p = S(\sigma)\tau(t)\,. \tag{44.1}$$

The hypothesis of similarity of creep curves is fairly well justified in the first parts of the curves. Indeed formula (43.9) is valid, at least over a certain stress range, and satisfies this condition. The similarity hypothesis may not be valid over a wider stress range. Bearing in mind that in formula (43.8) the functions $S(\sigma)$ and $v(\sigma)$ differ only by a constant factor, the creep curves described by this formula are also similar. In practice creep curves with only one similarity factor for both the first and second parts of the curve are by no means always similar. There is, in general, no similarity whatever between total creep curves which include the third section and no reasonably reliable formulae are available for describing the third section. The commencement of the third section immediately precedes fracture, it is generally not considered permissible for the material to enter this section and, therefore, it is not considered in creep calculations. Later on we shall see that in predicting the time to fracture on the basis of certain hypotheses that relate to the process of fracture it may be possible to approximate the creep curves up to the actual instant of fracture.

A series of creep curves at constant stress is a graphical representation of the functional relationship

$$e = e(\sigma, t)\,;$$

here the variables e and t are laid out on the coordinate axis and the value of σ is marked against the corresponding curves. However, the same functional relationship may be expressed otherwise, for instance, by plotting σ and e on the coordinate axes, as is usual in tensile-test curves. Then the time t is marked against the corresponding curve. Curves constructed in this way are termed isochronous creep curves. The method of construction calls for no special explanation, the point of intersection between a creep curve and a vertical line that corresponds to a fixed time value gives coordinates e and σ of the corresponding isochronous curve. The time $t = 0$ corresponds to the curve of instantaneous deformation.

In working out a large number of isochronous creep curves for various grades of steel it was found [98, 99] that the isochronous curves are often similar and consequently, their equation may be written

$$\sigma = \varphi(e)\theta(t)\,. \tag{44.2}$$

Note that the conditions of similarity expressed by formula (44.2) are quite different from those of (44.1). They coincide only when elastic strain can be neglected, that is if we consider $p \approx e$ and when the function $S(\sigma)$ is a power function.

If we define the function $\theta(t)$ in such a way that $\theta(0) = 1$ then $\sigma = \varphi(e)$ is the equation of an instantaneous deformation curve. In analysing published experimental data on creep it is usually found that information about the initial sections is unreliable and the instantaneous deformation that occurs immediately after applying the load is not sufficiently accurately recorded. Therefore, almost always the curve that corresponds to $t = 0$ or to values of t near to but different from 0 is not available. By selecting a suitable approximation to the function $\theta(t)$ we can restore the curve $\sigma = \varphi(e)$ by extrapolation, but of course we cannot be sure that this is indeed the actual curve of instantaneous deformation. A fairly good approximation to the function $\theta(t)$ is given by the formula [98, 99]

$$\theta(t) = \frac{1}{1 + at^{\beta}}\,. \tag{44.3}$$

Here it is found that the value of β is close to 0.3.

A series of isochronous creep curves for chrome-nickel-molybdenum steel plotted from Robinson's [342] test results, which lasted for 100 000 hours (from 27 March 1931 to 8 October 1942), are plotted in fig. 38. Selecting a function $\theta(t)$ in the form of (44.3) when $\beta = 0.3$ we obtain the curve of $\sigma = \varphi(e)$ shown in the figure. The scatter of points corresponding to individual specimens which are plotted on this curve in accordance with eq. (44.2), is not greater than the scatter of the moduli of elasticity of specimens.

Turner and Blomquist [372] used conditions of similarity of isochronous creep curves to extrapolate creep curves of aluminium alloy 24S-T3. They used Guarnieri's [223] test results for the temperature of 232°C, they had a tensile test curve of the material. The rate of extension is not indicated; however, at the temperature in question creep is not great enough to influence the tensile-test diagram at rates which are usual in testing machines. Therefore, the tension curve was taken as a graph of the function $\varphi(e)$. In formula (44.3)

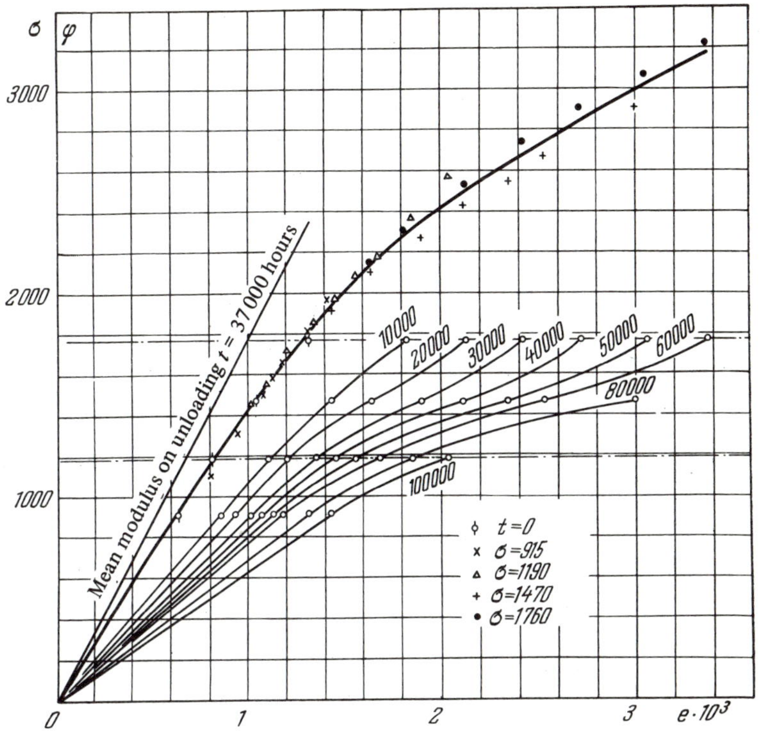

Fig. 38.

it was taken that $\beta = \frac{1}{3}$, the constant a was determined from the condition that at a stress of 10.54 kg/mm^2 the total deformation of 0.5% was achieved in 86 hours. The table below gives both experimental values and values calculated by formula (44.2) of the times to reach this deformation of 0.5% with various applied stresses.

σ, kg/mm^2	t, hours	
	experimental	calculated
11.95	37.0	41.1
14.06	16.0	14.0
15.46	6.0	6.5
17.57	3.5	1.9
21.09	0.15	0.10

Test results of Dorn and Tietz [192] were worked out for a temperature of 205°C. Two isochronous creep curves, corresponding to times of 10 and 20 hours, were used to determine the parameter a and a curve of $\varphi(e)$ was constructed. On comparing results of calculation by formula (44.2) with results of fairly long tests it was found that the actual deformation after 1000 hr was 0.7% with a stress of 10.54 kg/mm^2, whilst according to calculations based on short-term test results this deformation should be achieved at a stress of 10.27 kg/mm^2.

In comparing calculated and experimental creep results it should always be borne in mind that very small changes in stress have a great influence on the creep rate and consequently on the time required to achieve a particular deformation. Minor differences in structure or chemical analysis of the specimen also greatly affect the rate. Therefore, in tests to determine the time required to achieve a particular deformation at fixed stress the scatter of experimental results for individual specimens is usually great. If we determine the stress at which a particular deformation is achieved in a given time the scatter is very small. Therefore, the results given in the table, which relate to tests at a temperature of 232°C, must be considered to provide very good confirmation of formula (44.2), no worse than in the second case when the stress difference was only 2.7%.

However, the experimental results that have been given do not provide convincing confirmation of the rule of similarity of isochronous creep curves. In some cases this similarity is clearly not observed, but where it does exist it can be used to simplify calculations.

Goldhoff [213] gave a more fundamental verification of the conditions of similarity of isochronous curves. He worked out experimental results obtained in the General Electric Co. and other laboratories for the following materials:

a. ferritic steel containing 1% chromium, 1% molybdenum and 0.25% vanadium at a temperature of 562°C;

b. steel with 12% chromium at temperatures of 427, 538 and 649°C;

c. austenitic steel at temperatures of 649 and 704°C;

d. magnesium alloy at temperatures of 149 and 93°C;

e. two kinds of plastics;

f. dispersion hardened steel grade 17-7 at 482°C.

For each material the constants α and β and functions $\varphi(e)$ were found from the creep curves. The calculations were made on a computer in the following way. Values of β were chosen, a system of linear equations was found for a and $\varphi(e)$ and then values of a and $\varphi(e)$ were found by the method of least squares. The calculations were repeated for other values of β until the best combination of a, β and $\varphi(e)$ was obtained. Of course, the curve of $\varphi(e)$

calculated in this way differed somewhat from the tensile test curve, though the difference was not very great, and values of β were also somewhat different. The author then calculated the stress required to cause a particular deformation in a given time when $\beta = 0.33$, on the basis of the tensile test curve and one isochronous curve. The difference between calculated and experimental results usually lay within ± 5%, but was occasionally greater. The exceptions were low alloy steel (a) at relatively small deformations, magnesium alloy (d) in sheets and steel 17-7 (f). The author concludes that creep of structurally stable materials is sufficiently well described by eqs. (44.2) and (44.3). In all cases the value of β was 0.33 except for plastics for which a value of $\beta = 0.1$ had to be taken.

45. Temperature dependence

Hitherto we have considered creep curves at constant temperature. As a rule when the temperature is raised the creep rate increases and at a given stress a particular deformation is achieved in less time. To provide a detailed description of the influence of temperature on the creep curve it would be necessary to study the temperature dependences of all the parameters in any of the equations that satisfactorily describe the shape of the creep curve. Unfortunately, there are too many of these equations and the test results are too scattered. The main value of physical theories of creep in which the creep curves are obtained on the basis of certain assumptions concerning the internal mechanism of the process, is rather that they can predict, and sometimes indeed really do predict, the temperature dependence of the creep curve. In all creep theories it is assumed that plastic deformation is associated with the motion of certain structural elements caused by thermal motion. At a given instant of time, each of these elements has an energy u, and the distribution of energies between individual elements follows Maxwell's law.

Let $p\,du$ be the probability that the energy of a particular element at a particular instant of time lies between u and $u + du$. According to Maxwell's law

$$p = \frac{1}{RT}\exp\left(-\frac{u}{RT}\right)$$

where R is the gas constant.

The factor $1/RT$ is determined from the condition

$$\int_0^\infty p\,\mathrm{d}u = 1\,.$$

If the energy of a certain element exceeds a value of u_0, which is termed the energy of activation, the element is displaced and some deformation occurs. The probability of activation of an element, or the relative number of activated elements, which is the same thing, is determined as follows:

$$\int_{u_0}^\infty p\,\mathrm{d}u = \exp\left(-\frac{u_0}{RT}\right).$$

It is natural to assume that the rate of creep is proportional to the number of activated structural elements, therefore, with a given stress and given structural state of the material

$$v \approx \exp\left(-\frac{u_0}{RT}\right) f(\sigma)\,. \tag{45.1}$$

In all these considerations we have deliberately not considered the exact nature of the structural elements in question. These structural elements are different in different creep theories, they may be hypothetical or real. Accordingly we speak of different creep mechanisms. The energy of activation u_0 can be determined experimentally, if it is found that for one and the same material the energy of activation varies as function of certain conditions (for example in different ranges of load and temperature) it may be concluded that one or other particular creep mechanism predominates. This point of view was developed systematically by Dorn [192, 193]. The reservation that expression (45.1) is valid for a particular structural state of the material is also due to this author. If the material is structurally stable, the change in it that tends to reduce the creep rate (in the first section of the creep curve) results from an accumulation of deformations, and not from structural changes that are time-dependent irrespective of the applied load. Therefore, the following conclusion follows from eq. (45.1). Assume that the temperature is time-dependent and introduce a time function

$$\theta = \int_0^t \exp\left(-\frac{u_0}{RT}\right) \mathrm{d}t\,. \tag{45.2}$$

The quantity θ can be considered as the referred time. It follows from expression (45.1) that the creep rate relative to this referred time, i.e. $\mathrm{d}e/\mathrm{d}\theta$, depends only on the stress and on the structural state, which depends on the magnitude of the accumulated deformation e. Therefore, the creep curve at constant stress but variable temperature is given by

$$e = e(\theta)\,. \tag{45.3}$$

Therefore, given the creep curve at constant temperature the creep curve at variable temperature is obtained by appropriate change in the time scale. To check the statement that the state of the material which governs the instantaneous creep rate at a given stress depends only on the deformation and consequently, in accordance with eq. (45.3), on the value of the modified time θ, Dorn made metallographic and X-ray investigations of specimens undergoing creep and also carried out tensile tests. He found that the actual structure that results from creep depends on the parameter θ and in practice it can be considered the same for identical values of θ at different temperatures. Tests were made on pure aluminium, a creep curve in coordinates of e–θ was first constructed for a stress of 14.06 kg/mm^2 by combination of three creep curves at temperatures of 422, 478 and 528° Kelvin. The experimental points for the different temperatures lie sufficiently well about a single curve. In order to check the validity of eq. (45.3) at variable temperatures a specimen of the same pure aluminium was tested at the same stress of 14.06 kg/mm^2, but during a 100 hour test the temperature was varied in a rather arbitrary way from about 420 to 480°K. The creep curve was calculated from the variable temperature curve by means of eqs. (45.2) and (45.3) and the agreement with experimental results was found to be very good.

It should be noted that Dorn's tests and the results that he worked out relate to pure metal. As we shall see later dependences can be more complicated for alloys and particularly heat-resistant alloys.

Other physical theories lead to formulae which are somewhat different from the simplest form of temperature dependence of the type of (45.1). Numerous authors [269, 290], starting from various hypothetical schemes, have derived expressions for the creep rate as function of stress and temperature in the form

$$v \approx \exp\left(-\frac{u_0 - \gamma\sigma}{RT}\right). \tag{45.4}$$

Here u_0 is a constant energy activation, γ is a structure-sensitive coefficient.

The difference between this expression and (45.1) is that the energy of activation is stress dependent. A temperature dependence of the type of (45.4) was determined on a wide range of materials by Zhurkov and others [32–35]. Apparently a formula of the type of (45.4) gives the best agreement at moderate temperatures but expression (45.1) should be used at higher temperatures.

In the low temperature region Mott and Nabarro [297] proposed a model for which the creep rate is proportional to $T^{\frac{2}{3}}$. It was assumed that the crystal lattice contains a large number of uniformly dispersed inclusions on which dislocations are retarded. Thermal motion causes individual sections of the dislocation line between these inclusions to transfer to a position of stable equilibrium. The creep process is highly damped and disappears when all the loops of low activation energy are transferred to stable positions.

On the other hand, at very high temperatures close to the melting point, a completely different creep mechanism of purely diffusional character is, in principle, possible. A directed flow of diffusing atoms leads to the formation of new atomic planes and so to a visible deformation of the crystals. This scheme was considered in the articles [298, 229]. Livshits [68] developed a theory of diffusional-viscous flow of polycrystalline bodies and found the creep rate to depend on stress and temperature according to an expression of the type

$$\epsilon \sim \sigma \frac{D}{T}. \tag{45.5}$$

Here D is the autodiffusion coefficient which in its turn depends on temperature as follows:

$$D = D_0 \exp\left(-\frac{u_0}{RT}\right).$$

It should be remembered that each of these expressions relates to some particular mechanism which is assumed to take place in an ideal object; systematic deviations are usually observed in technical alloys. However, a unique explanation of direct experimental results is not always possible. Thus, it is found that formula (45.1) gives satisfactory results if the energy of activation u_0 is considered to be a function of temperature. Change in the function $u_0(T)$ can provide an indication of whether one mechanism or another predominates in a particular range; in this sense the corresponding standpoint is useful to the physicist but the value of formula (45.1) for extrapolation or interpolation is lost, any empirical formula of sufficiently simple structure is more convenient for technical theory.

Other methods of empirical or semi-empirical origin also exist for establishing relationships between creep results obtained at different temperatures. Various authors have proposed temperature-time parameters, i.e. combinations of temperature and time, as independent variables in describing creep. These parameters are mainly used to extrapolate long-term strength results and, therefore, they will be considered in detail in the appropriate chapter.

In relation to creep, empirical temperature-time parameters are mainly used to determine the conditions of the creep limit, i.e. the stress σ' at which a given arbitrary selected deformation e' is achieved in a given time t'. Creep test results may be represented in the following form (fig. 39).

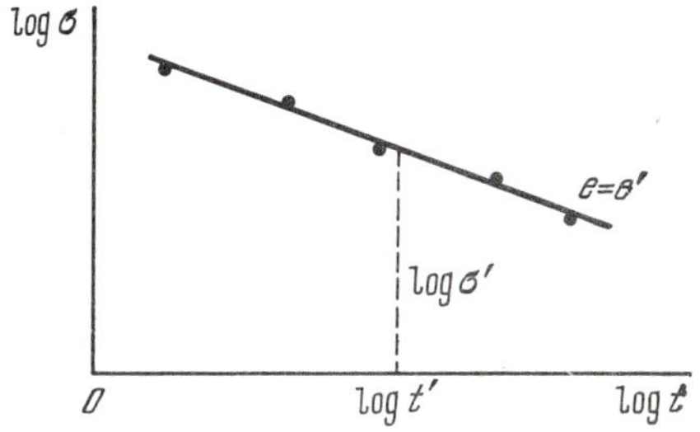

Fig. 39.

Stress is plotted on the ordinate and on the abscissus the time required for the deformation (total or creep deformation) to reach a value e'. Thus only one point on each creep curve is used. For convenience this curve is constructed in logarithmic coordinates. The quantities t' and e' are selected according to service conditions of the product, e' is usually 1% whilst t' can vary from minutes to thousands or tens of thousands of hours. Obviously each temperature has its own curve. As is shown in the figure, tests of finite duration can be used to find the creep limit only when the duration of some of the tests is greater than t', otherwise it is necessary to extrapolate the curve and this always gives unreliable results. If the temperature-time parameter can really be used to describe the creep process, by plotting this parameter on the abscissus in place of time we obtain a single curve for all temperatures. A single value of the parameter corresponds to a long-duration test at low temperature and to a short-duration test at high temperature, and so the test time can be shortened by raising the temperature. Larson and Miller [276] proposed the following quantity as temperature-time parameter:

$$(c + \log t)T \, .$$

Here T is the temperature in degrees Kelvin, c a constant which the authors considered the same for all materials, such that if we use logarithms to the base 10 of time in hours, $c = 20$. Recent checks have shown that extrapolation of creep limits by the Larson-Miller parameter is permissible only over a limited temperature range and does not give reliable results. Somewhat better results can be achieved by better selection of the constant c. Manson and Haferd [291] gave the following parameter:

$$\frac{T - T_a}{\log t - \log t_a} .$$

This includes two constants T_a and t_a and so use of this criterion can give better correlation of results.

The problem of extrapolating test results to other times and temperatures is outside the scope of the mechanical theory of creep. In the present state of knowledge creep calculations on products can only be considered reliable provided that direct experimental creep results are available for the material under consideration throughout the working range of stress and temperature. Formulae for the temperature dependence of creep are required not so much for extrapolation of available data as for interpolation and that usually over quite a narrow temperature range. Therefore, we are justified in preferring a simple formula even though it may have no direct physical meaning to a more well-founded formula which involves difficulties of calculation. The unavoidable scatter of experimental results and the complexity of the behaviour of technical alloys with which the designer has to deal provide still further justification for this point of view.

Strictly speaking physical theories that underlie the formulae that we have given relate to the phase of steady-state creep; the position is still more complicated in the first section of the creep curve.

46. **Stress relaxation**

The information given above relates only to one particular manifestation of creep, namely, tensile deformation of a specimen by a constant force or constant stress. This state of affairs is realised in standard creep tests but much more complicated effects are encountered in practical service. The load that acts on the body varies with time; a component that is subject to creep is associated with other components which may deform only elastically, therefore, even with constant external load the forces between components

Fig. 40.

of the system are redistributed with time. This can be illustrated on simple models. A rod of unit length and unit cross-sectional area is connected in series with a spring of stiffness c (fig. 40).

Initially the rod is subjected to an elongation e_0 and the stress in it is σ_0. The elongation of the spring is then σ_0/c. Thereafter the sum of the elongations of the specimen and of the spring remains constant, consequently,

$$e + \frac{\sigma}{c} = e_0 + \frac{\sigma_0}{c} = \text{const} . \tag{46.1}$$

If the material of the rod creeps its elongation is accompanied by a reduction in the stress because the energy of the system cannot increase. This process of stress diminution with time is termed relaxation. If the spring is not very stiff the displacement of its end that is associated with elongation of the specimen has little influence on the tensile force of the spring, and, therefore, the stress in the specimen remains constant and the conditions are those of a typical creep test. In the other extreme case the spring is extremely stiff and the second terms in the left- and right-hand sides of conditions (46.1) become negligible. The total length of the specimen remains constant and it follows from eq. (46.1) that

$$e = \text{const} .$$

We shall call this case pure relaxation.

At each instant, the total elongation of the specimen e is the sum of the elastic and creep elongations; therefore, the elasticity of the system illustrated in fig. 40 is not solely that of the spring shown in the diagram, there is, as it were, a second spring contained within the actual specimen and pure relaxation consists of creep at diminishing stress which takes place in such a way that the increase in creep deformation compensates the decrease in elastic deformation that results from diminution of the stress.

The general scheme which is represented by the model shown in fig. 40 is frequently realised in practice. Any statically indeterminate system under creep conditions reduces to it when the possible deformation of the component is restricted by the constraint of other components which remain elastic, or which reach a state of creep but can at the same time deform elastically. A flanged joint in pipework, illustrated in fig. 41, is a typical example of this. The bolts are tightened to a certain initial stress which decreases with time. The elongation of the bolt and the associated stress are related by a formula of the type of (46.1) because the flanges cannot be considered absolutely rigid. In this case it is difficult to determine the constant of rigidity c because the compressive stress is not distributed uniformly in the flange.

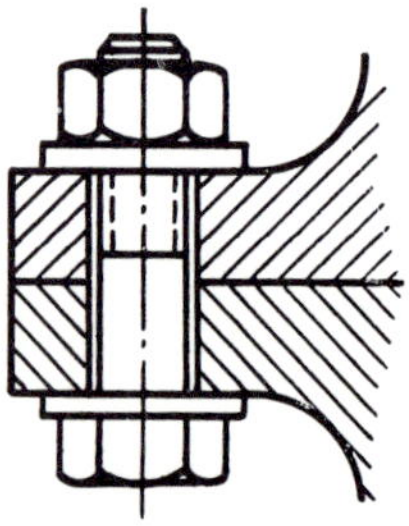

Fig. 41.

As we have shown, the general case of relaxation corresponding to condition (46.1) is typical of the operating conditions of structural elements that are subjected to constant load for a long time. The fundamental question, therefore, arises whether the relaxation process can be calculated from creep characteristics obtained from normal tests. If this is possible it is necessary to formulate a theory of unidimensional creep, that is to formulate equations capable of describing creep under variable loads. First, however, experimental results on stress relaxation are required. Tests of this kind have been made for a long time but the technical difficulties of relaxation testing are considerably greater than those of creep testing, and, therefore, not much relaxation testing has been done. The object of relaxation tests is not so definite as that of creep tests, since relaxation characteristics tend rather to provide a comparative rating of materials and they are not directly used in mechanical calculations. Relaxation tests interest us only from the standpoint of verifying particular creep theories, therefore, we are not concerned for example with such a widely used method of assessing relaxation stability as the Oding annular specimen method [93].

In the first place it is practically impossible to perform tests on metals in pure relaxation by the scheme shown in fig. 40. The spring should act as a dynamometer and consequently the force can be measured only from the elongation of the spring and this is contrary to the requirement that c is very great. The stiffness of the spring and of all the other intermediate links in the equipment, which are not shown in the diagram but which are essential in an actual construction, should be much greater than the stiffness of the specimen. In relaxation tests on plastics, which have a low modulus of elasticity and fairly high elastic deformation, a scheme of this kind is possible in principle if the requirements relating to fulfilment of the condition $e = \text{const}$ are not too severe. For metals this is quite unrealistic. From time to time, relaxation test methods have been proposed in which condition (46.1) has been deliberately achieved with some fixed stiffness c, and not a condition of pure relaxation, but these methods have not been widely accepted.

Another principle is employed in existing relaxation testing equipment using cylindrical specimens. The load applied to the specimen can be varied as required by a servomotor. An extensometer that is fitted to the specimen is provided with a contact device that operates the relay which controls the servomotor. In this way the load is varied and the elongation of the specimen is maintained constant. The actual contact device has a certain sensitivity ($1-2\ \mu$ in the best machines) and, therefore, the relaxation process consists of a series of creep processes at stepwise diminishing loads. The smaller the steps the more closely the scheme approximates to that of pure relaxation.

Here we shall not describe relaxation curves or the various formulae for these curves, whether empirical or based on theoretical considerations; the main thing that interests us is the possibility of predicting the relaxation curve from creep curves.

47. Elementary creep theories

The object of uniaxial creep theory is to relate measured values of stress, deformation, temperature and time by an equation or system of equations of universal character. The equations may contain implicit, internal parameters that characterise the state of the material. In constructing a mechanical theory we should in principle admit the possible existence of equations from which internal parameters can be excluded so that all the creep characteristics can be found by macro-experiments, that is by measuring mechanical quantities such as forces and deformations. We can imagine a theory that includes internal parameters which are not determined by macro-experiment, such as characteristics of the state of the material that can be measured by X-rays, chemical analysis, or optical or electronic metallography. But this possibility should be excluded in advance from mechanical theory, the object of which should be formulated in the following way.

We shall suppose that the time dependences of temperature and of one of the mechanical quantities, either σ or e are given. It is required to determine the time dependence of the second mechanical quantity. To be specific we shall suppose that we are given the functions $T(t)$ and $\sigma(t)$ and require to determine the time function $e(t)$. As the number of experiments is finite it is always possible to construct an expression for $e(t)$ that describes the experimental results correctly for all given values of $T(t)$ and $\sigma(t)$. It may be objected that it is always possible to carry out a new experiment that is essentially different from the previous ones. Functions $T(t)$ and $\sigma(t)$ might then have been selected for which the method of calculating $e(t)$ leads to an incorrect result. However, the experimental possibilities are in fact limited. The range of practical problems for which the theory is specifically required to obtain a solution is quite narrow. The investigators' task is to select means of describing the effect that reproduce its fundamental characteristics with sufficient accuracy, he must select an appropriate degree of approximation and rely on reasonably simple tests which are not too different from one another. By elementary creep theories we understand theories that cover all the necessary results from creep tests at constant stress described above.

The presence of the steady-state part of the creep curve on which the rate of creep is constant for a given stress suggests a comparison between creep in metals at high temperature and quasi-viscous flow of liquids. Unlike ordinary viscous flow, quasi-viscous flow occurs when the flow rate is not linearly dependent on the stress. This scheme is particularly attractive because, as we have seen, the most reliable test results relate to the steady-state part of the creep curve, that is to the dependence of the rate of steady-state creep v on stress σ. Let us then neglect elastic deformation, instantaneous plastic deformation and the presence of a first part of the creep curve and consider that $e = p$, and put

$$\dot{e} = \epsilon = v(\sigma, T)\,. \tag{47.1}$$

In this way the creep curves are replaced by straight lines that pass through the origin of coordinates (fig. 42).

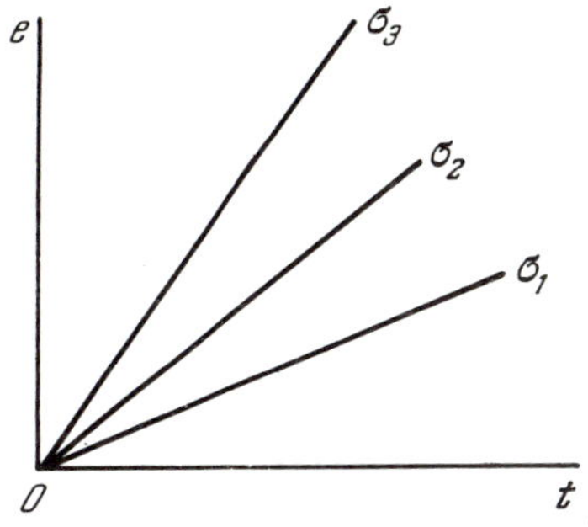

Fig. 42.

The creep theory that is based on eq. (47.1) is termed the steady-state creep theory. The applicability of this theory to practical objects and materials depends on the extent to which the creep curves in e–t coordinates can be approximated by straight lines. This is possible in two extreme cases. Firstly, it applies to products that operate for a very long time in which most of the deformation takes place at constant rate. Under conditions of long-term service, many materials are structurally unstable. Certain structural changes occur in them which sometimes influence the course of the creep curve and give rise to anomalies, which make a strict approach to description of the process impossible in the present state of understanding of its mechanism. Here the scatter of test results for different specimens is quite great. Therefore, an apparently greater precision of creep theory may be fictitious. In drawing straight lines through the origin of coordinates, we are within the limits of the scatter of individual curves that correspond to a given stress. In any case, creep calculations on products must be carried out along a particular experimental curve and so eq. (47.1) may be considered as taking an average and in principle no worse than any other, even though this equation provides a poor approximation to each individual creep curve.

The other extreme case is short-term creep at very high temperature and high stress. Fig. 43 shows creep curves for aluminium alloy D-16T at a temperature of 250°C. It will be seen that the primary creep period is practically absent, the rate of deformation is constant from the start and later accelerates.

Obviously we cannot describe the relaxation process by means of the theory of steady-state creep. Therefore, Söderberg proposed to allow for the initial elastic deformation, that is instead of eq. (47.1) he proposed

$$\dot{e} = \frac{\dot{\sigma}}{E} + v(\sigma, T)\,. \tag{47.2}$$

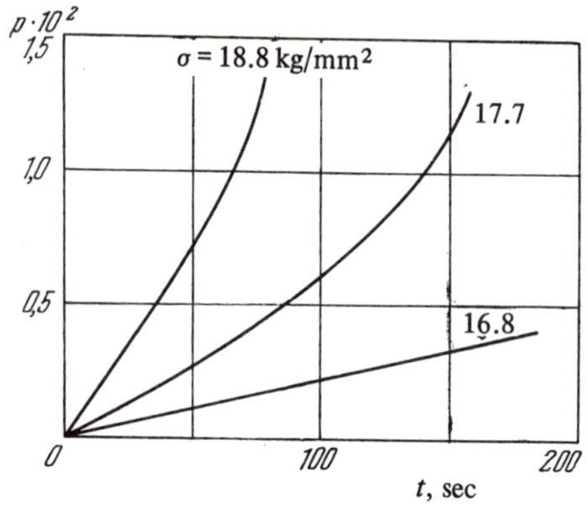

Fig. 43.

Putting $e = \text{const}$ we obtain a differential equation for σ and then the time dependence of stress on relaxation is

$$t = \frac{1}{E} \int_{\sigma}^{\sigma_0} \frac{d\sigma}{v(\sigma, T)} . \tag{47.3}$$

As a rule agreement between experimental results and eq. (47.3) is poor. This is understandable because deformation in the primary stage of creep is at least of the same order as elastic deformation and is often much greater. Therefore, eq. (47.3) does not describe the rapid diminution in stress in the initial period of relaxation. However, at high temperatures and stresses when the short-term creep curves are similar to those shown in fig. 43, eq. (47.3) reproduces fairly accurately at least the initial section of the relaxation curve in which there is a considerable reduction in stress in a short time.

For the same reason, eq. (47.2) must also be used with great care in describing creep at a given load. Allowing for elastic deformation whilst neglecting transient creep deformation often introduces only a small correction which does not justify the greater complexity of the theory that results when eq. (47.2) is extended to the complex stressed state.

Odquist [314] has proposed an important improvement in steady-state creep theory. If the straight line that corresponds to steady-state creep is prolonged to intersection with the e axis (fig. 44) its intercept on this axis is e_0, the value of which is a function of σ. The creep curve is now replaced by the broken line OAB, where $OA = e_0(\sigma)$. The creep law can now be written

$$\dot{e} = e_0'(\sigma)\dot{\sigma} + v(\sigma) . \tag{47.4}$$

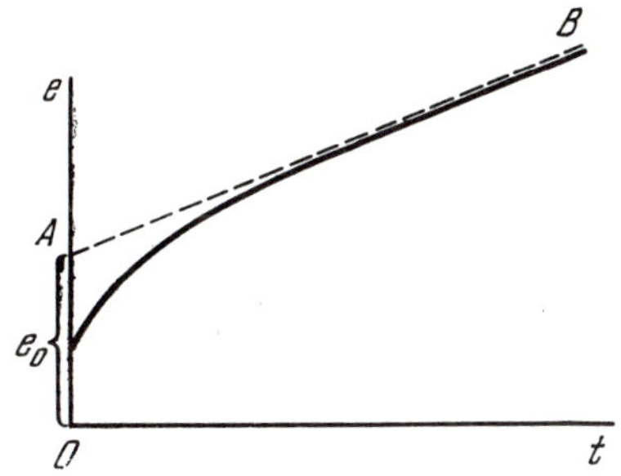

Fig. 44.

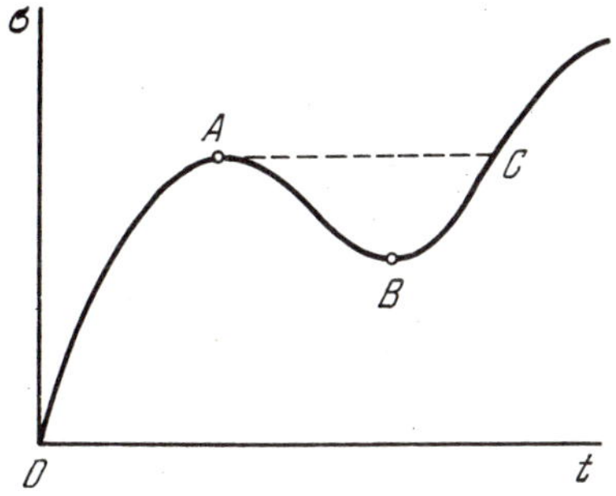

Fig. 45.

The quantity e_0 consists of three parts: it includes the elastic deformation, the instantaneous plastic deformation and the creep deformation that corresponds to the first part of the curve. The third component is combined with the second and is considered as the initial plastic deformation. Denote it $g(\sigma)$ then $e_0 = \sigma/E + g(\sigma)$. The deformation $g(\sigma)$ is irreversible and, therefore, the correct formulation of the creep law is

$$\dot{e} = \frac{1}{E}\dot{\sigma} + \kappa g'(\sigma)\dot{\sigma} + v(\sigma)\,. \tag{47.5}$$

The factor κ is unity when $\dot{\sigma} > 0$ and the value of σ is greater than any of its previous values. Otherwise $\kappa = 0$. A graph of change of stress with time in some hypothetical experiment is shown in fig. 45. On the segment OA the stress increases and creep deformation is accompanied by a plastic deformation $g(\sigma)$ at any instant. At point A a stress σ_A is reached, after which the stress diminishes. The creep continues but the accumulated plastic deformation $g(\sigma_A)$ remains unchanged whereas the elastic deformation diminishes in proportion to the diminution in stress. Starting from point B the stress σ again increases but as before κ is zero, the new yield point is σ_A and additional plastic deformation occurs only beyond point C when $\sigma > \sigma_A$. Strictly speaking this behaviour relates only to plastic deformation in the strict sense and not to the deformation $g(\sigma)$. We shall see later that creep tests with step-wise change of load are not completely in agreement with this hypothesis. However, Odquist's scheme is so simple that eq. (47.5) is successfully applied to the solution of quite complicated problems, in particular to problems of the stability of rods under creep conditions. Odquist proposed a power relationship with exponent $m + 1$ for $g(\sigma)$ and for $v(\sigma)$ he also recommends a power relationship with exponent n. Here $m + 1 < n$. By way of example of the use of eq. (47.5) we give an equation which was used by Hoff [237] to describe the behaviour of aluminium alloy. As the author was interested in problems of bending, the equation is written in such a form that it can be used for either tension or compression, namely

$$\dot{e} = \frac{\dot{\sigma}}{E} + k_1 \left(\frac{\sigma}{\mu}\right)^m \frac{\dot{\sigma}}{\mu} + k_2 \left(\frac{\sigma}{\lambda}\right)^n . \tag{47.6}$$

Here m and n are whole numbers. Only the case of monotonic loading is considered and so k_1 and k_2 are given by the following conditions:

when $\sigma > 0$, $\dot{\sigma} > 0$ $k_1 = 1$,
when $\sigma > 0$, $\dot{\sigma} < 0$ $k_1 = 0$,
when $\sigma < 0$, $\dot{\sigma} < 0$ $k_1 = 1$ for even values of m and $k_1 = -1$ for odd values of m ,
when $\sigma < 0$, $\dot{\sigma} > 0$ $k_1 = 0$.

For even values of n $k_2 = 1$ when $\sigma > 0$ and $k_2 = -1$ when $\sigma < 0$, and for odd values of n $k_2 = 1$ always. Numerical values of the constants in the above equation for aluminium alloy 24S-T4 at a temperature of 314° are

$$m = 1 , \quad n = 3 , \quad E = 5200 , \quad \mu = 562 , \quad \lambda = 61.5 .$$

Here σ is measured in kg/mm^2 and time in hours.

Note that extension of a creep law based on tensile tests to compressive stress requires special investigation. Certain experimental results that relate to this question will be considered later.

The problem of stress relaxation is solved by Odquist's theory in the same way as by using eq. (47.2) because in (47.6) we here take $k_1 = 0$.

48. Ageing (time-hardening) theories

If the equation of a family of creep curves at a given temperature is written in the form

$$e = f(\sigma, t) , \tag{48.1}$$

in attributing a universal validity to this relationship, the simplest assumption is to consider it valid not only for constant stress but also when the stress varies with time. In this way we arrive at a formulation of ageing theory in which stress, deformation and time are related by a finite relationship of the type of (48.1).

Ageing theory is very convenient because it is very simple. In using it there is no need to assign any analytical expression for the function $f(\sigma, t)$, in calculations direct use can be made of creep curves constructed either in the usual $e-t$ coordinates or, for greater convenience, isochronous curves constructed as described in section 44.

There is a third possible way of constructing creep curves: stress-time curves are plotted, each corresponding to a fixed deformation e. According to ageing theory creep curves constructed in this way should coincide with relaxation curves.

In structurally stable materials the use of eq. (48.1) leads to obvious contradictions. Indeed this equation is not invariant relative to the time that is counted as the initial time and the creep deformation at the instant t does not depend on whether the load has acted for the whole time starting from $t = 0$ or whether it has just been applied. For structurally unstable materials we can hypothesise that the relationship between stress and deformation depends on time, and that this dependence is governed by structural changes that depend only on certain internal processes that take place irrespective of the load that acts on the body or of the deformation that it acquires. Such processes actually occur in alloys. In steel spheroidising of cementite, dispersion hardening, and phase separation in complex alloys are apparently not associated with deformation. It is, however, difficult to suppose that these processes really govern creep. In form and in fact eq. (48.1) is not a creep equation, it is a law of non-linear elasticity of a material with time-dependent elastic properties. The following model may be considered for such a material. A system of elastic elements consisting of different kinds of springs with non-linear characteristics is located in a corrosive medium which attacks and dissolves the springs. Because of the reduction in the cross-sectional area of the springs, their stiffness diminishes and the deformation at constant load increases. Externally this corresponds to creep. The deformation remains constant, the force diminishes and relaxation occurs. But in actual fact creep deformation is essentially irreversible and the analogy between creep in metal and the behaviour of this system is quite superficial. It is easy to give a simple example which shows the unsuitability of eq. (48.1) to describe creep under varying load. Take two creep curves for stresses σ_1 and σ_2 (fig. 46). Suppose that at first $\sigma_1 < \sigma_2$; at the instant of t_1 when the deformation corresponds to point A, the stress is suddenly increased to σ_2. Then according to eq. (48.1) there is instantaneous deformation of an amount AB, and starting from

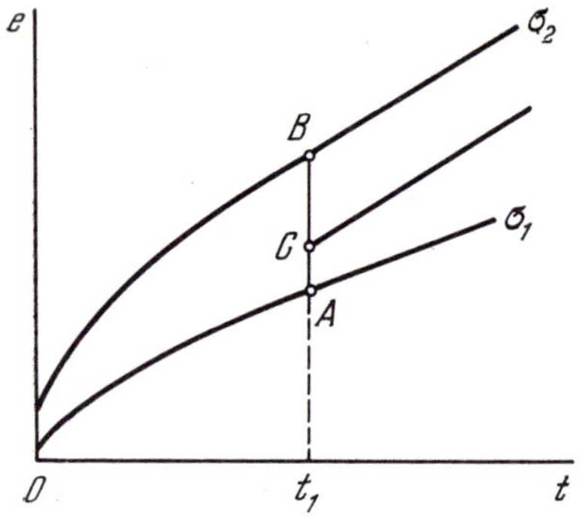

Fig. 46.

point B creep follows a curve corresponding to the new stress σ_2. After a fairly long time, the creep curves for stresses σ_1 and σ_2 become indefinitely widely separated and the instantaneous deformation AB is indefinitely great. In practice, nothing of the kind occurs. The statement that when the stress is reduced from σ_2 to σ_1, according to formula (48.1) the deformation should diminish by the same amount BA, is still more decisively contradicted by all known experimental facts.

In practical creep calculations on machine parts that operate under steady-state conditions, such a sharp change in stress does not occur. Typical conditions correspond to the scheme shown in fig. 40 for which the extreme cases are pure creep and pure relaxation. Any theory that satisfactorily predicts relaxation curves from creep curves will probably also be suitable for intermediate states. Relaxation curve predictions based on ageing theory are rather poor but not so bad as to be in severe disagreement with practice. Therefore, ageing theory can be justified as an approximate device for slightly changing loads. This particularly applies when the material is structurally unstable and it is difficult to attribute special features of the creep curves to some particular mechanism which can provide a basis for a more accurate and logical theory. Because primary creep curves with all their individual peculiarities can be used directly, ageing theory is often preferred to other better based but more complicated theories which cannot be used without making a mathematical approximation to the creep curves that inevitably distorts the experimental results.

In another variant of ageing theory the rate of creep deformation is considered a function of stress and time. The fundamental equation of this theory is

$$\dot{p} = \dot{e} - \frac{\dot{\sigma}}{E} = f(\sigma, t)\,. \tag{48.2}$$

Now the behaviour of material that is subject to creep conditions is similar to that of a non-linearly viscous liquid with time-dependent viscosity. Eq. (48.2) is obviously more logically acceptable than eq. (48.1). Creep deformation described by eq. (48.2) is irreversible, the instantaneous change in stress does not involve an instantaneous change in creep deformation, only the rate is altered. However, here too we must consider the possible existence of structural change mechanisms which govern the rate of creep at a given stress and which take place irrespective of the actual creep process. Modern physical theories of creep lead to the conclusion that these processes are not governing; this result follows from direct experimental verification of the theory. In the case of stepwise stress change from σ_1 to σ_2 the position is as follows (fig. 46). The deformation acquires an increment AC, equal to $(\sigma_2 - \sigma_1)/E$ in accordance with Hooke's law. From point C creep commences at a rate that corresponds to the time t_1 and stress σ_2, that is at the same rate as creep occurs under stress σ_2 at point B. Eq. (48.2) is of first order and, therefore, its integral depends on a single constant; consequently, the creep curve from point C is parallel to the upper curve commencing from point B. In fact at point C the creep rate is considerably greater than at point B as will be seen from the experimental curves given later. Relaxation curve predictions from the creep curve that are based on the second ageing theory are somewhat better than those based on the first ageing theory.

The second ageing theory, sometimes termed flow theory, has become fairly well known as a result of Kachanov's [52] work, it gives fairly satisfactory results for loads that change only slightly. Numerous methods of calculation have been based on this theory. It is particularly convenient to apply when the creep curves are similar, that is they are described by eq. (44.1). Then eq. (48.2) assumes the form

$$\dot{p} = \dot{e} - \frac{\dot{\sigma}}{E} = S(\sigma)\tau(t) .$$

Take as independent variable, not the time t but a function of time $\tau(t)$. This equation can then be written

$$\frac{\mathrm{d}p}{\mathrm{d}\tau} = \frac{\mathrm{d}e}{\mathrm{d}\tau} - \frac{1}{E}\frac{\mathrm{d}\sigma}{\mathrm{d}\tau} = S(\sigma) .$$

We have obtained the equation of the theory of steady-state creep with change in time scale. Indeed eq. (48.3) coincides in form with eq. (47.2).

The relaxation problem is solved very simply by means of (48.3), indeed putting e = const, we obtain

$$\tau(t) = \frac{1}{E} \int_{\sigma}^{\sigma_0} \frac{d\sigma}{S(\sigma)} . \tag{48.4}$$

The use of the name "flow theory" is justified in this case where eq. (48.3) is used as a basis.

Another variant of the ageing theory was proposed by Malinin [70], developing an idea of Belyaev [5]. The following relationship is taken between deformation and stress:

$$e = \sigma \left[\frac{1}{E} + \int_0^{\tau} \frac{S(\sigma)}{\sigma} \, d\tau \right] . \tag{48.5}$$

Belyaev's idea in formulating this theory was to make creep equations analogous to equations of the deformation theory of plasticity, which in their turn are the equations of the theory of elasticity with variable moduli. The expression in brackets in the right-hand side of eq. (48.5) represents a quantity which is the reciprocal of the creep modulus. Differentiating (48.5) with respect to τ and eliminating the integral we obtain another form of the equation

$$e' = e \frac{\sigma'}{\sigma} + S(\sigma) . \tag{48.6}$$

Here the primes denote differentiation with respect to τ. In using eq. (48.6) account must be taken of the initial condition that the instantaneous deformation follows Hooke's law. This condition follows from (48.5) but not (48.6). In the case of relaxation we must take $e' = 0$, $e = \sigma_0/E$. After separating variables and integrating we obtain

$$\tau(t) = \frac{1}{E} \int_{\sigma}^{\sigma_0} \frac{\sigma_0}{\sigma} \frac{d\sigma}{S(\sigma)} . \tag{48.7}$$

As $\sigma_0/\sigma > 1$, relaxation according to Malinin's theory is slower than according to flow theory. In fig. 47, curve 1 shows an experimental relaxation curve for copper at a temperature of 165° and an initial stress of 9.49 kg/mm^2 from the results of Davis [189], relaxation curves calculated by Malinin [108] by

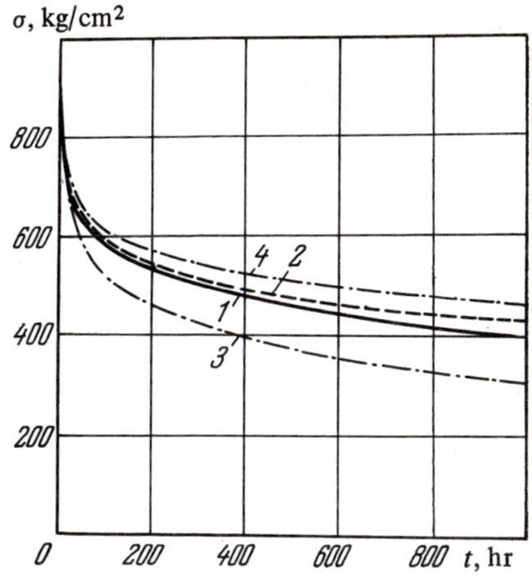

Fig. 47.

three variants of the ageing theory, are shown, curve 2 by eq. (48.7), curve 3 by eq. (48.4) and curve 4 by the ageing-hypothesis eq. (48.1). Until the early 1950's Davis' very carefully formulated experiments were probably the only reliable data for comparing creep and relaxation results. In the Soviet Union investigations on the mechanical theory of creep mainly commenced in the post-war years (Belyaev's first work relates to 1943) and theoretical work proceeded experimentally so that Davis' tests were widely used by all, including the present author, to check various hypotheses. A special feature of these tests was the very great care that was taken to ensure uniformity of

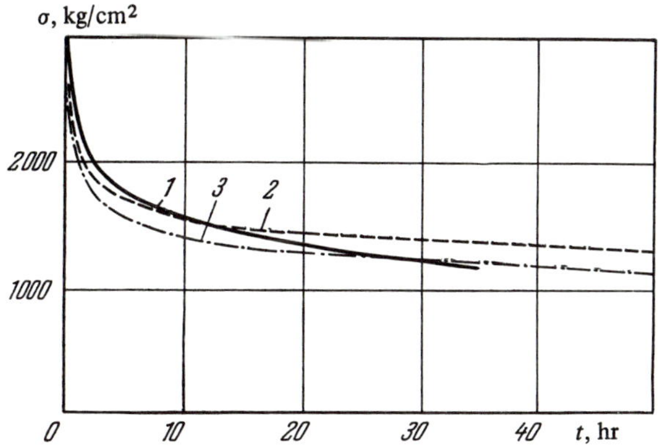

Fig. 48.

the creep and relaxation of test specimens. The copper was freed of oxygen which affected its properties and the tests were carried out in a neutral atmosphere.

Fig. 48 shows an experimental relaxation curve for steel 30KhMA at a temperature of 500° and an initial stress of 30.8 kg/mm^2, curve 1 from the results of [24], curve 2 calculated by eq. (48.7) and curve 3 calculated by eq. (48.4).

49. **Hypothesis of the equation of state**

In the case of structurally stable materials, i.e. materials whose structure and properties do not alter on prolonged exposure to the test temperature without load, it is most natural to suppose that the reduction in creep rate in the first part of the curve to a minimum value that corresponds to the second part is an effect analogous to that of work-hardening in ordinary plastic deformation. To a first approximation it may be considered that the degree of strain-hardening depends uniquely on the amount of plastic deformation that has accumulated. Therefore, the creep rate at any instant depends on the applied stress and the amount of accumulated creep-deformation. The corresponding equation is of the form

$$\Phi(\dot{p}, \sigma, p, T) = 0 . \tag{49.1}$$

and is termed the equation of state. The equation of state hypothesis was apparently first formulated by Ludvik [88], then by Nadai [300] and it was developed in an article by Davenport [188], who introduced the term strain-hardening theory as distinct from the time-hardening (or ageing) theory. This is the term that we shall use for preference. Returning to the presentation of Dorn's point of view in section 45, we note that in essence the work-hardening theory that he formulated is independent of temperature and of the order of the applied load, because the structural state of the material depends only on the amount of creep deformation that has accumulated. This structural state not only determines the rate of creep but can be recorded by various other independent methods.

The simplest method of checking the strain-hardening theory is by creep tests with stepwise change of load. Tests of this kind were carried out systematically on various materials in the laboratory of the Institute of Mechanics of the Academy of Sciences of the USSR, in Moscow State University and in the Institute of Hydrodynamics of the Siberian Division of the Academy of of Sciences of the USSR. These tests are simpler than relaxation tests and so give more contrasting results. Indeed, as we have seen from the examples given in section 48, the prediction of relaxation curves from creep curves on the basis of various time-hardening theories leads to results that are not too far from the truth. Qualitative predictions about the shape of the relaxation curve based on this theory are satisfactory and the quantitative discrepancies can always be related to scatter the results from variability in the properties of specimens; as we have mentioned several times, the scatter of creep test results is fairly large. Now consider creep tests with stepwise change in load. It is most convenient to construct creep curves in $p-t$ coordinates where p is the creep deformation (we shall discuss later the sense in which the quantity p should be understood). Two creep curves for stresses σ_1 and $\sigma_2 > \sigma_1$ are shown in fig. 49. Suppose that the stress σ_1 is maintained in the specimen and that creep follows the lower curve. At the instant t_1 we are at point A on the curve. Now let the stress be increased suddenly to σ_2. The additional elastic deformation is not plotted on the graph because p is only the creep deformation; consequently, the creep curve leaves point A at a rate which, according to eq. (49.1), depends on the deformation p_A and the stress σ_2. This rate is the angular coefficient of the tangent to the upper creep curve at point B, which has the same ordinate as point A. Since eq. (49.1) is a first order equation its integral depends on a single arbitrary constant and the creep curve from point A is a repetition of the creep curve with constant stress σ_2 commencing from point B. To obtain the creep curve that corresponds to stepwise change in load we must shift the upper curve parallel to itself to the right until point B coincides with point A. According to the second ageing theory the upper curve should be displaced downwards parallel to itself till point C coincides with point A. The difference between the predictions of the two

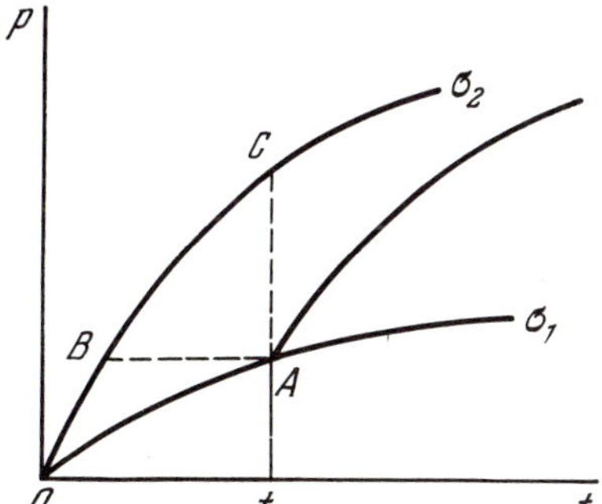

Fig. 49.

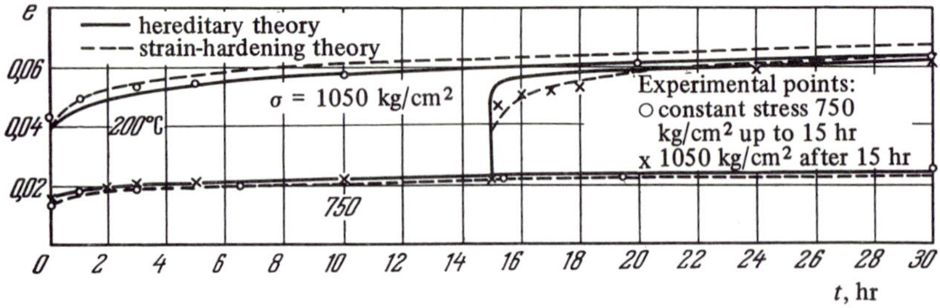

Fig. 50.

theories is very marked, and the strain-hardening theory agrees best with practice. Creep curves with stepwise loads for copper taken from the work [30] are compared with the predictions of the strain-hardening theory in fig. 50. Later we shall give similar results for the alloy D-16T from the results of Namyestnikov and Khvostunkov. It will be seen that the strain-hardening theory gives very satisfactory results whilst the time-hardening theory greatly underestimates the creep rate after change of load.

At the same time, the predictions of the strain-hardening theory are not entirely accurate, there are systematic deviations which are particularly marked when there are several stages of stress change. Later on, we shall consider in more detail possible improvements in the equation of state.

The strain-hardening theory does not describe certain secondary effects which are observed in creep tests, but for structurally stable materials it is a reliable means of describing the process to a first approximation on the basis of a physically clear and self-consistent hypothesis, which reflects the fundamental characteristics of the effect.

The equation-of-state hypothesis formulated in eq. (49.1) has been used by numerous authors to describe various processes in the deformation of metals. The quantity p is usually considered as the total plastic deformation and $\dot{p}$ as the rate of plastic deformation. If the temperature T is fixed, eq. (49.1) may be treated as an equation of a surface in space with coordinates $\dot{p}$, p and σ; the intersection of this surface with the plane $\dot{p} = \infty$ should give the instantaneous deformation curve. As we shall see later, this point of view is not justified and the problem of the instantaneous plastic deformation must be solved in another way. Therefore, we assume for the moment that there is no instantaneous plastic deformation but only elastic deformation and creep deformation p. Then evidently $e = \sigma/E + p$.

If creep curves are known the relaxation curve can be predicted by a graphical method due to Davenport [188]. Suppose that we have a fairly close network of creep curves constructed in p–t coordinates for different stresses (fig. 51).

We shall consider creep with stepwise change of stress instead of a continuous relaxation process. Suppose that at a time t_i the stress is σ_i, that is we have a point M_i on the relaxation curve (fig. 52). If this stress is maintained constant, creep follows the curve that corresponds to the stress σ_i and starts from point M_i (fig. 51). In a time Δt_i the creep deformation Δp_i accumulates. Now reduce the stress by an amount $\Delta\sigma_i = E\Delta p_i$, then the total deformation of the specimen due to elastic contraction again assumes the value that it had at the initial instant t_i. The stress becomes $\sigma_{i+1} = \sigma_i - \Delta\sigma_i$; creep again occurs at this stress along the curve that corresponds to the stress σ_{i+1} commencing from the point at which the creep deformation is $p_{i+1} = p_i + \Delta p_i$. In order to find this point we must draw horizontal lines through the point N_i which is the end of the creep section on the curve for the stress σ_i. The point of intersection between the curve σ_{i+1} and this line is the point M_{i+1} which corresponds to the point on the relaxation curve with coordinates σ_{i+1} and t_{i+1}. Continuing this process we construct a relaxation curve to any required accuracy depending on the closeness of the creep curves. In fact this method of calculation actually reproduces relaxation test conditions (see section 46).

In practice the grid of creep curves is not very dense and it is difficult to apply Davenport's method directly. A more reliable numerical method of constructing the relaxation curve was developed by Popov [328] who used it to work out Davis' results mentioned above.

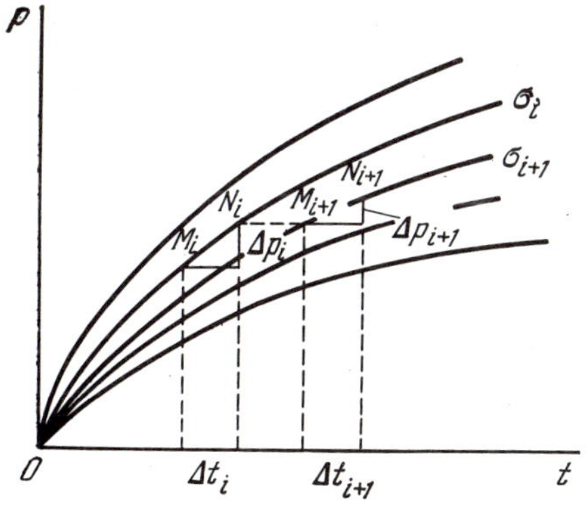

Fig. 51.

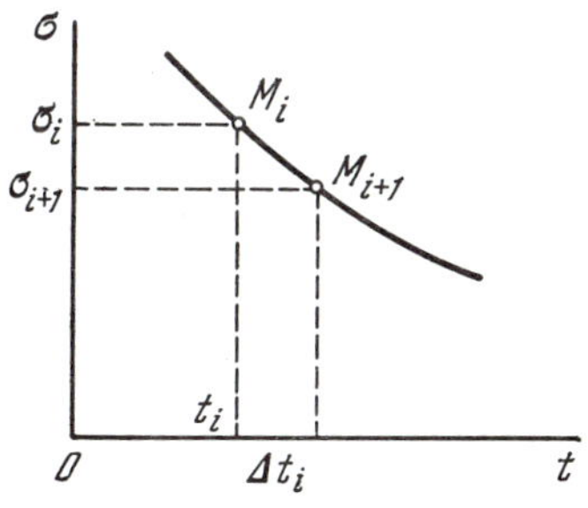

Fig. 52.

50. Analytical expressions for the strain-hardening law

Eq. (49.1) can be plotted as a series of curves in certain coordinates, but it is then difficult to use it for calculations. A more convenient method is to select a suitable analytical expression for the function $\Phi(\dot{p}, p, \sigma, T)$ which is sufficiently simple and which reproduces satisfactorily, for instance, creep curves at constant stress. The condition of similarity of creep curves limits the class of possible representations of the equation of state; at constant temperature this equation should necessarily be of the following form:

$$\dot{p} = S(\sigma)F\left(\frac{p}{S(\sigma)}\right) . \tag{50.1}$$

This will readily be seen if we represent the creep curve equation in the form $p = S(\sigma)\tau(t)$ then differentiate and eliminate time. If F is a power function we arrive at the following strain-hardening law:

$$\dot{p}p^{\alpha} = f(\sigma) . \tag{50.2}$$

Eq. (50.2) describes very well the first part of the creep curve. Indeed integrating it with σ constant and putting $m = 1/(1 + \alpha)$, we obtain

$$p = \left[\frac{f(\sigma)}{m}\right]^m t^m . \tag{50.3}$$

As we saw in section 43 the power approximation of the time dependence gives satisfactory results for the first part of the curve. A power function can also be selected for $f(\sigma)$,

$$f(\sigma) = B\sigma^n , \tag{50.4}$$

or an exponential function

$$f(\sigma) = k \exp \frac{\sigma}{A} . \tag{50.5}$$

In the first case, as follows from (50.3) we should have $n > (1 + \alpha)$; otherwise the creep rate is proportional to the stress to a fractional power. Materials that behave like this are not known and moreover, convincing considerations can be adduced to show that they cannot exist [149]. Just as in application to rate of steady-state creep, eq. (50.5) is unsuitable for small stresses σ, however, it cannot be corrected by one of the methods described in section 43. At small values of σ the function $f(\sigma)$ must tend to zero as $\sigma^n (n > (1+\alpha))$ and not as σ. One possible improvement in eq. (50.5) is for example

$$f(\sigma) = k\left(2 \operatorname{sh} \frac{\sigma}{A}\right)^n .$$

This expression is ill-suited for practical application and the low stress region is usually of little interest; in problems where the low stress region must be considered (for instance in bending about the neutral axis) it can be excluded by a device which will be explained later.

As we desire to use the equation of state to describe steady-state creep as well as the initial part of the curve we must alter the form of eq. (50.2), for example, to the following:

$$\dot{p} h(p) = f(\sigma) . \qquad (50.6)$$

The function $h(p)$ should behave like p^α for small values of p and should tend to a constant value with increase in the creep deformation. Shorr [154] proposed to define the function $h(p)$ as follows:

$$\begin{aligned} h(p) &= p^\alpha \qquad (p < p_c) , \\ h(p) &= p_c^\alpha \qquad (p > p_c) . \end{aligned} \qquad (50.7)$$

The quantity p_c depends on stress and temperature, it is the amount of creep deformation at the instant of transition from the first part of the creep curve to the second. The boundary between the parts of the creep curve is only nominal, but a satisfactory approximation to experimental creep curves can be obtained in this way.

In order to obtain continuous transition to steady-state creep we can select the function $h(p)$ as, for instance,

$$\frac{1}{h} = p^{-\alpha} + c ,$$

where c is a constant.

Then eq. (50.6) assumes the form

$$\frac{\dot{p}}{c + p^{-\alpha}} = f(\sigma) . \qquad (50.8)$$

The case when $\alpha = 2$ is particularly interesting, consequently Andrade's law is obeyed in the first part of the curve and creep deformation is proportional to time to the power $\frac{1}{3}$. Eq. (50.8) is then easily integrated at constant stress and we obtain the following creep curve equation:

$$p\sqrt{c} - \arctan(p\sqrt{c}) = \sqrt{c}\, v(\sigma) t\,. \tag{50.9}$$

Here $v(\sigma)$ is the rate of steady-state creep. Creep curves have asymptotes which intersect at a single point on the ordinate axis, namely at a distance of $\pi/2\sqrt{c}$ from the origin. Therefore, creep curves are extremely easy to work out by expression (50.9). Asymptotes are drawn to the creep curves, the slope of these asymptotes determines the function $v(\sigma)$. The ordinate of the point of intersection between the asymptotes and the p axis immediately gives the value of c. Individual creep curves are usually adequately represented by expression (50.9), but when a family of curves is worked out it is usually found that there is no common point of intersection and the value of c is a function of stress. We can, of course, consider that in eq. (50.8) $c = c(\sigma)$ but the theory becomes unsuitable in this form. If we allow for the scatter of experimental results and the relatively slight dependence of c on σ the use of this equation can be justified for a certain narrow stress range.

A somewhat different method of allowing for transient creep is that the creep deformation is considered as consisting of two parts: transient creep deformation described by the equation of state and steady-state creep deformation of the quasiviscous flow type. Thus we retain the equation of state and in it we understand by p the quantity

$$p = e - e_0(\sigma) - \int_0^t v(\sigma)\,\mathrm{d}t\,.$$

Systematic development of this point of view leads to a fairly complicated result even in the simplest cases but it is useful in interpreting certain experimental results.

Later we shall use the simpler creep eq. (50.2) which is suitable for describing transient creep. The form of (50.5) will be used preferentially for the function $f(\sigma)$ so that the strain-hardening law is written

$$\dot{p}p^{\alpha} = k \exp\frac{\sigma}{A} \qquad \left(p = e - \frac{\sigma}{E}\right). \tag{50.10}$$

Then the creep curve equation at constant stress is

$$p = e - \frac{\sigma}{E} = \left(\frac{k}{m}\right)^m \exp\left(\frac{\sigma m}{A}\right) t^m \qquad \left(m = \frac{1}{1+\alpha}\right). \qquad (50.11)$$

We shall now give an example of the procedure for determining the constants in eq. (50.10). The curves plotted in solid lines in fig. 53 are creep curves of steel grade 30KhMA at a temperature of 500° [24]. Each curve is an average of three experimental curves. Replotting these curves in logarithmic coordinates we obtain the graph shown in fig. 54. The experimental points that correspond to each stress lie fairly closely about a straight line; the slope of this line gives a value of m, which as will be seen, is not constant but stress-dependent. In the range 20 to 25 kg/mm^2 it may be considered constant, on increasing the stress to 27–28 kg/mm^2 m increases sharply.

The intercept made by each of the straight lines of fig. 54 on the line corresponding to $t = 1$ is $m \ln k/m + \sigma m/A$. If we plot a graph of the dependence of this quantity on σ the points lie about a straight line with an angular coefficient of m/A. Since m is already known A is determined. The intercept of the straight line on the σ axis is $m \ln k/m$, consequently, we find k. If we use the mean value of m then k and A are constants. The following mean values were found in this case: $m = 0.53$, $k = 2.77 \times 10^{-10}$ 1/hr, $A = 2.94$ kg/mm^2.

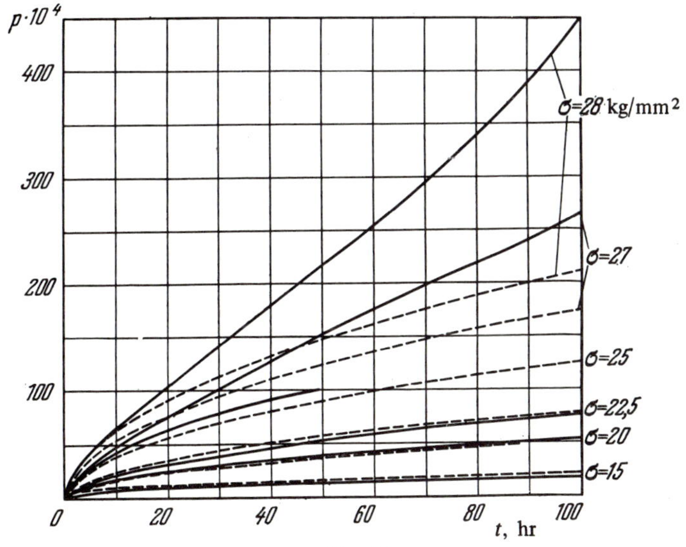

Fig. 53.

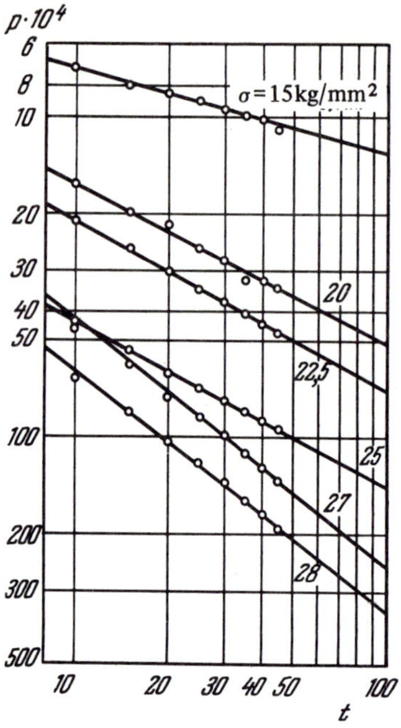

Fig. 54.

The curves shown dotted in fig. 53 are creep curves calculated by eq. (50.11) for these values of the constants. Up to a stress of $\sigma = 25$ kg/mm^2 the agreement between calculated and experimental curves is fully satisfactory. At stresses of 27 and 28 kg/mm^2 second and third parts are clearly visible on the curves and consequently, the equation that we have adopted is clearly unsuitable. However, even here in the initial parts agreement is satisfactory. It may be concluded that up to deformations of the order of 1%, eq. (50.10) with constant values of α, k and A is suitable for describing creep of this material, although creep tests at constant load cannot be considered a convincing demonstration, and experiments should be performed under other conditions, for example, under conditions of pure relaxation. We shall discuss this in the next section. For the time being note that if we consider the quantities m and A as functions of stress we achieve much better agreement with test results. These experimental results were worked out on the assumption that m and A are linear functions of σ. The experimental curves shown in fig. 55 are the same as those in fig. 53 and the dotted curves show the approximations with the assumptions mentioned above. It will be seen that agreement is very good.

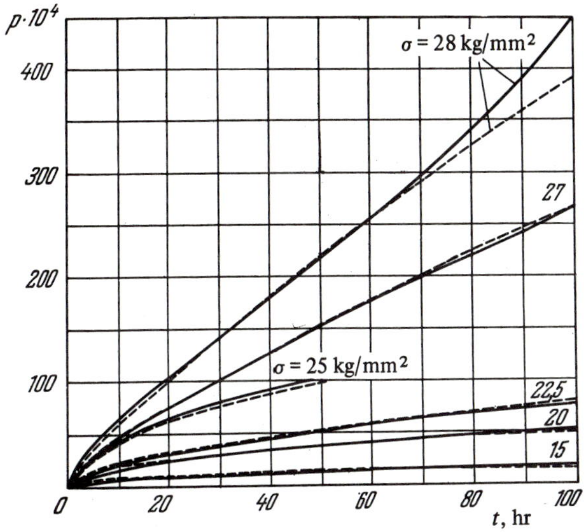

Fig. 55.

The creep law constants, for example, α, B and n in formula (50.4) or k and A in (50.5) generally depend on temperature. However, α is only slightly dependent on temperature. Also in some fairly narrow range of temperature we may consider that n or A is constant. On the other hand, the factor in the expression of the function $f(\sigma)$, i.e. quantity B in formula (50.4) or quantity k in formula (50.5), depends very much on temperature. In accordance with what we have said in section 45 we may take

$$B = B_0 \exp\left(-\frac{u_0}{RT}\right), \quad k = k_0 \exp\left(-\frac{u_0}{RT}\right).$$

Generally speaking the simplest general form of the strain-hardening law with allowance for the influence of temperature is

$$pp^{\alpha} = f(\sigma) \exp\left(-\frac{u_0}{RT}\right). \tag{50.12}$$

Here the constant a and the function $f(\sigma)$ are independent of temperature.

51. The relation between creep and relaxation according to the strain-hardening theory

Let us express eq. (50.10) in dimensionless variables in order to eliminate the constants k and A. Put

$$\hat{\sigma} = \frac{\sigma}{A}, \qquad \hat{e} = \frac{Ee}{A}, \qquad \hat{p} = \frac{Ep}{A}, \qquad \hat{t} = k\left(\frac{E}{A}\right)^{1+\alpha} t .$$

This equation then assumes the form

$$\dot{\hat{p}}\hat{p}^{\alpha} = \exp \hat{\sigma} \qquad (\hat{p} = \hat{e} - \hat{\sigma}) , \tag{51.1}$$

the point above a symbol denotes differentiation with respect to dimensionless time $\hat{t}$.

In the case of relaxation $\hat{e} = \hat{e}_0 = \text{const}$, consequently

$$\hat{p}_0 + \hat{\sigma}_0 = \hat{p} + \hat{\sigma} .$$

Whence

$$\hat{\sigma} = \hat{\sigma}_0 + \hat{p}_0 - \hat{p} .$$

Substitute this into (51.1) and separate variables, we obtain

$$\hat{p}^{\alpha} \exp \hat{p}\, \mathrm{d}\hat{p} = \exp (\hat{\sigma}_0 + \hat{p}_0) \mathrm{d}\hat{t} . \tag{51.2}$$

Introduce the notation

$$\int_0^z z^{\alpha} \exp z\, \mathrm{d}z = \chi(z) .$$

Integrating (51.2) with the initial condition $\hat{p}(0) = \hat{p}_0$ we find

$$\chi(\hat{p}) = \hat{t} \exp (\hat{\sigma}_0 + \hat{p}_0) + \chi(\hat{p}_0) .$$

Whence

$$\hat{\sigma} = \hat{\sigma}_0 + \hat{p}_0 - \chi^{-1} [\chi(\hat{p}_0) + \hat{t} \exp (\hat{\sigma}_0 + \hat{p}_0)] \tag{51.3}$$

(χ^{-1} is a function which is the reciprocal of the function χ).

The function $\chi(\hat{p})$ is easily calculated and tabulated. Equation (51.3) includes not only the initial stress but the initial creep deformation p_0, therefore, the case considered is the most general one. The point is that the material can reach the state of commencement of relaxation in various ways. In the usual tests the specimen is loaded to a stress σ_0 so rapidly that creep deformation cannot accumulate during the loading process and we may consider that $\hat{p}_0 = 0$. Eq. (51.3) is then much simpler and we find

$$\hat{\sigma} = \hat{\sigma}_0 - \chi^{-1}(\hat{t} \exp \hat{\sigma}_0) . \tag{51.4}$$

In section 50 we gave results of working out creep curves for steel 30KhMA at 500°C and found mean values of the constants m, k and A. For these values of the constants we calculated relaxation curves from initial stresses of $\sigma_0 = 14.1$; 20; 25.92 and 30.8 kg/mm^2. These curves are shown by dotted lines in fig. 56. The same figure shows experimental relaxation curves in solid lines. It will be seen that agreement is satisfactory for all curves except the uppermost which corresponds to $\sigma_0 = 30.8$ kg/mm^2. This is understandable for we have seen that for high stresses of 27 and 28 kg/mm^2 eq. (50.10) becomes unsuitable and for stresses of the order of 30 kg/mm^2 it will probably give entirely incorrect results. It is important to note that relaxation is accompanied by very little creep deformation and, therefore, satisfactory approximation of only the initial sections of the creep curve by the selected values of the constants ensures good prediction of relaxation curves. The work [24] also gives relaxation curves calculated on the assumption that m and A are functions of σ. As we have seen (fig. 55) this complication of the

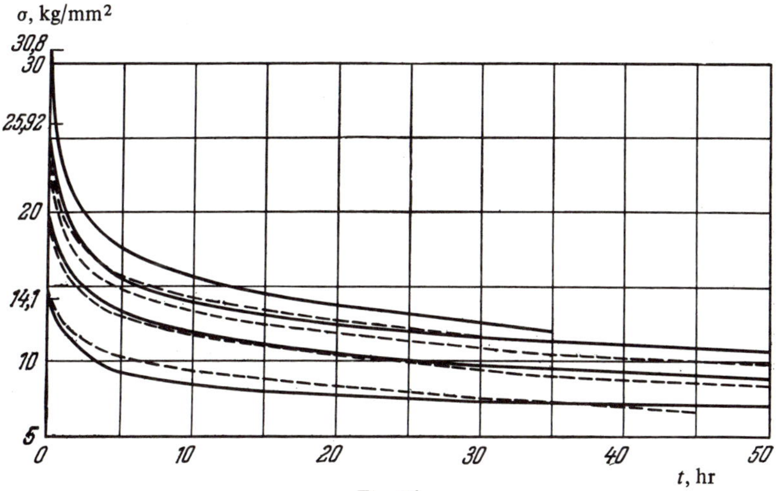

Fig. 56.

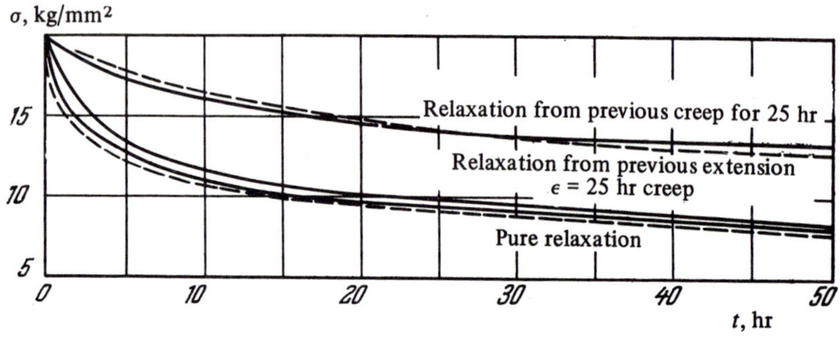

Fig. 57.

form of the equation of state gives a satisfactory description of creep curves over a wide range of stress and deformation, but in practice the agreement between calculated and experimental relaxation curves is not improved by this device. The following test was made to check the general relaxation eq. (51.3). The specimen was submitted to constant load for 25 hours after which the creep deformation p_0 had accumulated. The total deformation $(p_0 + \sigma_0/E)$ was maintained constant for the next 75 hours. Relaxation curves for a specimen that had not been strain-hardened and for a specimen that had been hardened by preliminary creep are shown in fig. 57. This figure also gives curves calculated by formulae (51.4) and (51.3) respectively. The values of the creep constants were determined independently beforehand in creep tests at

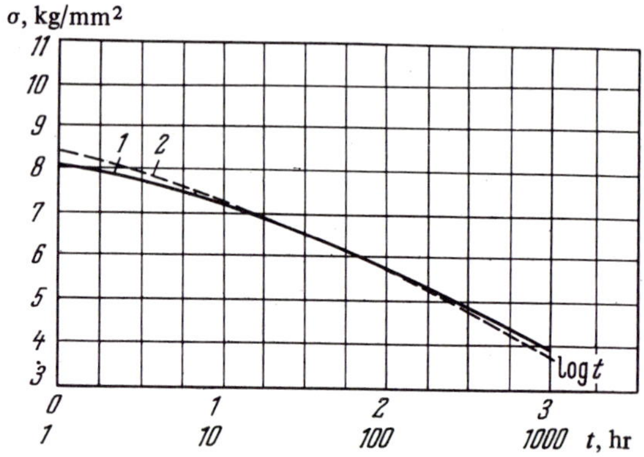

Fig. 58.

constant stress. The agreement between theory and experiment was exceedingly good and this convincingly confirms the strain-hardening hypothesis.

Similar verifications of the agreement between creep and relaxation tests based on the strain-hardening theory have frequently been carried out on various materials. With small test durations of the orders of tens or hundreds of hours the agreement was always very satisfactory. Churikov [145] worked out Davis' test results [189] for copper and his graph is shown in fig. 58 (1. experimental curve, 2. calculated). Namestnikov and Khvostunkov [85] made a series of tests on aluminium alloy D-16T. This test programme was more extensive and will be described later. Other comparisons between creep and relaxation on the basis of the strain-hardening hypothesis are given in the work [369].

52. **Hereditary theory of creep**

Volterra's viscoelasticity theory that we considered in the third chapter describes essentially reversible processes and assumes a linear relationship between stress and deformation. Therefore, it is quite unsuitable for describing creep in metals, even crudely. However, various generalisations of the hereditary theory are possible which make it suitable for studying processes of this kind. The idea behind this kind of consideration can be found by the study of isochronous creep curves given in section 44. Consider first the usual equation of the viscoelasticity theory

$$Ee = \sigma + \int_0^t K(t-\tau)\sigma(\tau)\,d\tau\,.$$

From this, if the stress is constant, we obtain the equation of a family of creep curves in the form

$$Ee = \sigma[1 + G(t)]\,. \tag{52.1}$$

Plotting this relationship in e–σ coordinates, for fixed values of t we obtain a series of straight lines which pass through the origin and which correspond to isochronous creep curves. The straight line $Ee = \sigma$ is the instantaneous deformation line, it also forms part of the series of lines. The condition of similarity (44.2) is fulfilled for the isochronous creep curves that we have obtained, the expression $1/(1 + G(t))$ plays the part of the function $\theta(t)$ and the linear function Ee that of the function $\varphi(e)$. In order to obtain an expression of the form of (44.3) for $\theta(t)$ we must take a power expression with exponent $\beta-1$. We have noted that very frequently isochronous creep curves for metals are near enough similar. Rewriting expression (44.2) in a form analogous to (52.1) we obtain

$$\varphi(e) = \sigma[1 + G(t)] \; .$$

This corresponds to the integral equation

$$\varphi(e) = (1 + K^*)\sigma \, . \tag{52.2}$$

Here $\varphi(e)$ is the ordinate of the instantaneous deformation curve, $K(t-\tau)$ is the kernel found from the creep curves. As we have seen the following approximation gives a satisfactory result for the function $G(t)$,

$$G(t) = at^{\beta} \; ,$$

and therefore, the creep nucleus should be of the form

$$K(t-\tau) = a\beta(t-\tau)^{\beta-1} \, . \tag{52.3}$$

However, most of the creep deformation is irreversible, therefore, the applicability of eq. (52.2) is limited by the condition that deformation does not diminish. Suppose that a deformation e' has been reached which corresponds to the ordinate of the instantaneous plasticity curve φ' (fig. 59) after which unloading occurs. On unloading the change in instantaneous deformation fol-

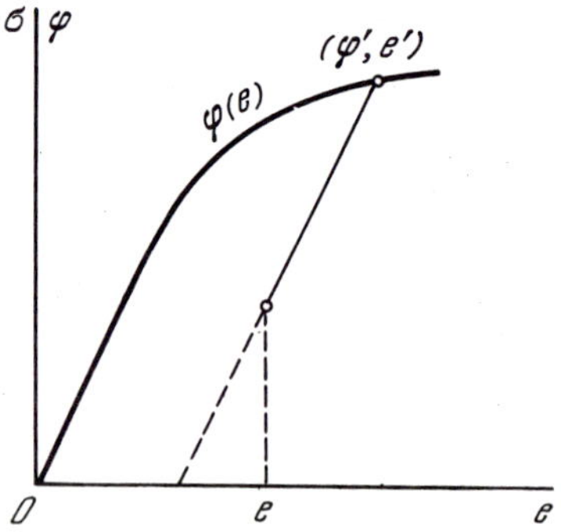

Fig. 59.

lows the elasticity law. Draw a straight line through the point (φ', e') with a slope corresponding to the modulus of elasticity; the equation of this instantaneous unloading line is

$$\varphi = E(e - e') + \varphi' .$$

Substitute this expression in place of the function $\varphi(e)$ in the left-hand side of eq. (52.2) and we obtain the creep law on unloading

$$E(e - e') + \varphi' = (1 + K^*)\sigma . \tag{52.4}$$

The case of relaxation is extreme in the sence that when $e = e'$, $\varphi = \varphi'$ and we remain on the same point of the instantaneous deformation curve. The relaxation law is

$$\sigma = [1 - R(t)]\,\sigma_0 . \tag{52.5}$$

Here $R(t)$ is the integral of the resolvent of the nucleus K, σ_0 is the initial stress. It is assumed that this initial stress is applied instantaneously and that the specimen length is immediately recorded.

It follows from formula (52.5) that relaxation curves for various initial stresses are similar. This consequence of the hereditary theory, like many others, is not verified in practice, nevertheless the theory qualitatively predicts certain effects which do not fit for example into the strain-hardening theory. In particular, eqs. (52.2) and (52.4) explain the effect of creep recovery, which is that after the load has been removed part of the deformation is recovered. Indeed suppose that a stress σ is applied to the specimen; for sim-

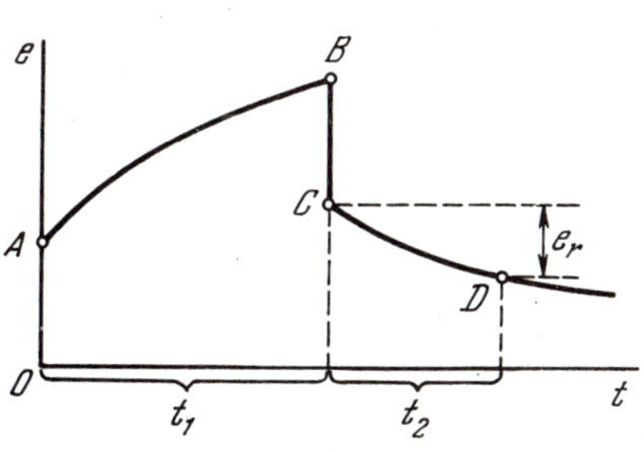

Fig. 60.

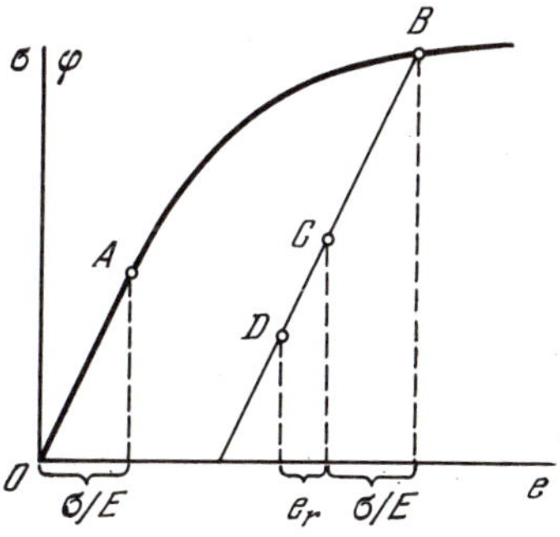

Fig. 61.

plicity we shall consider that it corresponds to the elastic part of the instantaneous curve; it causes an instantaneous deformation σ/E (point A in fig. 60) and for a time t_1 creep follows the curve AB. When $t = t_1$ the stress is removed and then the elastic deformation $BC = OA = \sigma/E$ disappears. The specimen continues to contract and the relationship between elongation and time is represented by the curve CD; at a time t_2 after removing the load the deformation e_r (point D) is recovered. Let us follow this process in the φ–e plane (fig. 61). Instantaneous application of load corresponds to movement from point O to point A; after time t_1 the deformation increases which corresponds to motion upwards along the curve to point B. Here the creep process is described by eq. (52.2) and for point B we obtain

$$\varphi(e_B) = [1 + G(t_1)]\,\sigma\,. \tag{52.6}$$

As a result of instantaneous unloading we pass down the straight line to point C, the length $BC = OA$; creep recovery corresponds to motion along the line to point D at the instant $t_1 + t_2$. To describe creep recovery we must use eq. (52.4) in which the integral in the right-hand side is $\sigma = \text{const}$ when $t < t_1$ and $\sigma = 0$ when $t_1 < t < t_1 + t_2$. We obtain

$$E(e - e_B) + \varphi_B = G(t_1 + t_2) - G(t_2)\,. \tag{52.7}$$

It will be seen from fig. 61 that

$$e_r = e_B - \frac{\sigma}{E} - e\,.$$

Combining (52.6) and (52.7) we find that

$$e_r = \frac{\sigma}{E}\,[G(t_1) + G(t_2) - G(t_1 + t_2)]\,. \tag{52.8}$$

The linear dependence of e_r on stress is well confirmed by Johnson's [253] tests, and the symmetrical form of the dependence of e_r on the times t_1 and t_2 is also approximately confirmed. The data given in this work are inadequate for quantitative verification. Other published works on creep recovery give unsystematic results and they have not been worked out successfully.

The method of generalising the hereditary theory equations to the case of linear creep that we have given is not, of course, the only one. Arutyunyan [4] and Rozovskiy [117] use another equation, namely

$$e = \psi(\sigma) + \int_0^t L(t - \tau) F[\sigma(\tau)]\,\mathrm{d}\tau\,. \tag{52.9}$$

This equation is somewhat more general because it includes not two but three functions that are determined from experimental results. Here the equation of the curve of instantaneous deformation is $e = \psi(\sigma)$ and the creep characteristics are not associated with this curve. As we have mentioned above the justification for selecting the simpler expression (52.2) was the similarity of isochronous creep curves and the possibility of extrapolating creep results by means of this similarity. The functions included in (52.2) are simpler to determine than the corresponding functions in (52.9). The quantitative and qualitative results obtained by each of the variants of the hereditary theory are approximately the same. The extent to which an equation of the type of (52.9) is applicable to processes that are accompanied by stress reduction and how it should be modified in this case have not been considered so far as we know. Apparently the hereditary type of creep theory is better suited to polymers than to metals. Creep tests with stepwise change of load for aluminium alloy [83] showed that in fact recovery on unloading is less than is predicted by any of the variants of the theory that we have mentioned.

Besides the variants of the linear hereditary theory that we have described a more general equation was considered by Volterra [376], namely

$$Ee = \int_{-\infty}^{t} \kappa_1(t-\tau)\mathrm{d}\sigma(\tau_1) + \int_{-\infty}^{t}\int_{-\infty}^{t} \kappa_2(t-\tau_1, t-\tau_2)\mathrm{d}\sigma(\tau_1)\mathrm{d}\sigma(\tau_2)$$

$$+ \int_{-\infty}^{t}\int_{-\infty}^{t}\int_{-\infty}^{t} \kappa_3(t-\tau_1, t-\tau_2, t-\tau_3)\mathrm{d}\sigma(\tau_1)\mathrm{d}\sigma(\tau_2)\mathrm{d}\sigma(\tau_3) + \ldots \tag{52.10}$$

The inconvenience of this variant is that the dependence of deformation on strain is highly non-linear so that the series in eq. (52.10) must contain a fairly large number of terms and this automatically gives more and more complicated hereditary kernels.

In recent years attempts have been made to apply equations of the type of (52.10) to describe the behaviour of polymeric materials [218, 301, 279, 381]. It was in fact difficult to find kernels, certain qualitative consequences of eq. (52.10) and which incidentally also follow from (52.2), were confirmed experimentally. Indeed eq. (52.2) is a particular case of (52.10). Put

$$\kappa_1(t_1) = \kappa(t_1),\ \kappa_2(t_1, t_2) = a_2\kappa(t_1)\kappa(t_2),\ \ldots,\ \kappa_n(t_1, t_2, \ldots, t_n)$$

$$= a_n\kappa(t_1)\kappa(t_2)\ \ldots\ \kappa(t_n),\ \ldots$$

Putting $\kappa'(t) = K(t)$, rewrite (52.10) in the form

$$Ee = s + a_2 s^2 + a_3 s^3 + \ldots + a_n s^n + \ldots \tag{52.11}$$

Here

$$s = \sigma + \int_{-\infty}^{t} K(t-\tau)\sigma(\tau)\mathrm{d}\tau .$$

Denoting by φ the conversion of the series in the right-hand side of (52.11), we obtain eq. (52.2).

53. Experimental verification of the strain-hardening hypothesis with variable loads

We have already given certain information on the experimental verification of the strain-hardening hypothesis by comparing theoretical and experimental relaxation curves. Creep tests with stepwise change in load provide a simpler method of verification. Moreover, stepwise change in stress more clearly reveals deviations from the predictions of theory and the results obtained provide better contrast. Indeed, as we have seen, calculation of the relaxation curve by various time-hardening theories gives the same qualitative result as by strain-hardening theory and even the curve calculated by strain-hardening theory is not always a better approximation to the experimental curve than the other theoretical curves. Therefore, comparison of creep and relaxation results does not provide sufficient data to confirm the time-hardening theory. With stepwise loading the difference between the predictions of the various theories is very great and is of qualitative character. However, this case is extreme for practical products. A theory that satisfactorily predicts the shape of the creep curve with sudden changes in stress state is certainly valid when the stress changes smoothly and continuously.

Very convincing verification of the strain-hardening hypothesis is given in the work of Namestnikov and Khvostunkov [85] for aluminium alloy D-16T at temperatures of 150 and 200°C. In all cases the test duration was about 100 hours, and altogether 130 specimens were tested so that random scatter of the results was unimportant. The tests were made on bar material in the state of delivery, that is work-hardened and naturally aged. The hardness of the specimens lay within a range of 5 Rockwell units. The heating conditions prior to starting the tests were strictly controlled in order to obtain reproducible results. Mean creep curves obtained at a temperature of 150° are shown

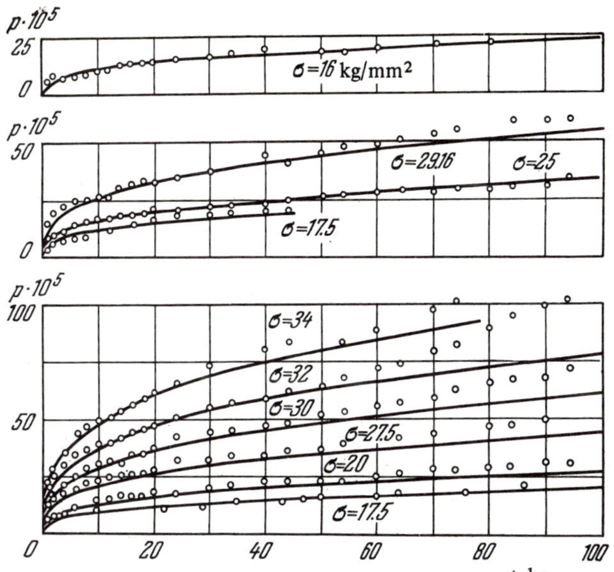

Fig. 62.

in fig. 62. The results plotted in logarithmic coordinates form parallel straight lines and so α is independent of stress. It was found that $m = 0.33$ in accordance with Andrade's law. Thus $\alpha = 2$. Eq. (50.11) was used in subsequent working-out. According to this equation the relation between $\log p$ and σ for a fixed time interval should be represented by a straight line, the actual graphs obtained are shown in fig. 63. The point of intersection between the two straight lines corresponds to a value of $\sigma = 25.3$ kg/mm^2, therefore, two pairs of values were obtained for the constants k and A:

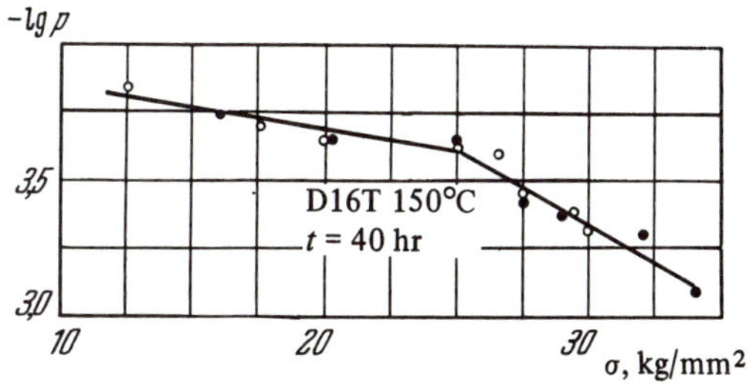

Fig. 63.

$$\sigma \leqslant 25.3\,, \qquad k = 5.68 \cdot 10^{-15}\,, \qquad A = 8.77\,\mathrm{kg/mm^2}\,,$$

$$\sigma \geqslant 25.3\,, \qquad k = 5.60 \cdot 10^{-18}\,, \qquad A = 2.52\,\mathrm{kg/mm^2}\,.$$

Creep curves calculated by eqs. (50.1) with the above values of the constants are shown in fig. 62. For stresses up to 25 kg/mm^2 agreement is very good. At higher stresses the creep curves contain second and third parts which are not described by formula (50.11) but again the approximation is fully satisfactory for the initial parts of the curves.

Ten relaxation tests were made on the same material with initial stresses of 20.0; 22.5 (two tests), 25.0 (two tests), 26.5; 27.5 (two tests), 30.0 and 32.0 kg/mm^2. The experimental points are plotted in figs. 64 and 65 in which the

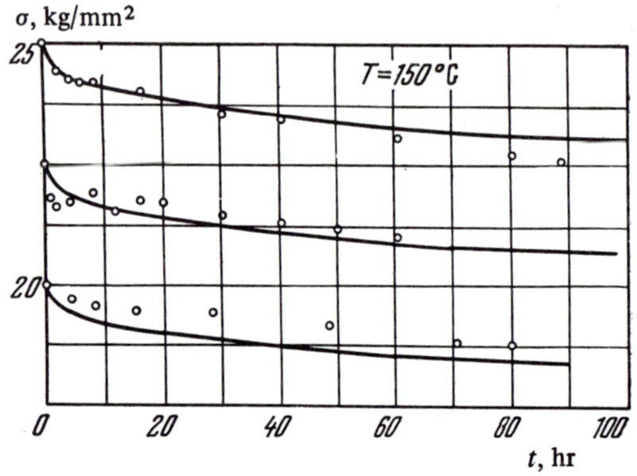

Fig. 64.

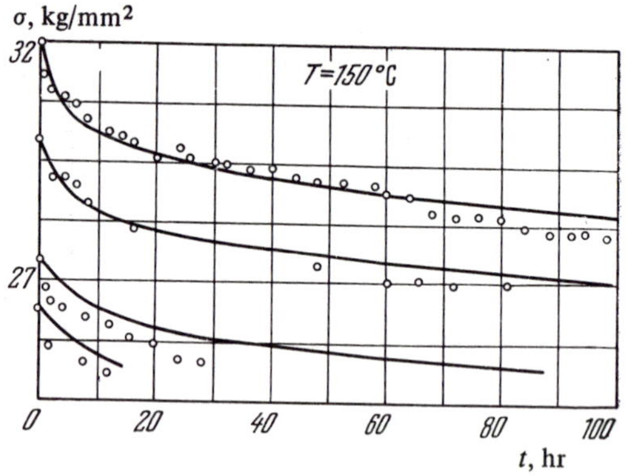

Fig. 65.

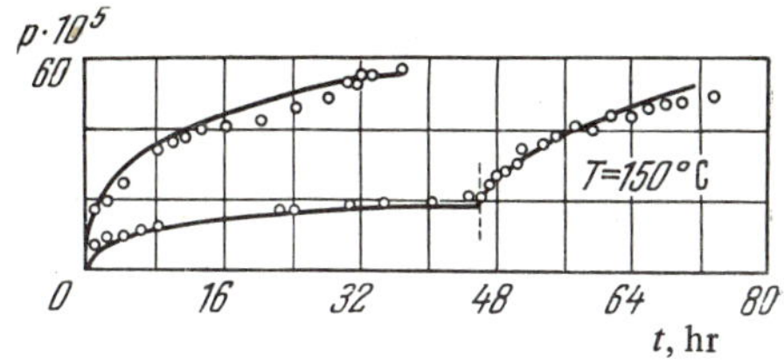

Fig. 66.

solid lines correspond to curves calculated by eq. (51.4) with the values of the creep constants given above, found from creep tests at constant stress. With the very smooth stress change that occurs during relaxation the strain-hardening hypothesis is sufficiently accurate. However, tests with stepwise load change immediately reveal certain systematic deviations from predictions of the hypothesis. A creep curve corresponding to a stress of 17.5 kg/mm^2 for 45 hours and a stress of 32 kg/mm^2 for the next 27 hours is shown in fig. 66. Here the agreement between experiment and theory is satisfactory. However, for higher stresses, for example, the tests shown in the upper part of fig. 67 for which the stress was 29.2 kg/mm^2 for the first 22 hours and 34 kg/mm^2 in the next 14 hours, it is clearly seen that the experimental points lie above the theoretical curve. This divergence between theory and experiment becomes still clearer when the load is altered several times. The lower part of fig. 67 shows a creep curve for the same material at a temperature of 200° for stress steps of 4.8 and 12 kg/mm^2. On the third step the difference between theory and experiment is already very great. The general conclusion can be drawn that the strain-hardening hypothesis satisfactorily describes creep at

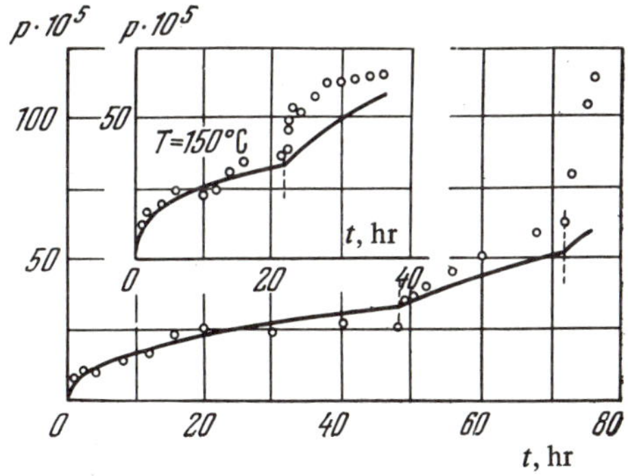

Fig. 67.

constant and diminishing loads but with increasing loads creep takes place somewhat faster than predicted by theory particularly in the first minutes and hours after changing loads if the stress level is such that the instantaneous deformation is mainly elastic. At higher stresses which cause considerable plastic deformation, use of the strain-hardening hypothesis gives considerable deviations from experimental results and the actual creep rate is greater than predicted by theory.

A very vivid experimental scheme to illustrate this was proposed by Odquist [94]. If the equation of state is of the form

$$\dot{p}h(p) = f(\sigma)\,, \tag{53.1}$$

then the following law of loading applies. Suppose that a stepwise change in load is applied to the specimen, the stress in each stage is σ_i and its duration is t_i. Then the total deformation after n load stages is independent of the order of application of the loads provided that the stress and duration of each stage remain unaltered. Indeed, separating variables in eq. (53.1) and integrating we obtain

$$\int_0^p h(p)\mathrm{d}p = \int_0^t f(\sigma)\mathrm{d}t = t_1 f(\sigma_1) + t_2 f(\sigma_2) + \ldots + t_n f(\sigma_n)\,.$$

The left-hand side is a single valued function of p and the right-hand side is independent of the order of numeration of the loading stages. A simple experiment to check the loading law was carried out by Odquist with only two

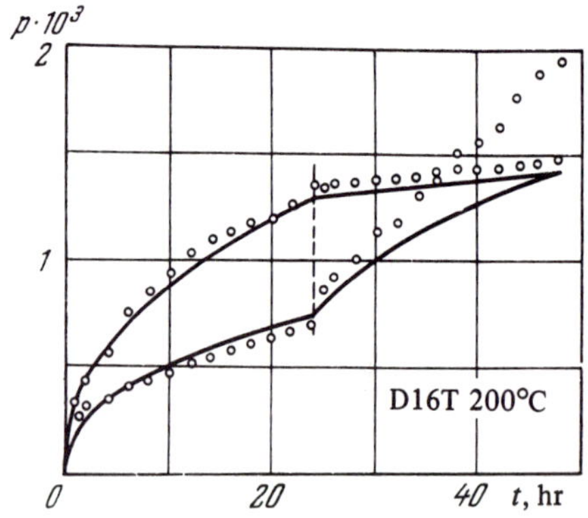

Fig. 68.

loading stages. It was found that the law is not observed, if $\sigma_2 > \sigma_1$ the total deformation is greater than if $\sigma_1 < \sigma_2$. Odquist's experiment was repeated by Namestnikov and Khvostunkov on several specimens at various temperatures. A typical result is plotted in fig. 68 in which first $\sigma_1 = 12$, and $\sigma_2 = 16$, and afterwards $\sigma_1 = 16$ and $\sigma_2 = 12$ kg/mm^2 and in both cases $t_1 = t_2 = 24$ hours. The solid line represents calculated curves which satisfactorily describe the experiment when $\sigma_2 < \sigma_1$. In the case of increasing load when $\sigma_2 > \sigma_1$ the deformation after 48 hours is much greater. Of course, the failure to observe the law cannot be attributed to a special selection of the type of equation of state (53.1) for which this law is valid, it is possible that the error here may be of the same order as the error in approximating creep curves for constant stress.

54. Other creep investigations with variable loading on the basis of the strain-hardening hypothesis

We shall not give a complete review of the experimental work that confirms the validity of the strain-hardening hypothesis within certain limits but shall describe a cycle of investigations carried out by Japanese authors [365]. The same systematic deviations from the theoretical predictions that were mentioned above are also observed during stepwise change of load. Creep was studied with stepwise cyclic load changes. The stress σ_1 acted on the specimen for a time t_1, then the stress was suddenly increased to σ_2 and creep continued to time t_2, then the stress was again reduced to σ_1 and the cycle was repeated (fig. 69). If the ratio t_2/t_1 is small the difference between the experimental results and the predictions of the strain-hardening theory is considerable, the actual total deformation is greater than would be expected from the hypothesis of the equation of state. A typical curve taken from the work [365] is shown in fig. 70, the author explains it as follows. Creep recovery

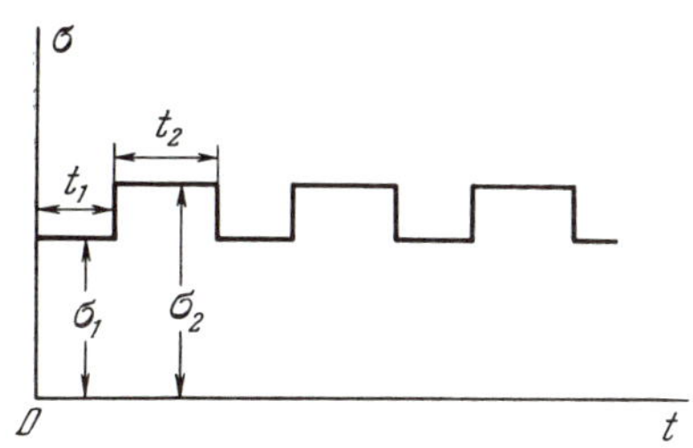

Fig. 69.

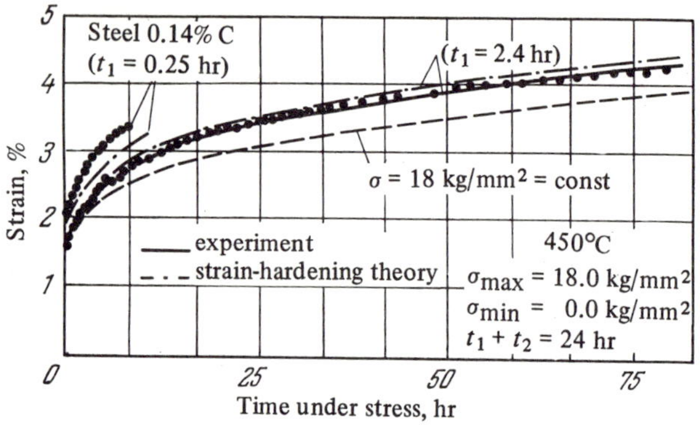

Fig. 70.

occurs during the time of action of the smaller stress, reverse-creep deformation is itself insignificant, the main effect is that when the stress is increased there is an instantaneous increase in creep rate, which is much greater than that to be expected from the equation of state for the given stress and deformation. If t_2 is small the total deformation in this time is much greater than would be expected in theory.

The influence of rapid change of temperature is similar to that of change of stress. In the work [309] it was found that if the temperature is raised rapidly during the creep test at first the deformation increases more rapidly than is to be expected from the equation of state. On the other hand, if the temperature is lowered, creep is temporarily delayed. During cyclic temperature variations these two effects compensate one another and overall the hypothesis of the equation of state is satisfactorily justified. Corresponding results have already been given in section 45 in discussing Dorn's work.

Let us take the law of creep with variable temperature in the form of (50.12). As we have already mentioned this simple form of temperature dependence gives satisfactory results provided the range of temperature variation is sufficiently narrow. The result of integrating the equation

$$\dot{p}p^{\alpha} = f(\sigma) \exp\left(-\frac{u_0}{RT}\right),$$

where σ is constant and T is a time function, can be represented as

$$p = \left[\frac{f(\sigma)}{m}\right]^m \exp\left(-\frac{mu_0}{RT_e}\right) t^m .$$

Here

$$\exp\left(-\frac{u_0}{RT_e}\right)=\frac{1}{t}\int_0^t \exp\left(-\frac{u_0}{RT}\right)\mathrm{d}t\,. \tag{54.1}$$

If $T(t)$ is a periodic function of time, that is the temperature varies cyclically, the integral in the right-hand side of (54.1) contains a term that is linearly dependent on time and confined to the periodic part. In calculating the equivalent temperature we shall consider that the time t is long enough and that the influence of the second part is insignificant. Then, more strictly

$$\exp\left(-\frac{u_0}{RT_e}\right)=\lim\left[\frac{1}{t}\int_0^t \exp\left(-\frac{u_0}{RT}\right)\mathrm{d}t\right]_{t\to\infty}\,.$$

Consider the following simple cases.

1. *The temperature varies according to a sinusoidal law* $T=T_0+\Delta T\sin\omega t$, the ratio $\Delta T/T$ is small. In this case

$$\exp\left(-\frac{u_0}{RT}\right)\approx\exp\left(-\frac{u_0}{RT}\right)\exp\left(\lambda\sin\omega t\right)\,.$$

Here we take

$$\lambda=\frac{u_0\Delta T}{RT^2}\,.$$

Therefore, we have to calculate the integral

$$\int_0^t \exp\left(\lambda\sin\omega t\right)\mathrm{d}t\,.$$

It is assumed that λ is small compared with unity. Expand the exponential function into a series and integrate each term individually. Note that $\int\sin^{2n+1}x\,\mathrm{d}x$ is a periodic function of x, whereas $\int\sin^{2n}x\,\mathrm{d}x$ is a sum of a periodic function and a term $\frac{1}{2^{2n}}\binom{2n}{n}x$ which increases indefinitely with time. Here, as usual the symbol $\binom{k}{l}$ denotes the binomial coefficient. It will now readily be seen that

$$\lim\left[\frac{1}{t}\int_0^t \exp(\lambda \sin \omega t)\,\mathrm{d}t\right]_{t\to\infty} = \sum \frac{1}{(2n)!}\frac{1}{2^{2n}}\binom{2n}{n}\lambda^{2n}.$$

Consequently, the equivalent temperature is given by the formula

$$\frac{1}{T_e} = \frac{1}{T_0} - \frac{R}{mu_0}\ln \Sigma \frac{1}{(2n)!}\frac{1}{2^{2n}}\binom{2n}{n}\lambda^{2n}. \tag{54.2}$$

The series in formula (54.2) converges very rapidly. With a periodic stress the result is obtained in an exactly similar way [103, 365] but the limitation that deviations from the medium should be small is unnecessary.

2. *Stepwise temperature change.* The duration of each cycle is $t_1 + t_2$, during the time t_1 the temperature is T_1 and during the time t_2 it is T_2. Obviously the equivalent temperature is given by

$$\exp\left(-\frac{u_0}{RT_e}\right) = \frac{1}{t_1+t_2}\left[t_1 \exp\left(-\frac{u_0}{RT_1}\right) + t_2 \exp\left(-\frac{u_0}{RT_2}\right)\right]. \tag{54.3}$$

Creep tests with sinusoidal temperature change were carried out by Ivanova [38] on chrome-nickel alloy EI-437B. These tests relate to the transient stage of creep when the creep curve is sufficiently well described by the equation of state (50.10). A special device was provided to ensure that the temperature varied according to the sinusoidal law. The creep curves were taken

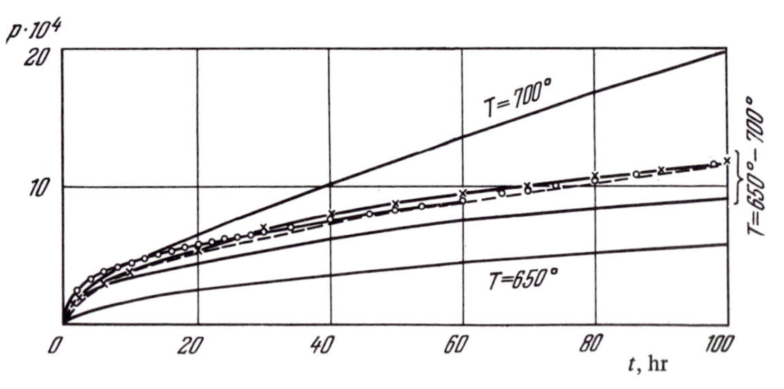

Fig. 71.

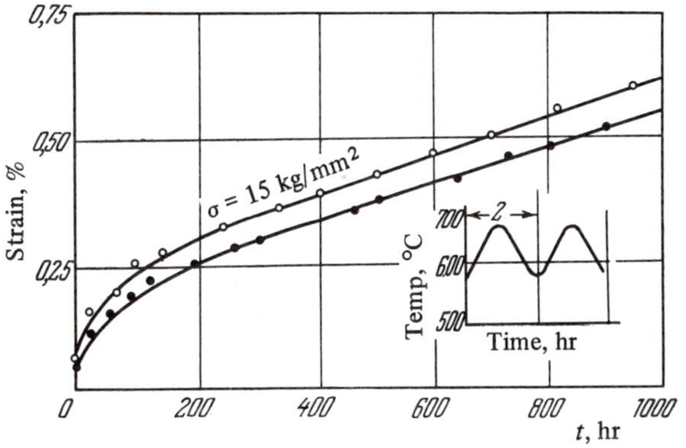

Fig. 72.

for a given value of stress at the upper and lower temperatures of the cycle so that the energy of activation u_0 could be determined. The calculated creep curve with variable temperature coincided satisfactorily with the experimental points (fig. 71). Creep curves at $T = 650°$ and $T = 700°$ with the temperature ranging from 650 to 700° with periods of 30 minutes, 1 hour and 2 hours are shown in fig. 71, and the calculated curve for variable temperature is shown dotted.

Similar tests on the same grade of stainless steel were carried out by Nishihara, Taira, Tanaka and Ohnami [310] and their test results are given in fig. 72. The temperature ranged from 575 to 675° as shown in the graph. The equivalent temperature calculated by formula (54.1) was 650°. The points denoted by circles on the graph correspond to the creep curve at variable temperature and the dots to tests at a constant temperature $T = 650°$. The difference between these curves is within the limits of experimental error. These tests were of fairly long duration and a steady-state part is clearly visible on them.

The authors use the scheme mentioned in section 50, namely, they consider separately creep deformation accompanied by strain-hardening and subject to the equation of state and steady-state creep deformation determined by the expression $\int v(\sigma)\mathrm{d}t$. The rate of steady-state creep or quasi-viscous flow depends on temperature, just like the function $f(\sigma)$ in the strain-hardening law, i.e.

$$v = v_0(\sigma) \exp\left(-\frac{u_1}{RT}\right) .$$

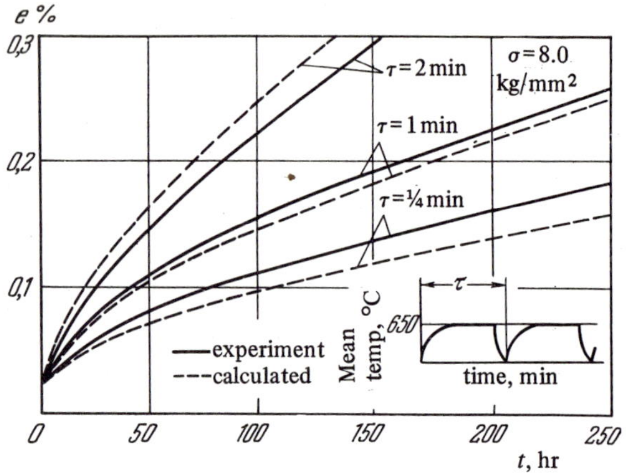

Fig. 73.

The corresponding energy of activation u_1 differs from u_0 and, therefore, the equivalent temperature, which can be calculated in this case also by formula (54.1), will be different.

The authors neglect this difference and use a formula which in our notation is

$$p = \left[\frac{f(\sigma)}{m}\right]^m \exp\left(-\frac{mu_0}{RT_e}\right) t^m + v_0(\sigma) \exp\left(-\frac{u_1}{RT_e}\right) t \, .$$

Similarly, Taira and Ohnami [362] studied creep with sharp temperature changes. Hollow specimens were heated by passing current and cooled by a jet of air, the cycle durations were 15 sec, 1 min and 2 min. Experimental creep curves are shown in fig. 73 in solid lines and calculated in dotted, a graph of temperature change with time is also given. In these tests the temperature distribution over the length of the specimen was irregular and an average had to be taken in working out the results. Despite this, there is reasonable agreement between theory and experiment.

55. **Kinetic creep equations**

The simplest of all assumptions is that creep deformation is a measure of work-hardening, but it is an assumption of a very special kind and is not confirmed by serious physical considerations. Indeed there may be different work-hardening mechanisms, and, therefore, there is no justification for asserting that the plastic deformation alone, always corresponds to one particular structural state of the material. Later we shall show that in this sense the instantaneous plastic deformation and the creep deformation play different parts and should be separated to improve the equation of state. It seems fairly obvious that the structural state cannot in general be considered to be uniquely determined solely by the accumulated creep deformation irrespective of the manner and the sequence in which this deformation was achieved.

Within the framework of phenomenological theory we cannot indicate any features that characterise the structural state; the only possible statement is that there is a certain finite number of parameters $q_1, q_2, ..., q_n$, in terms of which the structural state can be given. The meaning of this statement is that if all the parameters q_i are equal for two materials, then they will both have the same creep rate at the same temperature and stress. It should be emphasised that we are only considering the structural characteristics that govern the creep process. Accordingly the structural parameters q_i are introduced in a purely formal way and here we shall not consider their physical meaning, although certain of the results given below can immediately be translated into the language of dislocation kinetics.

Therefore, a more general hypothesis is that the creep rate depends on stress, temperature and a certain number of structural parameters, q_i, so that the general creep equation is

$$\dot{p} = f(\sigma, T, q_1, q_2, ..., q_n) . \tag{55.1}$$

Structural changes described by changes in the parameters q_i occur during creep. A sufficiently general assumption is that these changes are described by the following kinetic equations:

$$\mathrm{d}q_i = a_i \mathrm{d}p + b_i \mathrm{d}\sigma + c_i \mathrm{d}t + f_i \mathrm{d}T . \tag{55.2}$$

Here a_i, b_i, c_i, f_i are certain functions of p, σ, t, T and also $q_1, q_2, ..., q_n$. In general, eqs. (55.2) cannot be integrated. In this way we obtain extensive freedom to construct creep theories which may be as complicated and as accurate as necessary, but as yet this theoretical possibility has been very little used. Consider certain simple parameters.

a. Take $n = 1$ and put $\mathrm{d}q = \mathrm{d}p$. Hence $q = p$ and we obtain the usual strain-hardening theory.

b. As measure of work-hardening we may take not the amount of creep deformation but, for example, the work performed by the stresses acting on the creep deformations. Limiting ourselves again to only one parameter of work-hardening we put

$$\mathrm{d}q = \sigma \mathrm{d}p \ ,$$

consequently

$$q = \int \sigma \mathrm{d}p \ . \tag{55.3}$$

Now consider a simple creep test with stepwise change of load (fig. 74). If creep takes place at constant stress σ_1, a creep deformation p_1 has accumulated at the instant of time t_1 and the value of q is $p_1\sigma_1$. Now suppose that the stress is suddenly changed to σ_2, creep is described by the eq. (55.1) and at the first instant $q = p_1\sigma_1$. On the creep curve that corresponds to the constant stress σ_2 we denote a point B with ordinate p_2, such that

$$p_2\sigma_2 = p_1\sigma_1 \ .$$

After changing the load the creep rate is the same as on curve σ_2 at point B, and the subsequent course of the curve is the same. Therefore, in order to predict the shape of the creep curve with stepwise change of load from σ_1 to σ_2 at the time when $p = p_1$, we must find on the creep curve for the stress σ_2 a point for which $p = p_2 = p_1(\sigma_1/\sigma_2)$ and move part of the curve parallel to itself relative to this point as shown in fig. 74. According to this variant of the strain-hardening theory the rate after change of load is greater than that given by the ordinary strain-hardening theory which, as we have already seen, is in line with experimental results.

Now consider prediction of the relaxation curve by this theory. Assume

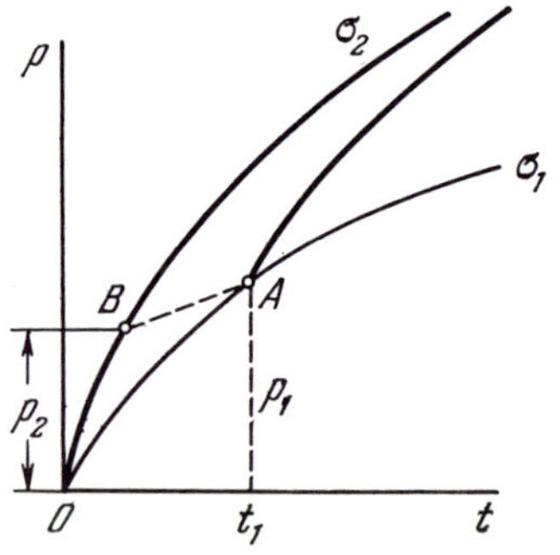

Fig. 74.

that a series of creep curves at constant stress is satisfactorily given by an equation of state of form

$$\dot{p}p^{\alpha} = \sigma^{n} . \tag{55.4}$$

Here the scales are selected in such a way that the equation does not contain dimensional constants and $p = e - \sigma$. In the case of relaxation $\sigma = \sigma_0, p = \sigma_0 - \sigma$ and $\dot{p} = -\dot{\sigma}$. Substituting these into (55.4) and separating variables we obtain

$$t_1 = \sigma_0^{-n+\alpha+1} \int_{\sigma/\sigma_0}^{1} \frac{(1-\zeta)^{\alpha}}{\zeta^{n}} \, d\zeta . \tag{55.5}$$

Now select the form of eq. (55.1) when $q = \int \sigma \, dp$ in such a way that for constant loads it coincides with (55.4). If $\sigma = \text{const}$, then $q = \sigma p$; and excluding p from (55.4) we obtain

$$\dot{p} = q^{-\alpha} \sigma^{n+\alpha} . \tag{55.6}$$

Now in the case of relaxation

$$q = \int \sigma \, dp = -\int_{\sigma_0}^{\sigma} \sigma \, d\sigma = \tfrac{1}{2} (\sigma_0^2 - \sigma^2) .$$

Substitute this into eq. (55.6), separate variables and integrate, we obtain

$$t_{\mathrm{II}} = \sigma_0^{-n+\alpha+1} \int_{\sigma_0/\sigma}^{1} \frac{(1-\zeta^2)^{\alpha}}{2^{\alpha} \zeta^{n+\alpha}} \, d\zeta . \tag{55.7}$$

The expression within the integral in the formula for t_{II} differs for the corresponding expression in the formula for t_{I} by the factor

$$\left(\frac{1+\zeta}{2\zeta}\right)^{\alpha} .$$

But when $\zeta < 1$ and $\alpha > 0$ this quantity is greater than unity; therefore, $t_{\mathrm{II}} > t_{\mathrm{I}}$ and this new variant of the strain-hardening theory predicts a slower relaxation curve than does the usual theory.

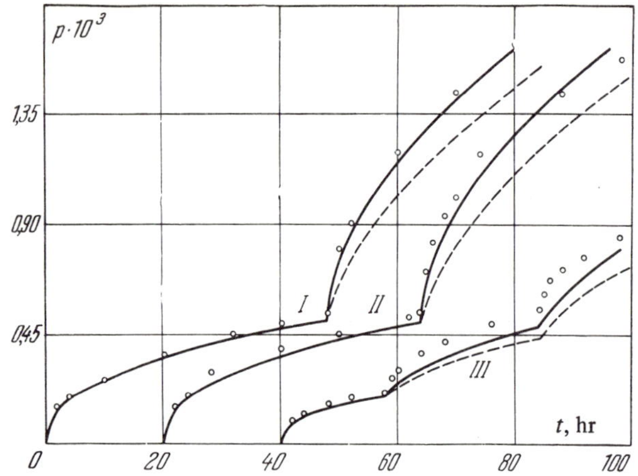

Fig. 75.

The variant of the strain-hardening theory in which expression (55.3) is taken as a measure of hardening was specially checked in the work [12]. The tests were carried out on aluminium alloy D-16AT at a temperature of 150°, the flat specimens were cut from sheet in the direction of rolling. Creep curves with stepwise change of load are shown in fig. 75, curves I and II correspond to steps of 15 and 25 kg/mm^2, curve III to steps of 10, 15 and 20 kg/mm^2. The results of calculation by equation (55.6) are shown by solid lines and of calculation by eq. (55.4), which is the usual strain-hardening theory, by dotted lines. In two cases the agreement between the new theory and the experimental results is extremely good, for the third curve the difference remains noticeable. Obviously we can, without great difficulty, achieve a very good description of stepwise loading tests if, as a measure of hardening, we take the quantity

$$q = \int a(\sigma)\,\mathrm{d}p\,. \tag{55.8}$$

The function $a(\sigma)$ can always be selected to give the best description of the course of a particular experiment, the simple expression (55.3) was taken to avoid introducing excessive complications. It should be borne in mind that stepwise load variations most clearly reveal deviation from theory; the case of smoothly varying loads is much easier and for these the usual strain-hardening theory provides an adequate description of the case of slowly varying loads and somewhat better results can be obtained by a parameter of the form

(55.3). Deformation curves for the same aluminium alloy at a rate of rise of stress of 0.002 kg/mm^2 per minute are plotted in fig. 76. The maximum stress does not exceed the limit of proportionality and, therefore, the apparent plastic deformation is creep deformation with increasing load. The solid line is calculated by eq. (55.6) and the dotted by eq. (55.4), in both cases the constants were determined from creep tests at constant load. The difference between the experimental points and the calculated curve is appreciable at low stresses, just as in the case of stepwise loading (fig. 75), and thereafter the agreement between the new variant of strain-hardening theory and the experimental results is fully satisfactory.

c. The work [84] attempts to improve the description of stepwise loads by means of two parameters of state; the usual measure of strain-hardening $q_1 = p$ and a parameter q_2 defined by the expression

$$dq_2 = p\, d\sigma\,. \tag{55.9}$$

If an instantaneous load is applied to the specimen and then maintained constant, $q_2 = 0$. With stepwise change in load from σ_1 to σ_2 at the time when $p = p_1$, the quantity q_2 acquires an instantaneous increment $\Delta q_2 = (\sigma_2 - \sigma_1)p_1$. The following creep law was taken in the work [84]:

$$\dot{p}p^{\alpha} = k \exp\left[\frac{\sigma}{A} + \frac{q_2}{B}\right]. \tag{55.10}$$

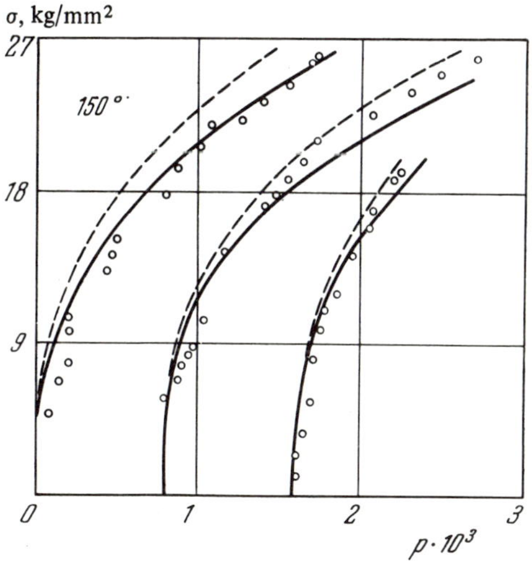

Fig. 76.

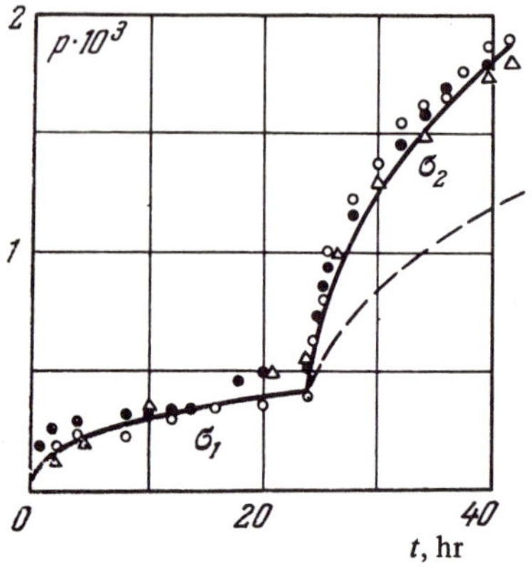

Fig. 77.

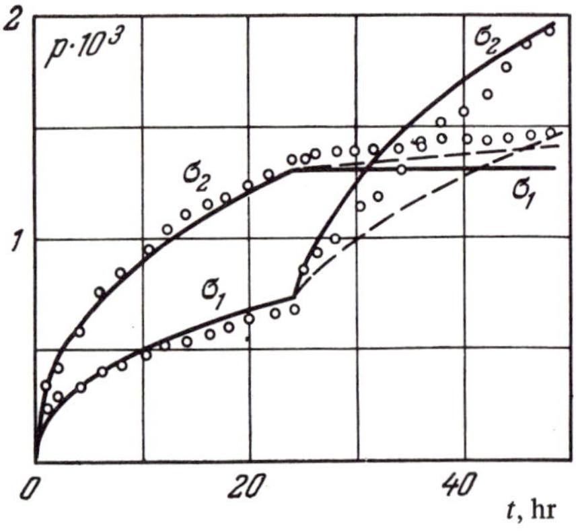

Fig. 78.

Experimental points for three specimens of alloy D-16T which were tested at a stress of 8 kg/mm^2 for the first 24 hours and then at a stress of 16 kg/mm^2 after 24 hours at $T = 200^\circ$ are plotted in fig. 77. The dotted curve is calculated by the usual equation of state, the solid line is a curve calculated by eq. (55.10). Results of similar tests for $\sigma = 12$ and 16 kg/mm^2 in which the stresses were applied in different sequence are shown in fig. 78. As before, the solid line corresponds to expression (55.10) and the dotted to the usual equation of state. It should be noted that if the load is diminishing the parameter q_2 makes things worse and not better. Calculation of relaxation curves by eq. (55.6) leads to results which are very little different from, and no better than, those of the ordinary theory. Therefore q_2 should be considered to alter only when the stress is increased. Accordingly

$$dq_2 = p\, d\sigma \quad (d\sigma > 0)\,, \qquad dq_2 = 0 \quad (d\sigma < 0)\,. \tag{55.11}$$

In addition to the simple variants of the kinetic creep equations which we have considered, various other assumptions may be made and more or less complicated variants of the theory derived. The available experimental data are still insufficient to enable the best variant to be chosen, but the point of view that we have described explains many of the observed facts, at least qualitatively, and offers possibilities for their quantitative description.

56. Softening during creep

So far we have considered creep only in structurally stable materials. In formulating the work-hardening law it was assumed that the state of hardness cannot be removed by holding the material at the test temperature. In real metals and alloys diffusion processes that take place at high temperatures can cause either softening or hardening, which in this case is termed ageing. These processes can be described formally by eqs. (55.1) and (55.2). Bailey has suggested that creep should be considered as the result of two simultaneous processes: deformation hardening and temperature softening. However, his suggestions were not presented in correct mathematical form, they remained purely qualitative and were subsequently criticised by some metallo-physicists and defended by others [93]. We shall not discuss this side of the matter but shall indicate how temperature softening can be allowed for. Suppose for simplicity that in eq. (55.1) $n = 1$, and take

$$dq = dp + c\,dt\,.$$

The coefficient c that characterises the rate of softening should be negative and increase in absolute value with increased hardening; the simplest assumption is that c is proportional to q and then

$$dq = dp - \mu q\,dt\,. \tag{56.1}$$

The factor μ is, of course, highly temperature dependent and will be considered constant at constant temperature. Suppose now that at a stress σ the specimen creeps until the deformation is p_1. The corresponding value of q is q_1, if p is known as a function of t (given the creep curve); q_1 may be calculated by integrating eq. (56.1). Now suppose that the specimen is unloaded and maintained at the same test temperature. Since when $\sigma = 0$ $\dot{p} = 0$, eq. (56.1) assumes the form

$$dq = -\mu q\,dt\,.$$

Integrate this considering $t = 0$ at the instant of unloading when $p = p_1$, we obtain

$$q = q_1 \exp(-\mu t)\,. \tag{56.2}$$

Eq. (56.2) describes temperature softening of a specimen that is not loaded.

If the stress σ is again applied to the specimen it creeps at a greater rate than that which preceded unloading. The constant μ has the dimensions of reciprocal time, therefore, $1/\mu$ can be considered as the characteristic weakening time. If this time is long enough, $q \approx p$ for the first sections of the creep curve and consequently, the equation of state may be written

$$\dot{\hat{p}}\hat{q}^{\alpha} = \exp \hat{\sigma} \tag{56.3}$$

(the dimensionless quantities $\hat{p}$, $\hat{\sigma}$ are defined as in section 51, $\hat{q} = Eq/A$). Eliminating $\dot{p}$ from eqs. (56.1) and (56.3) we obtain

$$\dot{\hat{q}}\hat{q}^{\alpha} + \hat{\mu}\hat{q}^{1+\alpha} = \exp \hat{\sigma}$$

$$\left(\hat{\mu} = \frac{1}{k}\left(\frac{A}{E}\right)^{1+\alpha} \mu\right).$$

Integrating with the condition σ = const and for $t = 0$ $q = 0$, we find

$$\hat{q}^{1+\alpha} = \frac{\exp \hat{\sigma}}{\hat{\mu}} \{1 - \exp [-\hat{\mu}(1+\alpha)\hat{t}]\} . \tag{56.4}$$

To determine $\hat{p}$ we must square once more, which gives a very complicated $\hat{p}(\hat{t})$ dependence and, therefore, we shall consider only the extreme cases.

a. The time t is short compared with the characteristic softening time $1/\mu$. Expand the exponent into a series and retain the first terms. From eq. (56.4) it follows that

$$\hat{q} = \exp (m\hat{\sigma}) \left(\frac{\hat{t}}{m}\right)^{m} .$$

Integrating (56.1) we find the creep deformation

$$\hat{p} = \exp (m\hat{\sigma}) \left(\frac{\hat{t}}{m}\right)^{m} \left[1 + \frac{\hat{\mu}}{1+m} \hat{t}\right]. \tag{56.5}$$

For small values of $\hat{\mu}t$ the second term in brackets is small and formula (56.5) gives the same result as (50.3). The constant m can be found by working out the first parts of the creep curve.

b. The time t is long. The quantity $\hat{q}$ given by eq. (56.4) tends to a limiting value $\hat{q} = \exp (m\hat{\sigma})/\hat{\mu}^{m}$, the creep rate becomes

$$\hat{p} = \hat{\mu}^{1-m} \exp(m\hat{\sigma}) . \tag{56.6}$$

The creep rate becomes constant when hardening and softening compensate one another, this idea was proposed by Bailey and later by Oding [91].

This scheme has not been thoroughly and quantitatively verified because of lack of experimental data; however, the fact of softening by holding at high temperature and unloading has been observed. For example, we mention certain results obtained on specimens of technically pure aluminium grade A-DIM at a temperature of 22.5°C in Petrusev's University diploma project. Creep took place in this specimen at a stress of 5.61 kg/mm^2 until the creep deformation came to 2.45%. Then the specimen was unloaded and maintained without load at the test temperature for a time t, after which it was again loaded to the same stress. Creep curves on repeated loading for various values of t are plotted in fig. 79 and the softening effect is clearly seen.

During creep of dispersion-hardening alloys separation of the hardening phase may be accompanied by creep so that the picture becomes confused. However, the test results on duralumin specimens can only be worked out by the equation of state provided that the material has first been aged. To a first approximation the ageing process may be considered to be independent of the state of stress and, therefore, in describing the properties of an ageing

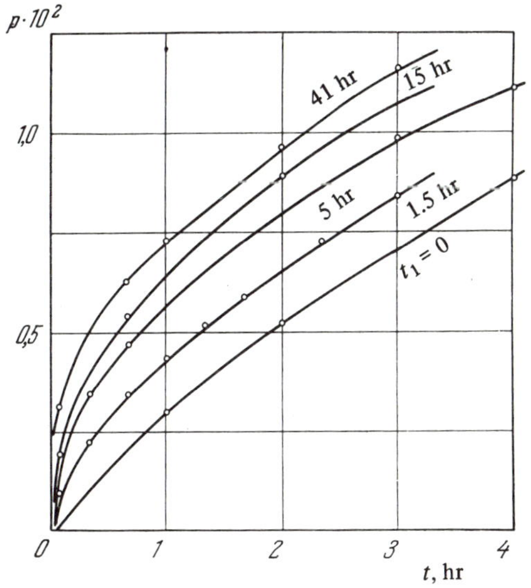

Fig. 79.

material it is convenient to introduce two parameters: q_1, which is a measure of deformation hardening and which is either the creep deformation or is defined by one of the methods mentioned above; and q_2 which determines the degree of ageing. If in the fully aged condition $q_2 = q^*$ it is natural to take

$$dq_2 = \kappa(q^* - q_2)dt .$$

Then one possible way of writing the equation of state is

$$\dot{p} = q_1^{-\alpha} q_2^{-\beta} f(\sigma) . \tag{56.7}$$

Japanese authors [307, 365] attempted to construct a creep theory on the basis of certain concepts developed in an earlier work of Orowan [319]. In effect these authors introduced a hardening parameter q, for which

$$dq = \left(\frac{\partial q}{\partial e}\right) de + \left(\frac{\partial q}{\partial t}\right) dt . \tag{56.8}$$

Here the total rate of plastic deformation is considered to be a function of stress, temperature and the parameter q. The quantity $\partial q/\partial e$ is interpreted as the rate of hardening, $\partial q/\partial t$ as the rate of softening. The form of eq. (56.8) is unsatisfactory as the existence of functions $q(e, t)$ is nowhere assumed and the quantities denoted $\partial q/\partial e$ and $\partial q/\partial t$ are independent functions determined experimentally. Of course, this generalisation of the equation of state can provide a satisfactory description of experimental results with variable stress, although the authors do not give explicit dependences of the parameters $\partial q/\partial e$ and $\partial q/\partial t$ (in our notation a_1 and c_1) in the corresponding arguments.

57. Creep and instantaneous plastic deformation

Older works on the equation of state (see section 49) did not differentiate between creep deformation and instantaneous plastic deformation. Denoting plastic deformation by e_p irrespective of its nature, we can write the equation of state in the form

$$\Phi(\dot{e}_p, \sigma, e_p, T) = 0 . \tag{57.1}$$

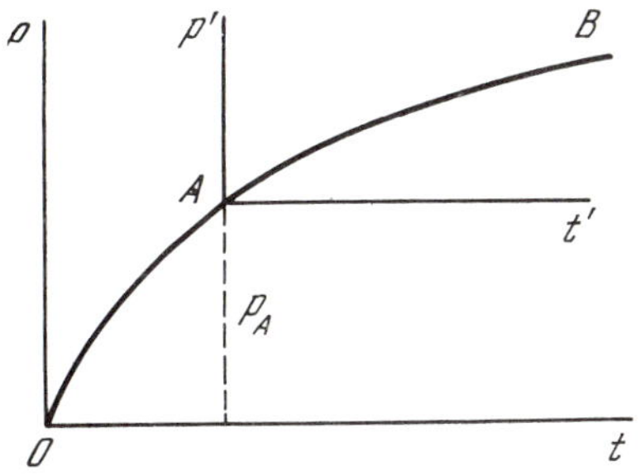

Fig. 80.

Universality is attributed to this expression. For example, when $\dot{e}_p = 0$ eq. (57.1) gives the relationship between e_p and σ, corresponding to extension at a fairly low rate. Numerous investigations of the temperature-time dependence of the properties of various materials have been based on eq. (57.1). However, this equation is incorrect for small deformations as can be shown by simple experiments. Apply to the specimen a load that sets up a stress σ such that there is no instantaneous deformation and allow the specimen to creep. At a certain instant of time the creep deformation is p_A, and we are at point A on the creep curve (fig. 80). Now move the origin of coordinates to the point A. The continuation of the curve to the right of the point A is the creep curve with initial hardening p_A. If eq. (57.1) is valid this applies whatever the method by which the initial hardening was achieved. Suppose now that we extend the specimen to a plastic deformation of p_A and apply a constant stress σ. The creep curve of the specimen should coincide with AB, but in fact short-term plastic deformation has no influence on creep and the creep curve of the deformed specimen is approximately the same as that of the undeformed. Fig. 75 gives relaxation curves of three specimens of steel grade 30KhMA, the first specimen was of undeformed material, the second underwent creep at $\sigma = 20$ kg/mm^2 for 25 hours and then was subject to relaxation tests, before the third specimen was subjected to relaxation testing it was extended at the test temperature until the plastic deformation in it was the same as the creep deformation of the second specimen. It will be seen that within the limits of experimental error the first and third specimens relax in exactly the same way and the second relaxes somewhat more slowly. Curves calculated from the equation of state on the basis of characteristics determined from creep tests at constant load are plotted in dotted lines. Accordingly, we conclude that short-term plastic deformation does not harden the material relative to creep and the equation of state should be written

$$\Phi(\dot{p}, \sigma, p, T) = 0 .$$

Now p is determined as follows:

$$\mathrm{d}p = \mathrm{d}e - \frac{\mathrm{d}\sigma}{E} - \kappa g'(\sigma)\mathrm{d}\sigma . \tag{57.2}$$

Here $\kappa = 1$, if $\mathrm{d}\sigma > 0$, and $\kappa = 0$ if $\mathrm{d}\sigma < 0$. The equation of the instantaneous deformation curve is

$$e = \frac{\sigma}{E} + g(\sigma) .$$

In formula (57.2) we use the same notation $g(\sigma)$ as in writing formula (47.5) in section 47 though here this is the instantaneous plastic deformation in the direct sense and previously it was a quantity that allowed for deformation during the first period of creep. However, in both cases the formal role of the function $g(\sigma)$ is the same and, therefore, we make no differentiation in notation.

This treatment of the equation of state is only valid to a first approximation and for sufficiently small instantaneous plastic deformations, apparently not more than 1%. Greater preliminary plastic deformation influences creep significantly. Kaptelin [49] made a detailed investigation of the influence of preliminary plastic deformation on the form of the equation of state for annealed copper, the picture that he obtained was complicated and confused and although he did find equations to describe his results they are hardly suitable for making calculations.

It is also possible to allow for hardening caused by instantaneous plastic deformation within the framework of the theory presented in section 55. Indeed, for this purpose, it suffices to take

$$\mathrm{d}q = \mathrm{d}p + \chi'(\sigma)\mathrm{d}\sigma .$$

If the function $\chi(\sigma)$ coincides with the function $g(\sigma)$, $\mathrm{d}q$ is the increment of the total plastic deformation and we return to the equation of state in the formulation of (57.1). Generally speaking, $\chi(\sigma)$ is a different function that must be found experimentally.

Returning to the usual form of the equation of state let us consider the possibility of determining the function $g(\sigma)$ experimentally, that is of finding the diagram of instantaneous plastic deformation. For this purpose a tensile test can be made at constant rate of loading or constant rate of deformation.

With extension at constant rate of loading the equation of state is easily integrated, if it is in suitable analytical form. It is particularly convenient to use the equation of state (55.4). Putting $\sigma = ct$ and integrating we obtain

$$p = \left[\frac{1}{m(n+1)}\right]^m \frac{1}{c^m} \sigma^{m(n+1)} . \tag{57.3}$$

If we use eq. (55.6) the result is

$$p = \frac{(1+m+mn)^{1-m}}{m(n+1)} \frac{1}{c^m} \sigma^{m(n+1)} . \tag{57.4}$$

It will readily be seen that the numerical factor in the right-hand side of formula (57.4) is always greater than the corresponding factor in formula (57.3). Given creep characteristics derived from working out short-term creep curves, we can calculate by eq. (57.3) or (57.4) the creep deformation that is accumulated during the extension test time and subtract it from the measured total deformation. The curves of fig. 76 were constructed from these formulae. A check is that the instantaneous deformation curve obtained in this way should be the same if the initial deformation curves are determined at different rates of loading.

Tension tests are more difficult to perform at constant rate than at constant loading, if we attempt to obtain a constant rate of deformation on a given basis instead of simply applying a constant rate to one of the machine clamps. Moreover it is much more difficult to work out the results of a test of this kind, it is possible only if we are given the form of the function $g(\sigma)$; we have not succeeded in finding any convenient device for establishing the form of the function $g(\sigma)$ from test results at constant rate of extension. Note that the tension curve always has a maximum, unlike a constant rate of loading curve. Here, of course, we do not consider the instability of the extension process and necking of the specimen. As $\dot{p} = v - g'(\sigma)\dot{\sigma}$, it follows from the creep equation that

$$\dot{\sigma} = \frac{v - f(\sigma, p)}{g'(\sigma)} .$$

The rate of stress increase $\dot{\sigma}$ becomes zero when $v = f(\sigma, p)$, that is if the rate of extension becomes equal to the creep rate for the particular stress and creep deformation that has been achieved.

Later we shall show that the instantaneous plastic deformation is quite easily separated from the creep deformation in the case of short-term, high-temperature creep.

The formulation of the work-hardening law that is given in this section relates to the case when the body is subject to a small plastic deformation (of the order of 1%) and subsequent creep takes place at stresses lower than that used to create the plastic deformation. The problem of creep in the plastic region is extremely confused. In the first place we have seen that on reaching the limit of proportionality there is a marked change in the values of the constants in the creep law. This circumstance was noted in section 53 for aluminium alloy at 150°C, the stress of 25.3 kg/mm^2 at which the two straight lines on the graph of fig. 63 intersect corresponds fairly accurately to the limit of proportionality of the material determined in tension at a sufficiently high rate. Very many similar results can be given but to avoid introducing new data specially for this purpose consider the graphs given in Chapter 5 (figs. 107–121) where there is also a significant change in the creep constants on transition to the plastic region.

The absence of work-hardening influence of a small instantaneous plastic deformation may be explained as follows. Instantaneous plastic deformation takes place as a result of localised shears in systems of glide planes which are thereby made much stronger, though the main volume of the grain is still undeformed. Creep deformation also results from shear, but these shears are more or less uniformly distributed throughout the volume and, therefore, the entire volume becomes work-hardened to the same extent. If a metal has been subjected to small instantaneous plastic deformations, subsequent creep takes place mainly in the undeformed volume.

On the other hand, there is no reason to assert that creep has no influence on the instantaneous deformation characteristics. The instantaneous deformation curve is of itself not an absolute characteristic of the material, its form

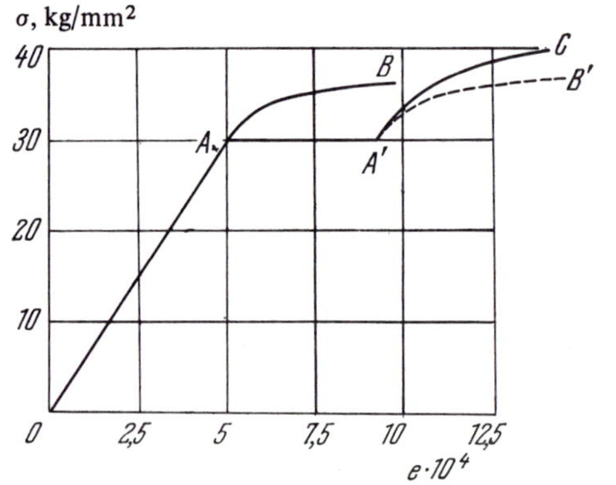

Fig. 81.

alters after repeated loading and unloading because of work-hardening and ageing. It will be noted that no allowance is made for the hardening effect in the usual theory of plasticity; under creep conditions matters are complicated by the circumstance that ageing occurs under load and is accompanied by creep. There is not sufficient information available to allow a phenomenological description of the interaction between creep and plastic deformation, but by way of illustration we give results of two tests. An instantaneous deformation curve of aluminium alloy D-16AT at 150° at loading rates above 6 kg/mm^2 · min is given in fig. 81, the deformation diagram is practically independent of the rate and may be considered as instantaneous, the creep accumulated during the deformation test is negligibly small. In the next experiment the specimen is loaded to $\sigma = 30$ kg/mm^2 (point A) and maintained under load. In this time, a creep deformation AA' is accumulated. On subsequent rapid loading, the relationship between stress and deformation is shown by curve $A'C$ and not by $A'B$, which repeats the initial deformation curve AB, and thus during creep there was hardening relative to instantaneous deformation.

58. **Short-term creep. The fundamental facts**

In recent years a number of works have been published on short-term creep, i.e. creep that results in the accumulation of considerable deformation in a time of some tens or hundreds of seconds. Magnesium alloys were studied in the works [185, 206], heat resistant austenitic steel in [332, 203, 331, 268] and high-melting point metals in [267]. Obviously short-term creep is of interest only at high temperature and stress, a region in which material is not usually tested. In the above works, for the most part, there is no attempt to analyse the creep curves on the basis of some theory or other, at best curves of stress against time to achieve a certain creep deformation were constructed or else isochronous creep curves were plotted. However, the following circumstances are immediately obvious from a brief review of this work.

1. The first part of the creep curve is not clearly expressed and is often altogether absent. Initially the creep rate is constant, later it increases and continues to increase up to the point of failure.

2. Instantaneous plastic deformation that occurs when load is applied is great.

Systematic investigations of short-term creep were carried out on a number of structural materials by Mileyko [74, 75] in the Institute of Hydrodynamics of the Siberian Division of the Academy of Sciences of the USSR. Our presentation is mainly based on results that he obtained by working out his own test results and those of others. Typical creep curves of the alloy D-16AT-2 at a temperature of 250° were given in fig. 43. This is a high temperature for this particular alloy, note the narrow stress range, the extreme curves correspond to stresses of 16.8 and 18.8 kg/mm^2, at lower stress there is no appreciable deformation in a time of 200 seconds and at higher stresses failure is instantaneous: at $\sigma = 16.8$ kg/mm^2 the creep rate is strictly constant and at $\sigma = 17.7$ and 18.8 kg/mm^2 the third part of the creep curve has already commenced at a deformation of about 0.5%. The first parts of the creep curves are absent, at the given temperature they did appear at lower stresses but then the creep deformation was negligibly small. In exactly the same way the first part appears at lower test temperatures. There is accordingly a certain boundary of temperature and stress beyond which creep takes place without being accompanied by hardening.

We have written the equation of transient creep in the form

$$p \sim [f(\sigma)t]^m \qquad \left(m = \frac{1}{1+\alpha}\right).$$

Creep at constant rate corresponds to $\alpha = 0$. Test results show that α diminishes with rising temperature and stress. In order to predetermine the possible limits of applicability of the short-term creep theory developed below, consider the schematic diagram shown in fig. 82, in which stress and temperature are plotted on the coordinate axes. The upper curve which represents the temperature dependence of the short-term strength is obviously the upper limit of possible use of the material under static conditions (the short-term strength may be exceeded under dynamic loading). In region A creep is

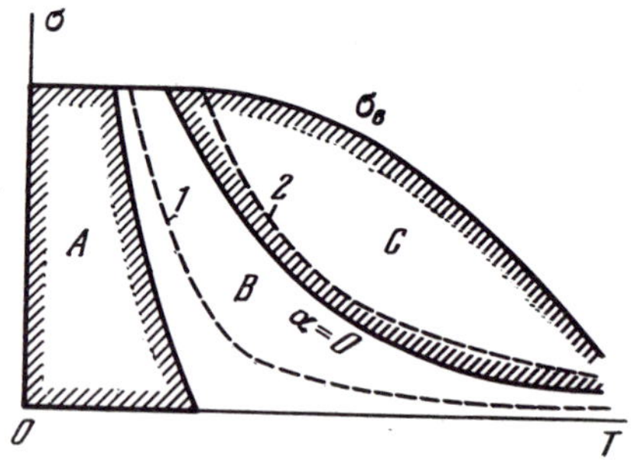

Fig. 82.

absent. The first curvilinear boundary of this region is not very clearly defined, it depends on the tolerances used in measuring the creep deformation. The region of stress and temperature at which creep occurs is in its turn divided into parts B and C by the curve $\alpha = 0$. In part B creep is accompanied by work hardening which can, of course, be terminated by moving into the steady-state region. No hardening occurs in region C, which is the one in which we are at present interested. The curves denoted 1 and 2 have nominal meanings. In application to any technical problem we can indicate a certain limiting creep deformation p' that need not be taken into account. If during the service life of the product the accumulated creep deformation is not greater than p' then creep can in general be ignored and the calculation is made by the methods of the theory of elasticity or of plasticity. Curve 1 in fig. 82 corresponds to the combination of stress and temperature which in a long time (say 100 000 hours) causes a creep deformation p'; the stress and temperature region in which creep must be allowed for in long-term service is located above curve 1. The upper boundary of this curve corresponds of course, not to the short-term ultimate strength σ but to the so-called long-term ultimate strength for the given service life. For our purpose, it is important that curve 1 is located in region B, therefore, creep calculations with long-term application of load should be made with allowance for the transient phase, that is in the work-hardening region.

If the product is under load for a short time (a few minutes) the occurrence of appreciable creep deformation during service corresponds to curve 2, which is higher up and is shown entirely in region C of the diagram. This means that short-term creep calculations relate to a stress and temperature region from which work-hardening is absent.

The scheme of fig. 82 is hypothetical, a large amount of experimental data, usually not available, would be required to construct these curves in practice, and to establish accurately the boundaries between regions. Curves of $\sigma_B(T)$ and of $\alpha = 0$ and $\alpha = 0.125$ for austenitic steel grade N-155 from the results of [167] are plotted in fig. 83 and so the boundaries of region C, which we shall term the region of short-term creep, are partially determined.

The fundamental equation that describes short-term creep is

$$\dot{e} = \frac{\dot{\sigma}}{E} + \kappa g'(\sigma)\dot{\sigma} + v(\sigma) . \tag{58.1}$$

Here it is assumed that there is an instantaneous deformation curve $e = \sigma/E + g(\sigma)$, whereas creep takes place at constant rate $v(\sigma)$ which, at the given temperature, depends only on stress and not on the previous deformation history. Eq. (58.1) coincides completely in form with eq. (47.5) but

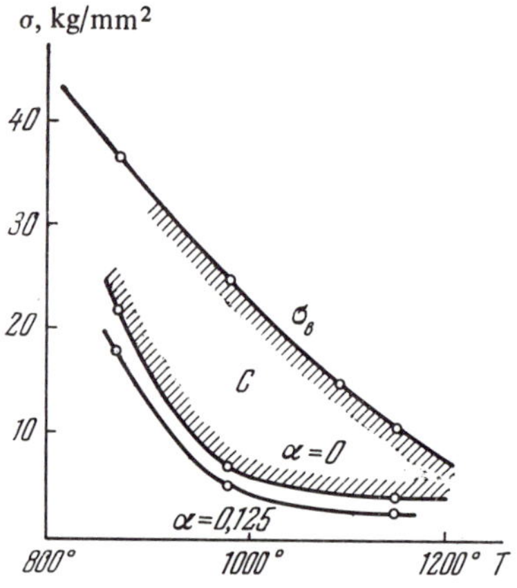

Fig. 83.

here $g(\sigma)$ represents the instantaneous plastic deformation in the direct sense.

The function $v(\sigma)$ is determined from creep tests, but the function $g(\sigma)$, which represents the instantaneous plastic deformation, can be found as follows. Carry out a series of tensile tests with constant rate of load increase, so that in each test $\sigma = ct$. Applying eq. (58.1) to this test we find

$$e = \frac{\sigma}{E} + g(\sigma) + \frac{1}{c} \int_0^\sigma v(\sigma)\,\mathrm{d}\sigma\,.$$

The quantity

$$p(\sigma) = \frac{1}{c} \int_0^\sigma v(\sigma)\,\mathrm{d}\sigma$$

is the creep deformation that accompanies extension. Knowing the function $v(\sigma)$ we can calculate $p(\sigma)$. Subtracting the quantity $p(\sigma)$ from the abscissa of the extension curve we determine the instantaneous deformation curve $e = \sigma/E + g(\sigma)$, shown schematically in fig. 84. A check on the correctness of the initial hypotheses is that the same instantaneous curve is found by starting from tension curves at various rates of loading. In fact, if the rate of loading is such that the entire deformation curve takes place in a time of about a second the accompanying creep deformation is small and the resultant curve may be taken as instantaneous. Now a check on the correctness of the initial hypothesis can be carried out in inverse order, namely add to abscissae of the

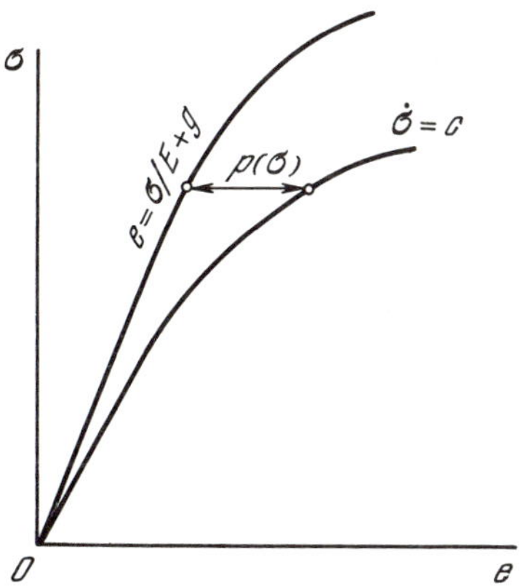

Fig. 84.

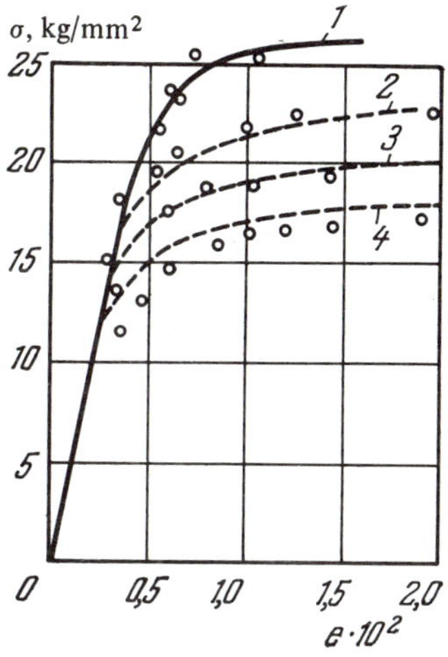

Fig. 85.

instantaneous curve values of $p(\sigma)$ calculated for various rates of loading c and compare the resulting tension test diagrams with experimental curves. The results of such a check are shown in fig. 85 for alloy D-16AT at a temperature of 250°, the upper curve was taken as instantaneous and the dotted curves were obtained by calculation for various rates of loading measured in kg/mm^2 per second, and experimental points are plotted against them: curve 1 was obtained at a rate of $c = 25$, curve 2 at $c = 0.289$, curve 3 at $c = 0.063$ and curve 4 at $c = 0.016$.

As a fundamental test method for determining short-term creep characteristics certain authors recommend testing at constant rate of deformation. But in practical structures the loads are always given and the rate of deformation cannot be controlled. In considering the design of a particular product we can always set real limits to the rate of change of load within which the mechanical characteristics should be determined. It would usually be impossible to state in advance the corresponding limits of rate of deformation. On the other hand, given curves of deformation at $\dot{e}$ = const, in working them out we encounter serious difficulties in separating plastic from creep deformation.

The remark concerning the possibility of taking the deformation curve that corresponds to a sufficiently high loading rate as the instantaneous curve is valid provided the temperature is not too high, otherwise the instantaneous curve must be constructed in the way that has been described. Results for titanium alloy VT-5-1 are given in fig. 86, the solid lines correspond to the reconstructed instantaneous curves, lines of dashes to deformation curves at a

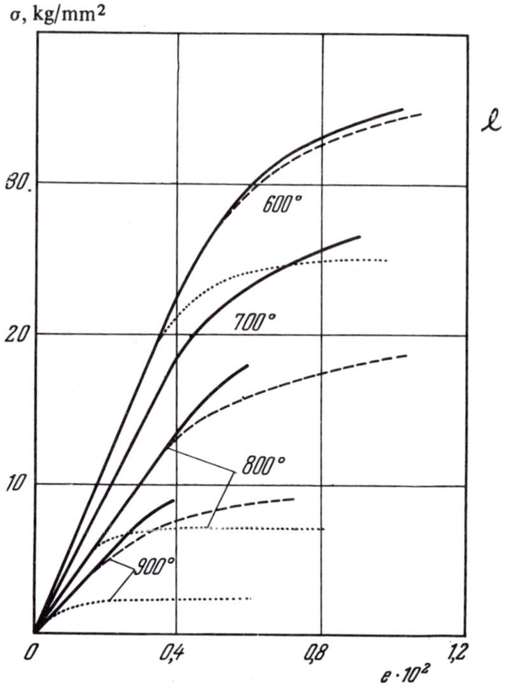

Fig. 86.

rate of change of stress of 20 kg/mm^2 per second, and the dotted lines to extension curves at constant rate of deformation corresponding to 1% elongation per minute. In the latter test the rate of deformation was checked directly from the extensometer readings, whereas it is the clamp rate that is measured in ordinary testing machines. A rate of deformation of the order of 1% per minute corresponds approximately to standard tensile test conditions; the comparative curves given in fig. 86 show the extent to which an ordinary testing machine diagram can give inadequate information about the true properties of material in short-term loading.

59. **Short-term creep. Dependence on rate**

As a result of working out test results on 15 different materials in the laboratories of the Institute of Hydrodynamics of the Siberian Division of the Academy of Sciences of the USSR it was found that the creep rate is satisfactorily represented as function of stress and temperature by the formula

$$\dot{p} = k \exp(\beta\sigma + \nu T) . \tag{59.1}$$

The form of the temperature dependence in formula (59.1) differs from that usually given in the physical literature and is somewhat more convenient for purposes of calculation. It was found that two temperature intervals must be distinguished for certain alloys: moderate temperatures where $T < T^*$ and high temperatures when $T > T^*$. In the moderate temperature region

$$k = k_1 = \text{const} , \quad \beta = \beta_1 = \text{const} , \quad \nu = \nu_1 = \text{const} \quad (T < T^*) . \tag{59.2}$$

In the high temperature region

$$k = k_2 = \text{const} , \quad \beta = \frac{\beta_2}{T_0 - T_1} , \quad \nu = \nu_2 = \text{const} , \tag{59.3}$$

$$T_0 = \text{const} \ (T > T^*) .$$

Instantaneous deformation curves for titanium alloy VT-5-1 have already been given in fig. 86, we now give values of the creep constants that were obtained for this material. Creep was studied in the temperature range 600–900°, which falls entirely within the high temperature range, consequently, the coefficients in eq. (59.1) assume the values of (59.3) where

$$k_2 = 7.76 \cdot 10^{-11} , \quad \beta_2 = 104 , \quad T_0 = 1020^\circ , \quad \nu_2 = 0.0140 .$$

Here the stress is measured in kg/mm^2, the time in seconds, and the temperature in degrees Celsius (Centrigrade).

In other titanium alloys and austenitic steels the region of intensive short-term creep lies entirely in the high temperature range whereas in aluminium alloys considerable creep is also observed in the moderate temperature range. Thus for material D-16T in bars, $T^* = 270°$, for alloy AMG-6M the range of practical interest, 150–250° is entirely in the moderate temperature range, and for alloy AMG-6N the region of practical importance, 200–300°, relates to the high temperature range.

Eq. (58.1) is applicable both when loads and temperatures are constant and when they vary. The error of calculations that are based on this assumption is not greater than the natural scatter of experimental results. Test results

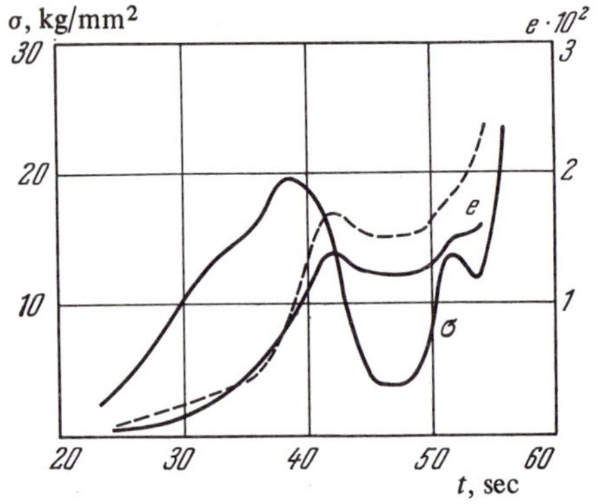

Fig. 87.

on alloy VT-5-1 at a temperature of 700° are given in fig. 87. The stress varies with time in a fairly complicated way, the experimental time dependence of deformation is shown by solid lines and that calculated by eq. (58.1) is shown by dotted ones.

In the absence of the first parts of the creep curves the relaxation curve should be calculated by eq. (47.3). Relaxation curves for alloy AMG-6M at 227° are shown in fig. 88, theoretical curves are plotted in solid lines and experimental curves dotted. To further verify that the creep rate is independent of previous loading, one specimen was tested several times in succession and after relaxation was again loaded to the initial stress and again relaxed, all the successive relaxation curves coincided accurately.

As we have seen, the diagram of instantaneous deformation also depends

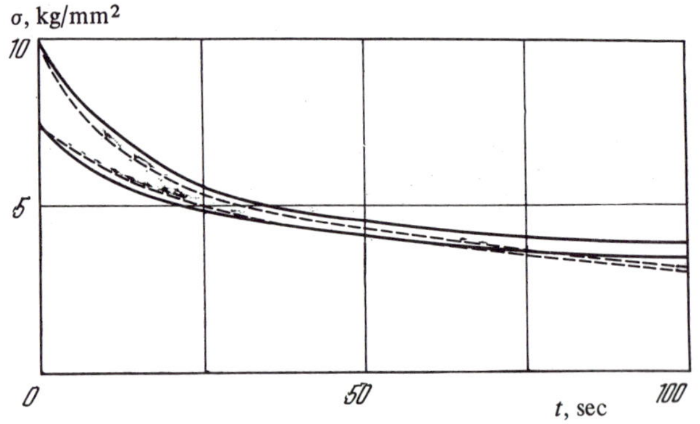

Fig. 88.

on temperature but the temperature dependence of the plastic properties is much less marked than that of creep characteristics. We shall consider that in eq. (58.1) the function $g(\sigma)$ is specific to each temperature and also that the modulus E is a function of T. Thus the total equation of short-term creep with allowance for temperature is

$$\dot{e} = \frac{\dot{\sigma}}{E(T)} + \kappa g'(\sigma, T)\dot{\sigma} + k \exp(\beta\sigma + \nu T)\,. \tag{59.4}$$

We shall not attempt to select approximate expressions for the function $g(\sigma, T)$, calculations should be based on a series of experimental curves of instantaneous deformation, corresponding to different temperatures.

In conclusion we shall briefly consider certain questions of short-term creep testing procedure. A very detailed account of test methods is given in the article [295] and in addition to the articles mentioned above this question is the subject of the publications [166, 205, 294, 347, 384]. Certain authors have made creep tests in compression, which is of direct interest for aviation structures.

Mileyko developed special equipment with electromagnetic application of force and automatic recording of deformation with time. Programmed loading is possible with this equipment, in particular the load can be varied in such a way that the total deformation is maintained constant (relaxation) or the rate of deformation is constant. Heating is by radiation at a maximum rate of 40°/sec. The method of heating by passage of current that is used by certain investigators is not to be recommended because it gives creep characteristics which are different from those obtained with radiation or convective heating.

60. **Creep in compression. Load reversal**

Creep testing in compression presents greater technical difficulties than testing in tension. When long specimens are used initial irregularities and eccentricity of load application play an important part. In tensile testing the influence of eccentricity diminishes during the test and the result is not greatly affected by initial irregularities and inaccuracy in application of load. This problem is considered in detail in the work [293]. In testing in compression initial deflections increase, the eccentricity of load application increases and the state of stress varies considerably during the course of the test. It is, therefore, difficult to obtain reliable compression curves and to compare them with tensile test results.

Sully and colleagues [124, 357, 358] have made numerous and systematic studies of creep in compression. The object of the investigations was to develop a simplified method of determining the comparative characteristics of heat-resistant alloys. They constructed simple machines for the purpose, which whilst not of the highest accuracy, were good enough for obtaining comparative data. Nevertheless considerable attention was devoted to the correspondence between creep characteristics obtained in tension and in compression because it was necessary to find out whether tests in compression could be replaced by standard tensile tests. Most of these investigations were on heat-resistant chrome-nickel alloys. The general conclusion is that for a given stress the absolute rate of steady-state creep in tension is the same as in compression, within the limits of experimental error. There is a difference in the first part of the creep curve, which may be related to experimental errors and the instability of creep characteristics in the first phase. It should be noted that the third phase of creep is also observed in compression and it commences at approximately the same amount of creep deformation as in tension. The difference between the two cases of creep in the third phase is that in tension the creep rate increases continuously to failure but in compression it first increases and then diminishes somewhat and then again increases, and so on; the resulting creep curve is wave shaped. The authors of the work [357] attribute the third phase to atomic rearrangement in the most deformed parts of the grains close to the boundaries, of the type found

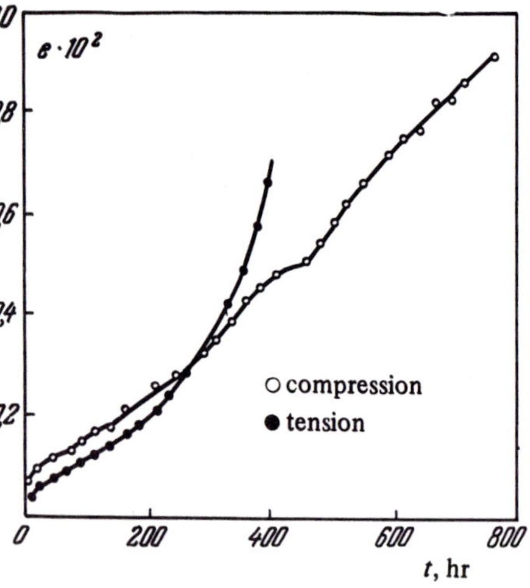

Fig. 89.

in the initial stages of recrystallisation. Creep curves in tension and in compression for chrome-nickel alloy at the same stress $\sigma = 1400$ kg/cm^2 and temperature 815°C, taken from the aforementioned article, are given in fig. 89.

Creep of aluminium alloys in compression was studied in the works [182, 183], in the first and second stages the creep curves approximately coincided and there was a difference in the third stage. In these works particular attention was paid to the third stage of creep, and, therefore, Torshenov [139] performed special tests to compare creep curves in tension and in compression in the first part of the creep curve and on transition to the second. The work was carried out on aluminium alloy D-16-T at a temperature of 200°. In order to produce compressive loads a special reversing device was attached to the tensile testing machine and special measures were taken to reduce the eccentricity of load application. The influence of the initial eccentricity on the creep curve in compression was studied both theoretically and experimentally on specimens of known initial eccentricity. The author found that in the test results that he worked out, the influence of eccentricity of load application

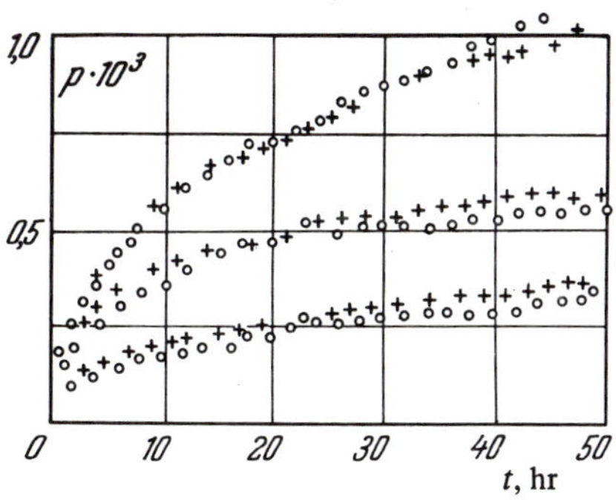

Fig. 90.

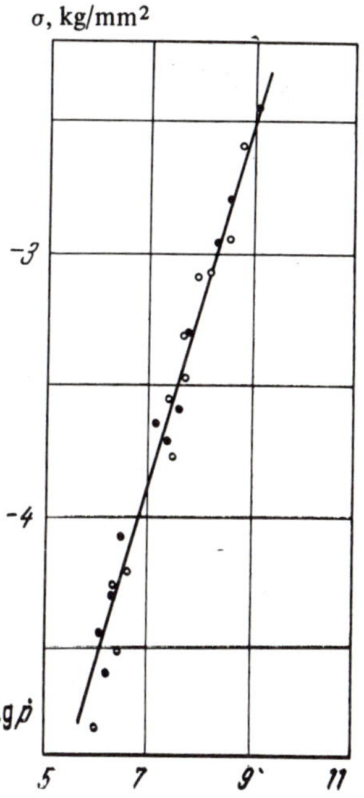

Fig. 91.

was within the limits of experimental error. Test results for stresses 8, 12, and 16 kg/mm^2 are shown in fig. 90, points corresponding to creep curves in tension are shown by crosses and those in compression by circles. The difference between the curves is within the limits of scatter of the test results.

The short-term creep behaviour in the sense of section 58, is apparently the same in tension and in compression for most materials. The graph of the creep rate against stress for alloy D-16-T taken from Zhigalkin's University diploma project is given in fig. 91, the circles correspond to tensile tests and the dots to compressive. Plotted in log-linear coordinates the points group about a straight line, which confirms formula (59.1), the deviations are of the nature of random scatter. The behaviour of magnesium alloys in compression is anomalous, at normal temperatures the usual deformation curves in tension and in compression are very different for these alloys [109]. For a given stress the creep rates in tension and in compression are also very different, as has been noted for example in the work [206]. Short-term creep test results for magnesium alloy VM-65-1 obtained in the Institute of Hydrodynamics of the Siberian Division of the Academy of Sciences of the USSR are plotted in fig. 92, the circles correspond to tensile tests and the points to compressive ones, the ordinates are the stress and the logarithm of the creep rate. Creep is found to be much more rapid in tension than in compression.

Creep under variable loading when the load changes sign has been little studied. Certain tests have been carried out in alternating torsion [2, 80] and it was found that deformation in the forward direction softens the material

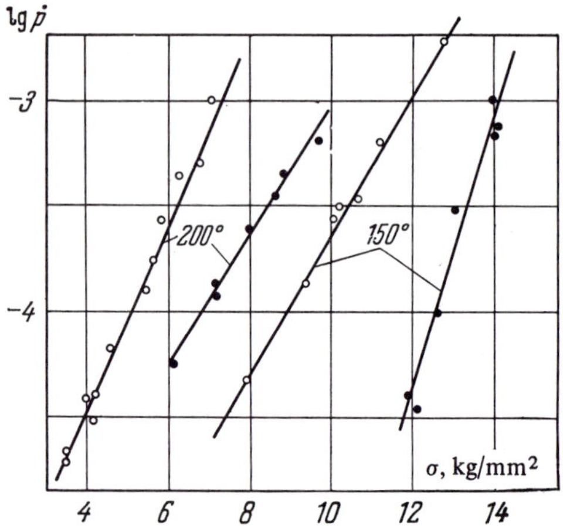

Fig. 92.

relative to creep in the opposite direction, after the load has changed sign creep is faster than on the first application of a load of the same intensity. In principle this effect can be described by the kinetic equations of section 55 but there are not sufficient experimental data available for us to do this.

Shorr [154, 155] proposed a hypothesis by which to assess the effect of work-hardening under alternating load, he took the creep law in the form

$$\dot{p}h(q) = \left|\frac{f(\sigma)}{\sigma}\right| \sigma .$$

The essentially positive quantity q is defined as follows:

$$q = p^+ + \lambda p^- \qquad \text{when } \sigma > 0 ,$$

$$q = p^- + \lambda p^+ \qquad \text{when } \sigma < 0 .$$

Here p^+ is the deformation accumulated in tension and p^- that accumulated in compression; the coefficient λ is an empirical constant of the material. We know of no experimental verification of this hypothesis or examples of determination of λ.

61. Small deviations from the basic state

In some applications of creep theory, for instance in considering stability problems, it is necessary to consider small deviations from a certain basic state of creep, for instance, the influence of a small increase or decrease in stress on creep at constant load. Taking any creep theory as basis we can write linearised equations which describe small deviations from the basic state. Suppose for example that the material observes the equation of state in the usual form

$$\dot{p} = \varphi(\sigma, p) , \qquad p = e - \frac{\sigma}{E} . \tag{61.1}$$

In the basic state p and σ are known functions of time. We suppose that if for example σ is given as function of time, eq. (61.1) is integrated and p is found as function of t. Now suppose that as a result of change in external conditions the stress becomes $\sigma + \delta\sigma$ and the creep deformation $p + \delta p$. If $|\delta\sigma/\sigma| \ll 1$ and $|\delta p/p| \ll 1$, the following linear approximation follows from eq. (61.1),

$$\delta \dot{p} = \frac{\partial \varphi}{\partial p} \delta p + \frac{\partial \varphi}{\partial \sigma} \delta \sigma . \tag{61.2}$$

The quantities $\partial\varphi/\partial p$ and $\partial\varphi/\partial\sigma$ are known functions of time. If we are interested in the behaviour of the material immediately after a change in load, the slowly changing time functions $\partial\varphi/\partial p$ and $\partial\varphi/\partial\sigma$ may be considered as constants. In eq. (61.2) substitute the total deformation δe for δp and we obtain

$$\delta \dot{\sigma} + \lambda \delta \sigma = E(\delta \dot{e} + \mu \delta e) . \tag{61.3}$$

Here

$$\lambda = E \frac{\partial \varphi}{\partial \sigma} - \frac{\partial \varphi}{\partial p} , \qquad \mu = - \frac{\partial \varphi}{\partial p} .$$

Eq. (61.3) coincides exactly in form with eq. (23.5) of Chapter 3. Accordingly, when small deviations from the basic state occur in a material that obeys the equation of state and is undergoing creep, it behaves like a viscoelastic Kelvin body. The instantaneous changes in stress and deformation are interrelated by the usual Hooke's law: after a fairly long time (which must, however, be short enough for changes in λ and μ to be negligible) the stress and deformation remain proportional to one another but the relationship between them is given by the long-term modulus

$$E_c = \frac{- E \dfrac{\partial \varphi}{\partial p}}{E \dfrac{\partial \varphi}{\partial \sigma} - \dfrac{\partial \varphi}{\partial p}} .$$

We shall term this quantity the creep modulus, it is readily shown that it is the tangential modulus of the deformation curve of the material at constant rate of creep deformation. It should be noted that although it is possible in principle to make creep tests at constant rate of deformation, in practice it is very difficult.

In order to take account of simultaneous quasiviscous flow we take

$$p = e - \frac{\sigma}{E} - \int v(\sigma) \mathrm{d}\sigma .$$

Using this relationship and eliminating the creep deformation from eq. (61.2) we obtain

$$\delta \ddot{\sigma} + (Ev' + \lambda) \delta \dot{\sigma} + \mu v' \delta \sigma = E(\delta \ddot{e} + \mu \delta \dot{e}) . \tag{61.4}$$

With alternating loading, predictions made by the equation of state are not entirely accurate and it is, therefore, to be expected that the difference between experimental results and the predictions of the linearised eq. (61.3) or the more general eq. (61.4) should be greater than the error that results from using the initial eq. (61.1), which in general provides a good description of the accumulation of total creep deformation under alternating load, but which is not suitable for a time interval immediately after a sudden change of load. Therefore, the direct study of creep laws with small changes in the stressed state is of interest. Certain information about this subject is contained in the work of Onat and Wang [318] who assume that variations of σ and e are related by the following linear differential equation,

$$p_2 \delta\ddot{\sigma} + p_1 \delta\dot{\sigma} + p_0 \delta\sigma = q_2 \delta\ddot{e} + q_1 \delta\dot{e}\,, \tag{61.5}$$

which is of the same form as (61.4) but the values of the coefficients remain unspecified. Creep tests on lead wire at normal temperature showed that if the change in stress is not more than 4% of the basic stress the additional deformation is proportional to the change in stress, as follows from the linearity of eq. (61.5). Further, if at the instant $t = t_0$ the stress is suddenly increased by an amount $\delta\sigma$ and is then maintained constant the following time dependence follows from eq. (61.5):

$$\delta e = \left\{ \frac{p_2}{q_2} + \frac{p_0}{q_1}(t - t_0) + \left(\frac{p_1}{q_1} - \frac{p_2}{q_2} - \frac{p_0 q_2}{q_1^2} \right) \left[1 - \exp\left(-\frac{t - t_0}{\tau} \right) \right] \right\} \delta\sigma\,. \tag{61.6}$$

Here

$$\tau = \frac{q_2}{q_1}\,.$$

Experimental curves are satisfactorily approximated by an expression of this type but the delay time is much less than is given by eq. (61.4). In one of the examples $\tau = 37.5$ minutes whereas the corresponding value determined from the initial creep law is 30 hours. This was to be expected from the creep test results under alternating loads given above. Taking a more general form of the equation of state, that is introducing the strain-hardening parameters q_i, we obtain more complicated equations for variation of σ and e. Onat and Wang's results give no basis for selecting a more accurate form of the equation of state. It can be shown that the introduction of a strain-hardening parameter of time $q = \int a(\sigma)\,dp$ does not improve the position: the delay time τ remains unchanged and a greater creep rate results from including a term that is linearly dependent on time in eq. (61.6).

62. Dynamic creep

We shall define dynamic creep as that which takes place when a periodic alternating stress component of fairly high frequency is superposed on a constant mean stress. Dynamic stress is observed in various machine components subject to alternating loads, in turbine blades and other structures and, therefore, numerous works have been devoted to the study of dynamic creep. Depending upon the material and the test temperature the cyclic component of the load may increase or decrease the creep rate. For example, in Greenwood's [220] tests on lead, it was found that the creep rate increased tens of times on application of an alternating stress component. Cyclic loading greatly accelerated the recrystallisation of lead. The influence of alternating load on structural change in the material has been noted by many other authors [226, 251]. In the works of Guarnieri [223, 224] for example, it was shown that in chrome-nickel steel and aluminium alloy 24-S-T3 alternating load accelerates creep, whereas in Inconel X the creep rate remains unaltered and in alloy N-155 it even diminishes. The change in structure during creep with constant and alternating loads was also studied in these works. If creep is not accompanied by recrystallisation and phase conversions alternating loads apparently have no specific effect and creep deformation can be calculated with reasonable accuracy by one of the theories used to describe creep under slowly varying loads, for example, by the equation of state in some form or other. Thus in our unpublished work of 1948 carried out in the Institute of Mechanics of the Academy of Sciences of the USSR, it was found that creep in lead at normal temperature and in copper at a temperature of 150° under a load consisting of a constant part and a relatively small component that varied sinusoidally with time, took place in the same way as creep at a constant stress which was somewhat greater than the mean stress of the cycle but less than the maximum stress. The tests were comparatively short, about a day, and the stresses were not high so that, for example, no recrystallisation was observed in lead. Similar results have been obtained by numerous other authors [248, 350, 353]. A detailed review of numerous investigations on dynamic creep was made by Isaksson [249].

If we assume that the equation of state is valid in the form

$$\dot{p}h(p) = f(\sigma)$$

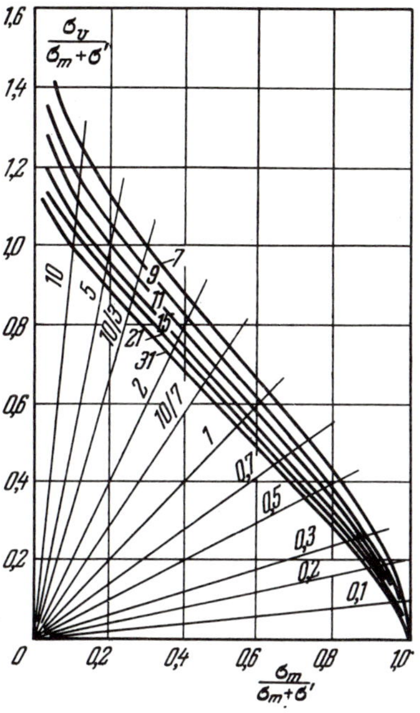

Fig. 93.

and take

$$\sigma = \sigma_m + \sigma_v \sin \omega t \, ,$$

creep takes place as if the material were subject to a constant effective stress $\sigma = \sigma_m + \sigma'$; here

$$f(\sigma_m + \sigma') = \lim \left[\frac{1}{t} \int_0^t f(\sigma_m + \sigma_v \sin \omega t) \mathrm{d}t \right]_{t \to \infty} . \tag{62.1}$$

If $f(\sigma) = \epsilon_e \exp \sigma/\sigma_e$ it follows from formula (62.1) that

$$\sigma' = \sigma_e \ln \sum \frac{1}{(2n)!} \frac{1}{2^n} \binom{2n}{n} \left(\frac{\sigma_v}{\sigma_e} \right)^{2n} . \tag{62.2}$$

An analogous formula was obtained for the case of alternating temperature

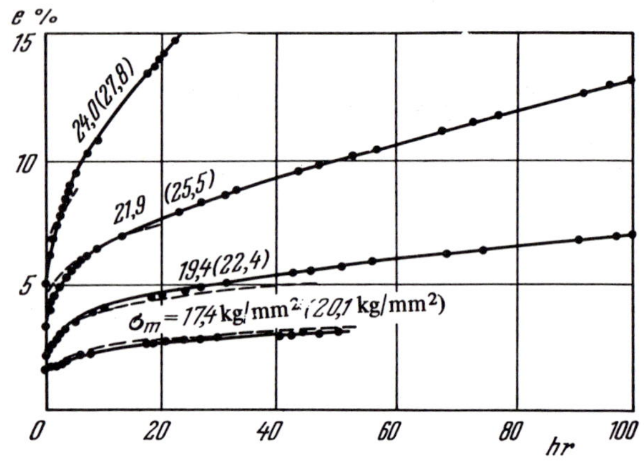

Fig. 94.

(section 54), but it differs from formula (62.2) in that the latter is suitable for any value of $\sigma_v < \sigma_m$. In deriving it the only assumption was that the stress remains positive. It is still unknown what happens when the stress changes sign. Calculations could of course be made by the strain-hardening theory but their real meaning is obscure in the absence of experimental data.

For the power law of creep the integral in formula (62.1) can be calculated for various values of n. Curves from the work [365] that can be used to determine the equivalent static stress $\sigma_m + \sigma'$, given the constant component σ_m and the alternating component σ_v, are plotted in fig. 93. In this graph the ratios $\sigma_m/(\sigma_m + \sigma')$ and $\sigma_v/(\sigma_m + \sigma')$ are plotted on the coordinate axes and the lines drawn from the origin of coordinates correspond to constant ratios σ_v/σ_m. The curves are constructed for $n = 7, 9, 11, 15, 21$ and 31, working from top to bottom. Experimental and theoretical curves of carbon steel at 450° are compared in fig. 94 for the case $\sigma_v/\sigma_m = 0.23$, theoretical curves are shown by dashes. The figures in brackets on the curves are equivalent stresses $\sigma = \sigma_m + \sigma'$. The authors note that the agreement between experiment and theory is satisfactory if $\sigma_v/\sigma_m < 0.7$. For higher values of the ratio of the alternating component of the stress to the mean value, creep is accompanied by fatigue damage which influences the creep rate and particularly the time to failure.

5

Creep in a State of Complex Stress

In most creep calculations on engineering structures the creep conditions are complex. Laboratory creep tests with complex stress conditions present technical difficulties, the experiments must be performed very carefully if the results are to be reasonably reliable. Therefore, the available experimental data is slender and does not provide a reliable basis for a creep theory that is capable of describing the behaviour of material under complex stresses. Moreover, tests can only be made with plane stresses and we have no information about creep performance with stresses on three axes.

In the theory of plasticity the state of affairs is much the same. Reasonably reliable information is available concerning the behaviour of material with proportional loading but with arbitrary loading paths different theories predict different results and the experimental data is not decisively in favour of any particular theory.

Like plasticity theory, the theory of creep under complex stress is based on certain speculative considerations which are only partially confirmed experimentally. The simplest case of creep is that in which the stressed state is uniform and constant with time, and this is the case to which the great majority of experimental results relate. Creep of this kind can be described fairly simply as a kind of viscous flow. There are various ways in which the theory can be extended to varying stressed states and the number of possibilities is much greater than in the theory of plasticity. In real objects the nature of the stressed state usually varies comparatively little with time, and, therefore, the different theories lead to results which are not very different from one another.

This chapter presents the fundamental concepts of steady state and quasi-steady-state creep for both isotropic and anisotropic media. The concepts of rate and stress potentials in creep are introduced. The possibility of constructing a theory of transient creep is considered. Then almost all the published experimental material on creep is analysed. It is concluded that quasi-steady-state creep is satisfactorily described by means of the rate potential which depends only on the second invariant of the deviator (Mises type theory), sometimes a third invariant has an influence, the nature of which is that the true equipotential surface lies between a cylinder $\tau_0 = \text{const}$ and a prism $\tau_{max} = \text{const}$.

63. Steady-state creep

In the preceding chapter we considered the possibility of describing creep with uniaxial stress solely on the basis of tests performed in pure tension. This is because tensile tests are easy to perform, a uniform stressed state is achieved in a fairly large volume and the deformation is measured on a large base length so that the accuracy is satisfactory. Other methods of achieving an approximately uniform stressed state, such as tensile, compressive and torsion tests on thin-walled tubes, are more difficult to perform and are rarely used. The correlation of tensile tests results with those in say torsion can only be resolved by means of a creep theory which is based on the general case of an arbitrary stressed state. Generally speaking, the theories with which we were concerned in application to tension in Chapter 4 can be extended to the complex stressed state in various ways and it is possible to describe a very large number of variants of spatial creep equations on the basis of simple schemes of steady-state creep, creep with strain-hardening and so on. The experimental possibilities of achieving more or less pure types of complex stress, and measuring the resulting deformation are limited. And therefore, inevitably the theory is only partially verified experimentally and confirmation of any particular variant is indirect. In this chapter we shall consider, from the general standpoint, several fairly extensive possibilities of describing creep and we shall make a selection among these possibilities, on the basis of agreement with test results on the one hand and of simplicity and convenience for practical application on the other. In creep in tension we have seen that the simplest possible scheme is that of steady-state creep in which the creep rate is constant at constant stress. Two cases of steady-state creep must be distinguished; in the first strain-hardening is absent from the start, in the second the rate becomes constant because the material becomes fully strain-hardened and loses the ability to become further strain-hardened as deformation continues. We have seen that strain-hardening is practically absent from short-term creep tests at sufficiently high temperatures and stresses. At relatively low temperatures the rate becomes constant only after a fairly long time and steady-state creep is preceded by a first phase during which strain-hardening occurs. The difference between the two cases of steady-state creep is observed with variable loading. If the material does not strain-harden the instantaneous creep rate depends only on the instantaneous stress and is independent of the previous history of loading; if steady-state creep is preceded by a first phase, and, during steady-state creep there is a stepwise change in load, the creep rate does not immediately become constant at a value corresponding to the new stress but rather the transient section of the creep curve is repeated. Accordingly, at high temperatures the region of short-term creep is the one to which the theory applies in its purest form. Creep investigations in this region are of great theoretical and practical interest because here the fundamental characteristics of the effect are observed most clearly, uncontaminated by auxiliary complications.

With complex stressing the presence or absence of strain-hardening is still more important. The point is that strain-hardening which results from creep or short-term plastic deformation creates anisotropy in the material relative to subsequent creep. Creep characteristics are particularly sensitive to anisotropy. For example, rolled metals such as steel or aluminium alloy have very similar elastic properties along and across the direction of rolling, the difference between plastic properties such as the nominal yield point or short-term strength is no more than 10–15% in the two directions. However, the creep rate determined at a given temperature and stress can differ by a factor of 2 or 3 between specimens cut along and across the direction of rolling. Further on, we shall see that anisotropy that develops as a result of strain-hardening during creep is still more marked.

In the first place we shall consider the possibility of describing steady-state creep with complex stressing. We must distinguish between isotropic and anisotropic creep, here anisotropy may be governed by the following factors:

1. The initial state of the material, which is associated with the method of manufacture of the specimen or product, and primarily with the texture that results from working the metal by pressure.

2. Strain-hardening or change of texture caused by instantaneous plastic deformation. This occurs either at the instant of loading, if the load is high

enough, or during creep if the stressing is not uniform and stresses become redistributed, this is the more difficult case to analyse.

3. Strain-hardening that takes place in the first phase of creep. In principle, it might be supposed that anisotropy also arises during steady-state creep. This means that the accumulated deformation in steady-state creep also causes structural changes in the material which are not observed as a strain-hardening effect in simple uniaxial tests but influence redistribution of the components of the creep deformation-rate tensor. We shall not consider this possibility here but shall consider steady-state creep to be that in which with a constant state of stress all the components of the creep rate tensor remain constant.

64. **Isotropic creep**

The equations of steady-state creep theory are relationships between components of the rate of creep deformation tensor ϵ_{ij} and components of the stress tensor σ_{ij}, namely: $\epsilon_{ij} = f_{ij}(\sigma_{kl})$. Generally the relationship between the symmetrical tensors $x_{ij} = \varphi_{ij}(y_{kl})$ is isotropic provided it does not change when the reference system is changed. Suppose that for a new system of coordinates the tensor components are respectively x'_{ij} and y'_{kl}. Then by definition of an isotropic relationship

$$x'_{ij} = \varphi_{ij}(y'_{kl}) .$$

We satisfy this condition by taking φ_{ij} in the form

$$\varphi_{ij}(y_{kl}) = \varphi\delta_{ij} + aA_{ij} + bB_{ij} + cC_{ij} + \dots . \qquad (64.1)$$

Here, φ, a, b, c are scalars, δ_{ij} is the unit tensor, A_{ij}, B_{ij}, C_{ij}, ... are arbitrary tensors which have the same principal axes as the tensor y_{kl}. Indeed if one of the tensor A_{ij}, B_{ij}, ... had different principal axes from those of y_{kl} it would mean that there were certain preferential directions in the material which are not related to the nature of the state of stress and which characterise properties of the material (if we return to the physical problems as initially formulated).

In the expansion of eqs. (64.1) there is point in retaining only linearly independent tensors A_{ij}, B_{ij}, ...; the set of linearly independent tensors is limited, as the number of their components is finite. Moreover, in this particular case we are considering tensors which have the same principal directions and which are constructed from the tensor y_{ij}. Put $A_{ij} = y_{ij}$, $B_{ij} = y_{ik}y_{kj}$, $C_{ij} = y_{ik}y_{ks}y_{sj}$ and so on, that is expand eqs. (64.1) as powers of the tensor y_{ij}. It can readily be shown that this expansion can be terminated at the second term. Indeed let x_{ij} represent the third power of the tensor y_{ij}, that is $x_{ij} = y_{ik}y_{ks}y_{sj} = C_{ij}$. We shall show that it is always possible to select scalars φ_0, a_0 and b_0 so that the following equation is valid:

$$C_{ij} = \varphi_0\delta_{ij} + a_0A_{ij} + b_0B_{ij} . \tag{64.2}$$

All the tensors in this equation are coaxial and, therefore, it can be written for the principal axes, giving

$$\begin{aligned} y_1^3 &= \varphi_0 + a_0y_1 + b_0y_1^2 , \\ y_2^3 &= \varphi_0 + a_0y_2 + b_0y_2^2 , \\ y_3^3 &= \varphi_0 + a_0y_3 + b_0y_3^2 . \end{aligned} \tag{64.3}$$

This is a system of three equations for the determination of three quantities φ_0, a_0 and b_0. Moreover, in order to solve this system directly, note that y_1, y_2 and y_3 are roots of one and the same cubic equation which can be written (see section 2)

$$y^3 - I_{\mathrm{I}}y^2 + I_{\mathrm{II}}y - I_{\mathrm{III}} = 0 . \tag{64.4}$$

Here

$$I_{\mathrm{I}} = y_1 + y_2 + y_3 ,$$

$$I_{\mathrm{II}} = y_1y_2 + y_2y_3 + y_3y_1 ,$$

$$I_{\mathrm{III}} = y_1y_2y_3 .$$

In section 2 we give formulae that express the invariants I_{I}, I_{II} and I_{III} in terms of invariants constructed in the usual way. Comparing formulae (64.3) and (64.4) we find that $b_0 = I_{\mathrm{I}}$; $a_0 = -I_{\mathrm{II}}$, $\varphi_0 = I_{\mathrm{III}}$; therefore, eq. (64.2) can be written

$$C_{ij} - I_{\mathrm{I}}B_{ij} + I_{\mathrm{II}}A_{ij} - I_{\mathrm{III}}\delta_{ij} = 0 .$$

Since A_{ij}, B_{ij} and C_{ij} are successive powers of the tensor y_{ij}, then using the notation $y_{ij}^{(n)}$ for power n of tensor y_{ij} we obtain

$$y_{ij}^3 - I_{\mathrm{I}}y_{ij}^{(2)} + I_{\mathrm{II}}y_{ij} - I_{\mathrm{III}}\delta_{ij} = 0 . \tag{64.5}$$

This is termed the Caley-Hamilton identity. It is not difficult to show that any higher powers of the tensor y_{ij} are expressed in terms of the same tensor, its second power and unit tensor in a linear way, and the coefficients of this linear dependence are invariants. Therefore, in eq. (64.1) the term with the factor c and all subsequent terms should be excluded, this exclusion is compensated by appropriate selection of φ, a and b. The most general form of the isotropic creep law can now be written

$$\epsilon_{ij} = P\delta_{ij} + Q'\sigma_{ij} + R\sigma_{ik}\sigma_{kj}\,. \tag{64.6}$$

Here P, Q' and R are functions of the invariants S_{I}, S_{II} and S_{III}. Further investigation will consist in rational limitation of the generality of eqs. (64.6) on the basis of physical or experimental considerations. In the first place it follows from both experimental and physical considerations that creep deformation is not accompanied by change of volume. Later we shall give experimental results to confirm this for the phase of creep in which there is not yet appreciable embrittlement. In Chapter 6 we shall show that microcracks appear in the third phase of creep that precedes fracture. The opening of these microcracks is of course accompanied by change in volume, which has been observed experimentally. Assuming that there is no change in volume, in the steady-state stage of creep, multiply eq. (64.6) by δ_{ij} and contract, obtaining

$$\epsilon_{ij}\delta_{ij} = 3P + S_{\mathrm{I}}Q' + S_{\mathrm{II}}R = 0\,.$$

Whence

$$P = -\,\sigma Q' - \tfrac{1}{3}S_{\mathrm{II}}R\,.$$

Now eliminate P from eq. (64.6) giving

$$\epsilon_{ij} = Q'(\sigma_{ij} - \sigma\delta_{ij}) + R(\sigma_{ik}\sigma_{kj} - \tfrac{1}{3}S_{\mathrm{II}}\delta_{ij})\,.$$

It will now be convenient to introduce into consideration the stress tensor deviator components $\bar{\sigma}_{ij} = \sigma_{ij} - \sigma\delta_{ij}$ it is readily shown that the second term in brackets can be written

$$\sigma_{ik}\sigma_{kj} - \tfrac{1}{3}S_{\mathrm{II}}\delta_{ij} = \bar{\sigma}_{ik}\bar{\sigma}_{kj} - \tfrac{1}{3}\bar{S}_{\mathrm{II}}\delta_{ij} + 2\sigma\bar{\sigma}_{ij}\,.$$

Therefore

$$\epsilon_{ij} = Q\bar{\sigma}_{ij} + R(\bar{\sigma}_{ik}\bar{\sigma}_{kj} - \tfrac{1}{3}\bar{S}_{\mathrm{II}}\delta_{ij})\,. \tag{64.7}$$

Here $Q = Q' + 2\sigma R$ and R is a function of the invariants. As basis for the system of invariants we can select the quantities σ, $\bar{S}_{\mathrm{II}}$ and $\bar{S}_{\mathrm{III}}$, though another selection is more convenient in some cases. The assumption of incompressibility is often supplemented by the firmer assumption that the first invariant of the stress tensor σ does not influence creep. Then Q and R depend only on $\bar{S}_{\mathrm{II}}$ and $\bar{S}_{\mathrm{III}}$.

Creep involves dissipation of mechanical energy and, therefore, the power dissipation should be introduced into consideration,

$$L = \sigma_{ij}\epsilon_{ij} = \bar{\sigma}_{ij}\epsilon_{ij} \, .$$

The stress tensor can be replaced by the tensor deviator because $\epsilon_{ij} = 0$ and hydrostatic stress performs no work. By means of eq. (64.7) we find

$$L = Q\bar{S}_{\mathrm{II}} + R\bar{S}_{\mathrm{III}} \, .$$

Obviously the functions Q and R should satisfy the condition that $L > 0$ for any value of σ_{ij}.

65. Creep potential

We shall present certain considerations that favour the creep potential hypothesis. First let us define energy dissipation in creep

$$L = \sigma_{ij}\epsilon_{ij} = L(\sigma_{ij}) \, .$$

Putting $L(\sigma_{ij}) = \text{const}$ we obtain a surface of constant dissipation in the stress space; the power dissipation is k for all states of stress that are represented by the points on this surface. Obviously surfaces of constant energy dissipation cannot intersect, for otherwise the energy dissipation would not be uniquely expressed in terms of stress; these surfaces enclose one another and are shown schematically in fig. 95, here $k_1 > k^*$. Now consider a certain loading path from the point M^* on the surface $L = k^*$ to the point M_1 on the surface $L = k_1$. To be specific we shall consider that this path intersects each of the constant dissipation surfaces $L = k$, $k^* \leqslant k \leqslant k_1$ only once, so that there is no diminution in the power dissipated during loading.

We shall require that on a certain loading path when $\mathrm{d}L \geqslant 0$, the power of additional actions is not negative, this means that for each point M of the

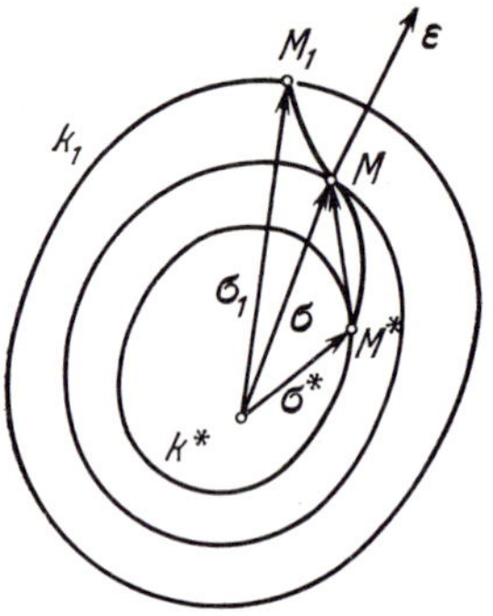

Fig. 95.

path M^*M_1

$$(\sigma_{ij} - \sigma_{ij}^*)\epsilon_{ij} \geqslant 0 \,. \tag{65.1}$$

In particular the points M^* and M can lie on a single surface $L = k$ and then the path M^*M lies entirely on this surface and it follows from eq. (65.1) that the deformation rate vector is in the direction of the normal to the surface $L(\sigma_{ij}) = k$. Therefore

$$\epsilon_{ij} = h(\sigma_{ij}) \frac{\partial L}{\partial \sigma_{ij}} \,. \tag{65.2}$$

In effect eq. (65.1) is a rephrasing of Drucker's postulate (section 14) to the case of steady-state creep. Just as in the theory of plasticity, condition (65.1) does not follow from the laws of thermodynamics and is not a thermodynamic postulate, this is a limitation which makes it possible to distinguish a class of materials that have certain stability properties; for these materials the solution of the creep theory problem is unique, as we shall see in Chapter 7, and certain variational principles are valid for them. However, it remains an open question whether condition (65.1) is a necessary condition of uniqueness, just as in the theory of plasticity. There may exist a function $\Phi(\sigma_{ij})$ such that

$$h(\sigma_{ij}) \frac{\partial L}{\partial \sigma_{ij}} = \frac{\partial \Phi}{\partial \sigma_{ij}} \,. \tag{65.3}$$

This function Φ is termed the creep rate potential and expressions (65.2) assume the form

$$\epsilon_{ij} = \frac{\partial \Phi}{\partial \sigma_{ij}} \,. \tag{65.4}$$

In particular, condition (65.3) is fulfilled if $h(\sigma_{ij}) = h(L)$. Indeed in this case

$$\Phi = \int_0^L h(L)\mathrm{d}L \,.$$

Multiplying eq. (65.2) by σ_{ij} we find

$$L = h(\sigma_{ij}) \frac{\partial L}{\partial \sigma_{ij}} \sigma_{ij} \,. \tag{65.5}$$

It follows from eq. (65.5) that the form of the function $h(\sigma_{ij})$ is fully determined by the structure of the function L; determining h from expression (65.5) and substituting in expression (65.2) we find

$$\epsilon_{ij} = \frac{L}{\sigma_{ks} \dfrac{\partial L}{\partial \sigma_{ks}}} \frac{\partial L}{\partial \sigma_{ij}} \,. \tag{65.6}$$

If L is a uniform function of stress of power $n + 1$ then $h = 1/(n + 1)$ and $\Phi = [1/(n + 1)]L$. If L is an arbitrary function of some uniform first power function s of σ_{ij} then $h = L(s)/(sL'(s))$ and $\Phi = \int_0^s L(s) \frac{\mathrm{d}s}{s}$.

We have thus shown that in consequence of condition (65.1) expression (65.3) for the creep rate is valid when certain fairly wide limitations are applied to the form of the functions $L(\sigma_{ij})$. The converse formulae follow immediately from these expressions; it is readily shown that there is a function $U(\epsilon_{ij})$ of the components of the creep rate tensor which can be used to determine the stress by the formula

$$\sigma_{ij} = \frac{\partial U}{\partial \epsilon_{ij}} \,. \tag{65.7}$$

Indeed, put

$$U = \sigma_{ij}\epsilon_{ij} - \Phi = L - \Phi \,.$$

Here we consider that the stresses σ_{ij} are expressed in terms of ϵ_{kl}, which is possible in principle, since the formulae (65.4) establish a mutually unique relationship between σ_{ij} and ϵ_{ij}. Therefore, the new function U depends on ϵ_{ij}. Differentiating both sides of the above equation with respect to ϵ_{ij} we obtain

$$\frac{\partial U}{\partial \epsilon_{ij}} = \sigma_{ij} + \epsilon_{kl} \frac{\partial \sigma_{kl}}{\partial \epsilon_{ij}} - \frac{\partial \Phi}{\partial \sigma_{kl}} \frac{\partial \sigma_{kl}}{\partial \epsilon_{ij}} .$$

But because of eq. (65.4) the last two terms on the right-hand side cancel one another and we obtain eq. (65.7).

In creep theory the function $L(\sigma_{ij})$ plays the same part as the flow function in the theory of plasticity. If condition (65.1) is fulfilled it follows that the surface L = const is convex. In the same way the surfaces $\Phi(\sigma_{ij})$ = const are also convex in the space of stresses and $U(\epsilon_{ij})$ = const in the space of rates of deformation.

It is evident that in an isotropic medium the creep potential can depend only on invariants of the stress tensor

$$\Phi = \Phi(S_{\mathrm{I}}, S_{\mathrm{II}}, S_{\mathrm{III}}) .$$

The condition of incompressibility applies some limitation to this relationship. Indeed it follows from the condition of incompressibility and (65.4) that

$$\frac{\partial \Phi}{\partial \sigma_{ij}} \delta_{ij} = 0 .$$

This is a first-order equation with partial differential coefficients that is equivalent to the following system of normal equations:

$$\mathrm{d}\sigma_{ij} = \delta_{ij} \mathrm{d}\sigma .$$

Hence it follows that the potential Φ depends not on the stress tensor components, but on the deviator components $\sigma_{ij} - \sigma\delta_{ij}$. Therefore, we may take

$$\Phi = \Phi(\bar{S}_{\mathrm{II}}, \bar{S}_{\mathrm{III}}) .$$

Eq. (65.4) is now rewritten

$$\epsilon_{ij} = 2 \frac{\partial \Phi}{\partial S_{\mathrm{II}}} \bar{\sigma}_{ij} + 3 \frac{\partial \Phi}{\partial S_{\mathrm{III}}} (\bar{\sigma}_{ik} \bar{\sigma}_{kj} - \tfrac{1}{3} \bar{S}_{\mathrm{II}} \delta_{ij}) . \tag{65.8}$$

Comparing with (64.7) we find that

$$Q = 2\frac{\partial \Phi}{\partial \bar{S}_{\mathrm{II}}}, \qquad R = 3\frac{\partial \Phi}{\partial \bar{S}_{\mathrm{III}}}.$$

Allowing for compressibility, if this is necessary, introduces no fundamental changes into creep theory, it is only a question of the degree of complexity of the resulting equations. By taking particular partial hypotheses it is possible to obtain more or less simple equations. Thus, a creep theory was developed in a recent work [324] in which the potential depends on a certain specially constructed invariant of the stress tensor, namely, on the quantity $S_{\mathrm{II}} + \alpha^2 S_{\mathrm{I}}^2$ where α is an experimentally determined constant associated with the coefficient of transverse deformation in tension, that is with the ratio of the rate of change of the transverse dimension of the extended specimen to the rate of axial extension.

The form of the relationship between stresses and rates that is obtained in creep theory on the assumption that the potential exists, strongly resembles the relationship between stress and deformation in the theory of non-linear elasticity. In the uniform case the equations of creep theory and of non-linear elasticity theory simply coincide, irrespective of the potential hypothesis and consequently, theorems of the type of Lagrange and Castigliano are valid in creep theory. Indeed consider a system consisting of unidimensional elements. This may be a statically determinate truss of n rods. The rate of elongation of each rod $\dot{q}_i$ is a function of the stress Q_i in this rod so that $\dot{q}_i = v_i(Q_i)$. Assume that the truss is in equilibrium when subject to a system of m external forces P_s. The external force P_s performs work on the generalised displacement, the rate of which is $\dot{u}_s$. Let us formulate the Lagrange equilibrium equation considering practically possible creep rates. We obtain

$$\sum_{s=1}^{m} P_s \dot{u}_s = \sum_{i=1}^{n} Q_i \dot{q}_i \,. \tag{65.9}$$

Consider an infinitesimally adjacent state of equilibrium that corresponds to change in only one of the external forces P_s by an amount δP_s. It follows from eq. (65.9) that

$$\dot{u}_s = \sum_{1}^{n} \dot{q}_i \frac{\partial Q_i}{\partial P_s} \,. \tag{65.10}$$

Now introduce the creep potential in the following way:

$$\Phi = \sum_{i=1}^{n} \varphi_i(Q_i)\,.$$

Here the functions φ_i are determined so that

$$\varphi_i'(Q) = v_i(Q)\,.$$

Differentiating Φ with respect to P_s

$$\frac{\partial \Phi}{\partial P_s} = \sum_{i=1}^{n} v_i(Q_i)\frac{\partial Q_i}{\partial P_s} = \sum_{i=1}^{n} \dot{q}_i \frac{\partial Q_i}{\partial P_s}\,.$$

Comparing with eq. (65.10) we find that

$$\dot{u}_s = \frac{\partial \Phi}{\partial P_s}\,.$$

The rods that form the unidimensional elements of the system may be in a state of bending or torsion as well as tension or compression. Therefore, the creep potential exists if the body can be modelled by unidimensional elements. One of the simplest models of this kind is considered in section 16. Consider a tube or rod of circular cross-section under conditions of bending. The individual fibres of the rod are in either tension or compression and, therefore, the foregoing analysis retains its force and creep equations can be written in terms of the potential for any direction of bending moment. The considerations that have been given do not, of course, serve to demonstrate that potential exists for a spatially stressed state, but they do make it very probable.

66. Special forms of creep law

It is found experimentally that the dependence of creep rate on stress is of the same nature for various types of stressed state. For instance, if the creep rate in tension is proportional to the tensile stress to power n, then in torsion the shear rate is also proportional to the tangential stress to this power. This suggests a particular selection of the form of the dependence of the potential on the invariants $\bar{S}_{II}$ and $\bar{S}_{III}$; the potential should depend on some uniform function of stress. For an incompressible material the following is a sufficiently general form of this dependence:

$$\Phi = \Phi\left[\bar{S}_{II} g_1\left(\frac{\bar{S}_{III}}{\bar{S}_{II}^{\frac{3}{2}}}\right)\right].$$

In place of $\bar{S}_{II}$ and $\bar{S}_{III}$ it is convenient to introduce the following quantities (see section 8)

$$\sigma_0^2 = \tfrac{3}{2}\bar{S}_{II}, \qquad \zeta = \sqrt{6}\,\frac{\bar{S}_{III}}{\bar{S}_{II}^{\frac{3}{2}}}.$$

Then the creep potential can be considered as a function of the product of σ_0 and $g(\zeta)$, namely

$$\Phi = \Phi[\sigma_0 g(\zeta)] .$$

Using the obvious relationships

$$\frac{d\sigma_0}{d\bar{S}_{II}} = \frac{3}{4\sigma_0}, \qquad \frac{\partial\zeta}{\partial\bar{S}_{II}} = -\frac{9}{4}\frac{\zeta}{\sigma_0^2}, \qquad \frac{\partial\zeta}{\partial\bar{S}_{III}} = \frac{9}{2}\frac{1}{\sigma_0^3},$$

we can rewrite eq. (65.8) as follows:

$$\epsilon_{ij} = \tfrac{3}{2}\Phi'\left\{(g-3\zeta g')\frac{\bar{\sigma}_{ij}}{\sigma_0} + 9g'\frac{\bar{\sigma}_{ik}\bar{\sigma}_{kj} - \frac{2}{9}\sigma_0^2\delta_{ij}}{\sigma_0^2}\right\}. \tag{66.1}$$

If the stress tensor is referred to the principal axes expression (66.1) can be written in another form representing stress by means of the formulae (8.9)

$$\bar{\sigma}_1 = \tfrac{2}{3}\sigma_0\cos\theta,$$

$$\bar{\sigma}_2 = \tfrac{2}{3}\sigma_0\cos\left(\theta - \frac{2\pi}{3}\right),$$

$$\bar{\sigma}_3 = \tfrac{2}{3}\sigma_0\cos\left(\theta - \frac{4\pi}{3}\right).$$

By formula (8.8)

$$\bar{S}_{\rm III} = \bar{\sigma}_1^3 + \bar{\sigma}_2^3 + \bar{\sigma}_3^3 = \frac{1}{\sqrt{6}} \bar{S}_{\rm II}^{\frac{3}{2}} \cos 3\theta \ .$$

Therefore

$$\zeta = \cos 3\theta \ .$$

Now write, for example, an expression for ϵ_1. The factor that enters into this expression is of the form

$$\sigma_{1k}\sigma_{k1} - \tfrac{2}{9}\,\sigma_0^2 = \tfrac{2}{9}\,\sigma_0^2(2\cos^2\theta - 1) \ .$$

Substituting this expression and also $\bar{\sigma}_1$ and ζ into eq. (66.1), after regrouping terms we obtain

$$\begin{aligned} \epsilon_1 &= \Phi' \left\{ g\cos\theta + 3g' \sin 3\theta \sin\theta \right\} , \\ \epsilon_2 &= \Phi' \left\{ g\cos\left(\theta - \frac{2\pi}{3}\right) + 3g' \sin 3\theta \sin\left(\theta - \frac{2\pi}{3}\right) \right\} , \\ \epsilon_3 &= \Phi' \left\{ g\cos\left(\theta - \frac{4\pi}{3}\right) + 3g' \sin 3\theta \sin\left(\theta - \frac{4\pi}{3}\right) \right\} . \end{aligned} \qquad (66.2)$$

Two other formulae are obtained by substituting $\theta - 2\pi/3$ and $\theta - 4\pi/3$ respectively for θ. Here $\sin 3\theta$ and also $\zeta = \cos 3\theta$, the argument of the function g, remain unaltered.

Instead of considering g as function of the variable ζ we may consider that $g = g(\theta)$. It is obvious that $g(\theta)$ is a periodic function of period $2\pi/3$; so far this function has been determined to within an arbitrary factor. We shall consider that $g(0) = 1$. Then in uniaxial tension the argument of the function $\Phi(\sigma_0 g)$ becomes equal to the magnitude of the tensile stress and consequently

$$\Phi'(\sigma_0 g) = v(\sigma_0 g) \ .$$

Here v is a function that gives the relationship between creep rate in tension and tensile stress. Now we can write formulae (66.2) in the following simpler form:

$$\epsilon_1 = v\left[g \cos\theta - \frac{dg}{d\theta}\sin\theta\right],$$

$$\epsilon_2 = v\left[g \cos\left(\theta - \frac{2\pi}{3}\right) - \frac{dg}{d\theta}\sin\left(\theta - \frac{2\pi}{3}\right)\right], \tag{66.3}$$

$$\epsilon_3 = v\left[g \cos\left(\theta - \frac{4\pi}{3}\right) - \frac{dg}{d\theta}\sin\left(\theta - \frac{4\pi}{3}\right)\right].$$

In the octahedral plane the stress intensity vector is at an angle θ to a straight line from the centre of the octahedral plane to the point of intersection with axis 1, as shown in fig. 3 and repeated in fig. 96.

It follows from formulae (66.3) that the vector of rate of deformation intensity ϵ is at an angle $\theta + \omega$ to this same straight line and

$$\tan\omega = \frac{1}{g}\frac{dg}{d\theta}.$$

To verify the theory experimentally and to obtain expressions for the functions $g(\theta)$ and $\nu(\sigma_0 g)$ we proceed as follows. Multiply the first of the eqs. (66.3) by $\cos\theta$, the second by $\cos[\theta - (2\pi/3)]$, the third by $\cos[\theta - (4\pi/3)]$ and add. We obtain

$$\epsilon_s = \tfrac{2}{3}\left[\epsilon_1 \cos\theta + \epsilon_2 \cos\left(\theta - \frac{2\pi}{3}\right) + \epsilon_3 \cos\left(\theta - \frac{4\pi}{3}\right)\right] = g\nu(\sigma_0 g). \tag{66.4}$$

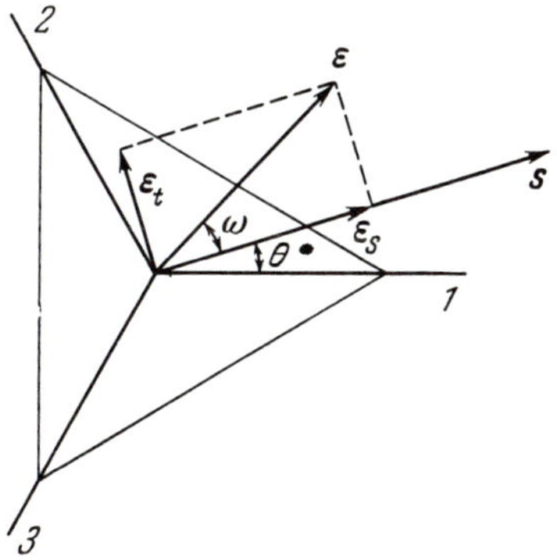

Fig. 96.

Similarly

$$\epsilon_t = -\tfrac{2}{3}\left[\epsilon_1 \sin\theta + \epsilon_2 \sin\left(\theta - \frac{2\pi}{3}\right) + \epsilon_3 \sin\left(\theta - \frac{4\pi}{3}\right)\right] = \nu(\sigma_0 g)\frac{dg}{d\theta}. \tag{66.5}$$

Dividing (66.5) by (66.4) we obtain

$$\frac{\epsilon_t}{\epsilon_s} = \frac{1}{g}\frac{dg}{d\theta} = \tan\omega .$$

In making tests with various states of stress we alter the quantities θ, and from measured values of ϵ_1, ϵ_2 and ϵ_3 we can always calculate ϵ_s and ϵ_t, so that the ratio of the one to the other is an experimentally determined function of θ. Then the foregoing expression is a differential equation for the function $g(\theta)$. Denoting the ratio $\epsilon_t/\epsilon_s = \kappa(\theta)$, here we have taken as argument the variable θ. Then obviously

$$g(\theta) = \exp \int_0^\theta \kappa(\theta)\, d\theta . \tag{66.6}$$

In using this formula it should be borne in mind that the function $g(\theta)$ is periodic, with period $2\pi/3$, therefore,

$$\int_{\theta_0}^{\theta_0 + \frac{2\pi}{3}} \kappa(\theta)\, d\theta = 0 .$$

Fulfilment of this condition for the experimentally determined function $\kappa(\theta)$ is a necessary condition of applicability of the theory. If the creep characteristics are the same in tension as in compression, then g depends on the modulus θ, consequently the period of $g(\theta)$ is $\pi/3$ and not $2\pi/3$.

In eq. (66.6) the constant of integration is taken as unity so that $g(0) = 1$. Now to find the function ν it suffices to use test results for any particular stressed state, the value of g for this stressed state is already known and in formula (66.4) the function ν is selected to give the best agreement with the test results. Note that it follows from (66.3) that the intensity of rate of deformation is

$$\epsilon_0 = [\tfrac{2}{3}\, \epsilon_{ij}\epsilon_{ij}]^{\frac{1}{2}} = \nu(\sigma_0 g)\sqrt{g^2 + g'^2} . \tag{66.7}$$

Hence if $g = 1$ we obtain an unchanged relationship between the intensity of creep rate and stress intensity. If $g(\theta)$ is not constant the expression differs for different values of θ in just the same way as the expression $\epsilon_s(\sigma_0 g)$ according to eq. (66.4). This second expression is more convenient in working out experimental results because formula (66.4), unlike formula (66.7), contains only the function $g(\theta)$ and not its derivative.

A very common method of making creep tests with complex stress is to test thin-walled specimens in combined tension and torsion or with combinations of tension or compression and internal pressure; therefore, we must consider the plane stressed state separately. For example, putting $\sigma_1 = 0$ in the first of the formulae (8.9), then $\sigma = -\frac{2}{3}\sigma_0 \cos\theta$; from the two successive formulae, after simplification we obtain

$$\sigma_2 = -\frac{2}{\sqrt{3}}\sigma_0 \sin\left(\frac{\pi}{3} - \theta\right), \quad \sigma_3 = -\frac{2}{\sqrt{3}}\sigma_0 \sin\left(\frac{\pi}{3} + \theta\right), \tag{66.8}$$

and also

$$\epsilon_s = -\frac{2}{\sqrt{3}}\left[\epsilon_2 \sin\left(\frac{\pi}{3} - \theta\right) + \epsilon_3 \sin\left(\frac{\pi}{3} + \theta\right)\right] = \frac{1}{\sigma_0}\left[\sigma_2\epsilon_2 + \sigma_3\epsilon_3\right],$$
$$\epsilon_t = \frac{2}{\sqrt{3}}\left[\epsilon_2 \cos\left(\frac{\pi}{3} - \theta\right) - \epsilon_3 \cos\left(\frac{\pi}{3} + \theta\right)\right]. \tag{66.9}$$

In particular it follows from (66.8) that

$$\sigma_2 + \sigma_3 = -2\sigma_0 \cos\theta = -3\sigma, \tag{66.10}$$

$$\sigma_2 - \sigma_3 = \frac{2}{\sqrt{3}}\sigma_0 \sin\theta. \tag{66.11}$$

The following formula that follows from formulae (66.10) and (66.11) is convenient for determination of θ with a given stressed state:

$$\sqrt{3}\,\text{ctan}\,\theta = -\frac{\sigma_2 + \sigma_3}{\sigma_2 - \sigma_3}. \tag{66.12}$$

In realising all possible plane stressed states we cover the range of angles from $\theta = 0$ to $\theta = 2\pi/3$. The following table gives certain values of the ratio σ_3/σ_2, corresponding values of θ, values of ζ and also of $\cos\theta$.

σ_3/σ_2	θ	ζ	$\cos\theta = 3\sigma/2\sigma_0$
$-\infty$	60°	-1	$-\frac{1}{2}$
-1	90°	0	0
0	120°	1	$\frac{1}{2}$
$\frac{1}{2}$	150°	0	$\sqrt{\frac{3}{2}}$
1	180°	-1	1
2	210°	0	$\sqrt{\frac{3}{2}}$
$+\infty$	240°	1	$\frac{1}{2}$

It should be noted that to establish the creep law we need experimental data for variation of the angle θ over a range of 120°. Indeed this is the period of the function $\zeta(\theta)$ and also of the factor $\sin 3\theta$ in eqs. (66.2). The range of the angle θ is shown schematically in fig. 97. A circle is constructed in the octahedral plane, the arcs are reckoned from axis (1); (2) and (3) denote the two other principal axes. The figures opposite corresponding points on the periphery at intervals of 30° represent the quantity ζ. The points where $\zeta = +1$ correspond to tension, and where $\zeta = -1$ to compression (with, of course, an arbitrary hydrostatic component superposed). Where $\zeta = 0$ the state of stress is pure shear, with superposed hydrostatic components. As has already been mentioned, to determine the function $g(\zeta)$ we

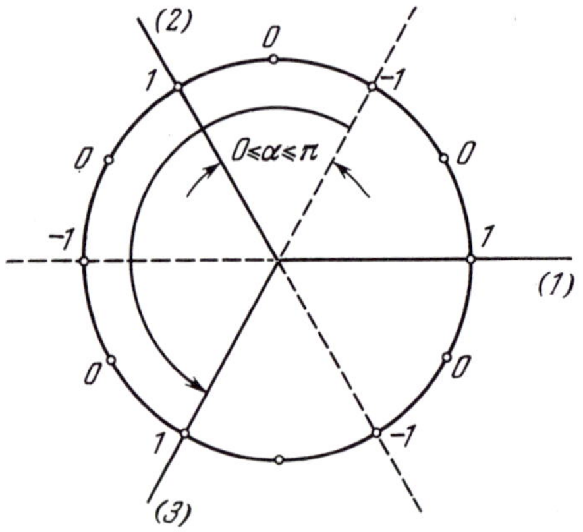

Fig. 97.

need only results for one sector in which the angle varies through 120°, demonstration that results in other sectors are identical can serve as a criterion for checking the validity of the theory. Deviations may result either from anisotropy of the material or from the influence of hydrostatic pressure which is associated with breakdown of the principle of the existence of creep potential. The columns in the table should be compared with points in the sector denoted by an arrow in fig. 97, the direction of the arrow corresponds to the order of the table reading from top to bottom. It will be seen that only the first point in which $\zeta = 0$ corresponds to pure shear, for the other two points the hydrostatic component is superposed on pure shear, such that $\sigma = (1/\sqrt{3})\sigma_0$.

In considering a plane stressed state, we arbitrarily put $\sigma_1 = 0$ and find expressions for σ_2 and σ_3. In applying other subscripts to the given stresses we can use the same formulae, first changing the origin for reading the angle θ. Thus if we take $\sigma_3 = 0$ we must write $\theta + 2\pi/3$ in place of θ in all the formulae. After this substitution the first formula (66.8) gives an expression for σ_1 and the second an expression for σ_2.

67. **Working out the results of tests in combined tension and torsion**

It is technically quite difficult to achieve the general case of a plane stressed state in creep tests, it is particularly difficult to measure peripheral deformation of thin-walled specimens and, therefore, most published results relate to combined tension and torsion. Measurements are made of the relative elongation e of the specimen and of the shear γ determined from the angle of twist. From the rates $\dot{e}$ and $\dot{\gamma}$ we still cannot determine the principal deformation rates ϵ_2 and ϵ_3 and, therefore, special consideration should be given to the case of combined tension and torsion. Let us put $\sigma_{11} = \sigma$, $\sigma_{12} = \sigma_{21} = \tau$, all the remaining components of the stress tensor are zero. The stress tensor deviator components are then

$$\bar{\sigma}_{11} = \tfrac{2}{3}\,\sigma\,, \qquad \bar{\sigma}_{22} = \bar{\sigma}_{33} = -\tfrac{1}{3}\,\sigma\,, \qquad \bar{\sigma}_{12} = \tau\,, \qquad \bar{\sigma}_{23} = \bar{\sigma}_{31} = 0\,.$$

The second invariant of the deviator $\bar{S}_{\mathrm{II}} = \frac{2}{3}(\sigma^2 + 3\tau^2)$. Consequently

$$\sigma_0^2 = \sigma^2 + 3\tau^2\,. \tag{67.1}$$

For brevity introduce the notation

$$s_{ij} = \bar{\sigma}_{ik}\bar{\sigma}_{kj} - \tfrac{1}{3}\,\bar{S}_{\mathrm{II}}\delta_{ij}\,.$$

We obtain

$$s_{11} = \tfrac{2}{9}\,\sigma^2 + \tfrac{1}{3}\,\tau^2\,, \quad s_{22} = -\tfrac{1}{9}\,\sigma^2 + \tfrac{1}{3}\,\tau^2\,, \quad s_{33} = -\tfrac{1}{9}\,\sigma^2 - \tfrac{2}{3}\,\tau^2\,,$$

$$s_{12} = \tfrac{1}{3}\,\sigma\tau\,, \quad s_{23} = s_{31} = 0\,.$$

Eqs. (66.1) assume the form

$$\begin{aligned}
\epsilon_{11} &= \tfrac{3}{2}\,v\left[\left(\frac{g}{3} - \zeta g'\right)\frac{2\sigma}{\sigma_0} + \frac{g'}{\sigma_0^2}(2\sigma^2 + 3\tau^2)\right],\\
\epsilon_{22} &= \tfrac{3}{2}\,v\left[-\left(\frac{g}{3} - \zeta g'\right)\frac{\sigma}{\sigma_0} + \frac{g'}{\sigma_0^2}(-\sigma^2 + 3\tau^2)\right],\\
\epsilon_{33} &= \tfrac{3}{2}\,v\left[-\left(\frac{g}{3} - \zeta g'\right)\frac{\sigma}{\sigma_0} + \frac{g'}{\sigma_0^2}(-\sigma^2 - 6\tau^2)\right],\\
\epsilon_{12} &= \tfrac{3}{2}\,v\left[\left(\frac{g}{3} - \zeta g'\right)\frac{3\tau}{\sigma_0} + 3\,\frac{g'}{\sigma_0^2}\,\tau\sigma\right].
\end{aligned} \tag{67.2}$$

Now put

$$\sigma = \sigma_0 \cos\alpha\,, \qquad \tau = \frac{1}{\sqrt{3}}\sigma_0 \sin\alpha\,;$$

then (67.1) becomes an identity. The experimental quantities that are measured directly are ϵ_{11} and ϵ_{12}, therefore, we convert only the first and last of eqs. (67.2). We obtain

$$\epsilon_{11} = \tfrac{3}{2}\,v[\tfrac{2}{3}\,g\cos\alpha + g'(-2\zeta\cos\alpha + 1 + \cos^2\alpha)]\,,$$

$$\epsilon_{12} = \frac{3\sqrt{3}}{2}\,v\,[\tfrac{1}{3}\,g\sin\alpha + g'(-\zeta + \cos\alpha)\sin\alpha]\,.$$

Comparing expressions for σ in terms of σ_0 and α with formula (66.10) we find

$$\cos\theta = -\tfrac{1}{2}\cos\alpha\,. \tag{67.3}$$

The circumstance that the numbering of the principal axes is different in formula (66.10) is unimportant, alteration of the numbering of the axes is equivalent to shifting the origin of reading of the angle θ through 120°, as was explained at the end of the previous section. We shall now express ζ in terms of α, namely

$$\zeta = \cos 3\theta = \tfrac{3}{2} \cos \alpha (1 - \tfrac{1}{3} \cos^2 \alpha) . \tag{67.4}$$

Eliminating ζ from the formulae for ϵ_{11} and ϵ_{12} we obtain the following result:

$$\begin{aligned} \epsilon_{11} &= v[g \cos \alpha + \tfrac{3}{2} g' \sin^4 \alpha] , \\ \epsilon_{12} &= \frac{\sqrt{3}}{2} v [g \sin \alpha - \tfrac{3}{2} g' \cos \alpha \sin^3 \alpha] . \end{aligned} \tag{67.5}$$

In these formulae g is considered as a function of ζ. If we consider g as function of α we have

$$g' = \frac{dg}{d\alpha} \frac{d\alpha}{d\theta} \frac{d\theta}{d\zeta} ,$$

but

$$\frac{d\alpha}{d\theta} = -2 \frac{\sin \theta}{\sin \alpha} , \qquad \frac{d\theta}{d\zeta} = - \frac{1}{3 \sin 3\theta} .$$

Consequently

$$g' = \tfrac{2}{3} \frac{1}{\sin \alpha (3-4 \sin^2 \theta)} \frac{dg}{d\alpha} = -\tfrac{2}{3} \frac{1}{\sin^3 \alpha} \frac{dg}{d\alpha} .$$

Eq. (67.5) can now be given the following simple form:

$$\begin{aligned} \dot{e} = \epsilon_{11} &= v \left[g \cos \alpha - \frac{dg}{d\alpha} \sin \alpha \right] , \\ \tfrac{1}{2} \dot{\gamma} = \epsilon_{12} &= \frac{\sqrt{3}}{2} v \left[g \sin \alpha + \frac{dg}{d\alpha} \cos \alpha \right] . \end{aligned} \tag{67.6}$$

Let us form the following combinations:

$$\epsilon_s = \dot{e}\cos\alpha + \frac{1}{\sqrt{3}}\dot{\gamma}\sin\alpha = v(\sigma_0 g)g\,,$$
$$\epsilon_t = -\dot{e}\sin\alpha + \frac{1}{\sqrt{3}}\dot{\gamma}\cos\alpha = v(\sigma_0 g)\frac{dg}{d\alpha}\,. \qquad (67.7)$$

Now divide the second of expressions (67.7) by the first, we obtain

$$\frac{1}{g}\frac{dg}{d\alpha} = \frac{\epsilon_t}{\epsilon_s} = v(\alpha)\,. \qquad (67.8)$$

Whence

$$g(\alpha) = \exp\int_0^{\alpha} v(\alpha)\,d\alpha\,. \qquad (67.9)$$

It will be seen from formulae (67.5) that if shear stresses are small compared with tensile, so that $\alpha \ll 1$, we have

$$\dot{e} = vg\,,$$

$$\dot{\gamma} = \sqrt{3}vg\alpha\,.$$

Note that the angle α varies in the range $\alpha = 0$ to $\alpha = \pi$, if the tests are made in both tension and compression. For these values the cosine of θ lies in the range $-\frac{1}{2}$ to $+\frac{1}{2}$, that is the angle θ varies through 60°, for example from 120° to 60° as shown in fig. 97.

68. **Quasilinear equations of steady-state creep**

As we have seen in Chapter 2, in the theory of plasticity it is not particularly important to allow for the influence of a third invariant, which is neglected in most theories. Experimental results which we shall consider later show that in the case of creep firm conclusions cannot always be drawn concerning the need to allow for a third invariant but its influence is clearly small for many materials under certain conditions. Considering that $\Phi = \Phi(\sigma_0)$, from formula (66.1) we obtain

$$\epsilon_{ij} = \tfrac{3}{2}\,v(\sigma_0)\frac{\bar{\sigma}_{ij}}{\sigma_0}\,,$$
$$v(\sigma_0) = \Phi'(\sigma_0)\,. \qquad (68.1)$$

Eqs. (68.1) are quasilinear in the sense that the relationships between components of creep rate and stress are linear; the non-linearity is entirely latent in the factor that depends on the invariant σ_0. In relating the rate of steady-state creep to the second invariant of the stress tensor deviator we, in essence, assume that creep is determined by tangential stresses; here σ_0 or $\overline{S}_{II}$ is a measure of tangential stress intensity. A still simpler hypothesis is that the creep process is pure shear caused by the greatest tangential stress. Let us put $\sigma_1 > \sigma_2 > \sigma_3$. Then the greatest tangential stress $\tau_{max} = \frac{1}{2}(\sigma_1 - \sigma_3)$. An element with faces parallel to the coordinate planes will be distorted in the plane 13 without change of volume; the rate of shear deformation $\dot{\gamma} = \epsilon_1 - \epsilon_3$ but $\epsilon_2 = 0$, consequently, $\epsilon_1 + \epsilon_3 = 0$, so that $\dot{\gamma} = 2\epsilon_1 = -2\epsilon_3$. It is postulated that $\dot{\gamma}$ is a function of τ_{max}, the form of this function is determined from a test in pure tension, consequently

$$\epsilon_1 = -\epsilon_3 = v(\sigma_1 - \sigma_3), \quad \epsilon_2 = 0. \tag{68.2}$$

The case when two principal stresses are equal to one another, for instance $\sigma_2 = \sigma_3$, requires special consideration. Let us superimpose on the stressed state a hydrostatic compression of intensity $-\sigma_2$, there then remains only one stress of $\sigma_1 - \sigma_2$ in the direction of the 1 axis and in this direction the creep rate is

$$\epsilon_1 = v(\sigma_1 - \sigma_2).$$

We can make no statement concerning creep rates ϵ_2 and ϵ_3 except that they satisfy the condition of incompressibility. Therefore, we take

$$\epsilon_2 = -\lambda\epsilon_1, \quad \epsilon_3 = -(1-\lambda)\epsilon_1. \tag{68.3}$$

The resultant indeterminacy is an unavoidable consequence of an oversimplified hypothesis. At first sight it might appear that considerations of symmetry oblige us to put $\lambda = \frac{1}{2}$ and this result follows from direct observations of creep in tension. However, this assumption unavoidably leads to contradictions in cases of non-uniform stress distribution. The condition $\sigma_2 = \sigma_3$ is too rigid and this is paid for by a somewhat arbitrary kinematic view of flow.

Eqs. (68.2) follow from the general equations if the potential is taken to depend on the maximum difference between the principal stresses. Indeed put

$$\Phi = \Phi(\sigma_1 - \sigma_3).$$

Then by formula (65.4)

$$\epsilon_1 = -\epsilon_3 = \Phi'(\sigma_1 - \sigma_3) = v(\sigma_1 - \sigma_3)\,, \qquad \epsilon_2 = 0\,.$$

These expressions retain their extreme simplicity only so long as we consider creep in the principal axes. Indeed in the general case we should express the quantity $\sigma_1 - \sigma_3$ in terms of invariants and use the general formulae (65.8). By means of eq. (8.9) we find

$$\sigma_1 - \sigma_3 = \bar{\sigma}_1 - \bar{\sigma}_3 = \tfrac{2}{3}\,\sigma_0 \left[\cos\theta - \cos\left(\theta - \frac{4\pi}{3}\right)\right]$$

$$= \frac{2}{\sqrt{3}}\,\sigma_0 \cos\left(\frac{\pi}{6} - \theta\right).$$

But $\sigma_1 - \sigma_3$ is the argument of the function Φ, consequently

$$g(\theta) = \frac{2}{\sqrt{3}} \cos\left(\frac{\pi}{6} - \theta\right). \tag{68.4}$$

Here the condition $g(0) = 1$ is fulfilled.

Now compare creep theories corresponding to eqs. (68.1) and (68.2). We shall call the first theory the Mises type and the second the Saint Venant type. For a Mises type theory the argument of the function Φ is σ_0 and for Saint Venant type $\sigma_0 g(\theta)$ where $g(\theta)$ is given by formula (68.4). Putting $\sigma_0 = 1$ we obtain a circle in the octahedral plane, shown in fig. 98, putting $\sigma_0 g(\theta) = 1$

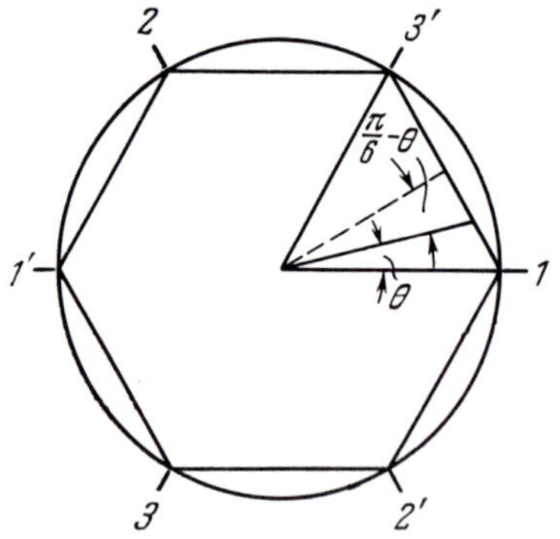

Fig. 98.

we obtain a chord with central angle 60°, passing through point 1. Indeed the height of the isosceles triangle that is shown is $\sqrt{3}/2$ and it follows from the condition $\sigma_0 g = 1$ that $\sigma_0 \cos(\pi/6 - \theta) = \sqrt{3}/2$. Obviously, formula (68.4) remains valid for $0 \leqslant \theta \leqslant \pi/3$ and in the range $\pi/3 \leqslant \theta \leqslant 2\pi/3$ we should take $g(\theta) = 2/\sqrt{3}\cos(\pi/2 - \theta)$ and so on. In principle, we could now write general creep equations for arbitrary axes using formulae (66.1). Solving the equation

$$\zeta = \cos 3\theta = 4\cos^3\theta - 3\cos\theta$$

relative to $\cos\theta$ we find θ as function of ζ, now eq. (68.4) defines g as function of ζ. The formulae that are obtained in this way are extremely complicated and hardly suitable for practical use.

On the other hand, combined tension and torsion is considered very simply. Putting $\sigma_{33} = 0$, as was done in section 67, we should take

$$g(\theta) = \frac{2}{\sqrt{3}}\sin\theta = \frac{1}{\sqrt{3}}\sqrt{3 + \sin^2\alpha}\,.$$

Then from formulae (67.6) we immediately obtain

$$\dot{e} = \epsilon_{11} = \sqrt{3}v\frac{\cos\alpha}{\sqrt{3 + \sin^2\alpha}}\,,$$

$$\tfrac{1}{2}\dot{\gamma} = \epsilon_{12} = 2v\frac{\sin\alpha}{\sqrt{3 + \sin^2\alpha}}\,.$$

The same result can be reached in an elementary way bearing in mind that in tension with torsion the principal axes are turned through an angle 2β relative to the tension axis, here

$$\tan 2\beta = \frac{2\tau}{\sigma} = \frac{2}{\sqrt{3}}\tan\alpha\,.$$

The principal rates of elongation are equal in magnitude and opposite in sign

$$\epsilon_1 = -\epsilon_2 = v(\sqrt{\sigma^2 + 4\tau^2})\,.$$

The principal axes of deformation coincide with the principal axes of the stress tensor, therefore

$$\epsilon_{11} = \epsilon_1 \cos 2\beta = \sqrt{3}v \frac{\cos\alpha}{\sqrt{3+\sin^2\alpha}},$$

$$\epsilon_{12} = \epsilon_1 \sin 2\beta = 2v \frac{\sin\alpha}{\sqrt{3+\sin^2\alpha}}.$$

It should be noted that the ratio of the creep rate to the shear rate is here

$$\frac{\dot{\gamma}}{\dot{e}} = \frac{2\epsilon_{12}}{\epsilon_{11}} = \frac{4\tau}{\sigma}.$$

According to Mises' type theory this ratio is

$$\frac{\dot{\gamma}}{\dot{e}} = \frac{3\tau}{\sigma}.$$

In essence the method of working out creep test results with combined tension and torsion that is described in section 67 can be used to determine the deviation of the behaviour of real material from the predictions of Mises' theory. If the material follows a creep law of Saint Venant type,

$$\nu(\alpha) = -\frac{\sin\alpha\cos\alpha}{4-\cos^2\alpha}, \tag{68.5}$$

$$g(\alpha) = \sqrt{\frac{4-\cos^2\alpha}{3}}. \tag{68.6}$$

Graphs of the functions $\nu(\alpha)$ and $g(\alpha)$ corresponding to formulae (68.5) and (68.6) are plotted in figs. 99 and 100. Experimental points plotted on the graph of fig. 99 should be grouped about the abscissus if the Mises' type flow law is valid, and grouped about the curve if the Saint Venant type law is valid.

The circle and hexagon shown in fig. 98 represent the section of the surface Φ = const on the octahedral plane. Unlike the case of plasticity we can select the constant arbitrarily. With the limitations that we have adopted the argument of the function Φ is $\sigma_0 g(\theta)$ and, therefore, we can select as standard equipotential surface that for which

$$\sigma_0 g(\theta) = 1. \tag{68.7}$$

The condition $g(0) = 1$ and the property of periodicity of the functions $g(\theta)$

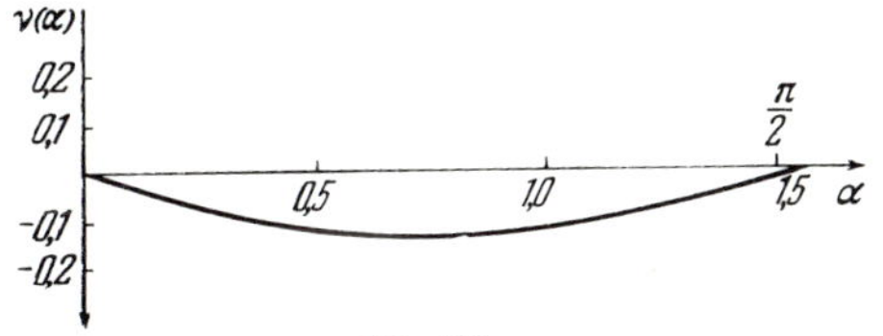

Fig. 99.

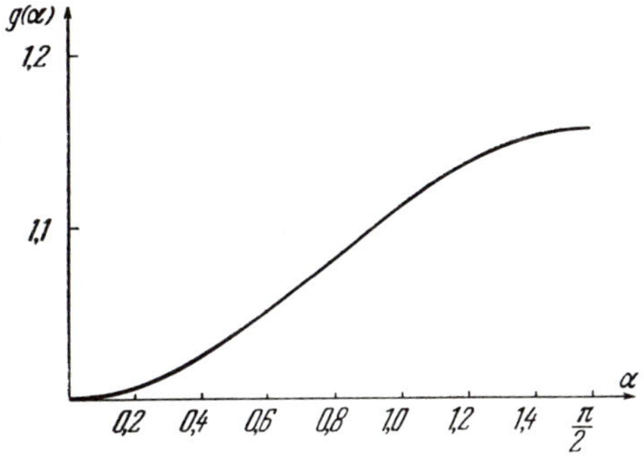

Fig. 100.

fix the points 1, 2 and 3; the assumption of identical creep laws in tension and in compression fixes the points 1′, 2′ and 3′; they all lie on a circle of unit radius. If the function Φ is convex all possible curves that represent condition (68.7) in the octahedral plane are convex curves passing through six fixed points. Obviously all these convex curves lie between the circumscribed and inscribed hexagons, as shown in fig. 101. It will be seen from the figure that the equation of the side of the circumscribed hexagon which passes through point 1 can be written

$$\sigma_0 \cos\theta = 1 .$$

Consequently, if $-\pi/6 \leqslant \theta \leqslant \pi/6$, then

$$g(\theta) = \cos\theta .$$

For the other sides of the hexagon we obtain

$$g(\theta) = \cos\left(\theta - \frac{k\pi}{3}\right) \qquad (k = 0, 1, ..., 5) . \tag{68.8}$$

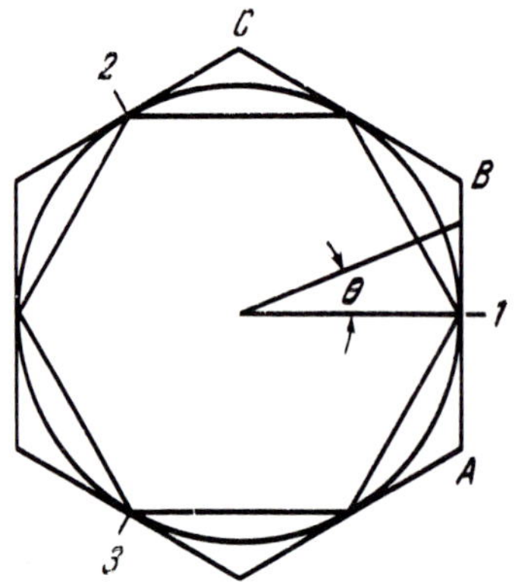

Fig. 101.

Now from formulae (66.3) we readily find

$$\epsilon_i = v \cos \frac{2i - 2 - k}{3} \pi . \tag{68.9}$$

If $k = 0$

$$\epsilon_1 = v , \quad \epsilon_2 = -\tfrac{1}{2} v , \quad \epsilon_3 = -\tfrac{1}{2} v . \tag{68.10}$$

This case is realised when $\sigma_1 > 0$ and is the principal stress of greatest absolute magnitude. If $\sigma_3 < 0$ and is greater in absolute magnitude than the other principal stresses, we are on the side BC of the polygon shown in fig. 101; then $k = 1$ and

$$\epsilon_1 = \tfrac{1}{2} v , \quad \epsilon_2 = \tfrac{1}{2} v , \quad \epsilon_3 = -v .$$

Special consideration is required when the two principal stresses are equal in absolute magnitude and we are at an angular point, for instance, at point B for which $\sigma_3 = -\sigma_1$. Just as with Saint Venant type flow law, we should admit indeterminacy in deformation rate distribution. If ϵ_i' is the rate determined from the condition that point B belongs to side AB and ϵ_i'' that from the condition that point B belongs to side BC, then

$$\epsilon_i = \lambda \epsilon_i' + (1 - \lambda) \epsilon_i'' \qquad (0 \leqslant \lambda \leqslant 1) ,$$

or

$$\epsilon_1 = v\,\frac{1+\lambda}{2}, \qquad \epsilon_2 = v\,\frac{1-2\lambda}{2}, \qquad \epsilon_3 = v\,\frac{-2+\lambda}{2}.$$

Note that equation

$$\sigma_0 \cos\left(\theta - \frac{k\pi}{3}\right) = \text{const}$$

is equivalent to

$$\max |\bar{\sigma}_i| = \text{const} .$$

In the theory of plasticity an analogous flow condition was termed the condition of maximum referred stress. Let the stress tensor deviator of greatest absolute magnitude be $\bar{\sigma}_1$, then

$$\Phi = \Phi\,[\sigma_1 - \tfrac{1}{2}(\sigma_2 + \sigma_3)] \ ,$$

and we immediately obtain expression (68.10). The criterion of maximum referred stress was used for creep in the work [86].

In order to compare results of predictions by the three elementary quasi-

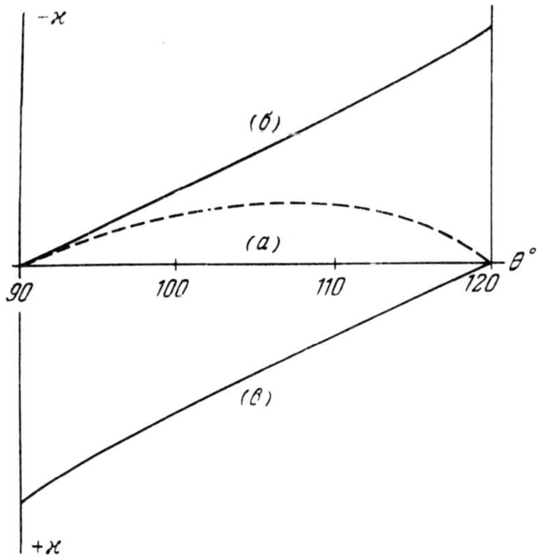

Fig. 102.

linear theories that we have considered, construct a graph of the function $\kappa(\theta)$ (fig. 102). Select a range of the angle θ from $\theta = \pi/2$ to $\theta = 2\pi/3$; referring to the diagram of fig. 97 we find that the point $\theta = \pi/2$ corresponds to tension and $\theta = 2\pi/3$ to torsion. For this region, by the criterion of maximum tangential stress

$$g(\theta) = \frac{2}{\sqrt{3}} \sin \theta \ ,$$

by the criterion of maximum referred stress

$$g(\theta) = \cos\left(\theta - \frac{2\pi}{3}\right).$$

We obtain: a) Mises' type criterion

$$\kappa = 0 \ .$$

b) the maximum tangential stress criterion

$$\kappa = \operatorname{ctan} \theta = - \tan\left(\theta - \frac{\pi}{2}\right).$$

c) the maximum referred stress criterion

$$\kappa = \tan\left(\frac{2\pi}{3} - \theta\right).$$

Determining the function $\kappa(\theta)$ experimentally, and plotting the experimental points on the graph we can immediately draw qualitative conclusions concerning the extent to which the behaviour of the material can be described by a theory of the type considered in section 66, and which of the simple schemes that we have described reproduces it best. If the experimental points are grouped around a certain smooth curve that lies entirely between curves (*b*) and (*c*), for instance as shown dotted, we shall say that the behaviour of the material is normal, otherwise the behaviour will be characterised as anomalous and associated either with anisotropy or with structural changes.

69. Anisotropic creep

Anisotropy probably affects creep tests more than any other kind of mechanical test. Specimens cut from rolled or forged material usually display negligible anisotropy of elastic properties. Anisotropy of the ordinary plasticity characteristics is more noticeable, the difference between the yield point for rolled steel specimens cut along and across the direction of rolling may be 10–15%. For a given stress and temperature the difference between the creep curves is sometimes enormous, elongation in a given time may be double or treble, depending on the orientation of the specimen relative to the grain. In some materials, for instance both pearlitic and austenitic steels, anisotropy can be completely removed by appropriate heat treatment. Anisotropy of aluminium alloys associated with the formation of texture in manufacture of the ingot or strip is apparently completely irremovable. Studies on the material of turbine discs manufactured by forging or stamping show that the standard heat-treatment which gives material of the best strength characteristics does not fully remove anisotropy and this should be borne in mind in accurate creep calculations. As already mentioned in section 63, anisotropy of creep characteristics is not necessarily associated with the initial state of the material, short-term plastic deformation that occurs during loading and previous creep already create anisotropy. Anisotropy also arises in the first stage of creep corresponding to strain-hardening. Therefore, consideration of anisotropic creep is a necessary part of the development of a general creep theory which allows for work-hardening. Generally speaking the state of material that is subject to plastic deformation in the neighbourhood of a certain point in the loading space or which is under certain creep conditions can be characterised by a number of tensors of different ranks. In this sense plastic anisotropy is considerably more complex than elastic anisotropy. Indeed, elastic anisotropy of the material is fully characterised by the fourth rank tensor of the elastic constants. In the most general case the number of constants is not more than 21. In speaking of elastic anisotropy we have in mind a linear relationship between forces and displacements and assume that the process is reversible; in this sense the properties of elements of the material and those of structures that consist of various elastic links is completely identical; we have no means of distinguishing between physical anisotropy of the material and so-called structural anisotropy. This latter is always described by means of 21 constants, the number of constants can be reduced from considerations of elastic symmetry in just the same way as in the other case. The position is much more complicated in plasticity theory, an example of this is the work [87], which considers conditions of limiting equilibrium of clamped plates and shells. The possibility that a structure may be transferred to the plastic state because of flow of only some of its components should be borne in mind in formulating the ultimate conditions of the system as a whole, and this makes it necessary to construct the surface of flow in parts, which may be fairly numerous. Considering the material as a totality of certain structural elements we should not, in principle, expect the plastic anisotropy to be described by simple expressions. Here we shall not consider the problem in all its generality, but only the theorems of anisotropic creep which lead to quasilinear relationships between rates and stresses. In this case also we assume the existence of creep potential and we should consider it as a function of combined invariants of the stress tensor and certain anisotropy tensors. On the other hand, if the stress tensor is referred to the principal axes, the creep potential should depend only on the three principal directions, and, therefore, the number of governing invariants is three. Selection of these invariants and anisotropy tensors respectively is somewhat arbitrary. Here any kind of general investigation becomes difficult, it is particularly difficult to apply complex theory even in working out experimental results. Similar difficulties arise in the theory of non-linear elasticity (see for example [219]). Therefore, the book to which we have referred uses another method of constructing the theory. It is assumed that the elastic potential is a polynomial function of the deformation components, and then certain types of elastic symmetry are considered. By requiring that the potential shall be invariant relative to particular groups of rotation it can be shown that potentials should depend on certain combinations of the deformation components.

Simpler assumptions, which are easily transferred to the case of creep, are generally used in theories of the plasticity of anisotropic media. From the very start we shall consider that the material is incompressible and assume, by analogy with the case of elasticity, that the creep potential depends only on the quadratic invariant that is formed from components of the stress tensor and components of the same constant, fourth rank, anisotropy tensor. Let us write this invariant

$$S = A_{ijkl}\sigma_{ij}\sigma_{kl}\,. \tag{69.1}$$

The anisotropy tensor is symmetrical relative to pairs of subscripts ij, kl and also relative to subscripts i and j, k and l so that

$$A_{ijkl} = A_{jikl} = A_{ijlk} = A_{klij}\,.$$

If the creep potential Φ depends on S, then

$$\epsilon_{ij} = \Phi'(S)\frac{\partial S}{\partial \sigma_{ij}} = 2\Phi' A_{ijkl}\sigma_{kl}\,.$$

Now use the condition of incompressibility $\epsilon_{ij}\delta_{ij} = 0$, we obtain

$$A_{ijkl}\delta_{ij} = 0\,. \tag{69.2}$$

Bearing in mind the symmetry of the anisotropy coefficients their number is 21, the six eqs. (69.2) reduce the number of independent coefficients to 15 and since it suffices to determine the function to within an arbitrary factor for complete characterisation of the anisotropic properties, on the assumption that the flow potential depends only on the quadratic invariant, it suffices to give 14 constants.

Note that because of expression (69.2)

$$S = A_{ijkl}\sigma_{ij}\sigma_{kl} = A_{ijkl}\bar{\sigma}_{ij}\bar{\sigma}_{kl}\,.$$

In an analogous way the creep law may be written

$$\epsilon_{ij} = \Phi'(S)2A_{ijkl}\bar{\sigma}_{kl}\,. \tag{69.3}$$

In real materials the presence of structure developed by working predetermines some particular form of symmetry of creep properties. The highest symmetry is found in bar or cylindrical specimens work-hardened by tension or compression along the axis. Let the specimen axis be axis number 3, it would then be expected that when axes 1 and 2 are rotated through an arbitrary angle β, the form of the invariant S remains unaltered and the coefficients of anisotropy remain as before. With this conversion the coordinate system of invariants becomes

$$\sigma_{33}\,,\quad \sigma_{31}^2 + \sigma_{32}^2\,,\quad \sigma_{11} + \sigma_{22}\,,\quad \sigma_{11}\sigma_{22} - \sigma_{12}^2\,.$$

The two latter expressions represent invariants of the plane tensor $(\sigma_{11}, \sigma_{22}, \sigma_{12})$; invariancy of the second expression is easily shown by noting that it is the square of the total tangential stress on an area perpendicular to the 3 axis. The quadratic function constructed from these invariants is necessarily of the form

$$S = a\sigma_{33}^2 + b\sigma_{33}(\sigma_{22} + \sigma_{11}) + c(\sigma_{31}^2 + \sigma_{32}^2)$$
$$+ d(\sigma_{22}^2 + \sigma_{11}^2 + 2\sigma_{11}\sigma_{22}) + e(\sigma_{11}\sigma_{22} - \sigma_{12}^2)\,.$$

We now write the developed expressions of the function S using formula (69.1) and compare coefficients. We obtain

$$A_{3333}=a\,,\quad A_{2222}=A_{1111}=d\,,\quad 2A_{1122}=2d+e=2f\,,$$

$$2A_{2233}=2A_{3311}=b\,,\quad 4A_{1212}=-e\,,\quad 4A_{2323}=4A_{3131}=c\,,$$

$$A_{1223}=A_{2331}=A_{3112}=A_{1112}=A_{1123}=A_{1131}=A_{2212}$$

$$=A_{2223}=A_{2231}=A_{3321}=A_{3323}=A_{3331}=0\,.$$

Using conditions (69.2) we find that the anisotropy coefficients should satisfy two further equations

$$d+f+\tfrac{1}{2}b=0\,,\qquad b+a=0\,.$$

Thus the number of independent anisotropy coefficients is three, and one of them may be fixed arbitrarily. For example let us require that in the case of tension in the direction of the 3 axis the quantity S coincides with $\bar{S}_{\mathrm{II}}$ determined for isotropic material. For this it is necessary that

$$A_{3333}=a=\tfrac{2}{3}\,.$$

Now put

$$d=\tfrac{1}{3}(2-\alpha)\,,\qquad c=\tfrac{4}{3}(\tfrac{3}{2}-\beta)\,,$$

then

$$b=-\tfrac{2}{3}\,,\qquad f=\tfrac{1}{3}(-1+\alpha)\,,\qquad e=\tfrac{4}{3}(-\tfrac{3}{2}+\alpha)\,.$$

The matrix of the anisotropic coefficients can now be written

$$|A_{ijkl}|=\frac{1}{3}\begin{vmatrix} 2-\alpha & -1+\alpha & -1 & 0 & 0 & 0\\ -1+\alpha & 2-\alpha & -1 & 0 & 0 & 0\\ -1 & -1 & 2 & 0 & 0 & 0\\ 0 & 0 & 0 & \tfrac{3}{2}-\beta & 0 & 0\\ 0 & 0 & 0 & 0 & \tfrac{3}{2}-\beta & 0\\ 0 & 0 & 0 & 0 & 0 & \tfrac{3}{2}-\alpha \end{vmatrix}\,. \tag{69.4}$$

In writing the matrix the following order of succession of pairs is taken: (11), (22), (33), (23), (31), (12). The case of isotropic material is obtained by putting $\alpha = \beta = 0$.

The requirement of positive determinacy of the quadratic form of S applies the following limitations to the quantities α and β:

$$\alpha < \tfrac{3}{2}, \quad \beta < \tfrac{3}{2}.$$

In plasticity theory there have been other attempts to construct systems of equations for the case of anisotropy which can also be transferred to creep. The theory that we have represented may be considered as a generalisation of Mises' type theory using quadratic stress tensor invariants. On the other hand, we could assume the properties of the material known in certain fixed directions and construct a piecewise-linear potential. The corresponding creep criterion is then a generalisation of the Tresca-Saint Venant criterion or of the maximum referred stress criterion. For ideal plasticity the corresponding equations are derived, for example, in the works [41, 346 and 244] and for creep in the work [137].

70. Determination of anisotropy parameters

The first possible type of test to determine the parameters of anisotropy and to check whether anisotropy actually exists and whether it has the assumed symmetry properties, is that the principal axes of the stress tensor are in the directions of the axes of symmetry. As before denoting the principal stresses by σ_1, σ_2 and σ_3 we write

$$S = A_{ijkl}\sigma_{ij}\sigma_{kl} = \tfrac{1}{3}\,[(2-\alpha)\sigma_1^2 + (2-\alpha)\sigma_2^2 + 2\sigma_3^2$$
$$+ 2(-1+\alpha)\sigma_1\sigma_2 - 2\sigma_2\sigma_3 - 2\sigma_3\sigma_1] = \bar{S}_{\mathrm{II}} - \frac{\alpha}{3}(\sigma_1 - \sigma_2)^2. \quad (70.1)$$

Obviously, in this formula the difference $\sigma_1 - \sigma_2$ may be replaced by the difference between the corresponding components of the deviators. Now representing the stress deviator in terms of the parameters σ_0 and θ, as in section 66, we obtain

$$S = \tfrac{2}{3}\sigma_0^2\left[1 - \tfrac{2}{3}\alpha\sin^2\left(\frac{\pi}{3} - \theta\right)\right].$$

Put

$$h(\theta) = \sqrt{1 - \frac{2}{\sqrt{3}} \alpha \sin^2 \left(\frac{\pi}{3} - \theta\right)}$$

and we shall consider the creep potential a function of the quantity $\sigma_0 h(\theta)$

$$\Phi = \Phi[\sigma_0 h(\theta)] \ .$$

After simple conversions we can write eqs. (69.3) in the form

$$\begin{aligned}
\epsilon_1 &= \tfrac{3}{2} \Phi'(\sigma_0 h) \frac{1}{\sigma_0 h} \left[\bar{\sigma}_1 - \frac{\alpha}{3}(\bar{\sigma}_1 - \bar{\sigma}_2)\right], \\
\epsilon_2 &= \tfrac{3}{2} \Phi'(\sigma_0 h) \frac{1}{\sigma_0 h} \left[\bar{\sigma}_2 + \frac{\alpha}{3}(\bar{\sigma}_1 - \bar{\sigma}_3)\right], \qquad (70.2) \\
\epsilon_3 &= \tfrac{3}{2} \Phi'(\sigma_0 h) \frac{1}{\sigma_0 h} \bar{\sigma}_3 ,
\end{aligned}$$

or turning to parameters σ_0 and θ,

$$\begin{aligned}
\epsilon_1 &= \frac{\Phi'}{h}\left[\cos\theta - \frac{\alpha}{\sqrt{3}} \sin\left(\frac{\pi}{3} - \theta\right)\right], \\
\epsilon_2 &= \frac{\Phi'}{h}\left[\cos\left(\theta - \frac{2\pi}{3}\right) + \frac{\alpha}{3} \sin\left(\frac{\pi}{3} - \theta\right)\right], \\
\epsilon_3 &= \frac{\Phi'}{h} \cos\left(\theta - \frac{4\pi}{3}\right).
\end{aligned}$$

Forming the same combinations of ϵ_1, ϵ_2 and ϵ_3 as in section 66 we obtain

$$\begin{aligned}
\epsilon_s &= \tfrac{2}{3}\left[\epsilon_1 \cos\theta + \epsilon_2 \cos\left(\theta - \frac{2\pi}{3}\right) + \epsilon_3 \cos\left(\theta - \frac{4\pi}{3}\right)\right] \\
&= \frac{1}{h} \Phi' \left[1 - \tfrac{2}{3} \alpha \sin^2\left(\frac{\pi}{3} - \theta\right)\right], \\
& \qquad\qquad\qquad\qquad\qquad\qquad (70.3) \\
\epsilon_t &= -\tfrac{2}{3}\left[\epsilon_1 \sin\theta + \epsilon_2 \sin\left(\theta - \frac{2\pi}{3}\right) + \epsilon_3 \sin\left(\theta - \frac{4\pi}{3}\right)\right] \\
&= \frac{1}{3h} \Phi' \alpha \sin\left(\frac{2\pi}{3} - 2\theta\right),
\end{aligned}$$

whence

$$\kappa(\theta) = \frac{\epsilon_t}{\epsilon_s} = \frac{\sin\left(\dfrac{2\pi}{3} - 2\theta\right)}{\dfrac{3-\alpha}{\alpha} + \cos\left(\dfrac{2\pi}{3} - 2\theta\right)} .$$

The quantity $\kappa(\theta)$ is determined experimentally, knowing it we can find α. Indeed it follows from the above expression that

$$\frac{3-\alpha}{\alpha} = \frac{\sin\left(\dfrac{2\pi}{3} - 2\theta\right)}{\kappa(\theta)} - \cos\left(\frac{2\pi}{3} - 2\theta\right) . \tag{70.4}$$

If the theory that we have presented is valid for the given material the right-hand side of formula (70.4) should be constant. Knowing α we can find $\Phi'(\sigma_0 h)$ by any of the formulae (70.3).

Deviations of test results from Mises' type theory can be attributed either to the influence of a third invariant or to anisotropy of the material. In the first case the experimental function $\kappa(\theta)$ has a period $2\pi/3$, and in the second π. If the tests are carried out over a narrow range of the angle θ it is not always easy to detect the difference and to draw a definite conclusion.

Now consider a test with combined tension and torsion. Put $\sigma_{33} = \sigma$, $\sigma_{23} = \tau$, all the remaining components of the stress tensor are zero. Also put

$$\epsilon_{33} = \dot{e} , \quad c_{23} = \tfrac{1}{2}\dot{\gamma} , \quad \epsilon_{11} = \epsilon_{22} = \dot{e}' .$$

Then it is easily shown that

$$S^2 = \tfrac{2}{3}\,[\sigma^2 + (3 - 2\beta)\tau^2] . \tag{70.5}$$

The creep rate is given by the following simple expressions:

$$\begin{aligned} \dot{e} &= 4\Phi'(S)\,\sigma , \\ \dot{\gamma} &= 4\Phi'(S)\,(3 - 2\beta)\tau , \\ \dot{e}' &= -\tfrac{1}{2}\dot{e} . \end{aligned} \tag{70.6}$$

Whence, in particular, it follows that

$$\frac{\dot{\gamma}}{\dot{e}} = (3 - 2\beta)\frac{\tau}{\sigma}. \qquad (70.7)$$

If for instance, it is found from tests that $\dot{\gamma}/\dot{e} = 4(\tau/\sigma)$ it may be concluded either that the material is isotropic and observes a Saint Venant type flow law or that it is anisotropic and $\beta = -\frac{1}{2}$. But if $\beta = -\frac{1}{2}$ the quantity S found by formula (70.5) becomes proportional to the square of the maximum tangential stress

$$S = \tfrac{2}{3}\,[\sigma^2 + 4\tau^2] = \tfrac{8}{3}\,\tau_{\max}^2 .$$

Formulae (70.6) here become indistinguishable from those of section 68 and tests in tension with torsion cannot be used to judge whether the material is really anisotropic or whether deformation is measured in a transverse direction. From the Saint Venant type theory we should have $\dot{\epsilon}_{22} = -\dot{e}$, $\epsilon_{11} = 0$, whereas for anisotropic material with quadratic creep potential $\epsilon_{22} = \epsilon_{33} = -\frac{1}{2}\,\dot{e}$. Wc have considered this case in order to show the difficulties that can arise in the interpretation of experimental results.

71. **Transient creep**

The transient creep theory, which is a generalisation to the special case of the simple assumption that the rate is constant at constant stress, is inadequate for solving many problems of practical interest. If elastic deformation is allowed for, the stresses in the body are redistributed with time from the initial state, which is determined by the methods of the theory of elasticity, to a limiting condition of steady-state creep, which cannot be fully realised in a finite time. The simplest theory with which steady-state creep can be considered is that, as before, the rate is stress-dependent and constant for constant stress, but elastic deformations are not negligible. The corresponding equations can, of course, be written for an anisotropic medium with allowance for the dependence of the creep potential on a third invariant, or finally for an anisotropic medium without any assumptions concerning the existence of potential. So as not to complicate matters, we shall commence from the quasilinear equations considered in section 68. Considering the material to be elastic-compressible we take $\sigma = 3Ke$ and find that

$$p_{ij} = \bar{e}_{ij} - \frac{1}{2G}\,\bar{\sigma}_{ij} .$$

Here p_{ij} is the creep deformation tensor. The creep rate is $\epsilon_{ij} = \dot{p}_{ij}$, therefore, it follows from (62.2) that

$$\dot{\bar{e}}_{ij} = \frac{1}{2G}\dot{\bar{\sigma}}_{ij} + \tfrac{3}{2}\frac{v(\sigma_0)}{\sigma_0}\bar{\sigma}_{ij}\,. \tag{71.1}$$

With allowance for bulk elasticity relationships eq. (71.1) can also be written in the form

$$\begin{aligned}
\dot{e}_{11} &= \frac{1}{E}[\dot{\sigma}_{11} - \nu(\dot{\sigma}_{22} + \dot{\sigma}_{33})] + \frac{v(\sigma_0)}{\sigma_0}[\sigma_{11} - \tfrac{1}{2}(\sigma_{22} + \sigma_{33})]\,,\\
&\cdots\cdots\cdots\cdots\cdots\cdots\cdots\cdots\cdots\cdots \qquad (71.2)\\
\dot{\gamma}_{12} &= \frac{1}{G}\dot{\sigma}_{12} + 3\frac{v(\sigma_0)}{\sigma_0}\sigma_{12}\,.
\end{aligned}$$

In a similar way we can derive equations for the Saint Venant type theory taking (68.2) as a basis, namely

$$\begin{aligned}
\dot{e}_1 &= \frac{1}{E}[\dot{\sigma}_1 - \nu(\dot{\sigma}_2 + \dot{\sigma}_3)] + v(\sigma_1 - \sigma_2)\,,\\
\dot{e}_2 &= \frac{1}{E}[\dot{\sigma}_2 - \nu(\dot{\sigma}_3 + \dot{\sigma}_1)]\,, \qquad (71.3)\\
\dot{e}_3 &= \frac{1}{E}[\dot{\sigma}_3 - \nu(\dot{\sigma}_1 + \dot{\sigma}_2)] - v(\sigma_1 - \sigma_3)\,.
\end{aligned}$$

The instantaneous deformation that occurs on application of load may be plastic as well as elastic and in this case, as before, by the tensor p_{ij} we must understand the tensor of creep deformation, consequently

$$p_{ij} = \bar{e}_{ij} - \frac{1}{2G}\bar{\sigma}_{ij} - e^{\mathrm{p}}_{ij}\,.$$

In order to combine the plasticity and creep equations we may start from different plasticity theories; the simplest assumption being that the initial plastic deformation is accompanied by isotropic strain-hardening, consequently

$$\mathrm{d}e^{\mathrm{p}}_{ij} = \tfrac{3}{2}\frac{g'(\sigma_0)}{\sigma_0}\bar{\sigma}_{ij}\mathrm{d}\sigma_0\,.$$

The equations that describe the initial plastic deformation and creep are

$$\dot{e}_{ij} = \frac{1}{2G}\dot{\bar{\sigma}}_{ij} + \tfrac{3}{2}\,[g'(\sigma_0)\dot{\sigma}_0 + v(\sigma_0)]\,\frac{\bar{\sigma}_{ij}}{\sigma_0} \quad \text{when } \dot{\sigma}_0 > 0\,,$$

$$\dot{e}_{ij} = \frac{1}{2G}\dot{\bar{\sigma}}_{ij} + \tfrac{3}{2}\,v(\sigma_0)\,\frac{\bar{\sigma}_{ij}}{\sigma_0} \quad \text{when } \dot{\sigma}_0 < 0\,. \tag{71.4}$$

In a similar way we could derive equations based on the creep law (68.3) but we shall not do so.

In practice, the first section of the creep curve is only absent at very high temperatures when creep is short-termed, calculations of the steady-state stage are usually based on the assumption that creep deformation is fairly large compared with elastic. In problems where elastic deformation must be allowed for, neglect of first phase creep usually introduces an error greater than the accuracy that results from allowing for elastic deformation. Therefore, eqs. (71.1)–(71.4) have a very limited field of application. This problem was considered in relation to the unidimensional case in section 47, where Odquist's [314] ideas were presented, namely that first phase creep deformation is replaced by a certain equivalent instantaneous deformation. Eq. (71.4) may be considered in just the same way when the function $g(\sigma_0)$ does not describe the initial plastic deformation but gives a correction to the first phase, namely as the function $g(\sigma_0)$ was introduced in section 47.

This point of view is useful when the second part of the creep curve is quite clearly expressed and we are interested in the behaviour of structures at times when the first part has certainly been passed through. The class of problem that can be solved by Odquist's scheme is unavoidably restricted to those in which the stressed state does not alter significantly.

In Chapter 4 we considered all possible unidimensional creep theories that establish relationships between the quantities p and σ, their derivatives and time; any of these theories can be extended to the spatial case, and in more than one way. The difficulty of constructing a system of equations for spatial creep is not so much a deficiency of analytical means as an excess. On the hypotheses of strain-hardening, ageing, heredity etc. we can write numerous systems of equations which lead to the same equations for the unidimensional case. Obviously the requirement that the formulation of the theory shall be invariant is difficult to fulfill in selecting between a large number of invariants which may be formed from say two tensors which are functions of time. The available experimental data is sparse, there are fundamental technical difficulties in formulating experiments and the properties of different specimens differ greatly and so we are often unable to give decisive preference to any particular creep theory. One saving grace is that in most cases the service conditions of a product that is subject to creep are relatively quiescent. As a rule external forces either remain unchanged or alter in such a way that the type of stressed state is unaltered. Elastic deformation in the first phase of creep causes some stress redistribution but it is quite smooth and usually does not involve rotation of the principal axes. By analogy with the theory of elasticity we are speaking of loading paths for which allowance for strain-hardening anisotropy does not introduce an important correction compared with calculations by the theory of isotropic work-hardening. For this kind of problem apparently the most important question is exactly what principle forms the basis of the theory which relates the variability of rate with time to time itself, to the amount of accumulated deformation, and the energy dissipated. Different formulations of the equations that describe creep on the basis of this principle are not too different from one another and we are justified in basing our first selection on considerations of simplicity and convenience.

72. **Theory of deformation type creep**

The Nadai-Hencky deformation theory of plasticity is, at least in principle, the simplest of the plasticity theories that can allow for work-hardening and elastic deformation of material. With the object of constructing a convenient technical theory of creep, Belyaev [5] made a suggestion that formed a basis for numerous variants of the theory. This idea is that in creep the stress and deformation are related by finite relationships that are fully analogous to the relationships of the deformation theory of plasticity (section 12), namely

$$\bar{e}_{ij} = \frac{3}{2}\frac{e_0}{\sigma_0}\bar{\sigma}_{ij}, \qquad e = \frac{\sigma}{3K}. \tag{72.1}$$

From (72.1) it follows (see section 12) that

$$G_s = \frac{\tau_0}{\gamma_0} = \frac{1}{3}\frac{e_0}{\sigma_0}.$$

It is now assumed that the quantities τ_0 and γ_0 or e_0 and σ_0 are related to one another by the same expressions that relate deformation and stress in the uniaxial case. Let us consider certain possibilities that result from this.

1. *Ageing theory.* Just as in the unidimensional case ageing theory assumes

the existence of relationships between deformation, stress and time, in the spatial case we shall consider that $e_0 = e_0(\sigma_0, t)$. This dependence may be given as isochronous creep curves [98], that is curves of e_0 against σ_0 for particular times. Here e_0 is the intensity of total deformation and eqs. (72.1) include total deformations so that elastic and initial plastic deformations are both automatically considered. The equations are written

$$\bar{e}_{ij} = \tfrac{3}{2}\, e_0(\sigma_0, t)\, \frac{\bar{\sigma}_{ij}}{\sigma_0}\,, \qquad e = \frac{\sigma}{3K}\,. \tag{72.2}$$

These are the equations of the deformation theory of plasticity which include time as a parameter. If the initial data are obtained from tensile tests, curves of e_0 against σ_0 should be reconstructed from curves of e against σ with correction for compressibility. There is usually no point in this refinement, because of the approximate nature of the ageing theory we can take $K = \infty$, that is consider that the material is also incompressible in the elastic region. Calculation by the ageing theory thus reduces to a series of calculations by the deformation theory of plasticity, a corresponding isochronous curve is selected for each instant of time and the state of the system at that instant is independent of its preceding states. We have seen that with slowly altering stresses the ageing theory gives results that are not so bad as would be expected from its obvious imperfections. For example, relaxation predictions from this theory are not too far from the truth (section 48). Moreover, all the difficulties that are associated with separating initial elastic deformation from plastic disappear and, which is most important, the primary creep curves are used directly without any approximation. At present nothing better than the ageing theory can be recommended in calculating products with a long life made from structurally unstable materials. Therefore, the variant of the ageing theory that we have presented has found wide practical application in strength calculations on turbines (see for example [73]).

We shall mention one analytical way of writing the ageing theory equations which were used by Marin [221]. In the principal axes these equations are written

$$e_1 = \frac{1}{E}[\sigma_1 - v(\sigma_2 + \sigma_3)] + \sigma_0^{n-1}\{k_1 + k_2[1 - \exp(-qt)] + k_3 t\}$$

$$\times\, [\sigma_1 - \tfrac{1}{2}(\sigma_2 + \sigma_3)]\,, \tag{72.3}$$

. . . .

It is evident that this is a particular case of eq. (72.2). The constants k_1, k_2, k_3 and q define a time function which should provide a satisfactory approximation to actual creep curves. It will be readily seen that eqs. (72.3) correspond to the case when the isochronous creep curves are similar and the dependence of e_0 on σ_0 is a power function with constant exponent n.

2. *Variants of the ageing theory.* The relationship between stress and deformation containing time explicitly can also be selected in a more complicated way. Thus, taking expression (48.2) as basis we may put

$$\dot{e}_0 - \frac{1}{3G}\dot{\sigma}_0 = f(\sigma_0, t) .$$

If the function on the right-hand side can be represented as a product of functions of stress $S(\sigma_0)$ and time $\dot{\tau}(t)$ by integration we obtain

$$e_0 = \frac{\sigma_0}{3G} + \int_0^{\tau} S(\sigma_0)\,\mathrm{d}\tau . \tag{72.4}$$

It would appear that nobody has examined such a combination of the deformation theory of plasticity and the second theory of ageing, it is doubtful whether it offers advantages over the first theory and for subsequent transfer of this point of view to creep, as to flow, the tensor relationships must also be formulated for rates.

3. *Belyaev's theory.* The unidimensional form of the creep theory proposed by Belyaev [5] and developed by Malinin [108] was presented in section 48. If eq. (48.5) is rewritten for intensities of stress and deformation we obtain

$$e_0 = \sigma_0 \left[\frac{1}{3G} + \int_0^{\tau} \frac{S(\sigma_0)}{\sigma_0}\,\mathrm{d}\tau\right]. \tag{72.5}$$

4. *Strain-hardening theory.* Various invariants may be selected as a measure of strain hardening which reduce to the same quantity, accumulated creep deformation, in the unidimensional case. If we follow the idea of not introducing other independent invariants except e_0 and σ_0 we should define the measure of strain-hardening p_0 as

$$p_0 = e_0 - \frac{\sigma_0}{3G} . \tag{72.6}$$

We must now assume that the quantities p_0 and σ_0 are related by the equation of state

$$\Phi(p_0, \dot{p}_0, \sigma_0, T) = 0 . \tag{72.7}$$

This variant of the strain-hardening theory in the three-dimensional creep problem was developed by Churikov [145]. It is readily shown that the quantity p_0 is related to the second invariant of the creep deformation tensor as follows:

$$p_0^2 = \tfrac{2}{3} P_{\mathrm{II}} = \tfrac{2}{3} p_{ij} p_{ij} .$$

5. *Hereditary theory.* The equations of the hereditary creep theory were introduced in a similar way in the article [99]. The relationship between invariants is

$$\varphi(e_0) = \frac{\sigma_0}{3G} + \int_{-\infty}^{t} K(t-\tau)\sigma_0(\tau)\mathrm{d}\tau . \tag{72.8}$$

Deformation type creep theories, if considered as certain physical relationships between stresses and deformation, are hardly adequate to describe actual creep processes in metals, but are rather of historical interest. However, the time-hardening theory as formulated in eq. (72.2) has not lost all its significance and because of the considerable experience that has been accumulated in research institutions and factories, it can still be recommended for use in engineering calculations.

73. Flow theory

Flow type theory establishes a direct relationship between the creep deformation rate tensor and the stress tensor. Taking a broad view of creep the existence of a potential can also be postulated for transient creep so that

$$\dot{p}_{ij} = \frac{\partial \Phi}{\partial \sigma_{ij}}$$

or

$$p_{ij} = H \frac{\partial \Phi}{\partial \sigma_{ij}} . \tag{73.1}$$

Here the functions Φ and H depend on a certain number of parameters that characterise the state of the material. Just as in steady-state creep, for an isotropic material the function Φ can depend on the second and third invariants of the stress tensor, but in their turn the parameters of state may depend on the stress. We shall return to this question in considering variants of the strain-hardening hypothesis and here we shall consider the simplest case. As parameter of state we may simply select time and we then obtain a variant of the flow theory. The corresponding equations will be termed the equations of the flow theory with ageing. If $\Phi(\sigma_{ij})$ can be represented as a product of functions of stress intensity and of time $\dot{\tau}(t)$ we have

$$\dot{p}_{ij} = \frac{3}{2}\frac{v(\sigma_0)}{\sigma_0}\bar{\sigma}_{ij}\dot{\tau}(t)\,. \tag{73.2}$$

Eqs. (73.2) are widely used in Kachanov's book [52], they are convenient because they coincide fully with the equations of the steady-state creep theory except that the rate is taken with respect to varied time $\tau(t)$. We have already shown in section 48 that non-invariance with respect to change in the location of the time origin makes the ageing theory equations physically unacceptable, but they can be used for slowly changing states of stress because the initial parts of the creep curves can be described adequately by appropriate selection of the function $\tau(t)$. In using eqs. (73.2) plastic deformations must be allowed for and, therefore, we can describe the process of stress redistribution.

Now proceed to equations that generalise the strain-hardening hypothesis. We shall first suppose that the only parameter of hardening is a certain scalar invariant p, which, in the case of uniaxial tension becomes the accumulated plastic deformation. Then $\Phi = \Phi(p, \sigma_{ij})$. Here and subsequently we assume that the stress dependence reduces to dependence on σ_0, that is the creep equations are quasilinear. We may make the more general assumption that the potential depends on a certain uniform function of zero power $s(\sigma_{ij})$, then within the framework of the same considerations we can consider a flow law of Saint Venant type and flow laws for anisotropic materials. We shall not consider these generalisations here. We introduce as invariant the intensity of creep deformation already used in section 72

$$p_0^2 = \tfrac{2}{3}p_{ij}p_{ij}\,.$$

However, this is not the only possibility. The intensity of rate of creep deformation defined in a similar way

$$\epsilon_0^2 = \tfrac{2}{3}\dot{p}_{ij}\dot{p}_{ij}\,,$$

is not the derivative of p_0 with respect to t, but an invariant quantity

$$q_0 = \int_0^t (\tfrac{2}{3}p_{ij}p_{ij})^{\frac{1}{2}}\,\mathrm{d}t \tag{73.3}$$

can be used in place of p_0 as a measure of strain-hardening. If we consider a nine-dimensional space of creep deformation, then p_0 is the length of the radius vector that represents the point whilst q_0 is the arc-length of its trajectory. For radial paths when all creep deformations alter in the same ratio, $p_0 = q_0$. In particular, for the unidimensional case we always have $p_0 = q_0$.

We have no serious reason to prefer one measure of strain-hardening to another, the use of q_0 as a measure of hardening is somewhat more convenient. Indeed if $\Phi = \Phi(\sigma_0, q_0)$

$$\dot{p}_{ij} = \frac{3}{2}\frac{\partial\Phi}{\partial\sigma_0}\frac{\bar{\sigma}_{ij}}{\sigma_0}\,.$$

Squaring and summing, allowing for (73.3), we obtain

$$\dot{q} = \frac{\partial\Phi}{\partial\sigma_0}\,.$$

In unidimensional creep $q_0 = p$, $\sigma_0 = \sigma$ and so the function Φ can be determined. If the law of unidimensional creep can be taken in the form of eq. (50.6)

$$\dot{q}_0 = \frac{f(\sigma_0)}{h(q_0)}$$

and

$$\dot{p}_{ij} = \frac{3}{2}\frac{f(\sigma_0)}{h(q_0)}\frac{\bar{\sigma}_{ij}}{\sigma_0}\,. \tag{73.4}$$

Eq. (73.4) gives the same results in the unidimensional case if p_0 is substituted for q_0; in the spatial case, in general, these are different equations.

Assuming that the degree of strain-hardening is defined by the creep deformation tensor, the creep potential should be a function of the stress tensor invariants, the creep deformation tensor and combined invariants. The total system of invariants is, of course, as follows (see for example [19])

$$\sigma_{ii}, \quad \sigma_{ij}\sigma_{ij}, \quad \sigma_{ij}\sigma_{jk}\sigma_{ki}, \quad p_{ii}, \quad p_{ij}p_{ij},$$

$$p_{ij}p_{jk}p_{ki}, \quad p_{ij}\sigma_{ij}, \quad p_{ij}p_{jk}\sigma_{ki}, \quad p_{ij}\sigma_{jk}\sigma_{ki}.$$

Introducing the creep deformation rate tensor into consideration, integral invariants must be added to these invariants; one example of such an invariant is the quantity q_0 defined above. A more general definition of this kind of invariant is

$$I = \int_0^t [\sigma_{ij}^{(m)} p_{jk}^{(n)} \dot{p}_{ki}^{(s)}]^{\frac{1}{s}} \, dt .$$

Here $\sigma_{ij}^{(m)}$, $p_{jk}^{(n)}$, $\dot{p}_{ki}^{(s)}$ are corresponding powers of the tensors. The possibilities become still wider when certain invariant integral operators act on the tensor components. For instance, we can construct tensors of form

$$e'_{ij} = K^* e_{ij}, \quad \sigma'_{ij} = \Gamma^* \sigma_{ij},$$

where K^* and Γ^* are integral Volterra operators and invariants are formed from these tensors to give certain equations of the type of the hereditary theory.

We shall not analyse the extremely extensive possibilities of constructing creep theories which are opened up by following this path, our problem is rather to find simple means of describing the effects. The tensor-linear relationships are very complicated even if we assume that the deformation rate tensor is a function of instantaneous values of the stress tensor and the creep deformation tensor, so that no allowance is made for the previous history of deformation. Examples of expressions constructed in this way may be found in the work [354].

74. Generalisation of strain-hardening theory

As in the unidimensional case (section 55) we shall assume that the instantaneous state of the material is defined by giving a certain number of parameters of state q_i such that $\Phi = \Phi(q_1, q_2, \ldots, q_n, \sigma_{ij}, T)$. For simplicity of analysis we shall assume that the dependence on stress reduces to dependence on a certain uniform function of zero power $s(\sigma_{ij})$, which may be the stress intensity, or the maximum tangential stress, or any other function of stress, for instance, piecewise-linear. The parameters of state q_i will be defined by the following linear differential equations [339]

$$dq_i = a^i_{rs} dp_{rs} + b^i_{rs} d\sigma_{rs} + c^i dt + d^i dT\,. \tag{74.1}$$

The coefficients a^i_{rs}, b^i_{rs}, c^i, d^i may depend on the stress and creep deformation tensors and on time, temperature and the parameters q_i. The fairly wide possibility of selecting a system of governing parameters given by expression (74.1) makes this point of view fairly flexible. Here the flow equations can be written

$$\dot{p}_{ij} = \frac{\partial\Phi}{\partial s}\frac{\partial s}{\partial\sigma_{ij}} + \frac{\partial\Phi}{\partial q_r} b^r_{ij}\,. \tag{74.2}$$

The additional terms that depend on b^r_{ij} can to some extent describe deformation, anisotropy and recovery.

Known forms of the flow theory equations and certain other variants follow from expressions (74.1) and (74.2). In particular one of the parameters q_i may be simply time and then we obtain flow theory with ageing. Let us consider certain other possibilities:

1. $a_{rs} = p_{rs}$, $b_{rs} = c = d = 0$. In this case

$$dq = p_{ij} dp_{ij}\,, \qquad q = \tfrac{1}{2} p_{ij} p_{ij}\,.$$

Putting $p_0^2 = 3q$ we introduce the usual measure of strain-hardening p_0, the intensity of creep deformation.

2. $a_{rs} = \sigma_{rs}$, $b_{rs} = c = d = 0$. Here

$$dq_1 = \sigma_{rs} dp_{rs}\,, \qquad q_1 = \int_0^t \sigma_{rs} dp_{rs}\,. \tag{74.3}$$

We have already seen in section 55 that selection of the energy q_1 dissipated in creep as a measure of strain-hardening has certain advantages over the usual variant of strain-hardening theory and hereafter we shall use the parameter q_1 defined in this way.

3.
$$\begin{aligned} &a_{rs} = 0\,, \quad b_{rs} = p_{rs}\,, \quad c = d = 0\,, \\ &dq_2 = p_{rs} d\sigma_{rs}\,, \quad q_2 = \int p_{rs} d\sigma_{rs}\,. \end{aligned} \tag{74.4}$$

In the unidimensional case introduction of the parameter q_2 permits satisfactory description of effects that are observed on sudden change of load. If we continue to assume that the potential exists, eqs. (74.2) assume the form

$$\dot{p}_{ij} = \frac{\partial \Phi}{\partial s}\frac{\partial s}{\partial \sigma_{ij}} + \frac{\partial \Phi}{\partial q_2} p_{ij} \, .$$

These equations describe total recovery on removing the load, which is never observed in metals.

4. $$a_{rs} = \sigma_{rs} \, , \qquad b_{rs} = p_{rs} \, , \qquad c = d \, ,$$

$$\mathrm{d}q_3 = \sigma_{rs}\,\mathrm{d}p_{rs} + p_{rs}\,\mathrm{d}\sigma_{rs} \, , \qquad q_3 = p_{rs}\sigma_{rs} \, .$$

The quantity q_3 was proposed as a measure of strain-hardening in one of Bailey's works [363] but he did not explore the corresponding equations in sufficient detail and he did not make a comparison with experiment or other theories.

5. Selection of the quantity q_0 defined by eq. (73.3) as a measure of strain-hardening does not fit into the scheme under consideration. Subsequently, in considering particular cases of transient creep, we shall use either q_0 or in some cases q_1 as a measure of strain-hardening. Assuming that creep curves are similar and defining s as the stress intensity we obtain

$$\dot{p}_{ij} = \frac{3}{2}\frac{f(\sigma_0)}{h(q)}\frac{\bar{\sigma}_{ij}}{\sigma_0} \, . \tag{74.5}$$

Here q is either q_0 or q_1. In a precisely similar way we obtain an equation for the case when we select as equivalent stress either the maximum tangential stress or the difference between the principal stresses. We shall not describe these equations now but will consider them in more detail in subsequent chapters when they are used to solve particular problems.

75. **Quasi-steady-state creep**

In some cases equations of the flow type of creep theory can be integrated to give finite relationships between deformations, stresses and time. In steady-state creep theory it is evident that if the stress is constant the rate of creep deformation ϵ_{ij} is also constant, and consequently $p_{ij} = \epsilon_{ij} t$; the equations can be written in finite form as follows:

$$p_{ij} = \frac{3}{2}\frac{v(\sigma_0)}{\sigma_0}\bar{\sigma}_{ij}t\,.$$

A completely similar situation arises with the equations of the theory of flow with ageing. When $\bar{\sigma}_{ij} = \text{const}$ it immediately follows from (73.2) that

$$p_{ij} = \frac{3}{2}\frac{v(\sigma_0)}{\sigma_0}\bar{\sigma}_{ij}\tau(t)\,. \tag{75.1}$$

A general definition of this case, called quasi-steady-state creep is that the stresses remain constant and the deformations are proportional to one and the same function of time $\tau(t)$. Strictly speaking quasi-steady-state creep is achieved only if the body is in a state of uniform stress and consequently, the conditions of compatibility are identically fulfilled for elastic deformations. In the general case of a body of arbitrary shape loaded by external forces the state of elastic stress differs from the state of stress during creep. Accordingly, we can consider creep as quasi-steady-state only when elastic deformations are negligibly small, and in particular bulk deformation is also negligibly small.

Quasi-steady-state creep is also possible in the strain-hardening theory, provided elastic deformations are again neglected, but with a special selection of the functions $h(q)$. Take as a basis eqs. (73.4) putting in them $h(q) = q^{\alpha}$, now select an Odquist measure of hardening, that is, put

$$q = q_0 = \int_0^t (\tfrac{2}{3}\dot{p}_{ij}\dot{p}_{ij})^{\frac{1}{2}}\,\mathrm{d}t\,.$$

Suppose that constant external forces act on the body and there are no external kinematic bonds. Then the equations of equilibrium are satisfied on the assumption that the stresses are independent of time, $\sigma_{ij} = \text{const}$. We shall seek creep deformation in the form $p_{ij} = p'_{ij}\tau(t)$ where p'_{ij} is a tensor that is independent of time. Then $q = \tau p'_0$, where p'_0 is the intensity of the tensor p'_{ij}. Substituting into eq. (74.5) we obtain

$$\tau = \left(\frac{t}{m}\right)^m, \quad \left(m = \frac{1}{1+\alpha}\right), \quad p'_{ij} = \tfrac{3}{2}\,p_0'^{-\alpha}\frac{f(\sigma_0)}{\sigma_0}\bar{\sigma}_{ij}\,.$$

Using the second equation to obtain an invariant p'_0 we obtain $p_0'^2 = p_0'^{-2\alpha}f^2(\sigma_0)$. Hence we find p'_0 as function of $f(\sigma_0)$, and we finally obtain

$$p'_{ij} = \frac{3}{2}\frac{f^m(\sigma_0)}{\sigma_0}\bar{\sigma}_{ij} \,. \tag{75.2}$$

When the directions of the principal axes of the stress tensor are known, simpler expressions can be obtained on the basis of Tresca type creep criteria. For this purpose we take the quantity $\sigma_1 - \sigma_3$ as function of s and as the measure of work-hardening we simply take the maximum shear, that is we put $2q = p_1 - p_3$. Then from the flow equations we obtain

$$\dot{p}_1 = -\dot{p}_3 = \varphi(q, s)\,, \quad \dot{p}_2 = 0\,. \tag{75.3}$$

The function $\varphi(q, s)$ is determined just as in the case of uniaxial tension; giving the strain-hardening law a special form we obtain

$$\dot{p}_1 = -\dot{p}_3 = q^{-\alpha} f(\sigma_1 - \sigma_3)\,, \quad \dot{p}_2 = 0\,. \tag{75.4}$$

The form of the right-hand side of the first of eqs. (75.4) is determined for the conditions of a tensile test, then only one principal stress remains and the other two are zero. However, when two of the principal stresses are equal to one another, Tresca type flow conditions do not lead to any conclusion concerning the distribution of deformations in these directions. Suppose for example $\sigma_1 > \sigma_2 = \sigma_3$. Applying hydrostatic compression of intensity $-\sigma_3$ we obtain uniaxial tension in direction 1, consequently

$$\dot{p}_1 = p_1^{-\alpha} f(\sigma_1 - \sigma_{2,3})\,. \tag{75.5}$$

All that we can say about the rate of flow in directions 2 and 3 is that from the condition of incompressibility their sum $\dot{p}_2 + \dot{p}_3$ equals $-\dot{p}_1$. The application of strain-hardening theory to materials with a Tresca type flow law is essentially simple because the measure of strain-hardening is expressed in terms of finite deformations. If the stresses are constant, putting $p_i = p'_i \tau(t)$ and substituting into (75.4) we obtain

$$\tau(t) = \left(\frac{t}{m}\right)^m ,$$

$$p'_1 = -p'_3 = [f(\sigma_1 - \sigma_{2,3})]^m \,.$$

If forces of constant magnitude are given everywhere on the boundary of the body, the equations of equilibrium can always be satisfied on the assumption that the stresses remain constant. Constant elastic deformations correspond to these constant stresses, but creep deformations are proportional to a definite time function $\tau(t)$. Therefore, generally speaking, equations of compatibility of deformations are not fulfilled. Quasi-steady-state creep can be achieved when creep deformations are sufficiently great and elastic deformations are negligible. Therefore, for materials that observe the equation of state with isotropic strain-hardening, quasi-steady-state creep is achieved provided that:

1. External forces are constant, and no kinematic bonds are applied to the body.
2. Creep curves are similar and are described by a power function of time.
3. Elastic deformations are negligibly small.

Note that quasi-steady-state creep can be realised with other forms of work-hardening law, for instance, when $q = q_1$ (74.3). This is a self-evident conclusion that we shall not consider further [339]. On the other hand, the possibility of quasi-steady-state creep occurring with less restrictive assumptions also follows from the equations of the hereditary theory of deformation of the type that the author has proposed elsewhere [99]. Strictly speaking in constructing the equations the author was trying to find fairly simple and general classes of solution. The starting point was the similarity of isochronous creep curves, noted in section 44. For the spatial case (section 72) the equations of this theory are of the form

$$e_{ij} = \frac{3}{2}\frac{e_0}{\sigma_0}\bar{\sigma}_{ij}\,, \qquad e = \frac{\sigma}{3K}\,.$$

Here e_0 depends on σ_0 in the following way:

$$(1 + K^*)\sigma_0 = \varphi(e_0)\,.$$

Now suppose that the material is incompressible, $\sigma_{ij} = \sigma'_{ij}\mu(t)$, $e_{ij} = e'_{ij}\lambda(t)$, σ'_{ij} and e'_{ij} are tensors that are independent of time. We also require that $\varphi(e_0)$ is a power function with exponent β. Then the time functions $\mu(t)$ and $\lambda(t)$ are related by the integral equation

$$(1 + K^*)\mu = \lambda^\beta\,, \tag{75.6}$$

whereas for σ'_{ij} and e'_{ij} we obtain a system of equations of the deformation theory of plasticity with power function strain-hardening.

By comparison with the previous conclusion, which is based on the hypothesis of the equation of state, we should not apply limitations to the form of the dependence of creep deformation on time, the nucleus $K(t-\tau)$ remaining arbitrary. In essence, elastic deformation is also rejected because no part of the graph of the function $\varphi(e_0)$ can be rectilinear. If the external forces are constant, $\mu = \text{const}$ and $\lambda(t)$ is determined from eq. (75.6). In the previous examination limitations were applied to the shape of the creep curves as function of time whereas here a definite stress dependence is taken. Creep will now be quasi-steady-state not only under constant loads but also under loads that vary with time in the same ratio. This case can be considered by strain-hardening theory only when $f(\sigma)$ is a power function.

Calculations by quasi-steady-state creep theory give satisfactory results when the creep curves are similar and the elastic deformations are negligibly small.

76. Experimental verification of creep theory with complex stressed state. Early work

The number of experimental works that have been published on creep under complex stress is fairly large. Almost all the tests were made on thin-walled tubes subject to various combinations of tension, torsion and internal pressure. The results of these tests do not always give a clear picture of the effect. Curves of creep in tension usually display considerable scatter. This also applies, of course, to tests with complex stress and moreover many investigators did not have sufficient testing machines and so did not collect enough data to make reliable averaging possible. In some of the works there was no evidence that the material was initially isotropic. In working out the results certain authors consider only the dependence of steady-state creep rate on stress and others compare entire creep curves including the first parts. Apparently, in all the tests without exception, the initial hypothesis that was subject to verification was that the creep rate components and the stress components are interrelated by expressions of the Mises' type theory, whilst the rates depend on the second invariant of the stress tensor deviator $\bar{S}_{\mathrm{II}}$ or the stress intensity σ_0. Therefore, the expression taken as the basis was either (68.1) or an analogous expression which may be written as follows for quasi-steady-state creep:

$$p_{ij} = \tfrac{3}{2}\, v(\sigma_0)\, \frac{\bar{\sigma}_{ij}}{\sigma_0}\, \tau(t)\,. \qquad (76.1)$$

It is, of course, only possible to work out results by eq. (76.1) when the creep curves are similar for various kinds of stressed state. In the first phase of creep the curves are indeed similar, but it is hardly possible to select a function $\tau(t)$ such that eq. (76.1) applies to both the first and second phases of creep. Therefore, strictly speaking, the results of working out short-term test data by eqs. (76.1) and steady-state creep studies on the basis of eqs. (68.1) are not strictly comparable.

Bailey [162] made the first experimental studies of creep with complex stress. The tests were made on lead pipes subject to internal pressure and tension or torsion and also on specimens of low-carbon steel at a temperature of 475°C and steel with an analysis of 0.45 C, 0.73 Mn, 0.20 Si, 0.19 Ni, 0.023 Cr, 0.032 S, 0.038 P at 480°C. The tests were continued up to the start of the straight-line part of the creep curve and the steady-state creep rate was measured.

A check on the suitability of any particular equation to describe test results should consist of two parts: on the one hand, the flow law should be checked, that is the relationship between the rate and stress ratios that follow from the theory, and on the other hand, the relationships between certain invariants of the rate and stress tensors or deformation and stress tensors should be checked. As we have seen in section 68, in tests with combined tension and torsion, we should have

$$\frac{\dot{\gamma}}{\dot{e}} = 3\,\frac{\tau}{\sigma}\,.$$

Tests on steel pipes generally confirmed this, but the deviations were sometimes fairly great. Agreement with theory was best at high stresses and temperatures. The results were generally worse for lead pipes than for steel. There was fairly large divergence between calculated and experimental creep rates. Bailey proposed fairly complicated equations to achieve the best agreement with the experimental results, namely

$$\begin{aligned}
\epsilon_1 &= A\sigma_0^{n-m}\,[(\sigma_1-\sigma_2)^m-(\sigma_3-\sigma_1)^m]\,,\\
\epsilon_2 &= A\sigma_0^{n-m}\,[(\sigma_2-\sigma_3)^m-(\sigma_1-\sigma_2)^m]\,,\\
\epsilon_3 &= A\sigma_0^{n-m}\,[(\sigma_3-\sigma_1)^m-(\sigma_2-\sigma_3)^m]\,.
\end{aligned} \tag{76.2}$$

When $m = 1$ these equations obviously coincide with eqs. (76.1). When $m = 3$ Bailey's theory also leads to eqs. (76.1). Indeed the validity of the identity

$$(\sigma_1 - \sigma_3)^3 - (\sigma_3 - \sigma_1)^3 = \tfrac{1}{2}(2\sigma_1 - \sigma_2 - \sigma_3)[(\sigma_1 - \sigma_2)^2 + (\sigma_2 - \sigma_3)^2 + (\sigma_3 - \sigma_1)^2]$$

is readily demonstrated. It is evident that a better description of the experimental data can always be achieved by introducing the redundant parameter m into the equation. However, when the deviations from eqs. (76.1) are great the much more complicated expressions (76.2) give little help. On the other hand, Bailey's theory is a partial form of the general expressions (64.7) and there is no special advantage in the choice of this particular form. Certain later authors attempted to apply expressions (76.2) in working out their experimental results but no further attempts are made now. When Bailey's equations are applicable the simplest theory of eqs. (76.1) gives a perfectly satisfactory approximation. We shall now mention the tests of Everett and Clark [202] in which creep in torsion and in tension were compared, but we shall not give a detailed review of their results. The test material was molybdenum steel. Torsion tests were made on thin-walled tubes and steady-state creep rates were compared with creep rates determined from tensile tests on cylindrical specimens which had been heat-treated in the same way. The degree of isotropy of the initial material was not defined and the influence of the method of making the tubular specimens is uncertain. Considerable divergences were found between the test results and predictions of Mises' type theory [69]. Attempts to work out Everett and Clark's test results were unsuccessful. The random scatter of torsion test results seemed to be very high, but it is not known why.

We shall also mention the tests of Norton [311] and Norton and Soderberg [312] which have frequently been discussed [52, 69]. They studied the behaviour of pipes subject to internal pressure when the axial stress was about half the hoop. Denoting the hoop stress by σ_1 we find from eq. (76.1) that

$$\epsilon_2 = 0\,, \qquad \epsilon_1 = \tfrac{1}{2}\, v\!\left(\frac{\sigma_1\sqrt{3}}{2}\right).$$

Note that this state of stress differs from that of pure shear only by a hydrostatic component. Indeed if $\sigma_3 = 0$ and $\sigma_2 = \frac{1}{2}\sigma$, we have

$$\bar{\sigma}_1 = \tfrac{1}{2}\sigma\,, \qquad \bar{\sigma}_2 = 0\,, \qquad \bar{\sigma}_3 = -\tfrac{1}{2}\sigma\,.$$

Norton's tests confirmed fairly well that there is no creep in the axial direction, that is indeed $\epsilon_2 = 0$. The results for the dependence of peripheral creep rate on stress confirmed the theory only approximately. Because of the considerable scatter of the experimental points their results cannot be used to improve the theory.

These experimental investigations and others made it clear that in studying physical creep relationships with complex stress the experiments must be formulated very carefully on material which is initially truly isotropic and they must cover a sufficiently wide range of stress and temperature. The structural changes that accompany creep in many alloys can greatly influence the observed relationships and, therefore, test results are required for a sufficiently wide class of materials. The behaviour of structurally stable materials, which are assuming ever greater importance in engineering, should be considered as a norm; when the behaviour of particular alloys deviates from this norm under particular external conditions the specific reason should be sought. Certain definite conclusions can be drawn from post-war investigations of complex stress carried out in numerous countries, and we shall now consider these investigations in more detail.

77. Creep with complex stress and constant load. Johnson's tests

Starting in 1932 systematic creep studies with complex stress were made under the leadership of Johnson, at first in the National Physical Laboratory in London and from 1952 in the National Engineering Laboratory. A brief review of the results was recently published separately [266]. The extensive experimental programme included creep tests in tension and in torsion of tubular specimens with constant and variable loads, relaxation tests with complex stress, long-term strength tests and also certain studies of instantaneous plastic deformation. The following materials were studied:

1. steel with a carbon content of 0.17% at temperatures of 350°C, 450°C, and 550°C;
2. steel with a carbon content of 0.2% at 450°C;
3. molybdenum steel (0.5% Mo) at 550°C;
4. aluminium alloy RR-59 at 150°C and 200°C;
5. magnesium alloy (2% Al) at 20°C and 50°C;
6. chrome-nickel alloy (Nimonic 75) at 550°C and 650°C;
7. technically pure copper at 250°C.

In the early work [252] on creep in carbon steel with 0.17% C and in molybdenum steel, the creep curves were found to be similar for different types of states of stress and the dependence of creep rate on stress could be represented approximately by a power function. However, because of anisotropy in the initial material, no clear conclusions could be drawn concerning creep relationships. Therefore, we consider in more detail the work published in 1949 on creep in steel with a carbon content of 0.17%, which was carried out very carefully taking all the necessary precautions [255], and also later work on other materials.

In order to check that the initial material was isotropic specimens were cut from different parts of an ingot and within the limits of experimental error all had the same analysis of 0.17% C; 0.19% Si; 0.026% S; 0.034% P and 0.64% Mn and so the material was chemically uniform. As a further check creep tests were made at a temperature of 450°C and stress of 945 kg/cm^2 on cylindrical specimens cut in various directions and no difference was found between the creep characteristics, neither did specimen orientation have any influence on subsequent tests on tubular specimens. The tests were made with combined tension and torsion of a tubular specimen with internal diameter of $\frac{1}{2}$ inch (12.7 mm) and a wall thickness of 0.020 inch (0.508 mm), the thickness tolerance of ± 0.0005 inch (about 12 microns) was maintained on the base of the specimen, which was 2 inches (50.9 mm). The test machine which was designed and constructed in the National Physical Laboratory was described in the work [252]. Altogether about 80 specimens were tested, each test lasted 150 hours. With the temperatures and stresses that were used the time of 150 hours almost always corresponded to the first stage of creep, reaching the second stage, and sometimes the start of the third, only at a temperature of 550°C and high stress. To economise on tubular specimens tensile tests were made on ordinary cylindrical specimens and no significant differences were observed in control tests on tubular specimens. At temperatures of 450°C and 550°C the surface of the specimen oxidised so that the effective cross-sectional area altered. Special tests were run to determine the influence of oxidation on change in area and an appropriate correction was made in working out the experimental results. The results were worked out on the basis of the observed fact that creep curves are similar at different stress levels with different components of the deformation tensor, for every possible state of stress. Creep curves in torsion depend less on temperature changes during the tests than do creep curves for other states of stress. Therefore, at the start the scales of the curves were altered so as to superpose all of the curves on one of them and afterwards all creep curves were reconstructed with change of scale so as to lie as close as possible to the torsion curve. The function $\tau(t)$ in eqs. (76.1) was determined in this way. Graphs of the derivative of this function for three temperatures are given in fig. 103. The function $\tau(t)$ was normalised so that $\dot{\tau}(150) = 1$ at 350° and 450° and $\dot{\tau}(100) = 1$ at

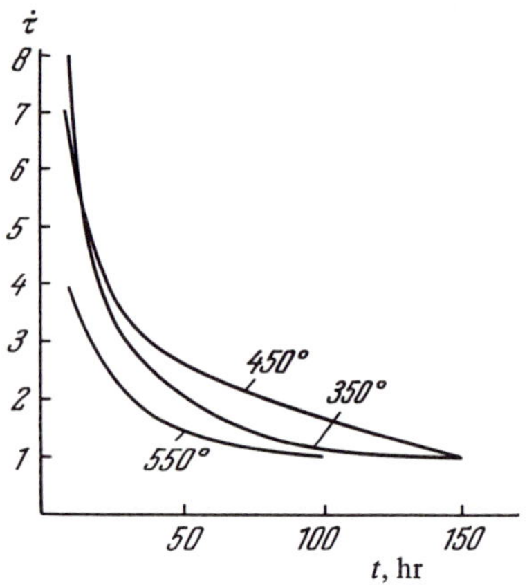

Fig. 103.

550°, all the experimental results were compared on the basis of the nominal creep rate at a fixed time, $t = 150$ hours for the two lower temperatures and $t = 100$ hours for 550°C, though in actual fact the first sections of the creep curves were compared as a whole; the values given in the article are not creep rates at the particular time, but scales that were used to refer the individual creep curve to the standard curve with the condition of minimum deviation throughout its length. At the higher temperature, 550°C, the nominal rates were compared for times of 100 hours instead of 150, because although similarity was satisfactory up to 100 hours during the next 50 hours creep was accompanied by second and even third stage effects and the curves were no longer similar. During the experiments the axial deformation and angle of twist were measured directly. The principal axes of the stress and deformation tensors were assumed to coincide, the material was also assumed incompressible and the author calculates the principal rates of deformation in the sense defined above. From the condition that the principal axes of the stress and deformation tensors coincide it follows that

$$\frac{2(\dot{e}_x - \dot{e}_y)}{\dot{\gamma}} = \frac{\sigma_x}{\tau}.$$

Here e_x is elongation in the axial direction, e_y elongation in the transverse direction, γ shear resulting from torsion, σ_x tensile stress, τ tangential stress. The quantities $\dot{e}_x$ and $\dot{\gamma}$ were measured directly and $\dot{e}_y$ was found from the equations above. Then the principal rates of deformation were calculated by the formulae

$$\epsilon_{2,3} = \frac{\dot{e}_x + \dot{e}_y}{2} \pm \tfrac{1}{2}\sqrt{(\dot{e}_x - \dot{e}_y)^2 + \dot{\gamma}^2}\,,$$

$$\epsilon_1 = -\epsilon_2 - \epsilon_3\,.$$

As the principal axes coincide these formulae assume the final form

$$\epsilon_2 = \dot{e}_x + \frac{\dot{\gamma}}{2}\left[\sqrt{1+\left(\frac{\sigma}{2\tau}\right)^2} - \frac{\sigma}{2\tau}\right],$$

$$\epsilon_3 = \dot{e}_x - \frac{\dot{\gamma}}{2}\left[\sqrt{1+\left(\frac{\sigma}{2\tau}\right)^2} + \frac{\sigma}{2\tau}\right], \qquad (77.1)$$

$$\epsilon_1 = -2\dot{e}_x + \dot{\gamma}\,\frac{\sigma}{2\tau}\,.$$

In combined tension and torsion one of the principal stresses is always positive and we shall denote it by σ_2, the second which is negative is denoted by σ_3, then the stress in the direction of the wall thickness, $\sigma_1 = 0$. To check the Mises' flow law, that is the relationship between the stress and rate ratios the author constructs a graph of $-\sigma_3/(\sigma_2-\sigma_3)$ against $-\epsilon_1/\epsilon_3$. Denoting the first ratio by x and the second by y, from eqs. (76.1) we find

$$y = \frac{1-2x}{2-x}\,. \qquad (77.2)$$

The quantity x becomes zero in tension and $\frac{1}{2}$ in torsion. Corresponding values of y are $\frac{1}{2}$ and 0. Fig. 104 shows a theoretical curve of the relationship between y and x and experimental points for a temperature of 350°C. It will be seen that there are systematic deviations from the theoretical predictions; the authors state that these deviations increase with stress intensity. The rate of octahedral shear $\dot{\gamma}_0$ was then calculated and a graph of $\dot{\gamma}_0$ against τ_0 was constructed as shown in fig. 105 on double logarithmic scale for a temperature of 350°C. The points lie fairly close around a single curve, a rather unexpected result is that points for pure torsion are all displaced to the left by about the same amount, at any rate for low stress levels, so that separate curves can be drawn through them as shown dotted. Points that correspond

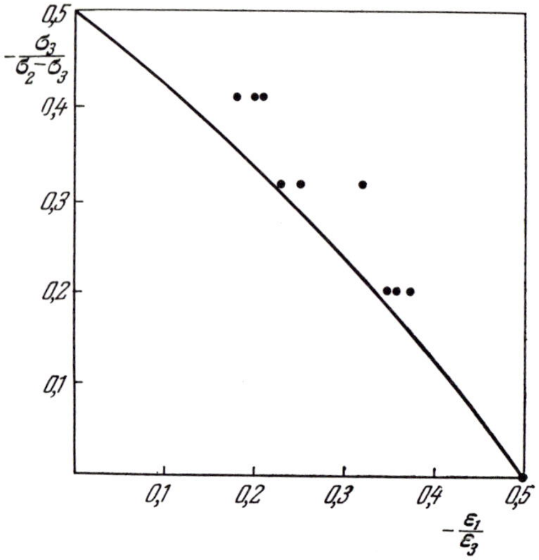

Fig. 104.

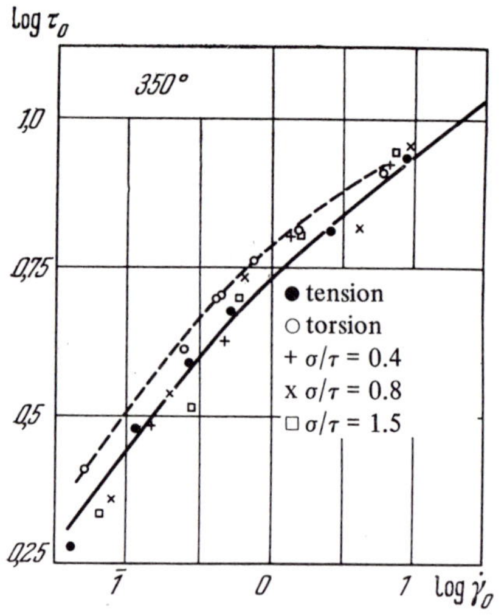

Fig. 105.

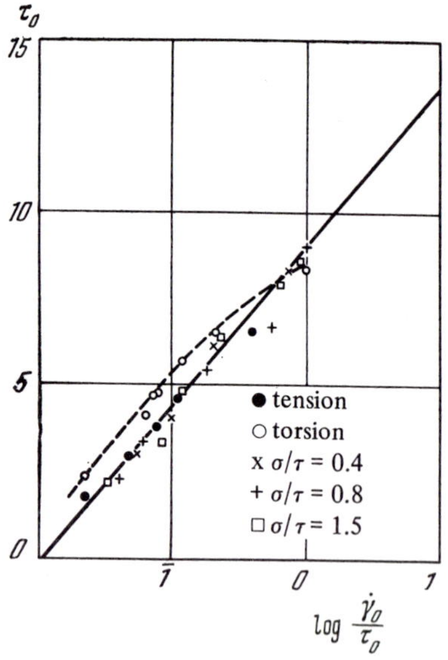

Fig. 106.

to stresses intermediate between tension and torsion do not tend to lie between the torsion and tension curves. A curve in log/log coordinates can be approximated by two straight-line sections, so that the power law of creep cannot be applied to the entire range of stress. The same results are plotted in other coordinates in fig. 106. The logarithm of the ratio $\dot{\gamma}_0/\tau_0$ is plotted on the abscissa and τ_0 on the ordinate. The points are grouped about a single straight line which corresponds to a relationship between $\dot{\gamma}_0$ and τ_0 of the form:

$$\dot{\gamma}_0 \sim \tau_0 \exp(c\tau_0) \, .$$

The author found the following approximation suitable for the function $\nu(\sigma_0)$:

$$\nu(\sigma_0) = A_1 \sigma_0^{n1} + A_2 \sigma_2^{n2} \, .$$

Johnson attributes the deviation of the behaviour of the material from the theoretical predictions at high stress intensities to anisotropy that is created in the initially isotropic material by the considerable plastic deformation that occurs at the instant at which load is applied. To describe this anisotropy it is assumed that the creep law is of the following form:

$$\begin{aligned} \epsilon_1 &= H(\sigma_0)[A_3(\sigma_1 - \sigma_2) - A_2(\sigma_3 - \sigma_1)] \, , \\ \epsilon_2 &= H(\sigma_0)[A_1(\sigma_2 - \sigma_3) - A_3(\sigma_1 - \sigma_2)] \, , \qquad (77.3) \\ \epsilon_3 &= H(\sigma_0)[A_2(\sigma_3 - \upsilon_1) - A_1(\sigma_2 - \sigma_3)] \, . \end{aligned}$$

Generally speaking, the anisotropy coefficients should depend on the system of stresses that is applied, that is they should depend on both the kind and the level of stress. As there were not sufficient data to establish these relationships the author selected constant coefficients for each type of stressed state to give the best description of the experimental results. It was found possible to take $A_1 = A_3 = 1$, whilst A_2 depended on the stress ratio, a quantity which is unity for pure torsion and increases as the stressed state approaches uniaxial tension. When τ/σ is 1.25 it was found that $A_2 = 5.75$. In fact with this kind of dependence of A_2 on the type of state of stress, eqs. (77.3) actually only take into account the redistribution of deformation rates between the directions 1 and 3. The creep rate in the direction of maximum stress σ_2 is independent of the constants of anisotropy and the rate of octahedral shear differs little from its usual value and, therefore, the use of expressions (77.3) does not, in practice, disturb the relationship between τ_0 and $\dot{\gamma}_0$.

Test results at temperatures of 450° and 550°C were worked out in a similar way and the relationship between τ_0 and $\dot{\gamma}_0$ was always independent of the type of state of stress, the points lay more or less satisfactorily about a single curve in the plane of log $\dot{\gamma}_0$–log τ_0 or log $\dot{\gamma}_0$–τ_0. No systematic separation of the experimental points into dependences on the ratio of τ/σ was found and so it appears that the scatter of the characteristics of individual specimens is greater than any systematic deviation from the assumed relationship. It should be noted that although the points lie sufficiently close about a common curve in logarithmic coordinates, the actual octahedral shear rate can sometimes differ by a factor of 1.5 for a given value of octahedral stress. As we have mentioned before, a considerable scatter of the experimental results is generally typical of creep tests.

Apparently the scatter is greater in thin-walled tubular specimens than in solid ones. Indeed, as Johnson indicates, the grain size is about $\frac{1}{6}$ of the specimen thickness and under these circumstances, it is quite arbitrary to describe the material as uniform. Preliminary tests to determine the uniformity of the material were carried out on normal cylindrical specimens and from these it was concluded that the creep characteristics were stable and the tests on tubular specimens were not duplicated.

The article [254] gives the results of similar investigations on aluminium alloy RR-59. The ingot was heat-treated by heating to 525°C for eight hours followed by quenching in hot water and holding at normal temperature for 24 hours and finally ageing for 16 hours at 170°C. Tests on specimens cut in different directions showed that the isotropy of the material was satisfactory. Creep tests in tension and torsion were made at temperatures of 150° and 250°C. Here, and in subsequent works, the method of working out the experimental data was the same as for steel with 0.17% C, and we shall not mention this point again.

Creep was tested in an exactly similar way on magnesium alloy (98% Mg, 2% Al) [256] after heat treatment by heating to 425°C for four hours and slow cooling to room temperature during ten hours, the creep test temperatures were 20 and 50°C. Finally, the chrome-nickel alloy Nimonic 75 was tested at 650°C. The specimens were cut from turbine disc forgings and it was found that the dependence of $\dot{\gamma}_0$ on τ_0 is always fulfilled for all temperatures and stresses to within the natural scatter of the results. At the same time, the condition that deviators should be similar can break down, the ratio of the rate ϵ_3 to ϵ_1 is sometimes significantly different from what would be expected from Mises' type theory. The deviation is insignificant for steel at a temperature of 350°C, at 450°C it is very significant; at this temperature the material behaves according to Mises' type theory at low and medium stresses; at high stresses deviations are significant. The boundary between low and medium stresses corresponds approximately to the inflection in the curve of log $\dot{\gamma}_0$ against log τ_0. At a temperature of 550°C the behaviour of the material becomes highly anomalous.

For aluminium alloy the condition that the deviators should be proportional is satisfactorily observed at 150°C, but high-stress tests were not made at this temperature. At a temperature of 200°C the deviators are proportional at low stresses and not at high, just as for steel with 0.17% C.

The authors draw similar conclusions for magnesium alloy [256]; finally, for the alloy Nimonic 75 [257] a Mises' type flow law was satisfactorily confirmed at all stresses. In coordinates of log $\dot{\gamma}_0$–log τ_0, the scatter was much greater for Nimonic 75 than for the other materials. This is possibly because the tests were made on material that had been subjected to the standard treatment, which gives a coarse grain size. It was required of all materials that the grain size should not exceed a sixth of the wall thickness, but this requirement was not fulfilled for Nimonic. Johnson ascribes deviation in the behaviour of material from the predictions of Mises' theory to the appearance of anisotropy and describes it in corresponding terms, mainly by means of equations similar to (77.3). Of course, it is possible to select the constants A_1, A_2 and A_3 for each individual specimen to give best agreement with experiment, but then the constants themselves are dependent on the type of stress, so that it becomes quite uncertain what kind of anisotropy is occurring in the specimen and with what it is associated. If it is assumed that anisotropy results from initial plastic deformation on loading, it is obvious that the method of applying load to the specimen is important. Apparently the best results are obtained by proportional loading of the specimen, maintaining a constant τ/σ ratio. It is by no means evident that the resultant anisotropy is exactly what Johnson supposes, namely that in the body there are three planes of symmetry which coincide with the planes of action of the principal stresses. Only on this assumption can we consider that formulae (77.3) are valid and that the principal axes of the stress tensor and creep rate tensors coincide. At the same time, Johnson's tests, which were very systematic and carefully performed, provide valuable material which throws light on the fundamental relationships of creep in a state of complex stress and, therefore, it is of interest to analyse them in more detail in the next few sections.

78. Further analysis of Johnson's test results

In working out his results Johnson assumed that the principal axes of the stress and rate of creep deformation tensors coincide. This is only possible for isotropic creep or for a particular type of anisotropic creep, namely, orthogonal isotropy of such a type that the axes of creep symmetry coincide with the principal axes of the stress tensor. The second possibility is of a rather particular kind, and, therefore, we shall not attempt to describe the true nature of the anisotropic behaviour of the material and only seek to determine the conditions to which anisotropy considerations do not apply. We shall use the method given in section 66. In combined tension and torsion tests the angle θ varies from $\theta = \pi/2$ in pure torsion to $\theta = 2\pi/3$ in pure tension.

It was explained in section 68 that if there is a creep potential and if the constant potential surface is convex the tangent of the angle between the vector of the octahedral tangential stress and the octahedral shear vector, namely $\kappa(\theta)$, should lie between $\kappa = -\tan[\theta-(\pi/2)]$ and $\kappa = \tan[(2\pi/3)-\theta]$. The use of the assumption that the creep potential is uniform as function of two invariants as a basis for our considerations, depends primarily on the observed similarity of creep curves for various states of stress; this fact was used to determine the characteristic creep rates that figure in all further working out of Johnson's results. The functions $g(\theta)$ should correspondingly lie between $g(\theta) = 2/\sqrt{3}\sin\theta$ and $g(\theta) = \cos[\theta-(2\pi/3)]$. Selecting $g(\theta) = 1$, we take tensile test results as initial data because the difference between the predictions of the dependence of creep rate on stress for several variants of the theory is greatest in pure shear when $\theta = \pi/2$. In this case by formula (66.4)

$$\epsilon_s = \frac{\dot{\gamma}}{\sqrt{3}} = g^* v(g^* \sqrt{3}\tau) .$$

Here $g^* = g(\pi/2)$. In octahedral coordinates

$$\dot{\gamma}_0 = g^* v(g^* \tau_0) .$$

With the Saint-Venant criterion $g^* = 2/\sqrt{3} > 1$, consequently for a given octahedral stress the rate of octahedral shear is greater in torsion than in tension. With the criterion of maximum referred stress $g^* = \sqrt{3}/2 < 1$, and consequently, for a given octahedral stress the rate of octahedral shear is less in torsion than in tension. If, for example, we plot the relationship between $\dot{\gamma}_0$ and τ_0 in logarithmic coordinates we should obtain a single curve if the creep potential depends on the invariant σ_0. If creep follows a theory of Saint-

Venant type the points should become separated, the curves for torsion and tension being different with the torsion curve displaced towards higher rates. If the maximum referred stress criterion applies, the points corresponding to torsion should be displaced in the opposite direction. Graphs of the dependence of ϵ_s on $\tau_0 = \sigma_0/\sqrt{3}$ plotted in log/log coordinates taken from the aforementioned work of Johnson and collaborators, and also graphs of $\kappa = \epsilon_t/\epsilon_s$ against the angle θ are given in figs. 107–122. The quantity ϵ_s is the creep rate expressed in nominal units as explained in section 77.

An obvious characteristic feature of curves of $\log \epsilon_s$ against $\log \tau_0$ is that they can always be approximated by two straight lines. The point of intersection between these lines separates the region of high stress from that of medium and low. Consequently, the values of κ which differ significantly from zero usually relate to high stresses. However, the way in which κ deviates from zero is different in different cases, so we shall consider these curves in more detail.

Steel 0.17% C, 350°C. The most obvious feature of fig. 107 is that points corresponding to pure torsion are displaced to the left in the lower stress region. This did not occur in other cases. Points on the graph of $\kappa(\theta)$ (fig. 108) lie close to the axis $\kappa = 0$, and below this axis values of κ are small and positive. Apparently, positive values of κ and displacement to the left of points corresponding to torsion in fig. 107 are in mutual agreement and are evidence that in the octahedral plane the true creep curve lies between a circle and the circumscribing hexagon. But in fig. 107 only points corresponding to

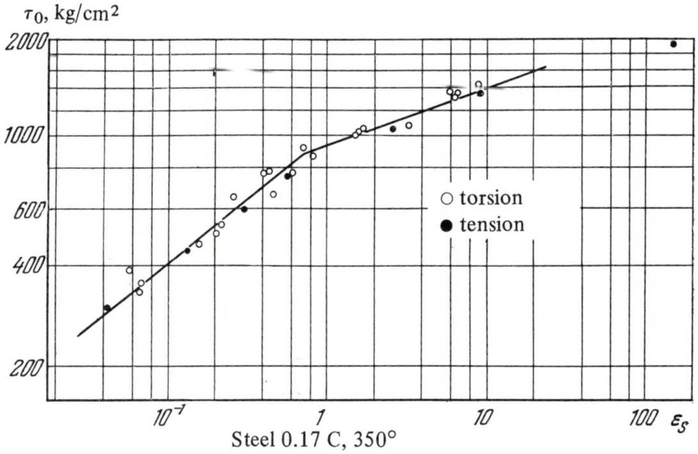

Steel 0.17 C, 350°

Fig. 107.

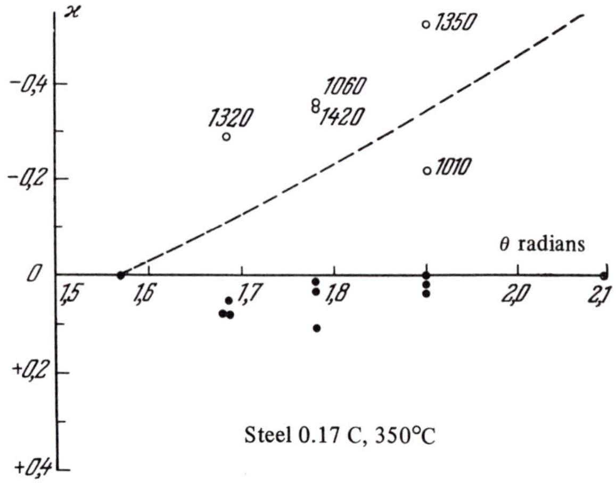

Fig. 108.

pure torsion stand out and for all the other states of stress the points lie on a single straight line, which confirms the validity of a Mises' type equation. The insignificant deviation of κ from zero at low stresses is probably a result of experimental error and random factors. At high values of τ_0 the position changes sharply, κ becomes negative and increases in absolute value with increase in θ. Values of τ_0 are shown near corresponding points on the figure. The curve of $\kappa = -\tan[\theta-(\pi/2)]$ which corresponds to Saint Venant theory, is shown dotted. Hence we may conclude that: steel of 0.17% C at 350°C follows a Mises' type theory satisfactorily at stress $\tau_0 < 1000$ kg/cm^2, a Saint Venant theory describes it better when $\tau_0 > 1000$ kg/cm^2. To confirm this conclusion it should be demonstrated that the dependence between ϵ_s and τ_0 also includes the corresponding quantity $g(\theta)$, however, the small number of experimental points at high stress and their considerable scatter render any such verification unconvincing.

Steel 0.17% C, 450°C (figs. 109, 110). The results at this temperature are similar to those at 350°C, and the picture is even clearer. At low stresses the relationship between ϵ_s and τ_0 is fulfilled very well, the scatter of the points is relatively small and obviously random, κ is small but everywhere positive. When $\tau_0 > 600$ values of κ lie close to the dotted curve that corresponds to the Saint Venant type flow theory. At higher stresses the scatter of ϵ_s becomes greater. Points 1 and 2 that indicate large positive values of κ should be noted. For point 2 for example, $\sigma_2 = 1080$ kg/cm^2, $\sigma_3 = -50$ kg/cm^2. The negative creep rate could be expected to be greater in absolute value than

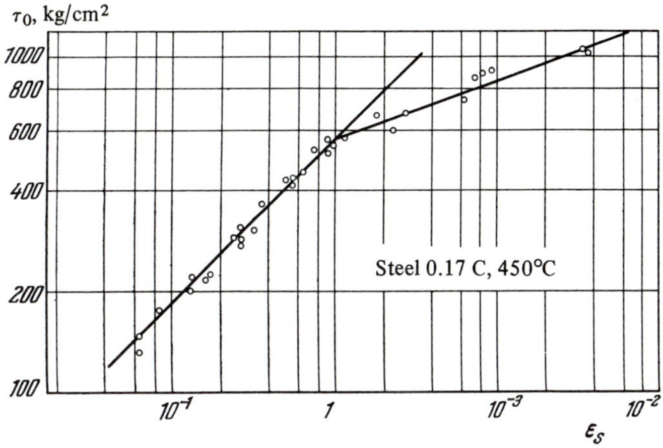

Fig. 109.

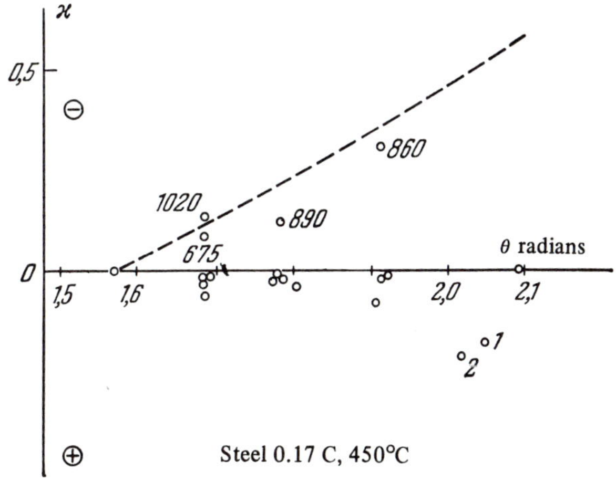

Fig. 110.

the rate ϵ_1, since $\sigma_1 = 0$. However, Johnson's figures for the creep rates are $\epsilon_2 = 0.62$, $\epsilon_3 = -0.21$, $\epsilon_1 = -0.41$. This result can be attributed either to anisotropy or to some unexplained experimental error. It is difficult to understand how anisotropy of such a strange kind could result from applying to the specimen a system of stresses which is close to pure tension in the direction 2 and why the properties are so different in the transverse direction. Exactly the same result was obtained for point 1. Here $\sigma_2 = 575$ kg/cm^2, $\sigma_3 =$

-52 kg/cm^2, $\epsilon_2 = 0.18$, $\epsilon_3 = -0.07$, $\epsilon_1 = -0.11$, the value of κ was almost the same. If the anisotropy results from plastic deformation during initial loading, the anisotropy should depend very markedly on the applied stress whereas in these two cases the octahedral stress differed by a factor of 1.8.

Steel 0.17% C, 550°C (figs. 111, 112). It will be noticed that there is considerable scatter of the points in coordinates of log ϵ_s against log τ_0, and it is difficult to see any kind of system in this scatter. In general, κ is near to zero, but when $\theta = 1.9$ there are three points that correspond to high positive values of κ. This behaviour of the material is anomalous and possible only if

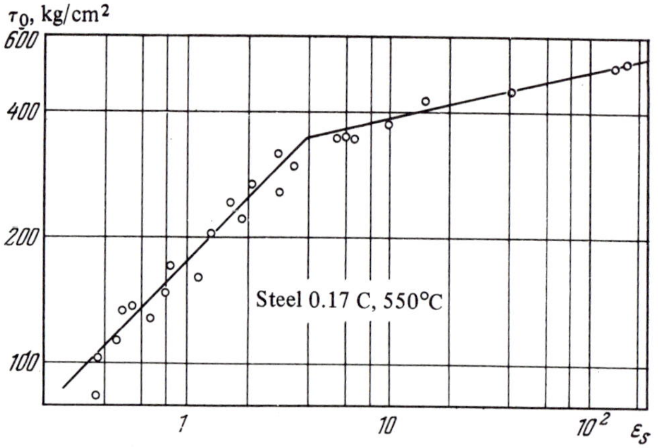

Fig. 111.

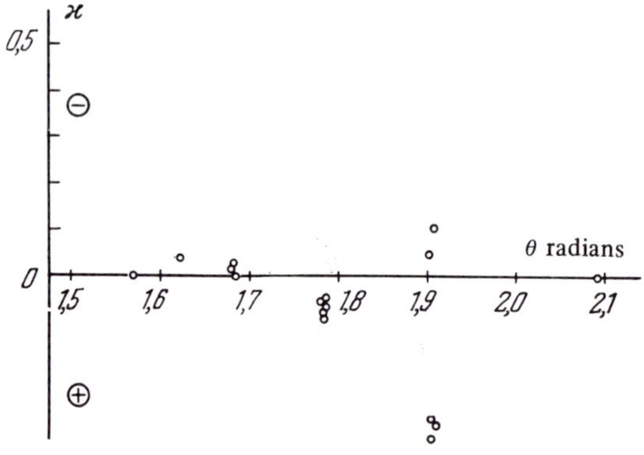

Steel 0.17 C, 550°C

Fig. 112.

there is marked anisotropy of properties in the transverse direction. Here again, as at 450°C, the anomaly occurs when the state of stress is close to uniaxial tension. Metallographic investigations showed that at 550°C creep is accompanied by structural changes. Spheroidalisation of cementite was noted together with separation of cementite within the ferrite grains and also on the grain boundaries. At high stresses polygonisation was observed in the ferrite grains.

Except for a few tests, and ignoring scatter, all the results indicate that here too the Mises' type theory provides the best basis for describing creep, there is no tendency to change to the Saint Venant flow law at high stresses.

Aluminium alloy 150°C (figs. 113, 114). As will be seen from the graphs, for this material the agreement between theory and experiment is exceedingly good.

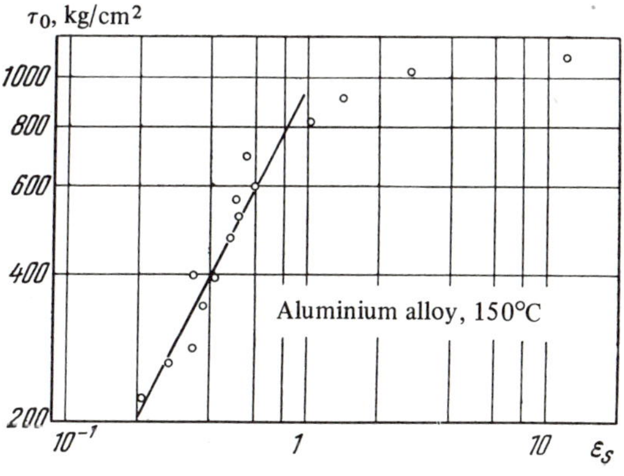

Fig. 113.

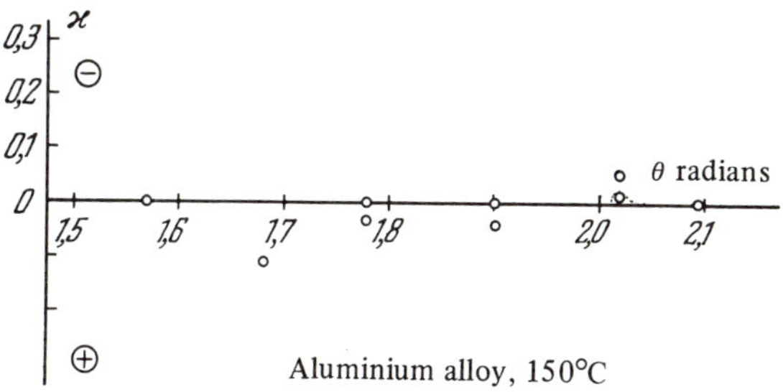

Fig. 114.

Aluminium alloy 200°C (figs. 115, 116). As with steel, the scatter of results for aluminium alloy is considerable at higher temperatures. At low and medium stresses $\kappa \approx 0$, at stresses exceeding approximately 500 kg/cm^2 the nature of the dependence of ϵ_s on τ_0 alters and a Saint Venant flow law is approached. Anomalous creep was not observed in aluminium alloy.

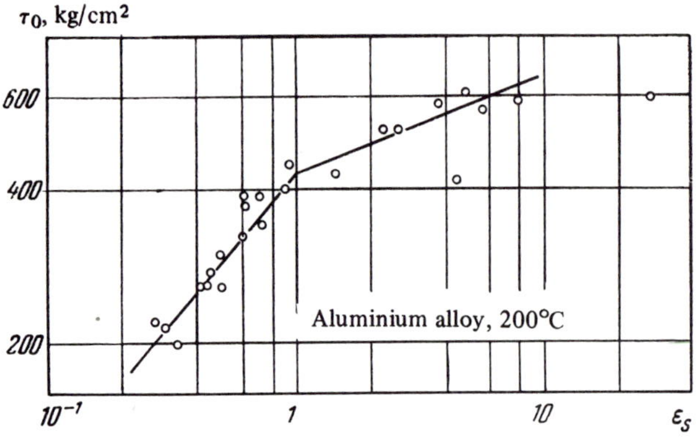

Fig. 115.

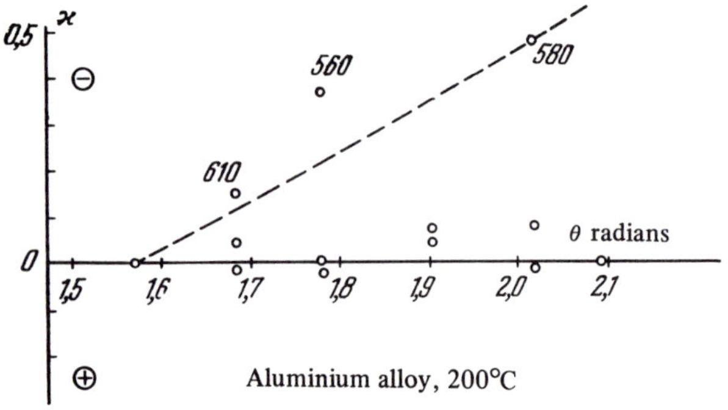

Fig. 116.

Magnesium alloy, 20°C (figs. 117, 118). At room temperature the behaviour of magnesium alloy is normal in the sense that the Mises' flow law is valid at low stress and the Saint Venant flow law at high. The scatter of experimental points is considerable in coordinates of log ϵ_s against log τ_0.

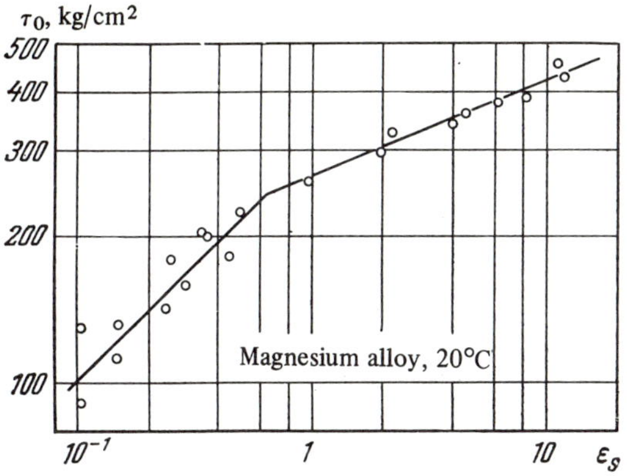

Fig. 117.

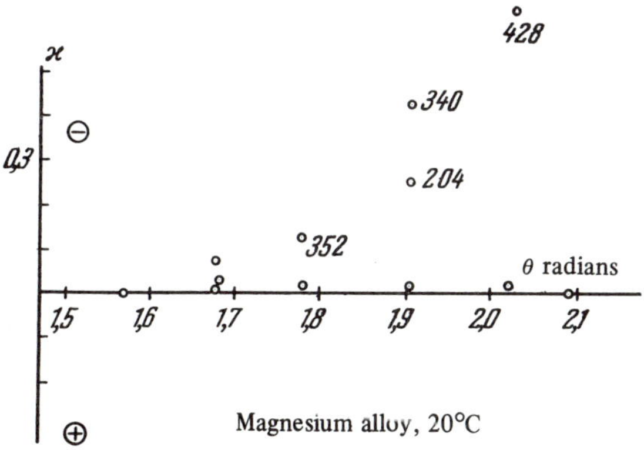

Fig. 118.

Magnesium alloy, 50°C (figs. 119, 120). At low and medium stress simple theory is in very good agreement with the experimental results and the scatter of experimental points is small. It is important that behaviour is anomalous at high stress, particularly when the state of stress is close to pure tension. For instance, the point $\kappa = -0.578$ corresponds to a state of stress $\sigma_2 = 411$, $\sigma_3 = -11$. Here it was found that

$$\epsilon_2 = 6.99\,, \qquad \epsilon_3 = -0.18\,, \qquad \epsilon_1 = -6.81\,.$$

Therefore, almost all flow in the transverse direction proceeds through the

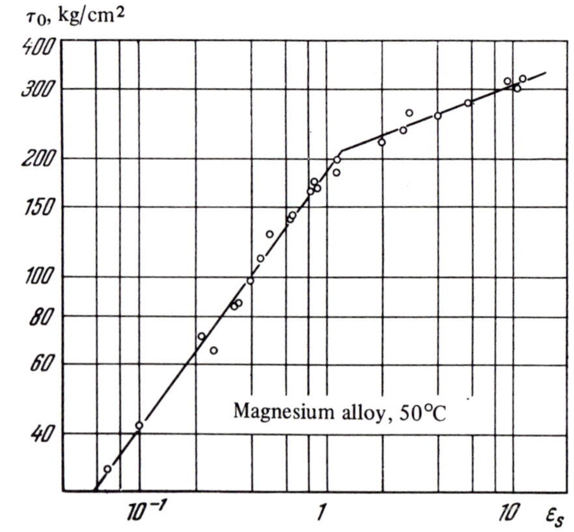

Fig. 119.

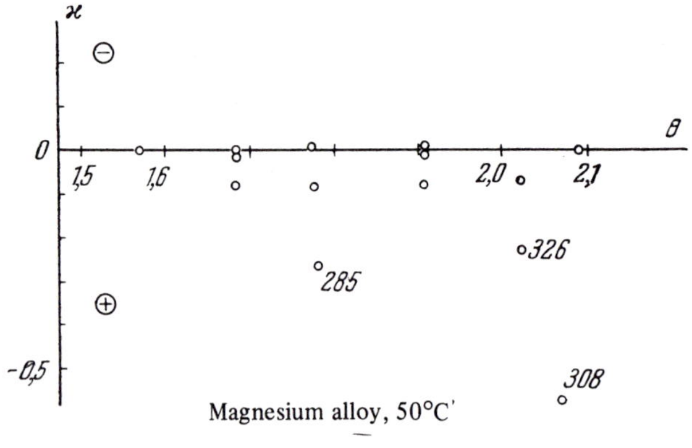

Fig. 120.

thickness of the specimen where no stress is applied; the flow picture is the exact opposite of that which follows from the Saint Venant theory.

Nimonic 75, 650°C (figs. 121, 122). Here $\kappa = 0$, Mises' type theory is valid throughout the range of stress, the scatter is fairly considerable but is apparently associated with the heat treatment of the material and the large grain size.

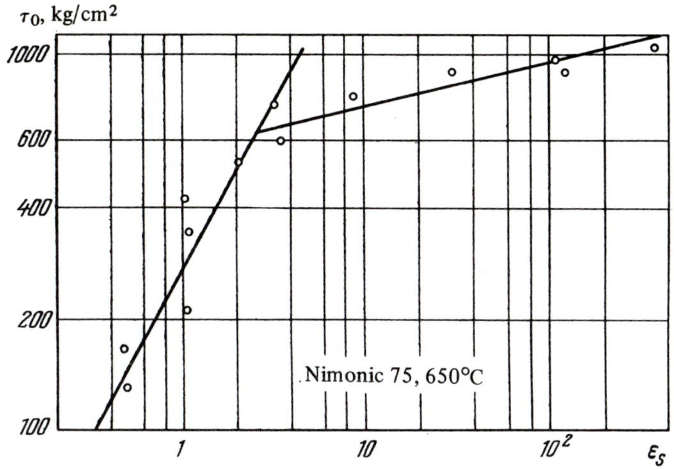

Fig. 121.

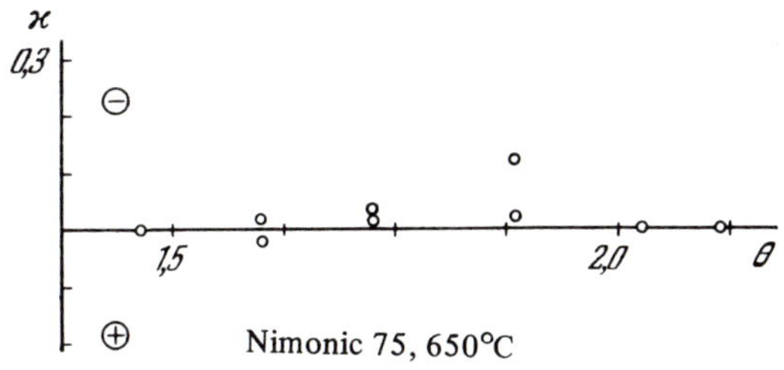

Fig. 122.

79. **Experimental studies of creep with complex stressed state** (continued)

We have considered Johnson's results in some detail because his numerous publications contain primary data that can be worked out in different ways. Other articles usually contain only certain final results. Without attempting to make an exhaustive review of all the available material we note certain results.

The work of Nishihara, Tanaka and Shima [303]. The tests were made on a specially constructed machine with the special feature that torsion and tension were both developed by the same load, so that loading was proportional. During the tests an additional load on the tension-application lever was displaced so as to compensate for the change in cross-sectional area, the torque was apparently not corrected. The rests were made on tubular specimens of 10 mm o.d. and 9 mm i.d. of steel with a carbon content of 0.100%, the only test temperature was 450°C. The test duration was always short, of the order of 500 minutes, the stress was fairly high, the octahedral tangential stress ranged from 8.5 to 11.3 kg/mm^2, and the corresponding value of the yield point at normal temperature was $\tau_0 = 11.3$ kg/mm^2. Creep was always preceded by considerable plastic deformation (of the order of 5%), after 500 minutes creep deformation was also quite large and in some tests the total deformation reached 15%. The authors checked fulfillment of the following relationship between octahedral shear and octahedral tangential stress:

$$\gamma_0 = \alpha\tau_0^k + \beta\tau_0^n t^m .$$

The best agreement with experimental results was achieved with the following values of the constants: $\alpha = 2.26 \times 10^{-5}, \beta = 1.17 \times 10^{-8}, k = 3.16, n = 5.95$, $m = 0.284$; here τ_0 is measured in kg/mm^2 and time in minutes. The results of all the tests were plotted in coordinates of γ_0 against t and for comparison curves were constructed from eq. (76.1). In all cases agreement was very satisfactory, the divergence was somewhat greater at high stress. The authors did not check the flow law, however from the curves given in the article, the ratio γ/e can be determined and compared with the value $3\tau/\sigma$ (section 68). The difference between these values is not more than 10% and divergences are on both sides. The use of proportional loading would be expected to exclude effects due to anisotropy caused by initial deformation. The article gives no information about the initial isotropy of the material, the specimens were apparently cut from a bar. The considerable plastic deformation makes the influence of the initial anisotropy insignificant and in this sense it is impossible to compare Nishihara's results with Johnson's.

Tests of Kennedy, Harms and Douglas on Niconel [274]. The tests were made on heat-resistant steel with a nominal analysis of 80% Ni, 15% Cr, 5% Fe at a temperature of 812°C. The object was to study the properties of the material with the types of stress that are realised in boilers and turbine discs where the principal stresses are usually positive. Therefore, the tests were made with internal pressure and axial force. The specimens were made from seamless tube, $\frac{3}{4}$ inch diameter, they were annealed for 2 hours in a hydrogen atmosphere at a temperature of 1060°C, the tests were conducted in argon, because the influence of an oxidising medium on creep and long-term strength of Niconel has been mentioned in the literature. The studies covered both the creep and long-term strength, certain of the tests at low stress lasted for 4000 hours. The first parts of the creep curves were absent and at a given stress the rate remained almost constant throughout and, therefore, the test results could be treated as true rates. Only the axial deformation was measured in the experiments, the transverse deformation was determined after failure, and on this basis certain conclusions were drawn concerning the character of deformation immediately before failure. We shall return to these conclusions later in considering the question of long-term strength. The data about deformation was incomplete as a result of the test programme and method of working out the results that were adopted. The Mises' type theory was checked and also the Saint Venant type theory, but with the flow law associated with the Mises' criterion. In this case the rate of axial deformation is given by

$$\epsilon_z = \frac{\nu(s)}{s}\left(\sigma_z - \tfrac{1}{2}\sigma_\theta\right).$$

Here σ_z is the axial stress, σ_θ the peripheral stress. In the first case $s = \sigma_0 = \sqrt{\sigma_z^2 + \sigma_\theta^2 - \sigma_z\sigma_\theta}$, in the second $s = \sigma_{max} - \sigma_{min}$; depending on the axial stress ratio the peripheral stress and the zero stress in the direction of the wall thickness appear in various combinations in the expression for s. In tensile tests the dependence of rate on stress was found to be a power function with exponent $n = 6$. Therefore, from any of the theories it follows that

$$\epsilon_z = K\sigma_z^n f_1\left(\frac{\sigma_z}{\sigma_\theta}\right)$$

or

$$\epsilon_z = K\sigma_\theta^n f_2\left(\frac{\sigma_z}{\sigma_\theta}\right).$$

For these theories the functions f_1 and f_2 are easily written in explicit form. In each series of tests either σ_z or σ_θ was maintained constant while the ratio σ_z/σ_θ assumed various values. Theoretical curves of the dependence of ϵ_z on σ_z/σ_θ with σ_z or σ_θ constant were constructed and experimental points were plotted. Fig. 123 gives results of tests with constant axial force; the solid lines correspond to the Mises type and the dotted to the Saint Venant type theory with the Mises flow law. The figures near the curves denote values of the axial stress σ_z in kg/cm^2. Theoretical and experimental results with constant peripheral force are compared in figs. 124 and 125. The experimental points lie quite close to the Mises curve, values of σ_θ are noted near the corresponding curves. Note that if the Saint Venant criterion is used, with the asso-

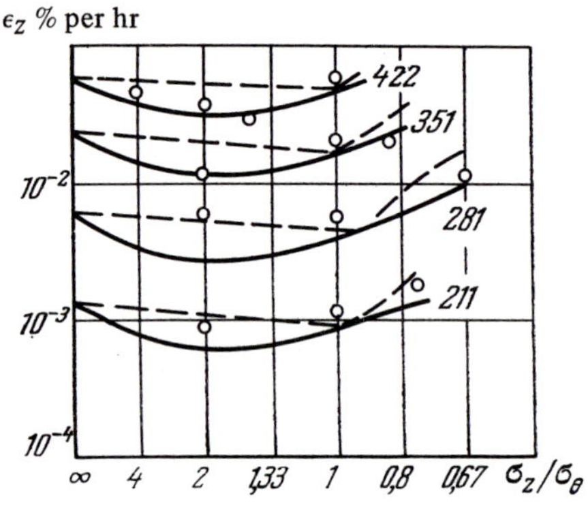

Fig. 123.

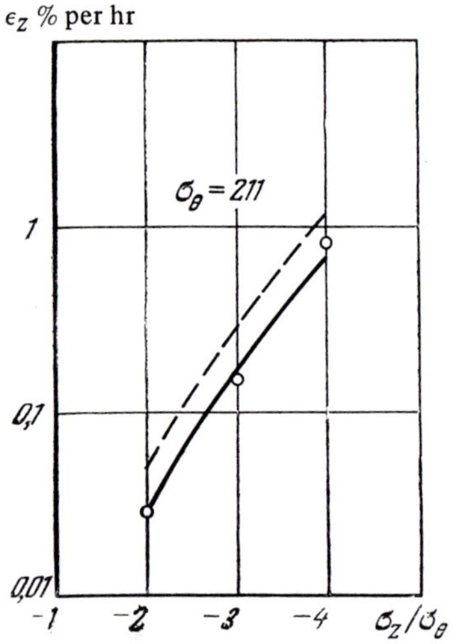

Fig. 124.

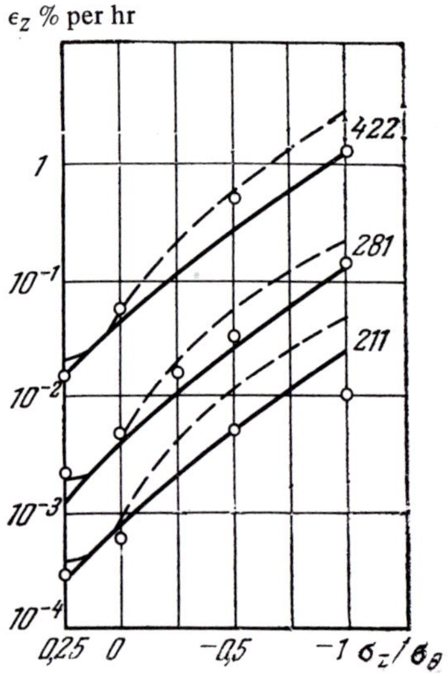

Fig. 125.

ciated flow law the difference between the curves increases. For example, in fig. 123, the linear part of the dotted curve which corresponds to the region where $\sigma_z/\sigma_\theta > 1$, becomes horizontal.

The authors conclude that Mises type theory is satisfactorily justified for their material with their temperature conditions and the maximum tangential stress criterion can be used.

Tests of Oding and Tulyakov [92]. The tests of these authors were unlike most of the others described above in being of fairly long duration, up to 2000 hours. The test material was austenitic steel 1Kh18N9T, subjected to austenisation at 1100°C and cooled in air. The test temperature was 600°C. Thin-walled tubular specimens were subjected to combined tension and torsion. The creep curves were of the same character for all kinds of states of stress, the first parts of the curves terminated quite quickly and for most of the time the creep rate was constant and the third part appeared at high stress.

Accordingly, the steady-state creep rate was determined quite reliably. The test results were presented as curves of the dependence of octahedral shear on octahedral tangential stress in logarithmic/logarithmic and logarithmic/linear coordinates, points were nearer to a straight line in the latter case than in the former. The authors conclude from this that the exponential creep law is to be preferred to the power law.

The test results of Oding and Tulyakov, prepared by the procedure described above, are plotted in fig. 126 in coordinates of $\log \epsilon_s - \log \sigma_0$. Double logarithmic coordinates were selected precisely because then the points do not lie about a straight line and there is a fairly clear boundary between low and high stress. In working out Johnson's results the nature of the creep process altered on transition from low to high stress. Values of $\nu(\alpha)$ determined

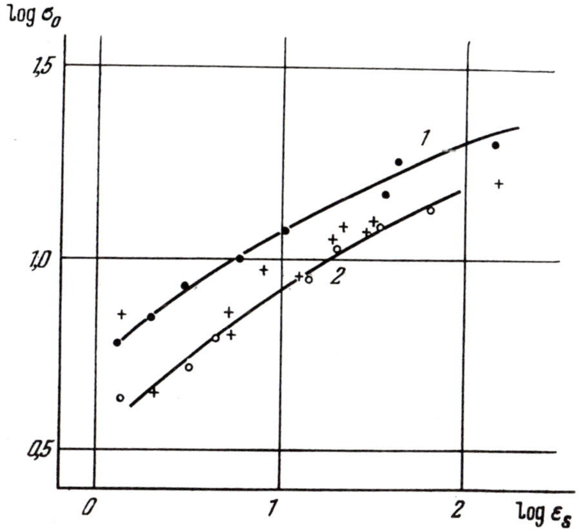

Fig. 126.

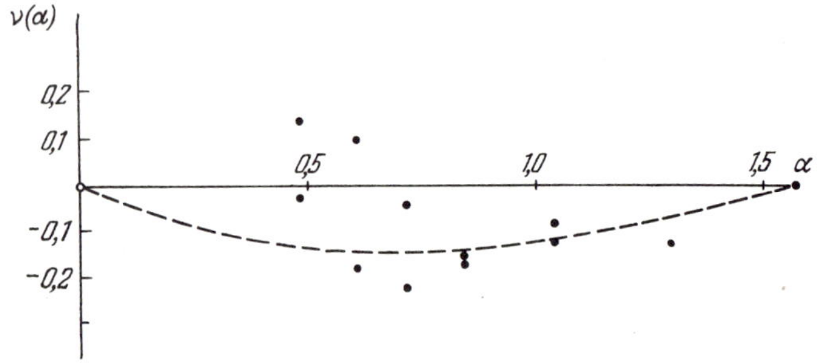

Fig. 127.

by formula (67.8) are plotted in fig. 127, the dotted curve corresponds to the Saint Venant type of flow criterion (68.5). The scatter of the experimental points is quite large, but all the points except two correspond to negative values of ν and are better grouped about the dotted line than about the abscissa. Therefore, the behaviour of the test material is normal and one would expect it to be described better by the Saint Venant than by the Mises type of theory. Better agreement between the theoretical and experimental results could be expected by making appropriate corrections to the relationship between ϵ_s and σ_0.

The values of ϵ_s/g are plotted against sg in accordance with the first of the formulae (67.7) in fig. 128 in the same log/log coordinates and to the same scale as before. Here the function $g(\alpha)$ is found by formula (68.6). The points correspond to tension tests, the circles to pure torsion; combined test results are denoted by crosses, ++. Curves 1 and 2 are plotted from test results in tension and torsion respectively.

In fig. 126 the tension and torsion curves differ quite markedly, the separation is similar if the results are plotted, as the author did, in coordinates of $\dot{\gamma}_0-\tau_0$. The points that correspond to combined tests lie between these curves; those that correspond to smaller values of τ/σ are further away from the torsion curve and nearer to that for tension. In fig. 128 all points for combined tests lie very close to the torsion curve and the deviations are not systematic, but random. As before the curve for pure tension is separate, but here curves 1 and 2 are much closer together. One point denoted by the figure 7 lies well off the curve, as in the preceding graph. It is difficult to understand why pure tension results do not lie on the common curve whereas points that correspond to a small ratio of τ/σ, where tension clearly predominates, nevertheless lie close to it.

In general the definite conclusion follows from Oding and Tulyakov's

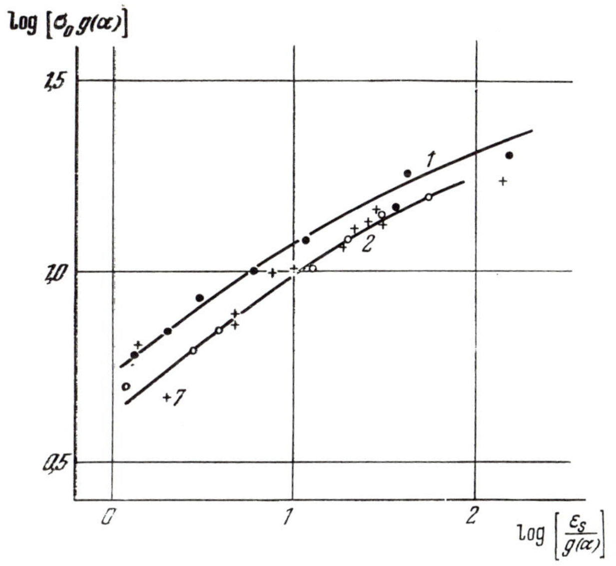

Fig. 128.

tests, that steady state creep of the particular material at the particular temperature is well described by the theory which takes the maximum tangential stress as the criterion of creep.

Soderquist's tests [356]. The tests were made on aluminium alloy and in general the results confirm eq. (76.1). The special feature of the work was that the specimens were disc-shaped and were extended in two mutually perpendicular directions.

80. Namestnikov's investigations

Namestnikov's experiments on austenitic steel grade EI-257 at temperatures of 600°C and 500°C relate to the period of transient creep. The author's object is to verify individually the principal hypotheses of the Mises type theory, namely the assumption that the deviators are similar and to determine the relationship between octahedral shear, octahedral stress and time [78]. The test machine, which was designed and made in the Institute of Mechanics of the Academy of Sciences of the USSR, permitted smooth proportional loading, the loads were tanks which were filled by water flowing through calibrated apertures from a tank at constant level. By altering the apertures any ratio between tensile force and twisting moments could be achieved. The specimens were made of bar material, they were annealed at 1150°C, cooled in water and normalised for 10 hours at 760°C followed by cooling in the furnace. Altogether 24 specimens with a mean diameter of 16 mm and a wall thickness of 0.4–0.8 mm were tested. Two series of tests were made at a temperature of 600°C at levels of 17 and 12 kg/mm^2, and in each series the ratio τ/σ was altered. At 500°C the stress was 14 kg/mm^2, and three individual tests were made at higher stresses. From Mises' type theory we should have

$$\frac{\dot{\gamma}}{\dot{e}} = 3\frac{\tau}{\sigma}.$$

Graphical differentiation, which introduces an error, is required to determine the rate on the steady-state creep section, alternatively it is assumed that the creep curve can be given analytically. But if the condition given above maintains its force for any instant in the creep process it follows that

$$\frac{\gamma^p}{e^p} = 3\frac{\tau}{\sigma}. \tag{80.1}$$

This is the expression that was checked during the tests from which it follows that

$$\frac{2\sigma}{3e^p} = \frac{2\tau}{\gamma^p}. \tag{80.2}$$

The left- and right-hand sides of (80.2) are functions of time which are

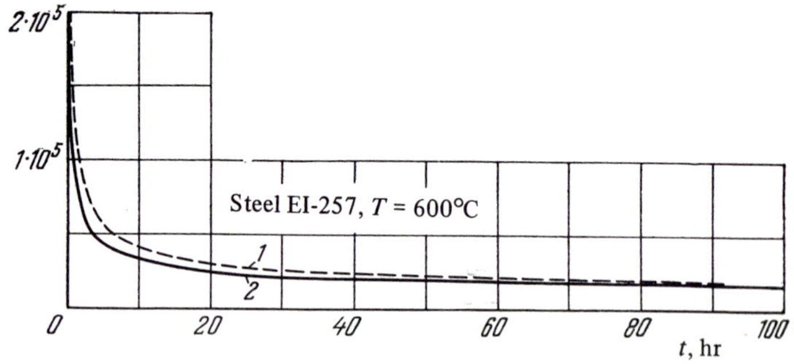

Fig. 129.

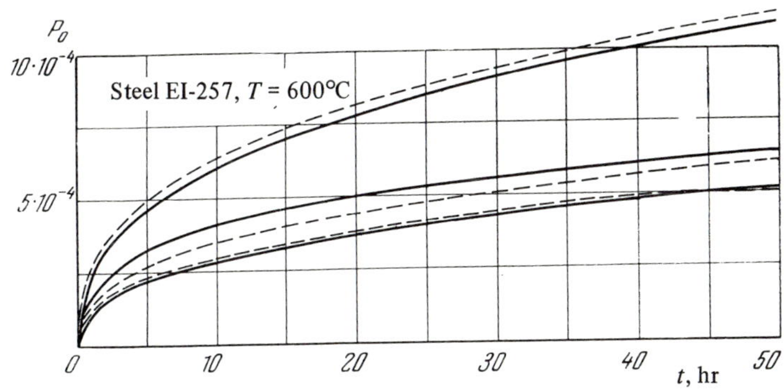

Fig. 130.

plotted in fig. 129 for one of the tests. Starting from about 60 hours condition (80.2) is fulfilled extremely well. The author states that an analogous picture is obtained in all other cases. The work [81] gives the working out of the data previously obtained on steel EI-257 and also new results for alloy D-16T at a temperature of 150°C. In all cases at the end of the tests it is found that $\gamma/e < 3(\tau/\sigma)$ and at 600°C this difference ranges from 3.3% to 19.5%, at 500°C it even reaches 28%; in one test only for alloy D-16T the equation $\gamma/e = 3(\tau/\sigma)$ is observed satisfactorily with little scatter on both sides. In explaining the deviations Namestnikov does not assume that the material is incompressible. For an elastic material with Poisson's ratio v

$$\frac{\gamma}{e} = 2(1+v)\frac{\tau}{\sigma}. \tag{80.3}$$

If it is assumed that the deviator of total deformations and the stress deviator are similar under creep conditions eq. (80.3) is also valid for creep but the quantity ν is the transverse deformation coefficient which usually varies with increasing deformation. At the initial time v is Poisson's ratio; if creep takes place without change of volume, v tends to $\frac{1}{2}$. Taking $v = 0.35$, the author obtained good agreement between formula (80.3) and experiment.

Creep curves in coordinates $e_0 - t$ at constant stress intensity did not coincide for different values of the ratio τ/σ, creep is faster in tension than in torsion. A typical series of such creep curves is plotted in fig. 130, where in all cases $\sigma_0 = 17$ kg/mm^2, the lower curve corresponds to pure torsion, the next to a ratio $\tau/\sigma = 1.5$, whilst the upper curve corresponds to pure tension. The author devised the following formula to describe these results:

$$p_0 = \left(\frac{kt}{m}\right)^m \exp\left(\frac{m\sigma_0}{A} - \frac{m|\tau_{max}|}{A_0}\right). \tag{80.4}$$

The values of the constants are

	500°	600°	Units of measurement
m	0.230	0.341	
k	1.12×10^{-35}	2.27×10^{-16}	hr^{-1}
A	0.258	0.659	kg/mm^2
A_0	0.398	0.578	kg/mm^2

Curves corresponding to eq. (80.4) are shown dotted in fig. 130. It is not difficult to interpret expression (80.4) from the general standpoint developed in section 66 and to obtain the corresponding associated flow law; however, in this case the agreement with test results is worse than for the Mises' flow law.

Data that confirm the isotropy of the material in the initial state were given in a later publication by the same author [302]. A characteristic feature of the results is that the existence of a creep potential is not confirmed, the condition of similarity of deviators is not associated with the invariant relationship (80.4).

81. Creep in a state of complex stress with variable loading

As we have seen, the problem of creep under conditions of complex stress, when the stress is constant with time, is far from simple, and an exhaustive solution has not yet been found. It will, therefore, be understood that the picture is much more complicated when the type of state of stress changes. The first fundamental question that arises is whether work-hardening is isotropic, that is can the degree of work-hardening be characterised by some single invariant parameter. In order to obtain a direct answer to this question Namestnikov [79] performed special tests by the following procedure. Consider two specimens, one of which is tested in tension and the other in torsion. In both the creep rate diminishes during the period of transient creep because of work-hardening. Similarity of creep curves in tension and in torsion indicates that the work-hardening law is the same in both cases; in tension, the creep deformation in tension is a measure of work-hardening, and in torsion the shear creep deformation. If work-hardening is isotropic the degree of hardening achieved on extension characterises the work-hardening of the material relative to creep in any state of stress. Now consider that the specimen is tested in creep with tension. At a certain time when considerable work-hardening has already occurred and the creep rate has diminished appreciably from its initial value, we remove the tensile load and apply a torque. The creep in torque should then be slower than that of a specimen which has not first been subjected to creep in tension. Tests were made on steel EI-257 at 500°C and 600°C under the same conditions as those at constant load on the material described in section 80. Each specimen was creep tested, for 50 hours with a single type of state of stress, then it was unloaded and a new system of loads was applied to it, with the same intensity but a different ratio of τ/σ. Preliminary tension had no work-hardening effect for subsequent torsion and vice versa. It was thus shown that work-hardening in creep is very anisotropic. Qualitatively this result may be understood in the light of the ideas that underlie the theories of slip in plasticity, see section 20. In the particular state of stress that is characterised by the ratio τ/σ creep takes place on the grains and in the planes where the tangential stresses are greatest. The dependence of creep rate on stress is very non-linear and, therefore, planes which are favourably located relative to the principal axes of the stress tensor work-harden considerably whilst on other planes there is practically no creep or work-hardening. When the principal axes are turned through a sufficiently large angle practically all the creep takes place in other grains and in other slip systems where there has been no work-hardening. If the test is made under less contrasting conditions, namely if the specimen is first obliged to creep with a stress ratio $(\tau/\sigma)_1$ and then with a stress ratio $(\tau/\sigma)_2$,

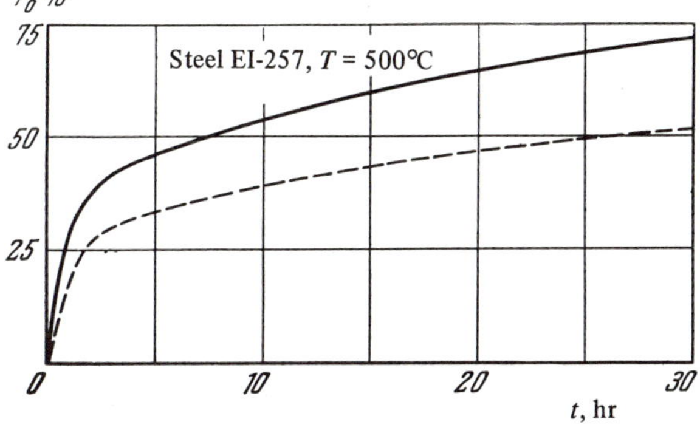

Fig. 131.

which is not too different, work-hardening due to the previous creep will be significant. Two creep curves taken at 500°C are shown in fig. 131, $\sigma_0 = 14$ kg/mm^2 and $\tau/\sigma = 1.5$. The intensity of creep deformation is plotted on the ordinate. The continuous line curve relates to an undeformed specimen and the dotted one to a specimen that has previously crept for 50 hours with the same stress intensity but with $\tau/\sigma = 0.28$.

Johnson and collaborators [262] carried out a fairly extensive test programme with complex stresses and variable loads. The materials tested were the same as were studied under constant load conditions (see section 77), namely aluminium alloy RR-59 at 150°C, magnesium alloy at 20°C and 50°C and steel containing 0.2% carbon at a temperature of 450°C (the latter was different from that which was tested at constant load). Axial loading was first applied in each test and then torque was applied in stages.

The results were compared with predictions by the time-hardening and the strain-hardening theories. The time-hardening theory was taken in the usual form

$$\dot{p}_{ij} = \tfrac{3}{2}\, v(\sigma_0)\, \frac{\bar{\sigma}_{ij}}{\sigma_0}\, \dot{\tau}(t)\,. \tag{81.1}$$

Two variants of the strain-hardening theory were considered, in both cases the intensity of creep deformation $p_0 = (\frac{2}{3}\, p_{ij} p_{ij})^{1/2}$ was taken as a measure of work-hardening. The equations were used in the forms

a) $$\dot{p}_{ij} = \frac{3}{2}\frac{f(\sigma_0)}{h(p_0)}\frac{\bar{\sigma}_{ij}}{\sigma_0}\,, \tag{81.2}$$

b) $$\dot{p}_{ij} = \tfrac{3}{2}\, [f(\sigma_0) - \chi(p_0)]\, \frac{\bar{\sigma}_{ij}}{\sigma_0}\,. \tag{81.3}$$

Finally, combined equations of time and strain-hardening were considered, the right-hand sides of (81.2) and (81.3) were multiplied by a certain time function similar to (81.1). In all cases the agreement between theory and experiment was poor. This was to be expected because eqs. (81.1)–(81.3) describe isotropic work-hardening.

Namestnikov's work [302] had the definite object of studying creep with proportionately varying loads. Tests were made in tension and torsion on steel grade EI-257 at 600°C, on technically pure copper at 150°C and on alloy D-16T at 150°C. In this work, unlike the earlier work, the initial isotropy of the material was checked. A special machine was designed and made

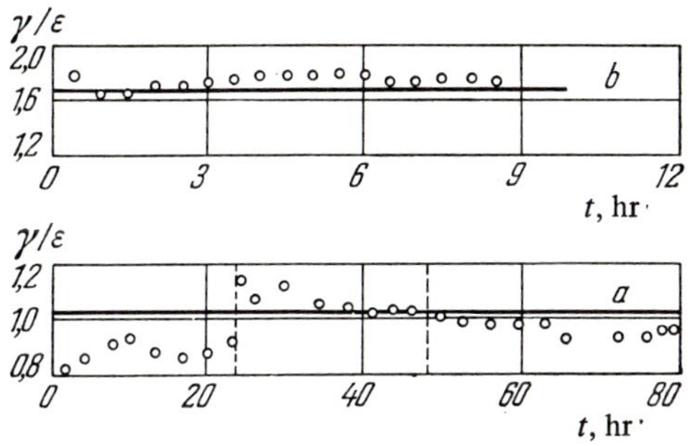

Fig. 132.

for creep testing small specimens with an overall length of 28 mm, which could be cut from bars in the transverse direction. Steel and copper were sufficiently isotropic but appreciable anisotropy was observed in alloy D-16T. To work out the results Namestnikov had to construct certain equations of the strain-hardening theory. In the new tests the condition of similarity of deviators was extremely well fulfilled, the ratio $\gamma/e : \tau/\sigma$ ranged from 2.8 for instantaneous elastic deformation to 3 for creep. Typical results are given in fig. 132, the points correspond to experimental values of γ/e, the straight line to the quantity $3(\tau/\sigma)$. Therefore, the assumption of similarity of deviators was taken as a basis. If the hypothesis of existence of a creep potential is retained in creep with work-hardening we must consider that the potential depends on σ_0. The author attributes the divergence of creep curves in coordinates $p_0 - t$ with different kinds of stress state and identical σ_0, not so much to the form of the state of stress as to the nature of the deformed state, he introduces a dimensionless parameter r defined as follows:

$$r^3 = \frac{4}{3} \frac{p_{ij}p_{jk}p_{ki}}{(\frac{3}{2}\, p_{ij}p_{ij})^{3/2}} \,,$$

and also defines a measure of work-hardening

$$q_3 = p_{ij}\sigma_{ij}$$

(see section 74).

He took the creep equations in the form

$$\dot{p}_{ij} = \tfrac{3}{2}\, h(q_3) \frac{\partial f(\sigma_0, r)}{\partial \sigma_0} \frac{\bar{\sigma}_{ij}}{\sigma_0} \,. \qquad (81.4)$$

The following approximations are proposed for the functions $h(q_3)$ and $f(\sigma_0, r)$:

$$h(q_3) = \kappa q_3^{-\alpha} ,$$

$$(n + \alpha + 1) f(\sigma_0, r) = \sigma_0^{n+\alpha+1} \exp [\sigma_0 \chi(r)] .$$

For alloy D-16T the values found were $\alpha = 1.57$, $n = 9$, $\kappa = 2.23 \times 10^{-22}$,

$$\chi(r) = -2.7 \times 10^{-3} \frac{r}{1.02 - r} .$$

Here force is measured in kilograms, length in millimeters and time in hours. The expression that was used in the calculations follows from eq. (81.4), namely

$$\dot{p}_0 p_0^{\alpha} = \kappa \sigma_0^n \left\{ 1 + \frac{\sigma_0 \chi(r)}{n + \alpha + 1} \right\} \exp [\sigma_0 \chi(r)] . \tag{81.5}$$

We have seen that the analytical possibilities of describing transient creep in the complex stressed state are almost limitless, the fairly complicated eqs. (81.4) represent only one possible variant. For the state of uniaxial stress this equation leads to the usual equation of the theory of strain-hardening considered in Chapter 4, tests in a state of complex stress with loads that are constant with time are satisfactorily described by this equation. In tests with stepwise change of load it is found that, with an arbitrarily fixed ratio τ/σ, the agreement between theory and experiment is the same as was found in working out tensile tests with stepwise change of load by the strain-hardening theory. When there is sudden increase in load the experimental points lie somewhat above the theoretical curve and in the next stage the divergence increases. Hence, we may consider that the case of proportionally varying loads is just as well described by eq. (81.4) as is the analogous uniaxial case. The following tests were made to give clearer confirmation of this conclusion. Two different states of stress were selected for which the right-hand sides of (81.5) were identical and varied only with time. Graphs of $p_0 - t$ were constructed and coincided accurately within the amount of scatter.

By way of example, two such curves are plotted in fig. 133, the specimen numbers are marked against the curves. Specimen 219 was subjected to torsion with tension of 12.7 kg/mm^2 in the first part and 10.4 kg/mm^2 in the second. Specimen 237 was first extended by a stress of 34 kg/mm^2 and then

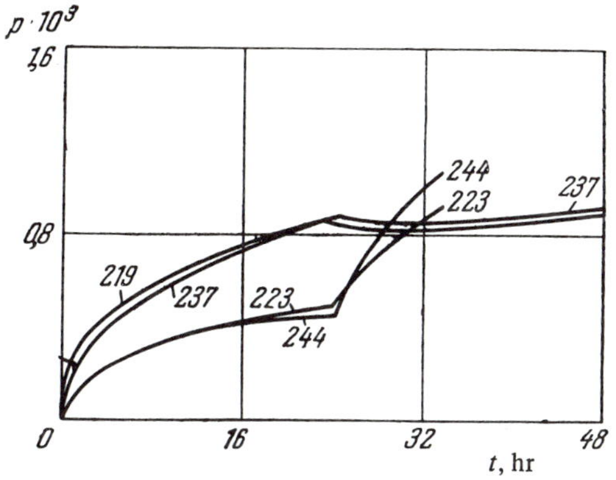

Fig. 133.

by one of 30.1 kg/mm^2. The curves practically coincided. Torsion tests on specimen 223 and combined tension and torsion with a ratio $\tau/\sigma = 0.56$ on specimen 244 were compared in just the same way. This experiment depends very little on the particular selection of the strain-hardening law and provides convincing confirmation of the general conclusion that there is complete analogy between the behaviour of the material in proportional and in uniaxial loading.

82. Stress relaxation in a state of complex stress

It follows from the results presented above that work-hardening anisotropy cannot be observed with proportional loading, therefore, any theory of creep with isotropic work-hardening that is satisfactorily confirmed by experiment may be used in this case. Similar conditions obtain in stress relaxation. Consider a body that is in a state of uniform stress; displacements, which are maintained constant, are given on part of its surface Σ_u, there are no forces on the remainder of the surface Σ_T. In any particular case Σ_T may be absent. We shall speak of general and special cases of relaxation respectively. Take as basis the equations of the theory of strain-hardening in the form

$$\dot{p}_{ij} = \frac{3}{2}\frac{f(\sigma_0)}{h(q_0)}\frac{\bar{\sigma}_{ij}}{\sigma_0}. \tag{82.1}$$

Here

$$\dot{q}_0^2 = \tfrac{2}{3}\dot{p}_{ij}\dot{p}_{ij}, \quad p_{ij} = \bar{e}_{ij} - \frac{1}{2G}\bar{\sigma}_{ij}.$$

Bulk deformation is considered elastic and subject to Hooke's law

$$e = \frac{\sigma}{3K} .$$

Now suppose that a state of uniform stress is set up in the body. In the special case of relaxation in which displacements of all points on the surface of the body are fixed, elastic deformations e^*_{ij} and elastic stresses σ^*_{ij} arise at the initial instant. Assume that the deformations e^*_{ij} also remain unchanged during relaxation. Because of Hooke's law, the hydrostatic component of the stress tensor also retains the constant value $\sigma = 3Ke^*$. Put $\bar{\sigma}^*_{ij} = \bar{\sigma}^*_{ij}\varphi(t)$, where $\varphi(0) = 1$. Substituting into (82.1) we obtain

$$\dot{\varphi} = -3G \frac{f(\sigma^*_0 \varphi)}{h\left(\dfrac{\sigma^*_0 \varphi}{3G}\right)} \frac{1}{\sigma^*_0} . \tag{82.2}$$

The quantity σ^*_0 is constant because the state of stress is assumed uniform and, therefore, the differential eq. (82.2) defines the function $\varphi(t)$. The stress is expressed as follows:

$$\sigma_{ij} = \bar{\sigma}^*_{ij}\varphi(t) + \sigma\delta_{ij} .$$

The forces on the surfaces are

$$T_i = (\bar{\sigma}^*_{ij}\varphi + \sigma\delta_{ij})\nu_j . \tag{82.3}$$

It will be seen that the forces do not relax to zero but to a certain value that corresponds to the limiting hydrostatic stressed state in the body.

The solution that has been obtained is not suitable for the general case of relaxation. Indeed if on the surface Σ_T the quantities T_i become zero when $t = 0$, then at later times, according to formula (82.3) they differ from zero.

The general case of relaxation is readily considered for an incompressible material. Indeed when $K = \infty$, $e = 0$, consequently σ can assume any value. Put $\sigma = \sigma^*\varphi(t)$ then

$$\sigma_{ij} = \sigma^*_{ij}\varphi(t) .$$

On the fixed part of the body surface stress and relaxation continue to zero and the conditions are always fulfilled on the free surface.

In relaxation tests with complex stress, displacements are always fixed on part of the specimen surface, and, therefore, a strict solution of the relaxation problem with allowance for compressibility of the material is far from simple. Even in the simplest cases use of the strain-hardening theory requires numerical integration of the corresponding equations.

The earliest experimental study of stress relaxation in a state of complex stress is probably that of Griffith and Marin [221]. The authors use the equations of the ageing (time-hardening) theory in the form of (72.3). The tests were made on specimens of aluminium alloy grade 3S-0 at room temperature. The constants in eq. (72.3) were first determined from creep tests on cylindrical specimens in tension. Relaxation tests were made: a) in tension; b) in torsion, and c) with combined tension and torsion. In the latter case constant deformation was maintained and the torque was varied so that the ratio of the principal stresses σ_2/σ_3 remained constant. Theoretical curves were constructed by eqs. (72.3), which were solved graphically. We have seen (section 48, Chapter 4) that relaxation curves predicted by the time-hardening theory do not deviate too seriously from practice in the uniaxial case and here again the agreement was fully satisfactory. Curves of axial stress relaxation and of the change in the amount of shear are shown in fig. 134. The

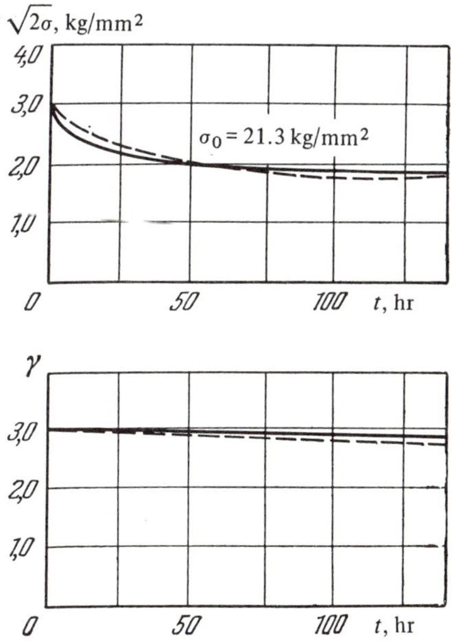

Fig. 134.

calculated relaxation curve in pure torsion gives somewhat worse agreement with experiment.

The next work performed on relaxation in a state of complex stress was by Johnson and colleagues [263]. The relaxation test machine used by the authors was described earlier [259], a special feature of its construction is that the ratio of the axial force to the torque is always constant, both torque and load diminish proportionately so that the angle of twist retains its initial, given value. Here the axial deformation remains uncontrolled. The authors indicate that in fact it also remains constant, but more accurate information about this point is not given. In the elastic state the ratio of shear to axial deformation is expressed as

$$\frac{\gamma}{e} = 2(1+v)\frac{\tau}{\sigma}.$$

The same expression with $v = \frac{1}{2}$ is also valid for creep deformation that takes place without change of volume. The process of relaxation is one of transition from elastic deformation to plastic, consequently, if τ/σ is given and γ is constant, e is not constant and the change in e corresponds to the change in v from its value for the elastic state to $\frac{1}{2}$. Accordingly, pure relaxation was not achieved in the tests. The specimens studied were aluminium alloy RR-59 at 200°C and magnesium alloy at 50°C (see section 77) and the tests were made in the stress range in which the material is reasonably isotropic and consequently, eqs. (81.1) are valid. If the compressibility of the material is neglected any mechanical theory may be used to solve the problem of relaxation. The authors compare the flow theory with the time-hardening theory, the strain-hardening theory in its usual formulation when the magnitude of the octahedral shear is taken as a measure of work-hardening and the combined theory. The time-hardening theory usually gave the best agreement with experiment although in some tests the strain-hardening theory gave the same or even better results. At the temperature of test it was found that for the aluminium alloy $v = 0.43$ and for the magnesium alloy $v = 0.39$; these values differ little from $\frac{1}{2}$ and therefore the authors considered that relaxation did occur in these experimental conditions. In calculating relaxation curves by the time-hardening and strain-hardening theories, elastic compressibility was allowed for, but not accurately. Namestnikov [302] states that accurate calculation of the relaxation curve (under conditions somewhat different from those used by Johnson and colleagues) gives results which are hardly distinguishable from those of the approximate solution.

Several tests on specimens of alloy D-16T have been published by Namest-

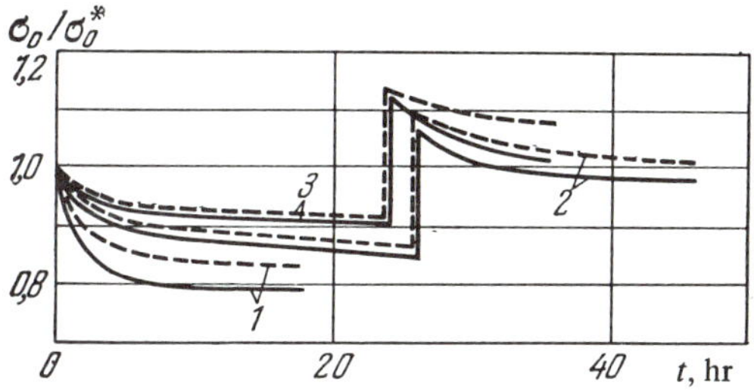

Fig. 135.

nikov [302]. In this work, the axial elongation and the angle of twist were maintained constant simultaneously, and, therefore, both force and torque relaxed together. In working out the results it was, from the start, assumed possible to neglect compressibility, therefore, it was decided in advance that the stresses should diminish in a given ratio. In fact the ratio τ/σ varied slightly during the process, it diminished during the first hours, and then remained practically constant, the difference between the maximum and minimum values of this ratio being not more than 14%. Stress intensity and relaxation curves are given in fig. 135, σ_0^* is the initial stress intensity. Results calculated by eq. (82.1) are plotted in dotted lines. For specimen 1 $\sigma_0^* = 33$ kg/mm^2, $\tau^*/\sigma^* = 0.58$; for specimen 2 in the first stage $\sigma_0^* = 18.73$ kg/mm^2, $\tau^*/\sigma^* = 1.45$, after 26 hours the stress was raised to $\sigma_0^* = 20.4$ to maintain the ratio τ/σ constant. In specimen 3 $\sigma_0^* = 28.1$, $\tau^*/\sigma^* = 0.242$ and at $t = 24$ hours the stresses were increased and for the next stage $\sigma_0^* = 32.2$, $\tau^*/\sigma^* = 0.214$. As in Johnson's tests relaxation was in fact somewhat faster than predicted by theory but the difference was not so great and the totality of published results on the experimental study of relaxation in a state of complex stress can be considered as serious confirmation of the validity of the equations of flow with work-hardening.

In many structures the question of stress redistribution with creep is one of primary importance and the occurrence of quite small creep deformations greatly reduces the peak stresses that occur at the initial instant when the behaviour of the material is still elastic. In problems of this kind we are concerned with processes very close to that of stress relaxation, and, therefore, direct verification of the theory in the region of small creep deformations that accompany relaxation is of direct interest.

6

Long Term Failure at High Temperatures

In assessing the strength of products that operate at high temperatures the possibilities that must be considered are that deformation may become too great and that failure may occur. There have been numerous investigations of long-term failure of metals and alloys at high temperatures and a great deal of experimental data has been accumulated, but the problem of extrapolating these long-term strength data cannot be considered solved. Here we shall review the fundamental experimental facts and present a phenomenological interpretation and also describe certain mechanical theories that can be used to predict the time of failure. The central idea of these theories is that there are two main types of failure, viscous and brittle. In actual materials these types may appear in pure form or may interact with one another giving a picture of mixed failure. Criteria of long-term strength in a state of complex stress are also considered.

83. Fundamental information concerning long-term strength

If the applied stress is high enough, creep of the specimen at constant load or constant stress terminates by transition to the third phase of creep which leads to failure after a certain time. The time to failure increases with diminution in the applied stress. In the standard method of long-term strength testing the specimen is subject to constant load and the dependence of the time to failure on the nominal stress referred to the initial cross-sectional area of the specimen is determined. The test results are presented in the form of curves of long-term strength, an example of which is shown in fig. 136. Given the curve of long-term strength we can determine the so-called long-term ultimate strength, that is the minimum stress at which failure occurs after a given time. Like the creep limit, the long-term ultimate strength is a conventional quantity; the life for which the long-term ultimate strength is determined depends upon the service conditions of the product and may range from a few minutes to tens of thousands of hours. The curve shown in fig. 136 apparently does not have a horizontal asymptote and so it is hardly possible to speak of the absolute, long-term ultimate strength. At a high stress level failure occurs relatively quickly and is accompanied by considerable elongation, which consists of an instantaneous plastic part and creep deformation. As a rule, when the stress is reduced and the time that the specimen carries load before failure is increased, the amount of deformation that takes place up to the instant of failure is less and failure tends to become brittle. Many alloys, particularly heat-resistant alloys of complex structure, display brittle long-term failure with hardly any plastic deformation. A typical graph of the relationship between elongation at failure and time to failure is shown in fig. 137. In many materials, such as certain austenitic steels and technically pure molybdenum, deformation at failure reaches a minimum at a certain value of t, longer term failure at lower stress giving greater elongation. A curve of this kind is shown dotted in fig. 137.

Quite small changes in stress have a great influence on the time to failure, and, therefore, it is convenient to construct long-term ultimate strength curves in logarithmic/logarithmic or logarithmic/linear coordinates and in both cases the logarithm of time is plotted on the abscissa. A typical diagram of

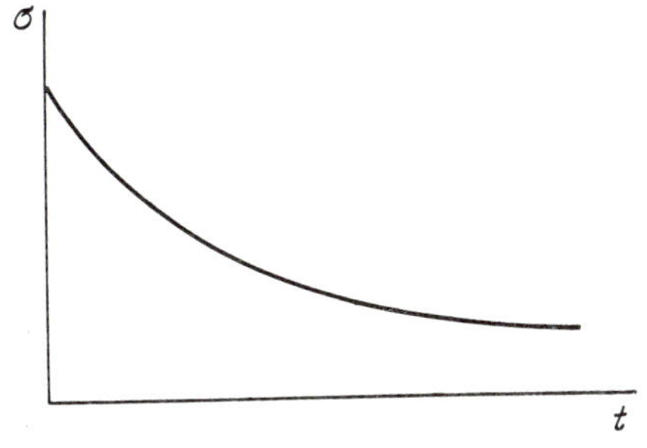

Fig. 136.

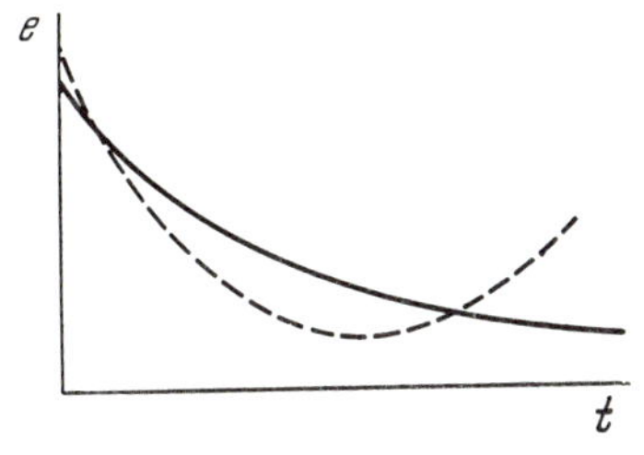

Fig. 137.

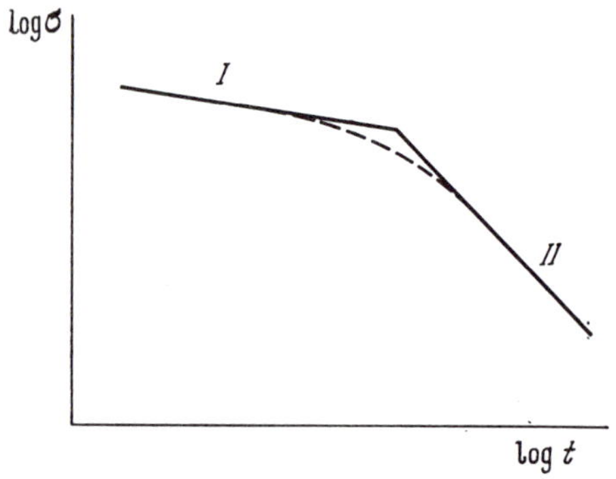

Fig. 138.

long-term strength in logarithmic/logarithmic coordinates, which is the form generally used, is shown in fig. 138. This diagram is usually satisfactorily approximated by two straight lines, one of lesser slope at high stresses and one of greater slope at low stresses. Sometimes the point of intersection between the lines is quite clear and sometimes there is a transitional curved section, as shown dotted. In structurally unstable materials the long-term strength diagram may be more complicated.

The inflection point on the long-term strength diagram corresponds to a change in the nature of failure. In part I failure is viscous, it is accompanied by considerable deformation and failure cracks pass through the body of the grains. In part II plastic deformation is small and failure is preceded by the formation of numerous cracks on the intergrain boundaries. These cracks grow and join together until the specimen fails.

Depending upon the material and the test temperature the transition from intra-grain to transgranular failure takes place at various times ranging from minutes to thousands of hours. Sometimes the inflection point is not found on the diagram which comprises only the first section. In calculating products intended for long service life it is necessary to know the long-term ultimate strength on a long basis, for instance, 100 000 hours. The test duration is usually selected so as clearly to include the second linear part of the diagram, which is extrapolated to the required value of t to give the long-term ultimate strength.

In the first part failure occurs with considerable plastic deformation and, therefore, the diagram is of a conventional nature; the true stress at the instant of failure is very different from the value plotted on the ordinate in constructing the diagram. In the second part matters are better in this respect, the deformation is small and so the nominal stress is little different from the true value.

Numerous studies have been made of the mechanism of brittle failure in creep. The observations indicate that intergrain cracks can have various origins. At relatively moderate temperatures joints between three neighbouring grains play an important part. Because of the different orientation of these grains it is difficult for the metal to flow in the neighbourhood of the joint so that there is a stress concentration which leads to cracking. At higher temperatures pores form on the boundaries and these unite to form wide cracks with rounded ends [187].

The relationship between crack formation and creep is far from clear but an empirical fact which has been observed, at any rate for certain materials, mainly pure metals and single-phase alloys, is that the product of the rate of steady-state creep and time to failure is approximately constant and independent of stress and temperature [35, 204]. Because of this a number of authors [33, 320, 190] have used the same formulae to extrapolate the creep and long-term strength results.

84. Temperature-time dependence of long-term strength

A great deal of experimental work is required to obtain complete long-term strength characteristics of a material throughout the range of operating temperatures, and the work becomes altogether impossible if the service life is very long. Naturally enough, therefore, there have been many attempts to find formulae by which to extrapolate the results of short-term tests or to replace long-term tests at low temperatures by shorter tests at higher temperatures. Apparently no such universal formulae exist, different alloys behave differently and the simple physical models that form the basis of the various attempts relate to an idealised material rather than to a real technical alloy. The literature on the extrapolation of long-term strength results is very great and the information given below is intended mainly to illustrate the complexity of the problem. Certain information about temperature dependences of creep rate was given in section 45, and in particular formula (45.4) was given for the creep rates as function of stress and temperature. We shall rewrite this formula in more general form, namely

$$v \sim \exp\left(-\frac{u(\sigma)}{RT}\right) \times \text{const}\,. \tag{84.1}$$

Formula (45.4) is now obtained as a particular case when $u(\sigma) = u_0 - \gamma_0$. As was mentioned at the end of section 83, the product of the creep rate v and the time to failure t_* is approximately constant, denoting it by e_*, we have

$$v = \frac{e_*}{t_*} . \tag{84.2}$$

From (84.1) and (84.2) it follows that

$$T(c + \log t_*) = F(\sigma) . \tag{84.3}$$

The left-hand side of eq. (84.3) is termed the Larson-Miller parameter [276] and we have already used it in section 45. If long-term strength test results are plotted by laying out stress on one axis and the Larson-Miller parameter on the other, according to eq. (84.3) points for all temperatures and stresses should lie on a single curve.

Somewhat greater flexibility results from using the Manson-Haferd [291] parameter which was developed empirically. These authors propose the following expression:

$$\frac{T - T_a}{\log t_* - \log t_a} = F(\sigma) . \tag{84.4}$$

Formula (84.4) contains two constants T_a and t_a and, therefore, it can, of course, give better agreement with experiment. If we take the dependence of creep rate on stress and temperature in the form of formula (45.1), namely

$$v = \exp\left(\frac{u_0}{RT}\right) f(\sigma) ,$$

test results at various temperatures are represented by a single curve in coordinates σ–θ_* where θ_* is the referred time to failure,

$$\theta_* = \int_0^{t_*} \exp\left(\frac{u_0}{RT}\right) \mathrm{d}t = F(\sigma) . \tag{84.5}$$

This method of presenting long-term strength results was proposed in an article by Orr, Sherby and Dorn [320].

The possibility of predicting long-term strength characteristics by means

Calculated failure stress for actual life and error, %.

T, °C	σ, kg/mm^2	t_*	Formula (84.3)	%	Formula (84.4)	%	Formula (84.5)	%
			Nimonic 90					
650	4090	4110	4090	0	4090	0	4160	+2
	3780	6440	3900	+3	3840	+2	3940	+4
	3620	5370	3960	+9	3940	+9	4060	+12
	3150	15290	3540	+23	3440	+9	3630	+15
700	2040	12090	2440	+19	2300	+12	2470	+21
	1970	20400	2230	+14	2040	+4	2230	+14
730	1570	7590	1570	0	1300	−5	1600	+2
	1100	22160	1260	+14	1040	−6	1210	+10
815	865	3340	895	+4	820	−5	880	+2
	630	6830	740	+18	645	+2	675	+8
870	315	16292	284	−10	220	−30	236	−25
			Nimonic 80					
650	3150	5270	3210	+2	3100	−1.5	3220	+1.5
	2830	8170	2990	+6	2850	+0.5	2980	+5
	2520	13390	2760	+10	2520	0	2770	+10
700	2040	4840	2140	+5	2010	−1.5	2180	+7
	1570	10900	1820	+16	1590	+1	1820	+16
	1160	34060	1390	+26	1130	+3	1370	+24
750	1260	4490	1260	0	1240	−1	1240	−1
	945	13090	975	+3	880	−7	945	0
	630	22660	835	+33	725	+15	790	+25

of formulae (84.3), (84.4), (84.5) and others has been checked by many authors on extensive data [190, 212, 215, 250]. In general the results are rather indeterminate. The authors note that as a rule, each method provides a first, fairly rough approximation, somewhat better results are obtained with the Manson-Haferd parameter. By way of illustration we give results obtained in long-term tests, up to 34 000 hours, on Nimonic 80 and Nimonic 90 [169].

Formulae (84.3), (84.4) and (84.5) were used to calculate the stress which corresponds to the actually determined time to failure. The columns of the table give the values of these stresses and the differences from the actual failure stress as percentages.

The work of Claus [186] should be considered in more detail, the results that it contains clearly illustrate the difficulties that arise in applying different criteria. The relationship between stress and the Larson-Miller parameter for an austenitic steel is shown in fig. 139. The stress is plotted on the ordi-

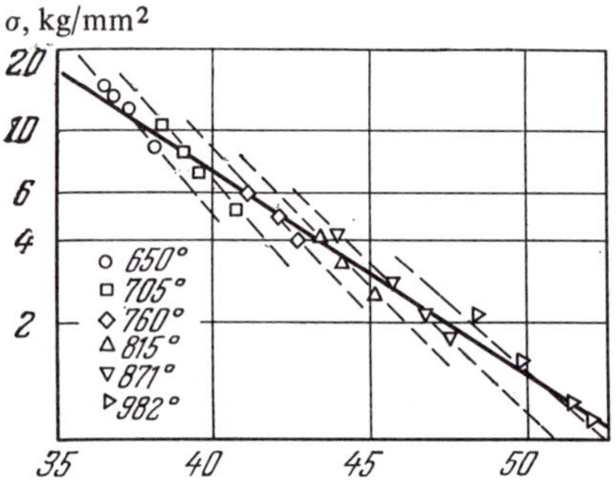

Fig. 139.

nate to logarithmic scale and the Larson-Miller parameter in conventional units on the abscissa. The points corresponding to each temperature lie on their own straight line, the straight lines are approximately parallel and are shown dotted. But a general straight line can also be drawn, as shown by the solid line, and the experimental points lie fairly closely around it. Given the experimental data for the extreme temperatures this curve can be drawn and can be used to give an approximate interpolation. But if we have experimental results for one temperature only, or for two neighbouring temperatures, a line of the required slope cannot be drawn through the points, and if the points corresponding to a single temperature are plotted and a line drawn through them, extrapolation to left or right will give considerable error.

Claus reaches the following conclusions from the analysis of extensive experimental material. In all cases the dependence of time to failure on stress and temperature could be written

$$\log t_* = a \log \sigma + b(T) . \tag{84.6}$$

Materials can be classified into four groups according to the behaviour of the quantity a.

I. $a = \text{const}$,

II. $a = a(\sigma)$,

III. $a = a(T)$,

IV. $a = a(T, \sigma)$.

Simple methods of representing the experimental results can be found for the first three groups. In particular, materials whose behaviour is described by Larson-Miller or Manson-Haferd criteria (84.3) and (84.4) belong to the third group. Austenitic steel, for which the graph of fig. 139 is constructed, clearly belongs to the first group. Indeed at a fixed temperature the Larson-Miller parameter is a linear function of $\log t_*$ as the points for different temperatures lie on parallel lines, the relationship between $\log \sigma$ and $\log t_*$ is linear with constant slope a. To combine all the dotted straight lines they must be shifted parallel to themselves, which involves appropriate selection of the function $b(T)$, the graph reconstructed in this way is given in fig. 140. It is obvious that when a = const formula (84.6) follows from formula (84.5) if we assume that $f(\sigma)$ is a power function. Here the energy of activation can be an arbitrary function of temperature.

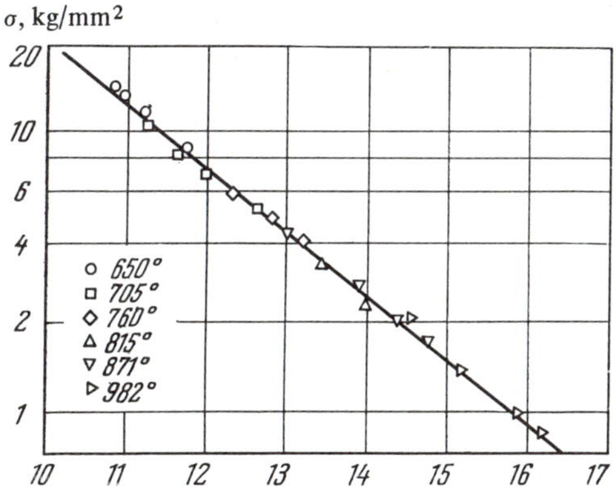

Fig. 140.

Materials of the second group differ from those of the first only in that the dependence of time to failure on stress can be arbitrary and is not necessarily a power function. Therefore, materials of the first and second groups observe a certain generalised Orr, Sherby and Dorn criterion.

The definition of the fourth group essentially says nothing more than that the time to failure is a function of temperature and stress. There are no established principles by which a material can be classed into any of these groups without performing long-term strength tests, for example, it does not depend on chemical analysis or structure and, therefore, the results that we have given are purely empirical.

85. **Viscous failure**

If the material is not embrittled by creep, failure will be described as viscous. In this case, with constant load the stress steadily increases because of reduction in the cross-sectional area. Initially the specimen constricts uniformly, but at a certain instant uniform flow becomes unstable and a neck is formed, if the stress in the neck is high enough, failure is of the same kind as in ordinary short-term tests. In some materials, for instance, pure niobium, plasticity is so great that the neck becomes so thin as to disappear completely and the halves of the failed specimen terminate in points. The time from the start of necking to the instant of failure is usually not very long compared with the total life of the specimen, and, therefore, necking need not be considered in assessing the life. Assume that uniform elongation of the specimen may be indefinitely great and that the creep law determined for small deformations remains valid throughout. Write this law in the form of eq. (50.6)

$$\dot{p}h(p) = f(\sigma)\,. \tag{85.1}$$

Here, of course, elastic deformation is unimportant and consequently, we may put

$$p = e - g(\sigma)\,. \tag{85.2}$$

Denote the length of the specimen during deformation by x, the initial length by x_0, and the cross-sectional areas by F and F_0 respectively. As measure of final deformation we take

$$e = \ln \frac{x}{x_0}\,.$$

From the condition of incompressibility

$$xF = x_0 F_0\,.$$

Hence

$$F = F_0 \exp(-e)\,.$$

Therefore, the true stress σ is related to the nominal stress σ_0, referred to the initial cross-sectional area by the equation

$$\sigma = \sigma_0 \exp(e)\,. \tag{85.3}$$

From eqs. (85.2) and (85.3) we obtain

$$p = \ln \frac{\sigma}{\sigma_0} - g(\sigma) .$$

Substitute this expression for p into (85.1):

$$\dot{\sigma}\left(\frac{1}{\sigma} - g'(\sigma)\right) h\left[\ln \frac{\sigma}{\sigma_0} - g(\sigma)\right] = f(\sigma) .$$

Separating the variables and integrating we obtain

$$t_* = \int_{\sigma_0}^{\sigma_*} \frac{d\sigma}{\sigma f(\sigma)} [1 - \sigma g'(\sigma)] h\left[\ln \frac{\sigma}{\sigma_0} - g(\sigma)\right] . \tag{85.4}$$

Here t_* is the time to failure and the instant of failure is defined as that at which the stress σ reaches a certain value σ_*.

If in eq. (85.4) we consider the upper limit of integration as a variable we obtain an equation for the creep curve which is valid for fairly large deformations and which describes the behaviour of the specimen up to the time of necking. This curve is shown schematically in fig. 141, it has a vertical asymptote. When $\sigma = \sigma_*$ we are at the point A of the creep curve; the equation ceases to be valid when $\sigma > \sigma_*$ because the extension ceases to be uniform and a neck forms on the specimen. However, the abscissa of point A differs

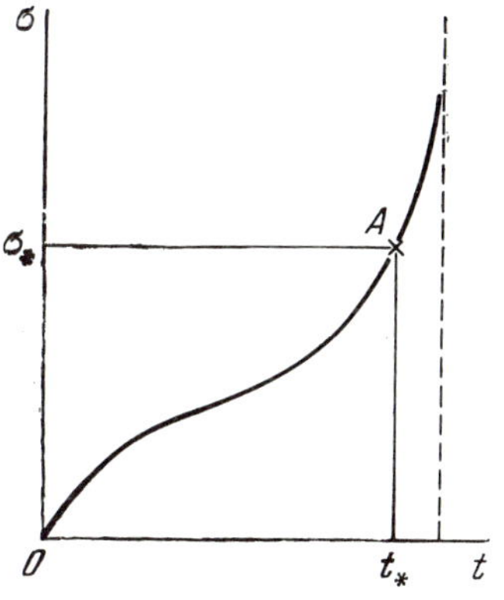

Fig. 141.

little from that of the asymptote and beyond point A the curve rises steeply. This gives ground for considering the time to failure to be that at which the creep curve reaches its asymptote, that is to take $\sigma_* = \infty$ in eq. (85.4).

The creep curve is not always of the shape shown in fig. 141, for this it is necessary that the sub-integral expression should not change sign; that is for any value of σ

$$1 - \sigma g'(\sigma) \geqslant 0 .$$

It may happen that this inequality breaks down at certain values of σ which we denote σ_*. This means that when $\sigma = \sigma_*$ where σ_* is given by

$$1 - \sigma_* g'(\sigma_*) = 0 ,$$

the creep rate becomes infinitely great. The integral curve given by formula (85.4) is of the shape shown in fig. 142. Only section OA of the curve has real meaning, when the point A is reached equilibrium becomes impossible and failure is instantaneous.

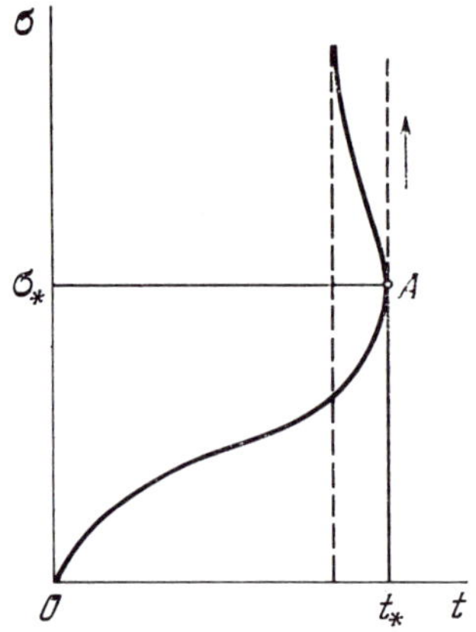

Fig. 142.

We shall now consider several particular cases of the application of eq. (85.4) to assess long-term strength.

a) *Viscous-plastic non work-hardening body.* For a body of this kind $h(p) = 1$, $f(\sigma) = \frac{1}{3}\mu(\sigma - \sigma_*)$, $g(\sigma) = 0$. We also take $\sigma_* = \infty$; therefore, t is the time at which the length of the extended specimen becomes infinitely great and the section infinitely small. Substituting into eq. (85.4) we obtain

$$t = 3\mu \int_{\sigma_0}^{\infty} \frac{d\sigma}{\sigma(\sigma - \sigma_T)} = \frac{3\mu}{\sigma_T} \ln \frac{\sigma_0}{\sigma_0 - \sigma_T} . \tag{85.5}$$

Formula (85.5) was derived by Hencky in 1925 [228]; Odquist [313] provided a generalisation for a work-hardening viscoelastic body.

b) *Creep without work-hardening.* The simplest method of assessing the time to viscous failure of a metal rod undergoing creep in tension was derived by Hoff [233]. Following this author put $\sigma_* = \infty$, $h(p) = 1$, $g(\sigma) = 0$. Then eq. (85.4) leads to the following result:

$$t_* = t_1 = \int_{\sigma_0}^{\infty} \frac{d\sigma}{\sigma f(\sigma)} . \tag{85.6}$$

If, as Hoff supposed, $f(\sigma) = A\sigma^n$ (in this chapter for convenience we shall use this method of writing the power law of creep instead of the standard form) then

$$t_* = \frac{1}{An\sigma_0^n} . \tag{85.7}$$

If $f(\sigma) = \epsilon_e \exp \sigma/\sigma_e$,

$$t_* = -\frac{1}{\epsilon_e} \operatorname{Ei}\left(-\frac{\sigma_0}{\sigma_e}\right). \tag{85.8}$$

These results can be made more precise by taking account of the instantaneous plastic deformation. The corresponding solution was derived by Odquist [316]. Following this author put

$$g = b\sigma^m$$

and retain Hoff's remaining assumptions, then

$$\sigma_*^n = \frac{1}{bm} , \qquad t_* = \int_{\sigma_0}^{\sigma_*} \frac{d\sigma}{A\sigma_0^{n+1}} (1 - bm\sigma^m) .$$

Noting that $b\sigma_0^m$ is the initial plastic deformation corresponding to the stress

σ_0, rewrite this formula as

$$t_* = \frac{1}{An\sigma_0^n}\left[1 - \frac{mn}{n-m}\, g(\sigma_0) + \frac{m}{n-m}\left(\frac{\sigma_0}{\sigma_*}\right)^n\right]. \tag{85.9}$$

(In the original work [316], it is assumed that $\sigma_* = \infty$, therefore, the formula for t_* differs from (85.9) because the third term is absent from the right-hand side.)

A simpler scheme for allowing for the possibility of plastic deformation was proposed by Rosenblyum [114]. He assumed that during instantaneous deformation the material is ideally plastic, it is of little importance whether it is elastically or rigidly plastic, because the elastic deformation is small. In eq. (85.6) he substituted the yield point σ_T for the upper limit of integration on the basis that when the stress reaches this value unlimited flow commences. With the power law of creep instead of eq. (85.7) we obtain

$$t_* = \frac{1}{An\sigma_0^n}\left[1 - \left(\frac{\sigma_0}{\sigma_T}\right)^n\right]. \tag{85.10}$$

Note that when $n = 1$ creep turns into viscous flow, however, Rosenblyum's scheme differs in principle from that of Hencky and, therefore, the result of eq. (85.5) cannot be obtained from eq. (85.10). If the corresponding body is modelled by elastic-plastic and viscous components, as described in section 23, but feedback is introduced which alters the plasticity and viscosity characteristics as functions of the elongation, which corresponds to change in cross-sectional area of the specimen, then with Hencky's scheme the components should be connected in parallel and with Rosenblyum's in series, as for eq. (85.4).

Turning again to Odquist's model, which, with the particular assumption concerning the form of the function $g(\sigma)$ leads to formula (85.9), note that the function $g(\sigma)$ can include both the instantaneous plastic deformation and deformation of the first stage of creep. The power approximation for the function $g(\sigma)$ is justified for small deformations but it is by no means obvious that it is satisfactory for large deformations where a logarithmic measure of deformation must be used. At the same time the qualitative result depends on the form of the function $g(\sigma)$ because this function determines the quantity σ_*.

Suppose for example that g is a power function of the conventional and not of the true stress. For an instantaneous deformation, $e = g$ and

$$g = b\sigma_0^m = b\sigma^m \exp(-mg)\,.$$

Differentiating both parts with respect to σ we find

$$\sigma \frac{dg}{d\sigma} = \frac{mg}{1+mg} < 1 .$$

Therefore, the dependence of σ on t is always as shown in fig. 141.

So far we have assumed that the loads acting on the specimen are constant. We shall now suppose that σ_0 is a given function of time, we neglect the instantaneous plastic deformation and take it that $f(\sigma)$ is a power function. Then $e \approx p$ and on the basis of eq. (85.3)

$$\dot{e} = f[\sigma_0 \exp(e)] = A\sigma_0^m \exp(ne) .$$

Separating variables and integrating we obtain

$$\int_0^\infty \exp(-ne) de = \frac{1}{n} = \int_0^t A\sigma_0^n dt . \tag{85.11}$$

If the stress σ_0 were constant, by formula (85.7) this would correspond to a time to failure t_* which we denote $t_*(\sigma_0) = 1/(An\sigma_0^n)$. Formula (85.11) can now be rewritten

$$\int_0^t \frac{dt}{t_*(\sigma_0)} = 1 . \tag{85.12}$$

Condition (85.12) expresses the so-called principle of linear summation of damage. If a stress σ_i is applied to the specimen for a time t_i it is considered that the specimen has acquired a degree of damage which is equal to the ratio of the time of action of the stress σ_i to the time required to cause failure at the given stress, $t_*(\sigma_i)$. With variable loads the damage is summated, and failure occurs when the amount of accumulated damage reaches a value of 1. This principle was proposed by Robinson [343]. It should be noted that for any viscous failure the principle of linear summation only follows from the equations that we have considered provided that very special assumptions are made concerning the power law of creep and the possibility of neglecting instantaneous plastic deformation.

c) *Creep with work-hardening.* If we do not allow for instantaneous plastic deformation the principle of summating damage is also valid for a material in which creep is accompanied by work-hardening, provided $f(\sigma)$ is a power function. Indeed, in this case eq. (85.1) gives

$$h(e)\exp(-ne)\mathrm{d}e = A\sigma_0^n\,\mathrm{d}t\,.$$

Denoting

$$\int_0^\infty h(e)\exp(-ne)\mathrm{d}e = \frac{1}{n}\,.$$

Then we obtain

$$1 = \int_0^t n'A\sigma_0^n\,\mathrm{d}t\,.$$

If σ_0^n = const, from this equation we obtain the time to failure

$$t_*(\sigma_0) = \frac{1}{n'A\sigma_0^n}\,.$$

This expression differs from formula (85.7) only by a numerical factor. Obviously, formula (85.11) remains valid for variable loads.

86. **Failure accompanied by embrittlement**

The scheme of viscous failure that we have described is realised only in certain pure metals and with short test durations. As we have already mentioned, at long durations failure occurs with much less deformation. Metallographic examination shows that failure is preceded by the formation of microcracks, principally on the grain boundaries. The preferential direction of these cracks is perpendicular to the line of action of the tensile stress. In the course of time the cracks grow and finally failure occurs. Failure is always preceded by an increase in creep rate, that is by the formation of a third part on the creep curve. We have seen in section 85 that the third part also occurs on the creep curve of fully 'viscous' material because of the reduction in cross-sectional area (see fig. 141). However, in actual fact the third part may commence at very low deformations and the quantitative result that is obtained by eqs. (85.4) does not explain the experimental facts.

We now give a rough, schematic description of the process of cracking.

Suppose that round cracks appear in the cross-section of the rods so that there is one crack of area $\omega = \pi\rho^2$ per unit cross-sectional area. The rate of propagation of the front of the crack is a function of the mean stress and the structural state of the material, which is characterised by its degree of work-hardening [104]. Then

$$\frac{d\omega}{dt} = 2\pi\rho \frac{d\rho}{dt} = \sqrt{\omega u}\,.$$

The quantity u is the rate of growth of the crack front to within a constant factor. If the crack is of elongated shape, for example, rectangular, and is propagated in only one direction

$$\frac{d\omega}{dt} = u\,.$$

In the general case it may be assumed that

$$\frac{d\omega}{dt} = \omega^{\beta} u \qquad (0 \leqslant \beta \leqslant \tfrac{1}{2})\,.$$

If the cross-sectional area of the uncracked specimen is F, the cross-sectional area that bears load is $F(1-\omega)$. On the other hand, because of creep the cross-sectional area is reduced in the ratio $\exp(-e)$. Therefore, the actual cross-sectional area at any instant is

$$F_0(1-\omega)\exp(-e)\,,$$

and the true stress is

$$\sigma = \frac{\sigma_0 \exp(e)}{1-\omega}\,.$$

We shall term the quantity ω the damage suffered by the material. The change in damage with time is described by the equation

$$\omega^{-\beta} \frac{d\omega}{dt} = u\left(\frac{\sigma_0 \exp(e)}{1-\omega}, p\right). \tag{86.1}$$

Besides eq. (86.1) we must consider the creep equation

$$\dot{p}h(p) = f(\sigma, \omega) . \tag{86.2}$$

In writing (86.2) we allow for the possibility that the extent of damage to the material influences the course of creep. Various assumptions may be made concerning the functions u and f, so there are extensive possibilities of describing experimental facts. We shall consider, for example, that creep strain is small at the instant of failure and the rate of crack propagation depends only on stress. It then suffices to consider only eq. (86.1) which assumes the form

$$\omega^{-\beta} \frac{d\omega}{dt} = u \left(\frac{\sigma_0}{1-\omega} \right) . \tag{86.3}$$

If $u(\sigma)$ is a power function so that

$$u(\sigma) = B\sigma^k ,$$

then in eq. (86.3) the variables are separated and we obtain

$$(1-\omega)^k \omega^{-\beta} d\omega = B\sigma_0^k \, dt .$$

Integrating we find

$$\int_0^1 (1-\omega)^k \omega^{-\beta} d\omega = \int_0^t B\sigma_0^k \, dt .$$

The integral in the left-hand side is taken from $\omega = 0$ to $\omega = 1$. Strictly speaking this scheme assumes that there is a certain number of incipient cracks, which remains unchanged. Therefore, at the initial time ω differs from zero. Representing the cracks by circles, we cannot permit the assumption that $\omega = 1$, which is geometrically impossible. On the other hand failure occurs at a certain value of $\omega = \omega_* < 1$ and, therefore, the upper limit of the integral should be ω_*, but we replace it by unity on the basis of the same considerations as those by which in formula (85.4) the upper limit σ_* is replaced by infinity. The integral in the left-hand side is always a finite number

$$\int_0^1 (1-\omega)^k \omega^{-\beta} d\omega = \frac{\Gamma(1+k)\Gamma(1-\beta)}{\Gamma(2+k-\beta)} = \frac{1}{1+k'}$$

(when $\beta = 0, k' = k$) .

Therefore, the condition of brittle failure is written

$$\int_0^t (1 + k')B\sigma_0^k \, dt = 1 \, . \tag{86.4}$$

The form of eq. (86.4) is no different from that of eq. (85.10). In exactly the same way if $\sigma_0 = \text{const}$, $t_*(\sigma_0) = t_2 = 1/[(1 + k')B\sigma_0^k]$ is the time to failure at a given stress. If σ_0 is variable, condition (86.4) leads to the law of linear summation of damage which is equally applicable to pure viscous and wholly brittle fracture. In logarithmic coordinates, $\log \sigma = \log t_*$, both eqs. (85.10) and (86.4) give straight lines but with different slopes. We saw in section 83 that parts of real long-term strength curves plotted in logarithmic coordinates are linear. Indeed the first part corresponds to transcrystalline fracture accompanied by considerable plastic deformation and the second to low loads and long durations when failure occurs through the grain boundaries and deformation is negligibly small. Between these extreme cases, there is an intermediate region of mixed failure to which the foregoing analysis is inapplicable, and the system of eqs. (86.1) and (86.2) should be considered with more general assumptions.

87. Mixed failure. Kachanov's hypothesis

In considering mixed failure, Kachanov [51] proposed the hypothesis that crack formation does not influence the creep process and consequently, eq. (86.2) does not include the quantity ω and this equation is integrated independently of eq. (86.1). The author presented the following considerations in support of this hypothesis. In general the mechanisms of creep and failure are different, brittle fracture develops along the grain boundaries and viscous through the body of the grains. On the other hand, if cracks do influence creep the creep curves from which the creep equations are established reflect the total effect. Limiting ourselves to a power law of creep without work-hardening we write eq. (86.2) in the form

$$\dot{e} = A\sigma_0^n \exp(ne) \, . \tag{87.1}$$

Hence, by integration we obtain

$$\exp(e) = \frac{F_0}{F} = (1 - nA\sigma_0^n t)^{-\frac{1}{n}} \, .$$

Note that by formula (85.7) the quantity $nA\sigma_0^n = 1/t_1$, where t_1 is the time to viscous failure for the given stress σ_0. Hence we may write

$$\exp(e) = \left(1 - \frac{t}{t_1}\right)^{-\frac{1}{n}}. \tag{87.2}$$

In writing eq. (86.1) Kachanov takes $\beta = 0$ and selects the function u in the form of a power function of σ alone. Bearing in mind (87.2) he obtains

$$\frac{d\omega}{dt} = B\frac{\sigma_0^k}{(1-\omega)^k}\left(1 - \frac{t}{t_1}\right)^{-\frac{k}{n}}. \tag{87.3}$$

Separating variables and integrating the left-hand side from $\omega = 0$ to $\omega = 1$ amd the right from $t = 0$ to $t = t_*$ which is the time to failure, we find

$$\frac{t_*}{t_1} = 1 - \left(1 - \frac{n-k}{n}\frac{t_2}{t_1}\right)^{\frac{n}{n-k}}. \tag{87.4}$$

Here t_2 is the time to brittle failure with a stress σ_0 in formula (86.4). Formula (87.4) is valid when the value of t_* that it gives is less than t_1 otherwise failure is purely viscous. If $t = t_1$

$$\frac{t_1}{n-k} = \frac{t_2}{n}.$$

Hence it follows that eq. (87.4) applies provided that the following condition is fulfilled:

$$\sigma_0 \leqslant \left(\frac{B}{A}\frac{k+1}{n-k}\right)^{\frac{1}{n-k}} = \sigma_0'.$$

Adopting the usual procedure for long-term strength curves, plot $\log\sigma$ against $\log t_*$ and draw straight lines corresponding to viscous failure $t = t_1(\sigma)$ and brittle fracture $t = t_2(\sigma)$ (fig. 143). At the point A $\sigma = \sigma'$; if $\sigma > \sigma'$ the long-term strength curve is a segment of the straight line $t_* = t_1$ to the left of the point A. For lower stresses this is a curve inscribed in the angle between the intersecting straight lines; this curve can be calculated by eq. (87.4); for low stresses $t_1(\sigma) \gg t_2(\sigma)$ and it follows from eq. (87.4) that $t_* \approx t_2$.

Kachanov considered in particular the case when $n = k$. The corresponding formula can be obtained as a limiting transition from eq. (87.3), but we shall

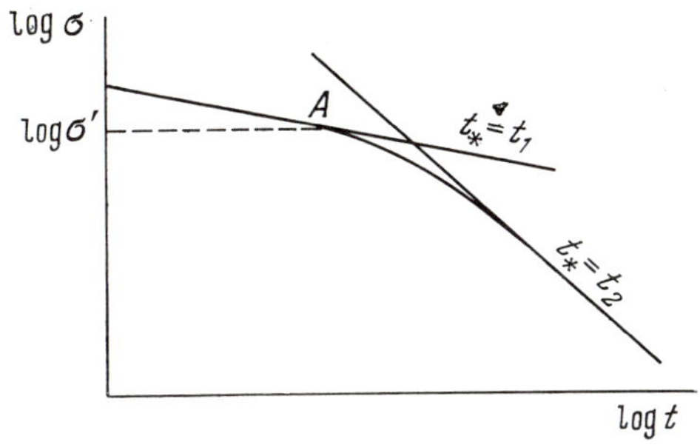

Fig. 143.

derive it independently. Rewrite eqs. (86.1) and (86.2) as follows:

$$\dot{\omega} = B\sigma_0^n (1-\omega)^{-n} \exp(ne) ,$$

$$\dot{e} = A\sigma_0^n \exp(ne) .$$

Dividing the second equation by the first we obtain

$$\frac{\mathrm{d}e}{\mathrm{d}\omega} = \frac{A}{B}(1-\omega)^n .$$

Integrating this equation we find

$$e = \frac{A}{B}\,\frac{1}{1+n}\,[1-(1-\omega)^{n+1}] .$$

At the instant of failure $\omega = 1$, consequently, $e = e_* = A/B[1/(n+1)]$. Thus if $n = k$ brittle failure always occurs at the same value of deformation. By formula (87.2)

$$\frac{t_*}{t_1} = 1 - \exp(-e_* n) = q = \text{const} . \tag{87.5}$$

Thus embrittlement reduces in constant ratio the life determined by the viscous failure scheme. This result explains to some extent the fact that life calculation by the Hoff scheme sometimes gives a satisfactory quantitative result (for example in the example given by Hoff) though the initial hypotheses are by no means confirmed.

Note that when $n = k$ the principle of linear summation of damage remains valid. This is easily confirmed by writing the following expression for the case of variable load instead of eq. (87.1):

$$\exp(ne) = 1 - \int \frac{dt}{t_1} .$$

Putting $e = e_*$ we find that $\int \frac{dt}{t_1} = q$, whence

$$\int \frac{dt}{t_*(\sigma_0)} = 1 .$$

Here as usual, t_* is the time to failure with constant stress σ_0.

Odquist [316] made further improvements in Kachanov's theory by allowing for instantaneous plastic deformation (or first-phase creep deformation) in the following way. Denote by $\alpha\sigma_0$ the stress at which failure occurs. It follows from formula (86.3) that $\alpha = \exp(e_*)$ where e_* is the deformation at the instant of failure. By formula (85.4) taking as in section 85, $f(\sigma) = A\sigma^n$, $g(\sigma) = b\sigma^m$, $h = 1$, we obtain

$$t = \int_{\sigma_0}^{\alpha\sigma_0} \frac{1 - bm\sigma^m}{A\sigma^{n+1}} \, d\sigma = t_3 - \frac{t_1}{\alpha^n} \left[1 + \alpha^m \left(\frac{t_3}{t_1} - 1 \right) \right] . \tag{87.6}$$

Here t_3 is the time given by formula (85.9). Now rewrite the equation of damage as

$$\dot{\omega}(1 - \omega)^k = B\sigma^k$$

or

$$(1 - \omega)^k \, d\omega = B\sigma^k \frac{dt}{d\sigma} \, d\sigma .$$

But from (87.6)

$$\frac{dt}{d\sigma} = \frac{1 - bm\sigma^m}{A\sigma^{n+1}} .$$

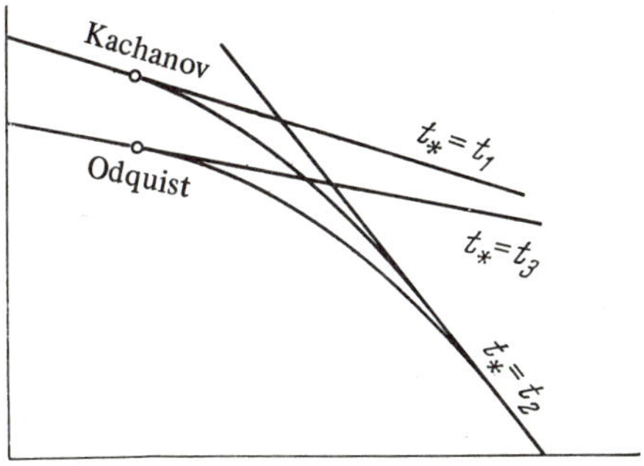

Fig. 144.

Therefore

$$(1-\omega)^k \, d\omega = \frac{B}{A} \frac{1 - bm\sigma^m}{\sigma^{n+1-k}} \, d\sigma \, .$$

Integrate the left-hand side from $\omega = 0$ to $\omega = 1$, and the right-hand from $\sigma = \sigma_0$ to $\sigma = \alpha\sigma_0$. After simplification we obtain

$$1 = \frac{t_2}{t_1}\left[\frac{n}{n-k}(1-\alpha^{k-n}) - \frac{n-m}{n-k-m}\left(1-\frac{t_3}{t_1}\right) \times (1-\alpha^{k+m-n})\right]. \tag{87.7}$$

Excluding the parameter α from (87.6) and (87.7) we obtain the long-term strength curve. Odquist's investigations showed that the long-term strength curve calculated by eqs. (87.6) and (87.7) is lower than the curve calculated by Kachanov's eq. (87.4) as shown in fig. 144. Certain further results in this direction are contained in the work [172].

88. More general hypothesis of failure. The brittle case

The embrittlement parameter ω, introduced in section 86, may, for clarity, be considered as the degree of reduction in effective cross-sectional area. However, this standpoint is not essential for the phenomenological theory. Returning to the general concept of kinetics of the creep process, section 55, we may consider the quantity ω as one of the structural parameters that define the state of the material. This quantity is a certain measure of cracking of the material, but we shall not attempt to establish a direct relationship between the number of cracks per unit volume, their dimensions and orientations on the one hand, and the parameter ω on the other. Therefore, our assumptions reduce to the following. Crack development takes place in a single way from the very start to the instant of failure. This means that if we have tests A and B in which crack development proceeds until failure, to each state of cracking in test A there corresponds an exactly similar state of cracking in test B and, in any experiment, these states follow one another in the same fixed sequence. This provides a basis for attributing a certain number ω to each state of cracking, however, conventionally we select this number in the range 0 to 1, here $\omega = 0$ corresponds to nominally undamaged material and $\omega = 1$ to the formation of macroscopic cracks. If we assume that actual crack development takes place differently under different conditions, for example, if we assume that cracks of two kinds exist, we shall be obliged to introduce two or more embrittlement parameters. Therefore, the hypothesis that we have proposed is the simplest [106].

Now assume, again for simplicity, that the quantity ω is the only structural parameter, so that, for example, work-hardening does not take place. Then in accordance with eq. (55.1) we should write

$$\dot{e} = f(\sigma, \omega) . \tag{88.1}$$

Here we have neglected the initial plastic and elastic deformations. We also make simple assumptions about the kinetics of change of ω, which follow as a particular case of eq. (55.2),

$$\dot{\omega} = \varphi(\sigma, \omega) . \tag{88.2}$$

Depending on whether we are in the low or high stress region we understand by σ the stress σ_0 referred to the initial cross-sectional area, or the true stress $\sigma = \sigma_0 \exp(e)$. In the same way, if deformations are small, e is normal deformation, otherwise e is the logarithmic deformation.

In considering simple schemes of failure, we approximated the functions that figure in the equations by means of power functions. In order to obtain comparable results we proceed in the same way in the general case, namely we take eqs. (88.1) and (88.2) in the form

$$\dot{e} = a\sigma^n (1-\omega)^{-q} , \qquad \dot{\omega} = c\sigma^k (1-\omega)^{-r} . \tag{88.3}$$

The coefficients a and c are functions of temperature, here in principle, the temperature dependences of creep and failure may be different, these processes being characterised by different energies of activation.

The case when the deformation at the instant of failure is small and consequently the change in cross-sectional area may be ignored, is considered to be that of pure brittle fracture. In eq. (88.3) put $\sigma = \sigma_0$, then from the second equation of system (88.3) it follows that

$$(1-\omega)^{r+1} = 1 - \frac{t}{t_2}. \tag{88.4}$$

Here t_2 is the time to brittle failure defined as in section 86:

$$t_2 = \frac{1}{c(1+r)\sigma_0^k}.$$

Note that it is not essential for the final result to introduce the parameter β which allows for the shape of the crack, therefore, we may take $\beta = 0$ in accordance with [51].

Now substitute expression (88.4) for ω in the first of eqs. (88.3) and integrate it. We then obtain the equation of the creep curve

$$e = \frac{\lambda}{n}\frac{t_2}{t_1}\left[1 - \left(1 - \frac{t}{t_2}\right)^{\frac{1}{\lambda}}\right]. \tag{88.5}$$

Here t_1 is the time to viscous failure defined in section 85:

$$t_1 = \frac{1}{an\sigma_0^n} \quad \text{and} \quad \lambda = \frac{r+1}{r-q+1}.$$

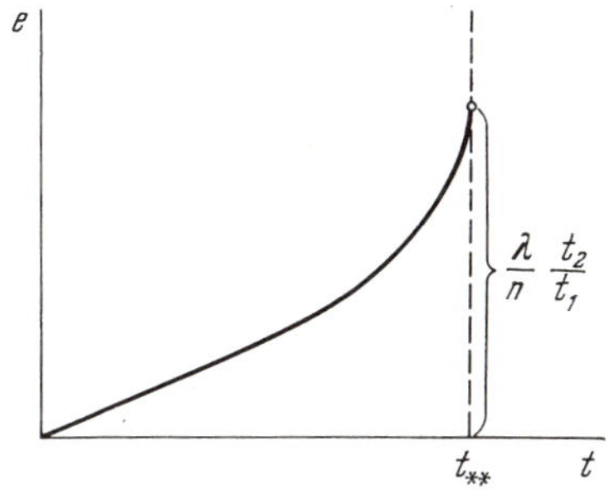

Fig. 145.

A typical example of a creep curve corresponding to eq. (88.5) is given in fig. 145. The value of the elongation at failure is

$$e_* = \frac{\lambda}{n}\frac{t_2}{t_1} = \frac{\lambda}{1+r}\frac{a}{c}\sigma_0^{n-k} . \tag{88.6}$$

Since everywhere $n \geqslant k$ the elongation at failure diminishes with diminishing stress.

It will be seen that use of the hypothesis expressed by eqs. (88.2) does not introduce any change into the definition of time to brittle failure but makes it possible to describe the third section of the creep curve that precedes failure. In statically indeterminate systems allowance for accelerated creep in the third stage permits more accurate description of stress redistribution in components of the system and more accurate assessment of life.

To determine the constants of eq. (88.3) experimentally we need data on creep in the steady-state parts of the curve, the long-term strength curve for the region of brittle failure and the dependence of elongation at failure on stress. The quantities a and n are found by working out creep curves, the constant k and the product $c(1+r)$ from working out the long-term strength curves. Moreover, if the relationship between e_* and σ_0 is plotted as a power dependence with exponent $n-k$, the coefficient λ can be found and is expressed in terms of r and q. It is not possible to determine the quantity r independently, therefore, some arbitrariness remains. Put for example $r = 0$, then

$$q = 1 - \frac{1}{\lambda} .$$

89. **Mixed failure. Short-term creep**

If the stress is high enough, failure occurs at large deformations and reduction of the cross-section must be allowed for. In eq. (88.3) putting $\sigma = \sigma_0 \exp(e)$ we obtain

$$\dot{e} = \frac{1}{nt_1}\exp(ne)(1-\omega)^{-q} ; \quad \dot{\omega} = \frac{1}{(1+r)t_2}\exp(ke)(1-\omega)^{-r} . \tag{89.1}$$

Dividing the one equation by the other we arrive at a differential equation for e as function of ω, namely

$$\frac{de}{d\omega} = \frac{1+r}{n} \frac{t_2}{t_1} \exp\left[(n-k)e\right](1-\omega)^{r-q} .$$

Integrating this equation with allowance for the initial condition $e(0) = 0$, we obtain

$$e = \frac{1}{n-k} \ln\left[1 - v + v(1-\omega)^{r-q+1}\right] . \tag{89.2}$$

Here

$$v = \frac{1+r}{n} \frac{n-k}{r-q+1} \frac{t_2}{t_1} = \frac{n-k}{r-q+1} \frac{a}{c} \sigma_0^{n-k} .$$

Now we shall consider two possibilities, namely:

1. $v < 1$, the condition of failure $\omega = 1$ is achieved at a certain finite value $e = e_{**}$ and

$$e_{**} = -\frac{1}{n-k} \ln(1-v) . \tag{89.3}$$

2. $v > 1$, the quantity e tends to infinity at a certain finite value $\omega = \omega_*$ and

$$\omega_* = 1 - \left(\frac{v-1}{v}\right)^{\frac{1}{r-q+1}} . \tag{89.4}$$

Now to determine the time to failure we substitute expression (89.2) for e into the second of the eqs. (89.1) and integrate. The result can be written

$$\frac{t}{t_2} = \lambda \int_{\zeta_*}^{1} (1 - v + v\zeta)^{\mu} \zeta^{\lambda - 1} d\zeta . \tag{89.5}$$

Here we have introduced the additional notation

$$\mu = \frac{k}{n-k} .$$

The lower limit of integration $\zeta_* = 0$ in the first case, when $v < 1$, and $\zeta_* = (v-1)/v$ in the second case when $v > 1$. In both cases the integral in formula (89.5) is expressed in terms of hypergeometric functions, for this purpose we use the known formula

$$\int_0^1 u^{b-1}(1-u)^{c-b-1}(1-uz)^{-a}\,\mathrm{d}u = \frac{\Gamma(b)\Gamma(c-b)}{\Gamma(c)}\,F(a;b;c;z)\,.$$

After simple conversions we obtain

$$t_* = t_2 F(-\mu, 1; 1+\lambda; v) \qquad \text{when } v<1\,, \tag{89.6}$$

$$t_* = t_1 F\left(1-v, 1; \mu+2; \frac{1}{v}\right) \qquad \text{when } v>1\,. \tag{89.7}$$

For small values of v we approximately obtain

$$t_* = t_2\left(1 - \frac{k}{n}\,\frac{\lambda}{1+\lambda}\,\frac{t_2}{t_1}\right).$$

The case when $k = n$ requires particular investigation. Here

$$\frac{\mathrm{d}e}{\mathrm{d}\omega} = \frac{1+n}{n}\,\frac{t_2}{t_1}(1-\omega)^{r-q}\,.$$

Whence

$$(1-\omega)^{r-q+1} = 1 - \frac{e}{e_*}\,.$$

Here

$$e_* = \frac{\lambda}{n}\,\frac{t_2}{t_1} = \frac{a}{c}\,\frac{1}{r+1-q} = \text{const}\,.$$

The value of $1-\omega$ that has been found is substituted into the first of eqs. (89.1), which is integrated giving

$$\int^{e} \exp(-ne)\left(1 - \frac{e}{e_*}\right)^{\frac{q}{r-q+1}} \mathrm{d}e = \frac{t}{nt_1}\,. \tag{89.8}$$

This is the creep curve equation. The integral (89.8) can be calculated approximately if we consider that $ne_* \ll 1$. Including only two terms of the expansion in the subintegral expression we obtain

$$\frac{t}{t_1} = \frac{ne_*}{\lambda}\left\{1-(1-ne)\left(1-\frac{e}{e_*}\right)^{\lambda} - \frac{ne_*}{1+\lambda}\left[1-\left(1-\frac{e}{e_*}\right)^{1+\lambda}\right]\right\}. \tag{89.9}$$

Hence we find the time to failure, putting $e = e_*$

$$t_* = t_1 \frac{ne_*}{\lambda}\left[1-\frac{ne_*}{1+\lambda}\right], \tag{89.10}$$

or

$$t_* = t_2\left[1-\frac{t_2}{t_1}\frac{\lambda}{1+\lambda}\right]. \tag{89.10'}$$

Short-term creep is particularly suitable for the application of this theory. As we have seen in section 58, short-term creep is characterised by the absence of work-hardening, and, therefore, the degree of embrittlement is the only structural parameter. In a stabilised material structural changes have no great influence on creep and failure. In working out Mileyko's results it was taken that $n = k = q = r$. The condition $n = k$ follows from the observation that in the time range from 3 seconds to 3 minutes, the elongation at failure was approximately constant. The condition that $q = r = n$ was taken for simplicity, which is natural if we consider that ω represents the reduction in effective sectional area. For sufficiently small deformations it follows from eq. (89.10) that

$$t_* = t_1 \frac{n}{n+1} e_* = \frac{e_*}{a(1+n)} \sigma_0^{-n} \qquad (\lambda = n+1). \tag{89.11}$$

The constants a and n are found by working out the initial parts of creep curves, the only constant on which the condition of failure depends is e_*. We treated this quantity as elongation at the instant of failure, but in practice, it is difficult to determine e_* in this way. Indeed, on the one hand, the deformation is localised at places of failure, and determination of the true deformation requires additional assumptions and, on the other, the development of macrocracks in the last stages of failure is not described by the scheme under consideration. However, knowing the time to failure t_* for any particular stress we can find e_* from formula (89.11). By way of a check we may construct the creep curve by eq. (89.9) or for small deformations by (88.5) which assumes the form

$$e = e_* \left\{ 1 - \left(1 - \frac{t}{t_*} \right)^{\frac{1}{1+n}} \right\} . \tag{89.12}$$

Experimental points on the creep curve of alloy D-16AT at a temperature of 250°C for a stress of 13.6 kg/mm^2 are plotted in fig. 146. The curve calculated by eq. (89.12) is shown dotted. The creep constants, and also the constant e_*, are determined by averaging the results of several creep curves, the points relate to an individual specimen and the divergence that is observed is typical of creep tests. The fact that the calculated curve and experimental points differ from the very start, where there is pure steady-state creep, is an indication that in this case the divergence is associated with individual scatter. The continuous line represents the calculated curve for $\sigma = 13.8$ kg/mm^2 with the same values of the constants, the experimental points coincide almost exactly with this curve. This example provides further illustration of the fact that we have frequently mentioned, that a small difference in stress (in this case only 1.5% which was within the error of the test machine) corresponds to considerable scatter in deformation or time.

When the influence of change of cross-sectional area at failure is insignificant the principle of summation of damage, established earlier for pure brittle failure, remains valid. In the work [74] this principle was verified for aluminium alloy in tensile tests with constant rate of loading. The creep constants were determined for creep tests at constant load, the mean value of the constant e_* was found from long-term strength tests, after which tests were performed at constant rate of loading $\dot{\sigma}$ until failure, the stress at the instant

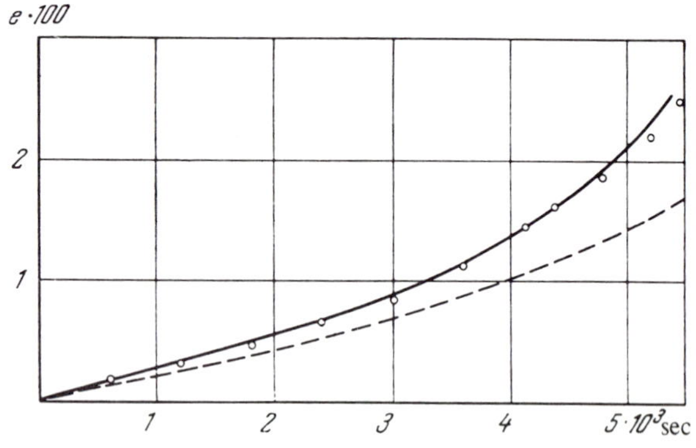

Fig. 146.

No.	T, °C	$\dot{\sigma}$ kg/mm^{2}sec	σ at failure, kg/mm^2		Error, %
			test	calculated	
1	230	0.35	27.6	29	+5
2	230	0.02	23.8	24.6	+3.5
3	275	1.4	26.8	26.9	+0.5
4	275	0.4	24.4	22.8	−6.5
5	275	0.1	20.6	21.3	+3.5
6	275	0.025	18.1	18.7	+3.5

of failure was recorded in the experiment and calculated by the law of linear summation of damage. Experimental and calculated results are compared in the table given above.

90. **Failure with cyclic loading**

Creep under cyclic loading was briefly considered in section 62. Failure under cyclic loading may result either from creep or from fatigue and strictly speaking it is impossible to separate the two. A good deal of work has been devoted to high temperature fatigue, but we cannot here include a detailed account of this subject. By way of illustration we give [277, 375] fatigue strength diagrams for different temperatures for the alloy S-816 (fig. 147).

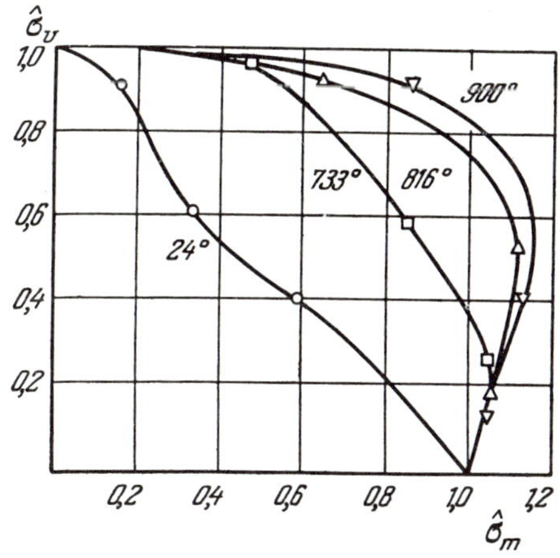

Fig. 147.

The amplitude of the cycle referred to the fatigue limit with a symmetrical cycle $\hat{\sigma}_v$ is plotted on the ordinate and the mean stress in the cycle referred to the ultimate long-term strength $\hat{\sigma}_m$ on the abscissa. This latter was determined on the basis of 100 hours and the fatigue limit on the basis of 2.16×10^7 cycles, the frequency was so chosen that this number of cycles was completed in 100 hours. It is evident that at the higher temperatures if $\hat{\sigma}_m < 0.5$ failure depends only on the alternating component of stress and may be considered to be pure fatigue. On the other hand, if $\hat{\sigma}_v < 0.3$ application of the alternating component has little influence on the long-term strength which moreover increases instead of diminishes. This phenomenon is not observed in all materials, it may result from the circumstance that application of the alternating component accelerates structural changes, which in this case have a strengthening effect.

There is a transitional region at 733° in which the relationship between $\hat{\sigma}_v$ and $\hat{\sigma}_m$ is of the same character as at normal temperature; at higher temperatures we can, to a first approximation, consider either that there is long-term failure which depends on the mean stress in the cycle, or that there is fatigue failure which depends only on the amplitude of the cycle.

Long-term failure under alternating loading may be considered from the standpoint of summation of damage. If failure is brittle, and the stress varies sinusoidally

$$\sigma = \sigma_m + \sigma_v \sin \omega t ,$$

then by formula (86.4)

$$kB \int_0^{t_*} (\sigma_m + \sigma_v \sin \omega t)^k \, \mathrm{d}t = 1 . \tag{90.1}$$

Failure under cyclic loading occurs in just the same way as under static loading at a stress σ_e, here

$$\sigma_e^k = \lim \left[\frac{1}{t} \int_0^t (\sigma_m + \sigma_v \sin \omega t)^k \, \mathrm{d}t \right]_{t \to \infty} . \tag{90.2}$$

The equivalent creep stress and equivalent failure stress are, generally speaking, different from one another, because the exponent k in formula (90.2) differs from the exponent n. Formula (90.2) permits of exactly the same graphical interpretation as the analogous formula for creep. In the work of Japanese authors [365] it is assumed that the rate of accumulation of damage depends on the absolute magnitude of the applied stress, therefore, in formula (90.2) an absolute value of the expression $(\sigma_m + \sigma_v \sin \omega t)$ is taken. According to their data this analysis gives satisfactory results if $\sigma_v/\sigma_m < 1.5$. If the amplitude of the cycle is increased relative to the mean amplitude, fatigue begins to occur. The authors [363] assume that fatigue damage takes place at the same time as damage by creep and independently of it, failure occurs when the sum of both kinds of damage reaches a critical value. The rate of accumulation of fatigue damage is proportional to some power of the stress amplitude, therefore the following condition is obtained:

$$t_* [kB\sigma_e^k + D\sigma_v^l] = 1 \, . \tag{90.3}$$

Here D and l are new constants that determine the kinetics of fatigue failure. Test results worked out by formula (90.3) [363] are shown in fig. 148. The ratios σ_v/σ_e and σ_m/σ_e are plotted on the coordinate axes, the circles and points correspond to experimental results for failure of aluminium alloy 24-S-T4 at a temperature of 260°C on the bases of 10^6 and 10^7 cycles respectively. The continuous line is the experimental curve. The chain-dotted lines correspond to failure due to creep alone and to fatigue alone. It will be seen that the region of mixed failures includes a comparatively narrow range

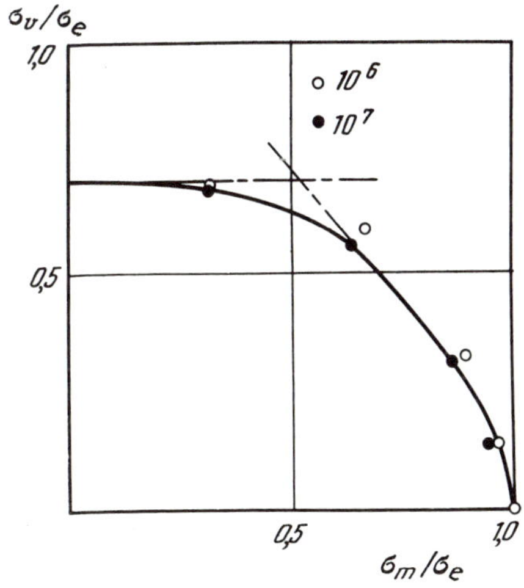

Fig. 148.

of the ratio σ_v/σ_m and to a first approximation it may be considered that failure occurs either only as a result of fatigue or only by creep. This point of view is confirmed in extensive experimental studies carried out under the leadership of Serensen [129, 130].

91. The experimental study of long-term strength in a state of complex stress

If a load acts for a long time under conditions of complex stress, just as in simple tension, the creep curves enter the third stage and creep is terminated by failure. Depending on the material, the temperature and the stress level, the deformation at failure may be either large or small, that is it may be viscous or brittle. Until recently, there were no reliable published data concerning criteria of failure under complex stress and hypotheses concerning the nature of failure were of a rather speculative kind. For example, Siegfried [352] suggested that at temperatures where intercrystalline failure occurs, the hydrostatic component of the stress tensor is a criterion of failure, and if transcrystalline failure occurs the governing quantity is the octahedral tangential stress, that is the deviator component of the stress tensor. It became clear from experimental results that this simple scheme is not justified. Before describing published experimental data we shall give certain general considerations based on analogy with the well-known uniaxial case. In tension we may speak of the time to viscous failure and the time to brittle failure. For rods the time to viscous failure is determined by the Hoff scheme. In a state of complex stress the time to viscous failure can also be calculated on the assumption that the material can creep indefinitely, observing a particular creep equation for finite deformations. In each case calculation of the time to viscous failure is a problem of creep theory that is soluble for a particular structure and tests on specimens are useful only to check and improve the creep law, but not to establish criteria of failure. In this sense the region of brittle failure is of much more interest. Indeed, in assessing the life of structures, it is usual to proceed from two standpoints. In the first place, it must be confirmed that deformation of the structure will not exceed a certain limit, beyond which the structural function of the product breaks down. Usually this limiting deformation is not very great and the service life of the product is limited by it and not by the possibility of viscous failure at large deformation. In the second place, it must be confirmed that brittle fracture or at any rate mixed type fracture does not occur before the limiting permissible deformation is reached. As concerns the criterion of brittle fracture it is logical to propose the rate of crack development, which is above all associated with the normal stresses on the planes in which cracks occur.

The maximum normal stress was the criterion used by Johnson and colleagues in working out their experimental results. The authors studied molybdenum steel at 550°C, copper at 250°C, aluminium alloy at 200°C, magnesium alloy at 20°C, steel with 0.2% C at 450°C and the alloy Nimonic 75 at 650°C [258, 260, 261, 266]. The different materials behaved very differently. Molybdenum steel, copper and Nimonic have a long and clearly expressed third stage of creep, i.e. accelerated creep before failure. Metallographic investigations showed that microscopic cracks, more or less uniformly distributed throughout the volume of the specimen, developed quite early. In the case of copper, for example, these cracks were clearly evident when the time elapsed was half that to failure. Johnson calls the time in which internal cracks accumulate and develop the failure incubation period; it is difficult to set a lower limit to this period, but it may be supposed that crack formation accompanies creep from the very start. The running together of microcracks in one section results in the development of a macroscopic crack which grows very quickly and leads to failure of the specimen, the time of growth of the macroscopic crack is much shorter than the incubation period. The authors represented results of long-term strength tests in coordinates of equivalent

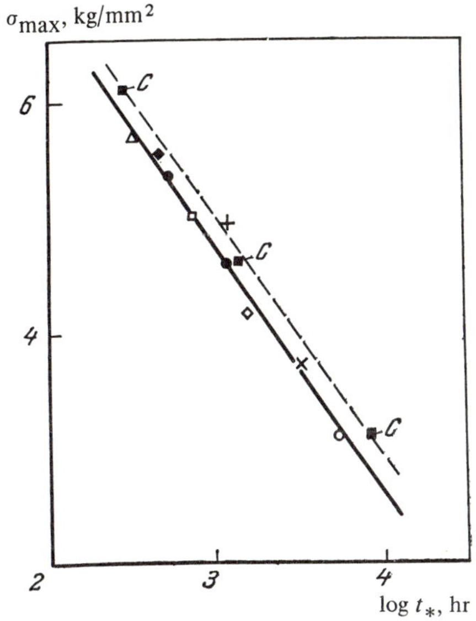

Fig. 149.

stress against time and found that points lie about a single curve if the maximum normal stress is taken as the equivalent stress. A typical long-term strength curve of copper at 250°C is given in fig. 149, the maximum normal stress is plotted on the ordinate and the logarithm of time to failure on the abscissa; the points correspond to various combinations of tension and torsion including pure tension and pure torsion. The three points denoted by the letter C were obtained with solid specimens in tension and a dotted straight line is drawn through them, all the remaining points are grouped very closely about the continuous straight line.

The third part of the creep curve must be associated with the formation of microcracks, as is confirmed by the following considerations. In all cases the dependence of deformation intensity on time could be described by an equation

$$p_0 = At + Bt^n + C \exp(kt) .$$

For copper, we may put $A = 0$. The value of k, which determined the creep rate in the third stage, was found to depend only on the maximum normal stress σ_1.

Aluminium alloy and carbon steel at 450°C behaved quite differently. Although in these the third stage of creep was clearly expressed, no crack

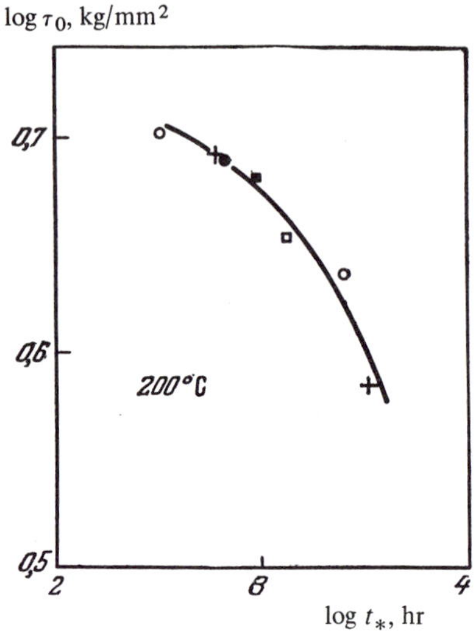

Fig. 150.

formation was observed in this stage and failure was of a local character. For example, in the case of aluminium alloy failure takes place at very small deformations of the order of 1% or less, and, therefore, here too we must speak of brittle failure, but of quite another nature. Long-term strength points for different states of stress lie about a single curve only when the stress intensity is taken as a measure of stress. The long-term strength curve for aluminium alloy is plotted in log/log coordinates in fig. 150, the points correspond to different states of stress ranging from pure torsion to pure tension. If the maximum normal stress is taken as a measure of the equivalent stress, as for materials of the first group, the points are orderless and do not lie around any curve.

The dependence of the intensity of deformation on stress during the third period is of the usual form for creep, p_0 depends on σ_0 but not on other combinations of stress tensor components. The experimental results are described by the following empirical expression:

$$p_0 = \sigma_0^{n_1}(A_1 t^{m_1} + A_2 t^{m_2}) + \sigma_0^{n_2}(B_1 t^{m_1} + B_2 t^{m_2}) .$$

Here $m_1 = 0.32$, $m_2 = 2.6$, A_1, A_2, B_1, B_2, n_1, n_2 are constants. Similar results were obtained for carbon steel, only a few tests were made on magnesium alloy and no certain conclusions could be drawn.

92. The simplest criteria of long-term strength

A large and systematic programme of long-term strength tests under complex stress was carried out by Sdobyrev [127]. The tests were made on specimens of alloy EI-437B made from bar material and from forged blanks for gas turbine discs. The bars were heat-treated by hardening at 1080°C with a heating time of 8 hours, ageing at 750°C for 16 hours and cooling in air. Control tests on specimens cut in a crosswise direction demonstrated satisfactory isotropy. Six series of tests were made, two on specimens from forgings at temperatures of 700°C to 750°C and the other three on bar material at 600, 650, 700 and 750°C. Some of the bar specimens were tested first and the corresponding results were published in the article [126]. The remaining specimens from the same batch of material and from the same melt were tested after about a year. It was found that the delay stabilised the properties of the material, the scatter of the experimental points was much less on the retained specimens than those which were tested immediately. Verification of various long-term strength criteria led Sdobyrev to the con-

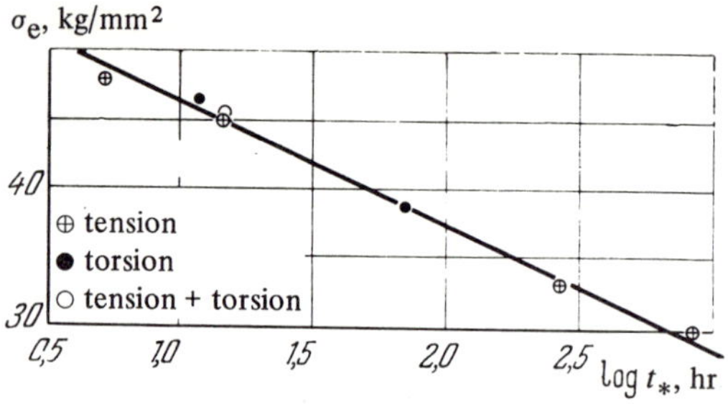

Fig. 151.

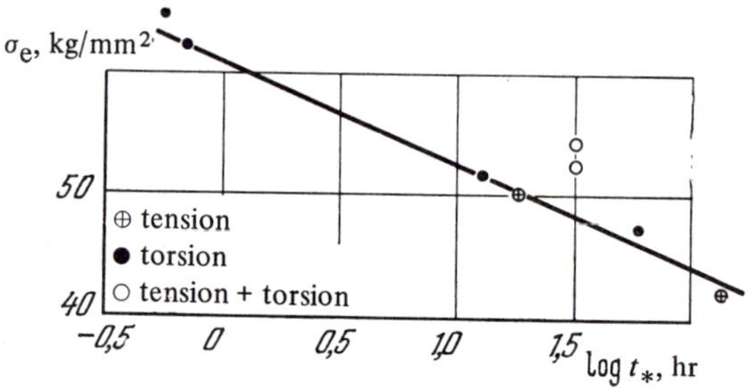

Fig. 152.

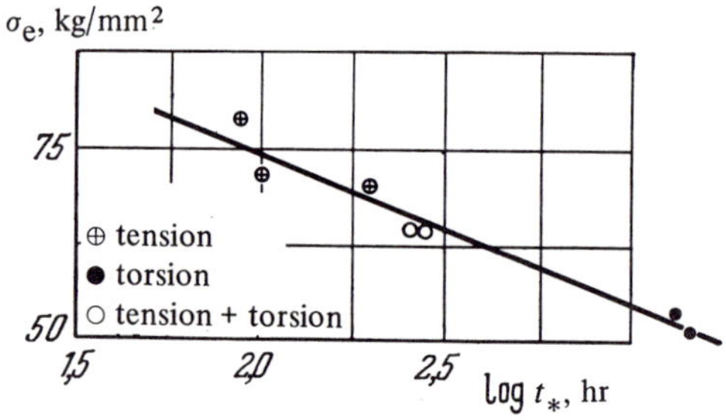

Fig. 153.

clusion that the maximum normal stress criterion is the most accurate of existing criteria. He then attempted to select an empirical equivalent stress that would make the experimental points lie most closely about a single curve. In failure with embrittlement the criterion of long-term strength is

σ_{max}, and for localised failure σ_0, therefore it is natural to take

$$\sigma_e = \beta\sigma_{max} + (1-\beta)\sigma_0 \,. \tag{92.1}$$

Sdobyrev put $\beta = \frac{1}{2}$, he then had

$$\sigma_e = \frac{\sigma_{max} + \sigma_0}{2}. \tag{92.2}$$

In working out the test results it was found that in coordinates σ_e–$\log t_*$ where t_* is the time to failure, the experimental points always lay close to a single straight line. Results of tests on specimens of bar material at 700°C are plotted in fig. 151, at 600°C in fig. 152, and for specimens cut from a disc at 700°C in fig. 153. The scatter is somewhat greater in the latter case.

Sdobyrev applied criterion (92.2) to work out the test results of other authors. He showed that Johnson's tests on molybdenum steel and copper [310] confirmed this criterion just as well as the maximum normal stress criterion. In these latter tests the number of points is small and the scatter of results is fairly considerable. Further he analysed the results of Kats [53, 54] who tested tubular specimens of carbon and austenitic steel in tension combined with internal pressure at temperatures of 500, 650 and 700°C, the tests of Zver'kov on alloy EI-496 at 700°C [36] and the tests of Kooistra and others [275] on carbon steel. In all cases the best results were obtained when σ_e determined by formula (92.2) was taken as the equivalent stress.

It should be noted that in cases of practical importance Sdobyrev's equivalent stress with a plane stressed state does not differ much from the maximum normal stress. It is easy to calculate that if the principal stresses are of the same sign the minimum value of σ_e is 0.936 σ_{max}, that is the difference is only 6.4%. Therefore, for example, in calculations on discs where both principal stresses are usually positive, it does not matter which criterion is used.

In concluding this section we return again to the results of Kennedy and others [274] for Niconel. Their creep results were considered in section 79, and it will be recalled that each series of tests was carried out with constant axial stress σ_z and the peripheral stress σ_θ was altered. Axial stress values that were used were 438, 292 and 219 kg/cm^2. From the test results, curves were constructed of the time to failure as function of the ratio σ_θ/σ_z, each curve corresponding to a particular fixed value of σ_z. The scatter of the points was rather great, therefore the curves are somewhat conventional. Now, using these curves, we can construct lines in the plane σ_θ–σ_z that correspond to failure

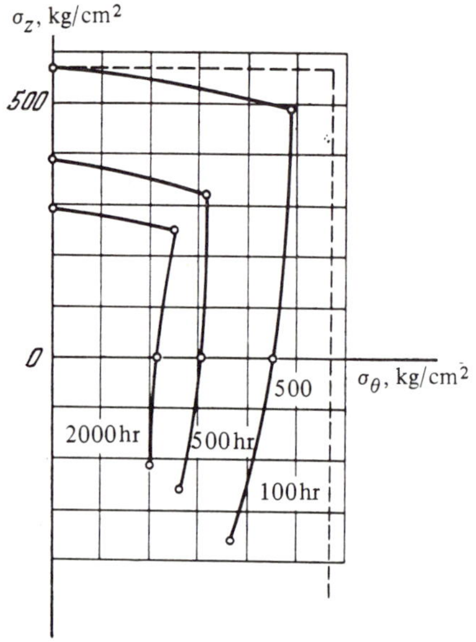

Fig. 154.

after a given time. This graph is reproduced in fig. 154, the three curves correspond to failure at times of 100, 500 and 2000 hours respectively. It is important to note that hoop stress gives more rapid failure than axial stress of the same value. Later we shall return to this question and show that viscous failure of tubes under internal pressure occurs sooner than the failure of rods extended by the same nominal stress; in rods the increase of stress at constant load results only from decrease in the cross-sectional area, in a tube at constant pressure the stress rises both because of reduction in the wall thickness and because of increase in the radius. The dotted lines show the condition of failure after 100 hours by the criterion of maximum normal stress on the assumption that this stress remains constant. In all cases failure of the specimens was accompanied by fairly large creep deformation and, therefore, the results can only be interpreted if allowance is made for the final deformation. These tests confirm qualitatively the condition of maximum tensile stress.

93. General concepts of long-term failure in a state of complex stress

From the experimental results which have been briefly considered above it can be stated that the nature of failure under complex stress essentially depends on the type of state of stress. Indeed viscous failure results from creep and consequently, depends on the stress intensity, or on some other equivalent stress, for example, τ_{max}. Brittle failure depends primarily on the maximum normal stress. The scheme shown in fig. 138 has meaning only in reference to extension of a specimen when the value of the stress is the only characteristic.

In a state of complex stress we shall take the stress intensity σ_0 as the quantity which governs the stress level. Construct a long-term strength diagram in coordinates of $\log \sigma_0 - \log t_*$ (fig. 155). The long-term failure part is constructed as in fig. 138 although in this case this has only conventional meaning. Indeed, in determining the time to viscous failure of a specimen in tension we go beyond the accepted geometrical picture of deformation; for other initial states of stress, even if they are uniform, the conditions of viscous failure must be formulated differently. Thus in section 92 in considering Kennedy's experimental data we noted that tubes that are subject only to hoop stress fail faster and, therefore, the increase in stress results not only from reduction in wall thickness but also from increase in diameter. On the other hand, in pure torsion, the cross-sectional area remains unchanged and viscous failure in Hoff's sense is impossible. In practice, when the danger of brittle failure does not arise, the suitability of structures is assessed in terms of some total deformation and the permissibility or otherwise of any particular condition is determined separately from the creep rate, which must not exceed a certain value. In any case we can locate the viscous failure part of the long-term strength diagram in such a way that it has a certain conventional meaning for all the states of stress.

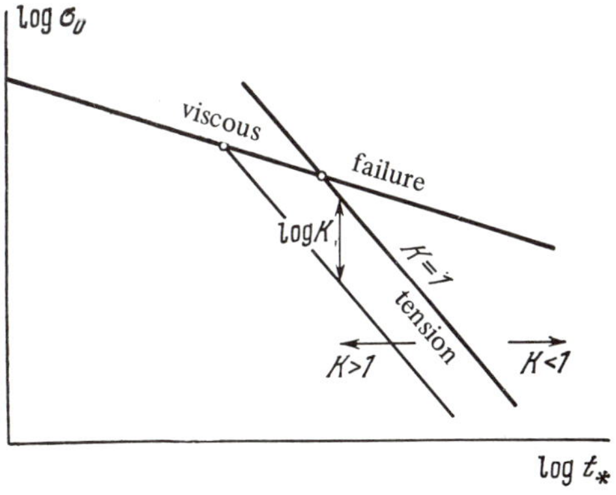

Fig. 155.

The position is quite different when brittle failure can occur, different types of states of stress give different curves in coordinates of $\sigma_0 - t_*$. As a characteristic of the state of stress let us take the quantity

$$K = \frac{\sigma_{max}}{\sigma_0}.$$

The straight line that corresponds to brittle failure in fig. 139, where σ is understood to mean σ_{max}, corresponds to a series of straight lines in fig. 155; the line $k = 1$ corresponds to tension. When $k < 1$ the time that corresponds to transition from viscous failure to brittle increases and when $k > 1$ it decreases. In a plane stressed state $k > 1$ if $\sigma_2 > 0$, and, therefore, with tension in two directions brittle failure commences sooner than with tension in one direction. The most unfavourable case occurs when $\sigma_2 = \frac{1}{2}\sigma_1$ and $k = 2/\sqrt{3}$. In a state of triaxial stress k may be indefinitely great, and, therefore, brittle failure is possible in materials where it does not occur in tension of any practical duration. Brittle failures of this kind are observed in zones of stress concentration. Mirkin and Trunin [76] studied the extension of cylindrical specimens with an annular notch, made of pearlitic steel EI-10 and austenitic steel EI-257. Intensive cracking along the grain boundaries was observed in the zone of stress concentration, the cracks were formed by the action of normal stresses, as may be judged both from the direction of the macroscopic cracks and from the circumstance that they did not originate on the surface, but at a certain depth at which the normal stresses are maximum. It was noted that intercrystalline failure commences in the zone of stress concentration much sooner than in smooth specimens.

We shall give certain results that relate to testing copper tubes of square section. The wall thickness was about 1 mm and the external dimension 4 mm. At a temperature of 650°C a tube of this kind failed on short term application of an internal pressure of 150–180 atm, there was considerable plastic deformation, the square section became almost circular, and failure resulted from local thinning of the walls. A photograph of part of the section of a failed tube close to the corner is shown in fig. 156 and the commencement of local wall thinning is clearly visible. Under an internal pressure of 100 atm the tube held the load for a short time and then failed in quite a different way without appreciable plastic deformation. A photomicrograph of a specimen which failed after 30 seconds is shown in fig. 157. The section illustrated is one where a through crack did not form. Intensive cracking along the grain boundary is visible throughout the extended zone and formation of macroscopic cracks from the internal angle can be seen. The similar

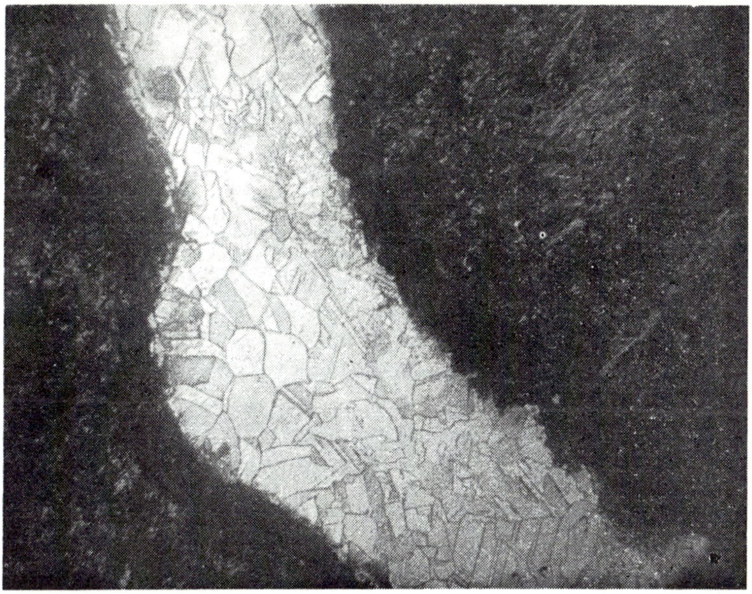

Fig. 156.

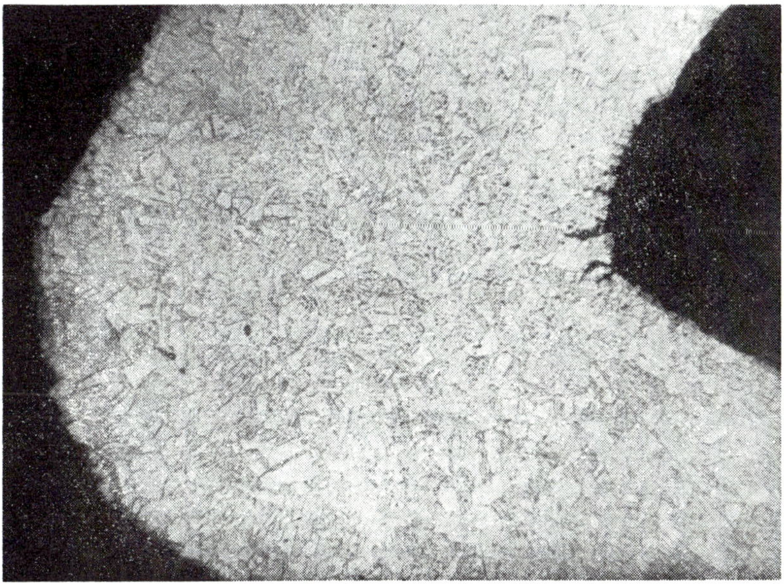

Fig. 157.

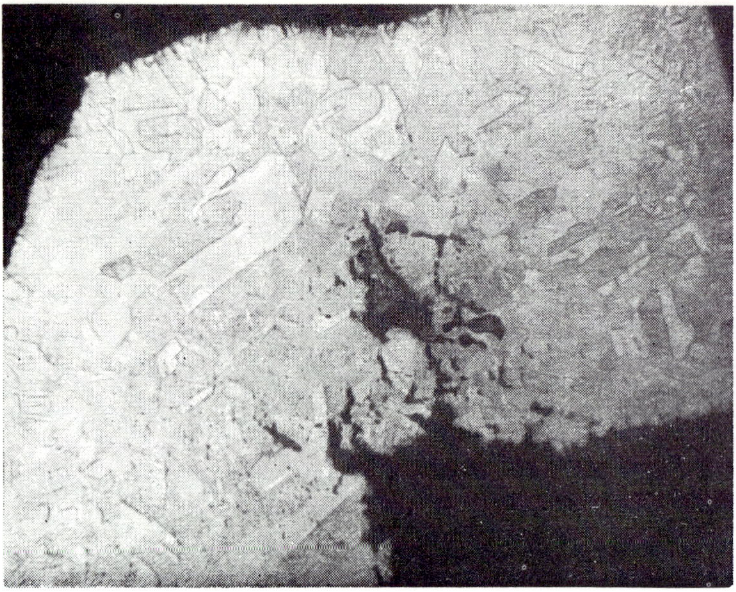

Fig. 158.

photograph shown in fig. 158 records a further stage in the development of a macroscopic crack. This photograph relates to another specimen which failed under the same conditions after 60 seconds.

This example shows how relative are the concepts of short-term and long-term strength, typical long-term failure can occur in a time of some tens of seconds.

94. The possibility of constructing a more general theory of long-term failure

The general considerations given above concerning long-term failure with complex stress are of a fairly schematic nature and only certain qualitative conclusions can be drawn from them concerning the causes of any particular type of failure in individual cases.

Generally speaking, to determine the life of a product of complicated shape under conditions of complex and non-uniform stress it is necessary to develop a general theory of creep which includes a description of the process of crack formation. The usual approach, which is currently widely used in design practice, is as follows. The stress distribution is determined on the basis of some creep theory. The value of the maximum normal stress is sought. This quantity is compared with the long-term strength curve, found by testing cylindrical specimens, and the nominal time to failure is determined from the long-term strength curve. This method of determining the life is conventional because of the following circumstances. In the first place, as we have seen, creep curves have a third phase in which the creep rate alters significantly with increase in crack formation. None of the theories of creep with complex stress described above take account of crack formation and they are unsuitable for the third phase of creep. Therefore, in calculating the state of stress, for example by the theory of steady-state creep, we cannot assess the stress at the instant preceding failure. Calculation of the state of stress by the time-hardening theory using isochronous curves has the advantage of using complete primary creep curves for the specimens, including the third section, although there is no reason to suppose that with complex stress the third phase develops in exactly the same way as in uniaxial stress and that tensor relationships of the deformation type continue to have a certain meaning.

Secondly, even if the stress distribution were accurately known, determination of the time to failure by the method indicated would only give the time for the first crack to appear at a danger point. Complete failure will not occur until the crack has passed right through the product and divided it into two parts.

Kachanov [52] uses the following procedure to determine life. It is considered that crack formation does not influence creep, consequently, the creep equations are taken in the usual form. It is assumed that crack formation occurs in areas which are perpendicular to the axis of maximum stress σ_1 and the equation of the kinetics of cracking is written in the form

$$\dot{\omega} = \varphi(\sigma_1, \omega) . \tag{94.1}$$

If certain constant loads act on the body, the stress distribution at first remains constant, the value of ω increases at each point in accordance with eq. (94.1) until at a certain time $t = t_1$, at some point $\omega = 1$. Thereafter two zones appear in the body; zone A where $\omega < 1$, and zone B where failure has already occurred and the material has lost its strength. The boundary between zones A and B, the failure front, moves in such a way that on it $\omega = 1$ and forces are not transmitted from zone B to zone A.

As we desire to assess the influence of crack formation on creep we make the extreme assumption that cracks only form on areas which are subject to the maximum normal stress σ_1, and its magnitude is the diminution in effective cross-sectional area for this direction. The corresponding effective principal stress is $\sigma_1/(1-\omega)$ whereas both other principal stresses remain unchanged. Taking the maximum tangential stress as the creep criterion we find

$$\epsilon_1 = -\epsilon_3 = v\left(\frac{\sigma_1}{1-\omega} - \sigma_3\right), \quad \epsilon_2 = 0, \tag{94.2}$$

where, as before, the change in the quantity ω is described by eq. (94.1).

In the case of short-term creep we must consider that φ is a power function of $\sigma_1/(1-\omega)$.

It follows from eq. (94.2) that creep accompanied by cracking takes place without change of volume. In fact this is not so, crack formation is always accompanied by increase in volume, but this effect is considered unimportant compared with the increased shear deformation that results from the presence of cracks.

This simple scheme is justified by the following considerations. When the degree of cracking is considerable it is indeed the stress σ_1 that determines the creep rate and in fact eq. (94.2) may be replaced by the equation

$$\epsilon_1 = v\left(\frac{\sigma_1}{1-\omega}\right), \tag{94.3}$$

which coincides with the first of the eqs. (88.3). Only a small correction is made by allowing for σ_3. In fact a kind of unidimensional creek takes place in the zone of cracking. Where ω is small, eqs. (94.2) are converted to the usual creep equations, the precise mode of transition from one extreme case to the other is not particularly important. Later on, we shall give certain examples of the application of eqs. (94.2) and (94.3).

The need or otherwise to allow for the influence of crack formation on creep rate is solved in different ways depending on the way in which cracking takes place in the particular material. In some cases, as we have already mentioned, cracks develop uniformly throughout the volume and are bound to influence the creep rate. The third phase of the creep curves is associated with uniform elongation of the specimen and can be predicted by theory, as will be seen from the example given in fig. 146. But in some materials failure is not accompanied by crack formation throughout the volume. Cracks, or groups of cracks, originate only in certain sections, they develop and lead to local failure. The third phase of the creep curve can also be observed in specimens of such materials, but these third phases are associated with local deformation close to the point of future failure and there is no justification for applying the theory presented above to materials of this kind.

7

Steady-State Creep. General Theory and Simple Examples

The simplest solutions of creep problems are obtained from the theory of steady-state creep. As was shown in Chapter 5, with certain assumptions, the equations of the theory of steady-state creep coincide with those of the non-linear theory of elasticity. Variational principles of the Lagrange and Catigliano type apply to geometrically linear problems of creep theory. A practical solution is possible with an arbitrary creep law but the power and exponential laws offer certain advantages. A recently demonstrated theorem of Calladine and Drucker is useful in obtaining approximate solutions with a power law of creep, it can be used to establish the limits within which the constant potential surface lies if the elastic state and the ultimate plastic state are known for the body. Later on the application of this theory to simple cases of designing girders is considered. The solution of statically indeterminate problems reduces to the solution of a certain system of non-linear equations and, therefore, various kinds of approximate devices are particularly useful. Satisfactory results can often be obtained if we are given the same stress distribution as is realised in the ultimate plastic state.

95. Uniqueness in the small and stability

As we have seen in Chapter 5 the assumption that creep potential exists severely limits the class of possible forms of the equations of steady-state creep theory in the case of a spatial stress, moreover this assumption is not strictly necessary and does not follow from any fairly general and reliable mechanical principles. However, this hypothesis permits the formulation of variational theorems for creep which are completely analogous with the known variational theorems of a linear elastic body. This opens the possibility of developing approximate methods of solving creep theory problems. Thus we assume that

$$\epsilon_{ij} = \frac{\partial \Phi}{\partial \sigma_{ij}}. \tag{95.1}$$

Since the quantities ϵ_{ij} are defined uniquely by stresses σ_{ij} the power dissipation

$$L = \sigma_{ij}\epsilon_{ij} \tag{95.2}$$

is a function of either only σ_{ij} or only ϵ_{ij} for any values of stress other than zero and for rates $L > 0$.

Let us introduce the new function $U(\epsilon_{ij})$ defined by the equation

$$U = L - \Phi\,. \tag{95.3}$$

It is assumed that L and Φ are functions of ϵ_{ij}. Differentiating (95.3) with respect to ϵ_{ij} we obtain

$$\frac{\partial U}{\partial \epsilon_{ij}} = \sigma_{ij} + \epsilon_{kl}\frac{\partial \sigma_{kl}}{\partial \epsilon_{ij}} - \frac{\partial \Phi}{\partial \sigma_{kl}}\frac{\partial \sigma_{kl}}{\partial \epsilon_{ij}}.$$

Because of eq. (95.1) the last two terms in the right-hand side cancel one another and we obtain

$$\sigma_{ij} = \frac{\partial U}{\partial \epsilon_{ij}}. \tag{95.4}$$

Eqs. (95.1) and (95.4) are fully analogous to the equations of the non-linear theory of elasticity, rates of creep deformation correspond to elastic deformations of the body in the non-linear theory of elasticity. The function U has a perfectly definite physical meaning, it is the elastic energy of the deformed body. The quantities L and Φ are introduced formally, the latter is termed the complementary energy. In creep theory the formal analogies are retained, but the essence of the matter is quite different. The only quantity that has a clear physical meaning is the energy dissipation L which is the work of plastic deformation per unit volume per unit time. Hereafter we shall call the function Φ the creep rate potential and the function U the stress potential.

Consider a body that is undergoing steady-state creep under the action of body forces F_i and surface forces T_i^e applied to part of the surface Σ_T. Rates of displacement of the surface points v_i^e are given on the remainder of the surface Σ_v. The state of stress which is developed in the body by the action of the forces and by the communication of forced rates to points on the surface, is defined at each point by the tensor σ_{ij} and the rate distribution v_i. Let us enumerate the conditions that σ_{ij} and v_i must satisfy. The stress tensor satisfies the differential equation of equilibrium

$$\sigma_{ij,j} + F_i = 0$$

and the boundary conditions

$$\sigma_{ij} \nu_j = T_i^e \qquad \text{or } \Sigma_T .$$

The rate distribution v_i should be continuous and differentiable, it satisfies the boundary conditions

$$v_i = v_i^e \qquad \text{or } \Sigma_v .$$

In calculating the deformation rate distribution by the formula $\epsilon_{ij} = \frac{1}{2}(v_{i,j} + v_{j,i})$ we should require that σ_{ij} and ϵ_{ij} are related by eqs. (95.1) or (95.4), which is the same thing. If all these conditions are fulfilled we have an exact solution of the problem of steady-state creep. The question whether a solution exists is fairly complicated and remains open, the question of uniqueness can be solved by using any particular form of the relationship between stress and rate of deformation. Here we consider the question of uniqueness in the small. Suppose that in addition to the known solution σ_{ij}, v_i there exists another infinitesimally close solution $\sigma_{ij} + \delta\sigma_{ij}$, $v_i + \delta v_i$ which satisfies the same boundary conditions. Then on the surface of the body $\delta\sigma_{ij} \nu_{ij} = 0$ on Σ_T and $\delta v_i = 0$ on Σ_v. The equations of equilibrium become: $\delta\sigma_{ij,j} = 0$. Multiply the left-hand side of the equilibrium equation by δv_i and integrate over the volume of the body, we obtain

$$\int_V \delta\sigma_{ij,j} \delta v_i \mathrm{d}V = 0 .$$

Convert this integral

$$\int_V \delta\sigma_{ij,j}\delta v_i \mathrm{d}V = \int_V [(\delta\sigma_{ij}\delta v_i)_{,j} - \delta\sigma_{ij}\delta v_{i,j}]\,\mathrm{d}V$$

$$= \int_\Sigma \delta\sigma_{ij} v_j \delta v_i \mathrm{d}\Sigma - \int_V \delta\sigma_{ij}\delta v_{i,j}\mathrm{d}V\,.$$

But the surface integral is zero because on Σ_T $\delta\sigma_{ij}v_j = 0$, and on Σ_v $\delta v_i = 0$. On the other hand $\delta\sigma_{ij}\delta v_{i,j} = \delta\sigma_{ij}\delta\epsilon_{ij}$ as follows from symmetry of the tensor $\delta\sigma_{ij}$ and the definition of $\delta\epsilon_{ij}$. Therefore

$$\int_V \delta\sigma_{ij}\delta\epsilon_{ij}\mathrm{d}V = 0\,. \tag{95.5}$$

There are two ways in which we may now proceed, either using eq. (95.1) from which it follows that

$$\delta\epsilon_{ij} = \frac{\partial^2\Phi}{\partial\sigma_{ij}\partial\sigma_{kl}}\delta\sigma_{kl}\,,$$

or finding $\delta\sigma_{ij}$ from (95.4), namely

$$\delta\sigma_{ij} = \frac{\partial^2 U}{\partial\epsilon_{ij}\partial\epsilon_{kl}}\delta\epsilon_{kl}\,.$$

In the first case it follows from (95.5) that

$$\int_V \frac{\partial^2\Phi}{\partial\sigma_{ij}\partial\sigma_{kl}}\delta\sigma_{ij}\delta\sigma_{kl}\mathrm{d}V = 0\,,$$

and in the second

$$\int_V \frac{\partial^2 U}{\partial\epsilon_{ij}\partial\epsilon_{kl}}\delta\epsilon_{ij}\delta\epsilon_{kl}\mathrm{d}V = 0\,.$$

If $\delta\sigma_{ij}$ and $\delta\epsilon_{ij}$ are different from zero these equations are impossible if the subintegral quadratic forms are of fixed sign, so that in this case an infinitely close state of creep that satisfies the same boundary conditions as the principal state does not exist. We shall show that for the principal state to be stable the quadratic form must be definitely positive. Indeed we have shown that if the quadratic forms are determinate, an infinitesimally close state of creep, that corresponds to the same boundary conditions and to the same mass forces, does not exist. A change in the surface forces $T_i^e + \delta T_i^e$ and change in the body forces $F_i + \delta F_i$ correspond to the changes $\sigma_{ij} + \delta\sigma_{ij}$, $v_i + \delta v_i$. Reproducing the considerations given above in place of eq. (95.5) we obtain

$$\int_\Sigma \delta T_i^e \delta v_i \, d\Sigma + \int_V \delta F_i \delta v_i \, dV = \int_V \delta\sigma_{ij} \delta\epsilon_{ij} \, dV .$$

The left-hand side represents the power developed by the additional forces on the additional rates. For a stable material this power should be positive, otherwise on applying an external action we extract work from the system instead of dissipating it. Therefore, for a stable material

$$\delta\sigma_{ij}\delta\epsilon_{ij} > 0 ,$$

or

$$\frac{\partial^2 \Phi}{\partial\sigma_{ij}\partial\sigma_{kl}} \delta\sigma_{ij}\delta\sigma_{kl} > 0 ,$$

$$\frac{\partial^2 U}{\partial\epsilon_{ij}\partial\epsilon_{kl}} \delta\epsilon_{ij}\delta\epsilon_{kl} > 0 . \tag{95.6}$$

Conditions (95.6) signify that the surface $\Phi = \text{const}$ in the stress space and $U = \text{const}$ in the deformation space are convex, as was mentioned in section 65.

96. The Lagrange variational principle

Suppose that, as in the preceding section, σ_{ij} and v_{ij} represent stresses and rates in the actual state determined from the creep equations with given boundary conditions. The material will be assumed stable, meaning that conditions (95.6) are fulfilled for it. Besides the actual rate distribution v_i consider a certain conceivable rate distribution v_i^* which we will assume to be continuous and differentiable so that the distribution of the deformation rate tensor ϵ_{ij}^* can be determined from it. Since the external forces F_i and T_i and the internal forces σ_{ij} are in equilibrium the work of the forces on the virtual rates is zero. Taking v_i^* as a distribution of virtual rates, the equations of equilibrium of the body, written in Lagrange form, are

$$\int_V F_i v_i^* \mathrm{d}V + \int_\Sigma T_i v_i^* \mathrm{d}\Sigma - \int_V \sigma_{ij} \epsilon_{ij}^* \mathrm{d}V = 0 \,. \tag{96.1}$$

It is easily shown that the differential equations of equilibrium and the boundary conditions indeed follow from eqs. (96.1). Because of symmetry of the tensor σ_{ij} and the definition of the quantity $\epsilon_{ij} = \frac{1}{2}(v_{i,j} + v_{j,i})$ $\sigma_{ij}\epsilon_{ij} = \sigma_{ij} v_{i,j}$. The latter integral may be converted as follows:

$$\int_V \sigma_{ij}\epsilon_{ij}^* \mathrm{d}V = \int_V \sigma_{ij} v_{i,j}^* \mathrm{d}V = \int_V [(\sigma_{ij} v_i^*)_{,j} - \sigma_{ij,j} v_i^*] \mathrm{d}V$$

$$= \int_\Sigma \sigma_{ij} \nu_j v_i^* \mathrm{d}\Sigma - \int_V \sigma_{ij,j} v_i^* \mathrm{d}V \,.$$

Substituting the converted value of the integral into eq. (96.1) we obtain

$$\int_V (F_i + \sigma_{ij,j}) v_i^* \mathrm{d}V + \int_\Sigma (T_i - \sigma_{ij}\nu_j) v_i^* \mathrm{d}\Sigma = 0 \,.$$

But this is an identity because the first integral becomes zero because of the equations of equilibrium and the second because of the boundary conditions. Now let us consider certain simple consequences of eqs. (96.1).

1. *Minimum power principle.* Besides the true rate distribution v_i consider another infinitesimally close to it, $v_i + \delta v_i$, which satisfies the same kinematic conditions, i.e. such that $\delta v_i = 0$ on Σ_v. Take as the imaginary rate distribution $v_i^* = \delta v_i$. Substituting into (96.1) we obtain

$$\int_V F_i \delta v_i \mathrm{d}V + \int_{\Sigma_T} T_i^e \delta v_i \mathrm{d}\Sigma - \int_V \sigma_{ij} \delta\epsilon_{ij} \mathrm{d}V = 0 \,. \tag{96.2}$$

Here the second integral does not extend over the entire surface Σ but only over the part of it Σ_T, the integral is zero for Σ_v because $\delta v_i = 0$ on Σ_v. By formula (95.4) $\sigma_{ij}\delta\epsilon_{ij} = (\partial U/\partial\epsilon_{ij})\delta\epsilon_{ij} = \delta U$. Now eqs. (96.2) can be rewritten

$$\delta \Big[\int_V F_i v_i \mathrm{d}V + \int_{\Sigma_T} T_i^e v_i \mathrm{d}\Sigma - \widetilde{U} \Big] = 0 \,.$$

Here, following Kachanov, we use the notation

$$\widetilde{U} = \int_V U \, \mathrm{d}V .$$

The sum of the first two integrals is the power developed by the external forces. Put

$$\int_V F_i v_i \mathrm{d}V + \int_{\Sigma_T} T_i^e v_i \mathrm{d}\Sigma = A .$$

Then we obtain

$$\delta(\widetilde{U} - A) = 0 . \tag{96.3}$$

Condition (96.3) expresses the theorem which is usually called the minimum total power theorem although the name is not precise in this case. Indeed A is the power developed by the external forces, that is the work done by these forces in unit time, whereas the power of the internal forces is $\widetilde{L}$ and not $\widetilde{U}$. Note that there is a similar imprecision in the name of the analogous theorem for elastic bodies. There $\widetilde{U}$ represents the actual work of internal forces, whereas A is not the work of external forces but only the sum of products of the final values of the forces and final values of displacements. But, just as in the theory of elasticity variation of the quantity A with constant external forces represents work on the additional displacement, that is an increment of work, so in creep theory variation of $\widetilde{U} - A$ represents change in power. It follows from eq. (95.1) that in this case there is indeed a minimum value of $\widetilde{U} - A$.

2. *Determination of forces that cause a given rate of displacement.* Suppose that on a certain part of the surface Σ' we are given the displacement rates v_i, which means that certain kinematic bonds are applied. The reactions of these bonds are surface forces T_i on part of the surface Σ'. Now put $v_i = v_i^0 q$. The power of the forces T_i on the rates v_i is

$$A = \int_{\Sigma'} T_i v_i \mathrm{d}\Sigma = q \int_{\Sigma'} T_i v_i^0 \mathrm{d}\Sigma = qQ .$$

The quantity q is termed the generalised rate, the quantity Q the corresponding generalised force. Now select the rate distribution variations such that $\delta v_i = \delta q v_i^0$ on Σ' and $\delta v_i = 0$ on $\Sigma - \Sigma'$, if $T_i \neq 0$. On that part of the remaining surface $\Sigma - \Sigma'$ where $T_i = 0$, δv_i may be of any value. Now assume that external forces are also absent. Then it follows from (96.3) that

$$\delta q Q - \delta \widetilde{U} = 0 ,$$

or

$$Q = \frac{\partial \widetilde{U}}{\partial q} . \tag{96.4}$$

Obviously to apply this formula we need an equation for $\widetilde{U}$ as function of q.

97. Castigliano's variational principle

As before, denote by σ_{ij}, v_i the actual stresses and rates, and consider a certain stress distribution σ^*_{ij} which satisfies the equations of equilibrium, $\sigma^*_{ij,j} + F^*_i = 0$, $\sigma^*_{ij}\nu_j = T^*_i$ on Σ. Since the system of external forces $F^*_i T^*_i$ is in equilibrium, with the internal forces, σ^*_{ij}, we can formulate the equation of equilibrium in the Lagrange form taking the true creep rates as virtual rates. We obtain

$$\int_V F^*_i v_i \mathrm{d}V + \int_\Sigma T^*_i v_i \mathrm{d}\Sigma - \int_V \sigma^*_{ij} \epsilon_{ij} \mathrm{d}V = 0 . \tag{97.1}$$

As in the previous paragraph we shall draw certain conclusions from this.

1. *The principle of minimum complementary energy dissipation.* In addition of the actual state of stress σ_{ij} consider an infinitesimally close statically possible state $\sigma_{ij} + \delta\sigma_{ij}$, i.e. one which satisfies the same equations of equilibrium and the same boundary conditions. This means that $\delta\sigma_{ij}$ form a self-equilibrating system of stresses, $\sigma_{ij,j} = 0$, $\delta\sigma_{ij}\nu_j = 0$ on Σ. In eqs. (97.1) put $\sigma^*_{ij} = \delta\sigma_{ij}$, then $F^*_i = 0$ and $T^*_i = 0$, consequently

$$\int_V \epsilon_{ij} \delta\sigma_{ij} = 0 .$$

But because of eqs. (95.1)

$$\epsilon_{ij} = \frac{\partial \Phi}{\partial \sigma_{ij}} , \qquad \epsilon_{ij}\delta\sigma_{ij} = \frac{\partial \Phi}{\partial \sigma_{ij}} \delta\sigma_{ij} = \delta\Phi .$$

Thus

$$\delta\widetilde{\Phi} = 0 . \tag{97.2}$$

The quantity $\widetilde{\Phi} = \int\limits_V \Phi \, \mathrm{d}V$ is termed the complementary energy dissipation.

Condition (97.2) expresses the principle of minimum complementary dissipation energy. It follows from eqs. (95.6) that there is indeed a minimum.

2. *Castigliano's theorem.* Forces T_i^e act on a certain part of the surface Σ'. Put $T_i^e = T_i^0 Q$. The power developed by these forces is

$$A = \int\limits_{\Sigma'} T_i^e v_i \, \mathrm{d}\Sigma = Q \int\limits_{\Sigma'} T_i^0 v_i \, \mathrm{d}\Sigma = Qq .$$

The quantity Q is termed the generalised force, q the generalised rate. These definitions are fully equivalent to those given in the previous section.

Now put $\sigma_{ij}^* = \delta\sigma_{ij}$, and $\delta\sigma_{ij,j} = 0$, $\delta\sigma_{ij}\nu_j = 0$ on $\Sigma - \Sigma'$, and $\delta\sigma_{ij}\nu_j = \delta Q T_i^0$ on Σ'. From (97.1) we obtain

$$q\delta Q = \delta\widetilde{\Phi} ,$$

or

$$q = \frac{\partial\widetilde{\Phi}}{\partial Q} . \tag{97.3}$$

98. Certain consequences of variational principles

Consider a body loaded by a certain combination of external forces in such a way that we can isolate a finite number r of generalised forces Q_i. The corresponding generalised rates are q_i. The instantaneous state of the system is fully defined by giving the quantities q_i or Q_i; thus if the solution of the creep theory problem is known we have $\widetilde{U} = \widetilde{U}(q_1, \ldots, q_r)$, $\widetilde{\Phi} = \widetilde{\Phi}(Q_1, \ldots, Q_r)$. It follows from Castigliano's theorem that

$$q_i = \frac{\partial\widetilde{\Phi}}{\partial Q_i} .$$

This result may be interpreted as follows. The equation $\widetilde{\Phi} = \text{const}$ is the equation of a surface in r-dimensional space, in which the generalised forces Q_i are plotted on the axes of a rectangular system of Cartesian coordinates.

We shall term this space a space of forces and the radius vector of points a force vector. Then the vector of rate of displacement with components q_i is always in the direction of the normal to the surface $\widetilde{\Phi}$ = const. In contrast to the theory of an ideal plasticity, a surface $\widetilde{\Phi}$ = const can pass through any point in the space of forces, therefore, there is an infinite number of such surfaces which correspond to different values of the constants. We shall term these surfaces of creep in the space of forces, the function $\widetilde{\Phi}(Q_i)$ will be called the rate potential. As was shown in section 95, the condition of uniqueness applies a limitation to the form of the functions $\Phi(\sigma_{ij})$, namely, it requires that the quadratic form whose coefficients are second partial differential coefficients of Φ is definitely positive. But it follows from this that the function $\Phi(Q_i)$ also satisfies an analogous condition. Indeed, consider a state of stress close to the basic one, $\sigma_{ij} + \delta\sigma_{ij}$, the corresponding external forces are $Q_i + \delta Q_i$, i.e. it is a solution of the creep problem for changed external forces. Then evidently,

$$\widetilde{\Phi}(Q_i) + \frac{\partial\widetilde{\Phi}}{\partial Q_i}\,\delta Q_i + \tfrac{1}{2}\,\frac{\partial^2\widetilde{\Phi}}{\partial Q_i \partial Q_j}\,\delta Q_i \delta Q_j + \cdots$$

$$= \int_V \left\{\Phi(\sigma_{ij}) + \frac{\partial\Phi}{\partial\sigma_{ij}}\,\delta\sigma_{ij} + \tfrac{1}{2}\,\frac{\partial^2\Phi}{\partial\sigma_{ij}\partial\sigma_{rs}}\,\delta\sigma_{ij}\delta\sigma_{rs} + \cdots\right\} \mathrm{d}V\,.$$

But if the forces and stresses of the basic system satisfy the conditions of equilibrium, the system of additional forces δQ_i and additional stresses $\delta\sigma_{ij}$ should also satisfy the conditions of equilibrium, consequently, the virtual power on the kinematically permissible rates is zero. As a permissible rate distribution we take the rate distribution of the basic state $q_i = \partial\widetilde{\Phi}/\partial Q_i$ and $\epsilon_{ij} = \partial\Phi/\partial\sigma_{ij}$, consequently

$$\frac{\partial\widetilde{\Phi}}{\partial Q_i}\,\delta Q_i = \int_V \frac{\partial\Phi}{\partial\sigma_{ij}}\,\delta\sigma_{ij}\,\mathrm{d}V\,.$$

We thus arrive at the following equation:

$$\frac{\partial^2\widetilde{\Phi}}{\partial Q_i \partial Q_j}\,\delta Q_i \delta Q_j = \int_V \frac{\partial^2\Phi}{\partial\sigma_{ij}\partial\sigma_{rs}}\,\delta\sigma_{ij}\delta\sigma_{rs}\,\mathrm{d}V\,. \tag{98.1}$$

Hence it follows that the quadratic form on the left-hand side of eq. (98.1) is of determinate sign. Geometrically this means that the creep surface in the space of forces is always convex.

We can consider the creep surface in the rate space q_i in a precisely similar way. The equation of this surface is $\widetilde{U}(q_i) = \text{const}$; the vector of forces with projections Q_s is in a direction normal to this surface, which is also convex. We shall call the function $\widetilde{U}(q_i)$ the force potential.

This latter remark relates to the construction of a dual assessment of the rate potential. The condition of minimum power has the following meaning. Select a certain kinematically possible deformation rate distribution ϵ^*_{ij} to which correspond generalised rates q^*_i. Calculate the quantity $\widetilde{U}(\epsilon^*_{ij})-Q_iq^*_i$. The minimum principle states that this expression assumes a minimum value if actual rates are substituted into it, subsequently

$$Q_iq^*_i - \widetilde{U}(\epsilon^*_{ij}) \leqslant Q_iq_i - \widetilde{U}(\epsilon_{ij}) .$$

But $Q_iq_i = L$, consequently, the right-hand side is $\widetilde{\Phi}(Q_i)$, thus we obtain

$$Q_iq^*_i - \widetilde{U}(\epsilon^*_{ij}) \leqslant \widetilde{\Phi}(Q_i) .$$

On the other hand, consider a certain statically possible state of stress that corresponds to given external forces Q_i, we will denote the corresponding stresses σ^{**}_{ij}. The quantities σ^{**}_{ij} are known functions of Q_i, therefore, calculating $\widetilde{\Phi}(\sigma^{**}_{ij})$ we obtain a certain function of Q_i, different from $\widetilde{\Phi}(Q_i)$. Denote this function $\widetilde{\Phi}^{**}(Q_i)$. Because of the principle of minimum complementary power dissipation

$$\Phi(Q_i) \leqslant \widetilde{\Phi}^{**}(Q_i) .$$

Uniting these inequalities we obtain the dual assessment for the rate potential

$$Q_iq^*_i - \widetilde{U}(\epsilon^*_{ij}) \leqslant \widetilde{\Phi}(Q_i) \leqslant \widetilde{\Phi}^{**}(Q_i) . \tag{98.2}$$

In formulating these inequalities we gave an expanded interpretation of the properties established above for the minima of functions $\widetilde{U}-A$ and $\widetilde{\Phi}$. Indeed in sections 96 and 97 we only established that a point exists for which these functions have a smaller value than at a sufficiently close neighbouring point. However, if the creep surface is strictly convex at all points this minimum is absolute and inequalities (98.2) are valid with any selection of permissible rate distribution and permissible stress distribution. The corresponding demonstration is geometrically obvious in the case of functions that depend on one or two variables. If at all points of the surface $f(x, y)$ the convexity faces downwards it cannot have more than one minimum point [230].

99. Particular forms of the equations of steady-state creep

In practical applications only those variants of steady-state creep theory that lead to linear tensor relationships are commonly used. The general formulation of a theory of this kind can be as follows. Select as equivalent stress a uniform first power function of the stress component $s(\sigma_{ij})$ defined so that in simple tension the stress σ is $s = \sigma$. Then the creep law is written in the form

$$\epsilon_{ij} = v(s)\,\frac{\partial s}{\partial \sigma_{ij}}\,. \tag{99.1}$$

The function $v(s)$ is the same function as that which gives the dependence of creep rate on stress in simple tension, when $\epsilon = v(\sigma)$. If the creep rate is defined as the second invariant of the stress tensor deviator then as shown in section 68:

$$s = \sigma_0 = \sqrt{\tfrac{3}{2}\bar{S}_{\mathrm{II}}}\,, \qquad \frac{\partial s}{\partial \sigma_{ij}} = \frac{3}{2}\,\frac{\bar{\sigma}_{ij}}{\sigma_0}\,. \tag{99.2}$$

If creep is defined by the maximum tangential stress and considered in the principal axes we have

$$s = \sigma_1 - \sigma_2\,, \quad \frac{\partial s}{\partial \sigma_1} = -\frac{\partial s}{\partial \sigma_2} = 1\,, \quad \frac{\partial s}{\partial \sigma_2} = 0 \qquad (\sigma_1 \neq \sigma_2)\,,$$

$$s = \lambda(\sigma_1 - \sigma_3) + (1 - \lambda)(\sigma_2 - \sigma_3) \qquad (\sigma_1 = \sigma_2)\,, \tag{99.3}$$

$$\frac{\partial s}{\partial \sigma_1} = \lambda\,, \quad \frac{\partial s}{\partial \sigma_2} = 1 - \lambda\,, \quad \frac{\partial s}{\partial \sigma_3} = -1\,.$$

Here λ is an arbitrary non-negative quantity, equal to or less than unity. In the case of anisotropy of the type described in section 69 we must take

$$s^2 = \sigma_{33}^2 - \sigma_{33}(\sigma_{22} + \sigma_{11}) + (3 - 2\beta)(\sigma_{31}^2 + \sigma_{32}^2)$$
$$+ (1 - \tfrac{\alpha}{2})(\sigma_{22} + \sigma_{11})^2 + (-3 + 2\alpha)(\sigma_{11}\sigma_{22} - \sigma_{12}^2)\,. \tag{99.4}$$

Assume that the expression $\epsilon = v(\sigma)$ is valid for creep in tension in the direction of the 3 axis. Then

$$\frac{\partial s}{\partial \sigma_{11}} = \frac{1}{2s}\left[(2-\alpha)\sigma_{11} - (1-\alpha)\sigma_{22} - \sigma_{33}\right] ,$$

$$\frac{\partial s}{\partial \sigma_{22}} = \frac{1}{2s}\left[-(1-\alpha)\sigma_{11} + (2-\alpha)\sigma_{22} - \sigma_{33}\right] ,$$

$$\frac{\partial s}{\partial \sigma_{33}} = \frac{1}{2s}\left[-\sigma_{11} - \sigma_{22} - 2\sigma_{33}\right] , \quad \frac{\partial s}{\partial \sigma_{23}} = \frac{1}{2s}(\tfrac{3}{2} - \beta)\sigma_{23} , \tag{99.5}$$

$$\frac{\partial s}{\partial \sigma_{12}} = \frac{1}{2s}(\tfrac{3}{2} - \alpha)\sigma_{12} , \quad \frac{\partial s}{\partial \sigma_{31}} = \frac{1}{2s}(\tfrac{3}{2} - \beta)\sigma_{31} .$$

It is sometimes convenient to consider rates of deformation as given and to determine the stresses. Since s is a uniform first power function of the stress components the quantities $\partial s/\partial\sigma_{ij}$ are uniform functions of zero order, that is functions of five ratios of the values of the tensor stress components to one of them. Six functions of five independent variables should satisfy a certain identical relationship, consequently, there exists a function F of six variables, such that

$$F\left(\frac{\partial s}{\partial \sigma_{ij}}\right) \equiv 0 .$$

Consequently, in view of eq. (99.1)

$$F\left(\frac{\epsilon_{ij}}{v}\right) = 0 .$$

From this equation we find v as function of ϵ_{ij}. Since the relationship does not alter when ϵ_{ij} and v are both multiplied by the same number, the function v is a uniform first-power function of ϵ_{ij}. Because $v = v(s)$ so $s = s(v)$. Multiply both sides of eq. (99.1) by σ_{ij} and contract. We obtain

$$\epsilon_{ij}\sigma_{ij} = vs .$$

In this equation we shall consider that $v = v(\epsilon_{ij})$, and $s = s(\sigma_{ij})$. Differentiating both sides with respect to ϵ_{ij} we obtain

$$\sigma_{ij} + \epsilon_{kl}\frac{\partial \sigma_{kl}}{\partial \epsilon_{ij}} = s\frac{\partial v}{\partial \epsilon_{ij}} + v\frac{\partial s}{\partial \sigma_{kl}}\,\frac{\partial \sigma_{kl}}{\partial \epsilon_{ij}} .$$

Because of eq. (99.1) the second terms on the left- and right-hand sides are equal and therefore we obtain

$$\sigma_{ij} = s(v) \frac{\partial v}{\partial \epsilon_{ij}} . \tag{99.6}$$

In the particular case when s^2 is a quadratic function of the stress components $s^2 = a_{ijkl}\sigma_{ij}\sigma_{kl}$, then

$$s \frac{\partial s}{\partial \sigma_{ij}} = a_{ijkl}\sigma_{kl} .$$

Denote the inverse matrix by a'_{ijkl}, then

$$\sigma_{ij} = a'_{ijkl} s \frac{\partial s}{\partial \sigma_{kl}}$$

If, as is usual in creep theory, we assume that creep is not accompanied by change in volume, then σ_{ij} is determined to within a spherical tensor and

$$\bar{\sigma}_{ij} = a'_{ijkl} s \frac{\partial s}{\partial \sigma_{kl}} .$$

Let us consider this case. Substituting $\partial s/\partial \sigma_{kl}$ from eqs. (99.1) we obtain

$$\bar{\sigma}_{ij} = \frac{s}{v} a'_{ijkl}\epsilon_{ij} .$$

Construct the quadratic form

$$h^2 = a'_{ijkl}\epsilon_{ij}\epsilon_{kl} . \tag{99.7}$$

Evidently

$$h \frac{\partial h}{\partial \epsilon_{ij}} = a'_{ijkl}\epsilon_{kl} .$$

Therefore

$$\bar{\sigma}_{ij} = \frac{s}{v} h \frac{\partial h}{\partial \epsilon_{ij}} . \tag{99.8}$$

Multiply by ϵ_{ij} and sum. Note that here h is a uniform first power function of ϵ_{ij}. We obtain

$$\bar{\sigma}_{ij}\epsilon_{ij} = \frac{sh^2}{v}.$$

On the other hand, multiplying (99.1) by σ_{ij} and summing we find that

$$\sigma_{ij}\epsilon_{ij} = \bar{\sigma}_{ij}\epsilon_{ij} = sv.$$

Therefore, we find that

$$h = v. \tag{99.9}$$

The quantity $h = v$, introduced by means of eq. (99.7), is the equivalent rate of deformation, which depends on the equivalent stress s in the same way that the rate of extension depends on the tensile stress. It follows from eq. (99.8) that $s = s(v)$ where the function $s(v)$ gives the dependence of stress on rate in the uniaxial case. Now eqs. (99.8) can be rewritten

$$\bar{\sigma}_{ij} = s(v)\frac{\partial v}{\partial \epsilon_{ij}}. \tag{99.10}$$

It is evident that in tension v is the rate of elongation in the direction of tension. If s is given by formulae (99.2) we have

$$v = \epsilon_0 = \sqrt{\tfrac{2}{3}H_{\text{II}}}, \qquad \frac{\partial v}{\partial \epsilon_{ij}} = \frac{2}{3}\frac{\epsilon_{ij}}{\epsilon_0}.$$

In using a Tresca type flow law the equivalent rate of deformation $v = \frac{1}{2}(\epsilon_1 - \epsilon_3)$, if $\sigma_1 \neq \sigma_2$, and $v = \epsilon_1 + \epsilon_2 = -\epsilon_3$ if $\sigma_1 = \sigma_2$.

The function $v(s)$ may be completely arbitrary, but in practical applications an analytical expression is usually given for it. We have seen that experimental results are satisfactorily described by a power function

$$v(s) = \epsilon_n\left(\frac{s}{\sigma_n}\right)^n$$

or an exponential function

$$v(s) = \epsilon_e \exp\left(\frac{s}{\sigma_e}\right).$$

Each form has its analytical advantages and disadvantages and, therefore, the type of expression is chosen to suit the particular type of problem.

100. **Modelling of steady-state and quasi-steady-state creep**

It is a property of the exponential creep law that, with appropriate choice of dimensionless parameters, the equations of steady-state and quasi-steady-state creep, in the sense given to these concepts in section 75, can be written in a form that does not contain any constants of the material. Consider first the equations of steady-state creep which can be of the form

$$\epsilon_{ij} = \epsilon_e \exp\left(\frac{s}{\sigma_e}\right) \frac{\partial s}{\partial \sigma_{ij}}.$$

Put

$$\hat{\sigma}_{ij} = \frac{\sigma_{ij}}{\sigma_e}, \qquad \hat{\epsilon}_{ij} = \frac{\epsilon_{ij}}{\epsilon_e},$$

then

$$\hat{s} = s(\hat{\sigma}_{ij}) = \frac{s(\sigma_{ij})}{\sigma_e}.$$

The previous equation can be written in the form

$$\hat{\epsilon}_{ij} = \exp(\hat{s}) \frac{\partial \hat{s}}{\partial \hat{\sigma}_{ij}}. \tag{100.1}$$

Now consider the equations of quasi-steady-state creep (75.2). In these equations p'_{ij} should be considered as a rate relative to the varied time $\tau = (t/m)^m$. If $f(\sigma) = k \exp(\sigma/A)$ then eq. (70.5) can be rewritten

$$\epsilon_{ij} = k^m \exp \frac{m\sigma_0}{A} \frac{\partial \sigma_0}{\partial \sigma_{ij}}.$$

Putting

$$\hat{\sigma}_{ij} = \frac{m\sigma_{ij}}{A}, \qquad \hat{\epsilon}_{ij} = \frac{\epsilon_{ij}}{k^m},$$

we obtain an equation which literally coincides with (100.1). If the elastic and instantaneous plastic deformations can be considered negligibly small compared with creep deformations, there follows a very extensive possibility of modelling creep of metallic parts at high temperature by means of models made of metals that creep at low temperature and even of plastics. Of course, the equations of steady-state and particularly quasi-steady-state creep are only valid when constant forces act on the body and there are no kinematic limitations.

Now consider the power law of creep. We shall only consider steady-state creep, quasi-steady-state creep is considered in an exactly similar way. Put

$$\hat{\sigma}_{ij} = \frac{\sigma_{ij}}{\sigma_n}, \qquad \hat{\epsilon}_{ij} = \frac{\epsilon_{ij}}{\epsilon_n};$$

then the creep equations are

$$\hat{\epsilon}_{ij} = \hat{s}^n \frac{\partial \hat{s}}{\partial \hat{\sigma}_{ij}}. \tag{100.2}$$

Unlike eqs. (100.1), (100.2) contain one dimensionless parameter n which characterises the properties of the material. Now consider modelling of creep in an arbitrary body. Suppose the external forces increase in proportion to some parameter λ and suppose that we are given the solution of the creep theory problem for a certain value of the parameter λ, for example, $\lambda = 1$. Let σ_{ij} be the corresponding stresses, and v_i the corresponding rates. Now change all the external forces in the ratio of λ. With the power law of creep, for a value of the loading parameter equal to λ the stresses are $\lambda\sigma_{ij}$ and the rates $\lambda^n v_i$. Indeed the equation of equilibrium, the boundary conditions, eqs. (100.2) and the conditions of compatibility of deformations are then fulfilled.

Such a simple result is not obtained with the exponential creep law. If σ_{ij} is a solution that corresponds to a certain load then $\lambda\sigma_{ij}$ is not, in general, a solution of the creep equations; this solution essentially depends on the value of the loading parameter λ which figures in the equations. With the power law, the parameter λ is excluded but the exponent n remains in the equations.

101. **Power law of creep. Theorem of Calladine and Drucker**

If $v(s)$ is a power function, for example $v(s) = \epsilon_n(s/\sigma_n)^n$ (section 43), then the expressions for the creep potential, the dissipative function and the stress potential are

$$\Phi = \frac{1}{n+1}\,\epsilon_n \sigma_n \left(\frac{s}{\sigma_n}\right)^{n+1}, \qquad L = \epsilon_n \sigma_n \left(\frac{s}{\sigma_n}\right)^{n+1},$$

$$U = \frac{n}{n+1}\,\epsilon_n \sigma_n \left(\frac{s}{\sigma_n}\right)^{n+1},$$

and eqs. (99.1) are rewritten

$$\epsilon_{ij} = \epsilon_n \left(\frac{s}{\sigma_n}\right)^n \frac{\partial s}{\partial \sigma_{ij}}.$$

In terms of dimensionless variables taking

$$\hat{\epsilon}_{ij} = \frac{\epsilon_{ij}}{\epsilon_n}, \qquad \hat{\sigma}_{ij} = \frac{\sigma_{ij}}{\sigma_n},$$

then $\hat{s} = s/\sigma_n$. We may now write

$$\hat{\epsilon}_{ij} = \frac{\partial \hat{\Phi}}{\partial \hat{\sigma}_{ij}} = \hat{s}^n \frac{\partial \hat{s}}{\partial \hat{\sigma}_{ij}}. \tag{101.1}$$

Here

$$\hat{\Phi} = \frac{1}{n+1}\,\hat{s}^{n+1}. \tag{101.2}$$

Because of the linearity of the equations of equilibrium and the boundary conditions for stress, the stresses are proportional to the applied forces, and, therefore, it is natural to introduce dimensionless loads $\hat{T}_i = T_i/\sigma_n$. If we can calculate the function $\widetilde{\Phi} = \int \Phi \, \mathrm{d}V$ in terms of the external forces we can, by Castigliano's theorem, find the rates of their points of application: in principle, to determine this function, we must be able to solve the problem of creep theory. A theorem recently demonstrated by Calladine and Drucker [178, 179] can be used in certain cases to select approximate solutions for functions $\widetilde{\Phi}$ and to assess the resultant error. Consider a body A loaded by forces T_i; we shall suppose that the material observes eqs. (101.1), the exponent in the power creep law is n. By altering the forces proportionally we alter the rates of deformation in the same ratio but to power n; therefore, if we have found a solution for the creep theory problem for some particular values of external forces we automatically obtain a solution for forces that are proportionally greater or less. Select as the standard system of loads a proportionally changed system $\hat{T}_i^n$ for which the power dissipation assumes a certain fixed value, suppose for example that the mean dimensionless power per unit volume is unity. This condition of normalisation is written

$$\frac{1}{V}\int_V \hat{L}\,\mathrm{d}V = 1\,.$$

Denote the dimensionless stresses that correspond to the normalised system of loads by $\hat{\sigma}_{ij}^n$, put $\hat{s}_n = \hat{s}(\hat{\sigma}_{ij}^n)$, then the conditions of normalisation can be written

$$\frac{1}{V}\int_V \hat{s}_n^{n+1}\,\mathrm{d}V = 1\,. \tag{101.3}$$

Now consider a second body B identical with the body A and loaded by the same forces T_i. The creep law for body B is also expressed by equations of type (101.1), the function s is determined in exactly the same way but the power exponent in formulae (101.1) is a certain value of $m \neq n$. For the body B determine a normalised system of loads T_i^m, which obeys the condition

$$\frac{1}{V}\int_V \hat{s}_m^{m+1}\,\mathrm{d}V = 1\,. \tag{101.4}$$

The stresses $\hat{\sigma}_{ij}^m$ satisfy the equations of equilibrium and boundary conditions corresponding to loads T_i^m. Alter these stresses in the ratio $\hat{T}_i^n/\hat{T}_i^m$, that is consider a stress system

$$\hat{\sigma}_{ij}^* = \hat{\sigma}_{ij}^m \frac{\hat{T}_i^n}{\hat{T}_i^m}\,.$$

Stresses σ_{ij}^* satisfy the equations of equilibrium and the boundary conditions corresponding to forces T_i^n, therefore they define a certain statically possible state of stress for the body A. By the theorem of minimum complementary energy, using eq. (101.2) we obtain

$$\int_V \hat{s}_n^{n+1}\,\mathrm{d}V \leqslant \left(\frac{\hat{T}_i^n}{\hat{T}_i^m}\right)^{n+1} \int_V \hat{s}_m^{n+1}\,\mathrm{d}V\,.$$

Dividing by V, note that the left-hand side is unity because of eq. (101.3), therefore from the inequality in the above expression it follows that

$$\left(\frac{\hat{T}_i^m}{\hat{T}_i^n}\right)^{n+1} \leqslant \frac{1}{V}\int\limits_V \hat{s}_m^{n+1}\,\mathrm{d}V = \frac{1}{V}\int\limits_V (\hat{s}_m^{m+1})^{\frac{n+1}{m+1}}\,\mathrm{d}V\,.$$

Now suppose that $m > n$. The integral on the right-hand side can be written

$$I(r) = \frac{1}{V}\int\limits_V x^r\,\mathrm{d}V\,.$$

Here $r = (n+1)/(m+1)$. It is evident that $I(0) = I(1) = 1$. We shall show that $I(r) < 1$ $(0 < r < 1)$. This is most easily done by using the Hölder inequality (see for example, [142]):

$$\int f^\alpha g^\beta\,\mathrm{d}V < (\int f\,\mathrm{d}V)^\alpha (\int g\,\mathrm{d}V)^\beta\,, \qquad (\alpha > 0, \beta > 0, \alpha + \beta = 1)\,.$$

Put $f = x, g = 1, \alpha = r, \beta = 1 - r$, then we obtain

$$\int x^r\,\mathrm{d}V < [\int x\,\mathrm{d}V]^r\,.$$

Returning to the initial inequality, we find

$$\frac{\hat{T}_i^m}{\hat{T}_i^n} \leqslant 1 \qquad (m > n)\,. \tag{101.5}$$

We shall now consider that the loads acting on the body are determined by a finite number of generalised forces Q_s, put $\hat{Q}_s = Q_s/\sigma_n$ and consider the space of the forces $\hat{Q}_s$ as was done in section 98.

The condition $1/V \int \hat{L}\,\mathrm{d}V = 1$ defines a certain surface of power dissipation in the space of forces, different surfaces are obtained for different values of n. The totality of generalised forces Q_s^n corresponds to a certain point on the constant power dissipation surface with exponent n. Forces Q^m are proportional to Q^n; consequently the point that represents the state Q^m lies on a line drawn from the origin of coordinates to the point Q^n. Here, in view of inequality (101.5) the radius vector of the point Q^m is shorter than or equal to the radius vector of the point Q^n. Therefore, if $m > n$, the constant power dissipation surface for the exponent m lies inside the surface that corresponds to the exponent n. By altering the value of n we obtain a series of surfaces enclosed within one another.

Usually, it is not difficult to construct a constant power dissipation surface for $n = 1$; if $s = \sigma_0$, for example, this surface corresponds to the stress distribution in an elastic incompressible body. On the other hand, consider that the material is ideally plastic and the condition of plasticity is of the form $\hat{s} = 1$. If the condition of plasticity is fulfilled at all points in the body so that the body is in a state of limiting equilibrium and subject to the action of a certain system of forces Q_i^∞, then condition (101.3) is fulfilled independently of n and consequently,

$$\frac{Q^\infty}{Q^n} \leqslant 1 \quad \text{for any value of } n.$$

Therefore, the limiting constant power dissipation surface when $n \to \infty$ is a surface of flow in a space of forces constructed on the assumption that the condition of plasticity $\hat{s} = 1$ applies. Note that for a given system of loads the state of limiting equilibrium for which $\hat{s} = 1$ at all points in the body is not always possible, the limiting condition may be achieved with the formation rigid regions within which $\hat{s} < 1$. Let the forces Q_i^∞ correspond to a state in which the condition $\hat{s} = 1$ is fulfilled in part of the body and in the remainder $\hat{s} < 1$. The stress distribution in the rigid parts can be selected in any way that is in accordance with the conditions of statics. Denote these stresses $\hat{\sigma}_{ij}^\infty$, $\hat{s}_\infty \leqslant 1$. Repeating the foregoing considerations take as a statically possible state of the body A a system of stresses

$$\hat{\sigma}_{ij}^{**} = \hat{\sigma}_{ij}^\infty \left(\frac{\hat{Q}_i^n}{\hat{Q}_i^\infty}\right).$$

We then obtain

$$\int \hat{s}_n^{n+1}\, \mathrm{d}V \leqslant \left(\frac{\hat{Q}_i^n}{\hat{Q}_i^\infty}\right)^{n+1} \int_V s_\infty^{n+1}\, \mathrm{d}V.$$

Hence, as the left-hand side is unity

$$\left(\frac{\hat{Q}_i^\infty}{\hat{Q}_i^n}\right)^{n+1} \leqslant \int_V \hat{s}_\infty^{n+1}\, \mathrm{d}V < 1.$$

The Calladine and Drucker theorem states that the entire series of constant power dissipation surfaces is enclosed between the two surfaces for $n = 1$ and for $n = \infty$. We shall call constant power dissipation surfaces in the space of generalised forces, Q-surfaces. Note that with a power law of hardening the functions L, Φ and U differ only by constant terms, therefore, the surface Q is a surface $\tilde{\Phi} = \text{const}$. According to Castigliano's theorem the generalised rates q_i are $\partial\Phi/\partial q_i$ so that the rate vector $\mathbf{q}(q_1, ..., q_s)$ is in the direction of the normal to the Q-surface, whereas because of the conditions of normalisation the scalar product of the vectors $\mathbf{Q}$ and $\mathbf{q}$ is equal to the volume of the body

$$\hat{Q}_i \hat{q}_i = V\,.$$

The condition

$$\frac{1}{V} \int \hat{L}\, \mathrm{d}V = 1$$

also corresponds to a surface of constant power dissipation in the space of rates $\hat{q}_i$. For different values of n we shall call these surfaces q-surfaces. It can be shown [178] that q-surfaces for different values of n also lie within one another but in inverse order: the surface that corresponds to the exponent $m < n$ lies inside the surface that corresponds to the exponent n.

Later we shall consider certain examples of application of the Calladine and Drucker theorem to construct approximate solutions for creep theory problems.

102. **Steady-state creep of bar type structures**

As usual, we shall define a bar type structure as a system of rods which are in tension or compression. It rarely happens in practice that a bar type structure is subject to creep so that only a few elementary schemes exist. Therefore, in contrast to the elastic and limit theories, practical examples of calculating relatively complicated rod systems have not been developed. The object of work on creep of bar structures has been to illustrate general principles and general theorems rather than to develop effective methods of calculation. This is the standpoint from which we shall consider steady-state creep in girders, we shall consider problems which illustrate by simple examples the application of general methods of deriving accurate or approximate solutions. We shall exclude the possibility of loss of stability of components in compression. If the force in a component of the girder is N_i, its cross-sectional area F_i, its length l_i, then the creep potential Φ is expressed by

$$\tilde{\Phi} = \sum_i l_i F_i \Phi\left(\frac{N_i}{F_i}\right). \tag{102.1}$$

The equations of statics are used to express the forces in the rods N_i linearly in terms of known generalised forces Q_r and 'redundant' unknowns X_k. Suppose that the number of redundant unknowns is p, then

$$N_i = \alpha_{ir} Q_r + \beta_{ik} X_k .$$

Summation extends over the subscripts r that correspond to external forces and over subscripts k from $k = 1$ to $k = p$. In consequence of the principle of minimum complementary energy dissipation we have

$$\frac{\partial \tilde{\Phi}}{\partial X_l} = 0 .$$

It follows from eq. (102.1) that

$$\sum_i l_i v\left(\frac{N_i}{F_i}\right) \beta_{ik} = 0 . \tag{102.2}$$

From these equations we find the redundant unknowns X_k as functions of the external forces Q_r, we can use eqs. (102.1) to determine the generalised displacements. Indeed by Castigliano's theorem

$$q_r = \sum_i l_i v\left(\frac{N_i}{F_i}\right) \alpha_{ir} . \tag{102.3}$$

It will be seen that calculation of statically determinate or indeterminate beams presents no fundamental difficulties. However, if in fact the degree of static indeterminancy is greater than unity, system (102.2) becomes very complicated to solve. Therefore, various authors have proposed different devices for approximate calculations on rod systems in which the components operate in tension and compression. In the simplest procedure the force distribution in the rods is simply guessed in advance. Suppose, for example, that for proportional change in the forces Q_r^0 the forces in the rods are N_i^0; if $Q_r = \lambda Q_r^0$, then $N_i = \lambda N_i$. Considering λ as a generalised force, by Castigliano's theorem we have

$$Q_r^0 q_r = \frac{\partial \tilde{\Phi}}{\partial \lambda} = \sum l_i v\left(\frac{\lambda N_i^0}{F_i}\right) N_i^0 . \tag{102.4}$$

This formula permits displacements to be determined very simply when only one force acts on the system, otherwise it is additionally necessary to give a surface

$$\widetilde{\Phi}(Q_1^0, Q_2^0, ..., Q_n^0) = H(Q_1^0, Q_2^0, ..., Q_n^0) = \text{const} .$$

Then $q_r = \mu(\partial H/\partial Q_r)$, the factor μ is found from eq. (102.4) after which the rate is readily calculated. In the work [231] it is proposed to take as the quantity N_i^0 the force that corresponds to the limiting state of a rigid-plastic system. To be specific, consider that the yield point is unity, then if the degree of static indeterminancy is p, in the limiting condition for $p+1$ rods $N_i^0 = \pm F_i$ and for the remainder $-F_i \leqslant N_i \leqslant F_i$. The corresponding forces Q_r^0 are proportional to the forces that actually act on the system; $Q_r^0 = (1/\lambda)Q_r$. In the limiting condition forces Q_r^0 are related by a certain expression, namely

$$H(Q_r^0) = \text{const} .$$

This expression is required to determine the rate distribution, but in writing it we assume something more than is contained in eq. (102.4). We shall consider some simple examples to illustrate this.

Example 1. A system of three rods, all of the same material and unit-cross-sectional area $\nu(\sigma) = \sigma^n$ is shown in fig. 159. Denoting by X the force in the centre rod, from the equation of equilibrium we find

$$N_1 = \frac{Q - X}{2 \cos \alpha} .$$

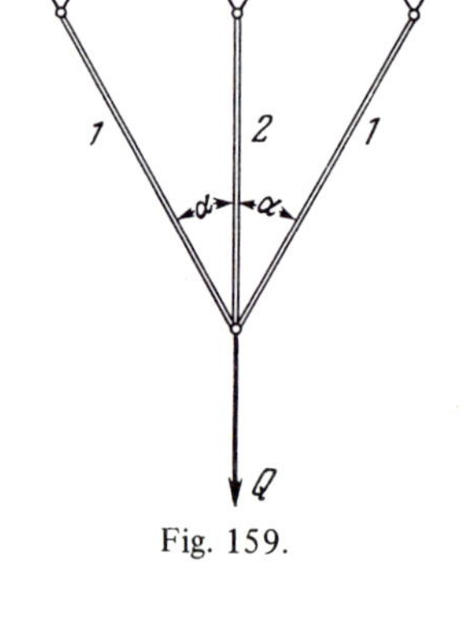

Fig. 159.

Consequently

$$\alpha_1 = \frac{1}{2\cos\alpha}, \quad \alpha_2 = 0, \quad \beta_1 = -\frac{1}{2\cos\alpha}, \quad \beta_2 = 1 .$$

Eq. (102.2) assumes the form

$$-\frac{1}{\cos^2\alpha}\left(\frac{Q-x}{2\cos\alpha}\right)^n + X^n = 0 .$$

Hence

$$X = \frac{Q}{1+\rho}, \qquad \rho = 2\left(\cos\alpha\right)^{1+\frac{2}{n}} .$$

Now from formula (102.3) we find the rate of displacement of the point of application of the force

$$q = l\left(\frac{Q}{1+\rho}\right)^n .$$

(Note that in this particular problem it is simpler to determine q as the rate of elongation of the vertical rod.)

Now solve the same problem approximately. Putting $N_1 = N_2 = 1$, we find $Q^0 = 1 + 2\cos\alpha = 1 + \rho_0$ where ρ_0 is the value of ρ when $n = \infty$. Now by formula (102.4)

$$(1+\rho_0)q^* = \frac{2l}{\cos\alpha}\lambda^n + l\lambda^n .$$

But $\lambda = Q/Q^0 = Q/(1+\rho_0)$, and finally

$$q^* = l\,\frac{4+\rho_0}{\rho_0(\rho_0+1)}\left(\frac{Q}{1+\rho_0}\right)^n .$$

The ratio q^*/q does not tend to unity when $n \to \infty$ and in determining rates the approximate method gives considerable error, but in creep problems we have to compare not rates but the forces that must be applied in order to obtain an identical rate of displacement. The force necessary to communicate a given rate of displacement as found by the accurate solution is denoted Q, and by the approximate solution Q^*. We find

$$\frac{Q^*}{Q} = \frac{1+\rho_0}{1+\rho}\left[\frac{\rho_0(1+\rho_0)}{4+\rho_0}\right]^{\frac{1}{n}}.$$

A graph of the dependence of $1+\rho$ on n is shown in fig. 160 and of the ratio Q^*/Q on n in fig. 161. The value of $\alpha = 60^\circ$ was taken in both cases. The force distribution in the rods, which is characterised by the quantity $1+\rho$, will be seen to remain fairly remote from that which corresponds to the limiting condition, whereas the ratio of the force that develops the given rate, calculated by the accurate method to that calculated by the approximate method, tends to unity more rapidly. It should be emphasised that in real structures the forces acting are rarely known with sufficient accuracy and, therefore, considerable error is possible in determining rates. The designer is not usually directly concerned with permissible rates and displacements but with permissible loads. Therefore, if when $n = 4$ for example the difference between Q and Q^* is 6.9%, this means that selecting the permissible load by the approximate method gives a 6.9% greater safety factor of load.

Example 2. The system of three rods considered in example 1 is loaded by two forces, a vertical force Q_1 and a horizontal force Q_2 (fig. 162). Put $N_2 = X$. In this case eq. (102.2) is fairly complicated, namely

$$\left(\frac{Q_1 - X}{2\cos\alpha} + \frac{Q_2}{2\sin\alpha}\right)^n + \left(\frac{Q_1 - X}{2\cos\alpha} - \frac{Q_2}{2\sin\alpha}\right)^n = 2\cos^2\alpha X^n .$$

If $Q_2 = 0$ we obtain the previous example, if $Q_1 = 0$ then obviously the solution is $X = 0$; consequently, $N_1 = -N_3 = Q_2/(2\sin\alpha)$ and the rate of displace-

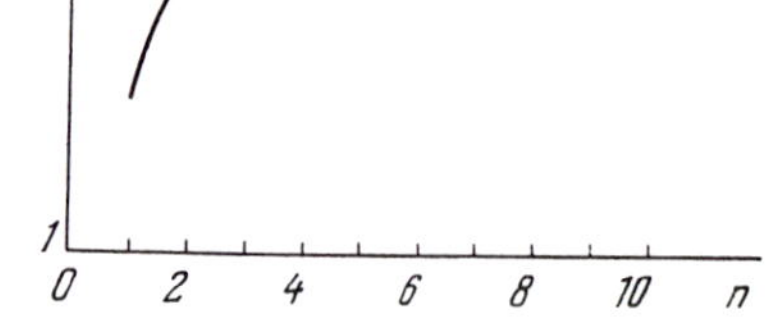

Fig. 160.

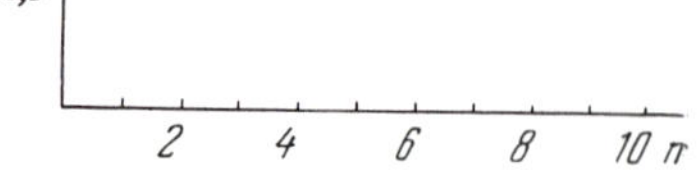

Fig. 161.

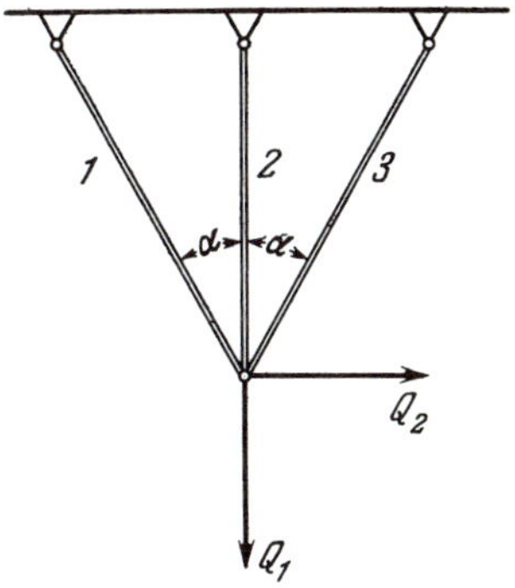

Fig. 162.

ment of the node in the direction of the force Q_2 is

$$q_2 = \frac{l}{\sin\alpha\cos\alpha}\left(\frac{Q_2}{2\sin\alpha}\right)^n .$$

In the general case when Q_1 and Q_2 are arbitrary the kinematic method that we shall describe later is more convenient and, therefore, we shall not consider the derivation of an accurate solution, but shall proceed immediately to the approximate method. From the equations of equilibrium

$$N_1 = \frac{Q_1 - X}{2\cos\alpha} + \frac{Q_2}{2\sin\alpha}, \qquad N_3 = \frac{Q_1 - X}{2\cos\alpha} - \frac{Q_2}{2\sin\alpha}, \qquad N_2 = X .$$

When the forces act in the directions shown in the sketch the following combinations are possible in the limiting state: $N_1 = N_2 = 1$ and $N_1 = -N_3 = 1$. In the first case

$$Q_1^0 + Q_2^0 \cot\alpha = 1 + 2\cos\alpha ,$$

$$N_1^0 = 1 , \quad N_2^0 = 1 , \quad N_3^0 = \frac{Q_0 - 1}{2\cos\alpha} - \frac{Q_2^0}{2\sin\alpha} .$$

In the second case

$$Q_2^0 = 2\sin\alpha , \quad N_1^0 = 1 , \quad N_3^0 = -1 , \quad N_2^0 = Q_1^0 .$$

The limiting state for positive values of Q_1^0 and Q_2^0 is represented by two straight lines in the plane Q_1, Q_2 as shown in fig. 163. Consider the first case.

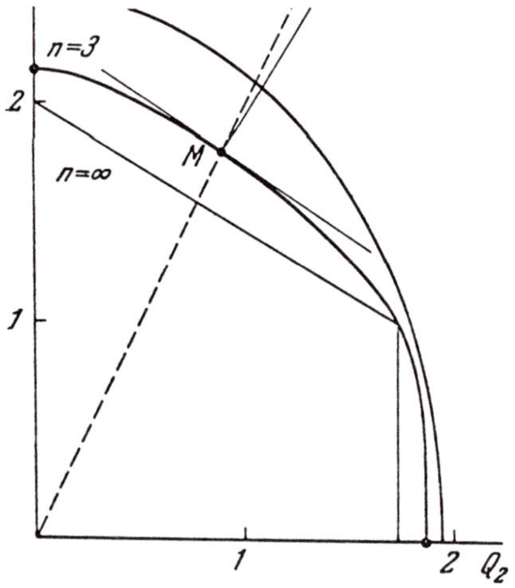

Fig. 163.

Substituting

$$Q_1^0 = \frac{1}{\lambda} Q_1 , \qquad Q_2^0 = \frac{1}{\lambda} Q_2$$

into the condition of the limiting state we obtain

$$\lambda = \frac{Q_1 + Q_2 \cot \alpha}{1 + 2 \cos \alpha} .$$

Formulate eq. (102.4):

$$Q_1^0 q_1 + Q_2^0 q = l\left(\frac{1}{\cos \alpha} + 1\right) \lambda^n + \frac{l}{\cos \alpha}\left(\frac{Q_1^0 - 1}{2 \cos \alpha} - \frac{Q_2^0}{2 \sin \alpha}\right) \lambda^n .$$

Substituting $1/\lambda(Q_1)$ and $1/\lambda(Q_2)$ for Q_1^0 and Q_2^0 and also the expression obtained for λ we find

$$Q_1^0 q_1 + Q_2^0 q_2 = l\left(\frac{1}{\cos \alpha} + 1\right) \times \left(\frac{Q_1 + Q_2 \cot \alpha}{1 + 2 \cos \alpha}\right)^n$$

$$+ \frac{l}{\cos \alpha}\left(\frac{1 + 2 \cos \alpha}{Q_1 + Q_2 \cot \alpha}\right) \times \left(\frac{Q_1 \cos \alpha - Q_2 \cot \alpha (1 + \cos \alpha)}{\cos \alpha (1 + 2 \cos \alpha)}\right)^{n+1} . \quad \text{(a)}$$

It follows from the equation of the limiting state that $q_2 = q_1 \cot \alpha$, therefore

$$Q_1^0 q_1 + Q_2^0 q_2 = q_1(Q_1^0 + Q_2^0 \cot \alpha) = q_1(1 + 2 \cos \alpha) .$$

Now consider the second case. Here

$$\lambda = \frac{Q_2}{Q_2^0} = \frac{Q_2}{2 \sin \alpha}, \qquad Q_1^0 = \frac{Q_1}{\lambda} = \frac{2Q_1 \sin \alpha}{Q_2} .$$

Formulate eq. (102.4)

$$Q_1^0 q_1 + Q_2^0 q_2 = \left(\frac{Q_2}{2 \sin \alpha}\right)^n \left[\frac{2}{\cos \alpha} + \left(\frac{2Q_1 \sin \alpha}{Q_2}\right)^{n+1}\right] l . \tag{b}$$

The corresponding limiting condition is represented by a straight-line segment parallel to the Q_1 axis, therefore $q_1 = 0$, consequently

$$Q_1^0 q_1 + Q_2^0 q_2 = Q_2^0 q_2 = 2 \sin \alpha q_2 .$$

When $Q_2 = 0$ we obtain an approximate result for the previous example, when $Q_1 = 0$ $q_2 = \dfrac{l}{\sin \alpha \cos \alpha} \left(\dfrac{Q_2}{2 \sin \alpha}\right)^n$ which coincides with the accurate solution. This is understandable when $Q_1 = 0$, the force in rod 2 is zero and the system behaves as statically determinate. In applying the approximate method the greatest error, of course, occurs near the angular point in fig. 163 where $Q_1^0 = 1$, $Q_2^0 = 2 \sin \alpha$. The quantity $Q_1^0 q_1 + Q_2^0 q_2$ given by eqs. (a) and (b) alters continuously on transition through this point, whereas the rate components q_1 and q_2 individually alter suddenly in value. If the rates q_1 and q_2 are given we have $q_1 \neq 1$ and $q_2 \neq q_1 \cot \alpha$ and we inevitably find that $Q_1 : Q_2 = 1 : 2 \sin \alpha$.

To judge the rate distribution we need to know the equation of the curve $\widetilde{\Phi} = \text{const}$ in the plane $Q_1 - Q_2$; the approximate location of this curve can be found by use of the Calladine-Drucker theorem.

Example 3. A rigid beam is suspended from four identical rods (fig. 164). We shall consider that $F_i = 1$ and $v = \sigma^n$. The load is applied at an arbitrary point on the beam and we replace it by forces Q_1 and Q_2 applied to the ends (in fig. 165 P and Q respectively). Put $N_1 = X_1, N_2 = X_2$, then

$$N_3 = Q_1 - \tfrac{2}{3} X_2 - \tfrac{1}{3} X_1 , \qquad N_4 = Q_2 - \tfrac{2}{3} X_1 - \tfrac{1}{3} X_2 .$$

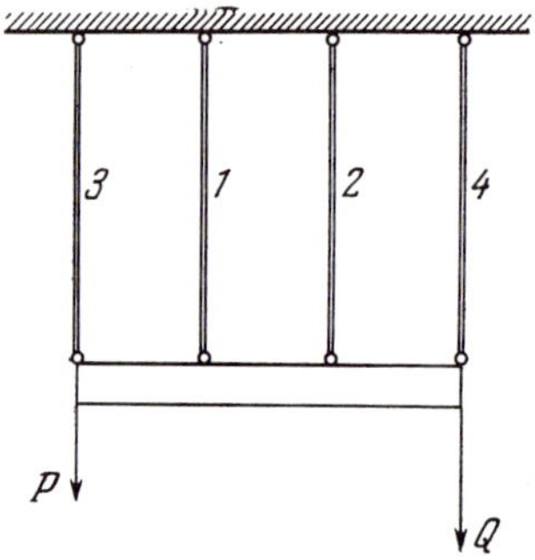

Fig. 164.

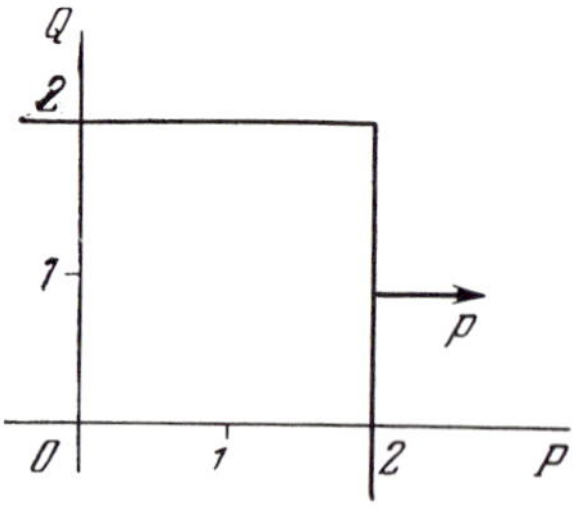

Fig. 165.

As in the previous example, eq. (102.2) is complicated and it is not possible to write the solution in explicit form. Let us use an approximate method. In the limiting state two cases are possible; either $N_1^0 = N_2^0 = N_3^0 = 1, Q_1^0 = 2$, $N_4^0 = Q_2^0 - 1$ or $N_1^0 = N_2^0 = N_4^0 = 1$, $Q_2^0 = 2$, $N_3^0 = Q_1^0 - 1$. The condition of the limiting state in the first quadrant of the plane Q_1, Q_2 is represented by two segments of straight lines parallel to the Q_1 and Q_2 axes whose intercepts on these axes equal 2 (fig. 165). Consider the first case when $Q_1^0 = \text{const}$ we should have

$$q_2 = 0\,, \qquad \lambda = \frac{Q_1}{Q_1^0} = \frac{Q_1}{2}\,, \qquad Q_2^0 = \frac{Q_2}{\lambda} = \frac{2Q_2}{Q_1}\,.$$

Now formulate eq. (102.4)

$$Q_1^0 q_1 + Q_2^0 q_2 = l[3\lambda^n + \lambda^n (Q_2^0 - 1)^{n+1}]\,.$$

Whence

$$q_1 = \frac{l}{2}\left(\frac{Q_1}{2}\right)^n \left[3 + \left(\frac{2Q_2}{Q_1} - 1\right)^{n+1}\right], \qquad q_2 = 0\,.$$

This equation is valid if $Q_1 > Q_2$. If $Q_2 > Q_1$ the same formulae are retained but Q_1 and Q_2, q_1 and q_2 change places. If $Q_1 \neq Q_2$ the rigid-plastic scheme admits only two possible forms of rate distribution; always either $q_1 = 0$ or $q_2 = 0$ so that one of the outermost rods remains unchanged in length and the beam rotates about the corresponding point. When $Q_1 = Q_2$ the beam may be displaced in any way provided that $q_1 Q_1, q_2 Q_2$ are positive. When $Q_1 = Q_2$

we obtain

$$q_1 + q_2 = 2l\left(\frac{Q_1}{2}\right)^n .$$

It is evident from the symmetry of the system that if $Q_1 = Q_2$ the beam must be displaced horizontally, the forces in the rods are then identical and it is readily shown that $q_1 = q_2 = l(Q_1/2)^n$. If Q_2 is less than Q_1, by however small an amount, then $q_2 = 0$ and $q_1 = 2l(Q_1/2)^n$; on the other hand if Q_2 is somewhat greater than Q_1 then $q_1 = 0$, and $q_2 = 2l(Q_2/2)^n$. This kind of paradox always arises when we seek an approximate solution of creep problems based on force distribution in limiting equilibrium and consider a state of loading in the neighbourhood of an angular point on the limiting surface.

103. Application of constant energy rate dissipation surfaces to the calculation of bar type structures

If for a rod system subject to the action of forces Q_r we know the creep potential expressed in terms of the external forces, the problem of determining the displacement is already solved because by Castigliano's theorem

$$q_r = \frac{\partial \tilde{\Phi}}{\partial Q_r} .$$

The Calladine-Drucker theorem, given in section 101, can be used to determine the potential $\tilde{\Phi}$ approximately, if creep follows a power law. For a power creep law the dissipative function and creep potential differ only by a factor and, therefore, if the surface $\tilde{L} = \text{const}$ is known then the function $\tilde{\Phi}$ is also known. It is usually found that with a certain combination of loads the creep problem of the system has a simple solution. Thus, in example 2 of section 102, extremely simple solutions were obtained when $Q_1 = 0$ and when $Q_2 = 0$; in example 3 the solution can be written without any difficulty for the case $Q_1 = Q_2$. Suppose that we are given the generalised force Q_1 that corresponds to the combination of loads. If all the loads are increased by a factor λ, then with a power creep law with exponent n the power dissipation increases by a factor λ^{n+1}, whence it follows that the power dissipation is a uniform function of power $n + 1$ of the generalised forces. Now the equation of the constant energy dissipation surface in the space of forces can be written

$$R(\hat{Q}_1, \hat{Q}_2, ..., \hat{Q}_n) = \text{const} .$$

Here R is a uniform first power function of the forces $\hat{Q}_r = Q_r/\sigma_n$. Obviously the function R is determined to within a factor, select this factor such that

$$R(\hat{Q}_1, 0, 0, ..., 0) = \hat{Q}_1 .$$

It will be seen that R plays the same part in relation to the external forces as the function s in relation to stresses; if s is a certain referred stress then R is a certain referred force. Supposing that the fundamental creep rate test results are obtained in tension, we have defined a uniform function $s(\sigma_{ij})$ such that in tension there is only one stress σ_{11}, that is $s = \sigma_{11}$. In just the same way if we consider that certain reliable information concerning the behaviour of the system is available when only one force Q_1 is acting, and the remaining forces are zero, we require that in this case $R = \hat{Q}_1$. Now the constant R = const in the equation of the constant dissipation surface assumes a definite meaning; it is the segment intercepted by the surface on the axis $\hat{Q}_1$. Denote this segment by a_1 and write the equation of the surface of constant power dissipation as

$$R(\hat{Q}_1, \hat{Q}_2, ..., \hat{Q}_n) = a_1 . \tag{103.1}$$

As standard constant power dissipation surface we select a surface normalised in the sense of section 101. The condition of normalisation is of the form

$$\sum_i l_i F_i \left(\frac{N_i}{F_i \sigma_n}\right)^{n+1} = \sum l_i F_i .$$

Here N_i are expressed in terms of external forces, consequently the condition of normalisation defines a surface with coordinates $\hat{Q}_1^0, \hat{Q}_2^0, ..., \hat{Q}_n^0$ in the space of forces. In accordance with eq. (103.1) we may write the equation of this surface as

$$R_0 = R(\hat{Q}_1^0, \hat{Q}_2^0, ..., \hat{Q}_n^0) = a_i^0 = \text{const} .$$

The creep potential is a function of R; $\widetilde{\Phi} = \widetilde{\Phi}(R)$; consequently

$$\hat{q}_2 = \frac{d\widetilde{\Phi}}{dR} \frac{\partial R}{\partial \hat{Q}_2} .$$

If $\hat{Q}_2 = \hat{Q}_3 = \dots \hat{Q}_n = 0$, then $\widetilde{\Phi} = \widetilde{\Phi}(Q_1)$ and $\hat{q}_1 = \mathrm{d}\widetilde{\Phi}/\mathrm{d}Q_1 = w(\hat{Q}_1)$. It is postulated that when only one load Q_1 acts the creep problem has a solution which can be written in explicit form, therefore, the function w is known and consequently

$$\hat{q}_i = w(R) \frac{\mathrm{d}Q}{\partial \hat{Q}_i}. \tag{103.2}$$

Suppose that in our space of forces $\mathbf{Q} = \hat{Q}\mathbf{n}$, and in the space of rates the vector $\mathbf{q} = \hat{q}\mathbf{v}$. Multiplying both sides of (103.2) by $\hat{Q}_i$ and summing, we find

$$\mathbf{qQ} = w(R)R\ ,$$

since by assumption R is a uniform first power function. Hence we obtain

$$\hat{q} = \frac{w(R)R}{\hat{Q}\cos(\widehat{\mathbf{Qq}})} \tag{103.3}$$

and

$$\hat{q}_i = \hat{q}v_i\ .$$

Formula (103.3) can be useful when the number of generalised forces is two and the functions $R(\hat{Q}_i)$ are given by a curve in the plane $\hat{Q}_1, \hat{Q}_2$ (fig. 166). If we are given the external forces $\hat{Q}_i$ we construct the point M in the space of forces and draw a line to it from the origin of coordinates; this line intersects a surface of constant normalised power dissipation at the point M_0 with coordinates $\hat{Q}_i^0$. Because of eq. (103.1) the quantity R is

$$R = a_1^0 \frac{\hat{Q}}{\hat{Q}_0}\ ,$$

where $\hat{Q}_0$ is the radius vector of the point M_0, a_1^0 is the radius vector of the point M, a_1^0 is the intercept of the surface of constant normalised power dissipation on the $\hat{a}_1$ axis, whilst the direction of the vector $\mathbf{q}$ is normal to the surface at the point M_0. To construct the constant normalised power dissipation surface, we have, by hypothesis, one point $\hat{Q}_1^0 = a_1^0$ on the $\hat{Q}_1$ axis, and two limiting surfaces, the surface of limiting states and the surface of elastic states. In individual problems it is usually found that the system permits of accurate solution for several other combinations of loads, for instance, for $\hat{Q}_2, \hat{Q}_3$. Then we know the points of intersection of the required surface with

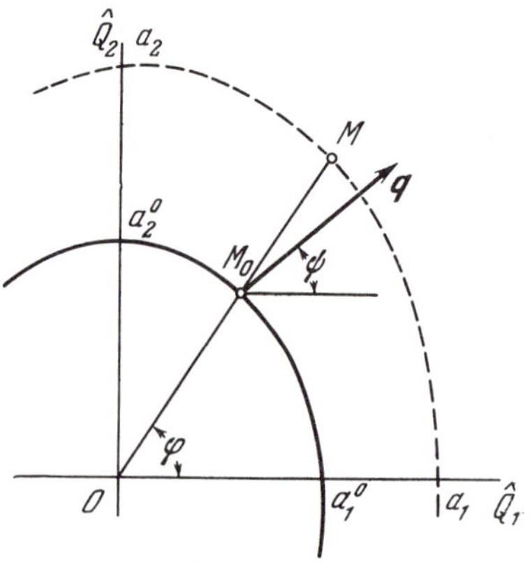

Fig. 166.

the $\hat{Q}_2$ and $\hat{Q}_3^0$ axes, denote these $\hat{Q}_2^0 = a_2^0$, $\hat{Q}_3^0 = a_3^0$. The problem of constructing surfaces R = const becomes somewhat more definite although much still remains arbitrary. Even when $n = 2$, selection of an appropriate analytical expression for the function $R(\hat{Q}_i^0)$ is difficult and when $n = 3$ or more it is completely hopeless; when $n = 2$ it is recommended simply to draw an arbitrary curve through the given points by eye and to seek a solution by means of eq. (103.3) using the drawing directly.

The method of construction is shown in fig. 166. Assume that we know the points $\hat{Q}_1^0 = a_1^0$ and $\hat{Q}_2 = a_2^0$, the corresponding normalised exact solutions for $Q_2 = 0$ and $Q_1 = 0$ respectively. Draw through these a smooth curve that is enclosed between the limiting curves (not shown in the drawings). Suppose that we are given the forces Q_1 and Q_2 and it is required to determine q_1 and q_2. Construct a point M with coordinates $\hat{Q}_1$ and $\hat{Q}_2$, draw a line OM, the coordinates of the point M_0 are $\hat{Q}_1^0$ and $\hat{Q}_2^0$ and they must be measured directly from the drawing. The length of the segment OM is the modulus of the force vector $\hat{Q}$. We assume that when $Q_2 = 0$ the solution is known, it is given by the formula $q_1 = w(\hat{Q}_1)$ so that w is a known function. Calculate the quantity R: $R = a_1^0(\hat{Q}_1/\hat{Q}_1^0) = a_1^0(\hat{Q}_2/\hat{Q}_2^0) = a_1^0(\hat{Q}/\hat{Q}^0)$ and draw a normal to the curve at the point M_0 and find the angle ψ between the normal and the $\hat{Q}_1$ axis, the angle φ between this axis and the line OM is known. By formula (103.3)

$$\hat{q} = \frac{w(R)R}{\hat{Q}\cos(\varphi - \psi)}.$$

Having found $\hat{q}$ determine $\hat{q}_1 = \hat{q}\cos\psi$ and $\hat{q}_2 = \hat{q}\sin\psi$.

104. The kinematic method of calculating bar type structures. Examples

The difficulty of obtaining accurate solutions of creep problems in bar structures by the method presented in section 102 is that of solving certain non-linear equations and in essence this is associated with the circumstance that it is required to find explicit expressions for internal forces in terms of external forces and for rates in terms of forces. This difficulty can sometimes be avoided if we consider the rates of the nodes as given. Let q_s be given rates of nodes of the structure; formulating kinematic equations we can always express the rate of elongation of each rod v_i in a linear way in terms of the rates of the nodes q_s, so that

$$v_i = \gamma_{is} q_s .$$

Formulate the expression for the force potential

$$\widetilde{U} = \sum l_i F_i U(v_i) .$$

By Lagrange's theorem

$$Q_s = \frac{\partial \widetilde{U}}{\partial q_s} = \sum l_i F_i U'(v_i) \gamma_{is} = \sum l_i F_i s(\gamma_{ik} q_k) \gamma_{is} . \tag{104.1}$$

The functions v and s are reciprocal; if $v = v(s)$ then $s = s(v)$.

Formulae (104.1) explicitly express the forces in terms of rates of nodes of the girder, it is by no means always possible to solve these equations relative to displacements, and in the general case the difficulties of the kinematic method are the same as those of the static. It is convenient to use the kinematic method when the number of possible displacements of the nodes is not great, for example, for the structure illustrated in fig. 167. Because of symmetry of the problem we are concerned with only one rate, the vertical rate q of the point of application of the force, therefore, formula (104.1) immediately gives the relationship between Q and q, the increase in degree of static indeterminancy, that is the increase in the number of sloping rods introduces no additional complications. On the other hand, there is no justification for using the kinematic method for the girder illustrated in fig. 168, where we have to consider four rates, two rate components of point A and two of point B. If we are interested only in the rate of the point of application of the force Q in its direction of action, the three remaining rates can be excluded from the equations. With the power creep law the kinematic method can be used to

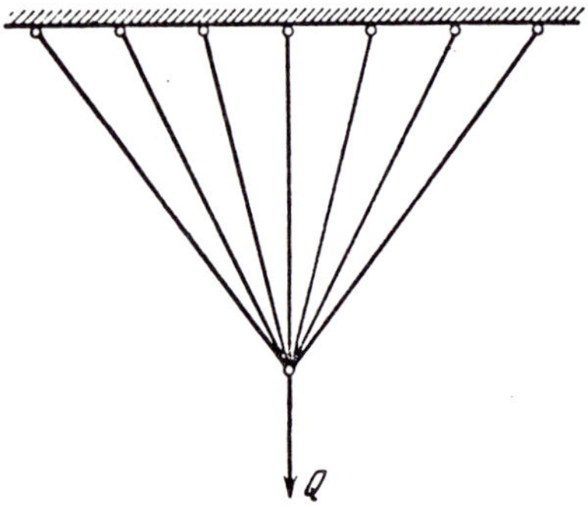

Fig. 167.

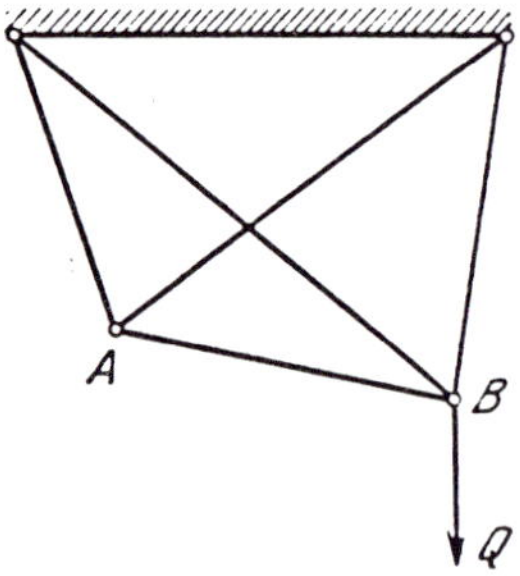

Fig. 168.

examine an approximate solution as was done in section 103 relative to the static method. Let us take

$$s = \sigma_n \left(\frac{v}{\epsilon_n}\right)^{\frac{1}{n}},$$

put as before $\hat{Q}_r = Q_r/\sigma_n$, $\hat{q}_r = q_r/\epsilon_n$. In exactly the same way as when the forces are given, it is required that the mean power dissipation per unit volume should be unity, so that

$$\sum l_i F_i \left(\frac{v_i}{\epsilon_n}\right)^{1+\frac{1}{n}} = \sum l_i F_i \, .$$

This condition defines the constant power dissipation surface in the space of rates. Write the equation of this surface in the form

$$r(\hat{q}_s^0) = b_1^0 \, . \tag{104.2}$$

Here the uniform first power function $r(\hat{q}_s)$ is defined such that $r = q_1$, if $q_2 = q_3 = \ldots = 0$. As we have seen in section 102 surfaces (104.2) for different values of n lie inside one another, but the order is the inverse of that for Q-surfaces; the inner boundary q-surface corresponds to the elastic state, that is $n = 1$ and the outer to the limiting rigid-plastic state. The Q-surfaces and q-surfaces are interrelated by reciprocity expressions. If we are given the vector $\overline{OM} = \mathbf{Q}$ in the space $\hat{Q}$ then the vector $\overline{OM}_0 = \mathbf{Q}^0$ is proportional to it, the vector $\mathbf{q}$ is in the direction of the normal to the Q surface at the point M_0. If we now construct the vector $\overline{ON} = \mathbf{q}$ in the space of rates and the vec-

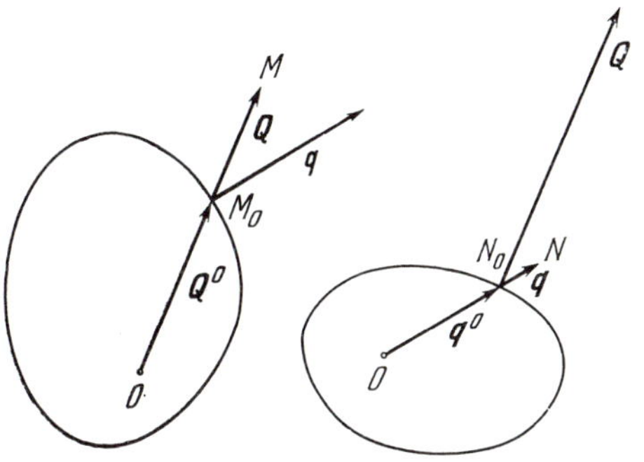

Fig. 169.

tor $\overline{ON}_0$ which is proportional to it, then the normal to the q-surface at the point N_0 gives the direction of the vector **Q**. For two forces when the limiting surface is converted into a limiting curve, this property can be used to construct Q curves if we are given q-curves and vice versa (fig. 169).

Now let us obtain more accurate solutions for examples 2 and 3 of section 102.

Example 2, section 102. We found that in the plane Q_1–Q_2 the limiting state of an ideal rigid-plastic system is represented by two straight lines $Q_1^0 + Q_2^0 \cot\alpha = 1 + 2\cos\alpha$ and $Q_2^0 = 2\sin\alpha$. These two lines are constructed for $\alpha = 60°$ in fig. 163. In the elastic state

$$N_2 = \frac{Q_1}{1 + 2\cos^3\alpha}, \qquad N_{1,3} = \frac{Q_1 \cos^2\alpha}{1 + 2\cos^3\alpha} \pm \frac{Q_2}{2\sin\alpha}.$$

The condition of normalisation

$$\frac{l}{\cos\alpha}(N_1^2 + N_3^2) + lN_2^2 = \frac{2l}{\cos\alpha} + l$$

leads to the following expression between Q_1^0 and Q_2^0:

$$\frac{Q_1^{02}}{1 + 2\cos^3\alpha} + \frac{Q_2^{02}}{2\sin^2\alpha\cos\alpha} = \frac{2 + \cos\alpha}{\cos\alpha}.$$

This is the equation of an ellipse, which is constructed in fig. 163 for $\alpha = 60°$. All the Q-curves for different values of n lie between the ellipse and the broken line for $n = \infty$ and the quantities Q_1^0 and Q_2^0, the intercepts of these curves on the corresponding axes, are easily found. Indeed the solutions are known when $Q_2 = 0$ and when $Q_1 = 0$, it remains to normalise them (with corresponding values of n). We obtain

$$a_1^0 = \left(1 + \rho\right)^{\frac{n}{n+1}} \left(\frac{2 + \cos\alpha}{\cos\alpha}\right)^{\frac{1}{n+1}},$$

$$a_2^0 = 2\sin\alpha \left(\frac{2 + \cos\alpha}{2}\right)^{\frac{1}{n+1}}.$$

Points corresponding to $n = 3$ are marked on the curve by circles, here $a_1^0 = 2.16$, $a_2^0 = 1.86$. Further we may attempt to guess how the Q-curve passes through these points but in this case the kinematic method gives more accurate solutions. Apply to the node a vertical rate q_1 and a horizontal rate q_2, then the rates of elongation of the rods are

$$v_{1,3} = \frac{\cos\alpha}{l}(q_1 \cos\alpha \pm q_2 \sin\alpha), \qquad v_2 = \frac{q_1}{l}.$$

By formula (104.1) we immediately obtain

$$Q_1 = \left[\frac{\cos\alpha}{l}(q_1 \cos\alpha + q_2 \sin\alpha)\right]^{\frac{1}{n}} \cos\alpha$$

$$+ \left[\frac{\cos\alpha}{l}(q_1 \cos\alpha - q_2 \sin\alpha)\right]^{\frac{1}{n}} \cos\alpha + \left(\frac{q_1}{l}\right)^{\frac{1}{n}},$$

$$Q_2 = \left[\frac{\cos\alpha}{l}(q_1 \cos\alpha + q_2 \sin\alpha)\right]^{\frac{1}{n}} \sin\alpha$$

$$- \left[\frac{\cos\alpha}{l}(q_1 \cos\alpha - q_2 \sin\alpha)\right]^{\frac{1}{n}} \sin\alpha.$$

Thus, forces are expressed in terms of rates.

In order to solve the inverse problem, that is to find the rates given the forces we must still construct Q curves, this may now be done as follows. Put $Q_1/Q_2 = \tan\varphi$, $q_1/q_2 = \tan\psi$; and from the solution that we have found it follows that

$$\tan\varphi = \cot\psi \frac{[\sin(\psi+\alpha)]^{\frac{1}{n}} + [\sin(\psi-\alpha)]^{\frac{1}{n}} + \dfrac{(\sin\psi)^{\frac{1}{n}}}{(\cos\alpha)^{1+\frac{1}{n}}}}{[\sin(\psi+\alpha)]^{\frac{1}{n}} - [\sin(\psi-\alpha)]^{\frac{1}{n}}}.$$

In calculating a table or constructing graphs of φ against ψ we start from a certain fixed point M_0 in the plane Q_1–Q_2 (fig. 170); the radius vector of this point forms an angle φ_0 with the abscissa. Through the point M_0 draw a straight line at an angle $\psi_0(\varphi_0)$ to the abscissa, the perpendicular to it at the point M_0 is in the direction of the tangent to the Q-curve. The small segment M_0M_1 of the tangent is taken as an element of arc, then the radius vector OM_1 slopes at an angle φ_1 to which corresponds the angle ψ_1 of slope of the normal; continuing this construction we obtain the point M_2 and so on. The construction could be made still more accurately but there is hardly need for this. The Q-curve obtained in this way for $n = 3$ is given in fig. 163.

Instead of constructing Q-curves we could have considered a space of rates, as will be shown in the next example.

Example 3 of section 102. We shall consider this example without referring to the concept of plane of forces. Apply to the ends of the rods downward rates q_1 and q_2, then the rates of elongation of the rods are: $v_3 = q_1$, $v_1 = \frac{2}{3}q_1 + \frac{1}{3}q_2$, $v_2 = \frac{1}{3}q_1 + \frac{2}{3}q_2$, $v_4 = q_2$. By formulae (104.1):

$$\begin{aligned} Q_1 &= q_1^{\frac{1}{n}} + \tfrac{2}{3}(\tfrac{2}{3}q_1 + \tfrac{1}{3}q_2)^{\frac{1}{n}} + \tfrac{1}{3}(\tfrac{1}{3}q_1 + \tfrac{2}{3}q_2)^{\frac{1}{n}}, \\ Q_2 &= q_2^{\frac{1}{n}} + \tfrac{1}{3}(\tfrac{2}{3}q_1 + \tfrac{1}{3}q_2)^{\frac{1}{n}} + \tfrac{2}{3}(\tfrac{1}{3}q_1 + \tfrac{2}{3}q_2)^{\frac{1}{n}}. \end{aligned} \qquad \text{(a)}$$

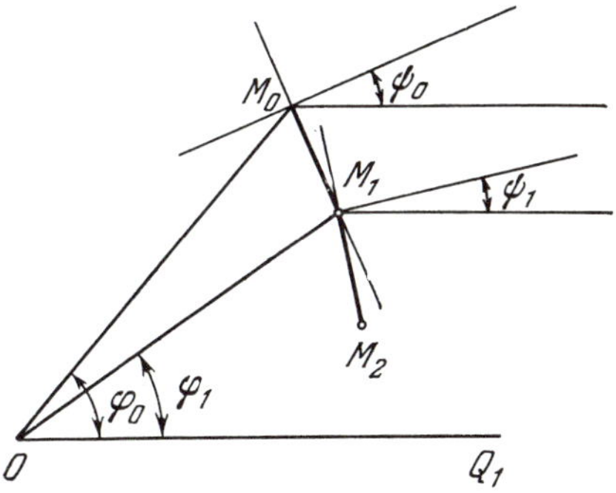

Fig. 170.

Now construct the corresponding q-curves. The condition of normalisation is written

$$q_1^{1+\frac{1}{n}} + \left(\tfrac{2}{3}q_1 + \tfrac{1}{3}q_2\right)^{1+\frac{1}{n}} + \left(\tfrac{1}{3}q_1 + \tfrac{2}{3}q_2\right)^{1+\frac{1}{n}} + q_2^{1+\frac{1}{n}} = 4\,.$$

When $n = \infty$ we obtain

$$q_1 + q_2 = 2\,.$$

When $n = 1$ the q-curve is an ellipse

$$7q_1^2 + 4q_1q_2 + 7q_2^2 = 18\,.$$

Its intercepts on the coordinate axes are $\sqrt{\frac{18}{7}} \approx 1.6$, and it touches the line of limiting state at the point $q_1 = q_2 = 1$. This always occurs for structures of minimum weight, i.e. so that with a certain combination of loads all the components become plastic simultaneously. If we now take $q_2 = 0$ or $q_1 = 0$, from the condition of normalisation we obtain

$$b_1^0 = b_2^0 = \left(\frac{4}{1 + \left(\frac{1}{3}\right)^{1+\frac{1}{n}} + \left(\frac{2}{3}\right)^{1+\frac{1}{n}}}\right)^{\frac{3}{4}}.$$

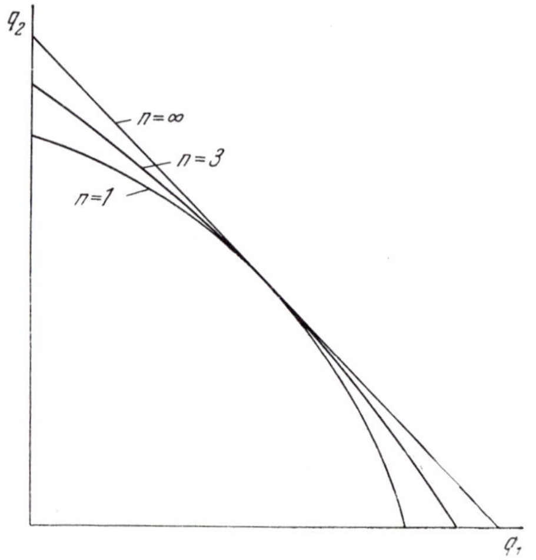

Fig. 171.

where b_1^0 and b_2^0 are intercepts of the q-curve on the coordinate axes. It is not difficult to construct the entire q-curve, the first quadrant is shown in fig. 171 for $n = 3$. Given Q_1 and Q_2 it is simplest to determine the rates as follows: on the graph of fig. 171 draw a line with slope Q_1/Q_2 and the tangent to the q-curve which is perpendicular to it. The abscissa and ordinate of the point of contact are in the ratio q_1/q_2. Knowing this ratio we can readily determine q_1 or q_2 by one of the formulae (a).

105. Kachanov's method

The general method of calculating statically indeterminate girders leads to non-linear equations for the redundant unknowns, however, these equations can only be solved for a system with one redundant unknown and if there are even two the difficulties become very great and the calculations laborious. In section 102 we considered the simplest device for obtaining approximate solutions, in which the distribution of the forces in the rods is given in advance and a certain system of forces λN_i^0 is selected which satisfies the equations of statics. More accurate results may be obtained as follows. Consider a self-equilibrating system of forces N_i' then the force distribution $\lambda(N_i^0 + kN_i')$ will also satisfy the equations of statics. Define the parameter k such that the rate potential is minimum. Substituting into eq. (102.1), differentiating with respect to k and equating to zero we obtain

$$\sum l_i v\left[\frac{\lambda(N_i^0 + kN_i')}{F_i}\right] N_i' = 0 .$$

With a power creep law, the quantity λ raised to the power n becomes a factor outside the summation sign and we arrive at the following equation for the parameter k:

$$\sum l_i v\left(\frac{N_i^0 + kN_i'}{F_i}\right) N_i' = 0 . \tag{105.1}$$

This non-linear equation for k is fairly complicated and is the only one which is independent of the degree of static indeterminancy. After k has been found proceed in the usual way as explained in section 102. Note that the factor k is determined from eq. (105.1) for a given system of loads, with other loads the factor will be different. Therefore, Castigliano's theorem for the determination of displacements can only be used in the form of (102.4) on the assumption that the loads alter proportionately. Kachanov proposed to take as the state N_i^0 the limiting state of a rigid-plastic structure with external loads $Q_i^0 = Q_i/\lambda$, and as the state N_i' the difference between forces calculated for the same force Q_i^0 on the assumption that the material is elastic and the forces are N_i^0. Some examples of the use of this method are given in a monograph [52] and we shall not consider them here.

8

Steady-State Creep. Bending and Torsion

The general methods of studying steady-state creep, described in Chapter 7, are applicable to the bending and torsion of rods. Approximate methods, based on replacing the true stress distribution by one which corresponds to the ultimate limiting state of an ideally plastic rod, are also useful, in particular they can be used to construct a fairly simple theory of tangential stresses in thin-walled rods (or tubes). The theory of torsion of thin-walled rods of open and closed profiles also presents no difficulties. In conclusion bending in the presence of a longitudinal force is considered.

106. Steady-state creep in pure bending

Here we consider creep in pure bending of a rod whose section has two axes of symmetry, x and y, the bending moment acting in the plane yOz. The height of the section is $2h$, its width b is given as function of the ratio y/h (fig. 172). Take the creep law in the form $\sigma = s(\epsilon)$, the rod sections remain plane on deformation, consequently $\epsilon = \kappa y$ where κ is the rate of change of curvature of the centre line of the rod. If the properties of the material are the same in tension and in compression the bending moment is

$$M = 2\int_0^h b\left(\frac{y}{h}\right) ys(\kappa y)\,\mathrm{d}y\,.$$

Take as variable of integration the quantity $\kappa y = \epsilon$, put $\kappa h = \epsilon'$, then

$$M = \frac{2h^2}{\epsilon'^2}\int_0^{\epsilon'} b\left(\frac{\epsilon}{\epsilon'}\right) \epsilon s(\epsilon)\,\mathrm{d}\epsilon\,.$$

Introduce the notation

$$m(\epsilon') = \frac{2}{\sigma^* b_0 \sigma'^2}\int_0^{\epsilon'} b\left(\frac{\epsilon}{\epsilon'}\right) \epsilon s(\epsilon)\,\mathrm{d}\epsilon\,.$$

Then

$$M = \sigma^* h^2 b_0 m(\kappa h)\,. \tag{106.1}$$

Here b_0 is a certain characteristic width, for example when $h = 0$, σ^* is an arbitrary constant with the dimensions of stress. For a power creep law

$$s(\epsilon) = \sigma_n \left(\frac{\epsilon}{\epsilon_n}\right)^{\frac{1}{n}},$$

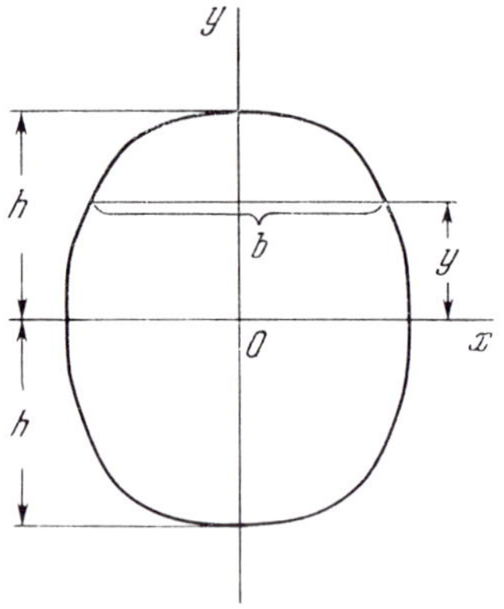

Fig. 172.

in the case of a rectangular section of width b and height $2h$ we obtain

$$m = \frac{2n}{2n+1}\left(\frac{\kappa h}{\epsilon_n}\right)^{\frac{1}{n}} \qquad (\sigma^* = \sigma_n)\,.$$

A feature of the power creep law is that the dependence of rate of change of curvature on torque is also a power law of exponent n. Indeed if

$$s(\epsilon) = \sigma_n \left(\frac{\epsilon}{\epsilon_n}\right)^{\frac{1}{n}},$$

then

$$M = \left(\frac{\kappa h}{\epsilon_n}\right) \frac{\sigma_n}{h^{\frac{1}{n}}} \int_F |y|^{1+\frac{1}{n}}\, \mathrm{d}F\,.$$

Denoting

$$\kappa_n = \frac{\epsilon_n}{h}, \qquad M_n = h\sigma_n \int_F \left|\frac{y}{h}\right|^{1+\frac{1}{n}} \mathrm{d}F\,.$$

Then

$$\kappa = \kappa_n \left(\frac{M}{M_n}\right)^n. \tag{106.2}$$

For a rectangle

$$M_n = \frac{2n}{2n+1}\, bh^2 \sigma_n\,.$$

The function $m(\kappa h)$ can also be calculated for other creep laws; if quadratures are not fulfilled it is always possible to construct the functions graphically or by numerical methods.

An extremely simple approximate solution of creep in bending is obtained if we assume a stress distribution which corresponds to the limiting state of a plastic rod. The cross-sectional area is F and the coordinate of the centre of gravity of half the section located in the region of positive values of y is $c/2$; we shall consider that when $y > 0$ the stress is a constant quantity σ and when $y < 0$, a quantity $-\sigma$. Then the bending moment is

$$M = \frac{cF}{2}\,\sigma\,,$$

consequently,

$$\sigma = \frac{2M}{cF}\,.$$

Formulate the expression for rate potential for unit length of the beam

$$\widetilde{\Phi} = \int_F \Phi(\sigma)\,\mathrm{d}F = F\Phi\left(\frac{2M}{cF}\right).$$

By Castigliano's theorem

$$\kappa = \frac{\partial\widetilde{\Phi}}{\partial M} = \frac{2}{c}\,v\left(\frac{2M}{cF}\right). \tag{106.3}$$

With a power law we can convert this formula to the form of (106.2) if we put

$$M_n = \tfrac{1}{2}\,cF\sigma_n\left(\frac{c}{2h}\right)^{\frac{1}{n}}.$$

Apply formula (106.3) to the case of a rectangular section with a power creep law. Here $F = 2bh$, $c = h$, consequently

$$\kappa = \frac{2\epsilon_n}{h}\left(\frac{M}{bh^2}\right)^n.$$

Hence

$$m = 2^{-\frac{1}{n}}\left(\frac{\kappa h}{\epsilon_n}\right)^{\frac{1}{n}}.$$

Comparing with the accurate solution for a rectangle we find it differs from the approximate only by a factor of $2n/(2n+1)$ in the first case and $2^{-1/n}$ in the second. For comparison the table on page 449 gives values of these factors for certain values of n.

The difference between the accurate and approximate solutions is comparatively small because of the nature of the stress distribution in bending. Stress diagrams are constructed in fig. 173 for different values of n and the

n	$\frac{2n}{2n+1}$	$2^{-\frac{1}{n}}$
2	0.800	0.707
3	0.858	0.795
4	0.889	0.840
6	0.923	0.892
8	0.940	0.917

same bending moment. It will be seen that the diagrams diverge for small values of y/h, but when y is near to h the stresses differ little from those that correspond to the plastic state. The principal contribution to the moment is made by stresses in parts of the section remote from the neutral axis.

Now suppose that the section has only one axis of symmetry, the y axis, and as before, bending takes place in the plane yOz. In contrast to the case of elastic bending the neutral axis of the rod does not pass through the centre of gravity of the section. Moreover, with an arbitrary creep law the position of the neutral axis depends on the magnitude of the bending moment, only with a power law is the neutral axis fixed for a given value of n. An approximate solution is very easily obtained in this case. In the limiting state the neutral

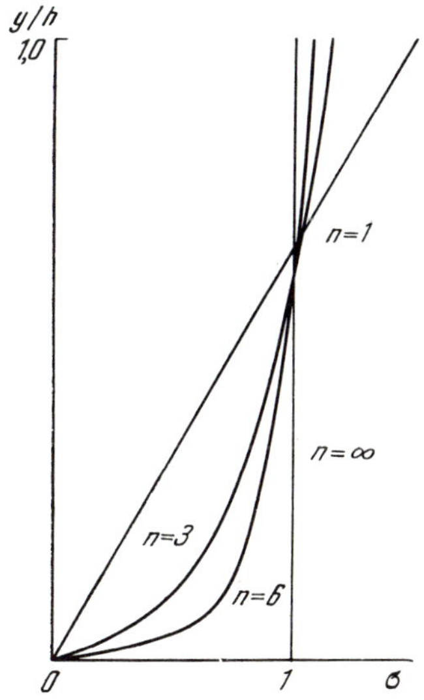

Fig. 173.

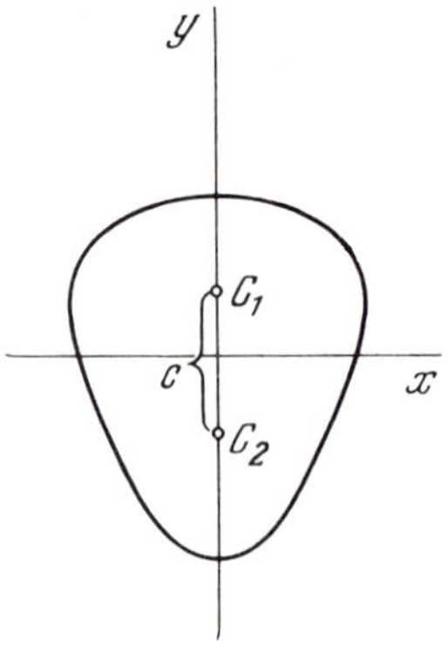

Fig. 174.

axis divides the cross-sectional area into two equal parts, suppose that the same applies during creep. Take this axis as the x axis (fig. 174), find the centres of gravity of the two halves of the section $C_1 - C_2$ and denote the distance C_1C_2 by c. Then as before the stress in the limiting state is $2M/cF$, the expression for potential is the same as in the case of symmetry about the x axis and expression (106.3) remains valid. With the power law, formula (106.2) is also true if there is only one axis of symmetry but the position of the neutral axis must be found before it can be applied.

Now consider bending of a rod of rectangular cross-section with the exponential creep law of eq. (43.3)

$$\sigma = \sigma_e \ln \frac{\epsilon}{\epsilon_e} . \tag{106.4}$$

This law ceases to be formally suitable when $\epsilon < \epsilon_e$, it is sufficiently accurate only when $\epsilon/\epsilon_e \gg 1$. Taking that eq. (106.4) is valid only when $\epsilon > \epsilon_e$, and when $\epsilon < \epsilon_e$, $\sigma = 0$. Then

$$M = 2b\sigma_e \int_{y_0}^{h} \ln \frac{\kappa y}{\epsilon_e}\, y\, \mathrm{d}y \qquad \left(y_0 = \frac{\epsilon_e}{\kappa}\right) .$$

Putting

$$M_e = \sigma_e h^2 b\,, \quad \kappa_e = \frac{\epsilon_e}{h}\,, \quad \hat{M} = \frac{M}{M_e}\,, \quad \hat{\kappa} = \frac{\kappa}{\kappa_e}\,,$$

we easily obtain

$$\hat{M} = \ln \hat{\kappa} - \tfrac{1}{2} + \frac{1}{2\hat{\kappa}^2}. \tag{106.5}$$

To avoid introducing regions for which $\sigma = 0$, in place of law (106.4) we use the hyperbolic sine law (section 43) which has better physical justification

$$\sigma = \sigma_e \text{ Arsinh } \frac{\epsilon}{2\epsilon_e}. \tag{106.6}$$

By simple calculation we find

$$\hat{M} = \left(1 + \frac{2}{\hat{\kappa}^2}\right) \text{Arsinh } \frac{\hat{\kappa}}{2} - \frac{\sqrt{\hat{\kappa}^2 + 4}}{2\hat{\kappa}}. \tag{106.7}$$

Graphs of $\hat{M}$ against $\hat{\kappa}$ by formula (106.5) and (106.7) respectively are plotted in fig. 175. It will be seen that when $\hat{\kappa} > 10$ the difference between these formulae is very small. This is because the low stress region is always close to the neutral axis of the rod and these low stresses are insignificant in calculating the bending moment.

A fairly large error is obtained by taking the stress distribution the same as

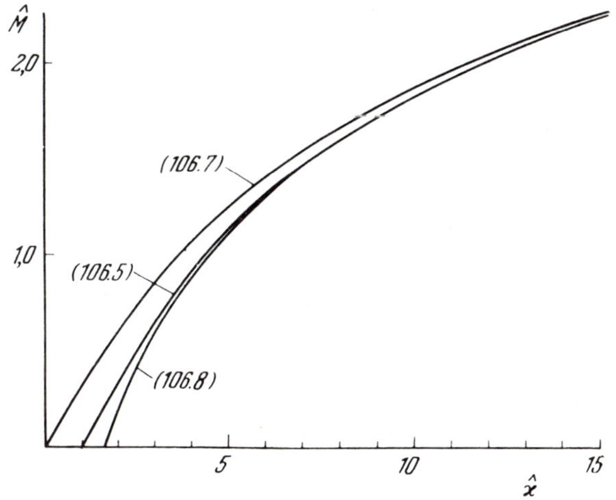

Fig. 175.

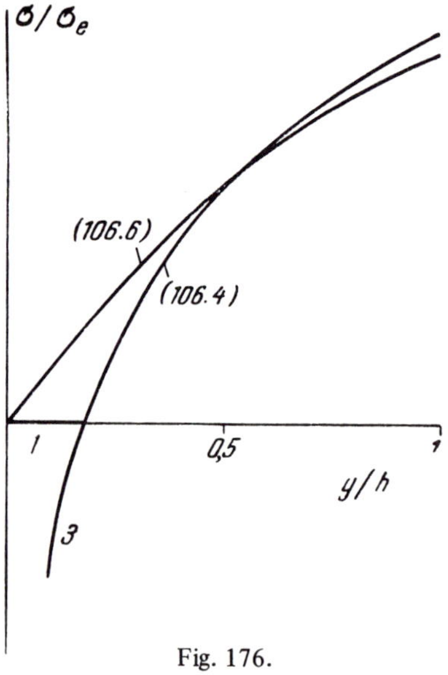

Fig. 176.

in the plastic state, that is applying formula (106.3) with an exponential creep law.

Stress diagrams across the section for the case $\kappa_e = 5$ are given in fig. 176; curve 1 corresponds to formula (106.4) with the stipulation mentioned, curve 2 to formula (106.6) and curve 3 to formula (106.4) without any limitations. In the latter case $\sigma = -\infty$ when $y = 0$. The moment calculated for this latter stress diagram remains finite and does not significantly differ from the value given by formula (106.5). If we make the lower limit of integration zero instead of y_0, we obtain

$$\hat{M} = \ln \hat{\kappa} - \tfrac{1}{2} . \tag{106.8}$$

Now consider bending of a rod of arbitrary cross-section having an axis of symmetry y with an exponential creep law. The example that we have given shows that, with a fairly large bending moment, the fact that the stress tends to infinity on the neutral axis has practically no influence on the relationship between moment and curvature and, therefore, there is no need to take special measures to improve the creep law near the neutral axis. Calculating the bending moment from the stresses given by formula (106.4) we obtain

$$M = \sigma_e \int_F \ln \frac{\kappa |y|}{\epsilon_e} |y| \, \mathrm{d}F = \sigma_e \left[\int_F \ln \frac{\kappa h}{\epsilon_e} |y| \, \mathrm{d}F + \int_F \ln \frac{|y|}{h} |y| \, \mathrm{d}F \right] .$$

Here h is an arbitrary linear dimension whence if follows that:

$$\frac{\kappa}{\kappa_e} = \exp \frac{M}{M_e} \,. \tag{106.9}$$

Here

$$\kappa_e = \frac{\epsilon_e}{h} \exp \left[-\frac{2}{Fc} \int_F |y| \ln \frac{|y|}{h} \, \mathrm{d}F \right],$$

$$M_e = \tfrac{1}{2} \sigma_e Fc \,,$$

where c is the distance between the centres of gravity of the parts of the section above and below the neutral axis. Generally speaking, in asymmetrical sections c is not constant, and depends on κ or M, but approximately we may take c = const.

For large values of $\hat{\kappa}$ the difference between eqs. (106.5), (106.7) and (106.8) becomes infinitesimally small and, therefore, there is no need to employ an artificial device to exclude the zone of the low creep rates. The curve corresponding to eq. (106.8) is also shown in fig. 175, as from $\hat{\kappa} = 5$ it no longer differs from the curve of eq. (106.5).

107. The general case of pure bending

Now let us consider the general case of an asymmetrical section. In the plane of the section, draw arbitrary axes x and y, consequently the law of plane sections is

$$\epsilon = \epsilon_0 + \kappa_x y - \kappa_y x \,.$$

For the neutral axis $\epsilon = 0$, consequently

$$\epsilon_0 + \kappa_x y - \kappa_y x = 0 \,.$$

The segments a and b that the neutral axis intercepts on the x and y axes are:

$$a = \frac{\epsilon_0}{\kappa_y} \,, \qquad b = -\frac{\epsilon_0}{\kappa_x} \,.$$

Now calculate the longitudinal force and bending moments:

$$N = \int_F \sigma\, \mathrm{d}F = \int_F s(\epsilon_0 + \kappa_x y - \kappa_y x)\, \mathrm{d}F\,,$$

$$M_x = -\int_F \sigma y\, \mathrm{d}F = -\int_F s(\epsilon_0 + \kappa_x y - \kappa_y x) y\, \mathrm{d}F\,, \qquad (107.1)$$

$$M_y = \int_F \sigma x\, \mathrm{d}F = \int_F s(\epsilon_0 + \kappa_x y - \kappa_y x) x\, \mathrm{d}F\,.$$

Requiring that $N = 0$ we obtain a relationship between ϵ_0, κ_x and κ_y; in the case of a power law ϵ_0 can be taken outside the integral sign and then the condition of absence of axial force is

$$\int_F s\left(1 - \frac{x}{a} - \frac{y}{b}\right) \mathrm{d}F = 0\,. \qquad (107.2)$$

The ratio of the moments M_x to M_y is:

$$\frac{M_x}{M_y} = -\frac{\int_F s\left(1 - \frac{x}{a} - \frac{y}{b}\right) y\, \mathrm{d}F}{\int_F s\left(1 - \frac{x}{a} - \frac{y}{b}\right) x\, \mathrm{d}F}\,. \qquad (107.3)$$

Eqs. (107.2) and (107.3) define a and b, thus with a power creep law the position of the neutral axis in the section depends on the ratio of the bending moments but not on their absolute values (see also [326]). If the neutral axis is known it should be taken as the x axis, and the beam curvature is then determined from the equation

$$M_x = -\int_F s(\kappa y) y\, \mathrm{d}F\,.$$

Even for sections of simple shape, both determination of the neutral axis and subsequent calculation of moment as function of curvature present considerable difficulties. An approximate device based on replacing the true state of stress by a state which corresponds to the limiting equilibrium of the rod is useful in this case. This was first proposed by Bailey [162] for bending calculations on turbine blades of symmetrical section. Suppose that arbitrary axes x and y are drawn in the section. Divide the plane of the section into two parts of equal size by a straight line nn, at an angle φ to the x axis

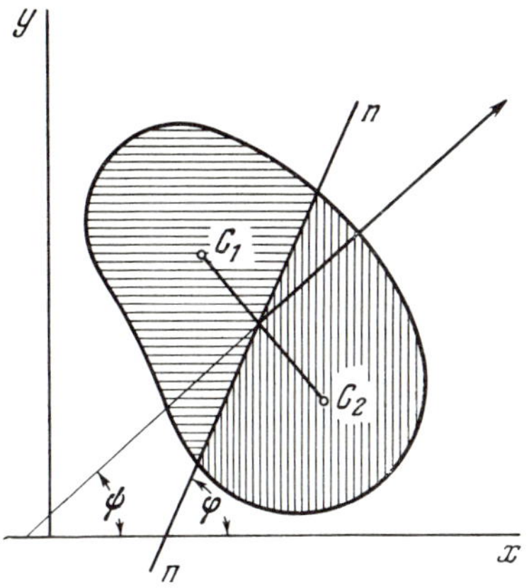

Fig. 177.

(fig. 177), denote the centres of gravity of these parts C_1, C_2. In the limiting state the direction of the moment is perpendicular to the straight line C_1C_2 and it forms an angle ψ with the x axis. Proceeding in exactly the same way as in the case of a symmetrical section we determine the rate of change of curvature of the beam, but in applying Castigliano's theorem it must be borne in mind that the generalised rate that corresponds to the torque M is $\kappa \cos(\varphi - \psi)$. Therefore

$$\kappa = \frac{2}{c \cos(\varphi - \psi)}\, v\left(\frac{2M}{cF}\right), \qquad c = C_1C_2 \,. \tag{107.4}$$

The principal difficulty in applying this method is to determine the relationship between φ and ψ or, which amounts to the same thing, to construct the relationship between M_x and M_y in the limiting state. Indeed if the limiting curve in the plane M_x, M_y is known the radius vector of the point M of this curve is at an angle ψ to the M_x axis and the normal at this point is at an angle φ to the axis. The absolute value of the moment is proportional to c, consequently, we can always construct the limiting curve in a scale such that $OM = c$.

Example. A thin-walled profile of rectangular section with sides $2a$ and $2b$ is shown in fig. 178, the wall thickness is assumed constant. Draw an axis nn, denote $\tan\varphi = \eta(b/a)$. The coordinates of the centre of gravity C_1 of the upper half of the section are

$$x_0 = -\frac{ab}{a+b}\eta\,, \qquad y_0 = \frac{b^2(1-\eta^2)+2ab}{2(a+b)}\,.$$

The quantity x_0 is proportional to M_y, and the quantity y_0 is proportional to M_x. To within an arbitrary factor we obtain

$$M_x = 2ab\eta\,, \qquad M_y = -b^2(1-\eta^2) - 2ab \qquad (0<\eta<1)\,.$$

If the neutral axis intersects the horizontal side, that is $\tan\varphi > b/a$ we denote the distance from the point of intersection to the y axis by ζa and obtain

$$M_x = a^2(1-\zeta^2) + 2ab\,, \qquad M_y = -2ab\zeta\,.$$

The limiting curve has two axes of symmetry, therefore, only one quadrant is shown in fig. 179. Values used in the construction were $a = 1$, $b = 2$. If $\varphi = 0$, then $c = b(b+2a)/(a+b)$ with the given values of a and b, $c = \frac{8}{3}$, alter the scale of the drawing so that the major semi-axis of the oval is $\frac{8}{3}$. Then the radius vector of the point on the limiting curve is c.

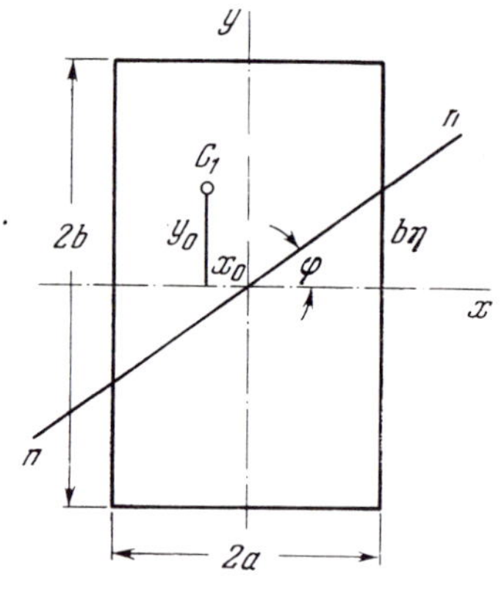

Fig. 178.

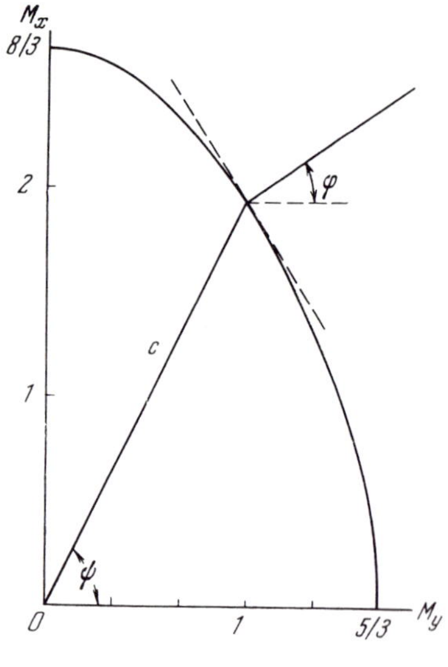

Fig. 179.

108. Thin-walled sections of open profile

In thin-walled sections of open profile, besides normal bending stresses it is necessary to consider shear stresses which equilibrate the shear force. These shear stresses are usually quite small but the important point is that under certain conditions they create an unbalanced couple of forces which twists the rod, and the torsional stiffness of a thin-walled profile is low, consequently as we shall see, torsional stresses may be considerable. The magnitude of the square of the stress intensity during the simultaneous application of normal and shear stresses is $\sigma^2 + 3\tau^2$; if τ is small compared with σ the error that results from substituting the normal stress for the intensity is of the order of τ^2/σ^2 compared with unity. Therefore, usually, in plasticity and creep theory, normal stresses in the beam are determined in the usual way, as in pure bending, and the tangential stresses are found separately from the conditions of statics. To explain the essence of the matter consider first bending of a rod of rectangular cross-section. The equation of equilibrium of an element in a state of plane stress is

$$\frac{\partial \sigma}{\partial z} + \frac{\partial \tau}{\partial y} = 0 .$$

Hence

$$\tau = -\int \frac{\partial \sigma}{\partial z}\, \mathrm{d}y + C . \tag{108.1}$$

The constant of integration is found from the condition that $\tau = 0$ when $y = \pm h$. With the power law $\sigma = \sigma_n(\kappa y/\epsilon_n)^{1/n}$, consequently

$$\tau = \sigma_n \epsilon_n^{-\frac{1}{n}} \frac{1}{n+1} \kappa^{\frac{1}{n}-1} \frac{\mathrm{d}\kappa}{\mathrm{d}z} [h^{1+\frac{1}{n}} - y^{1+\frac{1}{n}}] .$$

Using the relationship between moment and curvature by formula (106.2) for a rectangular section and noting that $\mathrm{d}M/\mathrm{d}z = Q$, where Q is the shear force we find

$$\tau = \frac{Q}{2bh} \frac{2n+1}{n+1} \left[1 - \left(\frac{y}{h}\right)^{1+\frac{1}{n}}\right] . \tag{108.2}$$

Here, as always when using a power law, when the stresses change sign it would be more correct to write an expression for σ as follows:

$$\sigma = \sigma_n \left(\frac{|\kappa y|}{h} \right)^{1/n} \operatorname{sign}(\kappa y) .$$

Then formula (108.2) is of the form

$$\tau = \frac{Q}{2bh} \frac{2n+1}{n+1} \left[1 - \left| \frac{y}{h} \right|^{1+\frac{1}{n}} \right].$$

For brevity we do not always stipulate this more accurate method of writing. In the limiting case when $n \to \infty$ we obtain from this a linear stress distribution, namely

$$\tau = \frac{Q}{bh} \left(1 - \frac{|y|}{h} \right) . \qquad (108.3)$$

The same linear shear stress distribution can be obtained if we are given the normal stress distribution corresponding to the limiting plastic state. Indeed if $\sigma = \sigma(z)$ it follows from (108.1) that

$$\tau = C - y \frac{\mathrm{d}\sigma}{\mathrm{d}z} .$$

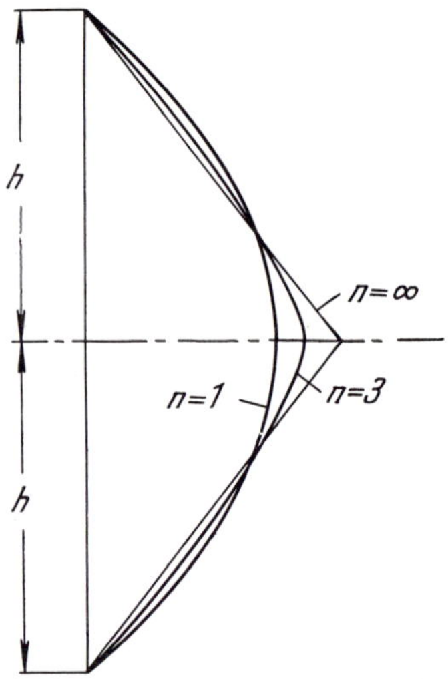

Fig. 180.

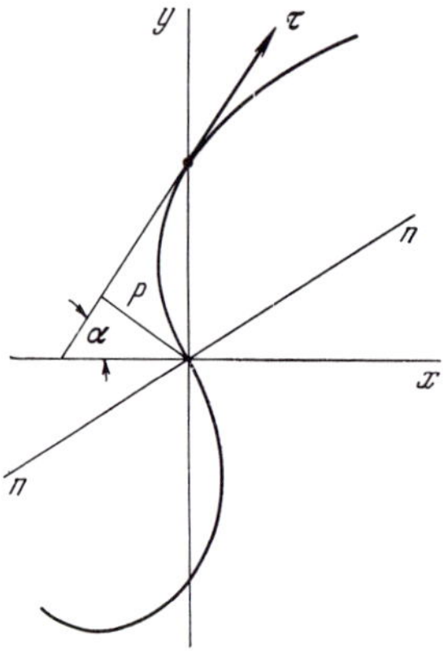

Fig. 181.

For a rectangle $\sigma = M/bh^2$, $d\sigma/dz = Q/bh^2$, therefore, we immediately obtain formula (108.3). Shear stress distribution diagrams in a beam of rectangular section are given in fig. 180 for several values of n.

It is interesting to note that a linear tangential stress distribution is also obtained with an exponential creep law. Using the statement made at the end of section 106 we pay no particular attention to the circumstance that the stress tends to $-\infty$ on the neutral axis, the logarithmic singularity disappears on integration. Since $\sigma = \sigma_e \ln \kappa y/\epsilon_e$, $d\sigma/dz = \sigma_e(d \ln \kappa/dz) = Q/bh^2$ when $y > 0$, consequently formula (108.3) again follows from formula (108.1).

The shear stresses in bending a thin-walled section of open profile can be determined in precisely the same way. For simplicity assume that the neutral axis is given in some way and intersects the mean line of the profile at only one point as shown in fig. 181. The profile thickness, which in the general case is variable, is denoted $\delta(l)$. Then from the equation of equilibrium, by the same method as was used to derive formula (108.1), we find

$$\tau = \frac{1}{\delta}\left[C - \int \delta(l)\frac{\partial\sigma}{\partial z}\,dl\right]. \tag{108.4}$$

It is readily shown that the resultant shear stresses in the section are equal to the shear force. Indeed, since $\sigma = s(\kappa_x y - \kappa_y x)$, we have

$$\frac{\partial}{\partial l}(\delta\tau) = -\delta\frac{\partial\sigma}{\partial z} = -\delta s'(\kappa_x y - \kappa_y x)\left(\frac{d\kappa_x}{dz}y - \frac{d\kappa_y}{dz}x\right).$$

Differentiating, with respect to z, the second of formulae (107.1), we obtain

$$Q_y = \frac{dM_x}{dz} = -\int \delta(l)s'(\kappa_x y - \kappa_y x)\left(\frac{d\kappa_x}{dz}y - \frac{d\kappa_y}{dz}x\right)y\,dl$$

$$= \int \frac{\partial}{\partial l}(\delta\tau)y\,dl.$$

Integrating by parts and bearing in mind that on the ends of the profile $\tau = 0$, we find

$$Q_y = -\int \delta\tau\cos\alpha\,dl.$$

Here α is the angle between the tangent to the mean line of the profile and the x axis. In an exactly similar way we find

$$Q_x = \int \delta\tau \sin\alpha \, dl \, .$$

If we take as the point of reference the origin of coordinates (or any other point in the plane of the section) then in addition to the resultant there is a couple of forces with moment

$$M_z = \int \delta\tau p \, dl \, .$$

Here p is the distance from the reference point to the tangent to the mean line of the profile, as shown in fig. 181. The reference point can always be selected in such a way that the moment M_z becomes zero. If through this point we construct a plane parallel to the plane of action of the bending moment the transverse forces applied in this plane cause only bending and not torsion of the rod. We call this plane the plane of bending, and the line of intersection between it and the plane of the cross-section is called the shear centre. It can be shown that for elastic rods all lines of bending that correspond to different directions of transverse load intersect at a single point, the shear centre. A shear centre does not exist in rods that are in a state of creep, this was shown by Phillips [326] in the case of the rod of semi-circular section illustrated in fig. 182.

In Phillips' work, general formulae are derived for shear centres with a power creep law, but they are very difficult to apply to particular cases. In order to show that there is not a common point of intersection of all the shear centres, the author considered a completely artificial case when $n = \frac{1}{3}$, which is less than unity, and only then did he succeed in filling all the

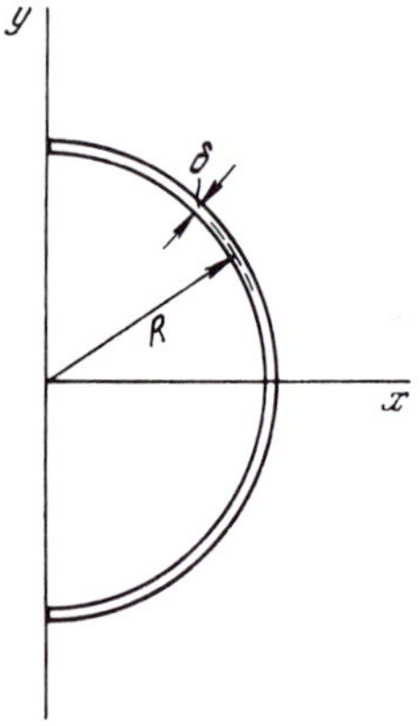

Fig. 182.

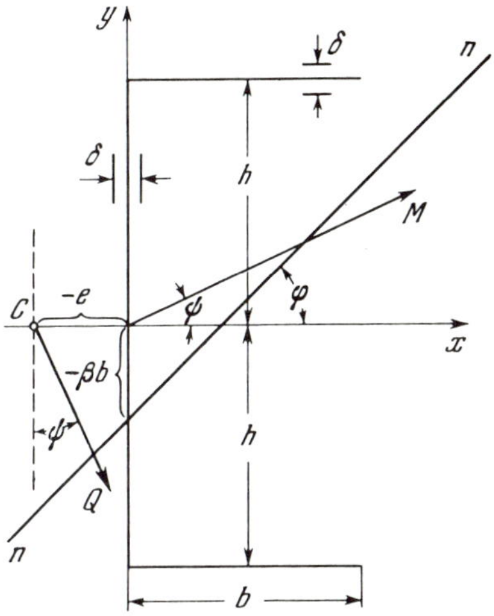

Fig. 183.

quadrants and writing the solution in explicit form. It was concluded from this example that, also when $n > 1$, a centre of bending should not exist. A second example, given in the articles of Patel and Venkatraman [322, 323], relates to a channel section of constant thickness such that $b = h$ (fig. 183). The x axis lies in the axis of symmetry of the profile, the y axis on the centre-line of the web. If the bending moment is at an angle ψ to the x axis, the neutral axis forms an angle φ with the same axis and its intercept on the y axis is $(-b\beta)$. The point C on the x axis with coordinate $(-e)$ has the property that if the plane of action of the force Q passes through this point, there should be no torsional stresses in the rod. In the table below we give results of

$\varphi°$	$n = 1$			$n = 3$		
	$\psi°$	β	e/b	$\psi°$	β	e/b
0	0°	0	0.3750	0°	0	0.3500
15	2°25′	0.067	0.3750	0°58′	0.034	0.3495
30	5°8′	0.144	0.3750	2°10′	0.072	0.3478
45	8°53′	0.250	0.3750	4°4′	0.134	0.3420
60	15°8′	0.433	0.3750	12°5′	0.323	0.2864
75	30°17′	0.933	0.3750	30°45′	0.670	0.2105
90	90°	–	–	90°	–	–

numerical calculations for $n = 3$ and also, for comparison, when $n = 1$, the elastic case.

It should be noted that when $\varphi = 45°$ the neutral axis intersects the beam flange, at lower values of φ when the neutral axis does not intersect the flange values of ψ are very small in the case of creep. This means that the small deviation of the moment vector from the x axis causes a considerable rotation of the neutral axis. The value of e then alters very little.

109. An approximate method of studying bending of thin-walled rods

As we have seen the fundamental difficulty in solving bending problems consists in finding the neutral axis. However, even after it has been found, to determine shear stresses it is necessary to fulfill certain quadratures and fairly cumbersome formulae are obtained. Here we shall use the approximate method given in section 107 in which the neutral axis is the same as that in the limiting condition of a rod of ideally plastic material. In parts of the rod that are subject to tension, where $0 < l < l_1$ (fig. 181) $\sigma = 2M/cF$, in the part under compression when $-l_2 < l < 0$, $\sigma = -2M/cF$. Here

$$\int_{-l_2}^{l_2} \delta(\operatorname{sign} l)\, \mathrm{d}l = 0\,.$$

From eq. (108.4) we find

$$\tau = \frac{2Q}{c\delta F} \int_{l}^{l_1} \delta\, \mathrm{d}l \qquad (0 < l < l_1)\,,$$

$$\tau = \frac{2Q}{c\delta F} \int_{-l_2}^{l} \delta\, \mathrm{d}l \qquad (-l_2 < l < 0)\,. \tag{109.1}$$

If $\delta = \text{const}$, then $l_1 = l_2$, $F = 2l_1\delta$ and $\tau = Q/c\delta(1 - l/l_1)$ when $l > 0$ and $\tau = Q/c\delta(1 + l/l_1)$ when $l < 0$. Formula (109.1) gives the maximum tangential stress when $l = 0$, it is

$$\tau_{\max} = \frac{Q}{c\delta}\,.$$

Now calculate the moment of shear stresses in the section about the reference centre O:

$$M = \int_0^l \tau \delta p \, dl .$$

Here, unlike the previous case, the arc is measured from one of the ends of the profile. As before, the quantity p is the length of the perpendicular drawn from the centre of reference O to the tangent to the mean line of the rod. Note that $p\,dl = d\omega$ is twice the area of the sector contained between the element dl and the radius vectors of the ends of this element. Integrating the expression for M by parts and bearing in mind that $\tau = 0$ on the ends of the profile we obtain

$$M = -\int \omega \delta \frac{\partial \sigma}{\partial z} \, dl .$$

Let the neutral axis nn intersect the axis of the profile several times as shown in fig. 184. The intercepts 01, 23 are in the region of compression and 12, 34 in the region of tension. In the region of tension, $\sigma = 2M/cF$, consequently, $d\sigma/dz = 2Q/cF$, whereas in the region of compression $d\sigma/dz = -2Q/cF$, therefore

$$M = -\frac{2Q}{cF} \{ \Sigma^+ \textstyle\int \omega \delta \, dl - \Sigma^- \int \omega \delta \, dl \} . \tag{109.2}$$

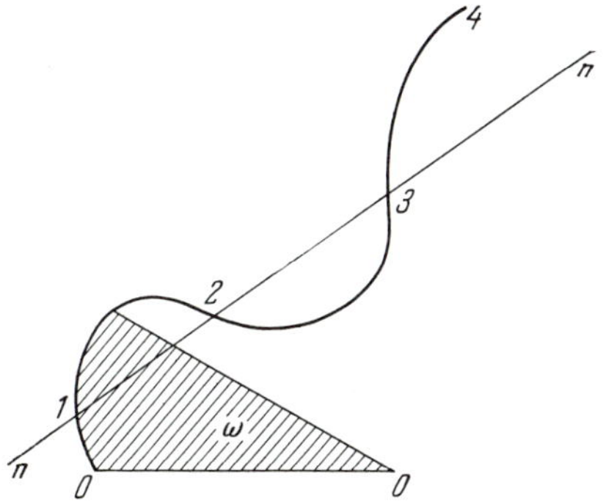

Fig. 184.

The symbol Σ^+ indicates that the summation extends over segments that are located in the region of tension and Σ^- denotes the sum of integrals over segments in the region of compression.

Example 1. Consider a channel of constant wall thickness, for simplicity we shall consider that $h = b = 1$, as in the example of Patel and Venkatraman considered above. If the neutral axis does not intersect the flanges it passes through the middle of the web, consequently, when $\varphi \leqslant 45°$, $\psi = 0$ and $\beta = 0$. A diagram of the sectorial areas for a pole O located in the centre of the web is shown in fig. 185. By formula (109.2), when $\varphi \leqslant 45°$, $M = -Q/3$, therefore the distance of the bending line from the point C, $e = \frac{1}{3}$. It will be recalled that for an elastic rod we found $e = 0.375$, when $n = 3$, $e = 0.350$. When $\varphi > 45°$ from the condition that the areas on the two sides of the neutral line are equal it follows that the neutral axis intersects the y axis at the point $-\beta$ and it cuts off an intercept β from the flange. Hence $\tan\varphi = (1+\beta)/(1-\beta)$. By simple calculations we find the coordinates of the centres of gravity of the halves of the section x_1, y_1 and x_2, y_2, namely

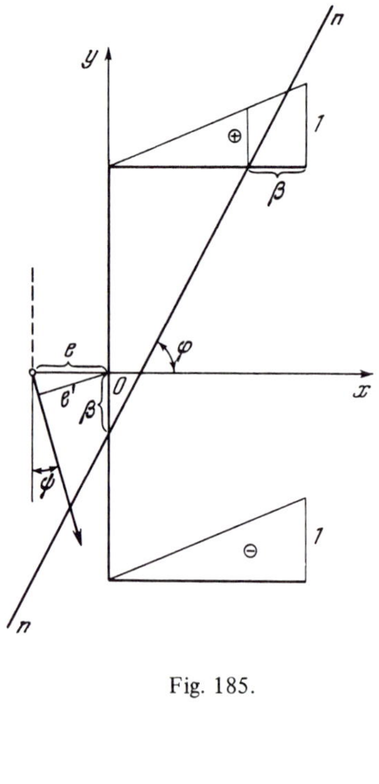

Fig. 185.

$$x_1 = \frac{(1-\beta)^2}{4}, \qquad y_1 = \frac{3-2\beta-\beta^2}{4},$$

$$x_2 = \frac{1+2\beta-\beta^2}{4}, \qquad y_2 = -\frac{3-2\beta-\beta^2}{4}.$$

We now find $\tan\psi$ and c:

$$\tan\psi = \frac{x_2-x_1}{y_2-y_1} = \frac{2\beta-\beta^2}{3-2\beta-\beta^2}, \qquad c = \tfrac{1}{2}\sqrt{(2\beta-\beta^2)^2+(3-2\beta-\beta^2)^2}.$$

The moment of the tangential stresses about the point O is found by formula (109.2),

$$M = -\frac{2Q}{cF}\left[\frac{(1-\beta)^2}{2} + \frac{1}{2} - \frac{(2-\beta)\beta}{2}\right] = -\frac{2Q}{cF}(1-\beta)^2.$$

The quantities in square brackets represent the areas of parts of the stress diagram ω that lie on the two sides of the nn axis, taken with appropriate signs. Dividing M by Q we determine the distance e' from the point O for the line of action of the force Q which does not cause twisting. The value of e is the absolute value of the coordinate of the point of intersection of the line of bending with the x axis equal to

$$e = \frac{e'}{\cos\psi}.$$

Calculations for $\varphi = 60^\circ$ and $\varphi = 75^\circ$ gave the following results: when $\varphi = 60^\circ$, $\beta = 0.267$, $\psi = 11^\circ$, $e = 0.246$; when $\varphi = 75^\circ$, $\beta = 0.578$, $\psi = 28^\circ 35'$, $e = 0.119$. Comparing these results with the exact solution for $n = 3$, it will be seen that when $\varphi = 60^\circ$ both ψ and e differ very little although the value of β differs quite appreciably. When $\varphi = 75^\circ$, the direction of the plane of bending for the elastic state, the creep state when $n = 3$ and the state of limiting equilibrium differ very little whereas the value of e is significantly different.

A note on this example is given in [181] in which the results of [322] are used to construct a curve of constant power dissipation for $n = 1$, 3 and ∞.

Example 2. A circular tube of radius R is cut on the axis as shown in fig. 186. The position of the neutral axis is given by an angle φ measured from the diameter that passes through the cut. The presence of the cut has no influence on determining the normal stresses, the distribution of normal stresses is sym-

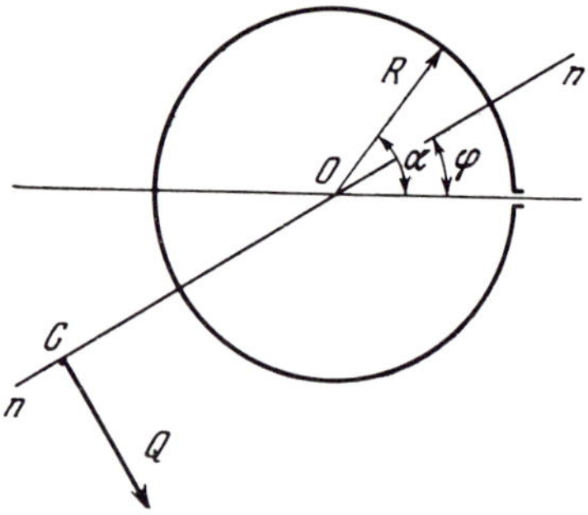

Fig. 186.

metrical and the plane of bending is perpendicular to the *nn* axis. The quantity c is independent of φ, $c = 4R/\pi$. Take as the reference origin the sectorial area when $\alpha = 0$, then $\omega = R^2\alpha$. By formula (109.2)

$$M = \frac{QR}{4}\left[\int_{\varphi}^{\varphi+\pi} \alpha \mathrm{d}\alpha - \int_{\varphi+\pi}^{2\pi} \alpha \mathrm{d}\alpha - \int_{0}^{\varphi} \alpha \mathrm{d}\alpha\right] = -QR\,\frac{\pi(\pi-2\varphi)}{4}\,.$$

The distance OC from the axis of the tube to the line of bending is

$$OC = R\,\frac{\pi(\pi-2\varphi)}{4}\,.$$

It is readily shown that there is no shear centre, i.e. there is not one point at which the lines of bending corresponding to different values of φ intersect.

In the example of a rod of rectangular section we saw that with an exponential creep law the shear stresses are distributed in the same way as when $n = \infty$. In the integral case, taking the neutral axis as the x axis, we should put

$$\sigma = \sigma_e \ln \frac{\kappa y}{\epsilon_e} \quad (y > 0)\,, \qquad \sigma = -\sigma_e \ln\left(\frac{-\kappa y}{\epsilon_e}\right) \quad (y < 0)\,.$$

Hence

$$\frac{\partial \sigma}{\partial z} = \sigma_e \frac{\partial \ln \kappa}{\partial z} \quad (y > 0)\,, \qquad \frac{\partial \sigma}{\partial z} = -\sigma_e \frac{\partial \ln \kappa}{\partial z} \quad (y < 0)\,.$$

But

$$\frac{\partial \ln \kappa}{\partial z} = \frac{\mathrm{d} \ln \kappa}{\mathrm{d}M}\, Q\,.$$

Therefore, a formula completely analogous to formula (109.2) is valid for the moment M:

$$M = -Q\sigma_e \frac{\mathrm{d} \ln \kappa}{\mathrm{d}M} \{\Sigma^+ \int \omega\delta \, \mathrm{d}l - \Sigma^- \int \omega\delta \, \mathrm{d}l\} . \tag{109.3}$$

Consider example 2 again, but now with an exponential creep law and in the exact formulation. The distance between a point on the circumference and the nn axis is $R \sin (\alpha-\varphi)$, and putting $\alpha-\varphi = \chi$, then when $0 < \chi < \pi$,

$$\sigma = \sigma_e \ln \frac{\kappa R \sin \chi}{\epsilon_e} .$$

The bending moment is

$$M = 4R^2\sigma_e \left\{ \ln \frac{\kappa R}{\epsilon_e} + \ln 2 - 1 \right\} .$$

Whence

$$\frac{\mathrm{d} \ln \kappa}{\mathrm{d}M} = \frac{1}{4R^2\delta\sigma_e} .$$

Applying formula (109.3) we again arrive at the previous result and therefore the approximate solution that we have found is accurate with an exponential creep law.

110. **Deformation of beams and frames in bending**

Knowing the relationship between the bending moment and the rate of change of curvature $\kappa = \kappa(M)$ we can determine the bending of beams by integrating the equation

$$\frac{\mathrm{d}^2 v}{\mathrm{d}z^2} = \kappa(M) . \tag{110.1}$$

Here v is the rate of displacement of the beam. If the moment is given as a function of the coordinate z, the integration of eq. (110.1) reduces to two quadratures which can be fulfilled either in the closed form or by numerical or graphical integration. The methods of integrating the bending equations which apply to elastic beams also apply fully to the case of creep. Indeed the equation of the displacement of an elastic beam is of the form

$$\frac{d^2w}{dz^2} = \frac{M}{EI}.$$

Here w is the deflection to the beam.

If we take a diagram of the function $\kappa(z)$ as the diagram of the bending moments divided by the stiffness, it is possible to use ordinary methods of structural mechanics. In particular a known graphical-analytical method is immediately generalised to the case of creep. In calculating the deflection the sole difficulty is to perform the integration. Simple statically indeterminate problems can evidently be solved in the same way. When it is required to determine the deflection at some particular points on the beam, or to determine the values of redundant unknowns in statically indeterminate systems, it is simpler to apply variational theorems. Consider κ as a generalised rate and M as a generalised force; we determine the rate potential as

$$\Phi = \int \kappa(M)\, dM$$

and the stress potential as

$$U = \int M(\kappa)\, d\kappa\ .$$

Consequently, for the beam or frame as a whole

$$\widetilde{\Phi} = \int_0^l \Phi(M)\, dz\ , \qquad \widetilde{U} = \int_0^l U(\kappa)\, dz\ .$$

Writing the corresponding formula for the power creep law we obtain

$$\kappa = \kappa_n \left(\frac{M}{M_n}\right)^n , \qquad \Phi = \frac{\kappa_n M_n}{n+1} \left(\frac{M}{M_n}\right)^{n+1} ,$$

$$U = \frac{n}{n+1} \kappa_n M_n \left(\frac{\kappa}{\kappa_n}\right)^{1+\frac{1}{n}} , \tag{110.2}$$

and for the exponential law

$$\kappa = \kappa_e \exp\left(\frac{M}{M_e}\right), \qquad \Phi = \kappa_e M_e \exp\left(\frac{M}{M_e}\right),$$
$$U = \kappa_e M_e \frac{\kappa}{\kappa_e}\left[\ln\frac{\kappa}{\kappa_e} - 1\right]. \tag{110.3}$$

Note that with the exponential law the second of formulae (110.3) gives a value of Φ that differs from zero when $M = 0$, but if the bending moments in the greater part of the beam are sufficiently large the error that this introduces is negligible. As in the case of bar systems in tension and compression the solution of statically indeterminate problems leads to transcendental equations for the redundant unknowns. Suppose that the redundant unknowns are $X_1, X_2, ..., X_n$, the generalised force for which we seek the displacement is $-Q$. The bending moments on the segments are linear functions of X_s and Q. Using the power law of creep the expression for $\widetilde{\Phi} = \int \Phi(M)\, \mathrm{d}z$ can always be written in the form

$$\widetilde{\Phi} = \frac{\kappa_n M_n l}{n+1}\left(\frac{Ql}{M_n}\right)^{n+1} R(\xi_1, \xi_2, ..., \xi_n).$$

Where ξ_s are dimensionless redundant unknowns, $\xi_s = X_s/P_s$, P_s are arbitrary quantities with the same dimensions as X_s, l is a certain characteristic length. The quantity ξ_s is determined from the conditions of minimum of the function R: $\mathrm{d}R/\mathrm{d}\xi_s = 0$. After values of ξ_s have been determined the generalised rate that corresponds to the force Q is determined by the Castigliano theorem

$$q = \frac{\partial\widetilde{\Phi}}{\partial Q} = \kappa_n l^2 \left(\frac{Ql}{M_n}\right)^n R(\xi_1, \xi_2, ..., \xi_n). \tag{110.4}$$

The principal difficulty is to solve the system $\mathrm{d}R/\mathrm{d}\xi_s = 0$; a simpler device is that values of ξ_s are given in advance, taking for example, the values of the redundant unknowns which are obtained for the corresponding elastic case or for a beam in the limiting plastic state. If we take $\xi_s = \xi_s^*$ the corresponding value of the function $R = R^*$ is greater than the true value that corresponds to fulfilment of the minimum condition, and, therefore, this device gives too high values of rate $q^* > q$.

Now let us consider some simple examples.

Example 1. A statically indeterminate beam is shown in fig. 187. It is required to determine the deflection at the point of application of the force Q. Consider the reaction of the support X as a redundant unknown. The bending

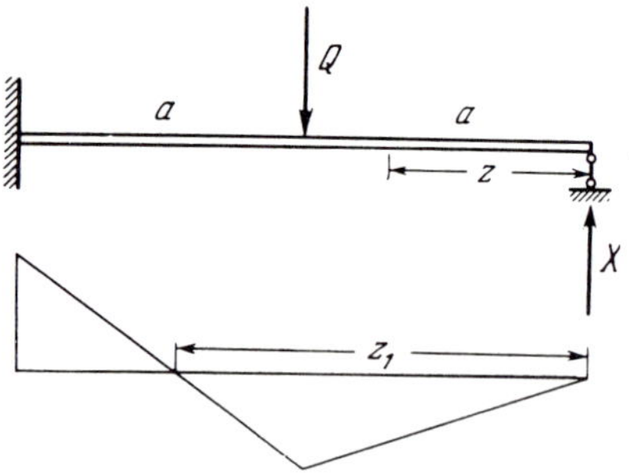

Fig. 187.

moment alters sign at the point with coordinate z_1. Divide the length of the beam into three parts as shown in the figure. The absolute values of the moments on various parts are expressed by

$$|M| = Xz \qquad (0 < z < a),$$
$$|M| = Xz - Q(z-a) \qquad (a < z < z_1),$$
$$|M| = -Xz + Q(z-a) \qquad (z_1 < z < 2a).$$

Consequently

$$\tilde{\Phi} = \frac{\kappa_n}{(n+1)M_n^n}\left\{\int_0^a X^{n+1} z^{n+1}\,\mathrm{d}z + \int_a^{z_1} [(X-Q)z + Qa]^{n+1}\,\mathrm{d}z \right.$$

$$\left. + \int_{z_1}^{2a} [-(X-Q)z - Qa]^{n+1}\,\mathrm{d}z \right\}.$$

Put $\xi = X/Q$, integrate and write the result in the form

$$\Phi = \frac{\kappa_n M_n a}{n+1}\left(\frac{Qa}{M_n}\right)^{n+1} R(\xi).$$

Here

$$R(\xi) = \frac{1}{n+2}\,\frac{(1-2\xi)^{n+2} + \xi^{n+1}}{1-\xi}.$$

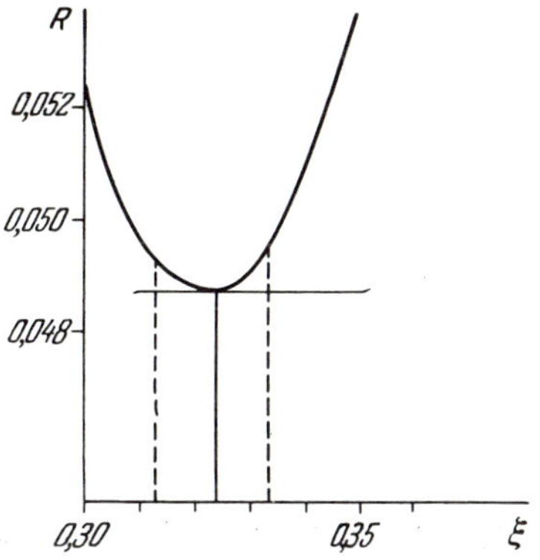

Fig. 188.

We now need to find the minimum of the function $R(\xi)$; this leads to a fairly complicated equation, the simplest way is to construct a graph of the function $R(\xi)$ and to find its minimum value. For an elastic beam $\xi = \frac{5}{16} = 0.313$; and for a beam in the limiting plastic state $\xi = \frac{1}{3} = 0.333$; it, therefore, suffices to investigate the function $R(\xi)$ in a very narrow range of values of the argument. A graph of the function $R(\xi)$ for $n = 3$ is shown in fig. 188, R reaches a minimum at $\xi = 0.324$, $R = 0.0487$. Therefore, by formula (110.4):

$$q = 0.0487\kappa_n a^2 \left(\frac{Qa}{M_n}\right)^n .$$

If we take the distribution of moments that corresponds to the elastic state we should put $\xi = 0.313$. The corresponding value of $R = 0.492$ is found from the graph and the rate is found to be too high by 1.03%. In the limiting plastic state $\xi = 0.333$ and $R = 0.494$ and here the error in determining the rate is 1.85%.

The two approximate solutions, one of which assumes a moment distribution as in an elastic beam whilst the other corresponds to the limiting condition of an elastic-plastic beam, both agree so well with the accurate solution because in the system under consideration the value of the redundant unknown differs very little in these two extreme cases. We shall consider two further examples which have frequently been considered in the literature [52, 231, 239], in a way they represent extremes of the application of approximate methods.

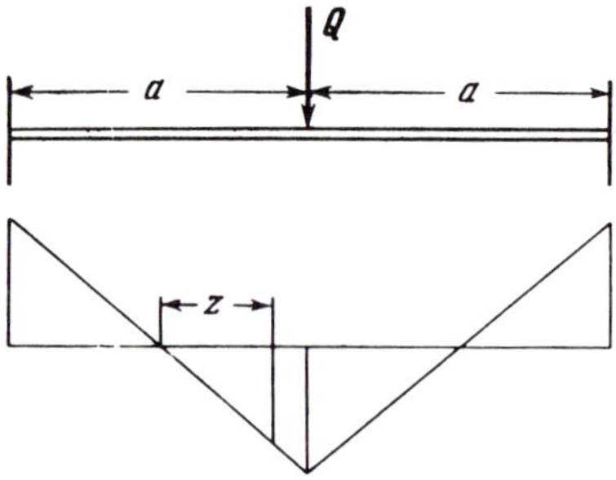

Fig. 189.

Example 2. A beam with built-in ends carries a concentrated load (fig. 189). The moment distribution diagrams are the same for both the elastic and limiting plastic states, the bending moments at the clamped end and at the point of application of the force are the same and equal to $Qa/4$. The moment distribution diagram is of course the same in creep. In each of the four triangles into which the moment distribution diagram is divided $|M|=Qz/2$, if the coordinate is reckoned from the point where $M=0$. Consequently,

$$\widetilde{\Phi} = \frac{4\kappa_n}{n+1}\,\frac{1}{M_n^n}\int_0^{a/2}\left(\frac{Qz}{2}\right)^{n+1} dz\,,$$

and

$$q = \frac{\kappa_n a^2}{2(n+1)}\left(\frac{Qa}{4M_n}\right)^n .$$

Example 3. The rectangular frame illustrated in fig. 190, is loaded by two forces. Consider as redundant the unknown bending moment in the strut X. By simple calculations we find

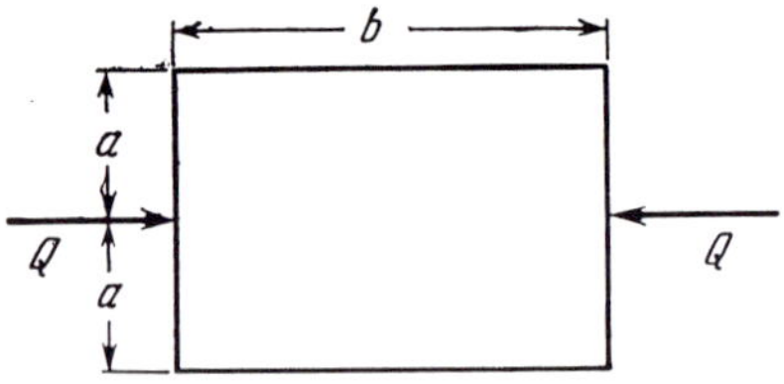

Fig. 190.

$$\widetilde{\Phi} = \frac{\kappa_n M_n a}{n+1} \left(\frac{Qa}{M_n}\right)^{n+1} R(\xi) .$$

Here

$$\xi = \frac{X}{Qa}, \quad R(\xi) = \frac{8}{n+2} [\xi^{n+2} + (\tfrac{1}{2} - \xi)^{n+2}] + 2\frac{b}{a}\xi^{n+1} .$$

If $b = 0$, we obtain

$$\xi = \tfrac{1}{4}, \quad q = \frac{\kappa_n a^2}{n+1} \left(\frac{Qa}{4M_n}\right)^n ,$$

that is twice the result of the previous example. This is to be expected, the short struts being equivalent to a rigid clamp so that this case is equivalent to the bending of two beams with clamped ends.

However, this frame behaves quite differently in the elastic and plastic states. If the frame is elastic $\xi = 1/4(1 + b/2a)$, when $b/a = 0$, $\xi = \frac{1}{4}$, when $b/a \to \infty$, $\xi \to 0$.

In the limiting state of an ideally rigid-plastic frame the moment at the point of application of the force is equal in absolute value to the moment at the strut and $\xi = \frac{1}{4}$, whatever the ratio b/a. It is evident that in a state of creep with increase in the ratio of b/a, ξ again tends to zero and therefore, an approximate calculation that is based on the assumption that the moments are distributed in the same way as in the limiting state can hardly be expected to give satisfactory results. Indeed assume that $b/a \gg 1$, then ξ is a small quantity. It is readily checked that the minimum value of R is reached when

$$\xi = \tfrac{1}{2}\left(\frac{2}{n+1}\,\frac{a}{b}\right)^{1/n} ,$$

and the corresponding value of R is

$$R = (\tfrac{1}{2})^n \left[\frac{2}{n+2} + \left(\frac{n+1}{2}\right)^{-\frac{n+1}{n}} \left(\frac{b}{a}\right)^{-\frac{1}{n}}\right].$$

When $b/a \to \infty$ R tends to the limit $(\frac{1}{2})^n [2/(n+2)]$, but it does so quite slowly because $1/n$ is a small quantity. In the elastic state for large values of b/a we have approximately $\xi \approx a/2b$. Substituting this value into the formula for R and neglecting small terms we obtain

$$R = (\tfrac{1}{2})^n \left[\frac{2}{n+2} + \left(\frac{b}{a}\right)^{-n}\right].$$

This expression tends to the same limit as the accurate solution but much faster. Finally, if we take, as in the limiting plastic state, $\xi = \frac{1}{4}$, we obtain

$$R = (\tfrac{1}{4})^n \left[\frac{1}{n+2} + \frac{1}{2}\frac{b}{a}\right].$$

This latter expression increases indefinitely with increase in the ratio b/a and, therefore, gives a completely unsatisfactory approximation.

For a square frame when $b = 2a$, the results of determining the redundant unknown for various values of n are given in the work [127], and the corresponding graph is plotted in fig. 191. For instance, when $n = 3$, $\xi = 0.175$ and $R = 0.0098$. Taking a value of $\xi = 0.125$ that corresponds to the elastic state we obtain too high a value of $R = 0.0128$; if we put $\xi = 0.250$ as for the limiting plastic state, we get a still higher value of $R = 0.0188$.

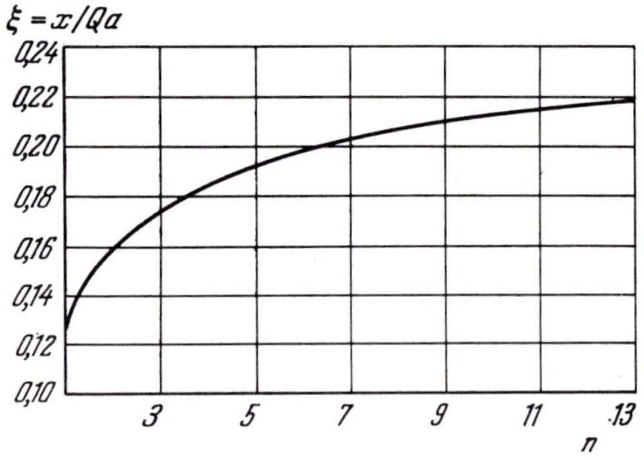

Fig. 191.

111. **Creep in torsion**

The theory of torsion of rods in steady-state creep is constructed in exactly the same way as the theory of torsion of elastic rods. Consider the z axis to be along the centre-line of the rod, locate the axes x and y arbitrarily in the plane of the cross-section and assume a kinematic picture of the effects bearing in mind that here, as in an elastic rod,

$$u = -\kappa zy\,, \qquad v = \kappa zx\,, \qquad w = \kappa\varphi(x,y)\,.$$

Here u, v and w are rates of flow in the directions of the x, y and z axes respectively, κ is the angular rate of torsion referred to unit length of rod. With this assumption it is found that all the components of the deformation rate tensor are zero except the components ϵ_{xz} and ϵ_{yz} which are given by the expressions

$$\epsilon_{xz} = \kappa\left(-y + \frac{\partial\varphi}{\partial x}\right), \qquad \epsilon_{yz} = \kappa\left(x + \frac{\partial\varphi}{\partial y}\right).$$

Taking the law of flow in accordance with eqs. (99.1) and (99.2) we find that all components of the stress tensor except τ_{xz} and τ_{yz} must be considered zero; for rates of deformation that differ from zero we find

$$\epsilon_{xz} = \tfrac{3}{2}v(\sigma_0)\frac{\tau_{xz}}{\sigma_0}\,, \qquad \epsilon_{yz} = \tfrac{3}{2}v(\sigma_0)\frac{\tau_{yz}}{\sigma_0}\,.$$

Here

$$\sigma_0^2 = 3(\tau_{xz}^2 + \tau_{yz}^2) = 3\tau^2\,.$$

The quantity τ is the modulus of the shear stress vector in the plane of the section. For convenience introduce the notation

$$\gamma(\tau) = \frac{\sqrt{3}}{2}v(\sqrt{3}\tau)\,. \tag{111.1}$$

Then

$$\epsilon_{xz} = \frac{\gamma(\tau)}{\tau}\tau_{xz}\,, \qquad \epsilon_{yz} = \frac{\gamma(\tau)}{\tau}\tau_{yz}\,. \tag{111.2}$$

Note that identical relationships are also obtained when we take as the basis a creep law based on the criterion of maximum shear stress. In this case, of course, the relationship between $\gamma(\tau)$ and the function $v(\sigma_0)$ determined from creep tests in tension is different, namely, $\gamma(\tau) = v(2\tau)$.

Differentiating the first of eqs. (111.2) with respect to y and the second with respect to z and subtracting the second from the first we obtain

$$\left[\frac{\partial}{\partial x}\left(\frac{\gamma}{\tau}\tau_{yz}\right) - \frac{\partial}{\partial y}\left(\frac{\gamma}{\tau}\tau_{xz}\right)\right] = 2\kappa\,. \tag{111.3}$$

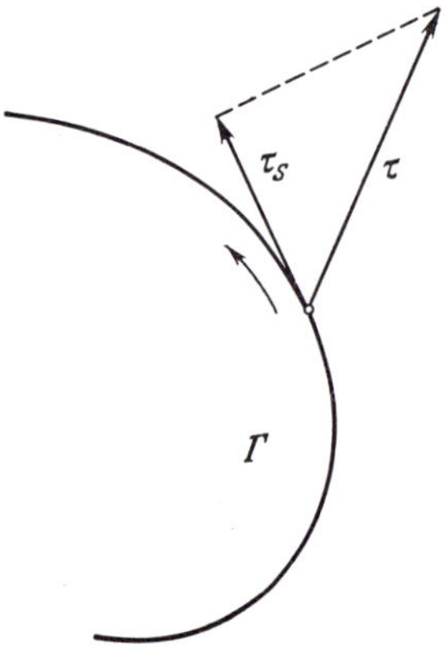

Fig. 192.

In the plane of the cross-section construct a closed contour Γ which does not intersect the boundary, but which can include part of the boundary of the region of the section. Denote the area bounded by this contour by Ω. Integrate both sides of eq. (111.3) over the region Ω. We obtain

$$\int_{\Omega} \left[\frac{\partial}{\partial x} \left(\frac{\gamma}{\tau} \tau_{yz} \right) - \frac{\partial}{\partial y} \left(\frac{\gamma}{\tau} \tau_{xz} \right) \right] \mathrm{d}\Omega = 2\kappa \Omega \, .$$

The integral in the left-hand side is converted by Riemann's formula into a contour integral which represents circulation of shear stress round the contour Γ and we obtain

$$\int_{\Gamma} \gamma \frac{\tau_s}{\tau} \, \mathrm{d}s = 2\kappa \Omega \, . \tag{111.4}$$

Formula (111.4) is a generalisation of the Bredt theorem concerning the circulation of shear stress. Here τ_s is the projection of the shear stress vector at a point of the contour Γ on the tangent to this contour, the counter-clockwise direction round the contour is considered positive and the sign of τ_s is selected accordingly (fig. 192).

It is necessary to add to eq. (111.3) a differential equation of equilibrium in the usual form

$$\frac{\partial \tau_{xz}}{\partial x} + \frac{\partial \tau_{yz}}{\partial y} = 0 \, .$$

We satisfy this equation if we introduce the stress function in such a way that

$$\tau_{xz} = \frac{\partial \psi}{\partial y}, \quad \tau_{yz} = -\frac{\partial \psi}{\partial x}, \quad \tau_2 = \left(\frac{\partial \psi}{\partial x}\right)^2 + \left(\frac{\partial \psi}{\partial y}\right)^2 .$$

Substituting into (111.3) we obtain

$$\frac{\partial}{\partial x}\left(\frac{\gamma}{\tau}\frac{\partial \psi}{\partial x}\right) + \frac{\partial}{\partial y}\left(\frac{\gamma}{\tau}\frac{\partial \psi}{\partial y}\right) = -2\kappa . \tag{111.5}$$

The boundary conditions are obtained in the usual way. Just as in the theory of elasticity, on the contour $\tau_{xz} \cos \widehat{\nu x} + \tau_{yz} \cos \widehat{\nu y} = 0$. Substituting $\cos \widehat{\nu x} = \mathrm{d}y/\mathrm{d}s$, $\cos \widehat{\nu y} = -(\mathrm{d}x/\mathrm{d}s)$ and the expressions for τ_{xz} and τ_{yz} in terms of the stress function, we can integrate the boundary condition round each of the contours Γ_i that form the boundary of the cross-section. Put

$$\psi = C_i \text{ on } \Gamma_i .$$

If the region is simply connected we can always take $C_1 = 0$; if the region is multiply connected the missing equation can be formulated by the circulation theorem to determine all constants except one which is fixed arbitrarily. A number of authors have obtained exactly the same equation for torsion of rods of work-hardening material [43, 52].

Note that if the stress function is known the twisting moment is calculated as

$$M = \int_F (x\tau_{yz} - y\tau_{xz})\,\mathrm{d}F = 2\int_F \psi\,\mathrm{d}F .$$

The integral extends over the cross-sectional area. Concerning the application of these equations an accurate solution of eq. (111.5) is only possible for a rod of circular section or for a tube, various approximate methods must be used in all other cases. It is comparatively easy to obtain a numerical solution of eq. (111.5) by the method of finite differences for any shape of cross-sectional contour giving a second order elliptical type of equation. However, the author has not encountered important technical problems in which it is necessary to consider torsion of a solid rod. Therefore the corresponding numerical methods have not been developed.

The second way of obtaining approximate solutions of torsion problems is to use Lagrange and Castigliano's variational principles. The corresponding functionals are fairly simple and in this sense the theory of torsion is extremely favourable for the application of general methods and comparing different devices with one another. Kachanov [52] devotes considerable space to this question in his book and he has obtained a number of effective results in this field, and therefore purely for purposes of illustration we shall only touch in passing on the application of variational methods to creep with torsion. We shall immediately consider the elementary problem of torsion of a circular rod as an example of the application of eq. (111.5) although it is hardly justified to do so in this simple case, the problem could be solved by an elementary method as in the case of elastic-plastic torsion in courses on the strength of materials. Accordingly, we shall assume that a circular rod of radius a is twisted by a torque M. We shall seek the stress as function of the radius $r = \sqrt{x^2 + y^2}$, then $\frac{\mathrm{d}\psi}{\mathrm{d}x} = \psi' \frac{x}{r}$, $\frac{\mathrm{d}\psi}{\mathrm{d}y} = \psi' \frac{y}{r}$, $\tau = -\psi'$ and eq. (111.5) is converted to the form

$$\frac{1}{r} \frac{\mathrm{d}}{\mathrm{d}r} (r\gamma) = 2\kappa \,.$$

Hence, bearing in mind that $\tau(0) = 0$, we obtain

$$\gamma(\tau) = \kappa r \,, \qquad \tau = \gamma^{-1}(\kappa r) \,.$$

The torsional moment is

$$M = 2 \int_F \psi \,\mathrm{d}F = 4\pi \int_0^a \psi r \,\mathrm{d}r \,.$$

Integrating by parts, bearing in mind that $\psi(a) = 0$, we obtain

$$M = 2\pi \int_0^a \tau r^2 \,\mathrm{d}r = 2\pi \int_0^a \gamma^{-1}(\kappa r) r^2 \,\mathrm{d}r \,.$$

The calculations for particular creep laws present no difficulties and we do not give them.

112. **Torsion of thin-walled rods of closed profile**

Tangential stresses in a thin-walled rod of closed profile must be considered to be in the direction of the mean line of the profile and constant through the thickness. The tangential stress flux $\tau\delta$ is constant. This follows from the equations of statics just as for elastic rods. The magnitude of the torque for a doubly connected section is found as follows. The force $\tau\delta\,\mathrm{d}l$ which is applied to a length $\mathrm{d}l$ of the mean contour line develops a moment about the point O with a lever arm p (fig. 193), $\mathrm{d}M = \tau\delta p\,\mathrm{d}l$, consequently,

$$M = \int \tau\delta p\,\mathrm{d}l\,.$$

But $\tau\delta$ = const, $p\,\mathrm{d}l$ is twice the area of a triangle of base $\mathrm{d}l$ and height p, that is $\mathrm{d}\omega$ (section 109). Therefore

$$M = 2\tau\delta F\,,$$

where F is the area bounded by the contour of the section. Hence

$$\tau = \frac{M}{2\delta F}\,. \tag{112.1}$$

Formula (112.1) is derived solely from the conditions of statics, it is valid whatever the physical nature of the material. To determine the angle of twist it is necessary to use the circulation theorem. In view of the assumptions that have been made $\tau_s = \tau$ and consequently, calculating the circulation for the mean line of the profile by formula (111.4) we obtain

$$\int_\Gamma \gamma(\tau)\,\mathrm{d}l = 2\kappa F\,.$$

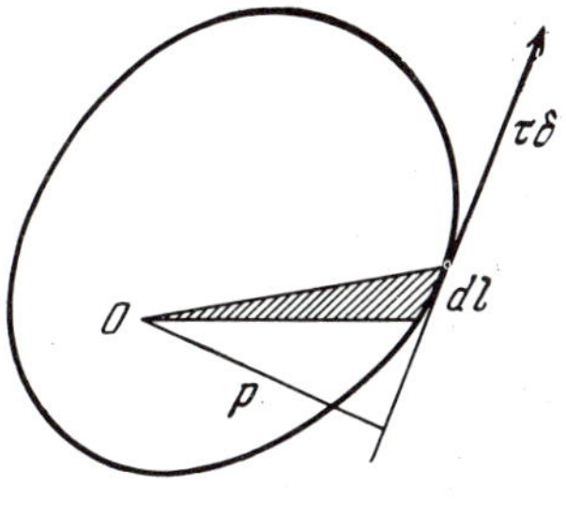

Fig. 193.

Substitute expression (112.1) for τ:

$$\int \gamma \left(\frac{M}{2\delta F} \right) \mathrm{d}l = \kappa F \,. \tag{112.2}$$

This eq. (112.2) is the equation that we require to connect the moment M and the current angular rate of twisting κ, to obtain further results we must particularise the form of the function $\gamma(\tau)$.

a) Power law. The function γ only differs by a factor from the function v and consequently we may take

$$\gamma = \gamma_n \left(\frac{\tau}{\tau_n} \right)^n \,.$$

Then

$$\kappa = \frac{\gamma_n}{2F} \left(\frac{M}{2F\tau_n} \right)^n \int_\Gamma \frac{\mathrm{d}l}{\delta^n} \,. \tag{112.3}$$

b) Exponential law $\gamma = \gamma_e \exp\,(\tau/\tau_e)$. In this case

$$\kappa = \frac{\gamma_e}{2F} \int_\Gamma \exp \left[\frac{M}{2F\tau_e\delta} \right] \mathrm{d}l \,.$$

No convenient formula has been obtained for the relationship between κ and M with an exponential creep law. In practical problems $\delta(l)$ is a piecewise-constant function and, therefore, the result is not very complicated. Indeed suppose that over the length l_i $\delta = \delta_i$, then

$$\kappa = \frac{\gamma_e}{2F} \sum l_i \exp \left[\frac{M}{2F\tau_e\delta_i} \right] . \tag{112.4}$$

As will be seen the solution of the problem of torsion of doubly-connected profiles is extremely simple and requires no explanation. Matters are somewhat more complicated when the cross-sectional region is multiply connected. Such a multiply connected thin-walled profile is illustrated schematically in fig. 194. Denote the external region by the index O and number the internal regions from 1 to k. Denote the arcs that form the mean line of the profile by two subscripts corresponding to the regions that are separated by this arc. Thus the arc AB has the subscript 01, the arc DC the subscript 23, and so on. The shear stress flux is constant on each arc and we denote $\tau\delta$ on the arc ij

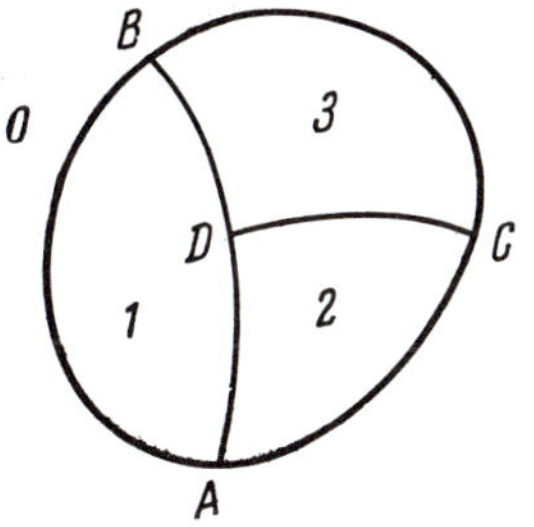

Fig. 194.

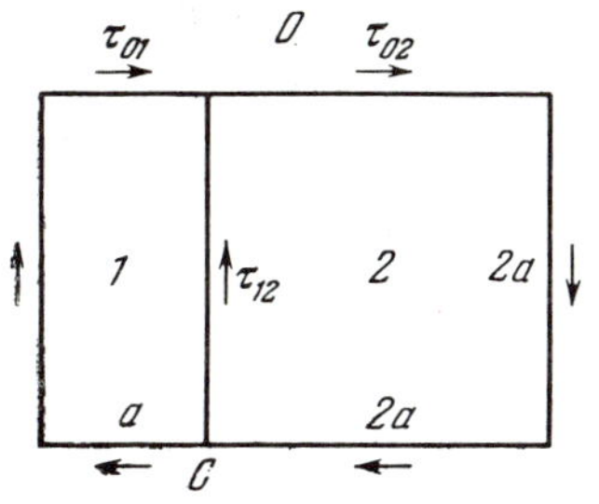

Fig. 195.

by t_{ij}. Then the moment about the centre O is

$$M = \int t \, \mathrm{d}\omega = \sum \int t_{ij} \, \mathrm{d}\omega \, . \tag{112.5}$$

Here the summation is extended over the entire arc l_{ij}, $\mathrm{d}\omega = p \, \mathrm{d}l$ is the sectorial area.

Now apply the circulation theory k times to all possible closed contours so that all the regions are covered. For example, we may go round each of the internal regions 1, 2 and 3 individually, other paths can also be selected. We obtain

$$\int_{\Gamma_i} \gamma \left(\frac{t_{ij}}{\delta} \right) \mathrm{d}l = 2\kappa F_i \, . \tag{112.6}$$

Moreover, the equations of equilibrium should be fulfilled at the nodes. Taking a longitudinal section of some node we find that the sum of tangential stress fluxes in the walls that are related to the particular node should be zero. If the boundaries of the region $r, q, s, \ldots$, converge at the node we should have

$$t_{rq} + t_{qs} + t_{sr} + \ldots = 0 \, . \tag{112.7}$$

We, therefore, obtain a complete system of equations for solution of the statically indeterminate problem.

Let us illustrate the method of calculation by a simple example. The section of a rectangular box beam with an internal barrier is shown in fig. 195. For simplicity assume the wall thickness constant and equal to δ, the calculations are not difficult when the wall thickness is different in different places. On the walls 01, 02, and 12 t_{ij} retains a constant value, and with constant wall thickness the tangential stresses τ_{01}, τ_{02} and τ_{12} are also constant.

Apply the circulation theorem, in the form of (112.6), once to the external contour and once to the right rectangle 2. With a power creep law we obtain

$$\gamma_n \left[\left(\frac{\tau_{01}}{\tau_n} \right)^n 4a + \left(\frac{\tau_{02}}{\tau_n} \right)^n 6a \right] = 2\kappa \cdot 6a^2 ,$$

$$\gamma_n \left[\left(\frac{\tau_{12}}{\tau_n} \right)^n 2a + \left(\frac{\tau_{02}}{\tau_n} \right)^n 6a \right] = 2\kappa \cdot 4a^2 .$$

Now formulate eq. (112.7) for point C. Stresses which are directed away from point C are denoted by the positive sign and those towards point C by the negative sign.

$$\delta(\tau_{01} + \tau_{12} - \tau_{02}) = 0 .$$

Put $\tau_{10}/\tau_{20} = \zeta$. Eliminating τ_{12} from the resulting equations we find

$$4\zeta^n - 3(1 - \zeta)^n = 3 , \tag{112.8}$$

$$\kappa = \frac{\gamma_n}{6a} \left(\frac{\tau_{02}}{\tau_n} \right)^n (3 + 2\zeta^n) . \tag{112.9}$$

The task is now reduced to the solution of the algebraic eq. (112.8). Assume that its root ζ is found, then

$$\tau_{01} = \zeta\tau_{02} , \qquad \tau_{12} = (1 - \zeta)\tau_{02} .$$

Formulate the equation for the moments (112.5) using the point C as a pole. Then the integral (112.5) assumes the form

$$M = \tau_{01}\delta \cdot 4a^2 + \tau_{02}\delta \cdot 8a^2 = 4\tau_{02}\delta a^2(2 + \zeta) .$$

Consequently

$$\tau_{02} = \frac{M}{4\delta a^2(2 + \zeta)} .$$

Substitute this value of τ_{20} into eq. (112.9) so as to obtain a relationship between the angular rate of torsion and the moment

$$\kappa = \frac{\gamma_n}{6a}(3 + 2\zeta^n)\left[\frac{M}{4\tau_n\delta a^2(2+\zeta)}\right]^n .$$

Eq. (112.8) can be solved numerically for any value of n. In our case it is very easy to obtain an approximate solution, when $n = 1$ it follows from (112.8) that $\zeta = \frac{6}{7}$. Consequently we should expect that $\zeta > \frac{1}{2}$ for any value of n, consequently, $(1-\zeta) < \frac{1}{2}$. As n increases the first term in the left-hand side of eq. (112.8) increases much faster than the second, and, therefore, if we write the equation in the form

$$\zeta = (\tfrac{3}{4})^{\frac{1}{n}}\,[1 + (1-\zeta)^n]^{\frac{1}{n}} ,$$

its solution by successive approximations should converge extremely rapidly. Indeed even when $n = 3$, to the third significant figure $\zeta = \frac{3}{4}{}^{1/n}$. Other somewhat more complicated examples of calculating box sections in torsion are given in the work [322].

113. **Torsion of thin-walled rods of open profile**

In thin-walled sections of open profile the shear stresses in each section of the profile are normal to its axial line, their direction is parallel to the tangent to the centre line and their distribution relative to the centre line is odd; if at the distance y on one side of the axial line (fig. 196) the stress is τ, then at the same distance on the other side it is $-\tau$. The two trajectories of the tangential stress unite at infinity; applying the theorem of tangential stress circulation we obtain (see [105]):

$$2\gamma(\tau) = 2\kappa y .$$

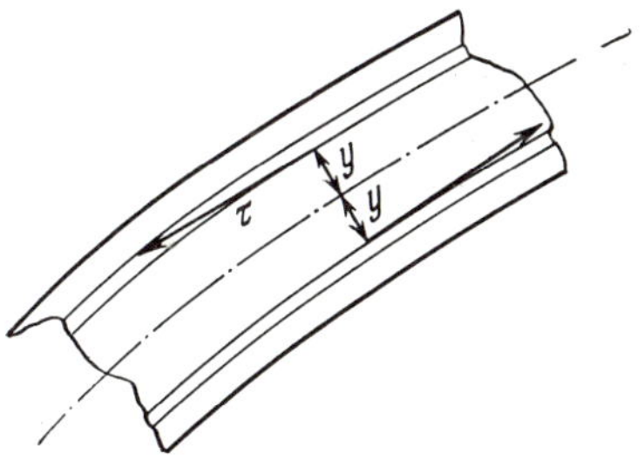

Fig. 196.

Consequently

$$\tau = \gamma^{-1}(\kappa y) .$$

A torque of $\mathrm{d}l \int_{-\delta/2}^{+\delta/2} \tau y \, \mathrm{d}y$ is applied to an element of the mean line arc, whilst the total moment for a rod of constant width with a mean line length l is

$$M = l \int_{-\delta/2}^{+\delta/2} \tau y \, \mathrm{d}y = l \int_{-\delta/2}^{+\delta/2} \gamma^{-1}(\kappa y) y \, \mathrm{d}y .$$

Hereafter it is convenient to assume a particular creep law, for example, the power law. Then

$$\tau = \tau_n \left(\frac{\kappa y}{\gamma_n}\right)^{\frac{1}{n}} , \qquad M = l \int_{-\delta/2}^{+\delta/2} \tau_n \left(\frac{\kappa y}{\gamma_n}\right)^{\frac{1}{n}} y \, \mathrm{d}y .$$

Put

$$\kappa_n = \frac{2\gamma_n}{\delta} , \qquad M_n = \frac{2n}{2n+1} \tau_n l \left(\frac{\delta}{2}\right)^2 .$$

The dependence of the angular rate of torsion on the torque is now given by

$$\kappa = \kappa_n \left(\frac{M}{M_n}\right)^n . \tag{113.1}$$

If $\delta = \delta(l)$ formula (113.1) may be retained; in calculating κ_n and M_n, δ is replaced by a certain, arbitrarily selected, characteristic width of profile δ_0, and l is replaced by the referred length

$$l_{\text{ref}} = \int_0^l \left(\frac{\delta}{\delta_0}\right)^{2+\frac{1}{n}} \mathrm{d}l .$$

As in the bending problem, a simple approximate solution can be obtained by assuming the stress distribution similar to that in an ideally plastic rod in the limiting state. Here the calculations are completely analogous with those in the problem of bending a rod of rectangular section. If the absolute value of τ is constant, the moment applied to an element of mean line of the rod is

$$\mathrm{d}M = \int_{-\delta/2}^{+\delta/2} \tau y \, \mathrm{d}y = \tfrac{1}{4} \tau\delta^2 .$$

The total torque is

$$M = \frac{\tau}{4} \int_0^l \delta^2 \, \mathrm{d}l ;$$

consequently $\tau = 4M/\delta_0^2 l_{\mathrm{ref}}$.

Here the referred length l_{ref} is determined as follows:

$$l_{\mathrm{ref}} = \int_0^l \left(\frac{\delta}{\delta_0}\right)^2 \mathrm{d}l .$$

The rate potential in pure torsion is given by

$$\Phi = \frac{\gamma_n \tau_n}{n+1} \left(\frac{\tau}{\tau_n}\right)^{1+n} .$$

For a constant absolute value of τ by integration over the cross-sectional area we obtain

$$\widetilde{\Phi} = \frac{\gamma_n \tau_n}{n+1} \left(\frac{\tau}{\tau_n}\right)^{1+n} F .$$

Substituting the expression derived for τ in terms of M, we again obtain formula (113.1) in which

$$\kappa_n = \frac{2F\gamma_n}{\delta_0^2 l}, \qquad M_n = 2^{-\frac{1}{n}} \tau_n l_{\mathrm{ref}} \left(\frac{\delta_0}{2}\right)^2 .$$

For a rod of constant thickness $\kappa_n = 2\gamma_n/\delta_0$. Then the exact and approximate solutions for M_n differ by a factor $2n/(2n + 1)$ in the first case and $2^{-1/n}$ in the second. When $n \to \infty$ these factors tend to unity so that the difference between the exact and approximate solutions disappears. However, this relates only to the magnitude of the torque that corresponds to a given rate; if the torque is fixed and the corresponding rate is determined the error of the approximate solution may be quite considerable, as was shown in section 106 on the example of bending a rod of rectangular section.

114. Variational methods of solving torsion problems

As we have already said, in studying torsion of solid rods variational methods are used to obtain approximate solutions. One of these variational devices has already been used in the preceding section. As we have shown the rate potential referred to unit length of rod is given by

$$\widetilde{\Phi} = \int \Phi(\tau)\, \mathrm{d}F\,.$$

Here

$$\Phi'(\tau) = \gamma(\tau)\,.$$

With a power law

$$\widetilde{\Phi} = \frac{\gamma_n \tau_n}{n+1} \int_F \left| \frac{\tau}{\tau_n} \right|^{n+1} \mathrm{d}F\,,$$

or if we introduce the stress function for torsion ψ,

$$\widetilde{\Phi} = \frac{\gamma_n \tau_n}{n+1} \int_F \left[\frac{|\operatorname{grad} \psi|}{\tau_n} \right]^{n+1} \mathrm{d}F\,. \tag{114.1}$$

Assume some statically possible stress distribution, that is define the function ψ which depends on a certain selection of parameters, c_i, in such a way that $\psi = \text{const}$ on the contour of the section, and $2 \int \psi\, \mathrm{d}F = M$. If the region is multiply connected we can fix a value of ψ on one of the contours, the values of the stress functions on the other contours are included in the parameters c_i. It is now necessary to minimise the function $\widetilde{\Phi}$ with respect to parameters c_i, and this leads to a system of non-linear equations for c_i. Substituting the values of c_i that are found into the expression for $\widetilde{\Phi}$ we can determine the angular rate of torsion of the rod $\kappa = \mathrm{d}\widetilde{\Phi}/\mathrm{d}M$. It is evident that in selecting the function ψ we must consider ψ proportional to M. If the function ψ is linearly dependent on the parameters c_i, the expression for $\widetilde{\Phi}$ can always be represented in the form

$$\widetilde{\Phi} = \frac{\kappa_n M_n}{n+1} \left(\frac{M}{M_n}\right)^{n+1} R(c_1, c_2, \ldots, c_k) .$$

Here κ_n and M_n are constants selected in an appropriate manner (compare section 110). Obviously the accurate expression for the rate potential with a power law of creep can be represented in this form and the true value of R is a certain constant. In view of the minimum principle, for a given value of M the true value of $\widetilde{\Phi}$ corresponds to the minimum value of R. Minimising R with respect to a limited number of parameters we always obtain a higher than true value of R. The rate of torsion that corresponds to the determined value of $\widetilde{\Phi}$ is

$$\kappa = \frac{\partial \widetilde{\Phi}}{\partial M} = \kappa_n \left(\frac{M}{M_n}\right)^n R , \tag{114.2}$$

and in view of the above it is always greater than the true rate; therefore, use of Castigliano's principle gives an upper limit for the angular rate of torsion, the rigidity of the rod is then less than it should be.

Now consider the general scheme of application of the Lagrange variational principle. With a power creep law the rate potential is expressed as

$$U = \frac{n}{n+1} \gamma_n \tau_n \left| \frac{\gamma}{\gamma_n} \right|^{1+\frac{1}{n}} .$$

Because of eq. (111.1)

$$\gamma^2 = \kappa^2 [x^2 + y^2 - 2y\varphi_{,x} + 2x\varphi_{,y} + \varphi_{,x}^2 + \varphi_{,y}^2] .$$

Therefore, the force potential per unit length of rod is

$$\widetilde{U} = \frac{n}{n+1} \kappa_n M_n \left(\frac{\kappa}{\kappa_n}\right)^{1+\frac{1}{n}} Q .$$

Here

$$Q = \frac{\tau_n}{M_n} \left(\frac{\kappa_n}{\gamma_n}\right)^{\frac{1}{n}} \int_F \left| x^2 + y^2 - 2y\varphi_{,x} + 2x\varphi_{,y} + \varphi_{,x}^2 + \varphi_{,y}^2 \right|^{\frac{n+1}{2n}} \mathrm{d}F .$$

Suppose that we are given the warping function of the section φ in the form of a linear expression that contains a certain number of parameters c_i. Minimisation of $\widetilde{U}$ with respect to these parameters reduces to minimisation of Q and with a finite number of parameters the value of Q that is obtained is equal to or greater than the true value. Determining a minimum value of Q for a given selection of c_i we determine the torque

$$M = \frac{\partial \widetilde{U}}{\partial \kappa} = M_n \left(\frac{\kappa}{\kappa_n}\right)^{\frac{1}{n}} Q \,. \tag{114.3}$$

In view of the minimum property of the rate potential, and consequently of the functional Q, the determined value of M is always greater than the true value, or in the extreme case equal to it. If we solve eq. (114.3) relative to κ the assessment of the angular rate of torsion is too high. Simultaneous application of a solution by means of the Lagrange and Castigliano variational principles gives a further assessment of the rate of torsion. The corresponding theorems for elastic torsion are well known [67], for work-hardening plastic media and for creep they were demonstrated by Kachanov [52] and examples of their application are given in his book.

115. **Bending and torsion of thin-walled rods of closed section**

Let us consider simultaneous bending and torsion of a rod on the assumption that the tangential stresses are small compared with the normal bending stresses, therefore we may consider that

$$\sigma_0 = (\sigma^2 + 3\tau^2)^{\frac{1}{2}} \approx \sigma \,.$$

Then from the creep equations it follows that

$$\epsilon = v(\sigma) \,, \qquad \gamma = 3v(\sigma)\frac{\tau}{\sigma} \,. \tag{115.1}$$

Here γ is the rate of shear. As the relationship between ϵ and σ does not include shear stresses and strains, the bending problem is solved separately and then the torsion problem is considered. Returning to the conclusion drawn in section 111, we should consider that the quantity γ in eqs. (112.2) and the subsequent equation is defined by the second of the formulae (115.1). Then eq. (111.4) reduces to

$$3\int_{\Gamma}\frac{\upsilon(\sigma)}{\sigma}\tau_s\,\mathrm{d}l = 2\kappa\Omega\,. \qquad (115.2)$$

The conditions of the circulation theorem for an elastic rod with variable shear modulus $G = \sigma/3\upsilon$ are written in an exactly similar way.

Now let us consider combined bending and torsion of thin-walled rods of closed profile. The shear stresses in a rod of closed profile are found by formula (108.4), i.e. in the same way as for a rod of open profile, but as yet the constant C remains indeterminate. Rewrite formula (108.4) as

$$\tau = \frac{1}{\delta}(C - t)\,.$$

Here

$$t = \int_0^l \delta(l)\frac{\partial\sigma}{\partial z}\,\mathrm{d}l\,.$$

By the circulation theorem

$$\int_{\Gamma}\frac{\upsilon(\sigma)}{\sigma}\tau\,\mathrm{d}l = C\int_{\Gamma}\frac{1}{\delta}\,\frac{\upsilon(\sigma)}{\sigma}\,\mathrm{d}l - \int_{\Gamma}\frac{t}{\delta}\,\frac{\upsilon(\sigma)}{\sigma}\,\mathrm{d}l = \tfrac{2}{3}\,\kappa\Omega\,. \qquad (115.3)$$

We can now calculate the torque relative to an arbitrary axis that intersects the cross-section at the point Q:

$$M_{\mathrm{k}} = \int_{\Gamma}\tau\delta p\,\mathrm{d}l = C\Omega - \int_{\Gamma}t\,\mathrm{d}\omega\,. \qquad (115.4)$$

Here as usual, p is the length of a perpendicular dropped from the point O onto the tangent of the mean line of the profile, $p\,\mathrm{d}l = \mathrm{d}\omega$, where ω is the sectorial area. If the rod is bent by forces in a plane of action at a distance e from the point O, then $M_{\mathrm{k}} = Qe$; the value of C is found from formula (115.4) and substituted into (115.3) giving the rate of change of the angle of torsion of the rod as a function of the shear force, the distance e and the bending moment in the section.

For a power creep law eqs. (115.3) and (115.4) assume the form

$$\frac{\epsilon_n}{\sigma_n}\left(\frac{M}{M_n}\right)^{n-1}\int_\Gamma (C-t)\left(\frac{\kappa_n y}{\epsilon_n}\right)^{\frac{n+1}{n}}\frac{\mathrm{d}l}{\delta}=\tfrac{2}{3}\kappa\Omega\,,$$

$$M_\mathrm{k}=C\Omega-\int_\Gamma t\,\mathrm{d}\omega\,,\qquad t=\frac{\sigma_n Q}{M_n}\int_0^l\left(\frac{\kappa_n y}{\epsilon_n}\right)^{\frac{1}{n}}\delta(l)\,\mathrm{d}l\,.$$

These formulae are very easily applied to the solution of problems, though the calculations are fairly cumbersome. The simplicity results from the circumstance that the dependence of shear rate on shear stress is linear and the problem is equivalent to the problem of the theory of elasticity with a variable, but predetermined shear modulus; the principle of superposition can be applied to problems of this kind and, therefore, the redundant unknown C has been removed from the formula in a linear manner and there is no need to solve the usual non-linear equations of creep theory.

Still simpler results are obtained if we use the approximate theory based on the assumption that the stress distribution is the same as in the plastic limiting state. Here the neutral axis nn divides the cross-sectional area into two equal parts. We shall reckon the arc of the mean line of the cross-section from the point O where the mean line intersects the neutral axis. The length of the arc of the upper half of the contour is l_1, and that of the lower half l_2. Here

$$\int_0^{l_1}\delta\,\mathrm{d}l=\int_{-l_2}^{0}\delta\,\mathrm{d}l=\tfrac{1}{2}F\,.$$

By formulae (109.1)

$$\tau=\frac{1}{\sigma}\left[C+\frac{2Q}{cF}\int_l^{l_1}\delta\,\mathrm{d}l\right],\qquad 0<l<l_1\,,$$

$$\tau=\frac{1}{\sigma}\left[C+\frac{2Q}{cF}\int_{-l_2}^{l}\delta\,\mathrm{d}l\right],\qquad -l_2<l<0\,.$$

The quantity $\upsilon(\sigma)/\sigma$ is constant for a given section, here $\sigma=2M/cF$. The circulation theorem gives the following result:

$$C\int_\Gamma \frac{dl}{\delta} + \frac{2Q}{cF}\left[\int_0^{l_1} dl \int_l^{l_1} \delta\, dl + \int_{-l_2}^{0} dl \int_{-l_2}^{l} \delta\, dl\right] = \tfrac{3}{2}\,\kappa\Omega\,\frac{\sigma}{v(\sigma)}.$$

The expression in square brackets may be neglected so that the final result is written

$$C\int_\Gamma \frac{dl}{\delta} + \frac{2Q}{cF}\left[\int_0^{l_1} l\delta\, dl - \int_{-l_2}^{0} l\delta\, dl\right] = \tfrac{3}{2}\,\kappa\Omega\,\frac{\sigma}{v(\sigma)}. \tag{115.5}$$

Now write the expression for the torsional moment

$$M_k = \int_\Gamma \tau\delta\, d\omega = C\Omega - \frac{2Q}{cF}\int \omega\delta\, dl. \tag{115.6}$$

116. **Bending of rods in the presence of a longitudinal force**

The problem of bending a rod in the presence of an axial force is particularly important in connection with the theory of stability under creep conditions. We shall consider this theory more closely later on, here we consider only some simple problems. Returning to the derivation of the general equations of the theory of bending in sections 106 and 107, it should be noted that the considerations are not altered by the presence of an axial force, except that the condition $N = 0$ should be replaced by the condition that N is equal to the given force. Here, however, practical calculations become fairly cumbersome, in order to elucidate the essential we consider here the problem of combined bending and extension of a rod of rectangular cross-section. Assuming that the bending takes place in the plane of symmetry of the rod, we use the following expressions in place of eq. (107.1):

$$\begin{aligned} N &= \int s(\epsilon_0 + \kappa y)\, dF, \\ M_x &= -\int s(\epsilon_0 + \kappa y) y\, dF. \end{aligned} \tag{116.1}$$

To calculate the integrals taken over the area of a rectangle of height $2h$ and width b it should be borne in mind that on the neutral axis when $y = y_0$, $\epsilon_0 + \kappa y_0 = 0$. The absolute value of y_0 may be less or greater than h, that is the neutral axis may intersect the section or may lie entirely outside it. In either case the following result is obtained with a power creep law:

$$\frac{N}{bh\sigma_n} = \frac{n}{n+1}\,\epsilon_0^{-1/n}\,\frac{1}{\kappa h}\,\{|\epsilon_0 + \kappa h|^{1+1/n} - |\epsilon_0 - \kappa h|^{1+1/n}\}\,,$$

$$\frac{M}{bh^2\sigma_n} = \frac{n}{2n+1}\,\epsilon_0^{-1/n}\,\frac{1}{(\kappa h)^2}\,\{|\epsilon_0 + \kappa h|^{2+1/n} - |\epsilon_0 - \kappa h|^{2+1/n}\} - \frac{\epsilon_0}{\kappa h}\,Nh\,. \qquad (116.2)$$

These equations cannot be solved for κh and ϵ_0, and therefore, it is difficult to use them for calculations. There are various ways of overcoming these difficulties. One is to select a certain approximating function for the creep rate potential which is represented in the form $\Phi = \Phi(R)$. Here by Φ we mean the potential per unit length of rod, whereas R is a function of M and N which is selected in such a way that when $N = 0$, $R = M$. Then

$$\kappa = \frac{\mathrm{d}\Phi}{\mathrm{d}R}\,\frac{\partial R}{\partial M}\,, \qquad \epsilon_0 = \frac{\mathrm{d}\Phi}{\mathrm{d}R}\,\frac{\partial R}{\partial N}\,.$$

If $N = 0$, by formula (106.2) we have

$$\kappa = \kappa_n \left(\frac{M}{M_n}\right)^n .$$

But here, as we agreed, $M = R$. Consequently

$$\Phi(R) = \frac{\kappa_n M_n}{n+1}\left(\frac{R}{M_n}\right)^{n+1} . \qquad (116.3)$$

To select the function R we use the Calladine and Drucker theory as was shown in section 103 for bar structures. In a rod of rectangular cross-section in the limiting conditions, loaded by a bending moment and a longitudinal force, the relationship between the force and moment is given by (see [105])

$$\hat{M} + \hat{N}^2 = 1\,.$$

Here

$$\hat{M} = \frac{M}{bh^2\sigma_\mathrm{T}}\,, \qquad \hat{N} = \frac{N}{2bh\sigma_\mathrm{T}}\,.$$

On the other hand in the elastic state $\sigma = N/2bh + \frac{3}{2}(My/bh^3)$. Formulating the condition of normalisation

$$b\int_{-h}^{+h} \hat{\sigma}^2 \,\mathrm{d}y = F = 2bh\,,$$

we obtain

$$\tfrac{3}{2}\hat{M}^2 + \hat{N}^2 = 1\,.$$

In $\hat{M}$–$\hat{N}$ coordinates the condition of the limiting state is a parabola and the condition of normalising of the power dissipation when $n = 1$ is an ellipse (fig. 197).

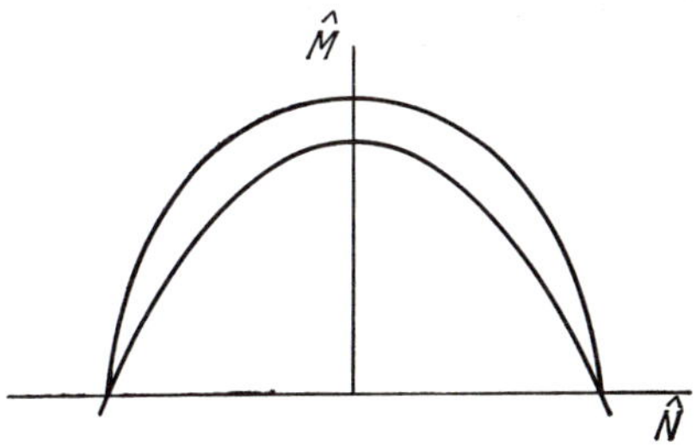

Fig. 197.

Obviously all curves of constant power dissipation pass through the point (1, 0) on the $\hat{N}$ axis. To determine the point of intersection between a curve corresponding to a given value of n and the M axis, we calculate the power dissipation in pure bending and subject it to the condition of normalisation. We obtain

$$L = \kappa_n M_n \left(\frac{M}{M_n}\right)^{n+1} = 2bh\,.$$

Whence

$$\hat{M} = \frac{M}{bh^2\sigma_n} = 2^{\frac{1}{n+1}} \left(\frac{2n}{2n+1}\right)^{\frac{n}{n+1}}\,.$$

Therefore, all curves of constant power dissipation intersect the $\hat{M}$ axis with an intercept between $\hat{M} = 1$ and $\hat{M} = 2/\sqrt{3}$. According to the Calladine and

Drucker theorem all of these curves that correspond to increasing values of n lie inside one another, all are located within an ellipse that corresponds to the case $n = 1$ and a parabola for $n = \infty$.

Assume that these curves are ellipses

$$\left(\frac{\hat{M}}{2k}\right)^2 + \hat{N}^2 = 1 .$$

Here

$$k = \left(\frac{n}{2n+1}\right)^{\frac{n}{n+1}} .$$

We must then take

$$R^2 = M^2 + k^2 h^2 N^2 . \tag{116.4}$$

The rates of change of curvature and elongation are written

$$\begin{aligned} \epsilon_0 &= \kappa_n \left(\frac{R}{M_n}\right)^n \frac{N}{R} (kh)^2 , \\ \kappa &= \kappa_n \left(\frac{R}{M_n}\right)^n \frac{M}{R} . \end{aligned} \tag{116.5}$$

It is easily checked that these formulae give correct results when $N = 0$ and when $M = 0$, that is in pure bending and in pure tension. It is difficult to compare formulae (116.4) and (116.5) with the exact solution because it is not possible to write an explicit expression for the potential Φ in terms of M and N. However, if we consider a space of rates the exact formulae (116.2) can be used to construct a curve of constant power dissipation. The conditions of power normalisation are

$$M\kappa + T\epsilon_0 = F = 2bh .$$

By formula (116.2) we obtain the relationship between ϵ_0 and κh:

$$\tfrac{1}{2} \frac{n}{2n+1} \frac{1}{\kappa h} \{ |\epsilon_0 + \kappa h|^{2+1/n} - |\epsilon_0 - \kappa h|^{2+1/n} \} = 1 . \tag{116.6}$$

When $n = 1$ it follows that

$$\tfrac{1}{3}(\kappa h)^2 + \epsilon_0^2 = 1\ .$$

When $n = \infty$, $1/2\kappa h[(\kappa h)^2 + \epsilon_0^2] = 1$ if $\kappa h > \epsilon_0$ and $\epsilon_0 = 1$ if $\kappa h < \epsilon_0$. It will be seen that curves of eq. (116.6) intersect the ϵ_0 axis at the point $\epsilon_0 = 1$ and the κh axis at the point $\kappa h = 1/k = [(2n + 1)/n]^{n/(n+1)}$.

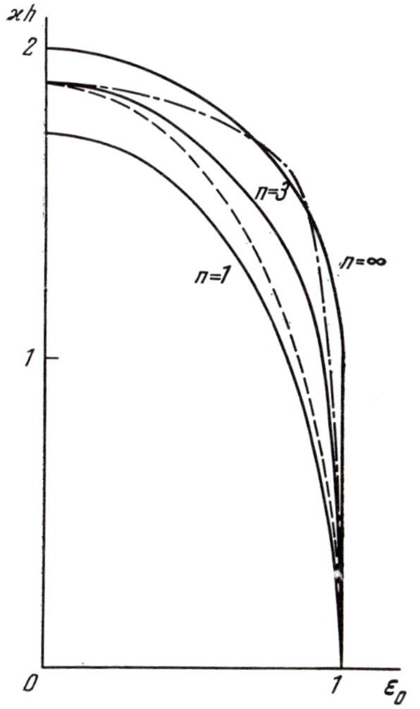

Fig. 198.

The corresponding curves of constant power dissipation and also the curve calculated for the value $n = 3$ are plotted in solid lines in fig. 198. With the approximate solution given by formulae (116.4) and (116.5) the curve of constant power dissipation in the plane of rates is also obviously an ellipse with intercepts on the axis of $\epsilon_0 = 1$ and $\kappa h = 1/k$. This ellipse, for $n = 3$, is plotted in fig. 198 as a line of dashes.

117. **Beam of ideal *I*-section**

In this paragraph we consider another approximate method of determining bending in the presence of a longitudinal force, it is based on replacing the

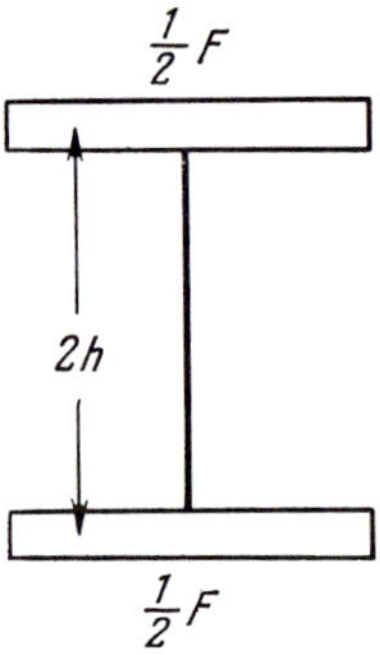

Fig. 199.

real beam by a beam of specially selected idealised cross-section. We shall call the section illustrated in fig. 199 an idealised *I*-section. Two flanges of equal cross-sectional area $\frac{1}{2}F$ are joined by a web which carries the intersecting load (force) but does not participate in the resistance to bending. We shall consider that the distance between the flanges $2h$ is great compared with their thickness and that the stress distribution in the flanges is uniform; $\sigma = \sigma^+$ in the upper flange and $\sigma = \sigma^-$ in the lower. Then

$$N = \frac{F}{2}(\sigma^+ + \sigma^-), \qquad M = \frac{hF}{2}(\sigma^+ - \sigma^-).$$

Put

$$\frac{N}{F\sigma^*} = \hat{N}, \qquad \frac{M}{Fh\sigma^*} = \hat{M}.$$

Then

$$\sigma^+ = (\hat{N} + \hat{M})\sigma^*, \qquad \sigma^- = (\hat{N} - \hat{M})\sigma^*.$$

If ϵ is the rate of deformation of the centre line of the rod, κ the rate of change of curvature, then the rate of deformation of the upper flange is $\epsilon^+ = \epsilon + \kappa h$, and the rate of deformation of the lower flange is $\epsilon^- = \epsilon - \kappa h$, then

$$\epsilon^+ = v(\sigma^+), \qquad \epsilon^- = v(\sigma^-),$$

or

$$\sigma^+ = s(\epsilon^+), \qquad \sigma^- = s(\epsilon^-).$$

Hence it follows that

$$\begin{aligned} \epsilon &= \tfrac{1}{2}\,\{v[(\hat{N}+\hat{M})\sigma^*] + v[(\hat{N}-\hat{M})\sigma^*]\}\,, \\ \kappa h &= \tfrac{1}{2}\,\{v[(\hat{N}+\hat{M})\sigma^*] - v[(\hat{N}-\hat{M})\sigma^*]\}\,, \end{aligned} \tag{117.1}$$

or

$$\begin{aligned} \hat{N} &= \frac{1}{2\sigma^*}\,\{s(\epsilon+\kappa h) + s(\epsilon-\kappa h)\}\,, \\ \hat{M} &= \frac{1}{2\sigma^*}\,\{s(\epsilon+\kappa h) - s(\epsilon-\kappa h)\}\,. \end{aligned} \tag{117.2}$$

With a power creep law

$$\begin{aligned} \hat{\epsilon} &= \frac{\epsilon}{\epsilon_n} = \tfrac{1}{2}\,[|\hat{N}+\hat{M}|^{n-1}(\hat{N}+\hat{M}) + |N-M|^{n-1}(\hat{N}-\hat{M})]\,, \\ \hat{\kappa} &= \frac{\kappa h}{\epsilon_n} = \tfrac{1}{2}\,[|\hat{N}+\hat{M}|^{n-1}(\hat{N}+\hat{M}) - |\hat{N}-\hat{M}|^{n-1}(\hat{N}-\hat{M})]\,, \end{aligned} \tag{117.3}$$

or

$$\begin{aligned} \hat{N} &= \tfrac{1}{2}\,[|\hat{\epsilon}+\hat{\kappa}|^{\frac{1}{n}-1}(\hat{\epsilon}+\hat{\kappa}) + |\hat{\epsilon}-\hat{\kappa}|^{\frac{1}{n}-1}(\hat{\epsilon}-\hat{\kappa})]\,, \\ \hat{M} &= \tfrac{1}{2}\,[|\hat{\epsilon}+\hat{\kappa}|^{\frac{1}{n}-1}(\hat{\epsilon}+\hat{\kappa}) - |\hat{\epsilon}-\hat{\kappa}|^{\frac{1}{n}-1}(\hat{\epsilon}-\hat{\kappa})]\,. \end{aligned} \tag{117.4}$$

It remains to relate the characteristics of the idealised *I*-section to those of a real beam. If F is the cross-sectional area of the real beam, in the absence of bending $\sigma^+ = \sigma^- = N/F = \hat{N}\sigma^*$ and it follows from eq. (117.1) that $\epsilon = v(\hat{N}\sigma^*)$, therefore the ideal beam behaves in just the same way as the real one. Now select the dimension h such that in the absence of longitudinal force the relationship between moment and curvature for the ideal *I*-section coincides with the corresponding expression for the real section. We shall assume that this latter has two axes of symmetry and then the relationship between moment and curvature with a power creep law is given by formula (106.2):

$$\frac{\kappa H}{\epsilon_n} = \left(\frac{M}{M_n}\right)^n, \qquad M_n = 2\sigma_n bHI ,$$

$$I = \frac{1}{2bH} \int_F \left|\frac{y}{H}\right|^{1+\frac{1}{n}} \mathrm{d}F .$$

Here H is the height of the real section which enters into the expression for M_n. On the other hand when $\epsilon = 0$ it follows from eq. (117.4) that:

$$\frac{\kappa h}{\epsilon_n} = \left(\frac{M}{Fh\sigma_n}\right)^n .$$

Comparing we find that

$$\frac{h}{H} = \left(\frac{2bH}{F} I\right)^n .$$

For a rectangular section

$$\frac{h}{H} = \left(\frac{n}{2n+1}\right)^{\frac{n}{n+1}} .$$

The condition of constant power dissipation is written in the new notation as

$$\hat{N}\hat{\epsilon} + \hat{M}\hat{\kappa} = \tfrac{1}{2}\left[\,|\hat{\epsilon} + \hat{\kappa}|^{1+\frac{1}{n}} + |\hat{\epsilon} - \hat{\kappa}|^{1+\frac{1}{n}}\right] = 1 \tag{117.5}$$

The chain dotted line in fig. 198 shows a curve corresponding to eq. (117.5) when $n = 3$.

Only with a power creep law is it possible to select the parameters of the model beam in such a way that its behaviour in pure tension and in bending is identical with that of the real beam. In other cases the ratio h/H depends on the moment or curvature. Note that in considering bending and tension or compression of a rod with only one plane of symmetry a model can be constructed in the form of an ideal I-section profile with flanges of unequal area.

118. **Simple problems of longitudinal force and transverse bending**

In the simplest problems of bending in the presence of a longitudinal force no allowance need be made for the bending moment due to the longitudinal

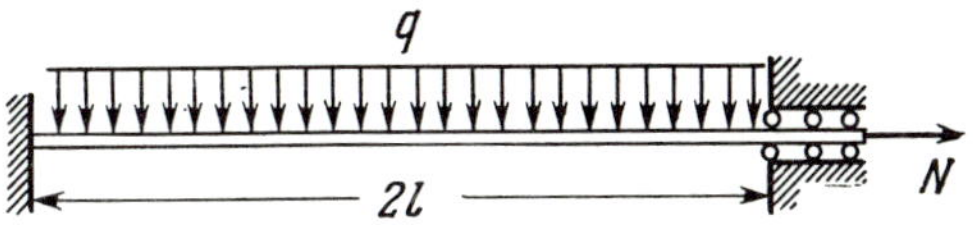

Fig. 200.

force, the magnitude of this force is given and remains constant. By way of example, consider a beam of length $2l$ which carries a uniformly distributed load q, the ends of the beam are fixed in such a way that they cannot move vertically or rotate, but can move axially so that the length of the beam can alter during bending (fig. 200). We shall solve this problem for an ideal I-section, we shall see that in this case the approximation (116.5) leads to an entirely analogous calculation. We first find an accurate solution for the particular value of $n = 3$, the calculations are more cumbersome for other whole-number values of n but in principle the course of the solution remains the same. Denote the unknown bending moment in the clamp by M_0, then in a section with the coordinate x

$$M = M_0 - qlx + \frac{qx^2}{2} .$$

Here we have not taken account of the bending moment due to the longitudinal force which arises only when the deflection is different from zero, which is indeed a specific feature of longitudinal-transverse bending. More complicated problems of this kind are considered in Chapter 12.

Introduce the notation

$$\frac{2M_0}{ql^2} = \zeta , \qquad \frac{ql^2}{2Fh\sigma_n} = p , \qquad \frac{x}{l} = \xi .$$

Then changing to the dimensionless quantities introduced in section 117 we find

$$\hat{M} = p(\zeta - 2\xi + \xi^2) .$$

Further put

$$\frac{2Nh}{ql^2} = \lambda .$$

Consequently

$$\hat{N} = \lambda p \ .$$

Denoting the deflection of the beam $w(\xi, t)$ the rate of change of curvature is

$$\kappa = \frac{\partial^2 \dot{w}}{\partial x^2} = \frac{1}{l^2} \frac{\partial^2 \dot{w}}{\partial \xi^2} = \frac{\epsilon_n}{h} \hat{\kappa}(\xi) \ ,$$

therefore

$$\frac{\partial^2 \dot{w}}{\partial \xi^2} = \frac{\epsilon_n l^2}{h} \hat{\kappa}(\xi) \ .$$

In an exactly similar way the rate of relative deformation of the axis of the beam $\epsilon = \epsilon_n \hat{\epsilon}(\xi)$. When $n = 3$ eqs. (117.3) lead to

$$\hat{\kappa} = (3\hat{N}^2 + \hat{M}^2)\hat{M} \ ,$$

$$\hat{\epsilon} = (\hat{N}^2 + 3\hat{M}^2)\hat{N} \ .$$

Substituting expressions for $\hat{M}$ and $\hat{N}$ we obtain

$$\hat{\kappa} = p^3 [3\lambda^2 + (\zeta - 2\xi + \xi^2)^2] (\zeta - 2\xi + \xi^2) \ ,$$
$$\hat{\epsilon} = \lambda p^3 [\lambda^2 + 3(\zeta - 2\xi + \xi^2)^2] \ . \qquad (118.1)$$

On expanding the brackets in the right-hand side of the first formula we obtain a sixth power polynomial relative to ξ and in the second formula, a fourth power polynomial. But $\hat{\kappa}$ is proportional to the second derivative of the deflection. When $\xi = 0$, $w = \partial w/\partial \xi = 0$, and because of symmetry also $\partial w/\partial \xi = 0$ at the centre of the beam, when $\xi = 1$. Integrating the expression for $\hat{\kappa}$ once and using this condition we obtain an equation for determination of the redundant unknown ζ,

$$\zeta^3 - 2\zeta^2 + (3\lambda^2 + \tfrac{8}{5})\zeta - (2\lambda^2 + \tfrac{16}{35}) = 0 \ .$$

Note that in the absence of an axial force, that is when $\lambda = 0$, the solution for a beam with clamped ends is given in [317], which also gives a calculated value of the real root of the cubic equation, namely $\zeta = 0.6041$. The conclusion drawn in this article should be noted: if the deflection at the centre of an elastic beam with clamped ends is a fifth that of a freely supported beam, then under creep conditions when $n = 3$ clamping the ends reduces the deflection rate by a factor of 24.5.

Integrating a second time and putting $\xi = 1$ we find the rate of deflection at the centre of the beam

$$\dot{w}(1) = \frac{\epsilon_n l^2}{h} \left(\frac{ql^2}{2Fh\sigma_n} \right)^3 \alpha .$$

Here

$$\alpha = \tfrac{1}{2}\zeta^3 - \tfrac{3}{4}\zeta^2 + (\tfrac{3}{2}\lambda^2 + \tfrac{1}{2})\zeta - \tfrac{3}{4}\lambda^2 - \tfrac{1}{8} .$$

Now we can determine the change in length of the beam that results from the axial force. Denoting this elongation Δ, we find:

$$\dot{\Delta} = \int_0^{2l} \epsilon \, dz = \epsilon_n l \int_0^1 \hat{\epsilon}(\xi) \, d\xi .$$

Substituting the value of $\hat{\epsilon}(\xi)$, given by the second of eqs. (118.1), we obtain

$$\dot{\Delta} = \epsilon_n l\lambda \left(\frac{ql^2}{2Fh\sigma_n} \right)^3 \beta .$$

Here

$$\beta = 6\zeta^2 - 8\zeta + 2\lambda^2 + \tfrac{16}{5} .$$

These formulae can be used for example to solve the following problem. A

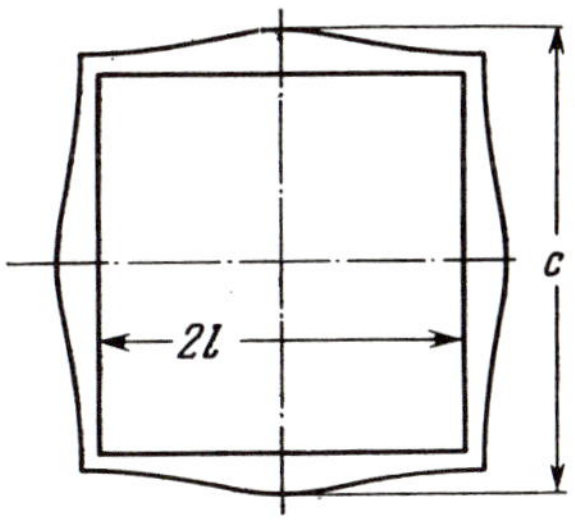

Fig. 201.

square frame is loaded by internal pressure (fig. 201). It is required to determine the rate of change of the dimension c shown in the figure.

It will be seen that each side of the frame is in the condition of the beam in the previous problem, here $N = ql$, consequently, $\lambda = 2h/l$. The rate of change of the quantity c is:

$$\dot{c} = 2\dot{w}(1) + \dot{\Delta} = 2\epsilon_n \frac{l^2}{h} \left(\frac{ql^2}{2Fh\sigma_n}\right)^3 \left(\alpha + \frac{h^2}{l^2}\beta\right).$$

This solution can be brought to a conclusion because we have taken $n = 3$. It is quite obvious how to proceed when n is any whole number. However, as we have seen the number n is determined experimentally and can assume any value. In the general case when n is arbitrary it is recommended to use a numerical method of solving problems of this kind. Maintaining the same notation we find that for any value of n

$$\frac{\hat{\kappa}}{p^n} = \tfrac{1}{2}\left[\,|\lambda + \mu|^{n-1}(\lambda + \mu) - |\lambda - \mu|^{n-1}(\lambda - \mu)\right],$$

$$\frac{\hat{\epsilon}}{p^n} = \tfrac{1}{2}\left[\,|\lambda + \mu|^{n-1}(\lambda + \mu) + |\lambda - \mu|^{n-1}(\lambda - \mu)\right]. \qquad (118.2)$$

Here $\mu = \mu(\xi) = \hat{M}/p$, and in our case $\mu = \zeta - 2\xi + \xi^2$, in other examples this may be a different function of ξ which contains a redundant unknown ζ in a linear manner. Given an arbitrary value of ζ we can construct a diagram of the function $\hat{\kappa}/p^n$, i.e. the right-hand side of the first of eqs. (118.2). On the basis of the usual considerations of the graphical-analytical method of the theory of deflection of beams we can always find out whether the kinematic condition which we have formulated is fulfilled for a particular value of ζ. In the example under consideration this kinematic condition leads to the equation $\int_0^2 \hat{\kappa}(\xi)\,d\xi = 0$, consequently the area of the diagram should be zero. Repeating the calculations for several values of ζ we easily select the required value of the unknown. After ζ has been determined the diagram of $\hat{\kappa}/p^n$ can be used to determine the coefficient of deflection at the centre of the beam; the quantity α is the bending moment at the centre section due to the imaginary load $\hat{\kappa}$, in accordance with the well-known rule for a beam with clamped ends the imaginary reactions and reactive moments are zero. We can now write

$$\dot{w}(1) = \frac{\epsilon_n l^2}{h} \left(\frac{ql^2}{2Fh\sigma_n} \right)^n \alpha .$$

In exactly the same way we find the rate of axial elongation

$$\dot{\Delta} = \epsilon_n l \left(\frac{ql^2}{2Fh\sigma_n} \right)^n \lambda\beta .$$

The factor $\lambda\beta$ is equal to the area of the diagram for the right-hand side of the second of eqs. (118.2), that is the function $\hat{\epsilon}/p^n$. Of course in constructing this diagram we must use a previously determined value of ζ.

We have already mentioned that approximation (116.5) leads to analogous calculations. Indeed with n odd, substituting the expression for the moment M into eq. (116.5) it is found that ϵ is a polynomial of power $(n-1)$ relative to M, and of power $2(n-1)$ relative to x, whereas κ is a polynomial of power n relative to M and of power $(2n-1)$ relative to x. The second equation is integrated once to determine the angle of slope, and a second time to determine the deflection, the condition that the tangent is horizontal at the centre of the span leads to an algebraic equation of power n for the unknown reactive moment M_0. For even values of n the analytical solution is more complicated, but in all cases the procedure described above to obtain a numerical solution is applied without any fundamental changes.

9

Plane Axially-Symmetrical Problems of the Theory of Steady-State Creep

Plane axially-symmetrical problems of creep theory are unidimensional, they are solved by integrating several systems of ordinary non-linear differential equations. A good deal of attention has naturally been paid to these problems and there is a large number of works on the subject. Moreover, they are of considerable technical interest and include for example calculation of thick-walled tubes subject to internal pressure and of a rotating disc. There is a particularly extensive literature on creep of rotating discs and Soviet scientists and engineers have made a substantial contribution to the development of practically convenient methods of calculating discs. The theory of steady-state creep is not always adequate in relation to actual structures and, therefore, manufacturers and designers generally use the theory of ageing which makes it possible to use primary creep curves of the material in all their variety. Calculations by the ageing theory are carried out in exactly the same way as by the theory of steady-state creep, and, therefore, the methods presented here will find application in Chapter 11. On the other hand, the steady-state creep theory is adequate for rough calculations and for determining the limiting state which is established after a sufficiently long time.

119. **Fundamental equations of the plane problem**

For convenience we first give the fundamental equations of creep theory for plane problems.

A plane strain state. This is the name given to the state that is realised in a cylindrical body when the velocity component $v_3 = 0$, the x_3 axis being taken on the centre-line of the cylinder. There is no very good justification for retaining the subscript notation in considering plane problems and, therefore, we shall consider the x and y axes as located in the plane of the section perpendicular to the centre line of the cylinder z. Since $\epsilon_z = 0$, it follows from the condition of incompressibility that

$$\epsilon_x + \epsilon_y = 0 .$$

Now consider various creep criteria:

a) Quadratic creep criterion (Mises) type. From eq. (68.1) it follows that:

$$\epsilon_{33} = \epsilon_z = \frac{3}{2}\frac{v(\sigma_0)}{\sigma_0}\bar{\sigma}_{33} = 0 .$$

Consequently, $\bar{\sigma}_{33} = \bar{\sigma}_z = 0$ and

$$\sigma_z = \tfrac{1}{2}(\sigma_x + \sigma_y) .$$

Eliminating the stress σ_z from the expression for σ_0 we obtain

$$\sigma_0 = \frac{\sqrt{3}}{2}\sqrt{(\sigma_x - \sigma_y)^2 + 4\tau^2} \tag{119.1}$$

or for the principal axes

$$\sigma_0 = \frac{\sqrt{3}}{2}(\sigma_1 - \sigma_2) . \tag{119.2}$$

Now calculate the quantity ϵ_0, bearing in mind that $\epsilon_z = 0$,

$$\epsilon_0 = \tfrac{2}{3}\sqrt{\epsilon_x^2 + \epsilon_y^2 - \epsilon_x\epsilon_y + \tfrac{3}{4}\gamma^2} = \frac{2}{\sqrt{3}}\sqrt{\epsilon_x^2 + \tfrac{1}{4}\gamma^2} , \tag{119.3}$$

and for the principal axes

$$\epsilon_0 = \frac{2}{\sqrt{3}}\epsilon_1 = -\frac{2}{\sqrt{3}}\epsilon_2 . \tag{119.4}$$

Now the flow equations are

$$\epsilon_x = \frac{3}{2}\frac{v(\sigma_0)}{\sigma_0}\bar{\sigma}_x = \frac{3}{4}\frac{v(\sigma_0)}{\sigma_0}(\sigma_x - \sigma_y),$$

$$\epsilon_y = \frac{3}{2}\frac{v(\sigma_0)}{\sigma_0}\bar{\sigma}_y = -\frac{3}{4}\frac{v(\sigma_0)}{\sigma_0}(\sigma_x - \sigma_y), \tag{119.5}$$

$$\gamma = 3\frac{v(\sigma_0)}{\sigma_0}\tau .$$

b) Tresca type creep criterion. Since the rate of creep in the direction of the z axis is zero this is the axis of mean stress and then the maximum tangential stress is always $\frac{1}{2}(\sigma_1 - \sigma_2)$, here the 1 and 2 axes lie in the plane x, y. Therefore

$$s = \sigma_1 - \sigma_2 , \tag{119.6}$$

or

$$s = \sqrt{(\sigma_x - \sigma_y)^2 + 4\tau^2} . \tag{119.7}$$

The flow equations are

$$\epsilon_x = -\epsilon_y = \frac{v(s)}{s}(\sigma_x - \sigma_y), \qquad \gamma = 4\frac{v(s)}{s}\tau , \tag{119.8}$$

or for the principal axes

$$\epsilon_1 = -\epsilon_2 = v(\sigma_1 - \sigma_2) . \tag{119.9}$$

c) Referred stress criterion. Put

$$s = \lambda[\sigma_1 - \tfrac{1}{2}(\sigma_2 + \sigma_3)] + \mu[\sigma_2 - \tfrac{1}{2}(\sigma_1 + \sigma_3)] . \tag{119.10}$$

Whence

$$\epsilon_1 = v(s)(\lambda - \tfrac{1}{2}\mu), \qquad \epsilon_2 = v(s)(-\tfrac{1}{2}\lambda + \mu),$$

$$\epsilon_3 = v(s)(-\tfrac{1}{2}\lambda - \tfrac{1}{2}\mu) .$$

We require that when $\epsilon_3 = 0$, $\mu = -\lambda$; from the condition that when $\sigma_2 = \sigma_3 = 0$, $s = \sigma_1$, we must put $\lambda = \frac{2}{3}$. Now

$$s = \sigma_1 - \sigma_2, \quad \epsilon_1 = v(\sigma_1 - \sigma_2), \quad \epsilon_2 = -v(\sigma_1 - \sigma_2).$$

We have obtained exactly the same result as by the maximum stress criterion.

Plane stress state. This is the state that occurs in a thin sheet loaded by forces that act in its plane. Putting the z axis in the direction of the normal to the plane of the sheet we find that $\sigma_z = 0$.

a) Quadratic creep criterion:

$$\sigma_0 = \sqrt{\sigma_x^2 + \sigma_y^2 - \sigma_x\sigma_y + 3\tau^2}. \tag{119.11}$$

Use the condition of incompressibility to eliminate ϵ_z from the expression for ϵ_0, we find

$$\epsilon_0 = \frac{2}{\sqrt{3}}\sqrt{\epsilon_x^2 + \epsilon_y^2 + \epsilon_x\epsilon_y + \tfrac{1}{4}\gamma^2}. \tag{119.12}$$

The flow law is

$$\epsilon_x = \frac{v(\sigma_0)}{\sigma_0}(\sigma_x - \tfrac{1}{2}\sigma_y), \quad \epsilon_y = \frac{v(\sigma_0)}{\sigma_0}(\sigma_y - \tfrac{1}{2}\sigma_x). \tag{119.13}$$

It is often necessary to express the stress in terms of deformation, this is done

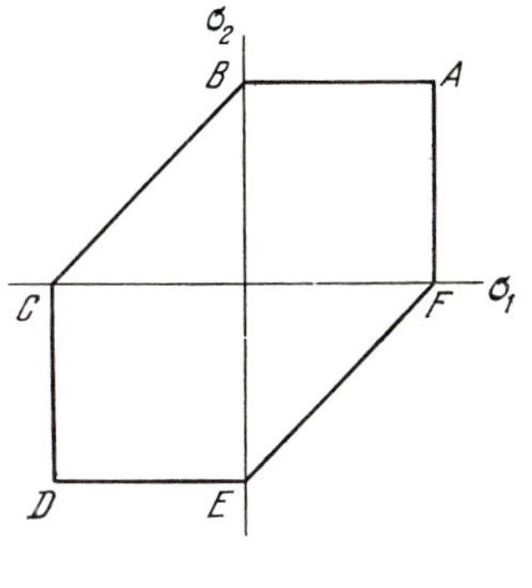

Fig. 202.

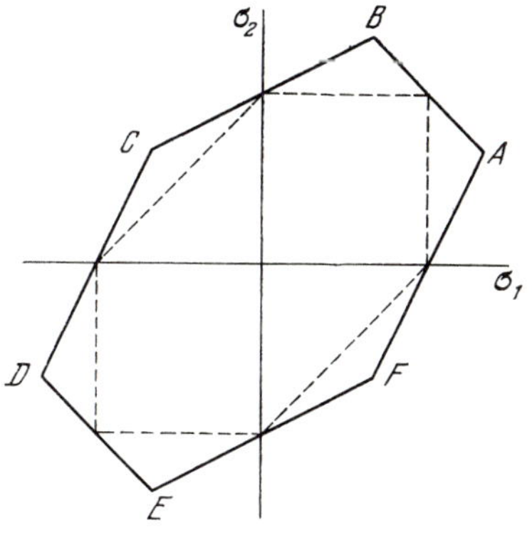

Fig. 203.

by the formulae

$$\sigma_x = \frac{4}{3}\frac{s(\epsilon_0)}{\epsilon_0}(\epsilon_x + \tfrac{1}{2}\epsilon_y)\,, \qquad \sigma_y = \frac{4}{3}\frac{s(\epsilon_0)}{\epsilon_0}(\epsilon_y + \tfrac{1}{2}\epsilon_x)\,,$$
$$\tau = \frac{1}{3}\frac{s(\epsilon_0)}{\epsilon_0}\gamma\,. \tag{119.14}$$

b) The maximum shear stress criterion. The condition τ_{max} = const is represented by a hexagon in the plane of principal stresses (fig. 202). The corresponding values of s and the flow law are

$$\begin{aligned}
&s = \sigma_1\,, \quad \epsilon_1 = v(s)\,,\ \epsilon_2 = 0\,,\\
&\qquad \epsilon_3 = -v(s)\ \text{(sides } FA \text{ and } CD)\,,\\
&s = \sigma_2\,, \quad \epsilon_1 = 0\,,\ \epsilon_2 = v(s)\,,\\
&\qquad \epsilon_3 = -v(s)\ \text{(sides } BA \text{ and } DE)\,,\\
&s = \sigma_1 - \sigma_2\,,\ \epsilon_1 = v(s)\,,\ \epsilon_2 = -v(s)\,,\\
&\qquad \epsilon_3 = 0 \quad \text{(sides } BC \text{ and } EF)\,.
\end{aligned} \tag{119.15}$$

It remains to consider the conditions represented by the angular points

$$\begin{aligned}
&s = \lambda\sigma_1 + (1-\lambda)\sigma_2 = \sigma_1\,,\\
&\epsilon_1 = \lambda v(s)\,, \quad \epsilon_2 = (1-\lambda)v(s)\,, \quad \epsilon_3 = -v(s)\,,\\
&\qquad 0 \leqslant \lambda \leqslant 1\ \text{(points } A \text{ and } D)\,,\\
&s = \lambda(\sigma_1 - \sigma_2) + (1+\lambda)\sigma_2 = \sigma_2\,,\\
&\epsilon_1 = \lambda v(s)\,, \quad \epsilon_2 = v(s)\,, \quad \epsilon_3 = -(1+\lambda)v(s)\,,\\
&\qquad 0 \leqslant \lambda \leqslant 1\ \text{(points } B \text{ and } E)\,,\\
&s = -\lambda(\sigma_1 - \sigma_2) + (1+\lambda)\sigma_1 = \sigma_1\,,\\
&\epsilon_1 = v(s)\,, \quad \epsilon_2 = \lambda v(s)\,, \quad \epsilon_3 = -(1+\lambda)v(s)\,,\\
&\qquad -1 \leqslant \lambda \leqslant 0\ \text{(points } F \text{ and } C)\,.
\end{aligned} \tag{119.16}$$

c) Referred stress criterion. The condition $\bar{\sigma}_i$ = const in the plane of principal stresses $\sigma_1 - \sigma_2$ is represented by the hexagon shown in fig. 203. The conditions on the sides of the hexagon are

$$
\begin{aligned}
s &= \tfrac{1}{2}(\sigma_1 + \sigma_2)\,, & \epsilon_1 &= \epsilon_2 = \tfrac{1}{2}\,v(s)\,, \\
& & \epsilon_3 &= -v(s) \quad \text{(sides } AB \text{ and } DE)\,, \\
s &= \sigma_1 - \tfrac{1}{2}\sigma_2\,, & \epsilon_1 &= v(s)\,, \\
& & \epsilon_2 &= -\tfrac{1}{2}\,v(s) = \epsilon_3 \quad \text{(sides } AF \text{ and } DC). \\
s &= \sigma_2 - \tfrac{1}{2}\sigma_1\,, & \epsilon_1 &= \epsilon_3 = -\tfrac{1}{2}\,v(s)\,, \\
& & \epsilon_2 &= v(s) \quad \text{(sides } CB \text{ and } EF)\,.
\end{aligned}
\tag{119.17}
$$

For states that correspond to the apexes we have

$$
\begin{aligned}
& s = \frac{\lambda}{2}(\sigma_1 + \sigma_2) + (1-\lambda)(\sigma_1 - \tfrac{1}{2}\sigma^2) = \tfrac{3}{4}\sigma_1\,, \\
& \epsilon_1 = \left(1 - \frac{\lambda}{2}\right)v(s)\,, \quad \epsilon_2 = \left(\lambda - \tfrac{1}{2}\right)v(s)\,, \quad \epsilon_3 = -\frac{1+\lambda}{2}v(s)\,, \\
& \qquad 0 \leqslant \lambda \leqslant 1 \text{ (points } A \text{ and } D)\,, \\
& s = -\frac{\lambda}{2}(\sigma_1 + \sigma_2) + (1-\lambda)(\sigma_2 - \tfrac{1}{2}\sigma_1) = \tfrac{3}{4}\sigma_2\,, \\
& \epsilon_1 = (\lambda - \tfrac{1}{2})v(s)\,, \quad \epsilon_2 = \left(1 - \frac{\lambda}{2}\right)v(s)\,, \quad \epsilon_3 = -\frac{1+\lambda}{2}v(s)\,, \\
& \qquad 0 \leqslant \lambda \leqslant 1 \text{ (points } B \text{ and } E)\,, \\
& s = \lambda(\sigma_1 - \tfrac{1}{2}\sigma_2) - (1-\lambda)(\sigma_2 - \tfrac{1}{2}\sigma_1) = \tfrac{3}{2}\sigma_1\,, \\
& \epsilon_1 = \frac{1+\lambda}{2}v(s)\,, \quad \epsilon_2 = \frac{-2+\lambda}{2}\,v(s)\,, \quad \epsilon_3 = \frac{1-2\lambda}{2}v(s)\,, \\
& \qquad 0 \leqslant \lambda \leqslant 1 \text{ (points } F \text{ and } C)\,.
\end{aligned}
\tag{119.18}
$$

120. Thick-walled tubes

Tubes subject to internal pressure and turbine discs were the first objects to which creep theory was applied. The number of publications on the subject of creep in thick-walled tubes is now quite immense. The fundamental interest of this problem is that it is the simplest possible unidimensional problem in which the stressed state is triaxial and no principal stress is zero. Therefore, it is quite natural that the case of a thick-walled tube subject to internal pressure served as a touchstone to assess and compare various creep theories, from the simplest to the most complicated.

The solution of this creep problem is particularly simple when the ends of the tube are sealed and the internal pressure causes an axial tensile force, the tube is then subject to plane deformation. The simplest method of solution is to assume in advance that the deformation is plane and then to calculate the axial force and to confirm that it is of the required magnitude. Suppose for example that a uniformly heated tube is loaded by an internal pressure q, its internal radius is a and its external b. Putting the z axis along the centre-line of the tube we have $\epsilon_z = 0$ whereas $\epsilon_\theta = u/r$, $\epsilon_r = \mathrm{d}u/\mathrm{d}r$ if $u(r)$ is the rate of radial displacement. From the condition of incompressibility

$$\epsilon_\theta + \epsilon_r = \frac{u}{r} + \frac{\mathrm{d}u}{\mathrm{d}r} = 0 ,$$

whence

$$u = \frac{\sqrt{3}}{2}\frac{c}{r}, \qquad \epsilon_r = -\frac{\sqrt{3}}{2}\frac{c}{r^2}, \qquad \epsilon_\theta = \frac{\sqrt{3}}{2}\frac{c}{r^2} .$$

Here c is a constant of integration, the factor $\sqrt{3}/2$ is introduced for convenience.

Now calculate ϵ_0 by formula (119.4). We find that $\epsilon_0 = c/r^2$. From the creep law eq. (119.5) it follows that:

$$\sigma_\theta - \sigma_r = \frac{4}{3}\frac{s(\epsilon_0)}{\epsilon_0}\epsilon_\theta , \qquad \sigma_z - \sigma = 0 ,$$

whence

$$\sigma_\theta - \sigma_r = \frac{2}{\sqrt{3}} s\left(\frac{c}{r^2}\right), \qquad \sigma_z = \tfrac{1}{2}(\sigma_r + \sigma_\theta) . \tag{120.1}$$

In cylindrical coordinates the equation of equilibrium of an element is

$$\frac{\mathrm{d}\sigma_r}{\mathrm{d}r} + \frac{\sigma_r - \sigma_\theta}{r} = 0 .$$

Using the first of eqs. (120.1) we obtain

$$\mathrm{d}\sigma_r = \frac{2}{\sqrt{3}} s\left(\frac{c}{r^2}\right)\frac{\mathrm{d}r}{r} = -\frac{1}{\sqrt{3}}\frac{s(v)}{v}\mathrm{d}v \qquad \left(v = \epsilon_0 = \frac{c}{r^2}\right) .$$

Integrating this equation with the initial condition that $\sigma_r(a) = -q$, we obtain

$$\sigma_r = -q - \frac{1}{\sqrt{3}}\int_{v_a}^{v}\frac{s(v)\,\mathrm{d}v}{v} . \tag{120.2}$$

We now easily find σ_θ and σ_z:

$$\sigma_\theta = -q - \frac{1}{\sqrt{3}} \int_{v_a}^{v} \frac{s(v)\,dv}{v} + \frac{2}{\sqrt{3}} s(v)\,, \tag{120.3}$$

$$\sigma_z = -q - \frac{1}{\sqrt{3}} \int_{v_a}^{v} \frac{s(v)\,dv}{v} + \frac{1}{\sqrt{3}} s(v)\,. \tag{120.4}$$

The quantity v contains an unknown constant c, which should be determined from the boundary condition $\sigma_r(b) = 0$, which leads to the equation

$$-q - \frac{1}{\sqrt{3}} \int_{v_a}^{v_b} \frac{s(v)\,dv}{v} = 0\,. \tag{120.5}$$

In the general case if $s(v)$ is some function given graphically, in a table, or as a formula that is inconvenient for calculation, the following procedure is recommended. Determine the function

$$\psi(x) = \int_0^x \frac{s(x)\,dx}{x}\,.$$

This integral is convergent for any reasonable creep law that suitably describes the region of small values of x. Then eq. (120.5) becomes

$$q\sqrt{3} = \psi(v_a) - \psi(v_b)\,. \tag{120.6}$$

On the other hand $v_a = c/a^2$, $v_b = c/b^2$, therefore $v_b/v_a = a^2/b^2$. If we construct a graph of the function $\psi(x)$ we can readily use a graphical method to construct a series of curves of the relationship between q and v_a for different values of a/b. Indeed, since

$$q\sqrt{3} = \psi(v_a) - \psi\left(\frac{a^2}{b^2} v_a\right),$$

by fixing a value of a^2/b^2 we establish a proportional circle such that the ratio of the measured segments is equal to the given value, a^2/b^2. Measuring

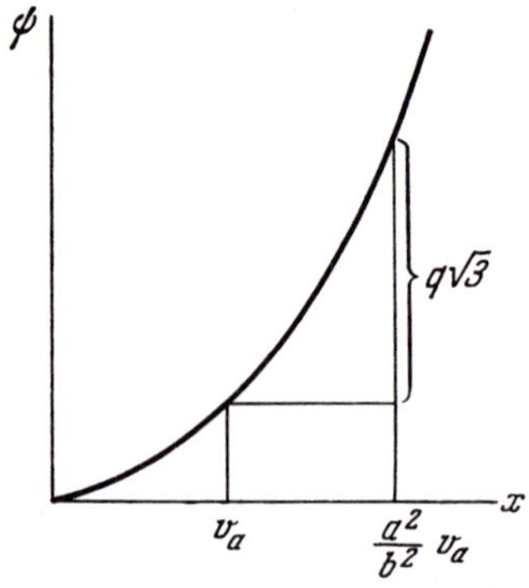

Fig. 204.

the ordinate that corresponds to any value of v_a (fig. 204), we immediately measure the ordinate that corresponds to the abscissa $a^2/b^2 \cdot v_a$; the difference between these ordinates is $q\sqrt{3}$, and therefore we construct a graph of q against v_a for the given ratio of the radii.

With a power creep law all the formulae are written in closed form. Put $s = \sigma_n(\epsilon/\epsilon_n)^{1/n}$, then $\psi(x) = n\sigma_n(x/\epsilon_n)^{1/n}$, consequently

$$\sigma_n \left(\frac{v_a}{\epsilon_n}\right)^{1/n} = \frac{q\sqrt{3}}{n\left[1 - \left(\frac{a}{b}\right)^{2/n}\right]}. \tag{120.7}$$

By formulae (120.2), (120.3), and (120.4) we readily find

$$\sigma_r = \frac{q}{\left(\frac{b}{a}\right)^{2/n} - 1}\left[1 - \left(\frac{b}{r}\right)^{2/n}\right],$$

$$\sigma_\theta = \frac{q}{\left(\frac{b}{a}\right)^{2/n} - 1}\left[1 - \left(1 - \frac{2}{n}\right)\left(\frac{b}{r}\right)^{2/n}\right], \tag{120.8}$$

$$\sigma_z = \frac{q}{\left(\frac{b}{a}\right)^{2/n} - 1}\left[1 - \left(1 + \frac{1}{n}\right)\left(\frac{b}{r}\right)^{2/n}\right].$$

Now check the statement that the stress distribution that has been found corresponds to a pipe with sealed ends. Calculate the axial stress

$$P = 2\pi \int_a^b \sigma_z r\,\mathrm{d}r = -\pi c \int_{v_a}^{v_b} \sigma_z \frac{\mathrm{d}v}{v^2}.$$

Substitute into this the expression for σ_z from formula (120.4) and note that in integrating the first term in the right-hand side which is $-q$, there is no need to change to the variable of integration v, because this term gives a component in the integral equal to $-\pi q(b^2 - a^2)$. The result of integrating the remaining parts is

$$\int_{v_a}^{v_b} \frac{\mathrm{d}v}{v^2} \int_{v_a}^{v} \frac{s(v)}{v}\,\mathrm{d}v - \int_{v_a}^{v_b} s(v)\frac{\mathrm{d}v}{v^2} = -\frac{1}{v_b}\int_{v_a}^{v_b} \frac{s(v)}{v}\,\mathrm{d}v$$

$$+ \int_{v_a}^{v_b} \frac{s(v)}{v^2}\,\mathrm{d}v - \int_{v_a}^{v_b} \frac{s(v)\,\mathrm{d}v}{v^2} = \frac{q\sqrt{3}}{v_b} = \frac{q\sqrt{3}}{c} b^2 .$$

Therefore

$$P = -\pi q(b^2 - a^2) + \pi q b^2 = \pi q a^2 ,$$

which demonstrates the assumption made. It will be seen that this elementary solution is a simple paraphrase of the problem of plastic deformation of tubes, solved on the basis of the deformation theory of plasticity. Consideration of more complicated cases, when the axial stress is zero (an open tube) or the axial stress is given in some other way, is based on the same principles but the axial rate of deformation is assumed different from zero. The solution of the general form of this problem is given in the book [43] and also in various forms by other authors.

In conclusion let us consider what results can be obtained by the use of Tresca's criterion. If $\sigma_\theta > \sigma_z > \sigma_r$ we must put $\sigma_\theta - \sigma_r = s(\epsilon_\theta - \epsilon_r)$ and the entire difference consists in the value of the constant c, which means that with the same creep law for tension the numerical value of the creep rate is different. As before the stress distribution is given by formulae (120.2) and (120.3), but the axial stress σ_z is not in any way fixed and it can assume any value between σ_θ and σ_r. This indeterminancy is immediately avoided if we assume that creep deformation is accompanied by elastic deformation, and within the framework of steady-state creep it is necessary to be satisfied with this remark.

121. **Pure bending of part of a circular cylinder**

Consider part of a circular cylinder cut out by two meridional sections and loaded by moments in a plane perpendicular to the centre line which are uniformly distributed along the generators. In the elastic case this is the problem of pure bending of a curved beam, under creep conditions the difference between the plane stressed and plane deformed states is important and the problem is no longer that of a curved beam. This latter problem for plasticity and creep has been considered by a number of authors with the usual assumptions of the approximate theory (see for example [52]); under conditions of plane deformation, that is for a sufficiently long cylinder, the procedure of calculation is of course identical and the final functions are analogous. However, there is a very simple accurate solution of this problem similar to Golovin's solution for bending an elastic rod of rectangular cross-section. This reveals certain typical features which are absent from the approximate solution. Let a be the internal radius of the cylinder and b the external radius. To be specific consider that the bending moment tends to straighten the beam (fig. 205). We seek a state of stress that does not depend on the angle θ. Here, of course, the displacements remain functions of both r and θ. Considering a state of stress that is independent of θ we take the tangential stress τ as zero, and consequently, the remaining stresses σ_r and σ_θ satisfy exactly the same equation of equilibrium as for axially symmetrical deformation of a tube:

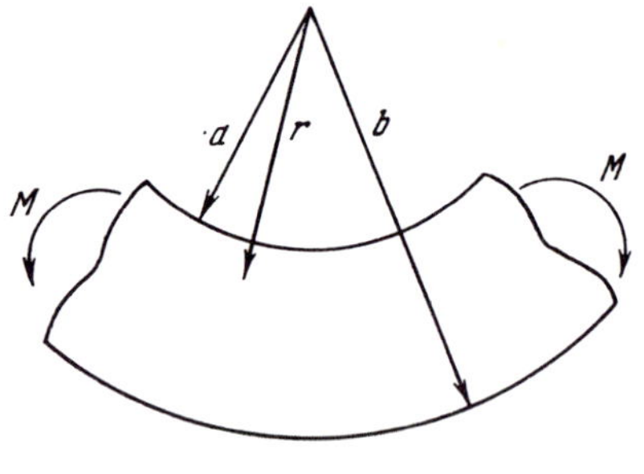

Fig. 205.

$$\frac{d\sigma_r}{dr} + \frac{\sigma_r - \sigma_\theta}{r} = 0 . \tag{121.1}$$

The usual expressions for the components of deformation in polar coordinates are

$$\epsilon_r = \frac{\partial u}{\partial r}, \quad \epsilon_\theta = \frac{u}{r} + \frac{1}{r}\frac{\partial v}{\partial \theta}, \quad \gamma = \frac{1}{r}\frac{\partial u}{\partial \theta} + \frac{\partial v}{\partial r} - \frac{v}{r}.$$

From the condition of incompressibility when $\epsilon_z = 0$ it follows that:

$$\epsilon_r + \epsilon_\theta = 0 .$$

On the other hand with the Mises criterion, and with the Saint Venant criterion, the difference between the stresses $\sigma_\theta - \sigma_r$ is a function of the difference between the rates of deformation $\epsilon_\theta - \epsilon_r$, and therefore, $\epsilon_\theta - \epsilon_r$ should depend only on r. Therefore, we may write

$$\epsilon_\theta - \epsilon_r = 2\varphi'(r) .$$

Bearing in mind the condition of incompressibility we find

$$\epsilon_r = \frac{\partial u}{\partial r} = -\varphi'(r) , \qquad \epsilon_\theta = \frac{u}{r} + \frac{1}{r}\frac{\partial v}{\partial \theta} = \varphi'(r) .$$

Integrating the first equation we obtain

$$u = -\varphi(r) + h'(\theta) .$$

Substituting this value of u into the second equation we find the function v:

$$v = \theta\,[r\varphi'(r) + \varphi(r)] - h(\theta) + g(r) .$$

Now substitute the expressions that have been derived for u and v into the formula for γ and use the condition $\gamma = 0$. We obtain

$$\frac{1}{r}\,[h''(\theta) + h(\theta)] + \theta\left[r\varphi''(r) + \varphi'(r) - \frac{\varphi}{r}\right] + g'(r) - \frac{g(r)}{r} = 0 .$$

Hence it follows that

$$h''(\theta) + h(\theta) = C\theta\,, \qquad \varphi''(r) + \frac{\varphi'}{r} - \frac{\varphi}{r^2} + \frac{C}{r^2} = 0\,. \tag{121.2}$$

The constant C can be made zero because in determining stress it suffices to know the function $\varphi(r)$ to within an additive constant. Integrating (121.2) we find

$$2\varphi'(r) = \frac{B}{r^2} - A\,.$$

Consequently

$$\epsilon_\theta - \epsilon_r = \frac{B}{r^2} - A\,. \tag{121.3}$$

To proceed further with the solution we use the criterion of maximum tangential stress and put

$$\sigma_\theta - \sigma_r = \sigma_n \left(\frac{\epsilon_\theta - \epsilon_r}{\epsilon_n}\right)^{1/n}.$$

Bearing in mind eq. (121.3) we can rewrite this equation as

$$\sigma_\theta - \sigma_r = \sigma_n \epsilon_n^{-1/n} \left(\frac{B}{r^2} - A\right)^{1/n}.$$

This latter formula is valid as long as $\epsilon_\theta - \epsilon_r > 0$. This applies in the region under tension near to the surface $r = a$. At a certain value of $r = c$ the quantity $\epsilon_\theta - \epsilon_r$ becomes zero and we have

$$\frac{B}{r^2} - A = 0\,.$$

Hence $B = Ac^2$. Therefore, in the region of tension

$$\sigma_\theta - \sigma_r = \sigma_n \left(\frac{A}{\epsilon_n}\right)^{1/n} \left(\frac{c^2}{r^2} - 1\right)^{1/n}. \tag{121.4}$$

In the region of compression $\epsilon_\theta - \epsilon_r < 0$, consequently we should put

$$\sigma_\theta - \sigma_r = -\sigma_n \left(\frac{A}{\epsilon_n}\right)^{1/n} \left(1 - \frac{c^2}{r^2}\right)^{1/n}. \tag{121.5}$$

Write the equation of equilibrium for the region of tension, substituting eq. (121.4) into eq. (121.1), we obtain

$$\frac{d\sigma_r}{dr} - \frac{2m}{nr}\left(\frac{c^2}{r^2} - 1\right)^{1/n} . \tag{121.6}$$

Here we use the notation

$$m = \frac{n\sigma_n}{2}\left(\frac{A}{\epsilon_n}\right)^{1/n} . \tag{121.7}$$

Change variables putting

$$\frac{c^2}{r^2} - 1 = u^2 .$$

We obtain

$$d\sigma_r = -m\,\frac{u^n}{1+u^n}\,du .$$

When $r = c$, $u = 0$, when $r = a$, $u_a = (\alpha - 1)^{1/n}$, where $\alpha = c^2/a^2 > 1$; therefore

$$\begin{aligned} \sigma_r &= \sigma^* - m\int_0^u \frac{u^n\,du}{1+u^n} , \\ \sigma_\theta &= \frac{2m}{n}u + \sigma^* - m\int_0^u \frac{u^n\,du}{1+u^n} . \end{aligned} \tag{121.8}$$

Here σ^* denotes the radial stress σ_r when $r = c$.

Calculations are analogous for the zone of compression and the equation of equilibrium is written

$$\frac{d\sigma_r}{dr} + \frac{2m}{nr}\left(1 - \frac{c^2}{r^2}\right)^{1/n} = 0 .$$

Now we put

$$1 - \frac{c^2}{r^2} = u^n ,$$

and obtain the following expressions for stress:

$$\sigma_r = \sigma^* - m \int_0^u \frac{u^n \, du}{1-u^n},$$

$$\sigma_\theta = -\frac{2m}{n} u + \sigma^* - m \int_0^u \frac{u^n \, du}{1-u^n}. \tag{121.9}$$

When $r=c$, $u=0$, when $r=b$, $u_b=(1-\beta)^{1/n}$ where $\beta=c^2/b^2<1$. In using these formulae for calculations we may proceed as follows. Introduce the notation

$$\varphi(u) = \int_0^u \frac{u^n \, du}{1+u^n},$$

$$\psi(u) = \int_0^u \frac{u^n \, du}{1-u^n}.$$

The stress σ_r becomes zero when $r=a$ and $r=b$, consequently, when $u=u_a$ and $u=u_b$. By formula (121.8) and (121.9) we obtain

$$\sigma^* - m\varphi(u_a) = \sigma^* - m\psi(u_b) = 0.$$

Hence it follows that

$$\varphi(u_a) = \psi(u_b) = \frac{\sigma^*}{m}.$$

Graphs are constructed for the functions $\varphi(u)$ and $\psi(u)$ and the values u_a and u_b are noted which correspond to the same value of the ordinate on both curves, each such ordinate is σ^*/m. In this way each value of σ^*/m corresponds to a pair of numbers u_a and u_b. But $u_a=(c^2/a^2-1)^{1/n}$, $u_b=(1-c^2/b^2)^{1/n}$; eliminating c we find that the given quantity σ^*/m corresponds to a certain ratio of b/a, namely

$$\lambda^2 = \frac{b^2}{a^2} = \frac{1+u_a^n}{1-u_b^n}.$$

We have thus found σ^* as a function of λ: $\sigma^* = mC(\lambda)$. Now the stress formulae may be rewritten:

for the zone of tension:

$$\frac{\sigma_r}{m} = C(\lambda) - \varphi(u)\,, \qquad \frac{\sigma_\theta}{m} = \frac{2}{n}u + C(\lambda) - \varphi(u)\,; \tag{121.10}$$

for the zone of compression:

$$\frac{\sigma_r}{m} = C(\lambda) - \psi(u)\,, \qquad \frac{\sigma_\theta}{m} = -\frac{2}{n}u + C(\lambda) - \psi(u)\,.$$

The functions $\varphi(u)$ and $\psi(u)$ can easily be written in closed form for whole-numbered values of n. However, the resulting expressions are not very convenient for calculations and therefore, in considering the particular example for $n = 3$ it was found convenient to use numerical integration, first separating certain terms. Accordingly, it was taken that

$$\varphi = u - \int_0^u \frac{du}{1 + u^3}\,,$$

$$\psi = -u - \tfrac{1}{3}\ln(1 - u) + \tfrac{1}{3}\int_0^u \frac{2 + u}{1 + u + u^2}\,du\,.$$

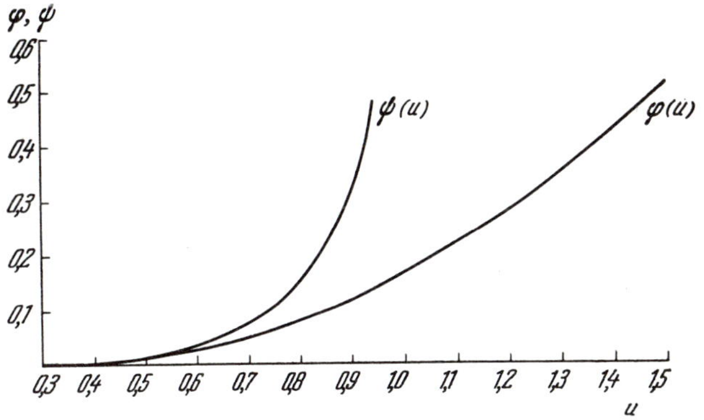

Fig. 206.

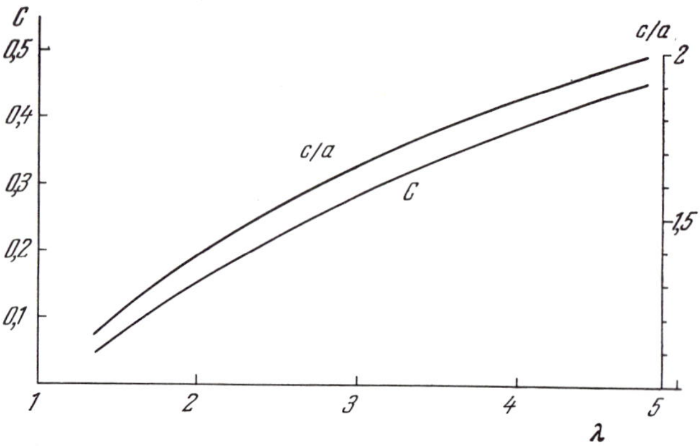

Fig. 207.

Graphs of the functions $\varphi(u)$ and $\psi(u)$ are shown in fig. 206. This figure and the procedure described above were used to determine the dependence of the function C on $\lambda = b/a$, this relationship is plotted in fig. 207, which also gives the ratio c/a as function of λ. Stress diagrams of σ_θ and σ_r for $\lambda = 4$ are shown in fig. 208. It is interesting to note that the maximum stress σ_θ is found at some inner point and not when $r = a$, as it was in the elastic case and as is always found by the approximate theory. This result is not unexpected, indeed with increase in n the stress distribution tends to that in a body of ideally plastic material. The corresponding solution for the case of ideal

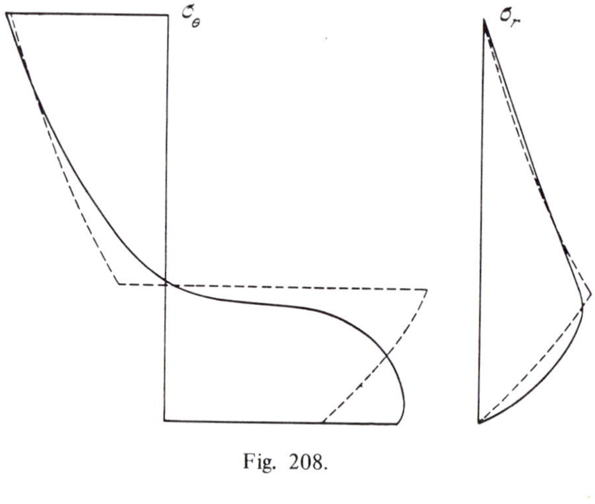

Fig. 208.

plasticity is immediately obtained. Putting $\sigma_\theta - \sigma_r = \sigma_T$ in the region of tension and $\sigma_\theta - \sigma_r = -\sigma_T$ in the region of compression we immediately find that in the first region

$$\sigma_r = \sigma_T \ln\frac{r}{a}, \qquad \sigma_\theta = \sigma_T\left(1 + \ln\frac{r}{a}\right).$$

In the compression region

$$\sigma_r = -\sigma_T \ln\frac{r}{b}, \qquad \sigma_\theta = -\sigma_T\left(1 + \ln\frac{r}{b}\right).$$

From the condition of continuity of σ_r on the boundary of the region when $r = c$ we obtain

$$c = \sqrt{ab}\,.$$

The bending moment is easily calculated. The dotted line in fig. 208 corresponds to the stress distribution when $n = \infty$ for the same value of the bending moment as when $n = 3$.

We now find the value of the radius r for which σ_θ is maximum. Differentiating the second of the formulae (121.8) with respect to r and equating the derivative to zero we obtain

$$\frac{2}{n} - \frac{u^n}{1 + u^n} = 0\,.$$

Whence

$$u^n = \frac{c^2}{r^2} - 1 = \frac{2}{n-2},$$

or

$$\frac{c}{r} = \sqrt{\frac{2}{n-2}}. \qquad (121.11)$$

From eq. (121.11) it first follows that a maximum exists only when $n > 2$. If this condition is fulfilled the maximum stress occurs at an internal point, when the value of r given by formula (121.11) is greater than a, that is for the case of sufficiently large curvature. For example, when $n = 3$ we obtain $c/r = 1.73$ and it will be seen from the graph of fig. 207 that this corresponds to $\lambda = 3.33$.

This result is interesting in that in bending cylindrical elements of this kind the region of high tensile stress penetrates to a considerable depth so that crack formation may be intensive.

122. **Rotating discs**

The case of a rotating turbine disc was the first in which serious allowance had to be made for creep, and considerable attention has been devoted to it since Bailey's early work. The special feature of this problem is that the stress in the disc results from acceleration in rotary motion but external loads are not great. Therefore, thickening the disc does not reduce the stress, the most favourable operating conditions for the material are only achieved by improving the profile and the possibilities of reducing the working stress are limited. The material unavoidably works at the limit and consequently the accuracy required of the calculations is very high. On the other hand, the configuration of a turbine disc is quite simple, the method of manufacture of discs has been worked out in great detail and technological factors have been well studied. As a result, the differences between calculations and real designs are minimal. Certain typical disc profiles that are used in steam and gas turbines are shown in fig. 209. The classic design of a disc is shown in fig. 209a, a heavy hub is provided to shrink it onto the shaft and in the design scheme, depending on the stiffness of the hub, the disc is either considered as extending to the shaft, as shown dotted, or the hub is considered to be absolutely rigid and displacements of points on the radius equal to the outer radius of the hub is assumed zero, or finally the actual conditions of contact between disc and hub are considered. They are usually assumed to be elastic. The justification for this

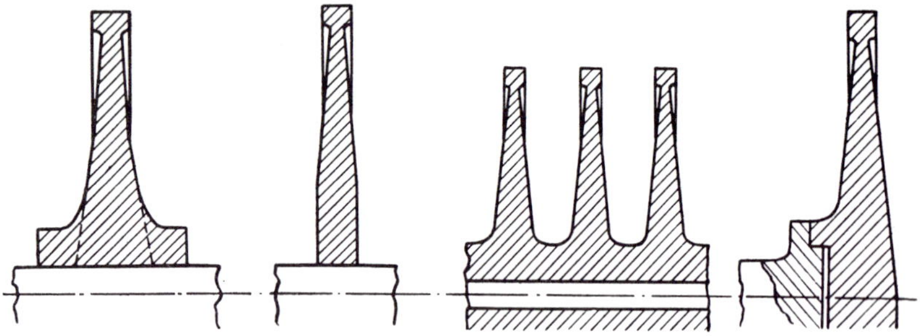

a) b) c) d)

Fig. 209.

latter scheme is that the disc is hottest close to the rim and the hub temperature is usually low. The tendency to reduce the size of machines and to improve the system of glands makes it necessary to avoid long hubs. A diagram of a design that is sometimes used for smaller steam turbines is shown in fig. 209b, the disc is mounted directly on the shaft with a shrink fit. In order to tighten the fit initial stresses are set up in the disc by plastic deformation.

A solid forged rotor is illustrated in fig. 209c, the discs and shaft form a single piece. The shaft usually has an axial bore which increases the working stress somewhat but which makes it possible to inspect for flaws in the central part of the rotor metal, where they are most probable because of the manufacturing conditions. Finally, a single-piece disc that is fixed to the shaft by a bracket flange is shown in fig. 209d.

The peripheral rim of the disc is always thickened and the blades are fixed to it in some way. Inertia forces of the blades create a distributed radial load on the rim. The rim is sometimes considered as a ring and the creep conditions must then be formulated separately for a disc subject to an unknown load transmitted from the rim, and for the rim considered as a rotating ring subject to an internal reaction from the disc and an external load from the blades. A simpler procedure is to neglect the load-carrying capacity of the rim and to consider the disc as loaded on its periphery by known centrifugal forces of inertia of the rim and blades.

Correct formulation of the boundary conditions for these schemes is not always possible within the framework of the theory of steady-state creep. For example, in case a) consideration of the interaction between the elastic hub and creep of the disc has meaning only if we allow for elastic deformation of the disc itself, that is we are concerned with transient creep. In case b) the firm fit of the disc on the shaft ensures that the disc is subject to elastic deformation and the problem of calculation is not so much to assess the stress in the disc or to determine its deformation after a given time, as to investigate relaxation of the fit, that is to determine the time at which the contact pressure ceases to ensure a reliable fit. It is probably simplest to apply the concepts of steady-state creep to case d).

The selection of disc profile essentially depends on the stress distribution in the elastic disc, and formerly when creep calculations were not made, particular attention was devoted to the question of optimum profile. Creep tends to equalise the stresses and, therefore, under creep conditions, the influence of profile is not so important and modern discs usually have a fairly simple profile, mainly conical, because this offers manufacturing advantages.

As a rule the temperature distribution in discs is irregular, the rim is heated much more than the central part and it is important to allow for temperature irregularities.

In calculating a disc the designer is usually concerned with materials that have been thoroughly studied and for which creep curves are available over a long time for various temperatures and stresses. The use of any kind of formula for creep rate as function of stress and temperature unavoidably distorts the experimental results; at the same time, all fairly complicated real problems must be solved by numerical methods. When numerical methods are used the advantages of simple creep laws, whether power or exponential, are not particularly great and, therefore, the most valuable numerical methods are those which use dependences of creep rate on stress and temperature determined directly from experiment. This is particularly important because the calculation of transient creep of discs on the basis of the theory of ageing using isochronous curves proceeds in precisely the same way as the calculation of steady-state creep, except that deformations are considered instead of rates and deformations against stress in a given time instead of the dependence of rate of deformation on stress (see Chapter 11).

The large number of methods of designing discs that have been proposed may be classified into two groups:

a) accurate solutions of differential equations (in closed form or by numerical integration);

b) various methods of successive approximation.

The second group of methods is most widely used in design practice and they are more convenient and simpler for manual calculation. Methods of the first group are becoming more promising because of the increasing use of computers.

123. **Creep of a solid disc of constant thickness**

In a series of works Wahl [377, 378, 379] developed methods of calculating rotating discs of both constant and variable thickness by means of the Tresca condition and associated flow laws. A fairly detailed presentation of the solution is contained in a book [52] and, therefore, we shall not repeat the presentation here; to illustrate the procedure, however, we consider certain fairly simple problems using the exponential creep law instead of the power law that Wahl used. As we have seen the exponential law offers some advantage in that the equations, which are presented in dimensionless form, contain no empirical constants and the solutions are quite universal. Consider first a solid disc of radius b rotating at an angular velocity ω. The blade load develops a radial stress $\sigma_r(b)$, which is proportional to ω^2, on the outer rim of the disc. The equation of equilibrium of a disc element is

$$\frac{\mathrm{d}}{\mathrm{d}r}(r\sigma_r) - \sigma_\theta + \rho\omega^2 r^2 = 0 .$$

Here ρ is the density of the disc material. As we shall use an exponential creep law

$$\epsilon = \epsilon_e \exp\left(\frac{\sigma}{\sigma_e}\right),$$

we determine the dimensionless stresses as follows:

$$\hat{\sigma}_r = \frac{\sigma_r}{\sigma_e}, \qquad \hat{\sigma}_\theta = \frac{\sigma_\theta}{\sigma_e} .$$

We also introduce the dimensionless radius $\xi = r/b$ and we shall consider the dimensionless velocity $\hat{u} = u/b\epsilon_e$ instead of the rate of radial displacement of points on the disc. The equation of equilibrium is rewritten

$$\frac{\mathrm{d}}{\mathrm{d}\xi}(\xi\hat{\sigma}_r) - \hat{\sigma}_\theta + m\xi^2 = 0 . \tag{123.1}$$

Here $m = \rho\omega^2 b^2/\sigma_e$. As the blade load is proportional to ω^2 we can always take $\sigma_r(b) = mq\sigma_e$, and therefore the boundary condition when $\xi = 1$ is $\hat{\sigma}_r(1) = mq$.

The dimensionless rates of deformation $\hat{\epsilon}_\theta = \epsilon_\theta/\epsilon_e$, and $\hat{\epsilon}_r = \epsilon_r/\epsilon_e$ are determined as follows:

$$\hat{\epsilon}_\theta = \frac{\hat{u}}{\xi}, \qquad \hat{\epsilon}_r = \frac{\mathrm{d}\hat{u}}{\mathrm{d}\xi} .$$

At the centre of the disc $\hat{\sigma}_\theta = \hat{\sigma}_r$, consequently in the region surrounding the centre we may have either $\hat{\sigma}_\theta > \hat{\sigma}_r > 0$, or $\hat{\sigma}_r > \hat{\sigma}_\theta > 0$, or $\hat{\sigma}_\theta = \hat{\sigma}_r$. The second possibility can be eliminated at once, indeed in this case we should have $\hat{\epsilon}_\theta = 0$, consequently, $\hat{u} = 0$ on the last segment, but then on this segment we should also have $\mathrm{d}\hat{u}/\mathrm{d}\xi = \hat{\epsilon}_r = 0$ and no creep takes place. The first possibility is eliminated for the same reason, if on a finite segment $\mathrm{d}\hat{u}/\mathrm{d}\xi = \hat{\epsilon}_r = 0$ the quantity $\hat{u}$ is also zero, since $\hat{u}(0) = 0$, and, therefore, also $\hat{\epsilon}_\theta = 0$. It remains to assume that in the region $0 \leqslant \xi \leqslant \alpha$, $\hat{\sigma}_\theta = \hat{\sigma}_r = \hat{\sigma}$. It immediately follows from the equation of equilibrium that

$$\hat{\sigma} = C - \tfrac{1}{2} m\xi^2 . \tag{123.2}$$

If the two principal stresses are equal, as was shown in section 119, the creep law must be taken as

$$\epsilon_\theta + \epsilon_r = v(\sigma) .$$

For an exponential law in dimensionless variables we obtain the following result:

$$\frac{d\hat{u}}{d\xi} + \frac{\hat{u}}{\xi} = \exp \hat{\sigma} = \exp C \cdot \exp\left(-\frac{m\xi^2}{2}\right). \tag{123.3}$$

Integrate this differential equation for the displacement $\hat{u}(\xi)$ putting $\hat{u}(0) = 0$, we find

$$\hat{u} = \frac{1}{m} \exp C \frac{1}{\xi}\left[1 - \exp\left(-\frac{m\xi^2}{2}\right)\right]. \tag{123.4}$$

Now calculate the rate of radial deformation

$$\hat{\epsilon}_r = \frac{d\hat{u}}{d\xi} = \frac{\exp C}{m} \left\{-\frac{1}{\xi^2}\left[1 - \exp\left(-\frac{m\xi^2}{2}\right)\right] + m \exp\left(-\frac{m\xi^2}{2}\right)\right\}. \tag{123.5}$$

The boundary of the region in which $\hat{\sigma}_\theta = \hat{\sigma}_r$ is determined from the condition that $\hat{\epsilon}_r$ becomes zero when $\xi = \alpha$, then when $\xi > \alpha$ we still have $\hat{\epsilon}_r = 0$ and therefore $\hat{\sigma}_\theta > \hat{\sigma}_r$. In eq. (123.5) putting $\xi = \alpha$ and $\hat{\epsilon}_r = 0$, we arrive at the equation

$$\exp x_0 = 1 + 2x_0 . \tag{123.6}$$

Here $x_0 = m\alpha^2/2$.

The root of this equation is readily calculated, namely $x_0 = 1.256$, therefore

$$m\alpha^2 = 2.512 .$$

We shall now distinguish two cases:

1. $m < 2x_0 = 2.512$. The stresses are equal throughout the disc when $0 \leqslant \xi \leqslant 1$, the constant C in formula (123.2) is determined from the condi-

tion $\hat{\sigma}_r(1) = mq$, consequently

$$\hat{\sigma}_r = \hat{\sigma}_\theta = m\left[q + \frac{1-\xi^2}{2}\right].$$

2. $m > 2x_0 = 2.512$, consequently, $\alpha^2 = 2.512/m < 1$. When $\xi > \alpha$ we should have $\hat{\sigma}_\theta > \hat{\sigma}_r > 0$, $\epsilon_r = 0$, consequently $\hat{u} = \hat{u}(\alpha)$ and $\hat{\epsilon}_\theta = \hat{u}(\alpha)/\xi$. The quantity $\hat{u}(\alpha)$ is determined from the condition of continuity of displacement when $\xi = \alpha$, that is from formula (123.4) when $\xi = \alpha$, namely

$$\frac{\hat{u}(\alpha)}{\alpha} = \exp(C - x_0).$$

Here we take account of condition (123.6).

Now the rate of peripheral deformation when $\xi > \alpha$ is expressed as

$$\hat{\epsilon}_\theta = \frac{\hat{u}(\alpha)}{\xi} = \frac{\hat{u}(\alpha)}{\alpha}\frac{\alpha}{\xi} = \frac{\alpha}{\xi}\exp(C - x_0).$$

In accordance with the creep law $\hat{\epsilon}_\theta$ depends only on $\hat{\sigma}_\theta$, namely $\hat{\epsilon}_\theta = \exp \hat{\sigma}_\theta$, therefore

$$\hat{\sigma}_\theta = C - x_0 + \ln\frac{\alpha}{\xi}.$$

Substituting this expression for $\hat{\sigma}_\theta$ into the equation of equilibrium (123.1) and integrating, we obtain

$$\hat{\sigma}_r = C - x_0 + \ln\frac{\alpha}{\xi} + 1 - \frac{m\xi^2}{3} + \frac{D}{\xi}.$$

To determine the constants of integration, C and D, we use the condition of continuity of the stress $\hat{\sigma}_r$ when $\xi = \alpha$ and the boundary conditions when $\xi = 1$. From the first condition it follows that

$$D = -\alpha(1 - \tfrac{2}{3}x_0).$$

From the second condition

$$C - x_0 = C' = m(q + \tfrac{1}{3}) - 1 + \alpha\left(1 - \frac{2x_0}{3}\right) - \ln\alpha =$$

$$= m(q + \tfrac{1}{3}) - 1 + 0.164\alpha - \ln\alpha.$$

Now the stress formulae can be rewritten as:

when $\xi < \alpha$:

$$\hat{\sigma}_r = \hat{\sigma}_\theta = C' + 1.256\left(1 - \frac{\xi^2}{\alpha^2}\right); \tag{123.7}$$

when $\xi > \alpha$:

$$\hat{\sigma}_r = C' - \ln\frac{\xi}{\alpha} + \left(1 - \frac{\xi}{\alpha}\right) + 0.836\left(\frac{\alpha}{\xi} - \frac{\xi^2}{\alpha^2}\right), \tag{123.8}$$

$$\hat{\sigma}_\theta = C' - \ln\frac{\xi}{\alpha}.$$

Check fulfilment of the condition $\hat{\sigma}_\theta > \hat{\sigma}_r$. The difference

$$\hat{\sigma}_r - \hat{\sigma}_\theta = \left(1 - \frac{\alpha}{\xi}\right)\left[1 - \frac{2x_0}{3}\left(1 + \frac{\xi}{\alpha} + \frac{\xi^2}{\alpha^2}\right)\right].$$

This expression is negative for any value of $\xi > \alpha$.

124. Creep of a disc with a central hole

Now consider the problem of creep in a disc of constant thickness with a central hole. Retaining the same dimensionless notation that was used in application to a solid disc we understand by α the relative internal radius of the disc, when $\xi = \alpha$ the radial stress is taken to be zero whereas when $\xi = 1$ as before, $\hat{\sigma}_r = mq$. Suppose that throughout the disc, that is when $\alpha \leqslant \xi \leqslant 1$, $\hat{\sigma}_\theta > \hat{\sigma}_r > 0$. Then the solution is written in exactly the same way as for a solid disc in the region $\xi > \alpha$, namely

$$\hat{\sigma}_\theta = C + \ln\frac{\alpha}{\xi},$$

$$\hat{\sigma}_r = C + \ln\frac{\alpha}{\xi} + 1 - \frac{m\xi^2}{3} - \frac{D}{\xi}. \tag{124.1}$$

In writing out these equations we have slightly altered the notation. From the boundary conditions $\hat{\sigma}_r(\alpha) = 0$ and $\sigma_r(1) = mq$ we find the constants

$$C = \frac{m}{1-\alpha}\left[q + \frac{1+\alpha+\alpha^2}{3}\right] - 1 - \frac{\ln\alpha}{1-\alpha},$$
$$D = m\left[q\,\frac{\alpha}{1-\alpha} + \frac{\alpha(1+\alpha)}{3}\right] - \frac{\alpha\ln\alpha}{1-\alpha}. \tag{124.2}$$

It remains to determine the limits of applicability of the solution. It is postulated that

$$\hat{\sigma}_r - \hat{\sigma}_\theta = 1 - \frac{m\xi^2}{3} - \frac{D}{\xi} < 0\,.$$

This inequality can only be infringed on the boundary or in the region $\alpha < \xi < 1$. In the first case it reduces to the conditions

$$D > 1 - \frac{m\alpha^3}{3}, \qquad D > 1 - \frac{m}{3}.$$

If we substitute expression (124.2) for D into these inequalities we find that the first inequality is always fulfilled whereas the second applies a certain limitation on m and q. When $q = 0$ the quantity D assumes a minimum value and, therefore, we need only find out from what values of m the solution that we have found is valid in the absence of a blade load; the presence of this load reduces the lower boundary for m. We obtain

$$m > 3\,\frac{1 + \dfrac{\alpha\ln\alpha}{1-\alpha}}{1+\alpha+\alpha^2}. \tag{124.3}$$

Now suppose that the maximum value of $\hat{\sigma}_r - \hat{\sigma}_\theta$ is achieved in the region $\alpha < \xi < 1$, and let us require that this value remains negative. As a result we arrive at the condition

$$\frac{\alpha(1+\alpha)}{3}\,m - \frac{\alpha\ln\alpha}{1-\alpha} > \frac{2}{3\sqrt{m}}. \tag{124.4}$$

In the range of values of α up to 0.30, condition (124.4) is found to be governing. The table on page 530 gives minimum values of m for which solution (124.1) remains valid for certain values of α.

α	0.05	0.10	0.15	0.20	0.25	0.30
m	6.28	3.22	2.17	1.65	1.27	1.05

These examples clearly show the advantages and disadvantages of applying the Tresca creep criterion to rotating disc problems. For certain conditions the solutions are simple, but it is a fairly complicated matter, requiring considerable calculation, to determine the limits of validity of these simple solutions. In other cases it may happen that the disc is divided into a number of regions such that each region corresponds to either a side or an apex of the Tresca hexagon. A solution, i.e. a formula for stresses and rates which contains certain constants, can be written for each such region. However, the boundaries of the regions are not known in advance, but are determined together with the constants of integration from the boundary conditions of conjugation so that we obtain fairly complicated non-linear equations which must be solved numerically.

A very detailed investigation of this kind in application to the power creep law was given by Wahl in the works that we have quoted. Besides a disc of constant thickness he considers a disc of thickness that varies according to the law $h = h_0 \xi^k$. In this case the integration is performed in finite form. However, despite the modern tendency to simplify disc profiles, in practice they still remain more complicated than those for which the graphs and tables have been constructed.

125. **Direct numerical integration of the rotating disc creep equations**

The rotating disc creep equations reduce to a second order system of quasi-linear equations, that is they are linear relative to arbitrary functions that it is required to determine; they are quite convenient for numerical integration. Sokolovskiy [133] applied this device to the problem of the plastic state of a rotating disc and Malinin [69] to the problem of creep. The method was found to be inconvenient in that the edge conditions are given on the two boundaries of the interval and, therefore, one of the initial conditions is not known in advance, and the calculation must be made by trial and error or a series of discs of the same type must be calculated. Since computers became available these difficulties are easily overcome and the method can unhesitatingly be recommended for practical use.

We shall assume, in the most general case, that a disc of external radius b is of variable thickness $h = h(r)$ and we shall also consider that the temperature varies across the radius. As is usual in disc calculations, we shall assume a generalised plane state of stress. In the case of an axially symmetrical problem this means that the stresses $\sigma_\theta = \sigma_1$ and $\sigma_r = \sigma_2$ are uniformly distributed through the thickness of the disc whereas $\sigma_3 = 0$. Using the stress representation according to formula (8.9) we exclude σ by means of the condition $\sigma_3 = 0$ and for convenience put

$$\varphi = \theta + \frac{\pi}{6}.$$

After simple conversions we obtain

$$\sigma_r = \frac{2}{\sqrt{3}}\sigma_0 \sin\left(\varphi - \frac{\pi}{6}\right), \quad \sigma_\theta = \frac{2}{\sqrt{3}}\sigma_0 \sin\left(\varphi + \frac{\pi}{6}\right). \qquad (125.1)$$

Now calculate the components of the rate of deformation by the Mises type theory, namely

$$\epsilon_r = -\tfrac{3}{2}\, v \cos\left(\varphi + \frac{\pi}{6}\right), \quad \epsilon_\theta = \tfrac{3}{2}\, v \cos\left(\varphi - \frac{\pi}{6}\right). \qquad (125.2)$$

When $\sigma_0 = \text{const}$ formulae (125.1) define an ellipse in the plane $\sigma_r - \sigma_\theta$, each point of the ellipse corresponds to a particular value of φ. Part of this ellipse is shown in fig. 210 with values of φ indicated against the corresponding

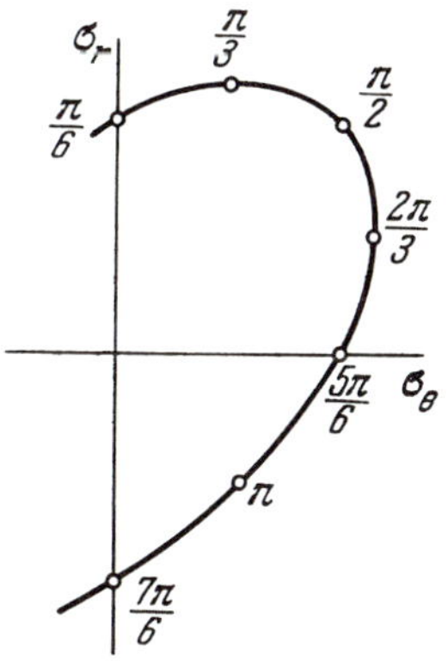

Fig. 210.

points, this sketch is useful in defining the region of change of φ in various particular cases.

The equation of equilibrium of an element of the disc is

$$\frac{\mathrm{d}}{\mathrm{d}r}(h\sigma_r) + h\,\frac{\sigma_r - \sigma_\theta}{r} + \rho\omega^2 hr = 0\,. \tag{125.3}$$

It is convenient to take

$$\zeta = \ln\frac{r}{b} = \ln\xi$$

as an independent variable. In place of the thickness h we are concerned with functions $\eta(\zeta)$ and

$$\eta = \ln\frac{h}{h_0}\,;$$

where h_0 is an arbitrary constant having the dimensions of length (the characteristic thickness of the disc). Substitute expression (125.1) for the stress into eq. (125.3). After simple conversions we obtain the equation

$$\frac{\mathrm{d}\varphi}{\mathrm{d}\zeta}\cos\left(\varphi - \frac{\pi}{6}\right) + \left(\frac{\mathrm{d}\eta}{\mathrm{d}\zeta} + \frac{\mathrm{d}\psi}{\mathrm{d}\zeta}\right)\sin\left(\varphi - \frac{\pi}{6}\right) - \cos\varphi + m\,\frac{\sqrt{3}}{2}\exp\left(2\zeta - \psi\right) = 0\,. \tag{125.4}$$

Here

$$\psi = \ln\frac{\sigma_0}{\sigma_*}\,, \qquad m = \frac{\rho\omega^2 b^2}{\sigma_*}\,,$$

where σ_* is an arbitrary constant with the dimensions of stress.

Now note that since $\epsilon_\theta = u/r$, $\epsilon_r = \mathrm{d}u/\mathrm{d}r$, the rates ϵ_θ and ϵ_r satisfy the following equation of compatibility

$$\frac{\mathrm{d}\epsilon_\theta}{\mathrm{d}r} + \frac{\epsilon_\theta - \epsilon_r}{r} = 0\,. \tag{125.5}$$

Put $z = \ln v/\epsilon_*$ where ϵ_* is an arbitrary constant with dimensions $1/t$. Substitute the values of ϵ_r and ϵ_θ from formulae (125.2) into eq. (125.5). We obtain

$$\frac{d\varphi}{d\zeta}\sin\left(\varphi-\frac{\pi}{6}\right)-\frac{dz}{d\zeta}\cos\left(\varphi-\frac{\pi}{6}\right)-\sqrt{3}\cos\varphi=0\,. \tag{125.6}$$

Therefore the problem has been reduced to integration relative to a simple non-linear system of eqs. (125.4) – (125.6). The quantity z, which figures in eq. (125.6) is a function of ψ. With a power creep law

$$v=\epsilon_n\left(\frac{\sigma_0}{\sigma_n}\right)^n .$$

Put $\epsilon_* = \epsilon_n$, $\sigma_* = \sigma_n$, then

$$z=n\psi\,. \tag{125.7}$$

With an exponential creep law

$$v=\epsilon_e \exp\frac{\sigma}{\sigma_e}\,.$$

Put $\epsilon_* = \epsilon_e$, $\sigma_* = \sigma_e$, then

$$z=\frac{\sigma_0}{\sigma_e}=\exp\psi\,. \tag{125.8}$$

Now in eqs. (125.4) and (125.5) it is convenient to retain a single variable z, eliminating ψ from eq. (125.4) by means of eq. (125.8) and $d\psi/d\zeta = 1/z(dz/d\zeta)$. Numerical integration is, of course, only possible when z is arbitrary, or is tabulated, or is given as a graphical function of ψ.

Now consider allowance for temperature variability. As was shown in section 45 the simplest assumption is that ϵ_* is a function of temperature and consequently of the radius. Thus $\epsilon_* = \epsilon_*(\zeta)$ is a known function. Now let ϵ_{**} be an arbitrary constant having the dimensions of rate of deformation

$$z=\ln\frac{v}{\epsilon_{**}}\,,\qquad \chi(\zeta)=\ln\left(\frac{\epsilon_{**}}{\epsilon_*(\zeta)}\right).$$

Then

$$\frac{\mathrm{d}}{\mathrm{d}\zeta}\ln\frac{v}{\epsilon_*}=\frac{\mathrm{d}z}{\mathrm{d}\zeta}+\frac{\mathrm{d}\chi}{\mathrm{d}\zeta},$$

and the following equation should be used in place of eq. (125.6):

$$\frac{\mathrm{d}\varphi}{\mathrm{d}\zeta}\sin\left(\varphi-\frac{\pi}{6}\right)-\left(\frac{\mathrm{d}\chi}{\mathrm{d}\zeta}+\frac{\mathrm{d}z}{\mathrm{d}\zeta}\right)\cos\left(\varphi-\frac{\pi}{6}\right)-\sqrt{3}\cos\varphi=0\,. \quad (125.9)$$

For a solid disc the point $r=0$ corresponds to $\zeta=-\infty$ which introduces inconvenience for numerical integration. In this case we should return to the variable $\xi=\exp\zeta$.

126. **Disc of equal strength**

A disc of equal strength is one in which the quantity σ_0 is the same at all points and consequently v is also constant. The problem is to determine the profile of this disc, that is to determine the function $h(r)$. Putting $\psi=\text{const}$ and $z=\text{const}$ in eqs. (125.4) and (125.6) we obtain

$$\sin\left(\varphi-\frac{\pi}{6}\right)\frac{\mathrm{d}\eta}{\mathrm{d}\zeta}+\cos\left(\varphi-\frac{\pi}{6}\right)\frac{\mathrm{d}\varphi}{\mathrm{d}\zeta}-\cos\varphi$$
$$+\frac{m\sqrt{3}}{2}\exp(2\zeta-\psi)=0\,, \quad (126.1)$$

$$\sin\left(\varphi-\frac{\pi}{6}\right)\frac{\mathrm{d}\varphi}{\mathrm{d}\zeta}-\sqrt{3}\cos\varphi=0\,. \quad (126.2)$$

Note first the existence of a trivial solution $\varphi=\pi/2$ for eq. (126.2). Substituting into (126.1) we obtain

$$\frac{\mathrm{d}\eta}{\mathrm{d}\zeta}=-m\exp(2\zeta-\psi)\,.$$

Without impairing generality we may take $\psi=0$, so that $\sigma_0=\sigma_*$. Now integrating we find

$$\eta=\ln\frac{h}{h_0}=C-\frac{m}{2}\exp2\zeta=C-\frac{mr^2}{2b^2}\,.$$

Whence

$$h = \exp\left(-\frac{\rho\omega^2 r^2}{2\sigma_0}\right) \times \text{const}\,. \tag{126.3}$$

This is the known solution of the problem of a disc of equal strength which is completely independent of the physical nature of the material of the disc; knowing for example the load applied to the outer periphery we determine the constant factor in eq. (126.3) and find the profile of a solid disc of equal strength.

If there is a hole in the disc and a different load is applied around its contour we cannot construct an equal strength profile by this method, because formula (126.3) does not contain sufficient constants. Therefore, we return to eqs. (126.1) and (126.2). The second of these is integrated and we obtain

$$\zeta = \ln\alpha - \tfrac{1}{2}\ln\frac{2\cos\varphi}{\sqrt{3}} - \frac{1}{2\sqrt{3}}\left(\varphi - \frac{\pi}{6}\right). \tag{126.4}$$

To be specific we have assumed that the radial stress is zero on the periphery of the hole when $r = a$ and $\zeta = \ln a/b = \ln\alpha$ and consequently as will be seen in the diagram of fig. 210, $\varphi = \pi/6$. Expression (126.4) can be rewritten as

$$r^2 = \exp 2\zeta = \frac{a^2\sqrt{3}}{2\cos\varphi}\exp\left(-\frac{\varphi - \frac{\pi}{6}}{\sqrt{3}}\right). \tag{126.5}$$

It will be seen that r tends to infinity when $\varphi = \pi/2$, therefore the range of change of the angle φ is from $\varphi = \pi/6$ to some value $\varphi^* < \pi/2$. Now turn to eq. (126.1). Considering the angle φ as the independent variable exclude ζ by means of eqs. (126.2) and (126.4). We obtain the following differential equation:

$$\frac{d\eta}{d\varphi} = -\operatorname{ctan}\left(\varphi - \frac{\pi}{6}\right) + \frac{1}{\sqrt{3}} - m\alpha^2\,\frac{\sqrt{3}}{4}\,\frac{\exp\left(-\dfrac{\varphi - \frac{\pi}{6}}{\sqrt{3}}\right)}{\cos^2\varphi}\,.$$

Before writing the integral of this equation let us determine the boundary conditions. Firstly, from eq. (126.4) we can find φ^* as function of α. It is readily shown that for real values of $\alpha < 0.3$ φ^* is very close to $\pi/2$. Therefore, we alter the variable of integration putting

$$\beta = \frac{\pi}{2} - \varphi\,.$$

Eq. (126.5) is now rewritten as

$$r^2 = a^2 \frac{\sqrt{3}}{2} \exp\left(-\frac{\pi}{3\sqrt{3}}\right) \frac{1}{\sin\beta} \exp\left(\frac{\beta}{\sqrt{3}}\right).$$

When $r = b$, $\varphi = \varphi^*$, $\beta = \beta^* = \pi/2 - \varphi^*$. If $b \gg r$ we need consider only the first powers of β in the expansion of the right-hand side. We obtain

$$\beta^* = \frac{a^2}{b^2} \frac{\sqrt{3}}{2} \exp\left(-\frac{\pi}{3\sqrt{3}}\right) \approx 0.473 \frac{a^2}{b^2}\,. \tag{126.6}$$

It is now convenient to write eq. (126.5) in the form

$$\frac{r^2}{b^2} = \frac{\beta^*}{\sin\beta} \exp\frac{\beta}{\sqrt{3}}\,. \tag{126.7}$$

Return to the differential equation that was derived from (126.1). Here also consider β as an independent variable and replace the combination of constants in the last term on the right-hand side by a new constant β^*, determined by formula (126.6). We obtain

$$\frac{d\eta}{d\beta} = \operatorname{ctan}\left(\frac{\pi}{3} - \beta\right) - \frac{1}{\sqrt{3}} + \frac{m\beta^*}{2} \frac{\exp\frac{\beta}{\sqrt{3}}}{\sin^2\beta}\,. \tag{126.8}$$

In the region of small values of β only the last term is significant and it becomes infinity when $\beta = 0$. Therefore, for small values of β

$$\eta \approx -\frac{m}{2} \frac{\beta^*}{\beta} = -\frac{mr^2}{2b^2}\,.$$

Hence we again obtain the result of eq. (126.3) for a solid disc. We shall consider that when $r = b$, $h = h_0$ and consequently, $\eta = 0$. Indeed the radial stress (the blade load) is given on the periphery of the disc and, therefore, the quantity h_0 may be selected such that the stress assumes a given value σ^*. We now write the integral of eq. (126.8) in the following way:

$$\eta = \ln \frac{\sin\left(\frac{\pi}{3} - \beta^*\right)}{\sin\left(\frac{\pi}{3} - \beta\right)} - \frac{\beta - \beta^*}{\sqrt{3}} + \frac{m\beta^*}{2} \int_{\beta^*}^{\beta} \frac{\exp\frac{\beta}{\sqrt{3}}}{\sin^2 \beta} d\beta .$$

In calculating the integral it is convenient to separate the part that increases indefinitely when $\beta^* \to 0$, therefore, we represent the final result in the form

$$\eta = \ln \frac{\sin\left(\frac{\pi}{3} - \beta^*\right)}{\sin\left(\frac{\pi}{3} - \beta\right)} - \frac{\beta - \beta^*}{\sqrt{3}}$$

$$+ \frac{m\beta^*}{2}\left[g(\beta) - g(\beta^*) - \operatorname{ctan}\beta + \frac{1}{\beta^*} + \frac{1}{\sqrt{3}} \ln \frac{2}{\beta^*} \tan \frac{\beta}{2}\right]. \quad (126.9)$$

Here

$$g(\beta) = \int_0^\beta \left[\exp \frac{\beta}{\sqrt{3}} - 1 - \frac{\sin \beta}{\sqrt{3}}\right] \frac{d\beta}{\sin^2 \beta}$$

is a function which does not have a singularity when $\beta = 0$. Values of this function are given in the following table:

β	0.1	0.2	0.3	0.4	0.5	0.6
$g(\beta)$	0.0175	0.0365	0.0571	0.0794	0.1036	0.1301
β	0.7	0.8	0.9	1.0	$\pi/3$	
$g(\beta)$	0.1588	0.1905	0.2260	0.2654	0.2846	

The profile of a disc of equal strength with a central hole, calculated by formula (126.9) when $a/b = 0.3$ and $m = 2.772$, is shown in fig. 211, it corresponds to a solid disc for which the thickness at the centre is four times that at the periphery. A disc profile determined by eq. (126.3) is also shown dotted. It will be seen that the difference is small throughout the disc except for the region near to the central hole. Here the thickness increases and tends

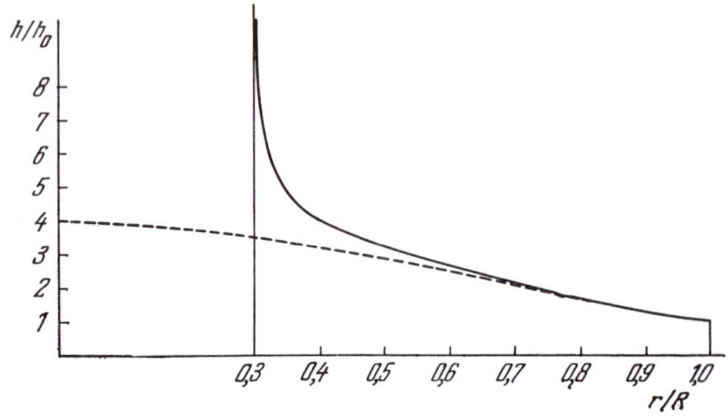

Fig. 211.

to infinity when $r=a$. Thus strictly speaking it is impossible to design a disc of equal strength with a central hole. In practice, there is a hub in the central part of the disc and for design reasons transition from the disc to the hub is made with a fairly large radius.

Note that the solution of a disc of equal strength is completely independent of the creep law, it also remains valid for elastic or plastic material when the actual deformation is considered instead of the rate of deformation. For a plastic disc only the condition of incompressibility is essential, the problem of a disc of equal strength made of elastic-plastic material with allowance for compressibility is considered in the work [100].

127. **Disc design by the method of successive approximations**

We have already noted that the use of direct numerical integration in disc calculations encounters the difficulty that the boundary conditions are given at the two ends of the interval, whereas for numerical integration it is necessary to know the values of the two constants at one end of the interval. Therefore, several variants of the method of successive approximations have been more widely used in design practice. One of these variants was first proposed by Bailey [162] and was improved by Popov [329] and later further improved in the works of Boguslavskiy [7], Kostyuk [61], Malinin [72, 108] and others. As a result of successive modifications which accompanied particular calculations it was possible to make the process very rapidly convergent.

Write the equation of equilibrium (125.3) for a disc, which in the general case is assumed to be of variable thickness, in somewhat different form, namely

$$\frac{\mathrm{d}}{\mathrm{d}\xi}(h\xi\sigma_r) - h\sigma_\theta + mh\xi^2 = 0 . \tag{127.1}$$

Here we have introduced the dimensionless radius $\xi = r/b$ but we have left the quantities σ_r, σ_θ and h dimensional. Therefore

$$m = \rho\omega^2 b^2 .$$

We shall now consider that in the general case the disc has a central hole of radius a, and the radial stress when $\xi = a/b = \alpha$ is given and equal to mp, and when $\xi = 1$, $\sigma_r = mq$. Integrating eq. (127.1), considering σ_θ as a given function of ξ we obtain:

$$\sigma_r = \frac{1}{h\xi}\int_\alpha^\xi h\sigma_\theta\,\mathrm{d}\xi - \frac{m}{h\xi}\left[\int_\alpha^\xi h\xi^2\,\mathrm{d}\xi - p\alpha h(\alpha)\right] .$$

The second integral in the formula is the moment of inertia relative to the axis of the part of the area of the section contained between the radii α and ξ; this function, which is always known, is denoted $I(\xi)$; rewrite the result as follows:

$$\sigma_r = \frac{1}{h\xi}\left[\int_\alpha^\xi h\sigma_\theta\,\mathrm{d}\xi - m\mathbf{I} + mp\alpha h(\alpha)\right] . \tag{127.2}$$

Expression (127.2) is the first equation that is used in all variants of Bailey's type method of successive approximations. The second equation is derived from the condition of compatibility, different authors using different methods. Write the equation of compatibility

$$\frac{\mathrm{d}\epsilon_\theta}{\mathrm{d}\xi} + \frac{\epsilon_\theta - \epsilon_r}{\xi} = 0 . \tag{127.3}$$

To be specific we shall take the usual Mises' type creep law, eq. (119.13)

$$\epsilon_\theta = \frac{v(\sigma_0)}{\sigma_0}(\sigma_\theta - \tfrac{1}{2}\sigma_r) , \qquad \epsilon_r = \frac{v(\sigma_0)}{\sigma_0}(\sigma_r - \tfrac{1}{2}\sigma_\theta) , \tag{127.4}$$

though a similar procedure of calculation is also possible with other laws. Our problem now is to replace eq. (127.3) with conditions (127.4) by some integral expression of the type

$$\sigma_\theta = F(\xi, \sigma_r, \sigma_\theta, C)\,. \tag{127.5}$$

Here F is a functional of some form that depends on σ_r, σ_θ, ξ and the constant of integration C. The temperature is assumed to be a known function of the radius and, therefore, does not explicitly enter into the functional. Evidently, a functional of type (127.5) can be constructed in many different ways, in the process of conversion we may also use the equation of equilibrium (127.1). In the work [61] the author replaces the rate of deformation in eq. (127.3) by the stress according to formula (127.4) and uses eq. (127.1) to eliminate the derivative $\mathrm{d}\sigma_r/\mathrm{d}\xi$ and integrates the resulting equation. Here the course of the calculations is as follows. Introducing values ϵ_r and ϵ_θ determined by eq. (127.4) into the equation of compatibility we obtain

$$\frac{\mathrm{d}\sigma_\theta}{\mathrm{d}\xi} - \frac{1}{2}\frac{\mathrm{d}\sigma_r}{\mathrm{d}\xi} = \frac{\mathrm{d}\ln\omega}{\mathrm{d}\xi}\left(\sigma_\theta - \tfrac{1}{2}\sigma_r\right) - \frac{3}{2}\frac{\sigma_\theta - \sigma_r}{\xi}\,.$$

Here $\omega = \sigma_0/v$.

On the other hand since $v = v(\sigma_0, T)$ we have

$$\frac{\mathrm{d}\ln\omega}{\mathrm{d}\xi} = \frac{1}{\sigma_0}\frac{\partial\ln\omega}{\partial\sigma_0}\left[\left(\sigma_r - \tfrac{1}{2}\sigma_\theta\right)\frac{\mathrm{d}\sigma_r}{\mathrm{d}\xi} + \left(\sigma_\theta - \tfrac{1}{2}\sigma_r\right)\frac{\mathrm{d}\sigma_\theta}{\mathrm{d}\xi}\right] + \frac{\partial\ln\omega}{\partial T}\frac{\mathrm{d}T}{\mathrm{d}\xi}\,.$$

Eliminate $(\mathrm{d}\ln\omega)/\mathrm{d}\xi$ from the first equation, regroup the terms and eliminate $\mathrm{d}\sigma_r/\mathrm{d}\xi$ by means of eq. (127.1). The following result is obtained:

$$(1+A)\frac{\mathrm{d}\sigma_\theta}{\mathrm{d}\xi} = -\left(1 - \tfrac{1}{2}B\right)\frac{\sigma_\theta - \sigma_r}{\xi} - \tfrac{1}{2}(1+B)\left(m\xi + \sigma_r\frac{\mathrm{d}\ln h}{\mathrm{d}\xi}\right) + \frac{\partial\ln\omega}{\partial T}\frac{\mathrm{d}T}{\mathrm{d}\xi}\left(\sigma_\theta - \tfrac{1}{2}\sigma_r\right)\,. \tag{127.6}$$

Here it is taken that

$$A = \Omega\left(1 - \tfrac{3}{4}\frac{\sigma_r^2}{\sigma_0^2}\right), \qquad B = \Omega\left(1 - \tfrac{3}{2}\frac{\sigma_r\sigma_\theta}{\sigma_0^2}\right),$$

$$\Omega = -\frac{\partial \ln \omega}{\partial \sigma_0}.$$

Therefore

$$\sigma_\theta = C + \int_\alpha^\xi H\,\mathrm{d}\xi\,, \tag{127.7}$$

where

$$H = \frac{1}{1+A}\left\{-\left(1-\tfrac{1}{2}B\right)\frac{\sigma_\theta - \sigma_r}{\xi} - \tfrac{1}{2}(1+B)\left(m\xi + \sigma_r\frac{\mathrm{d}\ln h}{\mathrm{d}\xi}\right)\right.$$
$$\left. + \frac{\partial \ln \omega}{\mathrm{d}T}\,\frac{\mathrm{d}T}{\mathrm{d}\xi}\left(\sigma_\theta - \tfrac{1}{2}\sigma_r\right)\right\}.$$

The constant C is determined from the second boundary condition for σ_r, namely $\sigma_r(1) = mq$. Using formula (127.2) we obtain

$$\frac{1}{m}\int_\alpha^1 h\sigma_\theta\,\mathrm{d}\xi - I(1) = qh(1) - p\alpha h(\alpha)\,. \tag{127.8}$$

The process of successive approximations is as follows. Assume a certain distribution of σ_θ as a first approximation, for instance put σ_θ = const. The quantity σ_θ in the first approximation is found by eq. (127.8). Substitute this value of σ_θ into formula (127.2) and find the corresponding radial stress distribution σ_r for the first approximation. Now use the values of σ_r and σ_θ of the first approximation to calculate the function $H(\xi)$ and find the second approximation of σ_θ by formula (127.7). The constant C is found from the condition (127.8), the second approximation of the radial stress distribution is found by formula (127.2). The entire process is repeated until two successive approximations practically coincide. The author indicates that the third approximation almost always gives acceptable accuracy. Bailey's original method and other modifications of it usually require five or six approximations.

With a higher temperature gradient the term $\partial T/\partial \xi$ is the most important in the expression for H. Kostyuk has shown that in this case the convergence of successive approximations is not so good, and, therefore, the method was somewhat modified. Put

$$-\frac{1}{1+A}\frac{\partial \ln \omega}{\partial T}\frac{\mathrm{d}T}{\mathrm{d}\xi}=L\,.$$

Then eq. (127.6) can be written in the form

$$\frac{\mathrm{d}\sigma_\theta}{\mathrm{d}\xi}+L\sigma_\theta=H+L\sigma_\theta\,.$$

The integral of this linear equation is

$$\sigma_\theta=\frac{1}{R}\left[C+\int_\alpha^\xi (H+L\sigma_\theta)R\,\mathrm{d}\xi\right]. \qquad (127.9)$$

Here

$$R=\exp\int_\alpha^\xi L\,\mathrm{d}\xi\,.$$

The entire process of successive approximation remains unaltered except that eq. (127.9) is used in place of (127.7).

This method is somewhat inconvenient in application to solid discs. Indeed in the expression for H there are terms that contain ξ in the denominator and if we do not apply the additional condition $\sigma_r(0)=\sigma_\theta(0)$, the integral for σ_θ is divergent. To overcome this difficulty isolate a certain region $0<\xi<\alpha$, where α is a sufficiently small quantity and seek the solution for this region in the form of series

$$\sigma_r=\sigma(0)(1+a_1\xi^2+\dots)\,,\quad \sigma_\theta=\sigma(0)(1+b_1\xi^2+\dots)\,. \qquad (127.10)$$

Now also assume that near to the point $\xi=0$ the disc thickness and temperature can be expressed in a similar way:

$$h=h_0(1-e_1\xi^2+\dots)\,,\qquad T=T_0+T_1\xi^2+\dots\,.$$

Substituting these series into the differential eqs. (127.1) and (127.6) we find

$$a_1 = -\frac{1}{4(4+\Omega_0)}\left\{6+(1+\Omega_0)\left(\frac{m}{\sigma(0)}-2e_1\right)-T_1\left(\frac{\partial \ln \omega}{\partial T}\right)_{\xi=0}\right\},$$

$$b_1 = -\frac{1}{4(4+\Omega_0)}\left\{6-(1+\Omega_0)\left(\frac{m}{\sigma(0)}-2e_1\right)-T_1\left(\frac{\partial \ln \omega}{\partial T}\right)_{\xi=0}\right\}. \tag{127.11}$$

Here we have taken $\Omega_0 = \Omega(0)$. Now in the region $0 < \xi < \alpha$ the stresses are calculated by formulae (127.9) and in the region $\alpha < \xi < 1$ by the method of successive approximations, just as for a disc with a hole.

Kostyuk's method is particularly convenient when the function $v(\sigma_0, T)$ is given numerically or graphically.

A similar method was proposed by Malinin [72] for the case of a power creep law. Put $v = \epsilon_n(\sigma_0/\sigma_n)^n$, here we shall consider σ_n as a constant and ϵ_n a known function of temperature and consequently, a known function of the radius. Dividing the equation of compatibility by ϵ_θ we obtain

$$\frac{\mathrm{d} \ln \epsilon_\theta}{\mathrm{d}\xi} = \frac{1}{\xi}\left(\frac{\epsilon_r}{\epsilon_\theta} - 1\right).$$

Integrate this equation

$$\epsilon_\theta = C \exp \int_\alpha^\xi \left(\frac{\epsilon_r}{\epsilon_\theta} - 1\right)\frac{\mathrm{d}\xi}{\xi}. \tag{127.12}$$

Denote the ratio σ_r/σ_θ by β, then $\sigma_0 = (1+\beta^2-\beta)^{\frac{1}{2}}\,\sigma_\theta$ and in accordance with the power creep law

$$v = \epsilon_n(\xi)\left(\frac{\sigma_\theta}{\sigma_n}\right)^n (1+\beta^2-\beta)^{n/2}.$$

The rate of deformation is expressed as

$$\epsilon_\theta = \epsilon_n\left(\frac{\sigma_\theta}{\sigma_n}\right)^n (1+\beta^2-\beta)^{\frac{n-1}{2}}\left(1-\tfrac{1}{2}\beta\right),$$

$$\epsilon_r = \epsilon_n\left(\frac{\sigma_\theta}{\sigma_n}\right)^n (1+\beta^2-\beta)^{\frac{n-1}{2}}\left(\beta-\tfrac{1}{2}\right).$$

Hence it follows that

$$\frac{\epsilon_r}{\epsilon_\theta} - 1 = -3\,\frac{1-\beta}{2-\beta}.$$

Now eq. (127.12) is rewritten

$$\epsilon_\theta = \epsilon_n \left(\frac{\sigma_\theta}{\sigma_n}\right)^n (1+\beta^2-\beta)^{\frac{n-1}{2}} (1-\tfrac{1}{2}\beta) = C\exp\left(-3\int_0^\xi \frac{1-\beta}{2-\beta}\,\frac{\mathrm{d}\xi}{\xi}\right).$$

Put

$$\psi(\xi) = (1+\beta^2-\beta)^{\frac{n-1}{2n}} (1-\tfrac{1}{2}\beta)^{\frac{1}{n}} \exp\left[-\frac{3}{n}\int_0^\xi \frac{1-\beta}{2-\beta}\,\frac{\mathrm{d}\xi}{\xi}\right].$$

Then

$$\sigma_\theta = \sigma_n \left(\frac{C}{\epsilon_n}\right)^{\frac{1}{n}} \psi(\xi)\,. \tag{127.13}$$

Eqs. (127.2) and (127.13) are now solved by the method of successive approximations. As before, the constant C is determined by eq. (127.7) which is written in the form

$$\frac{\sigma_n}{m}\,C^{1/n} \int_\alpha^1 \psi h \epsilon_n^{-1/n}\,\mathrm{d}\xi = I(1) + qh(1) - p\alpha h(\alpha)\,.$$

The work [72] recommends use of the stress distribution that corresponds to a disc in the elastic state as the initial approximation. It is, of course, possible to proceed otherwise, for example by assuming some distribution of the quantity β over the radius. For a solid disc free from peripheral load we always have $\beta(0) = 1$ and $\beta(1) = 0$, and, therefore, a curve of parabolic type is a satisfactory approximation. For further details concerning the method of successive approximations the reader is referred to the original work which also contains examples of calculations that we shall not reproduce here.

128. A further variant of the method of successive approximations

As is usual in calculating discs of elastic materials, here the system of the equations of equilibrium and compatibility may be replaced by a single non-linear second-order equation. Once again successive approximation methods may be used to solve this equation. One variant of this calculation proposed in the work [98] is given below.

In order to satisfy the equation of equilibrium (127.1) we introduce the stress function $z(\xi)$ such that

$$h\sigma_\theta = mh\xi^2 + \frac{dz}{d\xi}, \qquad h\sigma_r = \frac{z}{\xi}. \tag{128.1}$$

Put

$$\kappa = \ln \frac{\sigma_0 h}{v(\sigma_0)}.$$

Then the rate of deformation is expressed as

$$\epsilon_r = \exp(-\kappa)\left[\frac{z}{\xi} - \frac{1}{2}\frac{dz}{d\xi} - \frac{1}{2}mh\xi^2\right],$$

$$\epsilon_\theta = \exp(-\kappa)\left[\frac{dz}{d\xi} + mh\xi^2 - \frac{1}{2}\frac{z}{\xi}\right].$$

Substituting these expressions into the equation of compatibility of deformations, after simple modifications we obtain the following differential equation for $z(\xi)$:

$$(\xi z')' - \frac{z}{\xi} = -m\xi^2\left(\frac{7h}{2} + h'\xi\right) + \kappa' h\xi(\sigma_\theta - \tfrac{1}{2}\sigma_r). \tag{128.2}$$

To be specific consider the following boundary conditions:

$$\sigma_r(\alpha) = mp, \qquad \sigma_r(1) = mq.$$

Denoting the right-hand side of eq. (128.2) by $mf(\xi)$ we solve this equation with given boundary conditions by the method of variation of parameters. By the usual procedure we obtain

$$z(\xi) = m\xi[C_1 + f_1(\xi)] + \frac{m}{\xi}[C_2 - f_2(\xi)].$$

Here C_1 and C_2 are constants that must be determined from the boundary conditions, functions $f_1(\xi)$ and $f_2(\xi)$ are

$$f_1(\xi) = \tfrac{1}{2} \int_{\alpha}^{\xi} \frac{f(\xi)}{\xi} \mathrm{d}\xi \,, \qquad f_2(\xi) = \tfrac{1}{2} \int_{0}^{\xi} f(\xi)\xi \, \mathrm{d}\xi \,. \tag{128.3}$$

Now we determine the stress by formulae (128.1)

$$\begin{aligned} h\sigma_\theta &= m\left[h\xi^2 + C_1 + f_1(\xi) - \frac{C_2}{\xi^2} + \frac{f_2(\xi)}{\xi^2}\right], \\ h\sigma_r &= m\left[C_1 + f_1(\xi) + \frac{C_2}{\xi^2} - \frac{f_2(\xi)}{\xi^2}\right]. \end{aligned} \tag{128.4}$$

We use the boundary conditions to determine the constants C_1 and C_2 which are given by the formulae

$$\begin{aligned} C_1 &= \frac{1}{1-\alpha^2}\,[h(1)q - \alpha^2 h(\alpha)p] - \frac{1}{1-\alpha^2}\,[f_1(1) - f_2(1)] \,, \\ C_2 &= -\frac{\alpha^2}{1-\alpha^2}\,[h(1)q - h(\alpha)p] + \frac{\alpha^2}{1-\alpha^2}\,[f_1(1) - f_2(1)] \,. \end{aligned} \tag{128.5}$$

In our case $f(\xi)$ consists of two terms

$$f = -\xi^2\left(\tfrac{7}{2}h + h'\xi\right) + \frac{\kappa' h}{m}\left(\sigma_\theta - \tfrac{1}{2}\sigma_r\right)\xi \,.$$

Put

$$\tfrac{7}{2}h + h'\xi = H(\xi) \,, \qquad \kappa' h\left(\sigma_\theta - \tfrac{1}{2}\sigma_r\right) = g(\xi) \,.$$

The constants C_1 and C_2 also consist of two parts, one of which depends on the unknown function $H(\xi)$ and the boundary values of σ_r and the other depends on the unknown function $g(\xi)$. Separate these unknown parts by substituting $f(\xi) = -\xi^2 H(\xi)$ into the formulae for C_1 and C_2. Here we introduce the notation

$$H_1(\xi) = \tfrac{1}{2} \int_{\alpha}^{\xi} H(\xi)\xi \, \mathrm{d}\xi \,, \qquad H_2(\xi) = \tfrac{1}{2} \int_{\alpha}^{\xi} H(\xi)\xi^3 \, \mathrm{d}\xi \,.$$

By formula (128.3) we obtain

$$C_{10} = \frac{1}{1-\alpha^2}\,[h(1)q - \alpha^2 h(\alpha)p] + \frac{1}{1-\alpha^2}\,[H_1(1) - H_2(1)]\,,$$

$$C_{20} = \frac{1}{1-\alpha^2}\,[h(1)q - h(\alpha)p] + \frac{1}{1-\alpha^2}\,[H_1(1) - H_2(1)]\,.$$

In the same way we separate from the solution given by formulae (128.4) the part that corresponds to the case $f(\xi) = -\xi H(\xi)$ and the non-uniform boundary conditions of the problem. Introduce the notation

$$\begin{aligned}(h\sigma_\theta)^* &= m\left[h\xi^2 + C_{10} - \frac{C_{20}}{\xi^2} - H_1(\xi) - \frac{1}{\xi^2}H_2(\xi)\right],\\ (h\sigma_r)^* &= m\left[C_{10} + \frac{C_{20}}{\xi^2} - H_1(\xi) + \frac{1}{\xi^2}H_2(\xi)\right].\end{aligned} \tag{128.6}$$

Finally put

$$\tfrac{1}{2}\int_\alpha^\xi g(\xi)\,\mathrm{d}\xi = g_1(\xi)\,, \qquad \tfrac{1}{2}\int_\alpha^\xi g(\xi)\xi^2\,\mathrm{d}\xi = g_2(\xi)\,.$$

After simple conversions the stress formulae (128.4) can be rewritten as

$$\begin{aligned}h\sigma_\theta &= (h\sigma_\theta)^* - \frac{\alpha^2+\xi^2}{\xi^2(1-\alpha^2)}\,[g_1(1) - g_2(1)] + g_1(\xi) + \frac{1}{\xi^2}\,g_2(\xi)\,,\\ h\sigma_r &= (h\sigma_r)^* + \frac{\alpha^2+\xi^2}{\xi^2(1-\alpha^2)}\,[g_1(1) - g_2(1)] + g_1(\xi) - \frac{1}{\xi^2}\,g_2(\xi)\,.\end{aligned} \tag{128.7}$$

For numerical calculations it is convenient to introduce new unknowns, namely

$$2\sigma = \sigma_\theta + \sigma_r\,, \qquad 2\tau = \sigma_\theta - \sigma_r\,.$$

Then by formula (128.6) we obtain

$$\begin{aligned}(h\sigma)^* &= m\,[\tfrac{1}{2}h\xi^2 + C_{10} - H_1(\xi)]\,,\\ (h\tau)^* &= m\left[\tfrac{1}{2}h\xi^2 - \frac{1}{\xi^2}\,(C_{20} + H_2(\xi))\right].\end{aligned} \tag{128.8}$$

From formula (128.7) it follows that:

$$h\sigma = (h\sigma)^* - \frac{1}{1-\alpha^2}\,[g_1(1) - g_2(1)] + g_1(\xi)\,,$$
$$h\tau = (h\tau)^* - \frac{1}{1-\alpha^2}\,\frac{g_1(1) - g_2(1)}{\xi^2} - \frac{g_2(\xi)}{\xi^2}\,. \tag{128.9}$$

In the new variables the function $g(\xi)$ assumes the form

$$g(\xi) = \tfrac{1}{2}\,\kappa' h(\sigma + 3\tau)\,.$$

Here $\sigma_0^2 = \sigma^2 + 3\tau^2$. In the case of a solid disc the formulae are somewhat simpler, they are obtained from those given above by putting $C_{10} = 0$ in eq. (128.6) and $\alpha = 0$ in all the remaining equations.

Note that the left-hand side of eq. (128.2) is a Sturm-Liouville type operator, which we shall denote $L(z)$. If we define z^* as a function which satisfies the equation

$$L(z^*) = -m\xi^2 H(\xi)\,,$$

and the non-uniform boundary conditions, the equation can be replaced by the following integral equation:

$$z = z^* + \int_\alpha^1 G(\xi, t)\tan(t)\,\mathrm{d}t\,.$$

Here $G(\xi, t)$ is a Green function determined with the corresponding uniform boundary conditions. This is the presentation of the idea of the method that was given in the article [98]. It is evident that if we carry out all the calculations we arrive at the result given above.

We may take as initial approximation the quantities σ^* and τ^* and then the procedure is that, for each successive approximation, σ and τ are calculated by formulae (128.9) in which the function g is found in terms of the expressions for σ and τ of the previous approximation.

129. Stress concentration about a hole in a uniformly extended plate

One of the few plane problems of creep theory that permits of accurate solution is that of stress concentration about a circular hole in a plate that is subject to uniform tension. Winzer and Prager [383] obtained the corresponding result in application to the deformation theory of plasticity with a power law of work-hardening, it can also be obtained as a particular case from Kostyuk's [60] solution of the more general problem. Later the problem of stress concentration about a circular hole was studied with more general assumptions concerning the work-hardening law in the work [177]. Here the stress concentration coefficient can be determined by fairly simple means, determination of the stress distribution leads to fairly cumbersome formulae and, therefore, we shall here consider only the first problem, namely that of stress concentration, and subsequently, we shall give as an independent result the general solution of the more complicated problem given in the article [60].

Therefore, consider an infinite plane containing a hole of radius a. The origin of polar coordinates is located in the centre of the hole, we introduce the dimensionless coordinate $\zeta = \ln (r/a)$, the remaining notation is the same as in section 125. At infinity, $\sigma_r = \sigma_\theta = \sigma_1$, on the periphery of the hole when $\zeta = 0$, $\sigma_r(0) = 0$. It follows from formulae (125.1) that when $\zeta = \infty$, $\varphi = \pi/2$, when $\zeta = 0$, $\varphi = \pi/6$, consequently $\sigma_0(\infty) = \sigma_1$ and

$$\sigma_\theta(0) = \frac{2\sigma_0(0)}{\sqrt{3}} \sin\frac{\pi}{3} = \sigma_0(0) .$$

The stress concentration coefficient K is determined as follows:

$$K = \frac{\sigma_\theta(0)}{\upsilon_1} = \frac{\sigma_0(0)}{\sigma_0(\infty)} = \exp\left[\psi(0) - \psi(\infty)\right] .$$

By appropriate selection of the dimensionless constant σ_* we can always obtain $\sigma_0(\infty) = 1$ and $\psi(\infty) = 0$, consequently,

$$K = \exp \psi(0) . \tag{129.1}$$

Eqs. (125.4) and (125.6) which were derived for a rotating disc, remain valid in this case also provided we put in them $m = 0$ and $\eta = 0$, because the plate is considered to be of constant thickness. The quantity z in the second equation is an unknown function of ψ, therefore, $\mathrm{d}z/\mathrm{d}\zeta = z'(\psi)(\mathrm{d}\psi/\mathrm{d}\zeta)$. Rewriting eqs. (125.4) and (125.6) with these simplifications, we obtain

$$\frac{d\varphi}{d\zeta}\cos\left(\varphi-\frac{\pi}{6}\right)+\frac{d\psi}{d\zeta}\sin\left(\varphi-\frac{\pi}{6}\right)-\cos\varphi=0\,,$$

$$\frac{d\varphi}{d\zeta}\sin\left(\varphi-\frac{\pi}{6}\right)-z'\frac{d\psi}{d\zeta}\cos\left(\varphi-\frac{\pi}{6}\right)-\sqrt{3}\cos\varphi=0\,. \tag{129.2}$$

Multiply the first equation by $\sqrt{3}$ and subtract the second. The resulting equation does not contain the quantity ζ, and consequently, it relates the functions φ and ψ. This equation may be rewritten

$$2\cos\varphi\frac{d\varphi}{d\psi}+\sqrt{3}\sin\left(\varphi-\frac{\pi}{6}\right)+z'(\psi)\cos\left(\varphi-\frac{\pi}{6}\right)=0\,. \tag{129.3}$$

When $\varphi=\pi/2$, $\psi=0$; integrating eq. (129.3) to the point $\varphi=\pi/6$ we find the corresponding quantity $\psi(0)$ and consequently, the stress concentration coefficient in formula (129.1). It is evident that numerical integration is possible with any creep law. For a power creep law the integration is performed in finite form. Indeed with a power law, $z=n\psi$ and $z'=n$.

Now put

$$\varphi=\omega+\frac{\pi}{6}-\gamma\,.$$

If we select the angle γ in such a way that

$$\frac{\sqrt{3}}{\sqrt{3+n^2}}=\cos\gamma\,,\qquad \frac{n}{\sqrt{3+n^2}}=\sin\gamma\,,$$

then the sum of the second and third terms in eq. (129.3) assumes the form $\cos\gamma\sqrt{3+n^2}$ and the equation in which the independent variable was taken as ω instead of φ is rewritten in the form

$$\frac{d\psi}{d\omega}=-\frac{2}{\sqrt{3+n^2}}\,\frac{\cos\left(\omega+\frac{\pi}{6}-\gamma\right)}{\sin\omega}\,.$$

When $\varphi=\pi/2$, $\omega=\pi/3+\gamma$, $\psi=0$, when $\varphi=\pi/3$, $\omega=\gamma$, $\psi=\psi(0)$; by integration we obtain

$$\psi(0) = \frac{2}{\sqrt{3+n^2}}\left[\cos\left(\frac{\pi}{6}-\gamma\right)\ln\frac{\sin\left(\frac{\pi}{3}+\gamma\right)}{\sin\gamma} - \frac{\pi}{3}\sin\left(\frac{\pi}{6}-\gamma\right)\right].$$

Hence, eliminating the angle γ we find the following expression for the stress concentration coefficient

$$K = \left(\frac{n+3}{2n}\right)^{\frac{n+3}{n^2+3}} \exp\left(\frac{\pi}{\sqrt{3}}\,\frac{n-1}{3+n^2}\right). \tag{129.4}$$

This formula was derived in the work [177].

Now consider the same problem but on the basis of piecewise linear creep criteria. We shall use the same notation as in sections 123 and 124, putting $\hat{\sigma}_\theta = \sigma_\theta/\sigma_n$, $\hat{\sigma}_r = \sigma_r/\sigma_n$, $\xi = r/a$, $\hat{u} = u/a\epsilon_n$. Using a Tresca type criterion [115] we must assume that the condition corresponding to point A (fig. 212) is reached at infinity and point F when $\xi = 1$. The region that corresponds to side AF of the Tresca hexagon pertains to the periphery of the hole so that in this region $\hat{\sigma}_\theta > \hat{\sigma}_r > 0$, consequently, $\hat{\epsilon}_r = \mathrm{d}\hat{u}/\mathrm{d}\xi = 0$, and $\hat{u} = \text{const}$. The rate of peripheral deformation $\hat{\epsilon}_\theta = \hat{u}/\xi$, therefore, $\hat{\sigma}_\theta = (\hat{u}/\xi)^{1/n}$ and the equation of equilibrium is

$$\frac{\mathrm{d}}{\mathrm{d}\xi}(\hat{\sigma}_r\xi) = \left(\frac{\hat{u}}{\xi}\right)^{1/n}.$$

Its integral is

$$\hat{\sigma}_r = \frac{n}{n-1}\left(\frac{\hat{u}}{\xi}\right)^{1/n} + \frac{C}{\xi}.$$

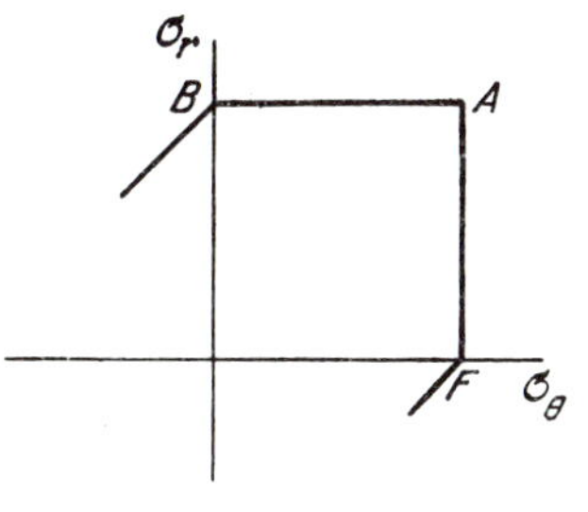

Fig. 212.

For a certain value $\xi = \alpha$ the inequality $\hat{\sigma}_\theta > \hat{\sigma}_r$ becomes an equation and we obtain

$$\sigma_r(\xi) = \sigma_\theta(\xi) = \sigma_1 .$$

From these conditions we determine the constants C and $\hat{u}$ after which the stress formulae assume the form

$$\hat{\sigma}_r = \frac{n\hat{\sigma}_1}{1-n}\left[\left(\frac{\alpha}{\xi}\right)^{1/n} - \frac{1}{n}\frac{\alpha}{\xi}\right],$$
$$\hat{\sigma}_\theta = \sigma_1\left(\frac{\alpha}{\xi}\right)^{1/n} . \tag{129.5}$$

We now determine the radius of the zone α from the condition $\sigma_r(1) = 0$. We obtain

$$\alpha = n^{\frac{n}{n-1}} .$$

According to the second of formulae (129.5) the stress concentration coefficient is

$$K = \alpha^{1/n} = n^{\frac{1}{n-1}} . \tag{129.6}$$

This simple formula gives greater values of the concentration coefficient than does formula (129.4). For example, when $n = 1$ formula (129.4) gives $K = 2$ and formula (126.9) gives $K = e$ (if the indeterminancy is developed).

In section 119 we gave equations based on a recently proposed piecewise-linear criterion of creep or plasticity, which has been termed the criterion of maximum referred stress, but we nowhere made actual use of it. Nemirovskiy [86] solved the problem of stress concentration about a hole by means of this criterion, and we give his solution below.

On the contour $\hat{\sigma}_r = 0$, and so near to the contour the condition represented by the segment AF on fig. 213 is achieved and for this state

$$2\hat{\epsilon}_r + \hat{\epsilon}_\theta = 2\frac{d\hat{u}}{d\xi} + \frac{\hat{u}}{\xi} = 0 .$$

Hence integrating we find

$$\hat{u} = \hat{\sigma}_m^n \xi^{-\frac{1}{2}} , \qquad \hat{\epsilon}_r = \hat{\sigma}_m^n \xi^{-\frac{3}{2}} .$$

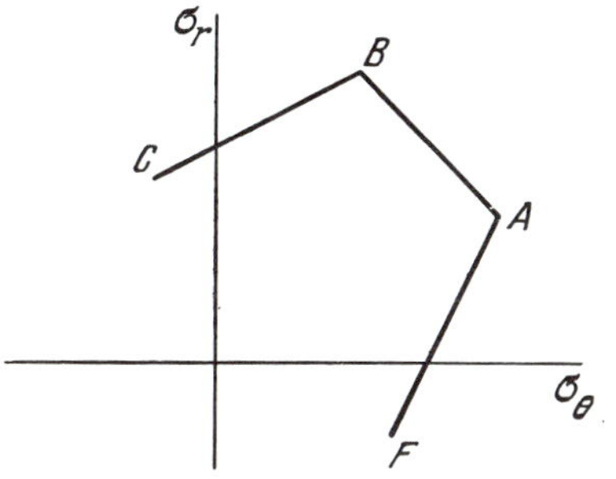

Fig. 213.

We denote the constant of integration $\hat{\sigma}_m^n$, the significance of this notation will become clear when we come to determine the stress. For the segment AF we have

$$\hat{s} = \hat{\sigma}_\theta - \tfrac{1}{2}\hat{\sigma}_r = \hat{\epsilon}_r^{\frac{1}{n}} = \hat{\sigma}_m \xi^{-\frac{3}{2n}} . \tag{129.7}$$

Use this expression to exclude $\hat{\sigma}_\theta$ from the equation of equilibrium and integrate it bearing in mind the boundary conditions $\hat{\sigma}_r(1) = 0$. We obtain

$$\hat{\sigma}_r = \hat{\sigma}_m \frac{2}{1-3v} \xi^{-\frac{1}{2}} (\xi^{\frac{1-3v}{2}} - 1) \quad \left(v = \frac{1}{n}\right) . \tag{129.8}$$

It follows from eq. (129.7) that when $\xi = 1$, $\hat{\sigma}_\theta = \hat{\sigma}_m$; therefore $\hat{\sigma}_m$ is the required maximum value of $\hat{\sigma}_\theta$ at the edge of the hole.

The equation $\hat{\sigma}_\theta = 2\hat{\sigma}_r$ is fulfilled for a certain value $\xi = \alpha$, that is we reach the point A of the hexagon shown in fig. 213. The value of α is found from this condition by means of eqs. (129.7) and (129.8), namely

$$\alpha = \left[\frac{3}{2+3v}\right]^{\frac{2}{1-3v}} . \tag{129.9}$$

Formulae (129.7) and (129.8) are valid provided that $0 < \xi < \alpha$, beyond this there is a certain zone that corresponds to point A of the hexagon, that is a zone in which $\hat{\sigma}_\theta = 2\hat{\sigma}_r$. The equation of equilibrium alone suffices to determine the stress, from which it follows that:

$$\hat{\sigma}_r = A\xi , \qquad \hat{\sigma}_\theta = 2A\xi . \tag{129.10}$$

From the condition of continuity of $\hat{\sigma}_r$ when $\xi = \alpha$ we find

$$A = \hat{\sigma}_m \frac{2}{2+3v} \alpha^{-\frac{3}{2}} . \tag{129.11}$$

The solution (129.10) is not suitable for satisfaction of the condition at infinity because the stress given by this formula increases indefinitely with increase in ξ and consequently the point A of the hexagon corresponds to a certain finite region $\alpha < \xi < \beta$. In this region the relationship between stress and strain is

$$2\hat{\epsilon}_\theta + \hat{\epsilon}_r = \tfrac{3}{2} (\tfrac{3}{4} \hat{\sigma}_\theta)^n ,$$

or

$$\frac{d\hat{u}}{d\xi} + \frac{2\hat{u}}{\xi} = \tfrac{3}{2} (\tfrac{3}{2} A\xi)^n .$$

Integrating with allowance for continuity of $\hat{u}$ when $\xi = \alpha$, we obtain

$$\hat{u} = (\tfrac{3}{2})^{n+1} \frac{A^n}{n+3} \frac{\xi^3 - \alpha^3}{\xi^2} + \hat{\sigma}_m^n \frac{\alpha^{\frac{3}{2}}}{\xi^2} . \tag{129.12}$$

When $\xi > \beta$ we realise the state represented by the side AB for which $\hat{\epsilon}_\theta - \hat{\epsilon}_r = 0$ or $d\hat{u}/d\xi - \hat{u}/\xi = 0$. Hence it follows that

$$\hat{u} = B^n \xi .$$

The constant B^n is found from the condition of continuity when $\xi = \beta$, namely

$$B^n = (\tfrac{3}{2})^{n+1} \frac{A^n}{n+3} \left(1 - \frac{\alpha^3}{\beta^3}\right) + \hat{\sigma}_m^n \frac{\alpha^{\frac{3}{2}}}{\beta^3} . \tag{129.13}$$

The relationship between stresses and rates in this region is

$$\hat{\sigma}_r + \hat{\sigma}_\theta = 2(\hat{\epsilon}_r + \hat{\epsilon}_\theta)^\nu .$$

Hence express $\hat{\sigma}_\theta$ and substitute into the equation of equilibrium, putting $\hat{\epsilon}_r = \hat{\epsilon}_\theta = B^n$ and integrating we obtain

$$\hat{\sigma}_r = \frac{1}{\xi^2}\,[\hat{\sigma}_1(\xi^2 - \beta^2) + A\beta^3]\,,$$
$$\hat{\sigma}_\theta = \frac{1}{\xi^2}\,[\hat{\sigma}_1(\xi^2 + \beta^2) - A\beta^3]\,. \tag{129.14}$$

Here we have taken

$$\hat{\sigma}_1 = 2^\nu B\,,$$

and allowed for the condition of continuity of $\hat{\sigma}_r$ when $\xi = \beta$. As will be seen σ_1 is the stress at infinity. It now remains to determine β. We require that the stresses given by formulae (129.14) correspond to point A for $\xi = \beta$ and obtain

$$\beta = \frac{2\hat{\sigma}_1}{3A}\,. \tag{129.15}$$

Now we can eliminate all quantities except $\hat{\sigma}_m$ and $\hat{\sigma}_1$ from eq. (129.13). Indeed B differs from $\hat{\sigma}_1$ only by a factor, A is expressed in terms of $\hat{\sigma}_m$ by formula (129.11), β is eliminated by eq. (129.15) and α is a known number. After all this simplification the following formula is obtained for the stress concentration coefficient $K = \hat{\sigma}_m/\hat{\sigma}_1$

$$K = (3)^{\frac{3}{n-3}}\left[\frac{2n+3}{n}\right]^{\frac{2(2n+3)}{9-n^2}}\,. \tag{129.16}$$

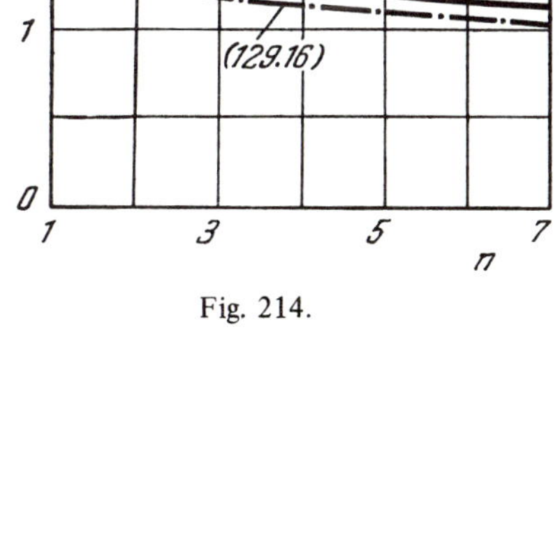

Fig. 214.

Graphs of the dependence of K on the exponent n by the three formulae (129.4), (129.6) and (129.16) are plotted in fig. 214. It should be mentioned that the difference is fairly considerable and is particularly great when $n = 1$, when the first formula gives $K = 2$, which is the known solution of the theory of elasticity, the second formula gives $K = e = 2.718$ and the third gives $K = 1.438$.

130. **Disc of hyperbolic profile loaded by radial forces**

The most general solution of the plane axially-symmetric problem of creep theory with a power law was obtained by Kostyuk [60]. He considered a disc of hyperbolic profile with an external radius b and internal radius a; here the dimensionless radius $\xi = r/a$. The thickness of the plate is a power function of the radius $h = h_0 \xi^{-\lambda}$. The equations of equilibrium and compatibility of deformation are

$$\frac{\mathrm{d}}{\mathrm{d}\xi}(\xi^{-\lambda}\sigma_r) + (\sigma_r - \sigma_\theta)\xi^{-1-\lambda} = 0\,, \tag{130.1}$$

$$\frac{\mathrm{d}\epsilon_\theta}{\mathrm{d}\xi} + \frac{\epsilon_\theta - \epsilon_r}{\xi} = 0\,. \tag{130.2}$$

We shall consider, as usual, that the dimensionless stresses and rates of deformation are defined such that in simple tension $\epsilon = \sigma^n$ or $\sigma = \epsilon^\nu$, consequently,

$$\epsilon_\theta = \sigma_0^{n-1}(\sigma_\theta + \tfrac{1}{2}\sigma_r)\,, \qquad \epsilon_r = \sigma_0^{n-1}(\sigma_r + \tfrac{1}{2}\sigma_\theta)\,. \tag{130.3}$$

Further we introduce a new variable $\omega = \sigma_r/\sigma_\theta$, then it follows from eq. (130.1) that

$$\sigma_\theta = C\frac{\xi^\lambda}{\omega}\exp\int \frac{1-\omega}{\omega}\frac{\mathrm{d}\xi}{\xi}\,. \tag{130.4}$$

Indeed, (130.1) can be rewritten

$$\frac{\mathrm{d}}{\mathrm{d}\xi}\ln(\xi^{-\lambda}\sigma_\theta\omega) - \frac{1-\omega}{\omega\xi} = 0\,,$$

whence we obtain eq. (130.4).

In an exactly similar way, if in eq. (130.2) we use rates of deformation by formulae (130.3) instead of stresses and introduce the parameter ω, then it follows from eq. (130.2) that

$$\sigma_\theta = D\,\frac{(1-\omega+\omega^2)^{-\frac{1-\nu}{2}}}{(2-\omega)^\nu}\exp\left(-3\nu\int\frac{1-\omega}{2-\omega}\,\frac{d\xi}{\xi}\right). \tag{130.5}$$

Equating the right-hand sides of (130.5) and (130.4) we obtain a certain integral expression for determination of ω; taking the derivatives of both sides of this equation we obtain the following differential equation:

$$\frac{1}{\xi}\frac{d\xi}{d\omega} = \frac{1}{2(1-\lambda-3\nu)}\,\frac{(\omega-2)^2+3\nu\omega^2}{(1-\omega+\omega^2)(\omega-p_1)(\omega-p_2)}. \tag{130.6}$$

Here

$$p_{1,2} = \frac{3(1-\nu)-2\lambda \pm \sqrt{9\nu^2+6\nu(1+\lambda)+(1-2\lambda)^2}}{2(1-\lambda-3\nu)}.$$

Eq. (130.6) is integrated and so we obtain the dependence of the radius ξ on the parameter ω:

$$\xi = A(1-\omega+\omega^2)^\alpha\,|\omega-p_1|^\beta\,|\omega-p_2|^\gamma \exp\left(\delta \arctan\frac{2\omega-1}{\sqrt{3}}\right). \tag{130.7}$$

Here A is a constant of integration, the remaining constants have the following values:

$$\alpha = \frac{3(1-\nu)(1-2\lambda-3\nu)}{4\Delta(1-\lambda-3\nu)^2}, \quad \beta = \frac{(p_1-2)^2+3\nu p_1^2}{2(1-\lambda-3\nu)(p_1-p_2)(1-p_1+p_1^2)},$$

$$\gamma = -\frac{(p_2-2)^2+3\nu p_2^2}{2(1-\lambda-3\nu)(p_1-p_2)(1-p_2+p_2^2)}, \quad \delta = \frac{3\sqrt{3}}{2}\,\frac{1-\nu^2}{\Delta(1-\lambda-3\nu)^2},$$

$$\Delta = (1-p_1+p_1^2)(1-p_2+p_2^2).$$

Now in expression (130.4) or (130.5) we can use eq. (130.6) to replace $d\xi/\xi$ by a certain function ω multiplied by $d\omega$. Squaring we obtain σ_θ in terms of ω, namely

$$\sigma_\theta = B(1-\omega+\omega^2)^k |\omega-p_1|^l |\omega-p_2|^m \exp\left(q \arctan\frac{2\omega-1}{\sqrt{3}}\right). \tag{130.8}$$

Here B is a constant of integration whilst

$$k = \frac{1-\nu}{2}\left[1-\tfrac{3}{2}\frac{\lambda(2\lambda-1)+6\nu(\nu+\lambda)}{\Delta(1-\lambda-3\nu^2)}\right], \quad l = -\beta\left[\lambda+3\nu\frac{p_1-1}{p_1-2}\right],$$

$$m = -\gamma\left[\lambda+3\nu\frac{p_2-1}{p_2-2}\right], \qquad q = -\frac{3\sqrt{3}}{2}\frac{(1-\nu)(\lambda+2\nu)}{\Delta(1-\lambda-3\nu)^2}.$$

Therefore, formulae (130.7) and (130.8) give a general solution of the problem represented in parametric form, σ_θ, $\sigma_r = \omega\sigma_\theta$, and ξ are expressed as functions of the parameter ω.

10

Steady-State Creep of Plates and Shells

In this chapter we consider the general theory of creep of plates with allowance for in plane stresses that act in the centre of the plane, the properties of the corresponding creep potentials and the possibility of partially linearising the equations. We then consider problems of symmetrically loaded circular plates which permit of relatively simple numerical solution, some of them being soluble in closed form.

The presentation of the theory of shells is limited to momentless shells, multi-layered shells and the development of an approximate theory based on replacing the real shell by a two-layer model. For axially symmetrical cylindrical shells in particular this approximate theory leads to fairly simple equations which are Euler equations of some functional.

131. **Bending of plates. Fundamental equations of the linear theory**

As usual, we shall define a plate as a body that is bounded by two parallel planes and a cylindrical surface normal to them. The distance between the planes, namely the thickness of the plate, $2h$, is assumed small compared with the least linear dimension of the figure which is formed by intersection between the boundary cylinder and the boundary plane. We shall define the mean plane of the plate as one which is equidistant from the boundary planes and we select the rectangular Cartesian coordinates such that the plane xOy coincides with the mean plane. Then, $z = \pm h$ is the boundary condition equation and $F(x, y) = 0$ is the equation of the boundary surface. Instead of the notation x and y we use x_1 and x_2 in accordance with the general system, but retain the notation z for the third coordinate because of its special role. Unit vectors in the directions of the coordinate axes are denoted $\mathbf{e}_1$, $\mathbf{e}_2$, $\mathbf{k}$ respectively. The radius vector of any point M in the plate can be written

$$\mathbf{r} = x_1\mathbf{e}_1 + x_2\mathbf{e}_2 + z\mathbf{k}\,. \tag{131.1}$$

Put $\boldsymbol{\rho} = x_1\mathbf{e}_1 + x_2\mathbf{e}_2$. The point P, with radius vector $\boldsymbol{\rho}$, is a projection of the point M on the mean plane, it is at a distance z from the point M. Now assume that the plate is bent and stretched but that displacements of points upon it are small. Within the meaning of the linear theory of elasticity, this means that the derivatives of the displacements with respect to the coordinates are negligibly small compared with unity. Kirchoff's hypothesis, which is generally used in the theory of bending plates, states that the relative location of points P and M is maintained during deformation. The segment PM, which is normal to the mean plane of the plate before deformation, remains normal to the surface into which the mean plane is converted and remains of unchanged length, z. Denote the positions of points P and M after deformation by P' and M' respectively. The radius vector of point P' is

$$\boldsymbol{\rho}' = (x_1 + u_1)\mathbf{e}_1 + (x_2 + u_2)\mathbf{e}_2 + w\mathbf{k}\,.$$

Here u_1, u_2 and w denote displacements of points on the mean surface which are functions of x_1 and x_2. Because of Kirchoff's hypothesis the radius vector of point M' is

$$\mathbf{r}' = \boldsymbol{\rho}' + z\mathbf{n}\,.$$

Here $\mathbf{n}$ is the unit vector normal to the deformed mean surface. The lines $x_1 = \text{const}$ and $x_2 = \text{const}$ form a coordinate grid on the deformed mean surface, the vectors that are tangential to these lines form the basis

$$\mathbf{e}_1' = \boldsymbol{\rho}'_{,1}\,, \qquad \mathbf{e}_2' = \boldsymbol{\rho}'_{,2}\,.$$

Here, as usual, the subscripts after the comma denote differentiation with respect to the coordinate x_i. We have

$$\mathbf{e}_1' = (1 + u_{1,1})\mathbf{e}_1 + u_{2,1}\mathbf{e}_2 + w_{,1}\mathbf{k}\,,$$
$$\mathbf{e}_2' = u_{1,2}\mathbf{e}_1 + (1 + u_{2,2})\mathbf{e}_2 + w_{1,2}\mathbf{k}\,. \tag{131.2}$$

The unit vector of the normal is orthogonal to the vectors $\mathbf{e}_1$ and $\mathbf{e}_2$ and consequently,

$$\mathbf{n} = \frac{\mathbf{e}_1 \times \mathbf{e}_2}{|\mathbf{e}_1 \times \mathbf{e}_2|}\,.$$

Once again, to within a second order of smallness we obtain

$$\mathbf{n} = -w_{,1}\mathbf{e}_1 - w_{,2}\mathbf{e}_2 + \mathbf{k}\,.$$

We can now write an expression for $\mathbf{r}'$:

$$\mathbf{r}' = (x_1 + u_1 - zw_{,1})\mathbf{e}_1 + (x_2 + u_2 - zw_{,2})\mathbf{e}_2 + (z + w)\mathbf{k}\,.$$

The quantities $u_1 - zw_{,1}$, $u_2 - zw_{,2}$, w are displacement components of an arbitrary point M relative to the initial system of Cartesian coordinates x_1, x_2, z, consequently, we can calculate the deformation components by the usual formulae:

$$e_{11} = u_{1,1} - zw_{,11}\,, \qquad e_{22} = u_{2,2} - zw_{,22}\,,$$
$$e_{12} = \tfrac{1}{2}(u_{1,2} + u_{2,1}) - zw_{,12}\,, \qquad e_{13} = e_{23} = e_{33} = 0\,. \tag{131.3}$$

Under creep conditions we are interested in rates of deformation which are obtained immediately by differentiating expressions (131.3), namely

$$\epsilon_{11} = \dot{u}_{1,1} - z\dot{w}_{,11}\,, \qquad \epsilon_{22} = \dot{u}_{2,2} - z\dot{w}_{,22}\,,$$
$$\epsilon_{12} = \tfrac{1}{2}(\dot{u}_{1,2} + \dot{u}_{2,1}) - z\dot{w}_{,12}\,. \tag{131.4}$$

Because of the linearity of eqs. (131.3) the rates of deformation are expressed in terms of displacement rates in exactly the same way as deformations are expressed in terms of displacements. We emphasise this point because circumstances are quite different in non-linear theory where the rates of deformation depend on the magnitudes of the displacements as well as on the rates. However, the problem of creep in a plate remains fairly complicated, even in the geometrically linear formulation. The next step is to introduce forces and moments and to use the conditions of equilibrium.

As usual the equations of equilibrium are written

$$\sigma_{11,1} + \sigma_{12,2} + \sigma_{13,3} = 0\,,$$
$$\sigma_{21,1} + \sigma_{22,2} + \sigma_{23,3} = 0\,, \qquad (131.5)$$
$$\sigma_{31,1} + \sigma_{32,2} + \sigma_{33,3} = 0\,.$$

Here we consider body forces absent and the boundary planes free from tangential stress, the loads acting normal to them are q^+ and q^- respectively, so that $\sigma_{33}(h) = q^+$, $\sigma_{33}(-h) = -q^-$. Integrate the equations of equilibrium (131.5) with respect to z between the limits $-h$ and $+h$ and introduce the notation

$$T_{ij} = \int_{-h}^{+h} \sigma_{ij}\,\mathrm{d}z \qquad (i, j = 1, 2)\,.$$

The first two equations give

$$T_{11,1} + T_{12,2} = 0\,, \quad T_{21,1} + T_{22,2} = 0\,. \qquad (131.6)$$

Integrating with respect to z terms of the form $\sigma_{13,3}$ we obtain the expression $\sigma_{13}(+h) - \sigma_{13}(-h)$ and a second analogous expression. These expressions are zero because of the assumption that the boundary planes carry only normal loads. Further, put

$$\int_{-h}^{+h} \sigma_{3i}\,\mathrm{d}z = N_i\,.$$

Then it follows from the third equation of equilibrium that

$$N_{i,i} + q = 0\,. \qquad (131.7)$$

Here $q = q^+ + q^-$ is the total transverse load on the plate. The quantities T_{ij} are termed forces and the quantities N_i shear forces.

Now introduce the moments M_{ij}, defining them as follows:

$$M_{ij} = -\int_{-h}^{+h} z\sigma_{ij}\,dz\,.$$

To derive the equations of the moments multiply each of eqs. (131.5) by z and integrate across the thickness. In the first equation for instance, the first and third terms are derivatives of the corresponding moments with respect to the coordinates, whilst the third term must be integrated by parts, namely

$$\int_{-h}^{+h} \sigma_{13,3}z\,dz = \sigma_{13}z\Big|_{-h}^{+h} - \int_{-h}^{+h} \sigma_{13}\,dz = -N_1\,.$$

Proceeding with the second equation in exactly the same way we obtain

$$\begin{aligned} M_{11,1} + M_{12,2} + N_1 &= 0\,, \\ M_{21,1} + M_{22,2} + N_2 &= 0\,. \end{aligned} \qquad (131.8)$$

It is unnecessary to consider the third equation of equilibrium obtained in this way. Indeed we should here introduce new characteristics of the type $\int_{-h}^{+h} \sigma_{31}z\,dz$ and $\int_{-h}^{+h} \sigma_{33}\,dz$.

132. Potentials of forces and moments. Fundamental equations of creep in plates

By definition the quantities T_{ij} and M_{ij} are two-dimensional symmetrical second rank tensors, with six components. Correspondingly, the state of creep of the plates is described by three components of rate of deformation of the mean surface and three quantities $\dot{w}_{,ij}$ which characterise the change in curvature. Denote the rate of deformation of the mean surface by ϵ_{ij}^0 so that

$$\epsilon_{ij}^0 = \tfrac{1}{2}(\dot{u}_{i,j} + \dot{u}_{j,i}) \qquad (i,j = 1, 2)\,. \qquad (132.1)$$

Correspondingly term the quantities $\dot{w}_{,ij}$ rates of change of curvature and put

$$\kappa_{ij} = \dot{w}_{,ij}\,. \qquad (132.2)$$

These notations are introduced because everything that follows will apply both to the system of Cartesian orthogonal coordinates that was originally selected for simplicity and to arbitrary components of the corresponding tensors, where the expressions of these components in terms of the corresponding displacement vectors are in general different. In the linear theory they are formally written in exactly the same way but the subscripts after the comma symbolise covariant differentiation. In non-linear theory the formulae must be replaced by others but the relationship between tensors ϵ_{ij}^0, κ_{ij} on the one hand and T_{ij} and M_{ij} on the other remain unchanged provided the deformations are small.

Now $\epsilon_{ij} = \epsilon_{ij}^0 - z\kappa_{ij}$, if $i, j = 1, 2$, $\epsilon_{ij} = 0$, if one of the subscripts is three.

To give a law of steady-state creep for a material means to give a potential $U(\epsilon_{ij})$, therefore the function U is assumed known. Here stresses are expressed as

$$\sigma_{ij} = \frac{\partial U(\epsilon_{ij})}{\partial \epsilon_{ij}}.$$

But the rate components that differ from zero are of the form $\epsilon_{ij}^0 - z\kappa_{ij}$, consequently, the function U may be considered as a function of ϵ_{ij}^0 and κ_{ij} containing z as a parameter. It is evident that

$$\frac{\partial U}{\partial \epsilon_{ij}} = \frac{\partial U}{\partial \epsilon_{ij}^0} = -\frac{\partial U}{\partial \kappa_{ij}} \frac{1}{z}.$$

Now write

$$\sigma_{ij} = \frac{\partial U}{\partial \epsilon_{ij}^0},$$

integrating both sides with respect to z between the limits $-h$ and $+h$, we obtain

$$T_{ij} = \int_{-h}^{+h} \frac{\partial U}{\partial \epsilon_{ij}^0} \, \mathrm{d}z = \frac{\partial U^*}{\partial \epsilon_{ij}^0}.$$

Here $U^* = \int_{-h}^{+h} U \, \mathrm{d}z$. On the other hand, write

$$\sigma_{ij} = -\frac{1}{z}\frac{\partial U}{\partial \kappa_{ij}}.$$

Multiply both sides of this equation by $-z$ and integrate, we obtain

$$M_{ij} = \int_{-h}^{+h} \frac{\partial U}{\partial \kappa_{ij}} \mathrm{d}z = \frac{\partial U^*}{\partial \kappa_{ij}}.$$

From now on the function U will not be marked with an asterisk but we shall use the notation ϵ_{ij} for the deformation of the middle surface, the function $U(\epsilon_{ij}, \kappa_{ij})$ may be termed the potential of forces and moments for the plate because

$$T_{ij} = \frac{\partial U}{\partial \epsilon_{ij}}, \qquad M_{ij} = \frac{\partial U}{\partial \kappa_{ij}}. \tag{132.3}$$

Note that the existence of two reciprocal formulae follows from eqs. (132.3):

$$\epsilon_{ij} = \frac{\partial \Phi}{\partial T_{ij}}, \qquad \kappa_{ij} = \frac{\partial \Phi}{\partial M_{ij}}. \tag{132.4}$$

Here $\Phi(T_{ij}, M_{ij})$ is the potential of rate of deformation of the middle surface and change of curvature. Here

$$\Phi = T_{rs}\epsilon_{rs} + M_{rs}\kappa_{rs} - U. \tag{132.5}$$

Indeed, differentiating expression (132.5) with respect to T_{ij} we obtain:

$$\frac{\partial \Phi}{\partial T_{ij}} = \epsilon_{ij} + T_{rs}\frac{\partial \epsilon_{rs}}{\partial T_{ij}} + M_{rs}\frac{\partial \kappa_{rs}}{\partial T_{ij}} - \frac{\partial U}{\partial \epsilon_{rs}}\frac{\partial \epsilon_{rs}}{\partial T_{ij}} - \frac{\partial U}{\partial \kappa_{rs}}\frac{\partial \kappa_{rs}}{\partial T_{ij}}.$$

All the terms on the right-hand side except the first cancel in pairs, consequently, expression (132.3) demonstrates the first of formulae (132.4). The demonstration of the second formula is analogous. It is readily shown that the expression

$$L = T_{ij}\epsilon_{ij} + M_{ij}\kappa_{ij}$$

represents power dissipation in creep. The relationships between forces and moments on the one hand and rates of deformation and changes of curvature on the other, are completely analogous in form to the relationships between stresses and rates of deformation in creep, and therefore we may term these quantities respectively the generalised stresses and generalised rates of deformation, and all the general theorems of steady-state creep theory can be simply paraphrased. In practice, it is difficult to determine the potential of stresses and moments, i.e. the functions $U(\epsilon_{ij}, \kappa_{ij})$, the corresponding integrals must be calculated for each particular creep law. It is still more difficult to write the function $\Phi(T_{ij}, M_{ij})$ in explicit form. Later we shall consider particular cases of creep law, but for the moment we shall proceed with a general formulation of the problem.

Firstly, note that by means of the equations of section 131 and formulae (132.3) we can obtain a complete system of equations for determination of three unknown rates: $\dot{u}_i$ and $\dot{w}$. The first two equations are obtained directly by substituting the expressions for T_{ij} by formulae (132.3) into the equations of equilibrium (131.6) and substituting for ϵ_{ij} and κ_{ij} the derivatives of the displacement given by eqs. (132.1) and (132.2). Further, it follows from eqs. (131.8) that:

$$N_i = -M_{ij,j} \, .$$

Substituting into (131.7) we obtain

$$M_{ij,ij} - q = 0 \, . \qquad (132.6)$$

The third equation is found by substituting the expression for M_{ij} in terms of the potential and replacing ϵ_{ij} and κ_{ij} by the derivatives of displacement.

Unlike the case of an elastic plate the equations of bending and the equations which describe deformation of the mean plane do not reduce to two independent systems. We shall not write these equations, because in any case they are of no practical use and in the general case we shall only consider problems in which they can be partially linearised.

133. **Partial linearisation of the equations of creep of a plate**

Here we consider two possibilities which can arise in solving particular classes of problems.

a) The principal state of stress of the plate is plane stress uniformly distributed across the thickness and bending stresses are small compared with the principal stresses. The stress potential depends on the arguments $\epsilon_{ij} - z\kappa_{ij}$. Considering that $|z\kappa_{ij}| \ll |\epsilon_{ij}|$, expand the function $U_0(\epsilon_{ij} - z\kappa_{ij})$ into a series of terms $z\kappa_{ij}$, and consider only first powers of κ_{ij}. (We have used the notation U_0 for the stress potential because in this chapter the potential of stresses and moments is denoted by U.)

$$U = U_0(\epsilon_{ij}) - z\kappa_{rs} \frac{\partial U_0}{\partial \epsilon_{rs}}.$$

Here U_0 depends only on ϵ_{ij}. Now the stresses are determined as

$$\sigma_{ij} = \frac{\partial U_0}{\partial \epsilon_{ij}} - z\kappa_{rs} \frac{\partial^2 U_0}{\partial \epsilon_{ij} \partial \epsilon_{rs}}. \tag{133.1}$$

Integrate both sides of this equation across the thickness, we obtain

$$T_{ij} = 2h \frac{\partial U_0}{\partial \epsilon_{ij}}.$$

Multiply both sides of eq. (133.1) by z and integrate across the thickness, we obtain

$$M_{ij} = \frac{2h^3}{3} \frac{\partial^2 U_0}{\partial \epsilon_{ij} \partial \epsilon_{rs}} \kappa_{rs}.$$

The function $U_0(\epsilon_{ij})$ multiplied by $2h$ will be called the potential $U(\epsilon_{ij})$. Further, introduce the notation

$$\frac{h^2}{3} \frac{\partial^2 U}{\partial \epsilon_{ij} \partial \epsilon_{rs}} = D_{ijrs}.$$

Then we obtain the formulae

$$T_{ij} = \frac{\partial U}{\partial \epsilon_{ij}}, \qquad M_{ij} = D_{ijrs} \kappa_{rs}. \tag{133.2}$$

The quantities T_{ij} are substituted into eqs. (131.6) which then assume the form

$$T_{ij,j} = \frac{\partial^2 U}{\partial \epsilon_{ij} \partial \epsilon_{rs}} \epsilon_{rs,j} = 0,$$

or

$$D_{ijrs}\dot{u}_{r,sj} = 0\,. \tag{133.3}$$

This system of equations suffices to determine the rates of deformation of the middle surface, which are determined independently of the deflection.

Now substitute the expression for moments into eq. (132.6) and obtain the following equation for the function w:

$$(D_{ijrs}\dot{w}_{,rs})_{,ij} - q = 0\,. \tag{133.4}$$

This is a fourth order linear equation with variable coefficients, which are known functions of the coordinates, because the solution of the first problem of determining the plane stressed state in a plate is considered known. It is readily shown that eq. (133.4) is a minimum condition of the functional

$$I = \iint \left[\tfrac{1}{2} D_{ijkl}\dot{w}_{,ij}\dot{w}_{,kl} - q\dot{w}\right] \mathrm{d}x_1\,\mathrm{d}x_2\,.$$

Because of this, approximate solutions of eq. (133.4) can be found by variational methods. Eq. (133.4) coincides exactly with the equation of deflection of an elastic anisotropic plate, the quantities D_{ijkl} serve as coefficients of anisotropy.

b) The principal state of stress is due to deflection, the rates of deformation of points on the mean surface being small compared with the rates of deformation of points on the boundary planes $z = \pm h$, i.e. $|\epsilon_{ij}| \ll |z\kappa_{ij}|$. The expansion for potential U_0 is:

$$U_0 = U_0(-z\kappa_{ij}) - \frac{1}{z}\frac{\partial U_0}{\partial \kappa_{ij}}\epsilon_{ij}\,.$$

Consequently

$$\sigma_{ij} = -\frac{1}{z}\frac{\partial U_0}{\partial \kappa_{ij}} + \frac{1}{z^2}\frac{\partial^2 U_0}{\partial \kappa_{ij}\partial \kappa_{rs}}\epsilon_{rs}\,.$$

Here U_0 is considered as function of the arguments $-z\kappa_{ij}$. Integrating both sides across the thickness we obtain

$$T_{ij} = -\frac{\partial}{\partial \kappa_{ij}}\int_{-h}^{+h} U_0(-z\kappa_{rs})\frac{\mathrm{d}z}{z} + \epsilon_{rs}\frac{\partial^2}{\partial \kappa_{ij}\partial \kappa_{rs}}\int_{-h}^{+h} U_0(-z\kappa_{rs})\frac{\mathrm{d}z}{z^2}\,.$$

Now multiplying both sides of the expression for σ_{ij} by z and integrating, we obtain

$$M_{ij} = \frac{\partial}{\partial \kappa_{ij}} \int_{-h}^{+h} U_0(-z\kappa_{rs})\,\mathrm{d}z - \epsilon_{rs} \frac{\partial^2}{\partial \kappa_{ij} \partial \kappa_{rs}} \int_{-h}^{+h} U_0(-z\kappa_{rs}) \frac{\mathrm{d}z}{z}\,.$$

It is now important to know whether the creep law is the same in tension and in compression. The only case that we shall consider is that in which these laws are identical and consequently $U(-z\kappa_{ij}) = U(z\kappa_{ij})$. Then, obviously, the first integral in the expression for T_{ij} and the second integral in the formula for M_{ij} disappear. We term the quantity

$$U = \int_{-h}^{+h} U_0(-z\kappa_{ij})\,\mathrm{d}z$$

the potential of moments and also put

$$C_{ijkl} = \frac{\partial^2}{\partial \kappa_{ij} \partial \kappa_{rs}} \int_{-h}^{+h} U_0(-z\kappa_{ij}) \frac{\mathrm{d}z}{z^2}\,.$$

Then

$$M_{ij} = \frac{\partial U}{\partial \kappa_{ij}}, \qquad T_{ij} = C_{ijkl}\epsilon_{kl}\,. \tag{133.5}$$

Now eq. (132.6) becomes

$$\left(\frac{\partial U}{\partial \kappa_{ij}}\right)_{,ij} - q = 0\,. \tag{133.6}$$

This equation contains one independent function $\dot{w}$. The second group of eqs. (133.5) expresses force in terms of rate of deformation by formulae which are completely analogous to the formulae that relate force and deformation in a linear-elastic anisotropic plate, and the coefficients of anisotropy are known functions of the coordinates.

134. Creep of plates in bending

If, from the boundary conditions of the problem, there are no forces on the contour of the plate and in its plane, then the forces T_{ij} are identically zero (so far we consider only the linear approximation) and creep in bending is completely described by eq. (133.6). Now let us determine the form of the creep law, and assume as in section 99 that the creep potential depends on some first power invariant function v of the rate-of-deformation tensor component ϵ_{ij}. If $\epsilon_{ij} = -z\kappa_{ij}$ then $v = zk$ because of uniformity and k is an expression which is formed in exactly the same way from the components of κ_{ij} as v is formed from ϵ_{ij}.

Now the potential of moments is given by

$$U(k) = \frac{1}{k} \int_{-hk}^{+hk} U_0(\xi)\, d\xi = \frac{2}{k} \int_0^{hk} U_0(\xi)\, d\xi \,. \tag{134.1}$$

The second expression can be written because creep in tension and in compression are assumed to follow the same law. Otherwise, as we have seen in section 133, bending is accompanied by deformation of the mean plane and these two forms of deformation must be considered simultaneously, stresses also arise in the mean surface with the corresponding zero boundary conditions. Now the expressions for moments are rewritten as

$$M_{ij} = U'(k) \frac{\partial k}{\partial \kappa_{ij}} . \tag{134.2}$$

It is not difficult to write the converse formulae. Proceeding as in section 99, determine the uniform first power function $m(M_{ij})$ which is expressed in terms of k. The potential of rate-of-change of curvature is

$$\Phi = U'(k)k - U(k) = \Phi(m) \,.$$

Consequently

$$\kappa_{ij} = \Phi'(m) \frac{\partial m}{\partial M_{ij}} . \tag{134.3}$$

Write the formula that relates moments to curvatures for creep criteria of Mises and Tresca types with power and exponential creep laws.

a) *Mises type criterion.* Put

$$m^2 = M_0^2 = M_{11}^2 + M_{22}^2 - M_{11}M_{22} + \tfrac{3}{2}(M_{12}^2 + M_{21}^2),$$

$$k^2 = \kappa_0^2 = \tfrac{4}{3}[\kappa_{11}^2 + \kappa_{22}^2 + \kappa_{11}\kappa_{22} + \tfrac{1}{2}(\kappa_{12}^2 + \kappa_{21}^2)] .$$

We then obtain the following formulae:

$$M_{11} = \frac{4}{3}\frac{U'(\kappa_0)}{\kappa_0}(\kappa_{11} + \tfrac{1}{2}\kappa_{22}), \quad M_{22} = \frac{4}{3}\frac{U'(\kappa_0)}{\kappa_0}(\kappa_{22} + \tfrac{1}{2}\kappa_{11}),$$

$$M_{12} = \frac{2}{3}\frac{U'(\kappa_0)}{\kappa_0}\kappa_{12} .$$

Substituting into the expression for M_0^2 we find that

$$U'(\kappa_0) = M_0 . \tag{134.4}$$

The following way of writing these formulae is more convenient in some respects:

$$M_{11} = \frac{4}{3}\frac{M_0(\kappa_0)}{\kappa_0}(\kappa_{11} + \tfrac{1}{2}\kappa_{22}), \quad M_{22} = \frac{4}{3}\frac{M_0(\kappa_0)}{\kappa_0}(\kappa_{22} + \tfrac{1}{2}\kappa_{11}),$$

$$M_{12} = \frac{2}{3}\frac{M_0(\kappa_0)}{\kappa_0}\kappa_{12} . \tag{134.5}$$

The converse formulae to (134.5) are

$$\kappa_{11} = \frac{\kappa_0(M_0)}{M_0}(M_{11} - \tfrac{1}{2}M_{22}), \quad \kappa_{22} = \frac{\kappa_0(M_0)}{M_0}(M_{22} - \tfrac{1}{2}M_{11}),$$

$$\kappa_{12} = \frac{3}{2}\frac{\kappa_0(M_0)}{M_0}M_{12} . \tag{134.6}$$

Calculate the function $M_0(\kappa_0)$ for the power law of creep. In this case

$$U_0(\epsilon) = \frac{n}{n+1}\sigma_n\epsilon_n\left(\frac{\epsilon}{\epsilon_n}\right)^{1+\frac{1}{n}}$$

Calculating $U(\kappa_0)$ by formula (134.1) we find

$$U = \frac{n}{n+1} M_n \kappa_n \left(\frac{\kappa_0}{\kappa_n}\right)^{1+\frac{1}{n}} . \tag{134.7}$$

In this formula we have

$$M_n = \frac{2n}{n+1} \sigma_n h^2 , \quad \kappa_n = \frac{\epsilon_n}{h} .$$

Hence, on the basis of eq. (134.4) it follows that

$$\frac{\kappa_0}{\kappa_n} = \left(\frac{M_0}{M_n}\right)^n . \tag{134.8}$$

Expression (134.8) coincides exactly with the fundamental equation that relates moment to rate of change of curvature in the bending of beams and the expressions for M_n and κ_n are the same as for a beam of square section of height $2h$ and unit width. Naturally, in considering bending of a narrow strip cut from a plate, M_0 and κ_0 become respectively the bending moment and the rate of change of curvature in the plane of bending.

Now for the exponential creep law we can simply take the relationship between moment and rate of curvature which was found for a beam in section 106. As we have already explained the singularity of the exponential law for the low stress region, that is for small values of z, has no significant influence on the relationship between M and κ, therefore we use formula (106.8):

$$\frac{M_0}{M_e} = \ln \left(\frac{\kappa_0}{\kappa_e}\right) - \tfrac{1}{2} .$$

Here

$$M_e = \sigma_e h^2 , \quad \kappa_e = \frac{\epsilon_e}{h} . \tag{134.9}$$

b) *Tresca type criterion.* This criterion only gives a simple result when the principal directions are known in advance. The corresponding bending moments are M_1 and M_2 and the rates of change of curvature κ_1 and κ_2. The rates of deformation at a point with coordinates x_1, x_2, z are $\epsilon_1 = -z\kappa_1$, $\epsilon_2 = -z\kappa_2$. The distribution of stresses that correspond to these rates of

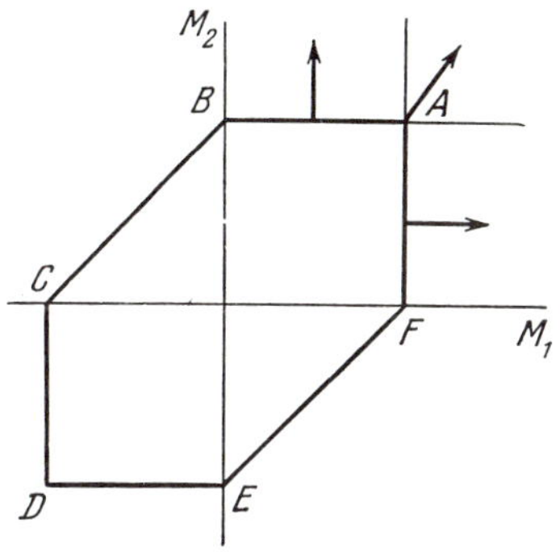

Fig. 215.

deformation depends on the rate distribution as shown in fig. 215, which illustrates a Tresca hexagon for moments. Consider each possible case in turn:

1. $\kappa_1 > 0$, $\kappa_2 > 0$, $\epsilon_1 > 0$, $\epsilon_2 > 0$ when $z < 0$, $\epsilon_1 < 0$, $\epsilon_2 < 0$ when $z > 0$. Both rates are positive simultaneously when and only when the state of stress is represented by point A of the Tresca hexagon, consequently, $\sigma_1 = \sigma_2 > 0$ when $z < 0$. In exactly the same way point D corresponds to positive values of z. Here the creep rates are related by the following condition (section 119):

$$\epsilon_1 + \epsilon_2 = v(\sigma_1) \quad \text{or} \quad \sigma_1 = \sigma_2 = s(\epsilon_1 + \epsilon_2) .$$

This expression is valid for both points A and D and, therefore, throughout the thickness of the plate, we have

$$\sigma_1 = \sigma_2 = s[-z(\kappa_1 + \kappa_2)] .$$

Multiplying by z and integrating we find

$$M_1 = M_2 = m(\kappa_1 + \kappa_2) . \tag{134.10}$$

Here the function s is the same as the function that defines the creep law with uniform tension, that is $s(\epsilon) = \mathrm{d}U_0/\mathrm{d}\epsilon$. By integration we obtain the function $U(k)$ and consequently $m(k) = U'(k)$, as in the previous case of eq. (134.4). The creep law can now either be written in the form of eq. (134.10) or in the form

$$\kappa_1 = \lambda k(M_1) , \quad \kappa_2 = (1-\lambda)k(M_1) , \quad 0 \leqslant \lambda \leqslant 1 . \tag{134.11}$$

2. $\kappa_2 = 0$, $\kappa_1 > 0$, $\epsilon_2 = 0$, $\epsilon_1 < 0$ when $z > 0$ and $\epsilon_1 > 0$ when $z < 0$. The state of stress is represented by the sides AF and CD of the Tresca hexagon and we readily obtain

$$M_1 = m(\kappa_1), \quad \kappa_1 = k(M_1). \tag{134.12}$$

3. $\kappa_1 > 0$, $\kappa_2 < 0$. The state of stress corresponds to points F and C, or B and E of the Tresca hexagon, in the first case $M_2 = 0$, whereas

$$M_1 = m(\kappa_1), \quad \kappa_1 = k(M_1). \tag{134.13}$$

The quantity κ_2 here remains indeterminate. In the second case

$$M_2 = m(\kappa_2), \quad \kappa_2 = k(M_2); \tag{134.14}$$

and the quantity κ_1 remains indeterminate.

4. $\kappa_1 = -\kappa_2$. The state of stress corresponds to the sides CB and EF,

$$M_1 - M_2 = m(\kappa_1 - \kappa_2), \quad \kappa_1 - \kappa_2 = k(M_1 - M_2). \tag{134.15}$$

In an exactly similar way we could consider creep of a plate using the maximum referred stress criterion.

135. **Creep of circular plates by Mises type theory**

Creep in circular plates with axially symmetric loading has been closely studied. These problems are unidimensional and therefore the differential equations that are obtained can be integrated, at least numerically. In general, when quadratic creep criteria are used only numerical solutions are possible and we shall consider how to obtain them.

In axially symmetric deformation of a round plate of radius b, M_r and M_θ respectively denote the radial and circumferential moments, and κ_r and κ_θ the corresponding rates of change of curvature. Here by formulae (134.5) of the preceding section we have

$$M_r = \frac{4}{3}\frac{M_0}{\kappa_0}(\kappa_r + \tfrac{1}{2}\kappa_\theta)\,,$$

$$M_\theta = \frac{4}{3}\frac{M_0}{\kappa_0}(\kappa_\theta + \tfrac{1}{2}\kappa_r)\,, \qquad (135.1)$$

$$M_0^2 = M_r^2 + M_\theta^2 - M_r M_\theta\,,$$

$$\kappa_0^2 = \tfrac{4}{3}(\kappa_r^2 + \kappa_\theta^2 + \kappa_r\kappa_\theta)\,.$$

The rates of change of curvature are found by formulae (132.2) in which we must now consider differentiation of the deflection w as covariant differentiation.

As the only independent variable is the radius r, we obtain

$$\kappa_r = \frac{d^2\dot{w}}{dr^2}\quad \kappa_\theta = \frac{1}{r}\frac{d\dot{w}}{dr}\,.$$

Eliminating $\dot{w}$ from these expressions we find the following equation of compatibility of rates of change of curvature:

$$\frac{d}{dr}(r\kappa_\theta) - \kappa_r = 0\,. \qquad (135.2)$$

Eq. (135.2) is fully analogous to the equation of compatibility of velocities (125.5) for axially symmetrical deformation.

Further write the equations of equilibrium (131.8) in polar coordinates. The second of these is fulfilled identically (if the subscript r is identified with the subscript 1), and the first assumes the form

$$\frac{d}{dr}(rM_r) - M_\theta + rN_r = 0\,. \qquad (135.3)$$

Moreover, because of eq. (131.7)

$$\frac{d}{dr}(rN_r) + rq = 0\,. \qquad (135.4)$$

In solving particular problems it is convenient to deal directly with the system of eqs. (135.2) – (135.4) without eliminating unknowns and to reduce the system to a single second order equation. Applying the idea that was used in the case of discs in section 125 we put

$$M_r = \frac{2}{\sqrt{3}} M_0 \sin\left(\varphi - \frac{\pi}{6}\right), \quad M_\theta = \frac{2}{\sqrt{3}} M_0 \sin\left(\varphi + \frac{\pi}{6}\right). \tag{135.5}$$

From these identities, by formulae (135.1), it follows that

$$\kappa_r = -\tfrac{3}{2}\kappa_0 \cos\left(\varphi + \frac{\pi}{6}\right), \quad \kappa_\theta = \tfrac{3}{2}\kappa_0 \cos\left(\varphi - \frac{\pi}{6}\right). \tag{135.6}$$

Now substitute expressions (135.5) for moments into the equation of equilibrium (135.3) and substitute equations (135.6) for rates of change of curvature into (135.2). We obtain

$$\begin{gathered}
\varphi' \hat{M} \cos\left(\varphi - \frac{\pi}{6}\right) + \hat{M}' \sin\left(\varphi - \frac{\pi}{6}\right) - \frac{\hat{M}}{\xi} \cos\varphi + \hat{N} = 0\,, \\
\varphi' \hat{\kappa} \sin\left(\varphi - \frac{\pi}{6}\right) - \hat{\kappa}' \cos\left(\varphi - \frac{\pi}{6}\right) - \hat{\kappa}\frac{\sqrt{3}}{2}\cos\varphi = 0\,.
\end{gathered} \tag{135.7}$$

Here

$$\hat{M} = \frac{M_0}{M_*}, \quad \hat{\kappa} = \frac{\kappa_0}{\kappa_*}, \quad \hat{N} = \frac{\sqrt{3}}{2}\frac{Nb}{M_*}, \quad \xi = \frac{x}{b}, \quad \varphi' = \frac{d\varphi}{d\xi}, \ldots .$$

M_* and κ_* are arbitrary constants with the dimensions of force and rate of change of curvature respectively, they are selected in such a way that the relationship between M_0 and κ_0 can be written in the form $\hat{\kappa} = \hat{\kappa}(\hat{M})$. Eqs. (135.7) are very similar to eqs. (125.4) and (125.6) for a disc of constant thickness, the only difference is in the last term of the left-hand side of the first equation, which corresponds to the load on the plate. This equation is written in somewhat different form because we have not introduced the logarithm of the radius as an independent variable and ln m is not one of the required functions, as was the case in section 125. For the purpose of numerical integration it is a matter of indifference which of these forms is used.

The shear force N for a solid plate is immediately found from eq. (135.4). Indeed

$$N = -\frac{1}{r}\int_0^r qr\,\mathrm{d}r\,.$$

When a concentrated force P is applied at the centre

$$N_r = \frac{P}{2\pi r}, \quad \hat{N} = \frac{\hat{P}}{\xi}, \quad \text{here } \hat{P} = \frac{\sqrt{3}}{4\pi}\,\frac{P}{M_*}\,.$$

With a distributed load

$$N_r = \frac{qr}{2}, \quad \hat{N} = \hat{q}\xi\,, \quad \text{here } \hat{q} = \frac{\sqrt{3}}{4}\,\frac{qb^2}{M_*}\,.$$

In the most general case we can always put

$$\hat{N} = \hat{q}\psi(\xi)\,,$$

where $\psi(\xi)$ is a known function of ξ, q is a dimensionless load parameter. Depending on the method of support or fixing, the function $\psi(\xi)$ may contain a certain number of indeterminate constants. Rewrite eqs. (135.7) as follows:

$$\begin{aligned} \frac{\mathrm{d}\varphi}{\mathrm{d}\xi} &= \frac{1}{\xi}\,\frac{D_\varphi(\varphi, \hat{M}, \xi)}{D(\varphi, \hat{M}, \xi)}\,, \\ \frac{\mathrm{d}\hat{M}}{\mathrm{d}\xi} &= \frac{1}{\xi}\,\frac{D_M(\varphi, \hat{M}, \xi)}{D(\varphi, M, \xi)}\,. \end{aligned} \tag{135.8}$$

Here

$$D = -\hat{M}\,\frac{\mathrm{d}\hat{\kappa}}{\mathrm{d}\hat{M}}\cos^2\left(\varphi - \frac{\pi}{6}\right) - \hat{\kappa}\,\sin^2\left(\varphi - \frac{\pi}{6}\right),$$

$$D_\varphi = (\hat{N}\xi - \hat{M}\cos\varphi)\,\frac{\mathrm{d}\hat{\kappa}}{\mathrm{d}\hat{M}}\cos\left(\varphi - \frac{\pi}{6}\right) - \sqrt{3}\hat{\kappa}\cos\varphi\sin\left(\varphi - \frac{\pi}{6}\right),$$

$$D_M = 2\hat{M}\hat{\kappa}\,\cos^2\varphi + \xi\hat{N}\hat{\kappa}\,\sin\left(\varphi - \frac{\pi}{6}\right).$$

Assume that the solution of these equations is found, i.e. we know $\hat{M}$ and φ as functions of ξ. Then the deflection is found by quadrature, indeed

$$\frac{1}{r}\frac{\mathrm{d}\dot{w}}{\mathrm{d}r} = \kappa_\theta = \tfrac{3}{2}\kappa_0 \cos\left(\varphi - \frac{\pi}{6}\right).$$

Introduce the dimensionless rate of deflection

$$\hat{w} = \frac{\dot{w}}{b^2\kappa_*}.$$

Then

$$\hat{w} = \tfrac{3}{2}\int \hat{\kappa}\cos\left(\varphi - \frac{\pi}{6}\right)\xi\,\mathrm{d}\xi. \tag{135.9}$$

In practical integration of the system (135.8) it is somewhat more convenient to take the angle φ as independent variable. Then the system becomes

$$\frac{\mathrm{d}\xi}{\mathrm{d}\varphi} = \xi\frac{D}{D_\varphi}, \qquad \frac{\mathrm{d}\hat{M}}{\mathrm{d}\varphi} = \xi\frac{D_M}{D_\varphi}.$$

With hand calculation the advantage is that the region of integration with respect to φ divides up into equal intervals and the trigonometric functions of the angle φ that enter into the expressions for D, D_φ and D_M are calculated first and then only very simple and easy operations are required to determine these quantities at each stage of the numerical calculations.

For example, the boundary conditions for a plate that is freely supported on the periphery are

$$M_r(0) = M_\theta(0), \qquad M_r(1) = 0.$$

We can write

$$\varphi(0) = \frac{\pi}{2}, \qquad \varphi(1) = \frac{\pi}{6}.$$

For plates that are clamped around the edge, $\kappa_\theta(1) = 0$, consequently,

$$\varphi(0) = \frac{\pi}{2}, \qquad \varphi(1) = -\frac{5\pi}{6}.$$

In making the calculations we assume a value of the function $\hat{M}(\xi)$ at the point from which calculations start, for a continuous plate at the point $\xi = 0$. Put $\hat{M}(0) = \mu$ and assume a certain value $\mu = \mu_1$. Now integrate numerically from $\xi = 0$ until the angle φ assumes the value that it should have on the second boundary. This does not, of course, occur when $\xi = 1$ as it would if we had guessed the initial value of μ accurately, but at a certain value $\xi = \xi_1$. This means that for the given value of the load parameter we have found a solution of the problem of bending a plate of which the radius is not b but $b_1 = \xi_1 b$. But the radius does not appear explicitly in the equations, it is only contained in the load parameter and the dimensionless rate of bending. To be specific, assume that the load is uniformly distributed. Then the solution that is found corresponds to the dimensionless load

$$\hat{q}_1 = \frac{\sqrt{3}}{4} \frac{qb^2}{\xi_1^2 M_*},$$

and by formula (135.9) we find the dimensionless rate of deflection

$$\hat{w}_1 = \frac{\dot{w}\xi_1^2}{b^2 \kappa_*}.$$

Repeating the calculation for different values of μ_1 we can construct a graph of the relationship between $\hat{w}_1$ and $\hat{q}_1$ which is universal for calculating plates of this type with the given creep law. The exponential creep law has the advantage that it contains no constants of the material except σ_e and ϵ_e, therefore, the graph constructed in this way is fully universal. By way of

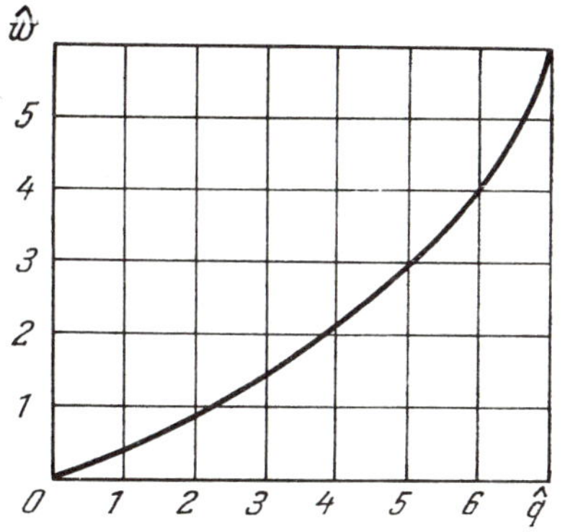

Fig. 216.

example, fig. 216 shows a graph of the dependence of the deflection when $\xi = 0$ on $\hat{q}$ for a continuous plate supported around the edge and with a uniformly distributed load [99]. With the exponential law of creep we should take $M_* = M_e = \sigma_e h^2$, $\kappa_* = \kappa_e = \epsilon_e / h$, therefore,

$$\hat{q} = \frac{\sqrt{3}}{4} \frac{qb^2}{\sigma_e h^2}, \qquad \hat{w} = \frac{\dot{w} b}{\epsilon_e h^2} .$$

With a power creep law the solution depends on the power exponent n but here other simplifications are possible. Indeed suppose

$$\hat{\kappa} = \hat{M}^n .$$

Assume that by the procedure described above we have found a solution for some arbitrary value $\hat{M}(0) = \mu_1$. The corresponding load parameter is $\hat{q}_1$, the corresponding rate of deflection at the centre is $\hat{w}_1$. Increase the load parameter by a factor λ. The first of the eqs. (135.7) is satisfied by taking that the moment $\hat{M}$ is also increased by a factor λ. The corresponding rate of change of curvature increases in the ratio λ^n, consequently, the second of the eqs. (135.7) is also satisfied. Hence it follows that when the load is increased in the ratio λ, the distribution of bending moments remains unchanged, and their values are altered in the ratio λ, the rates of deflection are altered in the ratio λ^n. We obtain

$$\hat{M} = \hat{M}_1 \left(\frac{\hat{q}}{\hat{q}_1} \right) , \qquad \hat{w} = \hat{w}_1 \left(\frac{\hat{q}}{\hat{q}_1} \right)^n . \tag{135.10}$$

Thus for a given value of the exponent n it suffices to make a single calculation starting with an arbitrary value of the moment $\hat{M}$ at the centre of the plate. Of course, the calculation must be repeated for each new value of n. This property of the power creep law has already been mentioned in section 100, and was used in the demonstration of theorems in section 101.

The construction of a universal solution of this kind for circular plates with simple boundary conditions is useful because for example, a circular, freely supported plate with a uniformly distributed load and with uniform temperature distribution is a fairly common structural element. The situation is quite different in the case of a rotating disc; a continuous disc of constant thickness with no load on the outer edge, rotating with a steady-state temperature distribution is comparatively rarely encountered as a structural element, and, therefore, problems in which universal graphs can be constructed are mainly of a model nature in creep theory and not of great practical importance.

Numerical integration of the equations of plasticity theory for plates with a power law of work-hardening and of the deformation theory of plasticity were developed by Sokolovskiy [134], the results can immediately be applied to the case of creep. Thus, for a freely supported plate carrying a uniformly distributed load when $n = 3$, Sokolovskiy obtained the following result:

$$\hat{w} = 0.87 \cdot 10^{-3} \left(\frac{qb^2}{2M_n}\right)^3 . \tag{135.11}$$

For a plate with a central hole, for example, the solution depends on one further parameter, the ratio of the radii, and a great deal of calculation is required to construct dimensionless graphs or tables that can be generally used in calculating plates of this type.

136. **The application of variational methods to the calculation of plates**

Creep calculations on plates, unlike those on discs, consist firstly in determining rates of deflection and in most technical applications the limiting factor is the maximum deflection and not the maximum stress. Therefore, particular attention has been paid to determining displacements rather than distribution of moments. The Ritz or Galerkin methods give sufficiently good results for rates of deflection, in the constrained single-parametric representation of the solution the difference between these two methods is insignificant. Kachanov [52] developed the Ritz method in application to bending of plates. On the basis of eq. (96.3) the corresponding variational Lagrange equation is

$$\int_a^b U(\kappa_0) r\,dr - \int_a^b q(r)\dot{w}(r) r\,dr = \min . \tag{136.1}$$

Here b and a are the outer and inner radii of the plate respectively. For practical solution by the Ritz method we write the deflection in the form

$$\dot{w}(r) = c_1\varphi_1(r) + c_2\varphi_2(r) + \dots .$$

Here the functions $\varphi_i(r)$ satisfy the geometrical boundary conditions. Calculate κ_0, substitute the result into (136.1); the left-hand side of this equation is a certain function of the constant c_i; minimising this function with respect to the argument c_i, we find the best approximation in Ritz terms.

The simplest way of applying the Ritz method to cases of this kind is to take as function $\varphi_1(r)$ the deflection in the corresponding problem of the theory of elasticity and to retain only one term in the expansion of $\dot{w}(r)$. It is particularly convenient to construct these approximate solutions for a power creep law. Indeed κ_0 is proportional to c_1, if we take only one term in the bending expression the function U_0 is proportional to κ_0^{n-1}, consequently, the first term of eq. (136.1) can be represented in the form $c_1^{n+1} I_1$, the second term is of the form $c_1 \hat{q} I_2$. Here I_1 and I_2 are constant numbers which are calculated as integrals of given and fully determinate functions, whereas $\hat{q}$ is a dimensionless load that is introduced by one means or another. Eq. (136.1) assumes the form

$$c_1^{n+1} I_1 - \hat{q} c_1 I_2 = \min .$$

Whence follows

$$c_1 = \left(\frac{\hat{q} I_2}{(n+1) I_1} \right)^{1/n} .$$

Kachanov used this method to calculate a continuous circular plate, supported or clamped at the edge and subject to uniform pressure. For a plate supported around the edge, when $n = 3$ the result differs from that given by formula (135.11), in that the factor is 0.85 instead of 0.87.

Malinin [71] used Galerkin's method to solve these problems. He substituted expressions (135.1) for moments in terms of rate of change of curvature into the equation of equilibrium (135.3) and obtained a second order differential equation for one unknown function $\theta(r) = \mathrm{d}\dot{w}/\mathrm{d}r$. With a power creep law and single-parametric form of the function $\theta(r)$, the parameter c_1 everywhere includes a factor, i.e. the same situation arises as was explained above in relation to the Ritz method. Every time the problem resolves to calculation of certain standard integrals of unknown functions. For the four simplest cases, a circular plate supported or clamped round the edge and bearing a uniformly distributed load or load concentrated at the centre, the relationship between deflection at the centre and the load has been calculated for different values of the exponent n. It is evident that the result can always be written in the form

$$\hat{w} = k\hat{q}^n \quad \text{or} \quad \hat{w} = k\hat{P}^n ;$$

the graph of the coefficient k against n for these cases is given in volume 3 of the book [108]. This book also gives solutions for an annular plate with internal radius a and external radius b and a plate resting on an inner contour and loaded with a distributed linear load Q on the outer edge. This solution is tabulated for two parameters: n and $\alpha = a/b$.

137. **Creep calculation of plates with the Tresca criterion**

As was shown in section 134, when using Tresca's criterion the relationships between rates of change of curvature and bending moments are exactly the same as those between the principal rates of deformation and the principal stresses in the plane state of stress. If we are given a point in the plane M_r, M_θ that represents the state of bending of an element of the plate we can always construct a Tresca hexagon for which the point lies either on one of the sides or on one of the apexes. The formula that expresses the relationship between moments and rates of change of curvature depends on the position of the point (see section 134 and fig. 215). Therefore, the problem of bending a plate is solved in the same way as that of creep of a disc except that in plates the loads and boundary conditions are more varied. In simple problems it is correspondingly possible to obtain simple solutions convenient for practical use. In more complicated problems use of the Tresca type criterion overcomes the difficulties that were encountered in using a quadratic creep criterion but leads to new difficulties which it is sometimes not easy to overcome. The essentials of the matter are illustrated by a few examples. Here the nature of the external load is important and accordingly we group the examples in order of increasing complexity. We commence directly from the equation of equilibrium; the kinematic freedom of deformation is severely limited for each side of the Tresca hexagon and so the parameters of change of curvature can be represented as known functions of the radius containing only a small number of constants. Suppose as before that the outer and inner radii of the disc are b and a respectively, put $\xi = r/b$, $\alpha = a/b$ and determine the dimensionless moments $\hat{M}_r = M_r/M_*$, $\hat{M}_\theta = M_\theta/M_*$ and the dimensionless rates of change of curvature $\hat{\kappa}_r = \kappa_r/\kappa_*$, $\hat{\kappa}_\theta = \kappa_\theta/\kappa_*$. Here M_* and κ_* are selected such that the relationship between the torque and the corresponding rate of curvature is of the form $\hat{M} = m(\hat{\kappa})$, where the function in the right-hand side does not contain dimensional constants and is of the simplest possible form. Thus, with a power creep law we require that $\hat{\kappa} = \hat{M}^n$. The dimensionless shear force $\hat{N}$ is defined by the ratio $\hat{N} = Nb/M_*$; unlike the case of section 135 we do not introduce the factor $\sqrt{3}/2$ into this definition, neither is this factor included in the dimensionless force $\hat{P}$ and the load $\hat{q}$ defined in that section. Therefore the equation of equilibrium is written in the form

$$\frac{d}{d\xi}(\xi\hat{M}_r) - \hat{M}_\theta + \xi\hat{N} = 0 . \qquad (137.1)$$

A plate loaded by a moment distributed round the periphery. Assume that $\hat{M}_\theta > \hat{M}_r > 0$, then $\dot{\hat{\kappa}}_r = d^2\dot{\hat{w}}/d\xi^2 = 0$. Here $\hat{w}$ is defined as in section 135; $\dot{\hat{w}} = \dot{w}/b^2\kappa_*$. The plate is deformed in such a way that $d\dot{\hat{w}}/d\xi = \theta = \text{const}$. Since with a load the moment of the shear force is zero, eq. (137.1) assumes the form

$$\frac{d}{d\xi}(\xi\hat{M}_r) = m\left(\frac{\theta}{\xi}\right).$$

With the power law

$$\frac{d}{d\xi}(\xi\hat{M}_r) = \left(\frac{\theta}{\xi}\right)^\nu \quad \left(\nu = \frac{1}{n}\right).$$

Integrating we find

$$\hat{M}_r = \frac{1}{1-\nu}\left(\frac{\theta}{\xi}\right)^\nu + \frac{C}{\xi}.$$

Consider particular cases of edge conditions.

a) *A plate loaded by a moment on the outer edge, the inner edge being free* (fig. 217). Then from the condition that $\hat{M}_r(1) = \mu$, $\hat{M}_r(\alpha) = 0$, we find

$$\theta^\nu = (1-\nu)\frac{\mu}{1-\alpha^{1-\nu}}. \qquad (137.2)$$

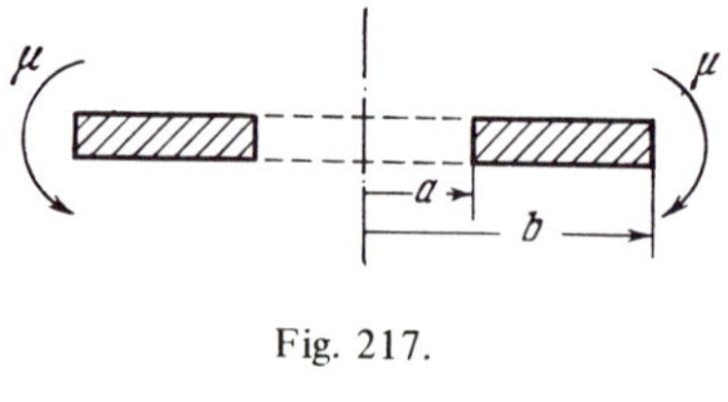

Fig. 217.

Here

$$\hat{M}_r = \frac{\mu}{1-\alpha^{1-\nu}} [\xi^{-\nu} - \alpha^{1-\nu}\xi^{-1}] .$$

We assume that $\hat{M}_\theta = (\theta/\xi)^\nu < \hat{M}_r$. Checking the condition for which these assumptions are fulfilled we obtain

$$\xi^{-\nu} > \frac{1}{1-\nu} [\xi^{-\nu} - \alpha^{1-\nu}\xi^{-1}] ,$$

or

$$1 - \alpha^{1-\nu}\xi^{-1+\nu} < 1 - \nu .$$

Hence it follows that we should have

$$\left(\frac{\alpha}{\xi}\right)^{1-\nu} > \nu .$$

This inequality is first infringed when $\xi = 1$, therefore, the solution that has been found is valid, provided that

$$\alpha > \nu^{\frac{1}{1-\nu}} . \tag{137.3}$$

If condition (137.3) is not fulfilled, for a certain value $\xi = \beta$ we have $\hat{M}_\theta = \hat{M}_r$. Assume that the condition $\hat{M}_\theta = \hat{M}_r$ is fulfilled in the region $\beta < \xi < 1$. In this region the equation of equilibrium is

$$\frac{d\hat{M}_r}{d\xi} = 0 . \tag{137.4}$$

Hence $\hat{M}_r = \text{const} = \mu = \hat{M}_\theta$, $\hat{\kappa}_\theta + \hat{\kappa}_r = \mu^n$.

To determine the deflection we may consider that either the outer or the inner edge is supported, it does not matter which. Put for example $w(\alpha) = 0$, then in the region $\alpha < \xi < \beta$, $\hat{w} = \theta(\xi - \alpha)$, where θ is given by formula (137.2). In the outer region $\beta < \xi < 1$

$$\frac{d^2\hat{w}}{d\xi^2} + \frac{1}{\xi}\frac{d\hat{w}}{d\xi} = \mu^n .$$

Whence

$$\hat{w} = \frac{\mu^n}{4}\xi^2 + C \ln \xi + D . \tag{137.5}$$

The constants C and D are determined from the condition of continuity of deflection when $\xi = \beta$. Finally a complete solution of the problem is written as follows [323]:

$$\hat{M}_r = \frac{\mu}{n-1}\left[n\left(\frac{\beta}{\xi}\right)^{1/n} - \frac{\beta}{\xi}\right],$$

$$\hat{M}_\theta = \mu\left(\frac{\beta}{\xi}\right)^{1/n}, \qquad \alpha < \xi < \beta,$$

$$\hat{M}_r = \hat{M}_\theta = \mu, \qquad \beta < \xi < 1,$$

$$\hat{w} = \frac{\mu^n}{4}\left[-1 - 3\beta^2 + 2\beta^2 \ln \beta + 4\beta\xi\right], \quad \alpha < \xi < \beta,$$

$$\hat{w} = \frac{\mu^n}{4}\left[\beta^2 - 1 + 2\beta^2 \ln \xi\right], \qquad \beta < \xi < 1.$$

Here

$$\left(\frac{\alpha}{\beta}\right)^{\frac{n-1}{n}} = \frac{1}{n}.$$

b) *A plate loaded with a moment on the outer edge with the inner edge clamped.* In this case the only solution is that in which $\hat{M}_r = \hat{M}_\theta = \mu$. Putting $\hat{w}(\alpha) = \hat{w}'(\alpha) = 0$ from formula (137.5) we find

$$\hat{w} = \frac{\mu^n}{4}\left[(\xi^2 - \alpha^2) - 2\alpha^2 \ln \frac{\xi}{\alpha}\right].$$

This latter example is also considered in the article of Patel and Venkatraman [323].

As we have seen, use of the piecewise-linear creep criterion greatly simplifies integration of the equation of equilibrium by parts but raises the new problem of the conjunction of the piecewise solutions on the boundaries of regions which are not known in advance. In the very simple examples which we have considered these difficulties do not arise, in example a) the quantity β, the radius of an unknown boundary, is very easily determined. However, difficulties do arise when the loading is more complicated.

Returning to example a) we find that in essence this is the problem of a

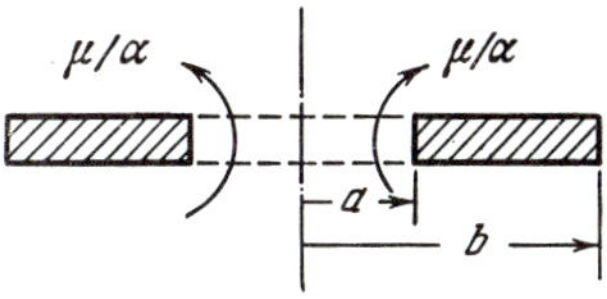

Fig. 218.

ring twisted by a moment uniformly distributed round the periphery, μ is the value of the moment referred to unit length of outer periphery in appropriate units. So long as condition (137.3) is fulfilled the resulting deformation is of extremely simple nature, every section of the ring remains rectangular on twisting. The elementary theory of this kind of deformation of a twisted ring is well known (see for example [69]); as might be expected formula (137.2) gives the relationship between the rate of increase of the angle of torsion and the moment μ by this theory. However, in the elementary theory the method of application of the turning moment is not limited in any way. In solving problems on the basis of the theory of bending of plates the method of application of the moment is vital and formula (137.2) is only valid if indeed throughout the plate $\hat{M}_\theta > \hat{M}_r > 0$. Suppose for example that the distributed turning moment is applied to the inner edge of the plate and not the outer (fig. 218). Let the intensity of this distributed moment in dimensionless units be μ/α, then the loads on the plates shown in figs. 217 and 218 are equivalent in the sense that elementary theory gives the same angle of torsion for each. Here, unlike the previous case, the middle surface of the plate is deformed in a surface of negative curvature, therefore, if $\hat{M}_r > 0, \hat{M}_\theta < 0$ and we are on side BC of the Tresca hexagon (fig. 215, $M_\theta = M_1, M_r = M_2$), consequently we have

$$\hat{\kappa}_\theta + \hat{\kappa}_r = \frac{1}{\xi}\frac{\mathrm{d}\hat{w}}{\mathrm{d}\xi} + \frac{\mathrm{d}^2\hat{w}}{\mathrm{d}\xi^2} = 0 .$$

Integrating we find

$$\frac{\mathrm{d}\hat{w}}{\mathrm{d}\xi} = \theta = \frac{k}{2\xi}, \qquad \kappa_\theta = -\hat{\kappa}_r = \frac{k}{2\xi^2}.$$

Now

$$\hat{M}_r - \hat{M}_\theta = (\hat{\kappa}_\theta - \hat{\kappa}_r)^\nu = \left(\frac{k}{\xi}\right)^\nu .$$

Substituting into the equation of equilibrium and integrating, we find

$$\hat{M}_r = \tfrac{1}{2}\left[-n\left(\frac{k}{\xi^2}\right)^{1/n} - C\right].$$

The constant C is found from the condition on the free edge, the constant k from the condition on the loaded edge, namely

$$k^{1/n} = \frac{2\mu}{\alpha n(1-\alpha^{-2/n})},$$

and

$$\hat{M}_r = \frac{\mu}{\alpha}\,\frac{\xi^{-2/n}-1}{\alpha^{-2/n}-1}.$$

This solution is valid for any value of α. Unlike the previous case in which the radius of the middle surface remained rectilinear it is here curved.

138. **More complicated types of loading**

Analysis of creep of circular plates on the basis of the Tresca type criterion becomes much more complicated if the plate carries a transverse load. Consider first of all a problem which is another variant of the problem solved above of a plate twisted by a distributed moment. Let the plate be loaded by forces distributed over the outer and inner peripheries (fig. 219). In order for it to be in equilibrium the forces applied to each of the contours must be the same. Denoting the force by P we find $\hat{N} = \hat{P}/\xi$. Assume that we remain on the side AF of the Tresca hexagon, then $\mathrm{d}\hat{w}/\mathrm{d}\xi = \theta = \text{const}$ and eq. (137.1) assumes the form

$$\frac{\mathrm{d}}{\mathrm{d}\xi}(\xi\hat{M}_r) = \left(\frac{\theta}{\xi}\right)^{\nu} - \hat{P}.$$

Whence

$$\hat{M}_r = \frac{1}{1-\nu}\left(\frac{\theta}{\xi}\right)^{\nu} - \hat{P} + \frac{C}{\xi}.$$

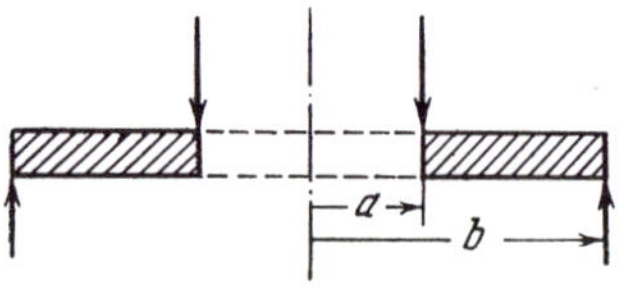

Fig. 219.

Now $\hat{M}_r(\alpha) = \hat{M}_r(1) = 0$, from these conditions we find θ and C and obtain

$$\theta^{\nu} = (1-\nu)\frac{1-\alpha}{1-\alpha^{1-\nu}}\hat{P} \tag{138.1}$$

and

$$\begin{aligned} \hat{M}_r &= \hat{P}\left[\frac{1-\alpha}{1-\alpha^{1-\nu}}\xi^{-\nu} - 1 - \frac{\alpha}{\xi}\frac{\alpha^{-\nu}-1}{1-\alpha^{1-\nu}}\right], \\ \hat{M}_\theta &= \hat{P}(1-\nu)\frac{1-\alpha}{1-\alpha^{1-\nu}}\xi^{-\nu}. \end{aligned} \tag{138.2}$$

Note that the load gives a moment $\hat{P}(1-\alpha)$ per unit length of outer edge. Putting $\hat{P}(1-\alpha) = \mu$ we confirm that formulae (138.1) and (137.2) give exactly the same dependence of the angle θ on the moment, in accordance with the result of elementary theory. However, in the latter case the limits of applicability of the solution are different and it is a fairly complicated matter to determine them. Comparing the expressions for $\hat{M}_\theta$ and $\hat{M}_r$ in eqs. (138.2) we confirm that if $\hat{M}_\theta > \hat{M}_r$ we have

$$\xi^{-\nu} < \frac{1-\alpha^{1-\nu}}{1-\alpha} + \frac{\alpha^{-\nu}-1}{1-\alpha}\frac{1}{\xi}.$$

The corresponding limiting value of α is found from the condition that if the inequality sign is altered to an equals sign the resulting equation in ξ has a double root. Making the appropriate calculations we confirm that solutions (138.1), (138.2) are valid if α satisfies the inequality

$$\left(\frac{\alpha^{-\nu}-1}{\nu(1-\alpha)}\right)^{\frac{\nu}{1-\nu}}(1-\nu) < \frac{1-\alpha^{1-\nu}}{1-\alpha}. \tag{138.3}$$

Inequality (138.3) is rather inconvenient, it must be solved numerically to determine the minimum value of α for a given value of ν.

These and similar problems can be solved with an exponential creep law. Study of the limits of applicability of the solutions is simpler in some cases than in others, with the power law the complication is about the same. The course of the solution is entirely analogous to that which was used for a rotating disc (section 124) and, therefore, we do not give the corresponding solutions.

139. **Deflection of plates by a distributed load**

The problem of bending a plate by a distributed load is identical with that of a rotating disc of constant thickness. Indeed with a distributed load $\hat{N} = \hat{q}\xi$. Substituting into the equation of equilibrium, (137.4), we obtain an equation that coincides exactly with eq. (123.1) for a disc of constant thickness, except that stresses are replaced by moments and the constant $\hat{q}$ replaces the constant m.

The condition of support round the periphery of the plate corresponds to the condition that the radial stress is zero on the periphery of the disc, that is there is no blade load.

For a freely supported plate, just as for a disc, we obtain two zones: when $0 < \xi < \alpha$, $\hat{M}_r = \hat{M}_\theta = \hat{M}$ whereas the rates of change of curvature may assume different positive values including those for which the sum is a given function of the moment. For the first zone the equation of equilibrium is

$$\frac{\mathrm{d}\hat{M}}{\mathrm{d}\xi} = -\hat{q}\xi \,. \tag{139.1}$$

The rate of change of the angle of slope θ is here determined by the equation

$$\hat{\kappa}_\theta + \hat{\kappa}_r = \frac{1}{\xi}\frac{\mathrm{d}}{\mathrm{d}\xi}(\xi\hat{w}) = \hat{\kappa}(\hat{M}) \,. \tag{139.2}$$

In the second zone when $\alpha < \xi < 1$, we have $\hat{M}_\theta > \hat{M}_r > 0$, consequently, $\kappa_r = \theta' = 0$ and $\theta = \text{const} = \theta_1$, therefore $\hat{\kappa}_\theta = \theta_1/\xi$. The bending moment $\hat{M}_r$ is a function of $\hat{\kappa}_\theta$, therefore, the equation of equilibrium assumes the form

$$\frac{\mathrm{d}}{\mathrm{d}\xi}(\xi\hat{M}_r) = m\left(\frac{\theta_1}{\xi}\right) - q\xi^2 \,. \tag{139.3}$$

The entire solution for a disc with an exponential creep law, that was derived in section 123 applies exactly to a plate. Consider here the problem of bending a plate according to a power law, like Wahl's [379] treatment of the case of a disc. Integrating eq. (139.1) for the first region, we obtain

$$\hat{M} = C\left[1 - \frac{1}{2}\frac{\hat{q}}{C}\xi^2\right]. \tag{139.4}$$

Substitute this value of the moment into eq. (139.2) and integrate, bearing in mind the condition that $\theta(0) = 0$. The result is

$$\theta = \frac{C^n}{n+1}\frac{C}{\hat{q}\xi}\left[1-\left(1-\frac{1}{2}\frac{\hat{q}}{C}\xi^2\right)^{n+1}\right].$$

Now determine the boundary of the first zone. Calculate the derivative $d\theta/d\xi$ and equate to zero, putting $\xi = \alpha$. Introducing the notation

$$\frac{1}{2}\frac{\hat{q}}{C}\alpha^2 = z\,, \tag{139.5}$$

we obtain the following equation for z:

$$(2n+1)(1-z)^{n+1} - 2(n+1)(1-z)^n + 1 = 0\,. \tag{139.6}$$

Therefore, the quantity z depends only on n and can be calculated for any value of n.

Now we can write expressions for $\hat{M}$ and θ as follows:

$$\hat{M} = \hat{M}_r = \hat{M}_\theta = C\left[1 - z\frac{\xi^2}{\alpha^2}\right],$$
$$\theta = \frac{C^n}{2(n+1)}\frac{1}{z}\frac{\alpha^2}{\xi}\left[1-\left(1-z\frac{\xi^2}{\alpha^2}\right)^{n+1}\right]. \tag{139.7}$$

Now proceed to the second zone, where $\theta = \theta_1 = \text{const}$, and the condition of continuity of θ when $\xi = \alpha$ must be fulfilled, therefore, the quantity θ_1 is found from the second of formulae (139.7) by substituting into it $\xi = \alpha$. Bearing in mind eq. (139.6) we obtain the following expression for θ_1:

$$\theta_1 = \alpha C^n (1-z)^n\,.$$

Eq. (139.3) is now written

$$\frac{d}{d\xi}(\xi\hat{M}_r) = C(1-z)\left(\frac{\alpha}{\xi}\right)^{1/n} - \hat{q}\xi^2\,.$$

Integrating we find

$$\hat{M}_r = C(1-z)\frac{n}{n+1}\left(\frac{\alpha}{\xi}\right)^{1/n} - \frac{\hat{q}\xi^2}{3} + \frac{D}{\xi}.$$

The moment $\hat{M}_r$ should satisfy the condition of continuity when $\xi = \alpha$ and also the condition of free support $\hat{M}_r(1) = 0$, consequently

$$C(1-z) = C(1-z)\frac{n}{n+1} - \frac{\hat{q}\alpha^2}{3} + \frac{D}{\alpha},$$

$$0 = C(1-z)\frac{n}{n+1} - \frac{\hat{q}}{3} + D\,.$$

Bearing in mind eq. (139.5), from the second equation we find

$$D = C\left[\frac{2}{3}\frac{z}{\alpha^2} - (1-z)\frac{n}{n+1}\alpha^{1/n}\right].$$

Substituting into the first boundary condition we obtain the following equation for determination of the unknown α:

$$\frac{n}{n+1}(1-z)\alpha^{1/n} - \frac{2z}{3}\left(\frac{1}{\alpha^3} - 1\right) - \frac{1-z}{n+1} = 0\,. \tag{139.8}$$

Hence we find the quantity α and consequently also C by the formula

$$C = \frac{q\alpha^2}{2z}\,.$$

Now we determine $\hat{M}_r$ and $\hat{M}_\theta$ from the first of formulae (139.7) when $0 < \xi < \alpha$ and by the formulae

$$\hat{M}_\theta = C(1-z)\left(\frac{\alpha}{\xi}\right)^{1/n},$$

$$\hat{M}_r = C\left\{\frac{n}{n+1}(1-z)\left(\frac{\alpha}{\xi}\right)^{1/n} - \frac{2z}{3}\left(\frac{\xi}{\alpha}\right)^2 + \left[\frac{2z}{3\alpha^3} - \frac{n}{n+1}(1-z)\alpha^{1/n-1}\right]\frac{\alpha}{\xi}\right\} \quad \text{when } \alpha < \xi < 1\,. \tag{139.9}$$

In the first zone the angle of rotation θ is given by formula (139.7) and in the second it remains of constant value, $\theta_1 = \alpha C^n(1-z)^n$. To determine the deflection quadrature is still required and when n is arbitrary the result cannot be expressed in terms of elementary functions.

This example is typical of the use of piecewise-linear creep criteria to solve problems of plates. Even in such a simple case as a plate supported round the periphery it is necessary to solve the algebraic eqs. (139.6) and (139.8) successively, which can only be done numerically. Numerical integration is again required to construct the deformed shape. Therefore, we decisively prefer the general method presented in section 135, in which the numerical solution is obtained by direct calculation to a standard programme. In this case the calculation of plates, for example clamped round the edge, proceeds in exactly the same way and no additional difficulties arise. Solution of the problem of a clamped plate by means of a Tresca type criterion is, of course, possible in principle, but hardly advisable, the resulting expressions are extremely cumbersome and there is hardly more than sporting interest in trying to overcome the purely algebraic and essentially simple difficulties.

Successive approximation methods are also applied with success to the calculation of plates, they are particularly suitable when the functions $m(\hat{\kappa})$ are given graphically or numerically. We shall not consider them specially because the methods that are used for rotating discs are immediately applicable to plates.

140. **Non-circular plates**

We have considered examples of circular plates with symmetrical fixing and carrying an axially symmetrical load. There are some elementary solutions for plates of different shapes.

a) *Pure bending of plates.* If a plate of arbitrary shape in plan is loaded round the edge with the same bending moment everywhere, the equations of statics are satisfied if we take $M_{11} = M_{22} = M$, $M_{12} = 0$. Then $M_0 = M$, and from eq. (134.6) it follows that:

$$\kappa_{11} = \kappa_{22} = \tfrac{1}{2}\kappa_0(M)\,. \qquad (140.1)$$

b) *Cylindrical bending.* Put $\kappa_{22} = \kappa_{12} = 0$, $\kappa_{11} = \kappa$; then by formulae (134.5)

$$M_{11} = 2M_{22} = M\,, \quad \frac{2}{\sqrt{3}}\kappa = \kappa_0\left(\frac{\sqrt{3}}{2}M\right), \qquad (140.2)$$

or

$$\frac{\sqrt{3}}{2}M = M_0\left(\frac{2\kappa}{\sqrt{3}}\right).$$

c) *Pure twist.* Hitherto we have not specially considered the boundary conditions because they were quite obvious in the simple examples that we have examined. It follows from the theory of elastic plates that in the general case one of the static conditions on the edge may be redundant. Thus for a free edge we might require that the bending moment, the twisting moment and the shear forces are zero. Formulated in this way the problem is indeterminate, generally speaking a solution cannot be derived for a biharmonic equation which satisfies three contour conditions. The well known way out of this contradiction (see for example [138]) is, after Kirchoff, instead of requiring that the torque H and the intersecting force N are zero, a single boundary condition is written

$$\frac{\partial H}{\partial s} + N = 0 . \tag{140.3}$$

Here s is an arc of contour of the plate. This follows from the fact that the distribution of torque round the arc AB is statically equivalent to a distributed linear load of intensity $\partial H/\partial s$ with the addition of concentrated forces at points A and B which are numerically equal to the moments H_A and H_B at these points.

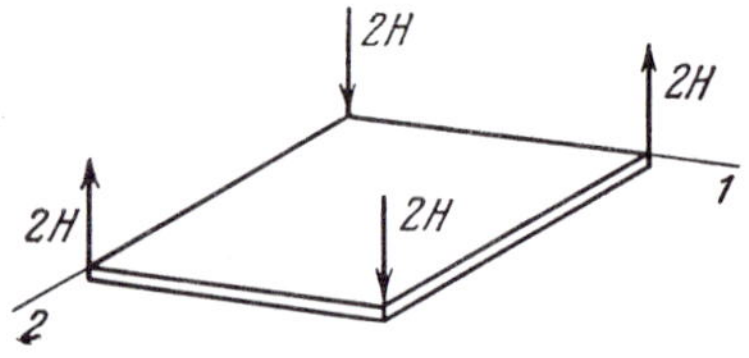

Fig. 220.

It is evident that these considerations, which derive purely from statics, also remain valid in creep theory, and therefore, we can consider pure twist of plates in creep by analogy with the corresponding problem of the theory of bending of elastic plates. Put $M_{11} = M_{22} = 0$, $M_{12} = H$. Assume that the plate is rectangular, then on the edge we obtain everywhere a torque of H. But $\partial H/\partial s = 0$, therefore, the contour load is statically equivalent to four forces each of $2H$ applied at the corners of the plate as shown in fig. 220. Now in view of the formulae of section 134, we have

$$\kappa_{12} = \frac{2}{\sqrt{3}} \kappa_0 \,(H\sqrt{3}) = \dot{w}_{,12} \, .$$

As in the elastic case the surface of the plate is a hyperboloid, with straight lines parallel to the 1 and 2 axes as generating lines.

Under more complicated conditions creep calculations on plates can, as yet, only be made by variational methods. Following the Ritz method we make a suitable approximation for the deflection of the plate. If we select the deflection of an elastic plate as an approximating function the solution is extremely complicated. Therefore, it is usual to select expressions which satisfy the kinematic boundary conditions, for example, either the first term of the double Fourier series, which represents solution for the elastic case, or some polynomial. Then the Lagrange equation is used, exactly as was explained in section 136 for a round plate; the fact that the function φ_1 now depends on two variables instead of one does not introduce fundamental complications.

In an exactly similar way we may be given a distribution of bending moments which satisfies the static boundary conditions and the equation of equilibrium (132.6). Suppose that this distribution is $M_{ij}^0(x_k)$. Find a second distribution of moments which satisfies the equation of equilibrium in the absence of external forces and with zero boundary conditions $M_{ij}^{00}(x_k)$. It is evident that

$$M_{ij} = M_{ij}^0 + cM_{ij}^{00}$$

is a statically possible state of the plate whatever the value of the constant c. We now formulate the Castigliano functional and minimise it with respect to the parameter c. Examples of the application of variational principles to the calculation of rectangular plates are contained in the book [52].

141. **Creep of shells**

The general equations of steady-state creep of plates and shells can be derived in the same way as the analogous equations of the theory of elastic shells. When this is done there are complications, similar to the theory of elasticity, which are associated with the need to limit the number of terms in expansions of the various functions in powers of the variable through the thickness of the shell. Variants of the fundamental equations can be constructed that differ from one another in the way of taking account of particular small terms. However, shell creep problems are so complicated that, as yet, solutions, or attempted solutions, are available only for the simplest cases, and, therefore, more general investigations could at best only suggest certain possibilities without indicating any way of realising them. Indeed in the simplest and generally accepted variant of the theory of elastic shells the relationship between forces and moments on the one hand, and rates of deformation of the mean surface and rates of change of curvature on the other, remain as in the theory of plates. For other shells the forces are related to deformations and moments to changes in curvature, whereas for shells undergoing creep we should take expressions (132.3) or their converses (132.4). Therefore, the first and fundamental difficulty in shell theory is to find an expression of the potential of forces and moments or the potential of the corresponding rates. This is only possible for a few fairly simple shells. In particular the shell may be subject to no torque or pure torque. In the absence of moment the forces are determined from the conditions of statics if the corresponding boundary conditions are given for forces. If the boundary conditions contain limitations on displacements it is sometimes possible to formulate statically indeterminate problems of the theory of moment-free shells for which the solutions are fundamentally little different from those of the corresponding problems of the theory of elastic shells. Examples of such statically indeterminate problems are creep of laminated shells with various assumptions concerning the nature of the bonds between layers.

Shells with pure moments are to be understood in the sense that the forces at the centre of the surface are given in advance, that is they are found from the equations of equilibrium and the boundary conditions, irrespective of the deflection. An independent system of equations is obtained for the bending moments and rates of change of curvature.

The presence of forces at the centre of the surface, even though they are known in advance, greatly complicates the relationships between moments and rates of change of curvature. Difficulties of this kind were encountered in considering the bending of beams in the presence of a longitudinal force. Practically soluble problems of shells with pure moments are of the same character.

The conditions of existence of moment-free and pure moment states in elastic shells have been adequately stated [13, 20], but we shall not refer to these general results or attempt to make a corresponding analysis for the case of creep, the cases considered below are sufficiently simple and obvious. As we have already confirmed, the effect of stress concentration is significantly less in creep than in the elastic state. Therefore, it would be expected that the edge effect in shells in creep is less clearly expressed than in elastic shells, it is less important and the equations of momentless theory can be applied with greater justification. This will become clearer later on in a few examples that relate to axially symmetrical deformation of cylindrical shells.

The problem of axially symmetrical deformation of cylindrical shells is one of the few problems for which at least a numerical solution is possible. Here the important circumstance is that the longitudinal force is determined immediately from conditions of statics and there is no change of curvature in the transverse direction. In this sense the state of stress of a cylindrical shell has certain properties of both moment-free and pure-moment states. To a certain extent this kind of circumstance is also encountered in certain other problems of shell theory and the simplification that results forms a basis for an approximate theory of elastic cylindrical shells of medium length subject to asymmetrical loading [13, 88]. In the theory of creep of shells the corresponding assumptions also introduce important simplifications and variants of the theory that are based on them are called technical theories of shells.

142. Theory of moment-free shells

Let us recall the fundamental facts of the theory of moment-free shells. Let a surface be referred to a curvilinear system of coordinates u^1, u^2 so that the radius vector of its points is $\mathbf{r}(u^1, u^2)$. The vector $\mathbf{r}_{,i}$, where the subscript after the comma denotes differentiation, forms a covariant basis on the surface, the vector of unit normal to the surface is denoted $\mathbf{n}$. The covariant metric tensor is $g_{ij} = \mathbf{r}_{,i}\mathbf{r}_{,j}$; the coefficients of the second quadratic form are $b_{ij} = \mathbf{r}_{,ij}\mathbf{n} = \mathbf{r}_{,i}\mathbf{n}_{,j}$. We recall the Gauss and Weingarten [110] formulae

$$\mathbf{r}_{,ij} = G^k_{ij}\mathbf{r}_{,k} + b_{ij}\mathbf{n}, \qquad \mathbf{n}_{,i} = -b^j_i\mathbf{r}_{,j}.$$

Here G^k_{ij} is the Christoffel symbol for the surface. The contravariant stress tensor in the shell T^{ij} may be formally introduced in the following way. Consider the shell as an inextensible surface. This means that if a point on the surface acquires a displacement $\mathbf{w}$ the deformation is zero. The radius vector of the curves shell is $\mathbf{r} + \mathbf{w}$, the condition of inextensibility may be written

$$d\mathbf{r}\, d\mathbf{w} = 0,$$

or

$$\mathbf{w}_{,1}\mathbf{r}_{,1} = 0, \quad \mathbf{w}_{,1}\mathbf{r}_{,2} + \mathbf{w}_{,2}\mathbf{r}_{,1} = 0, \quad \mathbf{w}_{,2}\mathbf{r}_{,2} = 0. \qquad (142.1)$$

Consider part of the shell S bounded by the contour Γ. The force acting per unit surface is $\mathbf{q}$, and the force $\mathbf{T}$ per unit length is applied on the contour. Equate to zero the virtual work of all forces that act on the segment of the shell bearing in mind that possible displacements are subject to the conditions of eqs. (142.1). For this purpose we introduce Lagrange factors which we denote T^{11}, T^{12} and T^{22}. If $T^{12} = T^{21}$, the conditions (142.1) are written as follows:

$$T^{ij}\mathbf{w}_{,i}\mathbf{r}_{,j} = 0 .$$

Equating to zero the virtual work of all forces, we obtain

$$\iint_S [\mathbf{q}\mathbf{w} - T^{ij}\mathbf{w}_{,i}\mathbf{r}_{,j}]\, \mathrm{d}S + \int_\Gamma \mathbf{T}\mathbf{w}\, \mathrm{d}s = 0 . \tag{142.2}$$

Let $\mathbf{v}$ be a vector normal to the contour Γ in the tangential plane so that $\mathbf{v} = \nu_k \mathbf{e}^k = \nu_k g^{sk}\mathbf{r}_{,s}$. Converting the surface integral, eq. (142.2) becomes

$$\iint_S [\mathbf{q} + T^{ij}_{,i}\mathbf{r}_{,j} + T^{ij}b_{ij}\mathbf{n}]\,\mathbf{w}\,\mathrm{d}S + \int_\Gamma [\mathbf{T} - T^{ij}\mathbf{r}_{,j}\nu_i]\,\mathbf{w}\,\mathrm{d}s = 0 . \tag{142.3}$$

Hence it follows that:

$$T^{ij}_{,i} + q^j = 0 , \quad T^{ij}b_{ij} + q_n = 0 , \quad T^i = T^{ij}\nu_j . \tag{142.4}$$

In the first of these expressions the subscript after the comma denotes the covariant derivative of the tensor T^{ij}.

The expansion of the first of eqs. (142.4) can be obtained in simpler and more symmetrical form. Put $\widetilde{T}_{ij} = T_{ij}\sqrt{g}$. Then, using the well-known relationships

$$\frac{\partial \ln g}{\partial u^i} = G^k_{ik} ,$$

we arrive at the result

$$\begin{gathered}
\frac{\partial \widetilde{T}^{11}}{\partial u^1} + \frac{\partial \widetilde{T}^{12}}{\partial u^2} + G^1_{11}\widetilde{T}^{11} + 2G^1_{12}\widetilde{T}^{12} + G^1_{22}\widetilde{T}^{22} + q^1\sqrt{g} = 0 , \\
\frac{\partial \widetilde{T}^{12}}{\partial u^1} + \frac{\partial \widetilde{T}^{22}}{\partial u^2} + G^2_{11}\widetilde{T}^{11} + 2G^2_{12}\widetilde{T}^{12} + G^2_{22}\widetilde{T}^{22} + q^2\sqrt{g} = 0 , \\
\widetilde{T}^{11}b_{11} + 2\widetilde{T}^{12}b_{12} + \widetilde{T}^{22}b_{22} + q_n\sqrt{g} = 0 .
\end{gathered} \tag{142.5}$$

The quantities $\widetilde{T}^{ij}$ do not form a covariant tensor; if we alter the sequence of the superscripts they may be considered as a mixed tensor. Indeed the quantity $\widetilde{T}^{ij}$ is formed by contraction of the contravariant tensor T^{ij} with a discriminant tensor (see for example, [47]). Further conversion of formulae (142.5) is obtained by taking the quantities $T^{ij}g$ as force components. Multiplying the components of the contravariant tensor by the quantity g is equivalent to twice lowering the superscript by means of the discriminant tensor, therefore in the result we obtain the covariant tensor whose components are

$$S_{11} = gT^{22}, \quad S_{12} = -gT^{12}, \quad S_{22} = gT^{11} .$$

The equations of equilibrium for the force tensor S_{ij} are written

$$\frac{\partial S_{11}}{\partial u^2} - \frac{\partial S_{12}}{\partial u^1} - S_{11}G^1_{21} + S_{12}(G^1_{11} - G^2_{12}) + S_{22}G^2_{11} + q^2 g = 0 ,$$
$$\frac{\partial S_{22}}{\partial u^1} - \frac{\partial S_{12}}{\partial u^2} + S_{11}G^1_{22} + S_{12}(G^2_{22} - G^1_{12}) - S_{22}G^2_{12} + q^1 g = 0 , \qquad (142.6)$$

$$S_{11}b_{22} + S_{22}b_{11} - 2S_{12}b_{12} = q_n .$$

The equations of moment-free theory in the form (142.6) were obtained in the work [97], it is convenient to introduce components S_{ij} of the force tensor because this makes it possible to introduce the stress function in a natural way so as to satisfy the first two of eqs. (142.6). For this purpose let us return to eq. (142.2) and convert the surface integral by integrating by parts. Note that here the element of surface $dS = \sqrt{g}\, du^1\, du^2$. This surface integral can be written in the form

$$\iint [\sqrt{g}\mathbf{q}\mathbf{w} - (\sqrt{g}T^{ij}\mathbf{r}_{,j})\mathbf{w}_{,i}]\, du^1 du^2 =$$
$$= \iint [\sqrt{g}\mathbf{q}\mathbf{w} - (\sqrt{g}T^{ij}\mathbf{r}_{,j}\mathbf{w})_{,i} + (\sqrt{g}T^{ij}\mathbf{r}_{,j})_{,i}\mathbf{w}]\, du^1 du^2 .$$

The integral of the second term in the sub-integral expression is converted to a contour integral. Put

$$\mathbf{S}_1 = T^{2i}\mathbf{r}_{,i}\sqrt{g} = \frac{1}{\sqrt{g}}[S_{11}\mathbf{r}_{,2} - S_{12}\mathbf{r}_{,1}] ,$$
$$\mathbf{S}_2 = -T^{1i}\mathbf{r}_{,i}\sqrt{g} = \frac{1}{\sqrt{g}}[S_{12}\mathbf{r}_{,2} - S_{22}\mathbf{r}_{,1}] . \qquad (142.7$$

Then the conditions of equilibrium become

$$\frac{\partial \mathbf{S}_1}{\partial u^2} - \frac{\partial \mathbf{S}_2}{\partial u^1} + \mathbf{q}\sqrt{g} = 0 \,, \tag{142.8}$$

and the boundary condition

$$\mathbf{T}\,\mathrm{d}s = \mathbf{S}_1\,\mathrm{d}u^1 + \mathbf{S}_2\,\mathrm{d}u^2 \,. \tag{142.9}$$

Eqs. (142.8) are the vector notation of the system of eqs. (142.6), which are readily derived from eq. (142.8) by Gaussian derivative formulae. Let $\mathbf{S}_1^*$ and $\mathbf{S}_2^*$ be a particular solution of eq. (142.8), the corresponding force components are S_{ij}^*. We obviously obtain a more general solution if to the particular solution $\mathbf{S}_1^*$ we add a solution of the uniform equation that is obtained from eq. (142.8) by putting $q = 0$, namely

$$\frac{\partial \mathbf{S}_1}{\partial u^2} - \frac{\partial \mathbf{S}_2}{\partial u^1} = 0 \,. \tag{142.10}$$

We satisfy this latter equation by taking $\mathbf{S}_1 = \mathbf{a}_{,1}$ and $\mathbf{S}_2 = \mathbf{a}_{,2}$ where $\mathbf{a}$ is an arbitrary vector:

$$\mathbf{a} = a^i \mathbf{r}_{,i} + \psi \mathbf{n} \,.$$

Now

$$\mathbf{a}_{,j} = \mathbf{S}_j = (a^i_{,j} - \psi b^i_j)\mathbf{r}_{,i} + (\psi_{,j} - a^i b_{ij})\mathbf{n} \,.$$

Since $\mathbf{S}_j$ and consequently also $\mathbf{a}_{,j}$ lie in the tangential plane we have

$$\psi_{,j} - a^i b_{ij} = 0 \,.$$

Whence

$$a^1 = \frac{1}{Kg}(\psi_{,2} b_{12} - \psi_{,1} b_{22}) \,, \quad a^2 = -\frac{1}{Kg}(\psi_{,2} b_{11} - \psi_{,1} b_{12}) \,. \tag{142.11}$$

Here K is the Gaussian curvature of the surface. Since S_{ij} is a symmetrical tensor we have $\mathbf{S}_i \mathbf{e}^i = 0$ or $\mathbf{a}_{,i}\mathbf{e}^i = 0$. Hence

$$a^1_{,1} - \psi b^1_1 + a^2_{,2} - \psi b^2_2 = 0 \,.$$

Writing the expressions for the covariant derivatives explicitly and using for simplicity the identity $(\partial \ln \sqrt{g})/\partial u^i = G^k_{ik}$, we obtain the following equation:

$$\frac{\partial}{\partial u^1}(a^1\sqrt{g}) + \frac{\partial}{\partial u^2}(a^2\sqrt{g}) - \psi\sqrt{g}(b^2_2 + b^1_1) = 0 .$$

Substituting eq. (142.11) into this we obtain the final equation for the stress function ψ:

$$\frac{\partial}{\partial u^2}\left(\frac{\psi_{,2}b_{11} - \psi_{,1}b_{12}}{K\sqrt{g}}\right) + \frac{\partial}{\partial u^1}\left(\frac{\psi_{,1}b_{22} - \psi_{,2}b_{12}}{K\sqrt{g}}\right)$$
$$+ \psi\sqrt{g}(b^1_1 + b^2_2) = 0 . \qquad (142.12)$$

The forces can now be written by means of the following symmetrical expressions:

$$S_{ij} = -\frac{\partial c_i}{\partial u^j} + c_k G^k_{ij} - \omega_{ik} b^k_j . \qquad (142.13)$$

Here

$$c_1 = -\sqrt{g}a^2 = \frac{1}{K\sqrt{g}}(\psi_{,2}b_{11} - \psi_{,1}b_{12}) ,$$

$$c_2 = \sqrt{g}a^1 = \frac{1}{K\sqrt{g}}(\psi_{,2}b_{12} - \psi_{,1}b_{22}) ,$$

$$\omega_{11} = \omega_{22} = 0 , \qquad \omega_{12} = -\omega_{21} = \psi\sqrt{g} .$$

This solution does not apply to the case of a developable surface, but if $K = 0$, it can easily be shown that we can introduce the normal stress function for the plane problem, namely, take

$$S_{ij} = \varphi_{,ij} . \qquad (142.14)$$

143. Theory of moment-free shells. Determination of deformations and displacements

The components of the stress tensor S_{ij} that we have introduced in a formal way have a simple mechanical meaning. In formula (142.9) put $\mathrm{d}u^2 = 0$,

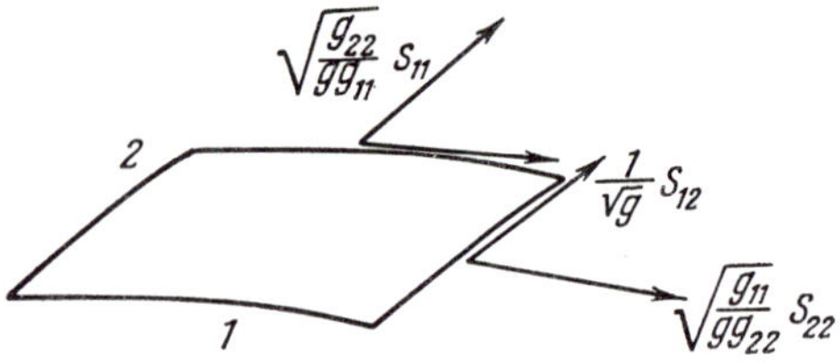

Fig. 221.

this means that we take as ds an element of arc of the line u^1, d$s = \sqrt{g_{11}}\,\mathrm{d}u^1$. Then from eqs. (142.7) and (142.10) it follows that:

$$\mathbf{T}\sqrt{g_{11}} = \frac{1}{\sqrt{g}}\,[S_{11}\mathbf{r}_{,2} - S_{12}\mathbf{r}_{,1}]\ .$$

The vectors $\mathbf{r}_{,2}$ and $\mathbf{r}_{,1}$ are in the directions of the tangents to the lines u^2 and u^1 respectively, their lengths are $\sqrt{g_{22}}$ and $\sqrt{g_{11}}$, whereas $\mathbf{T}$ is a force vector acting per unit length of the line u^1. From the expressions given above it follows that the quantities

$$S_{11}\frac{\sqrt{g_{22}}}{\sqrt{gg_{11}}} \quad \text{and} \quad -S_{12}\frac{1}{\sqrt{g}}$$

are "physical" components of force acting per unit length of the line u^1 resolved in the directions of the lines u^1 and u^2. In exactly the same way

$$-S_{21}\frac{1}{\sqrt{g}} \quad \text{and} \quad S_{22}\frac{\sqrt{g_{11}}}{\sqrt{gg_{22}}}$$

are physical components of force per unit length of line u^2. Fig. 221 shows an infinitesimally small element of surface with sides $\sqrt{g_{11}}\,\mathrm{d}u^1$ and $\sqrt{g_{22}}\,\mathrm{d}u^2$ and the forces which act on the element.

If the surface is described by orthogonal coordinates, for instance to lines of curvature as is usually done, its linear element is often taken in the form $\mathrm{d}s^2 = A^2(\mathrm{d}u^1)^2 + B^2(\mathrm{d}u^2)^2$, consequently, $g_{11} = A^2$, $g_{22} = B^2$ and the physical components of force are

$$\frac{1}{A^2}S_{11}\,, \quad \frac{1}{B^2}S_{22}\,, \quad -\frac{1}{AB}S_{12}\,.$$

Note that $(1/A^2)S_{11}$ is the normal force in the direction of u^2, whereas $(1/B^2)S_{22}$ is the normal force in the direction of u^1. If the usual notation: N_1, N_2 and N_{12} is taken for the physical components of the forces in orthogonal coordinates we obtain

$$N_1 = \frac{1}{B^2} S_{22}, \quad N_2 = \frac{1}{A^2} S_{11}, \quad N_{12} = -\frac{1}{AB} S_{12}. \tag{143.1}$$

The rate potential for the shell is expressed as

$$\Phi(T^{ij}) = \frac{1}{2h} \Phi(2h\sigma^{ij});$$

here the right-hand side includes the creep-rate potential expressed in terms of contravariant components of the stress tensor. The components of rate of deformation ϵ_{ij} are expressed in terms of potential in the following way:

$$\epsilon_{ij} = \frac{\partial \Phi}{\partial T^{ij}}.$$

Proceeding to the tensor S_{ij}, we obtain

$$\epsilon_{11} = g \frac{\partial \Phi}{\partial S_{22}}, \quad \epsilon_{22} = g \frac{\partial \Phi}{\partial S_{11}}, \quad \epsilon_{12} = -g \frac{\partial \Phi}{\partial S_{12}}. \tag{143.2}$$

In orthogonal coordinates when $g_{11} = A^2$, $g_{22} = B^2$, $g_{12} = 0$, the strain rates ϵ_1, ϵ_2 and the shear rate γ are expressed in terms of ϵ_{ij} as follows:

$$\epsilon_1 = \frac{\epsilon_{11}}{A}, \quad \epsilon_2 = \frac{\epsilon_{22}}{B}, \quad \frac{\gamma}{2} = \frac{\epsilon_{12}}{\sqrt{AB}}.$$

In physical stress components and rates of deformation expressions (143.2) are replaced by simpler ones which literally coincide with the expressions for the rates in terms of stresses in Cartesian coordinates for a plane state of stress. However, in shell theory, it is not always best to select lines of curvature or some other orthogonal network as the system of reference.

If we take a Mises type creep law the potential is a function of the force intensity which is expressed in general coordinates as follows:

$$T_0^2 = \tfrac{3}{2} \left[T^{ij} T^{kl} g_{ik} g_{jl} - \tfrac{1}{3} (T^{ij} g_{ij})^2\right].$$

Consequently,

$$\epsilon_{ij} = \frac{\Phi'(T_0)}{T_0} [3T^{kl} g_{ik} g_{jl} - (T^{kl} g_{kl}) g_{ij}] .$$

Proceed from components of force T^{ij} to quantities S_{ij}. By direct calculations we obtain the following form of the creep law:

$$\epsilon_{ij} = \frac{\Phi'(T_0)}{T_0} [(S_{kl} g^{kl}) g_{ij} - \tfrac{3}{2} S_{ij}] . \tag{143.3}$$

Here

$$T_0^2 = \tfrac{3}{2} [S_{ij} S_{kl} g^{kl} - \tfrac{1}{3} (S_{ij} g^{ij})^2] .$$

Solving (143.3) for S_{ij}, we find

$$S_{ij} = \frac{2T_0}{3\Phi'(T_0)} [2(\epsilon_{kl} g^{kl}) g_{ij} - \epsilon_{ij}] .$$

On the other hand, $\Phi'(T_0) = v(T_0)$ and, vice versa, $T_0 = s(v)$, the functions s and v are the same as in the fundamental creep law. Therefore, we rewrite the latter equation as

$$S_{ij} = \frac{2}{3} \frac{s(v)}{v} [2(\epsilon_{kl} g^{kl}) g_{ij} - \epsilon_{ij}] . \tag{143.4}$$

When the forces in the shell have been found, formulae (143.4) are used to determine the rates of deformation from which the rate of displacement of the shell can always be found. Indeed let the rate-of-displacement vector be given as follows:

$$\mathbf{w} = w^i \mathbf{r}_{,i} + w \mathbf{n} . \tag{143.5}$$

Then the components of rate of deformation are

$$\epsilon_{ij} = \tfrac{1}{2} (\mathbf{w}_{,i} \mathbf{r}_{,j} + \mathbf{w}_{,j} \mathbf{r}_{,i}) .$$

Calculate the derivatives of the vector $\mathbf{w}$, expression (143.5), by the Gauss and Weingarten derivative formulae. We obtain

$$\epsilon_{ij} = \tfrac{1}{2} (w_{i,j} + w_{j,i}) - w b_{ij} . \tag{143.6}$$

As is shown in the work [97] determination of the displacements depends on solving an equation of the same form as (142.12), but with the difference that the right-hand side differs from zero; when the solution has been found the displacement formulae are written in quadrature form. We shall not describe this method here because the problems that we shall consider are relatively simple.

144. **Moment-free shells of rotation**

A shell of rotation with an axially symmetric load is the simplest object of application of the general theory and is at the same time of great practical interest. A meridian of such a shell is shown in fig. 222, in which we take as the coordinate parameters the arc ξ of the meridian for direction 1 and the angle φ between neighbouring meridional sections (geographical-longitude) for direction 2. The radius of the point M is considered as a function of the arc ξ; $\mathbf{r} = \mathbf{r}(\xi)$. An element of arc of the shell surface is expressed as

$$\mathrm{d}s^2 = \mathrm{d}\xi^2 + r^2 \mathrm{d}\varphi^2 .$$

Therefore, $g_{11} = 1, g_{22} = g = r^2, g_{12} = 0$. The Christoffel symbols (see [110]) are

$$G_{11}^1 = 0 , \quad G_{12}^1 = 0 , \quad G_{22}^1 = -r \frac{\mathrm{d}r}{\mathrm{d}\xi} = -r \sin\theta ,$$

$$G_{11}^2 = 0 , \quad G_{12}^2 = \frac{1}{r} \frac{\mathrm{d}r}{\mathrm{d}\xi} = \frac{\sin\theta}{r} , \quad G_{22}^2 = 0 .$$

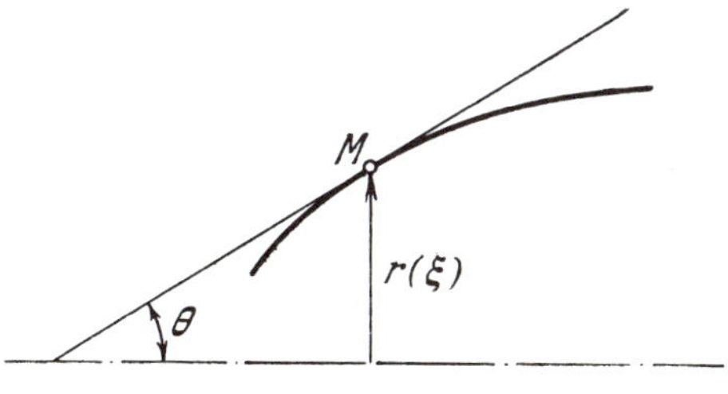

Fig. 222.

The coefficients of the second quadratic form are

$$b_{11} = \frac{1}{R_1}, \qquad b_{22} = \frac{r^2}{R_2}, \qquad b_{12} = 0 .$$

Here $R_1 = \mathrm{d}\xi/\mathrm{d}\theta$ is the radius of curvature of the meridian, $R_2 = r/\cos\theta$ is the radius of curvature of the normal section drawn through the parallel. With an axially symmetrical load the force does not depend on φ and $S_{12} = 0$, therefore, the first of eqs. (142.6) is satisfied identically and the second gives

$$\frac{\mathrm{d}S_{22}}{\mathrm{d}\xi} - \sin\theta \left(rS_{11} - \frac{1}{r} S_{22} \right) + q_1 r^2 = 0 .$$

The third equation becomes of the form

$$S_{11} \frac{r^2}{R_2} + S_{22} \frac{1}{R_1} + q_n r^2 = 0 .$$

In this particular simple case it is convenient to convert to physical components of force

$$N_1 = \frac{1}{r^2} S_{22}, \qquad N_2 = S_{11} .$$

We then obtain

$$\begin{aligned} &\frac{\mathrm{d}N_1}{\mathrm{d}\xi} + \frac{\sin\theta}{r}(N_1 - N_2) + q_1 = 0 , \\ &\frac{N_1}{R_1} + \frac{N_2}{R_2} + q_n = 0 . \end{aligned} \tag{144.1}$$

The system (144.1) has a well-known solution

$$\begin{aligned} &N_1 = -\frac{1}{r\cos\theta} \int r(q_n \sin\theta + q_1 \cos\theta)\, \mathrm{d}\xi , \\ &N_2 = -\frac{R_2}{R_1} N_1 - R_2 q_n . \end{aligned} \tag{144.2}$$

The components of rate of deformation are found by formulae (143.6) as follows:

$$\epsilon_{11} = \frac{dw_1}{d\xi} - \frac{w}{R_1}, \qquad \epsilon_{22} = r(w_1 \sin\theta - w\cos\theta).$$

The physical components of rate of deformation are

$$\epsilon_1 = \epsilon_{11} = \frac{dw_1}{d\xi} - \frac{w}{R_1},$$
$$\epsilon_2 = \frac{\epsilon_{22}}{r^2} = \frac{1}{r}(w_1 \sin\theta - w\cos\theta). \tag{144.3}$$

The relationships between rates of deformation and forces can now be written in the usual form as for plates. With the quadratic creep criterion we have

$$\epsilon_1 = \frac{v(N_0)}{N_0}(N_1 - \tfrac{1}{2}N_2), \quad \epsilon_2 = \frac{v(N_0)}{N_0}(N_2 - \tfrac{1}{2}N_1). \tag{144.4}$$

Here

$$N_0^2 = N_1^2 + N_2^2 - N_1 N_2.$$

If the fundamental creep law is given in the form $\epsilon = \epsilon_* v(\sigma/\sigma_*)$ then $v = v(N_0/N_*)$ where $N_* = 2h\sigma_*$.

When rates of deformation have been calculated displacements of points on the shell are found from eqs. (144.2) by squaring, namely

$$w_1 = \cos\theta \int \left(\epsilon_1 - \frac{R_2}{R_1}\epsilon_2\right)\frac{d\xi}{\cos\theta},$$
$$w = w_1 \tan\theta - R_2\epsilon_2. \tag{144.5}$$

145. Multi-layer shells with rigidly bonded layers

Shells that consist of layers of materials with different creep characteristics are frequently encountered as structural components. Moreover, a uniform shell with a temperature gradient across the thickness can be modelled by a multi-layer shell. In doing so it is assumed that within each layer the tempera-

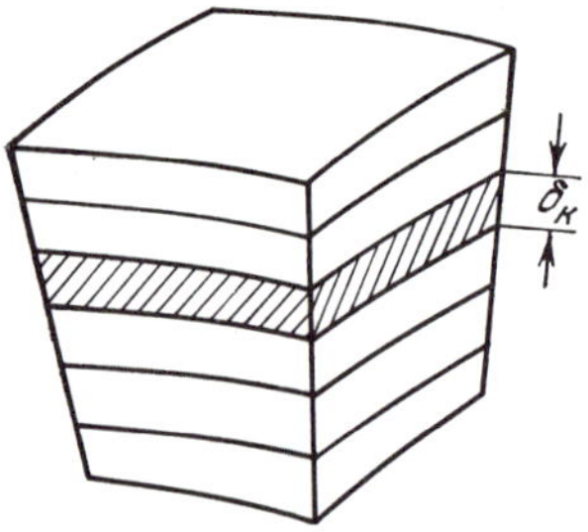

Fig. 223.

ture and consequently the creep characteristics are constant. We assume that the total thickness of the shell is small compared with its radii of curvature and, therefore, the coefficients of the first and second quadratic forms are taken the same for all layers.

A diagram of such a multi-layer shell is shown in fig. 223, layer number k is of thickness δ_k, and as before the total thickness of the shell is $2h$. Considering the shell as a whole we determine the forces S_{ij} by moment-free theory because the use of this theory is not associated with any assumptions concerning the structure and physical properties of the shell. The stress components in each layer are denoted σ_{ij}^k. Here σ_{ij} is not the ordinary covariant stress tensor but a tensor derived from the contravariant stress tensor by twice lowering the superscripts by means of the discriminant tensor; the quantities σ_{ij} are related to σ^{ij} in the same way as S_{ij} are related to T^{ij}. Each layer can be considered as a separate shell in a moment-free state. Therefore, by formulae (143.4)

$$\sigma_{ij}^k \delta_k = \frac{2}{3} \frac{s_k(v)}{v} \delta_k \, [2(\epsilon_{rs} g^{rs}) g_{ij} - \epsilon_{ij}] \, .$$

On the other hand it is evident that

$$\sum_k \sigma_{ij}^k \delta_k = S_{ij} \, .$$

Consequently,

$$S_{ij} = \frac{2}{3} \frac{\sum s_k(v) \delta_k}{v} \, [2(\epsilon_{rs} g^{rs}) g_{ij} - \epsilon_{ij}] \, .$$

We have assumed that the layers are rigidly bonded together and cannot slip relative to one another, therefore, the quantities ϵ_{ij} and consequently the invariant v, are the same for all layers. Therefore, the relationships between stresses and rates of deformation are given as before by formulae (143.4) in which

$$s(v) = \sum \delta_k s_k(v) . \tag{145.1}$$

Put

$$\eta_k = \frac{\delta_k s_k(v)}{\sum \delta_i s_i(v)} .$$

Then

$$\sigma_{ij}^k = \frac{\eta_k}{\delta_k} S_{ij} . \tag{145.2}$$

If the entire layer follows the power creep law with the same exponent n, such that

$$s_k = \sigma_n^k \left(\frac{v}{\epsilon_n^k}\right)^n ,$$

then

$$s(v) = s_n \left(\frac{v}{v_n}\right)^n .$$

Here

$$\frac{s_n}{(v_n)^n} = \sum \frac{\sigma_n^k \delta_k}{(\epsilon_n^k)^n} .$$

As usual, arbitrariness remains in selection of the quantities s_n and v_n, now

$$\eta_k = \frac{\delta_k \sigma_n^k}{s_n} \left(\frac{\epsilon_n^k}{v_n}\right)^{-n} .$$

In this case the quantities η_k are constants, in all other cases η_k is a function of v. Note that it follows from eq. (145.2) that

$$\eta_k = \delta_k \frac{\sigma_0^k}{S_0} . \tag{145.3}$$

Here σ_0^k and S_0 are any first power invariants formulated in the same way from tensors σ_{ij}^k and S_{ij}, in particular σ_0^k can be the stress intensity in the layer, S_0 the force intensity. Since in each case the metric is the same and identical with the metric of the shell as a whole, the physical components of stress are also expressed in terms of physical components of force by formulae (145.2).

If, as before, we understand by $\mathbf{q}$ the force vector applied to unit surface area of the shell, in order to consider interaction between layers we should consider the method of application of these forces more precisely. We shall consider that the force $\mathbf{q}^k$ acts on unit area of each layer such that

$$\mathbf{q} = \sum \mathbf{q}^k .$$

When the shell is deformed, in addition to the forces $\mathbf{q}^k$ each layer is subject to forces $\mathbf{p}^k$ which correspond to the reactions of the adjacent layers. Each layer can be considered as an individual moment-free shell and conditions of equilibrium can be formulated for it. For this purpose we introduce the notation

$$\boldsymbol{\sigma}_1 = \frac{1}{\sqrt{g}}(\sigma_{11}\mathbf{r}_{,2} - \sigma_{12}\mathbf{r}_{,1}) ,$$
$$\boldsymbol{\sigma}_2 = \frac{1}{\sqrt{g}}(\sigma_{12}\mathbf{r}_{,2} - \sigma_{22}\mathbf{r}_{,1}) . \tag{145.4}$$

Now the equation of equilibrium can be written in the form of eq. (142.8), namely

$$\frac{\partial}{\partial u^2}(\delta_k \boldsymbol{\sigma}_1^k) - \frac{\partial}{\partial u^1}(\delta_k \boldsymbol{\sigma}_2^k) + (\mathbf{q}^k + \mathbf{p}^k)\sqrt{g} = 0 . \tag{145.5}$$

Because of eq. (145.2)

$$\boldsymbol{\sigma}_1^k = \frac{\eta_k}{\delta_k}\mathbf{S}_1 , \quad \boldsymbol{\sigma}_2^k = \frac{\eta_k}{\delta_k}\mathbf{S}_2 .$$

Substituting these expressions into (145.5) we obtain

$$\mathbf{S}_1 \frac{\partial \eta_k}{\partial u^2} - \mathbf{S}_2 \frac{\partial \eta_k}{\partial u^1} + \eta_k \left(\frac{\partial \mathbf{S}_1}{\partial u^2} - \frac{\partial \mathbf{S}_2}{\partial u^1}\right) - (\mathbf{q}^k + \mathbf{p}^k)\sqrt{g} = 0 .$$

Hence with allowance for eq. (142.8),

$$\mathbf{p}^k = \eta_k \mathbf{q} - \mathbf{q}^k - \frac{1}{\sqrt{g}}\left(\mathbf{S}_1 \frac{\partial \eta_k}{\partial u^2} - \mathbf{S}_2 \frac{\partial \eta_k}{\partial u^1}\right),$$

or, in projections on the axes,

$$\begin{aligned}
p_1^k &= \eta_k q_1 - q_1^k - \frac{\sqrt{g_{11}}}{g}\left(S_{22} \frac{\partial \eta_k}{\partial u^1} - S_{12} \frac{\partial \eta_k}{\partial u^2}\right), \\
p_2^k &= \eta_k q_2 - q_2^k - \frac{\sqrt{g_{22}}}{g}\left(S_{11} \frac{\partial \eta_k}{\partial u^2} - S_{12} \frac{\partial \eta_k}{\partial u^1}\right), \\
p_n^k &= \eta_k q_n - q_n^k .
\end{aligned} \tag{145.6}$$

Here p_1, p_2, q_1, q_2 are physical components of the force vector in the direction of tangents to the coordinate lines u^1 and u^2. For a symmetrically loaded shell of rotation formulae (145.6) assume the form

$$p_1^k = \eta_k q_1 - q_1^k - \frac{\mathrm{d}\eta_k}{\mathrm{d}\xi} N_1, \quad p_n^k = \eta_k q_n - q_n^k . \tag{145.7}$$

In most cases the shell is loaded only by normal pressure q_n, applied to one of its outer layers. Then $q_1 = q_2 = q_1^k = q_2^k = 0$, $q_n^k = 0$ when $k \neq 1$, and $q_n^1 = q_n$. Nevertheless, according to formulae (145.6) each layer experiences tangential forces from the neighbouring layers and also bears part of the normal load; in this case formulae (145.6) give

$$\begin{aligned}
p_1^k &= -\frac{\sqrt{g_{11}}}{g}\left(S_{22} \frac{\partial \eta_k}{\partial u^1} - S_{12} \frac{\partial \eta_k}{\partial u^2}\right), \\
p_2^k &= -\frac{\sqrt{g_{22}}}{g}\left(S_{11} \frac{\partial \eta_k}{\partial u^2} - S_{12} \frac{\partial \eta_k}{\partial u^1}\right), \\
p_n^1 &= (\eta_1 - 1) q_n, \quad p_n^k = \eta_k q_n \quad (k \neq 1).
\end{aligned} \tag{145.8}$$

Tangential stresses are set up on the surface of junction between neighbouring layers and we shall denote them as follows. Let τ_i^k be the physical component in the direction of the vector $\mathbf{r}_{,i}$ of the tangential force that acts between the layers numbered k and $k+1$ (fig. 224). Then obviously, $p_i^k = \tau_i^k - \tau_i^{k-1}$, whence we obtain the following formula for τ_i^k:

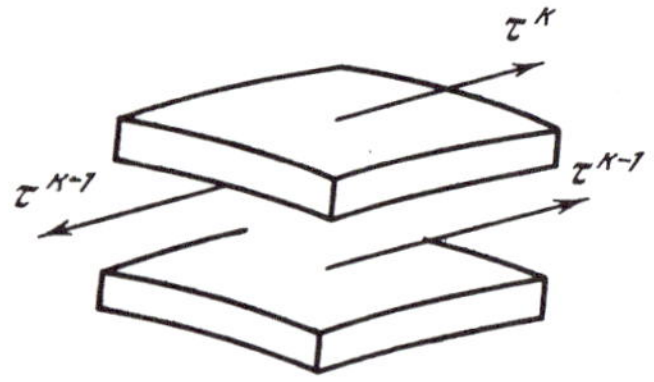

Fig. 224.

$$\tau_i^k = \sum_{s=1}^{k} p_i^s \, .$$

Knowing the quantities τ_i^k we can firstly assess the strength of the junctions between layers, and secondly judge the limits of applicability of the theory as tangential stresses ranging from τ_i^{k-1} to τ_i^k are set up in the thickness of the layer. These stresses should be small compared with the normal stresses, otherwise we should take account of shear occurring in the layer. The hypothesis that the shell behaves as a unit and that the tensor is constant in all layers will be incorrect if the quantities τ_i^k are significant. This is the case if the properties of the layers differ greatly, which occurs for example, when we are concerned with a layer of sandwich type in which metallic sheets are bonded together by a layer of light filler of low creep strength.

Note that if all layers observe the power creep law with the same exponent n, the quantities η_k are constant and consequently $p_1^k = p_2^k = 0$, and there are no tangential stresses in the shell.

146. **Two-layer model shell**

It is very difficult to construct a general theory of creep of thin shells with allowance for both tension and bending. Applying the Kirchoff-Love hypothesis of straight normals, even the first approximation leads to relationships between moments and forces on the one hand and rates of deformation and changes of curvature on the other, which repeat the corresponding expressions found for plates in section 132. Therefore, every one of the few existing solutions of the problem of creep of shells with allowance for bending is based on particular kinds of simplification, which ultimately reduce to the following. The rate potential or the potential of forces and moments are functions of six variables, M_{ij} and T_{ij} or ϵ_{ij} and κ_{ij} (see section 147). The condition U = const or Φ = const defines the surface in the corresponding six-dimensional space, the normals to the surface at each point represent the rate distribution for the given forces, or on the other hand force distribution for the given rates. In practice it is difficult to construct these surfaces, they always consist of several individual pieces of the analytical surfaces with more or less complicated boundaries. Real results can only be obtained in particular cases for certain subspaces with a smaller number of dimensions. However, certain points of these surfaces that correspond to simple states of stress can be found fairly simply. Without knowing the true form of the constant potential surface we draw a suitable surface through known points which provides a more or less successful approximation to the true surface.

This procedure is justified on the basis of the following considerations which apply to both creep and plasticity theories. The condition of plasticity that relates the values of the stresses in the limiting state is not essentially an expression of a certain definite physical fact, but only a more or less approximate method of describing the real behaviour of a certain set of specimens with adequately defined experimental conditions. Generally speaking, in simple tension tests the different specimens give a set of deformation diagrams which are replaced by a mean curve for purposes of calculation.

In particular the diagram of ideal plasticity without work-hardening may be selected as the mean diagram, this selection, though possible, is by no means the best for the great majority of materials, but it does permit a fully definite formulation of the limiting equilibrium. Now if we consider for example the bending of a rod, we can calculate the diagram for the variation of curvature with the bending moment on the basis of the idealised stress-strain diagram. On the other hand, we may consider that we are constructing a bending diagram for individual specimens by calculation or by experiment and then take a mean, being guided by a desire to achieve the simplest subsequent consideration of the bending problem. One possible approximation is that some particular curvature corresponds to constant limiting moment. This is the method used in constructing the theory of rigid-plastic beams to relate the idealised bending diagram with the concept of an ideal $\sigma - e$ diagram; the actual beam is replaced by a model, and either the elastic-plastic material is replaced by a rigid-plastic material, or the actual section of the beam is replaced by an ideal I-section.

Exactly the same state of affairs arises in creep theory, the elementary creep laws provide a more or less satisfactory approximation to true creep curves for a large number of specimens which differ quite markedly from one another, at any rate for technical alloys, unless special measures are taken to stabilise the properties of the material. In considering elements that are in a

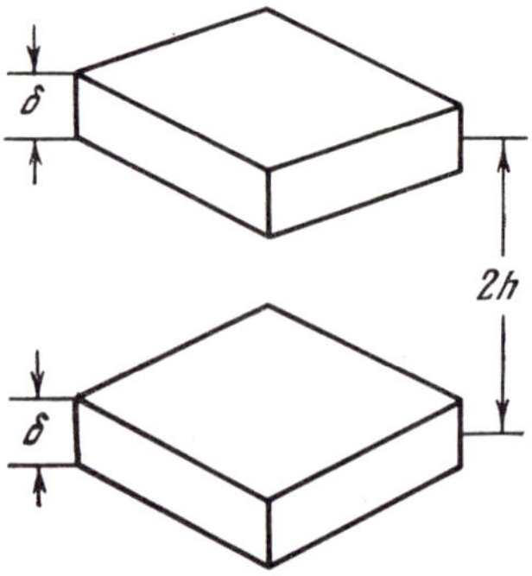

Fig. 225.

state of complex stress, we here add one further circumstance. The simplest laws of steady-state creep that provide a basis for the theory are only valid with constant or proportionately varying loads. These conditions are by no means always fulfilled in actual structures. Therefore, even if we proceed in a strictly accurate way in constructing a potential surface for a shell from some particular creep law, the practical value of the result should not be overestimated.

In constructing a simplified theory of shells we replace the real shell by a two-layer model, which is analogous to the known model of an ideal *I*-section for a beam. Consider a shell of two layers, each of thickness δ, the distance between the centre lines of the layers is $2h$ (fig. 225). We shall first consider that $\delta \ll h$. The layers are considered to be bonded together in such a way that the bond or filler transmits shear stresses but takes no part in carrying tensile forces and bending moments, therefore, the layers deform in accordance with the Kirchoff-Love hypothesis. The surface of reference will be considered to lie half way between the layers. In view of the thinness of each layer we can neglect the change in deformation across the thickness that follows from the hypothesis of straight normals and consequently, we can consider that the stress is uniformly distributed across the thickness of each layer.

The two-layer shell illustrated in fig. 225 serves as a model for a solid shell of thickness $2H$. Assuming that the properties of the materials of the real and model shells are the same, select the model parameters δ and h such that the behaviour of the real and model shells coincides in two cases; with a moment-free state of stress and in cylindrical bending.

By the formulae of section 119, in the plane state of stress we have

$$\sigma_{11} = \frac{4}{3} \frac{s(\epsilon_0)}{\epsilon_0} (\epsilon_{11} + \tfrac{1}{2} \epsilon_{22}) ,$$
$$\dots\dots\dots\dots\dots$$

In the moment-free state of stress the quantities ϵ_{ij} are constant across the thickness of a real shell, and identical in the top and bottom layers of the model shell. Therefore, the forces in the real shell are

$$T_{11} = \frac{4}{3} \frac{s(\epsilon_0)}{\epsilon_0} (\epsilon_{11} + \tfrac{1}{2} \epsilon_{22})\, 2H .$$

The forces in the model shell are

$$T_{11} = \frac{4}{3} \frac{s(\epsilon_0)}{\epsilon_0} (\epsilon_{11} + \tfrac{1}{2} \epsilon_{22})\, 2\delta ,$$
$$\dots\dots\dots\dots\dots$$

Comparing these two expressions we conclude that

$$\delta = H . \tag{146.1}$$

Now consider a state of bending of the shell in which the mean surface is not deformed. Because of the straight normals hypothesis, just as in the case of plates in bending, we obtain $\epsilon_{ij} = \kappa_{ij} z$; here κ_{ij} is the corresponding rate of change of curvature. In this case $\epsilon_0 = \kappa_0 z$, where κ_0 is an invariant and is derived from the rates of change of curvature in just the same way as ϵ_0 is derived from ϵ_{ij}. For simplicity all the considerations relate to physical components of forces, moments, rates of deformation and curvature in an orthogonal system of coordinates; the results are immediately transferable to an arbitrary system of reference. Now $\sigma_0 = s(\kappa_0 z)$. For the real shell the bending moment, for instance M_{11}, is expressed as follows:

$$M_{11} = \int_{-H}^{+H} \sigma_{11} z \, dz = \tfrac{4}{3} (\kappa_{11} + \tfrac{1}{2} \kappa_{22}) \frac{2}{\kappa_0} \int_0^H s(\kappa_0 z) z \, dz .$$

In the model shell the deformation of each layer is the same in value and opposite in sign, in the mean layer $\epsilon_{ij} = \pm h\kappa_{ij}$, the corresponding stresses, which are considered uniformly distributed across the thickness of each layer are

$$\sigma_{ij} = \pm \frac{4}{3} \frac{s(\kappa_0 h)}{\kappa_0} (\kappa_{11} + \tfrac{1}{2}\kappa_{22}) .$$

Consequently, the bending moment is

$$M_{11} = 2\sigma_{11}\delta h = \tfrac{4}{3}(\kappa_{11} + \tfrac{1}{2}\kappa_{22}) \frac{2}{\kappa_0} s(\kappa_0 h)\delta h .$$

Comparing the two expressions for M_{11}, we find

$$\delta h s(\kappa_0 h) = \int_0^H s(\kappa_0 z) z \, dz . \tag{146.2}$$

In particular with a power creep law

$$\delta h^{1+\frac{1}{n}} = \frac{n}{1+2n} H^{2+\frac{1}{n}} .$$

Hence, bearing in mind eq. (146.2) we obtain

$$h = \left(\frac{n}{1+2n}\right)^{\frac{n}{n+1}} H . \tag{146.3}$$

When $n = 1$, $h = (1/\sqrt{3})H \approx 0.577H$, when $n = \infty$, i.e. in the case of ideal plasticity, $h = H/2$. At first sight it might appear that the values found for δ and H are such that the initial assumptions are not fulfilled. Indeed we consider that $\delta \ll h$ whereas in actual fact these quantities are of the same order. However, for convenience we assumed that the creep laws of the material of the model of the real shell are the same. In actual fact we obtain exactly the same result if in the creep law

$$\sigma = \sigma_* s\left(\frac{\epsilon}{\epsilon_*}\right)$$

we use the same values of the constants ϵ_* for both the model and real shell, but arbitrarily increase the constant σ_* for the model, reducing the value of δ in the same ratio. It is evident that in this way we may make δ as small as we like relative to h. For convenience we use the actual characteristics of the material and ignore the circumstance that the inequality $\delta \ll h$ is not fulfilled.

Note that formula (146.2) establishes a relationship between h and H which generally contains the quantity κ_0, only with the power law is there a constant ratio of h to H. Thus, with the exponential creep law, depending on the value of κ_0 the ratio h/H varies within certain limits and tends to the value of $\frac{1}{2}$ with increase in curvature.

147. **Equations of the theory of shells**

The simplification that results from replacing the true shell by a two-layer model is not so great that we might think of constructing general equations in such a way that general methods of solution become realistic. Indeed the theory of elastic shells has mainly developed in the direction of investigating certain approximate systems of equations that describe a particular class of shells and particular types of deformations and states of stress that arise in them. We shall now describe in outline the considerations that underlie the selection of the corresponding simplified variants of the equations of the theory of shells in the case of an elastic shell. Assumptions are made concerning the nature of the decreases in certain quantities that characterise the states of stress and deformation of the shell along particular lines on the surface, which are usually taken as coordinates. Analysis of the order of magnitude of terms in the equations of the theory of shells inevitably leads to the result that if it is desired to limit the degree of accuracy in a rational way, for example, by neglecting terms of order h/R where R is the radius of curvature compared with terms of the order of unity or, with a coarser approximation to terms of the order of $(h/R)^{\alpha}$ $(0 < \alpha < 1)$, then a number of these terms must be rejected from the equations.

We shall not give the analysis in detail (see for example [20]), but simply mention that in addition to the accurate equations of the theory of shells a number of authors have proposed approximate variants based on mainly intuitive simplifications and assumptions. Subsequent stricter investigation usually provides some confirmation of the accuracy of the hypothesis. No such strict investigations are yet available for shells under conditions of creep or plasticity and, therefore, we shall consider several variants of the approximate equations of the theory, which are constructed by analogy with the corresponding variants of the theory of elastic shells which have proved to be well justified.

As usual, we shall refer the shell to lines of curvature of its mean surface. Put

$$ds^2 = A^2 d\xi_1^2 + B^2 d\xi_2^2 .$$

Here R_1 and R_2 denote the principal radii of curvature. Introduce the physical components of forces and moments T_1, T_2 and T_{12}, M_1 and M_2 and M_{12} and also the physical components of rates of deformation $\epsilon_1, \epsilon_2, \frac{1}{2}\gamma, \kappa_1, \kappa_2$ and τ. The rate-of-deformation vectors will again be given in terms of their physical components u, v and w in the directions of the coordinate lines ξ_1 and ξ_2 and the normal to the surface respectively. Let us write the fundamental equations of the theory of shells following [88].

a) The expressions for rates of deformation in terms of rates of displacement are:

$$\epsilon_1 = \frac{1}{A}\frac{\partial u}{\partial \xi_1} + \frac{v}{AB}\frac{\partial A}{\partial \xi_2} - \frac{w}{R_1},$$

$$\epsilon_2 = \frac{1}{B}\frac{\partial v}{\partial \xi_2} + \frac{u}{AB}\frac{\partial B}{\partial \xi_1} - \frac{w}{R_2}, \tag{147.1}$$

$$\gamma = \frac{B}{A}\frac{\partial}{\partial \xi_1}\left(\frac{v}{B}\right) + \frac{A}{B}\frac{\partial}{\partial \xi_2}\left(\frac{u}{A}\right),$$

$$\kappa_1 = \frac{1}{A}\frac{\partial}{\partial \xi_1}\left(\frac{1}{A}\frac{\partial w}{\partial \xi_1} + \frac{u}{R_1}\right) + \frac{1}{AB}\frac{\partial A}{\partial \xi_2}\left(\frac{1}{B}\frac{\partial w}{\partial \xi_2} + \frac{v}{R_2}\right),$$

$$\kappa_2 = \frac{1}{B}\frac{\partial}{\partial \xi_2}\left(\frac{1}{B}\frac{\partial w}{\partial \xi_2} + \frac{v}{R_2}\right) + \frac{1}{AB}\frac{\partial B}{\partial \xi_1}\left(\frac{1}{A}\frac{\partial w}{\partial \xi_1} + \frac{u}{R_1}\right), \tag{147.2}$$

$$\tau = \frac{1}{AB}\left(\frac{\partial^2 w}{\partial \xi_1 \partial \xi_2} - \frac{1}{A}\frac{\partial A}{\partial \xi_2}\frac{\partial w}{\partial \xi_1} - \frac{1}{B}\frac{\partial B}{\partial \xi_1}\frac{\partial w}{\partial \xi_2}\right)$$
$$+ \frac{1}{R_1}\left(\frac{1}{B}\frac{\partial u}{\partial \xi_2} - \frac{u}{AB}\frac{\partial A}{\partial \xi_2}\right) + \frac{1}{R_2}\left(\frac{1}{A}\frac{\partial v}{\partial \xi_1} - \frac{v}{AB}\frac{\partial B}{\partial \xi_1}\right).$$

b) The equations of equilibrium are:

$$\frac{1}{AB}\left[\frac{\partial}{\partial \xi_1}(BT_1) + \frac{\partial}{\partial \xi_2}(AT_{12}) + \frac{\partial A}{\partial \xi_2}T_{12} - \frac{\partial B}{\partial \xi_1}T_2\right] - \frac{N_1}{R_1} + q_1 = 0,$$

$$\frac{1}{AB}\left[\frac{\partial}{\partial \xi_1}(BT_{12}) + \frac{\partial}{\partial \xi_2}(AT_2) + \frac{\partial B}{\partial \xi_1}T_{12} - \frac{\partial A}{\partial \xi_2}T_1\right] - \frac{N_2}{R_2} + q_2 = 0, \tag{147.3}$$

$$\frac{1}{AB}\left[\frac{\partial}{\partial \xi_1}(BN_1) + \frac{\partial}{\partial \xi_2}(AN_2)\right] + \frac{T_1}{R_1} + \frac{T_2}{R_2} + q_n = 0,$$

$$\frac{1}{AB}\left[\frac{\partial}{\partial \xi_1}(BM_1)+\frac{\partial}{\partial \xi_2}(AM_{12})+\frac{\partial A}{\partial \xi_2}M_{12}-\frac{\partial B}{\partial \xi_1}M_2\right]+N_1=0\,,$$

$$\frac{1}{AB}\left[\frac{\partial}{\partial \xi_1}(BM_{12})+\frac{\partial}{\partial \xi_2}(AM_2)+\frac{\partial B}{\partial \xi_1}M_{12}-\frac{\partial A}{\partial \xi_2}M_1\right]+N_2=0\,. \tag{147.4}$$

In a strict and consistent theory of shells we should distinguish between the quantities T_{12} and T_{21} and M_{12} and M_{21}. Indeed calculating the force per unit length of mean surface we should bear in mind that the normals to the surface are not parallel and, therefore, for example

$$T_{12}=\int_{-h}^{+h}\sigma_{12}\left(1+\frac{z}{R_2}\right)\mathrm{d}z\,,$$

$$T_{21}=\int_{-h}^{+h}\sigma_{12}\left(1+\frac{z}{R_1}\right)\mathrm{d}z\,, \tag{147.5}$$

and $T_{12}\neq T_{21}$.

In the literature considerable attention has been paid to the construction of a consistent theory of elastic shells on the basis of the Kirchoff-Love hypothesis and various authors have proposed different systems of equations which are very similar to one another. If we take the expressions for forces in the form of eqs. (147.5) and in similar form for moments, fairly complicated expressions are obtained for these quantities in terms of deformations even for an elastic shell and they must be simplified for practical application. In creep theory we immediately neglect terms of the form z/R in formulae (147.5) and the corresponding expressions for M_{ij}; here it is found that T_{ij} and M_{ij} are expressed in exactly the same way as for plates, namely

$$T_{ij}=\int_{-h}^{+h}\sigma_{ij}\mathrm{d}z\,,\qquad M_{ij}=\int_{-h}^{+h}\sigma_{ij}z\,\mathrm{d}z\,. \tag{147.6}$$

Note that in our work [96] the quantities T_{ij} and M_{ij} were introduced directly as parameters that characterise the state of stress of the shell and which are not considered as forces and moments; these quantities do not figure in the theory at all. Then the equations of equilibrium of the shell do not coincide with eqs. (147.3) and (147.4) but have a more complicated structure, the theory is simplified by successive rejection of certain terms in the equations of equilibrium.

If formulae (147.6) are taken as expressions for forces and moments we fail to fulfil the further, sixth equation of equilibrium, which we have not written down, but, as many authors have shown, this contradiction is unimportant. Therefore, in the theory of creep of shells we consider that forces and moments are given by formulae (147.6) and consequently the creep law must be added to the equations that we have written down and we find, exactly as for plates (see section 132), either

$$T_{ij} = \frac{\partial U}{\partial \epsilon_{ij}}, \qquad M_{ij} = \frac{\partial U}{\partial \kappa_{ij}}, \tag{147.7}$$

or

$$\epsilon_{ij} = \frac{\partial \Phi}{\partial T_{ij}}, \qquad \kappa_{ij} = \frac{\partial \Phi}{\partial M_{ij}}. \tag{147.8}$$

148. Technical theory of shells

We shall use the general term "technical theory of shells" to combine various simplified systems of equations of shells for which the following conditions are fulfilled relative to the coordinate system that is formed by lines of curvature:

1) forces $T_{11} = 0$;

2) the rate of change of curvature $\kappa_{22} = 0$;

3) there are no tangential forces or twisting moments:

$$T_{12} = 0, \qquad M_{12} = 0.$$

These conditions may be fulfilled either accurately or approximately. In the latter case the essential thing is that by working to a particular degree of accuracy the corresponding terms of the fundamental equations of the problem can be neglected.

The following problems relate to technical theory:

a) *Theory of axially symmetrical deformation of a circular cylindrical shell.* Axis 1 is in the direction of the generator of the cylindrical shell, the corresponding coordinate is x and as coordinate y we take the arc of the contour of the cross-section. Here $A = B = 1$, $R_1 = \infty$, $R_2 = a$. In the axially symmetrical case there are two components of displacement u and w, $\epsilon_{12} = \kappa_{12} = 0$, $T_{12} = M_{12} = 0$, the remaining quantities depend only on x. From eqs. (147.1) and (147.2) it follows that:

$$\epsilon_{11} = \frac{du}{dx}, \quad \epsilon_{22} = -\frac{w}{a}, \quad \kappa_{11} = \frac{d^2w}{dx^2}, \quad \kappa_{22} = 0. \tag{148.1}$$

From the equations of equilibrium (147.3) and (147.4) we obtain

$$\frac{dT_{11}}{dx} + q_1 = 0, \quad \frac{dN_1}{dx} + \frac{T_{22}}{R_2} + q_n = 0, \quad \frac{dM_{11}}{dx} + N_1 = 0. \tag{148.2}$$

The first of the eqs. (148.2) is immediately integrated, therefore, T_{11} is a known function of x. If $q_1 = 0$, $T_{11} = \text{const}$; this constant becomes zero if there is no axial force on the end of the shell. In this case all the assumptions of the technical theory are fulfilled. Eliminating w, from the expressions for ϵ_{22} and κ_{11} by formulae (148.1) we obtain the following equation of compatibility:

$$\frac{d^2\epsilon_{22}}{dx^2} + \frac{1}{a}\kappa_{11} = 0. \tag{148.3}$$

Now eliminate N_1 from the second and third equations of the system (148.2). We obtain

$$\frac{d^2M_{11}}{dx^2} - \frac{1}{a}T_{22} - q_n = 0. \tag{148.4}$$

It is essential in the technical theory that the equations of compatibility and equilibrium are constructed symmetrically.

b) *Theory of long cylindrical shells.* In long cylindrical shells the state of stress varies slowly along the length. Longitudinal bending moments, twisting moments and forces in the transverse direction are negligible, the fundamental factors that determine the state of the shell are longitudinal forces and the transverse moments. Tangential forces exist and figure in the equations of equilibrium but they are not large and, therefore, the shears associated with them are negligibly small. The theory of long cylindrical shells has been developed by various authors using different approaches [21, 88]; either by studying the properties of certain solutions of the complete system of equations of the theory of shells or by explicitly formulated hypotheses. The first of these procedures was developed for elastic shells, in the case of creep it requires independent investigation; the second gives quicker results and, therefore, we give here the derivation of the fundamental equations of the theory of long cylindrical shells which is sometimes termed, following Vlasov, the semi-moment-free theory. We select a system of coordinates just as previously in considering axially symmetrical deformations of cylindrical shells. The difference is that the cylinder is no longer considered circular and, therefore, $R_2 = R(y)$. Rewriting expressions for ϵ_{ij} and κ_{ij} by formulae (147.1) for a cylindrical shell we obtain

$$\epsilon_{11} = \frac{\partial u}{\partial x}, \quad \epsilon_{22} = \frac{\partial v}{\partial y} - \frac{w}{R}, \quad 2\epsilon_{12} = \gamma = \frac{\partial v}{\partial x} + \frac{\partial u}{\partial y},$$

$$\kappa_{11} = \frac{\partial^2 w}{\partial x^2}, \quad \kappa_{22} = \frac{\partial}{\partial y}\left(\frac{\partial w}{\partial y} + \frac{v}{R}\right), \tag{148.5}$$

$$\kappa_{12} = \frac{\partial^2 w}{\partial x \partial y} + \frac{1}{R}\frac{\partial v}{\partial x}.$$

Vlasov's kinematic hypotheses are that the transverse elongation ϵ_{22} and the shear ϵ_{12} are zero. Therefore, the following conditions should be added to eqs. (148.5):

$$\epsilon_{22} = 0, \quad \epsilon_{12} = 0. \tag{148.6}$$

The meaning of the second of conditions (148.6) is that although the shear forces T_{12} on which ϵ_{12} primarily depend cannot be considered zero (we shall see later that the derivatives of T_{12} figure in the equations of equilibrium); they are small compared with the fundamental forces T_{11}, consequently, the shear deformation is small compared with the longitudinal deformation ϵ_{11}. A similar position arises in the elementary theory of bending of beams. Shear bending stresses appear in the equations of statics, their resultant for a given section is the shear force which cannot be made zero. Nevertheless, in considering the distribution of the fundamental stresses, i.e. the normal bending stresses, we use the plane-sections hypothesis and consequently, we do not take into account shear deformation associated with tangential stress.

The first condition is not so clear in nature. Indeed we are justified in assuming that $T_{22} \ll T_{11}$, but it does not follow from this that $\epsilon_{22} \ll \epsilon_{11}$, these quantities are of the same order. In an elastic shell, ϵ_{22} is infinitesimally small only when Poisson's ratio is zero, and this is the assumption that Vlasov used in developing the theory of long cylindrical shells. Subsequent investigations have shown that the first of conditions (148.6) can be rejected with very little change in the final result and the equations remain valid to the same degree of accuracy when $\nu \neq 0$. We retain Vlasov's condition for simplicity of presentation.

From eqs. (148.5), bearing in mind eqs. (148.6), we can obtain the equation of compatibility which relates the rate of deformation ϵ_{11} to the rate of change of curvature κ_{22}.

This equation is of the form

$$\Omega\epsilon_{11} + \frac{\partial^2 \kappa_{22}}{\partial x^2} = 0 . \tag{148.7}$$

Here the differential operator Ω is

$$\Omega = \frac{\partial^2}{\partial y^2}\left(R\,\frac{\partial^2}{\partial y^2}\right) + \frac{\partial}{\partial y}\left(\frac{1}{R}\,\frac{\partial}{\partial y}\right) . \tag{148.8}$$

Note that with the assumptions that have been made

$$\kappa_{11} = -R\,\frac{\partial^2 \epsilon_{11}}{\partial y^2} .$$

Consequently, the first term in eq. (148.7) is $\partial^2\kappa_{11}/\partial y^2$. Equations of the approximate theory have meaning only when terms of the same order are retained in them. The theory of long cylindrical shells should describe states of stress that vary slowly along the length of the shell and relatively quickly in the transverse direction. The characteristic length of shell for which the theory is valid is denoted l, and it is assumed that $a \ll l$, where a is a suitable radius of curvature of the cross-section. The order of the first term in eq. (148.7) is κ_{11}/a^2, the order of the last is κ_{22}/l^2 and as these orders should be the same (otherwise the corresponding term in the equation is rejected) we have

$$\kappa_{11} \sim \frac{a^2}{l^2}\,\kappa_{22} .$$

Thus we have found that the curvature of the generator is small compared with the curvature of the cross-section of the shell and we can take

$$\kappa_{11} = 0 \,. \tag{148.9}$$

It should be emphasised that condition (148.9) can be used to find forces and moments from the creep law; the error in the result is small but derivatives of κ_{11} can be considerable in the direction in which κ_{11} changes rapidly and, as we have seen, these derivatives cannot be neglected.

Now consider the equations of equilibrium. Write first the total system of equations for a cylindrical shell

$$\begin{gathered}\frac{\partial T_{11}}{\partial x} + \frac{\partial T_{12}}{\partial y} = 0 \,, \qquad \frac{\partial T_{12}}{\partial x} + \frac{\partial T_{22}}{\partial y} - \frac{N_2}{R} = 0 \,, \\ \frac{\partial N_1}{\partial x} + \frac{\partial N_2}{\partial y} + \frac{T_{22}}{R} + q_n = 0 \,, \\ \frac{\partial M_{11}}{\partial x} + \frac{\partial M_{12}}{\partial y} + N_1 = 0 \,, \qquad \frac{\partial M_{12}}{\partial x} + \frac{\partial M_{22}}{\partial y} + N_2 = 0 \,.\end{gathered} \tag{148.10}$$

In writing the equations of equilibrium it is assumed that only a normal load q_n acts on the shell. As we have established that $\kappa_{11} \ll \kappa_{22}$, we should expect the moments M_{11} and M_{22} to be of the same order, say M. Then the assessment of the orders of N_1 and M_{12} follows from the fourth equation

$$N_1 \sim \frac{M}{l} \,, \qquad M_{12} \sim \frac{a}{l} M \,.$$

If we accept this assessment of M_{12}, i.e. we assume that the derivative of M_{12} is important in the fourth equation, then we find that in the fifth equation the order of the first term is Ma/l^2, the order of the second M/a, consequently, the first term may be neglected with an error of a^2/l^2 compared with unity. This equation is rewritten as follows:

$$\frac{\partial M_{22}}{\partial y} + N_2 = 0 \,. \tag{148.11}$$

Hence follows the assessment for N_2:

$$N_2 \sim \frac{M}{a} \sim N_1 \frac{l}{a} \,.$$

Now consider the third equation of the system (148.10). Starting from the assessments which we have found for the orders of N_1 and N_2, we find that the order of the first term is M/l^2, and the order of the second is M/a^2, consequently the first term is negligibly small and we rewrite the equation in simplified form,

$$\frac{\partial N_2}{\partial y} + \frac{T_{22}}{R} + q_n = 0 . \tag{148.12}$$

The remaining equations are written without change:

$$\frac{\partial T_{11}}{\partial x} + \frac{\partial T_{12}}{\partial y} = 0 , \quad \frac{\partial T_{12}}{\partial x} + \frac{\partial T_{22}}{\partial y} - \frac{N_2}{R} = 0 . \tag{148.13}$$

Assuming that all terms in these equations are of the same order we find that the forces T_{11} are fundamental, compared with them forces T_{12} are small quantities of order a/l and forces T_{22} are small quantities of order a^2/l^2, thus justifying the hypothesis that shear, which depends mainly on T_{12}, is absent and we may approximately put

$$T_{22} = 0 . \tag{148.14}$$

Conditions (148.9) and (148.14) together with the conditions that shear is absent and that the moment M_{12} is small compared with M_{22} justify the statement made at the start of this section that this variant pertains to the technical theory of shells.

Now eliminate N_1, T_{12} and T_{22} from the equations of equilibrium (148.11) – (148.13). We find

$$\Omega M_{22} + \frac{\partial^2 T_{11}}{\partial x^2} - \frac{\partial^2}{\partial y^2} (R q_n) = 0 . \tag{148.15}$$

Comparing eqs. (148.7) and (148.15) we find the same symmetry as in the case of axially symmetrical loading of a cylindrical shell.

c) *Pure bending of a curvilinear tube.* An extensive literature (see [111]) has been devoted to solution of this problem for a tube of elastic material. An approximate solution of the creep problem is contained in [52]. Here we give a somewhat different variant of the fundamental equations of the problem [102].

Let R be the radius of the tube and a the radius of the section, the choice of coordinate system $\varphi = \xi_1$, and $\theta = \xi_2$ is clear from fig. 226. It is evident

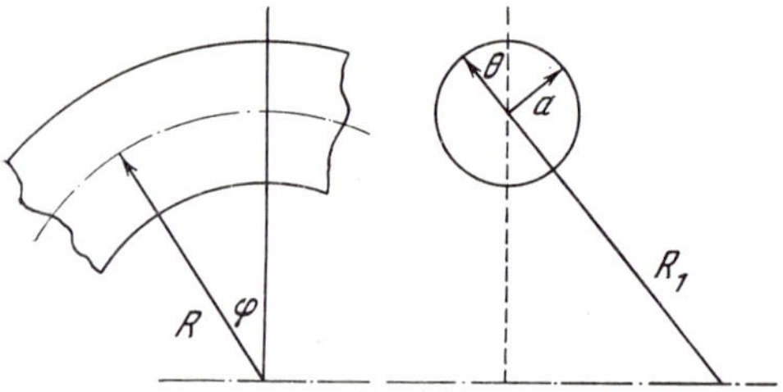

Fig. 226.

that

$$A = R + a\cos\theta\,, \quad B = a\,, \quad R_2 = a\,, \quad R_1 = \frac{R+a}{\cos\theta}\,. \tag{148.16}$$

Hereafter we shall consider that $a \ll R$, therefore, in place of expressions (148.16) we may write

$$A = R\,, \quad B = a\,, \quad R_1 = \frac{R}{\cos\theta}\,, \quad R_2 = a\,. \tag{148.17}$$

However, where for instance the quantity A must be differentiated with respect to θ we shall use the accurate expression (148.16).

If the tube is subject to pure bending and its section remains circular all the parameters of change of curvature and all the rates of deformation are zero except $\epsilon_1 = c\cos\theta$. In fact there are additional deformations associated with flattening of the tube. Denote the additional rates of displacement $v(\theta)$ and $w(\theta)$. By formulae (147.1) and (147.2) we obtain

$$\epsilon_2 = \frac{v' - w}{a}\,, \quad \epsilon_1 = c\cos\theta - \frac{v\sin\theta + w\cos\theta}{R}\,, \quad \kappa_2 = \frac{w'' + v'}{a^2}\,.$$

Since by assumption $a \ll R$ and in order for ϵ_2 and ϵ_1 to be of the same order the numerator in the expression for ϵ_2 should be small. It is required that ϵ_2 and ϵ_1 are of the same order in order to meet the condition $T_2 \ll T_1$. Therefore

$$w \approx v'\,.$$

Thus

$$\epsilon_1 = c\cos\theta - \frac{1}{R}(v\sin\theta + v'\cos\theta)\,,$$

$$\kappa_2 = \frac{1}{a^2}(v''' + v')\,.$$

Eliminating v we arrive at the following equation of compatibility:

$$\epsilon_1''' + \epsilon_1' + \frac{a^2}{R^2}\left[(\kappa_2\cos\theta)' - \kappa_2\sin\theta\right]\,. \qquad (148.18)$$

The equations of equilibrium (147.3) and (147.4) now lead to the following system:

$$\frac{1}{a}T_2' + \frac{1}{R}T_2\sin\theta - \frac{1}{a}N_2 = 0\,, \quad \frac{1}{a}M_2' + N_2 = 0\,,$$

$$\frac{1}{a}N_2' + \frac{1}{a}T_2 + \frac{\cos\theta}{R}\,T_1 = 0\,.$$

By eliminating T_2 and N_2 we obtain

$$M_2''' - M_2' - \frac{a^2}{R}\left[(T_1\cos\theta)' - T_1\sin\theta\right] = 0\,. \qquad (148.19)$$

Again, as in previous cases, the equations of compatibility and equilibrium are symmetrical. Similar equations in somewhat different form are obtained in [111].

149. **The technical theory of creep of two-layer shells**

Now consider a two-layer model shell with the assumptions formulated at the start of section 148. An element of the shell cut along the line of curvature is shown in fig. 225, the thickness δ of the layers and the distance $2h$ between them are selected in accordance with the formulae of section 146. The surface of reference is selected half way between the layers, quantities relating to the upper layer are denoted by the sign + and those relating to the lower layer by the sign −. Now let ϵ_1 and ϵ_2 be rates of elongation of the mean surface, κ_1 the rate of change of curvature in direction 1, $\kappa_2 = 0$ by hypothesis.

Because of the straight-normals hypothesis

$$\epsilon_{11}^{+} = \epsilon_1 + \kappa_1 h\,, \quad \epsilon_{11}^{-} = \epsilon_1 - \kappa_1 h\,, \quad \epsilon_{22}^{+} = \epsilon_{22}^{-} = \epsilon_{22}\,.$$

The stresses in the layers are expressed in terms of the rates of deformation as follows:

$$\begin{aligned}
\sigma_{11}^{+} &= \frac{4}{3}\frac{s^{+}}{v^{+}}\,[\epsilon_1 + \kappa_1 h + \tfrac{1}{2}\,\epsilon_2]\,,\\
\sigma_{11}^{-} &= \frac{4}{3}\frac{s^{-}}{v^{-}}\,[\epsilon_1 - \kappa_1 h + \tfrac{1}{2}\,\epsilon_2]\,,\\
\sigma_{22}^{+} &= \frac{4}{3}\frac{s^{+}}{v^{+}}\,[\epsilon_2 + \tfrac{1}{2}\,\epsilon_1 + \tfrac{1}{2}\,\kappa_1 h]\,,\\
\sigma_{22}^{-} &= \frac{4}{3}\frac{s^{-}}{v^{-}}\,[\epsilon_2 + \tfrac{1}{2}\,\epsilon_1 - \tfrac{1}{2}\,\kappa_1 h]
\end{aligned} \tag{149.1}$$

$$(s = \sigma_0\,, \quad v = \epsilon_0)\,.$$

Now the condition $T_{11} = 0$ should be fulfilled, this means that $\sigma_{11}^{+} = -\sigma_{11}^{-}$. From the stress formulae it will be seen that this occurs if

$$\epsilon_1 + \tfrac{1}{2}\,\epsilon_2 = 0\,, \tag{149.2}$$

and moreover

$$\frac{s^{+}}{v^{+}} = \frac{s^{-}}{v^{-}}\,.$$

As the material of both layers is assumed the same and $s = s(v)$ we should have $v^{+} = v^{-}$. We now confirm that this second condition is a consequence of eq. (149.2). Indeed if eq. (149.2) is fulfilled the stress formulae are rewritten as follows:

$$\begin{aligned}
\sigma_{11}^{+} &= \frac{4}{3}\frac{s^{+}}{v^{+}}\,\kappa_1 h\,, & \sigma_{11}^{-} &= -\frac{4}{3}\frac{s^{-}}{v^{-}}\,\kappa_1 h\,,\\
\sigma_{22}^{+} &= \frac{s^{+}}{v^{+}}\,[\epsilon_2 + \tfrac{2}{3}\,\kappa_1 h]\,, & \sigma_{22}^{-} &= \frac{s^{-}}{v^{-}}\,[\epsilon_2 - \tfrac{2}{3}\,\kappa_1 h]\,.
\end{aligned} \tag{149.3}$$

Now substitute the values of σ_{11}^{+} and σ_{11}^{-} into the expression

$$s^2 = \sigma_{11}^2 + \sigma_{22}^2 - \sigma_{11}\sigma_{22} \,.$$

As a result we obtain

$$(v^+)^2 = \epsilon^2 + \tfrac{4}{3}\kappa_1^2 h^2 \,. \qquad (149.4)$$

The stresses in the lower layer are expressed by the same formulae as those in the upper with $-h$ substituted for h, but expression (149.4) does not alter when h is replaced by $-h$, consequently, $v^+ = v^-$ and $s^+ = s^-$. Hereafter we shall not put signs against the quantities v and s.

Now calculate the force T_{22} and the bending moment M_{11}:

$$T_{22} = \delta(\sigma_{22}^{+} + \sigma_{22}^{-}) = 2\delta \frac{s}{v} \epsilon_2 \,,$$

$$M_{11} = \delta h(\sigma_{11}^{+} - \sigma_{11}^{-}) = \tfrac{8}{3}\delta \frac{s}{v}\kappa_1 h^2 \,.$$

Introduce the following dimensionless quantities:

$$\omega = \frac{\sigma_* v}{\epsilon_* s}, \quad u = \frac{\epsilon_2}{\epsilon_*}, \quad m = \frac{\sqrt{3}}{4}\,\frac{M_{11}}{h\delta\sigma_*} \,. \qquad (149.5)$$

Here σ_* and ϵ_* are arbitrary constants with the dimensions of stress and rate of deformation respectively. Now

$$T_{22} = 2\delta\sigma_* \frac{u}{\omega}, \quad \kappa_1 h = \frac{\sqrt{3}}{2}\epsilon_* m\omega \,. \qquad (149.6)$$

The first of formulae (149.5) defines ω as function of v; conversely we may represent v as a function of ω, then expression (149.4) assumes the form

$$\frac{1}{\epsilon_*^2} v^2(\omega) = u^2 + m^2\omega^2 \,. \qquad (149.7)$$

The function $v(\omega)$ is known for each particular creep law and, therefore, eq. (149.7) can be used to determine the quantity ω as function of the dimensionless variables m and u. It is somewhat inconvenient that ω is given as an implicit function of m and u.

The next step is to use eqs. (149.5) and (149.6) to eliminate from the equation of compatibility all other variables except u and m. We shall consider this in the next section and for the moment we shall examine eq. (149.7) in more detail.

Later on we shall require partial derivatives of ω with respect to m and u. Differentiating eq. (149.7) we obtain the following formulae for them:

$$(\psi' - 2m^2)\frac{\partial\omega}{\partial u} = \frac{2u}{\omega}, \quad (\psi' - 2m^2)\frac{\partial\omega}{\partial m} = 2m\omega . \tag{149.8}$$

The function $\psi(\omega)$ is defined by the expression

$$\psi = \frac{1}{\epsilon_*^2}\int\frac{\mathrm{d}(v^2)}{\omega} . \tag{149.9}$$

The primes in formulae (149.8) denote differentiation with respect to the parameter ω.

Now consider particular cases of the creep law.

a) *Power law* $\epsilon = \epsilon_*(\sigma/\sigma_*)^n$. Eq. (149.7) assumes the form

$$\omega^{\frac{2n}{n-1}} = u^2 + m^2\omega^2 . \tag{149.10}$$

A nomogram of eq. (149.10) is given in fig. 227. With a power creep law the function ψ defined by formula (149.9) is

$$\psi = \frac{2n}{n+1}\,\omega^{\frac{n+1}{n-1}} . \tag{149.11}$$

b) *Hyperbolic sine law.* As we shall explain later, difficulties sometimes arise in using the creep power law for the investigation of shells. These difficulties are associated with the circumstance that the derivative $\mathrm{d}s/\mathrm{d}v$ becomes infinite when $s = 0$. The hyperbolic sine law is free from this defect and it is probably the most suitable for numerical calculations. If it is replaced by the exponential law we are obliged to isolate the region of small stresses, which is inconvenient in calculations if they are to be put in standard form for use on a computer. Eq. (149.7) is then of the form

$$\varphi^2(\omega) = u^2 + m^2\omega^2 .$$

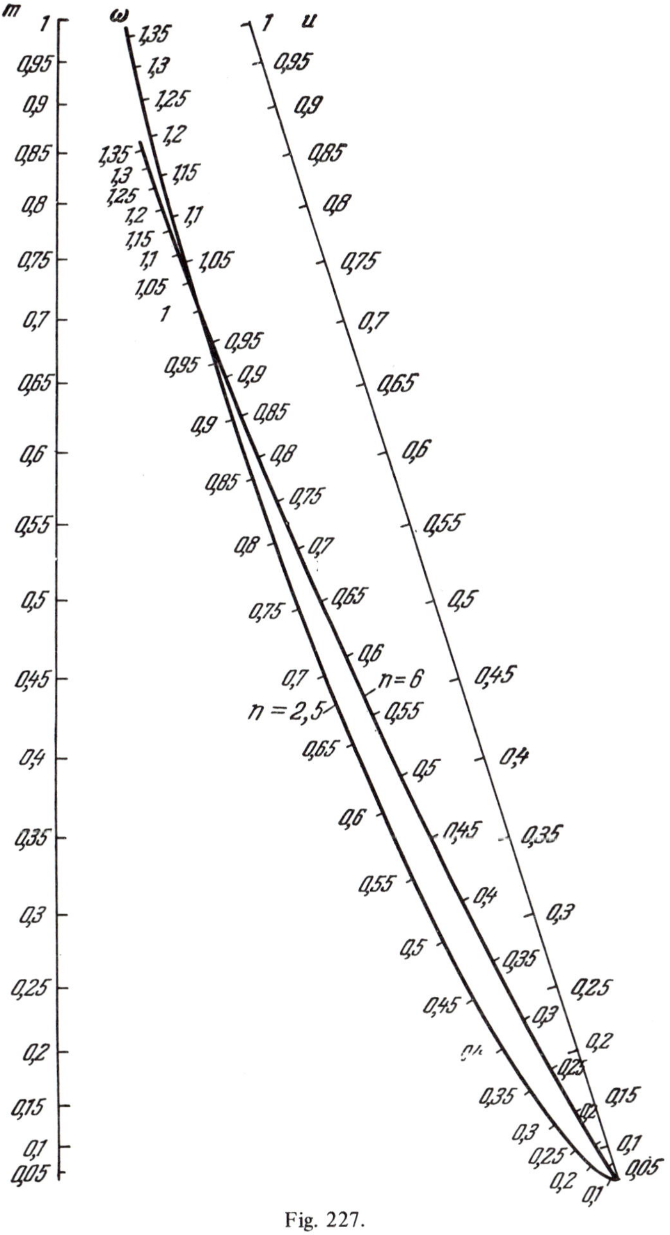

Fig. 227.

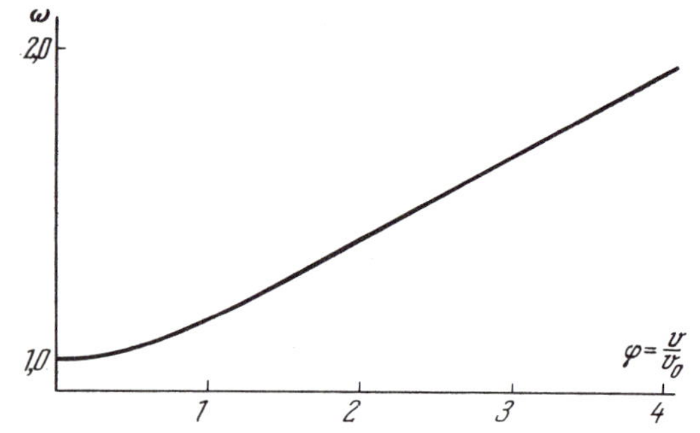

Fig. 228.

Here the function φ is the converse of the following ratio:

$$\omega = \frac{\varphi}{\operatorname{arsinh} \varphi} . \tag{149.12}$$

Here we consider that

$$\sigma_* = \sigma_e , \qquad \epsilon_* = 2\epsilon_e .$$

σ_e and ϵ_e are constants of the exponential law.

A graph of the function $\varphi(\omega)$ for the hyperbolic sine law is given in fig. 228. It should be borne in mind that with the hyperbolic sine law, as with the exponential law, the thickness of the model is not uniquely determined.

150. **Equations of axially symmetrical creep of a circular cylindrical shell**

We shall assume that the shell carries only a normal load q_n. In eqs. (148.3) and (148.4) convert to the variables u and m, that we introduced above. We obtain the following system:

$$u'' + 2m\omega = 0 , \tag{150.1}$$

$$m'' - \frac{2u}{\omega} + 2p = 0 . \tag{150.2}$$

Here

$$p = -\frac{q_n a}{2\delta\sigma_*},$$

the primes denote differentiation with respect to the dimensionless variable $\xi = x/b$, where $b^2 = (4/\sqrt{3})ah$. The function $\omega(u, m)$ that appears in eqs. (150.1) and (150.2) is given by expression (149.6).

Various methods of numerical integration may be used to solve this system, an approximate method of solution by the Galerkin method is given in the work [102], in which an analogous system is derived for elastic-plastic shells. Another possibility is to use variational methods. It is found that eqs. (150.1) and (150.2) express the condition that a certain functional is stationary and, therefore, direct methods of variational calculus can be applied to the solution of problems of the theory of shells. Eqs. (150.1) and (150.2) include two functions: $u(\xi)$ and $m(\xi)$; we shall vary them on the assumption that they are independent of one another and, therefore, the new variational principle is not a consequence of the Lagrange or Castigliano principles, on the contrary each of these principles follows from the new principle with certain specific assumptions. Therefore, we shall not refer to the general principles of creep theory but shall attempt to select a functional for which eqs. (150.1) and (150.2) are Euler equations. This is the following functional:

$$N = \int_0^l [u'm' + \psi(\omega) - 2m^2\omega - 2pu]\, d\xi\,. \tag{150.3}$$

Here $\psi(\omega)$ is a function introduced by expression (149.8). Calculate the variation of the functional N using formulae (149.7) for the derivatives $\partial\omega/\partial u$ and $\partial\omega/\partial m$. We obtain

$$\delta N = (u'\delta m + m'\delta u)_0^l$$

$$+ \int_0^l \left[\left(-m'' + \frac{2u}{\omega} - 2p\right)\delta u - (u'' + 2m\omega)\delta m\right] d\xi\,.$$

Therefore, $\delta N = 0$ if eqs. (150.1) and (150.2) are fulfilled together with the natural boundary conditions that follow from the equation

$$(u'\delta m + m'\delta u)_0^l = 0\,. \tag{150.4}$$

Formula (150.3) can be made more symmetrical by integrating the expression for ψ by parts:

$$\psi = \frac{v^2}{\epsilon_*^2} \frac{1}{\omega} - \int v^2 \mathrm{d}\left(\frac{1}{\omega}\right) .$$

Put

$$-\int v^2 \mathrm{d}\left(\frac{1}{\omega}\right) = \int \frac{v^2}{\omega^2} \mathrm{d}\omega = \chi(\omega) ;$$

then, bearing in mind expression (149.7), we obtain

$$\psi = \frac{u^2}{\omega} + m^2\omega + \chi(\omega) .$$

The expression for the functional N assumes the form

$$N = \int_0^l \left[u'm' + \chi(\omega) + \frac{u^2}{\omega} - m^2\omega - 2pu \right] \mathrm{d}\xi . \tag{150.5}$$

We shall now show that the Lagrange and Castigliano variational principles follow from the variational equation $\delta N = 0$.

Assume that the equation of compatibility (150.1) is satisfied in advance. This means that we give arbitrarily the function $u(\xi)$ which is varied; then the function $m(\xi)$ is given uniquely by means of eqs. (150.1) and (149.7). Convert the functional (150.3) by integrating the first subintegral term by parts:

$$N = mu'|_0^l + \int_0^l [-mu'' + \psi - 2m^2\omega - 2pu]\, \mathrm{d}\xi .$$

Because of expression (150.1) this expression becomes

$$N = mu'|_0^l + \int_0^l (\psi - 2pu)\, \mathrm{d}\xi .$$

Now let us find the mechanical meaning of the quantity $\psi(\omega)$. This is twice the stress potential per unit volume expressed in dimensionless units. Indeed

$$\psi = \frac{1}{\epsilon_*^2} \int \frac{1}{\omega} \, \mathrm{d}(v^2) = \frac{2}{\epsilon_* \sigma_*} \int s \, \mathrm{d}v \, .$$

Therefore, provided eq. (150.1) is fulfilled, the expression for N can be written

$$N = 2(\widetilde{U} - A) + mu' |_0^l \, . \tag{150.6}$$

This is in fact the Lagrange functional. Indeed $\widetilde{U}$ is the stress potential, $A = \int pu \, \mathrm{d}\xi$ is the work of external forces on the corresponding displacement, the last term is the work applied by the moments on the edges. But the variation of δN becomes zero if condition (150.4) is fulfilled. With this condition in mind, varying expression (150.6) we obtain

$$2\delta(\widetilde{U} - A) = (-m'\delta u + m\delta u')_0^l \, . \tag{150.7}$$

Eq. (150.7) is a Lagrange variational equation. If the comparison functions are selected in such a way that when they are varied the external forces applied to the ends of the specimens perform no work, eq. (150.7) reduces to

$$\delta(\widetilde{U} - A) = 0 \, . \tag{150.8}$$

We may proceed as follows for practical calculations of the functional $\widetilde{U} - A$. Formula (149.7) expresses ω as function of the argument $u^2 + m^2\omega^2$, therefore, $\frac{1}{2}\,\psi(\omega) = U(u^2 + m^2\omega^2)$, but we have assumed that eq. (150.1) is fulfilled from which $m^2\omega^2 = (u''/2)^2$. Therefore

$$\widetilde{U} - A = \int_0^l \left[U\left(u^2 + \frac{u''^2}{4}\right) - pu \right] \mathrm{d}\xi \, . \tag{150.9}$$

Now assume that the equation of equilibrium (150.2) is fulfilled, we are given the function $m(\xi)$ and from this equation we find the function u. Proceeding in exactly the same way as before, integrating by parts we convert the functional (150.3) to the form

$$N = um' |_0^l + \int_0^l \left[\psi - \frac{2v^2}{\epsilon_*^2 \omega} \right] \mathrm{d}\xi \, .$$

In the conversion we have taken into account that eq. (150.2) is fulfilled and have used expression (149.7). The subintegral quantity

$$\psi - \frac{2v^2}{\epsilon_*^2 \omega} = \frac{2}{\epsilon_* \sigma_*} \left[\int s \, dv - vs \right] = -\frac{2}{\epsilon_* \sigma_*} \int v \, ds \tag{150.10}$$

is twice the dimensionless rate potential referred to unit volume, consequently,

$$N = um' |_0^l - 2\widetilde{\Phi} \, . \tag{150.11}$$

If statically possible states that correspond to the given load are compared, in varying expression (150.1) terms that depend on conditions on the ends of the shell disappear and we obtain Castigliano's variational equation

$$\delta \widetilde{\Phi} = 0 \, . \tag{150.12}$$

The quantity $v^2/\epsilon_*^2\omega - \frac{1}{2}\psi(\omega)$ is a function of ω, consequently it may be considered as a function of the argument $u^2/\omega^2 + m^2$:

$$\frac{v^2}{\epsilon_*^2 \omega} - \tfrac{1}{2} \psi(\omega) = \Phi \left(\frac{u^2}{\omega^2} + m^2 \right) .$$

But from expression (150.2) it follows that

$$\frac{u}{\omega} = \tfrac{1}{2} (m'' + 2p) \, .$$

Therefore, the Castigliano functional can be calculated by the formula

$$\widetilde{\Phi} = \int_0^l \Phi \, [m^2 + \tfrac{1}{4} (m'' + 2p)^2] \, d\xi \, . \tag{150.13}$$

151. Power creep law. Approximate investigation of the edge effect

With the power creep law, by formulae (149.10) and (149.11) we have

$$\psi = \frac{2n}{n+1} \omega^{\frac{n+1}{n-1}} = \frac{2n}{n+1} \left[\frac{u^2}{\omega} + m^2 \omega \right] .$$

Therefore, expression (150.3) for the functional N becomes

$$N = \int_0^l \left[u'm' + \frac{2n}{n+1} \frac{u^2}{\omega} - \frac{2}{n+1} m^2\omega - 2pu \right] d\xi \, . \tag{151.1}$$

This form is more convenient as it does not contain the function ψ.

To formulate the Lagrange functional note that

$$\tfrac{1}{2}\psi = \frac{n}{n+1}\left(\omega^{\frac{2n}{n-1}}\right)^{\frac{n+1}{2n}} .$$

Therefore

$$\widetilde{U} = \frac{n}{n+1}\int_0^l \left(u^2 + \tfrac{1}{4}u''^2\right)^{\frac{n+1}{2n}} d\xi \, . \tag{151.2}$$

On the other hand from the equation

$$\omega^{\frac{2n}{n-1}} = u^2 + m^2\omega^2 ,$$

it follows that

$$\omega^{\frac{2}{n-1}} = \frac{u^2}{\omega^2} + m^2 .$$

Now

$$\frac{v^2}{\epsilon_*^2\omega} - \tfrac{1}{2}\psi = \frac{1}{n+1}\,\omega^{\frac{n+1}{n-1}} = \frac{1}{n+1}\left(\omega^{\frac{2}{n-1}}\right)^{\frac{n+1}{2}} .$$

Consequently,

$$\widetilde{\Phi} = \frac{1}{n+1}\int_0^l \left[m^2 + \tfrac{1}{4}(m'' + 2p)^2\right]^{\frac{n+1}{n}} d\xi \, . \tag{151.3}$$

Consider a semi-infinite cylindrical shell uniformly loaded with distributed pressure and clamped at the end. Two cases will be distinguished:

a) The edge of the shell is freely supported, consequently, if we locate the origin of coordinates at the end of the shell we have

$$u(0) = m(0) = 0\,.$$

b) The edge of the shell is firmly clamped, consequently

$$u(0) = u'(0) = 0\,.$$

At a sufficiently great distance from the end the shell is in a moment-free state. We shall consider that when $\xi = \infty$, $m = 0$, $u = u^*$, $\omega = \omega^*$. From expression (150.2) it follows that:

$$\frac{u^*}{\omega^*} = p\,,$$

and from expression (149.10)

$$(\omega^*)^{\frac{2n}{n-1}} = u^{*2}\,.$$

Whence

$$u^* = p^n\,, \qquad \omega^* = p^{n-1}\,. \tag{151.4}$$

Introduce functions U and V which satisfy the following system of linear equations:

$$U'' + 2\mu V = 0\,, \qquad V'' - 2\lambda(U - u^*) = 0\,, \tag{151.5}$$

and the boundary conditions of our problem are: $U(0) = V(0)$ in case a) and $U(0) = U'(0)$ in case b). Here λ and μ are constants which are still not determined. Excluding the functional V from eq. (151.5) we arrive at the equation

$$U^{\mathrm{IV}} + 4\alpha^4(U - u^*) = 0\,.$$

Here $\alpha^4 = \lambda\mu$. Find the solution of this equation which satisfies the condition $U(\infty) = u^*$:

$$U = u^* + e^{-\alpha\xi}(a \cos \alpha\xi + b \sin \alpha\xi)\,.$$

Now from the first of the eqs. (151.5) we obtain

$$V = \sqrt{\frac{\lambda}{\mu}}\,e^{-\alpha\xi}(b \cos \alpha\xi - a \sin \alpha\xi)\,.$$

The constants a and b are determined so as to satisfy the boundary conditions on the fixed edge. Introduce the additional notation

$$\sqrt{\frac{\lambda}{\mu}}u^* = \beta(u^*)^{1/n} ,$$

and put

$$U = u^*U_0(\alpha\xi) , \quad V = \beta(u^*)^{1/n}V_0(\alpha\xi) . \tag{151.6}$$

The functions U_0 and V_0 are defined as follows:
in case a):

$$U_0 = 1 - e^{-\alpha\xi} \cos \alpha\xi , \quad V_0 = e^{-\alpha\xi} \sin \alpha\xi ;$$

in case b):

$$U_0 = 1 - e^{-\alpha\xi}(\cos \alpha\xi + \sin \alpha\xi) , \quad V_0 = -e^{-\alpha\xi}(\cos \alpha\xi - \sin \alpha\xi).$$

We shall seek an approximate solution in the form

$$u = U , \quad m = V ;$$

and here the parameters α and β will be arranged in such a way that the function assumes a stationary value. Moreover put

$$\omega = \omega^*\omega_0 = \left(u^*\right)^{\frac{n-1}{n}} \omega_0 . \tag{151.7}$$

Substitute expressions (151.6) and (151.7) into eq. (149.10). Note that the factor that depends on u^* is cancelled and we obtain

$$\omega_0^{\frac{2n}{n-1}} = U_0^2 + \beta^2 V_0^2\omega_0^2 . \tag{151.8}$$

We must now calculate the functional (151.1). The integral is convergent for an infinite region because in consequence of expression (151.4) the quantity u^2/ω tends to the value $(u^*)^{[(n+1)/n]}$ when $\xi = \infty$ and the product pu tends to the same value. Substituting expressions (151.6) and (151.7) into eq. (151.1) and on the basis of eq. (151.4) substitute $(u^*)^{1/n}$ for p, and under the integral add the constant terms that are necessary to ensure its convergence. We obtain

$$N = \left(u^*\right)^{\frac{n+1}{n}} \int_0^\infty \left[\beta \frac{dU_0}{d\xi}\frac{dV_0}{d\xi} + \frac{2n}{n+1}\left(\frac{U_0^2}{\omega_0} - 1\right)\right.$$

$$\left. - \frac{2}{n+1}\beta^2 V_0^2 \omega_0 - 2(U_0 - 1)\right] d\xi \, .$$

Now convert from the variable ξ to the variable $x = \alpha\xi$. Here the prime denotes differentiation with respect to this variable. Also put

$$N = \left(u^*\right)^{\frac{n+1}{n}} N_0(\alpha, \beta) \, .$$

We obtain

$$N_0 = \alpha\beta \int_0^\infty U_0' V_0' \, dx + \frac{1}{\alpha}\int_0^\infty \left[\frac{2n}{n+1}\left(\frac{U_0^2}{\omega_0} - 1\right)\right.$$

$$\left. - \frac{2}{n+1}\beta^2 V_0^2 \omega_0 - 2(U_0 - 1)\right] dx \, . \tag{151.9}$$

It is evident that with this selection of functions of comparison the boundary conditions (150.4) are fulfilled for both the freely supported and clamped edges. Consequently N_0 considered as a function of α and β should assume a stationary value. Denoting

$$\int_0^\infty U_0' U_0' \, dx = A \, ,$$

$$\int_0^\infty \left[\frac{2n}{n+1}\left(\frac{U_0^2}{\omega_0} - 1\right) - \frac{2}{n+1}\beta^2 V_0^2 \omega_0 - 2(U_0 - 1)\right] dx = B(\beta) \, . \tag{151.10}$$

Then

$$N_0 = \alpha\beta A + \frac{1}{\alpha} B(\beta) \, .$$

The extremal conditions are

$$\beta A - \frac{1}{\alpha^2} B(\beta) = 0\,, \qquad \alpha A + \frac{1}{\alpha}\frac{\partial B}{\partial \beta} = 0\,.$$

Whence

$$\frac{\partial}{\partial \beta}(B\beta) = 0\,, \quad \alpha^2 = \frac{B}{A\beta}\,. \tag{151.11}$$

The quantity A can be written down immediately, in case a) $A = -\frac{1}{4}$, in case b) $A = \frac{1}{2}$. We must now select the parameter β for which the first of conditions (151.11) is fulfilled. For this purpose we must calculate $B(\beta)$ for a series of values of β by the second of formulae (151.10) and find the value for which βB is a minimum.

All these considerations remain valid for an elastic shell but the words "rate of deformation" are replaced by the word "deformation" and in all the above formulae $n = 1$. The expression for $B(\beta)$ becomes

$$B = \int_0^\infty [(U_0 - 1)^2 - \beta^2 V_0^2]\,dx\,.$$

Put

$$\int_0^\infty (U_0 - 1)^2\,dx = C\,, \qquad \int_0^\infty V_0^2\,dx = D\,.$$

Then

$$B = C - \beta^2 D\,.$$

From the first condition (151.11) we obtain

$$\beta^2 = \frac{C}{3D}\,.$$

The integrals denoted C and D are easily calculated. Thus for case a) we have

$$C = \tfrac{3}{8}\,, \qquad D = \tfrac{1}{8}\,.$$

Therefore $\beta^2 = 1$. If we put $\beta = -1$ by the second of formulae (151.11), $\alpha = 1$.

152. A semi-infinite shell with hinged edge fixing

Graphs of the functions $U_0(x)$ and $V_0(x)$ for a supported edge are constructed in fig. 229. The quantity V_0 is considerably less than U_0, particularly for large values of x; when $x = 0$ by formula (151.8) $\omega_0 = 0$, consequently, ω_0 is small for small values of x. Therefore, the first term in the right-hand side of eq. (151.8) is always considerably greater than the second. Therefore, we may take

$$\omega_0 = U_0^{\frac{n-1}{n}} (1+\delta),$$

where $\delta \ll 1$. Substituting into (151.8) and considering only the first power of δ we find

$$\delta = \frac{n-1}{2n} \beta^2 V_0^2 U_0^{-2/n}. \tag{152.1}$$

Now fix a certain value of $x = c$, commencing from which $|U_0 - 1| < \rho$, and $\rho \ll 1$. Represent the integral that gives an expression for $B(\beta)$ in the form of a sum of two integrals: B_1 from $x = 0$ to $x = c$ and B_2 from $x = c$ to $x = \infty$. By the second of formulae (151.10), bearing in mind the approximate expression for ω_0 which we have adopted and neglecting terms of the order of δ^2 we find

$$\tfrac{1}{2} B_1 = \int_0^c \left\{ \frac{n}{n+1} U_0^{\frac{n+1}{n}} - U_0 + \frac{1}{1+n} \right\} \mathrm{d}x - \tfrac{1}{2}\beta^2 \int_0^c V_0^2 U_0^{\frac{n-1}{n}} \, \mathrm{d}x .$$

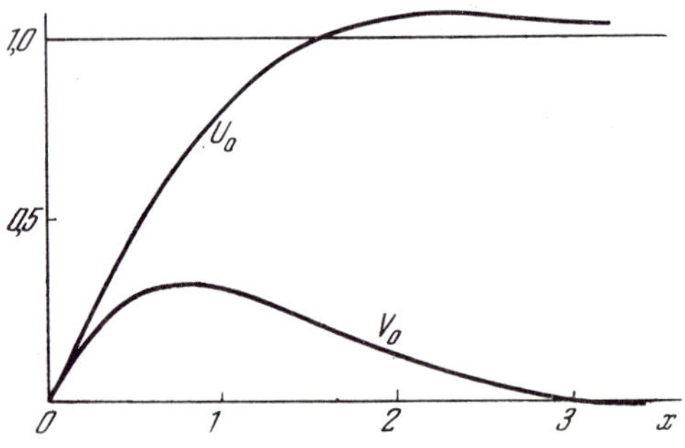

Fig. 229.

The quantity B_2 is calculated by the same formula with other limits of integration but terms of the order of ρ are neglected compared with unity, therefore

$$\tfrac{1}{2} B_2 = -\tfrac{1}{2}\beta^2 \int_c^\infty V_0^2 \,\mathrm{d}x \,.$$

This latter integral is easily calculated, namely

$$\int_c^\infty V_0^2 \,\mathrm{d}x = \int_c^\infty e^{-2x} \sin^2 x \,\mathrm{d}x = \tfrac{1}{8} e^{-2c}(4 + \cos 2c + \sin 2c) \,.$$

We give calculations for several values of n. Here we have taken $c = \pi$, therefore, $B_2 = -0.00093\beta^2$. The expression for B is given in the form

$$B(\beta) = 2I_1 - \beta^2 I_2 \,.$$

Here

$$I_1 = \int_0^\pi \left[\frac{n}{n+1} U_0^{\frac{n+1}{n}} - U_0 + \frac{1}{1+n}\right] \mathrm{d}x \,,$$

$$I_2 = \int_0^\pi V_0^2 U_0^{\frac{n-1}{n}} \,\mathrm{d}x + 0.00093 \,.$$

Now from the condition

$$\frac{\partial}{\partial\beta}(\beta B) = 0 \,,$$

we find

$$\beta^2 = \frac{2I_1}{3I_2} \,,$$

and by the second of formulae (151.11)

$$\alpha^2 = -4\left(\frac{2I_1}{\beta} - I_2\beta\right).$$

The table gives values of α and β for different values of n:

n	β	α
1	−1	1
2	−0.840	0.860
3	−0.728	0.763
4	−0.643	0.718
5	−0.598	0.678
6	−0.540	0.646

The approximation that results from the assumption that deflections and bending moments are distributed in the same way as in an elastic shell cannot be considered reliable for large values of n. Moreover, formulation of the problem of a shell loaded by pressure and hinged at the end leads to a contradiction when the power creep law is used. Indeed when $x = 0$, the deflection and moment are zero, consequently, also $\omega = 0$ and eqs. (150.1) and (150.2) are without a solution that satisfies the conditions $u(0) = m(0) = 0$ in the neighbourhood of the point $x = 0$. This is readily understood from the following considerations. Consider a shell of rigid-plastic work-hardening material corresponding to the diagram shown in fig. 230. When $\sigma < \sigma_T$ the plastic modulus is infinity and ω is the reciprocal of the plastic modulus. If the edge of this rigid-plastic shell is not loaded and the stresses are zero when $x = 0$, the stresses remain below the yield point in a certain region $0 < x < \delta$ and consequently part of the shell of length δ behaves as a rigid ring. The rest of the shell is in a plastic state, but when $x = \delta$ the condition of rigid clamping is fulfilled.

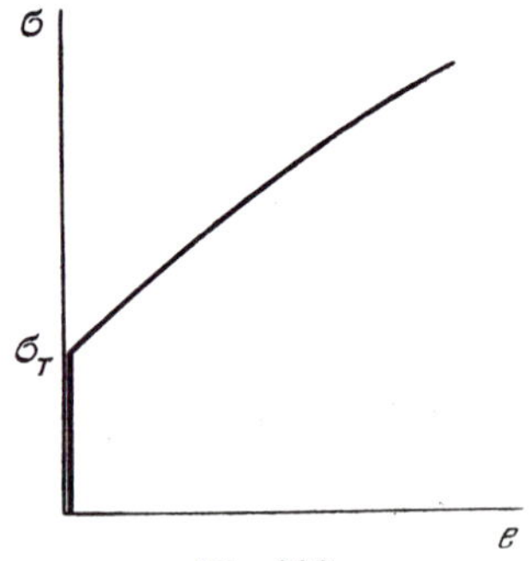

Fig. 230.

With a power creep law the $\sigma - e$ diagram is tangential to the σ axis at the origin of coordinates and at this point the plastic modulus is infinity. Therefore, even though δ tends to zero, a rigid ring remains, though its length is infinitesimally short, and a continuous solution is impossible.

153. A semi-infinite shell with clamped edge

In this case, as we have seen in section 151, the functions U_0 and V_0 are selected as follows:

$$U_0 = 1 - e^{-x}(\cos x + \sin x)\,,$$

$$V_0 = -e^{-x}(\cos x - \sin x)\,.$$

It is evident that in the case of a clamped edge there are no simplifications associated with the relative smallness of V_0. Write expression (150.10) for $B(\beta)$ in the form

$$\tfrac{1}{2}B(\beta) = I_1 - \frac{\beta^2}{n+1} I_2\,. \tag{153.1}$$

Here

$$I_1 = \int_0^\infty \left[\frac{n}{n+1}\left(\frac{U_0^2}{\omega_0} - 1\right) - U_0 + 1\right]\mathrm{d}x\,, \quad I_2 = \int_0^\infty V_0^2 \omega_0\,\mathrm{d}x\,.$$

For practical calculation of these integrals proceed as before. Select a certain value $x = c$ such that when $x > c$ $|U_0 - 1| < \rho$ and $\rho \ll 1$. Then as before when $x > c$, $\omega_0 = U_0^{[(n-1)/n]}(1+\delta)$ where δ is given by formula (152.1). Neglecting small quantities of higher order we obtain

$$I_1 = \int_0^c \left[\frac{n}{n+1}\left(\frac{U_0^2}{\omega_0} - 1\right) - U_0 + 1\right]\mathrm{d}x - \beta^2 \frac{n-1}{2n} K\,. \tag{153.2}$$

Here

$$K = \int_0^\infty V_0^2\,\mathrm{d}x\,. \tag{153.3}$$

In exactly the same way

$$I_2 = \int_0^c V_0^2 \omega_0 \, dx + K . \tag{153.4}$$

Now calculate the derivative $\partial B/\partial\beta$. Carry out differentiation under the integral sign and on the other hand differentiate eq. (151.8) with respect to β. As a result we obtain

$$(nU_0^2 + \beta^2 V_0^2 \omega_0^2) \frac{\partial \omega_0}{\partial \beta} = (n-1)\beta V_0^2 \omega_0^3 .$$

By means of this equation eliminate the subintegral derivative $\partial\omega_0/\partial\beta$ and we find

$$\frac{1}{2}\frac{\partial B}{\partial \beta} = -\beta I_2 . \tag{153.5}$$

Now the first of conditions (151.11) reduces to

$$I_1 - \beta^2 \frac{n+2}{n+1} I_2 = 0 . \tag{153.6}$$

The quantity K is calculated directly, namely,

$$K = \tfrac{1}{4} e^{-2c}(2 - \sin 2c - \cos 2c) .$$

Thus when $c = \pi$, as was assumed in the calculations, $K = \frac{1}{2} e^{-2\pi} = 0.00093$.

The calculation proceeds as follows. The quantities I_1 and I_2 are determined for several values of β that are selected in the region where the true solution is expected. The results of the preceding section can be used for guidance. Then the root of eq. (153.6) is determined graphically and B is calculated by formula (153.1) and α by the second of formulae (151.11) with $A = \frac{1}{2}$.

In practical calculations the position is as follows. Unlike the case of a simply supported shell the quantity β differs comparatively little from unity, which corresponds to an elastic shell. Indeed the bending moment in the real shell is proportional to h which is related to the actual thickness H by formula (146.3). Therefore, taking $\beta = 1$, we find that the moment in the clamp is

$$M_n \approx \left(\frac{n}{1+2n}\right)^{\frac{n}{n+1}}$$

By this formula the ratio of the moment in the limiting state when $n = \infty$ to the moment when $n = 1$ is $\sqrt{3}/2$. But when $n = \infty$ the limiting state is reached in the shell and if this limiting state is σ_* then $p = -1$. There is a hinged periphery in the clamp with a bending moment equal to the limiting value, but in the limiting state $m = \sqrt{3}/2$ as follows from the definition of M_n when $\sigma_{11}^{+} = -\sigma_{11}^{-} = \sigma_*$. Calculation for $n = 3$ shows that the value of β calculated by eq. (153.6) is indeed very close to unity. Putting $\beta = 1$ and making the calculation we find that when $n = 3$, $\alpha = 0.716$.

It should be noted that this problem was solved by the Lagrange variational equation in the work [170]. For simplicity it was assumed that the bending moment depends only on the change of curvature and peripheral forces only on deformation, therefore, a simpler functional was varied instead of functional (150.9). As in our case the form of deflection was assumed given and the parameter α, which is the moment calculated from the deflection, was sought. The results were very simple, for $n = 3$ the value was $\alpha = 0.659$. As always when the displacement rates are subject to variation it is difficult to judge the reliability of the resulting moment distribution. The problem of edge effect in a semi-infinite clamped shell was also considered by Volchkov [14], who integrated eqs. (150.1) and (150.2) directly by successive approximations. The idea of the method is that the linear parts of these equations were separated and the non-linear terms were transferred to the

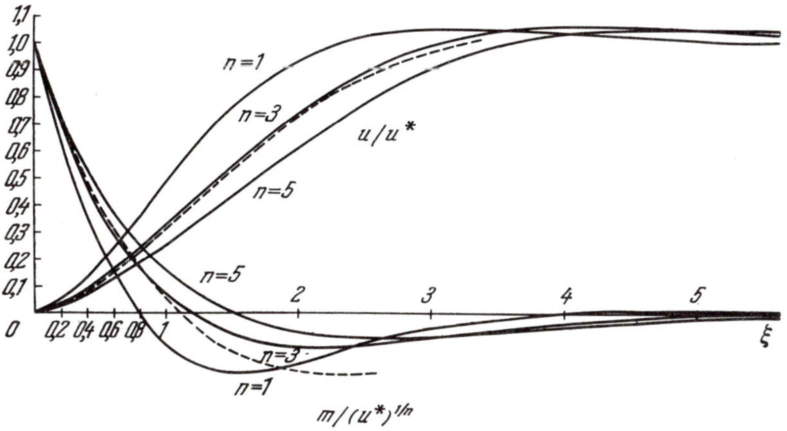

Fig. 231.

right. Knowing the functions u and m in the kth approximation the right-hand side of the equation was calculated and linear operators remained in the left-hand side. Functions u and m in approximation $k+1$ were found by integration of the resulting linear equations. The calculations were made on a computer and the fifth approximation was practically the same as the fourth. Graphs of the functions u and m for $n = 1$, 3 and 5 (solid lines) are given in fig. 231. The result for $n = 3$ obtained by the variational method, namely $\beta = 1$, is confirmed with very high accuracy for $n = 3$ and also for $n = 5$. For comparison graphs of $u(\xi)$ and $m(\xi)$ which correspond to the approximate solution given above are shown dotted.

Returning to this solution let us consider how it can be used to assess the stress in a real shell. Knowing U_0 and V_0 as functions of the argument x, the stresses can be calculated by the formulae of section 149. As stresses in the shell we shall take stresses that are found in the layers of the model. It is possible to proceed more accurately by obtaining formulae for stresses in the real shell for given values of u and m which were calculated for the model, but this improvement is likely to be of little value because of the approximate nature of the investigation. It is readily shown that the stress formulae are of the form

$$\frac{\sigma_{11}}{\sigma_*} = p\frac{2}{\sqrt{3}}\beta V_0(\alpha\xi)\,, \quad \frac{\sigma_{22}}{\sigma_*} = p\left[\frac{U_0}{\omega_0} \pm \frac{1}{\sqrt{3}}\beta V_0(\alpha\xi)\right]. \tag{153.7}$$

The quantity p is the dimensionless hoop stress at infinity

$$\left(\frac{\sigma_{22}}{\sigma_*}\right)_{\xi=\infty} = p\,.$$

154. Extension of the technical theory to the case $T_{11} \neq 0$

Generalisation of the above theory presents no special difficulties. We follow the presentation of section 149 and refer to the corresponding formulae, making changes as necessary. The expressions for deformation given at the start of that paragraph and the stress formulae (149.3) remain valid but now condition (149.2) is not fulfilled. As before

$$M_{11} = \delta h(\sigma_{11}^+ - \sigma_{11}^-)\,,$$

but

$$T_{11} = \delta(\sigma_{11}^+ + \sigma_{11}^-)\,.$$

Whence follows

$$\sigma_{11}^{+} = \frac{M_{11} + hT_{11}}{2\delta h}, \qquad \sigma_{11}^{-} = -\frac{M_{11} - hT_{11}}{2\delta h}. \tag{154.1}$$

On the other hand σ_{11}^{+} and σ_{11}^{-} are expressed in terms of deformations by formulae (149.1), therefore,

$$\epsilon_1 \pm \kappa_1 h = \frac{3\epsilon_*}{8\delta h\sigma_*}(M_{11} \pm hT_{11})\omega^{\pm} - \tfrac{1}{2}\,\epsilon_2\,. \tag{154.2}$$

Here, as before, $\omega^{\pm} = \dfrac{\sigma_*}{\epsilon_*}\dfrac{v^{\pm}}{s^{\pm}}$. Now introduce the notation

$$\frac{\sqrt{3}}{4}\frac{M_{11}}{\delta h\sigma_*} = m\,, \qquad \frac{\sqrt{3}}{4}\frac{T_{11}}{\delta\sigma_*} = \tau\,, \qquad \frac{\epsilon_2}{\epsilon_*} = u\,.$$

The notations for m and u coincide with those given in section 149. Now formula (154.2) can be rewritten

$$\frac{1}{\epsilon_*}(\epsilon_1 \pm \kappa_1 h) = \frac{\sqrt{3}}{2}(m \pm \tau)\omega^{\pm} - \tfrac{1}{2}\,u\,. \tag{154.3}$$

The expression for intensity of rate of deformation is

$$(v^{\pm})^2 = \tfrac{4}{3}\,[\epsilon_2^2 + (\epsilon_1 \pm \kappa_1 h)^2 + \epsilon_2(\epsilon_1 \pm \kappa_1 h)]\,.$$

Bearing in mind eq. (154.3) and the notation introduced above we obtain

$$\left(\frac{v^{\pm}}{\epsilon_*}\right)^2 = u^2 + (m \pm \tau)^2(\omega^{\pm})^2\,. \tag{154.4}$$

Expression (154.4) coincides accurately with expression (149.7), v^+ is a function of ω^+, v^- is a function of ω^-, ω^+ and ω^- are found by the same nomogram as was used to find ω except that the moment must be replaced by a known quantity $\pm\,\tau$ respectively.

Now calculate the force T_{22}:

$$T_{22} = \delta(\sigma_{22}^{+} + \sigma_{22}^{-}) = \tfrac{4}{3}\,\delta\sigma_*\left[u\left(\frac{1}{\omega^+} + \frac{1}{\omega^-}\right) + \tfrac{1}{2}(\epsilon_1 + \kappa_1 h)\frac{1}{\omega^+} + \tfrac{1}{2}(\epsilon_1 - \kappa_1 h)\frac{1}{\omega^-}\right].$$

By eq. (154.3) we rewrite this expression as

$$T_{22} = \delta\sigma_* \left[u \left(\frac{1}{\omega^+} + \frac{1}{\omega^-} \right) + \frac{2}{\sqrt{3}} \tau \right] . \tag{154.5}$$

It now remains to write the expression for the rate of change of curvature, which immediately follows from eq. (154.3):

$$\kappa_1 = \frac{\sqrt{3}}{4} \frac{\epsilon_*}{h} [(m+\tau)\omega^+ + (m-\tau)\omega^-] . \tag{154.6}$$

155. A cylindrical shell compressed by an axial force

Here we shall take into account only the influence of the axial force on creep of the shell but not the part that it plays in creating a bending moment. If the axial force is allowed for in the equation of equilibrium we must introduce the curvature of the generator into it and the problem becomes essentially non-stationary; we shall return to problems of this kind in Chapter 12. For the time being we retain eqs. (148.3) and (148.4) for the axially symmetrical deformation of a cylindrical shell. Substituting into these equations the expressions found above for T_{22} and κ_1, and also replacing ϵ_2 and M_{11} by their expressions in terms of u and m we arrive at the following equations:

$$m'' - u \left(\frac{1}{\omega^+} + \frac{1}{\omega^-} \right) - \frac{2}{\sqrt{3}} \tau + 2p = 0 , \tag{155.1}$$

$$u'' + (m+\tau)\omega^+ + (m-\tau)\omega^- = 0 . \tag{155.2}$$

Differentiation is with respect to the dimensionless coordinate ξ defined in the same way as in section 150. The quantity p retains the same meaning.

Following the method given in section 150, it is readily shown that eqs. (155.1) and (155.2) are Euler equations for the functional

$$N = \int_0^l \Big[u'm' + \tfrac{1}{2}\psi(\omega^+) + \tfrac{1}{2}\psi(\omega^-) - (m+\tau)^2\omega^+$$

$$- (m-\tau)^2\omega^- + 2\left(\frac{\tau}{\sqrt{3}} - p\right) u \Big] \mathrm{d}\xi . \tag{155.3}$$

With a power creep law this functional can be written in a somewhat more convenient form, namely

$$N=\int_0^l\left[u'm'+\frac{n}{n+1}u^2\left(\frac{1}{\omega^+}+\frac{1}{\omega^-}\right)-\frac{1}{n+1}\right.$$

$$\left.\times\left\{(m+\tau)^2\omega^+ +(m-\tau)^2\omega^-\right\}+2\left(\frac{\tau}{\sqrt{3}}-p\right)u\right]\mathrm{d}\xi\,, \quad (155.4)$$

which corresponds to expression (151.1) in the absence of axial force.

We shall not demonstrate this statement, because the demonstration is just the same as that given in section 150, the natural boundary conditions at which the variations of functional (155.3) become zero, which is equivalent to fulfilment of eqs. (155.1) and (155.2), remain the same as in the absence of axial force.

If we assume that the equation of equilibrium (155.1) or the equation of compatibility (155.2) is fulfilled, the requirement that the functional N should be stationary is equivalent to the condition that either the Castigliano or Lagrange functional is stationary. However, unlike the simple case considered earlier, these functionals cannot be written explicitly. Indeed suppose that eq. (155.1) is fulfilled, integrating by parts the first term of eq. (155.3) and bearing in mind eq. (155.1), we obtain

$$N=um'|_0^l+\widetilde{\Phi}\,.$$

Here

$$\widetilde{\Phi}=\int_0^l\left\{\tfrac{1}{2}\psi(\omega^+)+\tfrac{1}{2}\psi(\omega^-)-\frac{1}{\epsilon_*^2}\left[\left(\frac{v^+}{\omega^+}\right)^2+\left(\frac{v^-}{\omega^-}\right)^2\right]\right\}\mathrm{d}\xi\,. \quad (155.5)$$

This expression is the Castigliano functional and provided the static boundary conditions are fulfilled, $\delta N=0$ is equivalent to $\delta\widetilde{\Phi}=0$. However, in this case it is not possible to formulate a simple expression for $\widetilde{\Phi}$ in terms of the function $m(\xi)$, similar to expression (150.13). For example, with a power creep law formula (155.5) is rewritten

$$\widetilde{\Phi}=\frac{1}{n+1}\int_0^l\left[\left(\omega^+\right)^{\frac{n+1}{n-1}}+\left(\omega^-\right)^{\frac{n+1}{n-1}}\right]\mathrm{d}\xi\,.$$

From eq. (155.1) it follows that

$$u = \left(m'' - \frac{2}{\sqrt{3}}\tau - 2p\right)\frac{\omega^+\omega^-}{\omega^+ + \omega^-}\,.$$

The expression for u that has been found must now be substituted into eq. (154.4), giving a system of two equations, which are solved to determine ω^+ and ω^- in terms of the function $m(\xi)$; the values obtained are then substituted into the formula for $\tilde{\Phi}$. However, explicit expressions for ω^+ and ω^- are not obtained. Similar difficulties arise on transition to the Lagrange variational principle.

156. Edge effect in a cylindrical shell loaded with distributed pressure and an axial force

The procedure for solving this problem by variational principles with the power creep law is not essentially different from that developed in section 151. Let us put

$$\tau = \sigma c^{1/n}\,,\quad u = \gamma c U_0\,,\quad m = \beta c^{1/n} V_0\,,\quad \omega^{\pm} = c^{1-1/n}\omega_0^{\pm}\,. \tag{156.1}$$

Here U_0 is a function that satisfies the boundary conditions in bending and is unity at infinity, V_0 is a function that satisfies the boundary condition for moment and which disappears at infinity. It is natural to select these functions in the same way as in section 151, there they were functions of the argument $\alpha\xi$; the quantity α, which defines the rate of damping of the edge-effect, is one of the parameters that is subject to variation, the second such parameter is β. At a sufficient distance from the edge m becomes negligibly small, therefore, ω^+ and ω^- tend to the same limit when $\xi \to \infty$. Let us require that

$$\omega_0^+(\infty) = \omega_0^-(\infty) = 1\,.$$

Then from eq. (154.4) it follows that

$$1 = \gamma^2 + \sigma^2\,. \tag{156.2}$$

Moreover, when $\xi = \infty$, from eq. (155.1) we obtain

$$p = c^{1/n}\left(\gamma + \frac{\sigma}{\sqrt{3}}\right). \tag{156.3}$$

As the value of τ is given, the first of expressions (156.1) and eqs. (156.2) and (156.3) can be used to determine the values of all the parameters introduced, namely

$$\sigma = \frac{1}{\sqrt{1 + \left(\frac{p}{\tau} - \frac{1}{\sqrt{3}}\right)^2}}, \qquad \gamma = \frac{\frac{p}{\tau} - \frac{1}{\sqrt{3}}}{\sqrt{1 + \left(\frac{p}{\tau} - \frac{1}{\sqrt{3}}\right)^2}}, \tag{156.4}$$

$$c^{1/n} = \frac{\tau}{\sigma}.$$

Now substitute expressions (156.1) into eq. (154.4) which is written for a power creep law, we obtain

$$(\omega_0^{\pm})^{\frac{2n}{n-1}} = \gamma^2 U_0^2 + (\beta V_0 \pm \sigma)^2 (\omega_0^{\pm})^2. \tag{156.5}$$

Here the constant c cancels out. This constant to the power $1 + (1/n)$ is also taken outside the integral sign as a factor in the expression for N and the problem reduces to finding a stationary value of the following function of the variables α and β:

$$N_0 = \alpha\beta A + \frac{1}{\alpha} B(\beta).$$

Here

$$A = \gamma \int_0^\infty U_0' V_0' \, dx,$$

$$B(\beta) = \int_0^\infty \left\{ \frac{n}{n+1}\gamma^2 \left[U_0^2 \left(\frac{1}{\omega_0^+} + \frac{1}{\omega_0^-} \right) - 2 \right] \right. \tag{156.6}$$

$$\left. - \frac{1}{n+1} [(\beta V_0 + \sigma)^2 \omega_0^+ + (\beta V_0 - \sigma)^2 \omega_0^- - 2\sigma^2] - 2\gamma^2 (U_0 - 1) \right\} dx.$$

Here, as in section 151, U_0 and V_0 are functions of the argument x and there are additional terms under the integral sign to ensure its convergency. In this case the conditions that the expressions should be stationary with respect to the parameters α and β literally coincide with conditions (155.11) and the procedure of calculation is the same. The following remarks may be helpful in selecting a region in which to seek the value of β. Put $n = 1$, which corresponds to a linear-viscous shell or shell of elastic-incompressible material, if we are considering displacements instead of rates. In this case $\omega_0^+ = \omega_0^- = 1$ and the function $B(\beta)$ assumes the following simple form:

$$B(\beta) = \int_0^\infty [\gamma^2(U_0 - 1)^2 + \beta^2 V_0^2]\,\mathrm{d}x\,.$$

Now consider the case of a freely supported edge, then as we have seen

$$U_0 = 1 - e^{-x}\cos x\,, \qquad V_0 = e^{-x}\sin x\,,$$

$$\int_0^\infty (U_0 - 1)^2\,\mathrm{d}x = \tfrac{3}{8}\,, \qquad \int_0^\infty V_0^2\,\mathrm{d}x = \tfrac{1}{8}\,,$$

and consequently,

$$B = \tfrac{3}{8}\gamma^2 - \tfrac{1}{8}\beta^2\,.$$

The quantity $B\beta$ is minimum when $\beta = \gamma$. Under creep conditions when $n > 1$ the bending moments are less, and, therefore, the required value of β is always less in absolute magnitude than γ.

Notice that if $p/\tau = 1/\sqrt{3}$, $\gamma = 0$ and there is no bending. It is easily shown that in this case the axial stress σ_{11} is twice as great as the hoop stress σ_{22} and, therefore, the rate of creep in the peripheral direction is zero and fixing of the edge has no influence on the shell.

157. Approximate expressions for rate potential for an arbitrary shell

As we have seen, the principal difficulty in creep theory of shells is that, in the general case, rate potential cannot be written explicitly and, therefore, all the above results relate to particular cases and were obtained by various simplifying assumptions.

One rational method of obtaining an approximate expression for potential is to apply to a shell the same idea as was used in section 116 for a bent rod in compression. Suppose that

$$\Phi = \Phi(R),$$

where

$$\begin{aligned} R^2 &= M_0^2 + \alpha^2 T_0^2, \\ M_0^2 &= M_{11}^2 + M_{22}^2 - M_{11}M_{22} + \tfrac{3}{2}(M_{12}^2 + M_{21}^2), \\ T_0^2 &= T_{11}^2 + T_{22}^2 - T_{11}T_{22} + \tfrac{3}{2}(T_{12}^2 + T_{21}^2). \end{aligned} \tag{157.1}$$

Hence it follows that:

$$\begin{aligned} \epsilon_{11} &= \frac{\Phi'(R)}{R}(T_{11} - \tfrac{1}{2}T_{22}), \quad \epsilon_{22} = \alpha^2\frac{\Phi'(R)}{R}(T_{22} - \tfrac{1}{2}T_{11}), \\ \epsilon_{12} &= 3\alpha^2\frac{\Phi'(R)}{R}T_{12}, \\ \kappa_{11} &= \frac{\Phi'(R)}{R}(M_{11} - \tfrac{1}{2}M_{22}), \quad \kappa_{22} = \frac{\Phi'(R)}{R}(M_{22} - \tfrac{1}{2}M_{11}), \\ \kappa_{12} &= 3\frac{\Phi'(R)}{R}M_{12}. \end{aligned} \tag{157.2}$$

For a power creep law

$$\Phi'(R) = \kappa_n\left(\frac{R}{M_n}\right)^n,$$

because when $T_0 = 0, R = M_0$. Requiring that formulae (157.2) also give a correct result with a pure moment-free state of stress we take, as in section 116,

$$\alpha = h\left(\frac{n}{2n+1}\right)^{\frac{n}{n+1}}.$$

11

Transient Creep

Considerable difficulties are encountered in making transient creep calculations on the basis of strain hardening theory. The most effective, but still laborious method is numerical calculation in steps of time. This is illustrated by several unidimensional examples relating to bending of beams, to rotating discs and cylinders.

Some relatively simple transient creep problems can be solved approximately by variational principles. Methods of calculating transient creep based on ageing (time-hardening) theory are far simpler and are widely used.

In other transient creep problems it is required to determine the time to failure with allowance for crack formation and the influence of crack formation on rate and consequently stress redistribution. One such problem is considered at the end of the chapter.

158. Application of the strain-hardening hypothesis to transient creep calculations

In the problems considered in this chapter creep is not steady-state or quasi-steady-state and stress redistribution during creep is of fundamental importance. This particularly applies when creep deformations are comparatively small and of the same order of magnitude as elastic deformations. How-

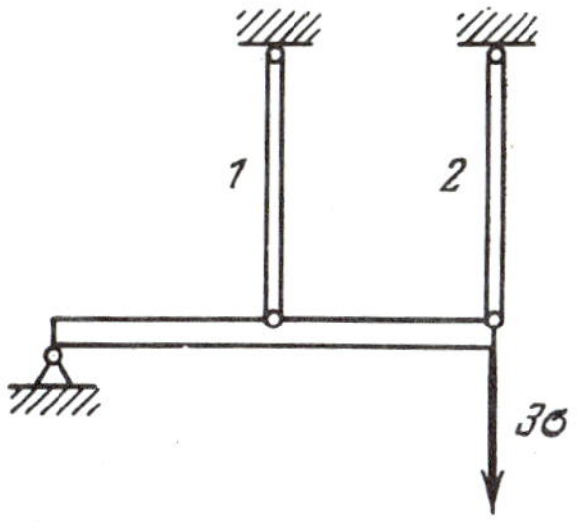

Fig. 232.

ever, in allowing for elastic deformation as a rule the first section of the creep curve cannot be neglected and since the stress varies with time the creep theory that is used should be suitable with variable stress. It was shown in Chapter 4 that the simplest variant of the strain-hardening hypothesis satisfactorily describes creep with variable stress provided that the stress changes sufficiently smoothly. It is hardly worthwhile to use the more accurate equations because the cost of the extra accuracy is too high, and we shall see later that even the simplest variant of the strain-hardening theory leads to very great difficulties in solving real problems. Therefore, at present, few problems are solved on the basis of strain-hardening theory and they involve immense calculations. Practical calculations are usually based on simpler theories of transient creep even though they may not have sufficient physical justification.

To illustrate the point, consider the simple rod system illustrated in fig. 232.

The equation of equilibrium of this system is

$$\sigma_1 + 2\sigma_2 = \sigma .$$

The equation of compatibility of deformations is

$$e_2 = 2e_1 .$$

Here σ_1, σ_2, σ are appropriately selected dimensionless stresses and e_2 and e_1 are deformations. Here the creep law is written

$$\dot{p}_i p_i^{\alpha} = \sigma_i^n , \quad p_i = e_i - \sigma_i .$$

To be specific we have chosen a power dependence on σ.

It is very difficult to solve this system, but certain conclusions can be drawn concerning the character of the solution. Indeed at the initial instant of time $t = 0$, $p_1 = p_2 = 0$, and consequently, $\sigma_1 = e_1$, $\sigma_2 = e_2$ and we find

$$\sigma_1(0) = \tfrac{3}{5}\,\sigma\,, \qquad \sigma_2(0) = \tfrac{6}{5}\,\sigma\,.$$

On the other hand when the time t is sufficiently long, $p_1 \approx e_1, p_2 \approx e_2$, the stresses remain practically unaltered and, therefore, $\dot{p}_1 = \dot{e}_1, \dot{p}_2 = \dot{e}_2$, consequently

$$\left(\frac{\sigma_1}{\sigma_2}\right)^n = \left(\frac{e_1}{e_2}\right)^{1+\alpha},$$

or

$$\sigma_2 = \sigma_1 2^{\frac{1+\alpha}{n}}\,.$$

It now follows from the equation of equilibrium that

$$\sigma_1 = \frac{3}{1+2^{1+\frac{1+\alpha}{n}}}\,\sigma\,, \qquad \sigma_2 = \frac{3\cdot 2^{\frac{1+\alpha}{n}}}{1+2^{1+\frac{1+\alpha}{n}}}\,.$$

Thus, if $n/(1+\alpha) = 3$, then $\sigma_1 = 0.854\sigma$, $\sigma_2 = 1.076\sigma$.

Accordingly, during creep the stress σ_1 increases from its initial value of $\sigma_1 = 0.600\sigma$ and tends to a limit of 0.854σ and the stress σ_2 diminishes from its initial value of 1.200σ and tends to a limit of 1.076σ. The limiting condition corresponds to quasi-steady-state creep and is found quite easily, in the general case all the methods described in Chapter 7 are used for this purpose. In transient creep investigations σ_1 and σ_2 are found as functions of time. If we are interested in displacements of points in the system, for instance displacement of points of application of external forces, then at the first instant the magnitude of this displacement is governed by elastic deformation of the rods and subsequently, the displacement varies with time.

Even the extremely simple example which has been given cannot be solved in closed form and numerical methods must be used. The simplest and most natural method is to calculate by finite differences with respect to time. Suppose that at an instant t_k the stresses in the rods are $\sigma_{1,k}$ and $\sigma_{2,k}$, the deformations are $e_{1,k}$ and $e_{2,k}$ and the corresponding creep deformations are $p_{1,k}$ and $p_{2,k}$. Then at a neighbouring instant of time, $t_{k+1} = t_k + \Delta t_k$,

$$p_{i,k+1} = p_{i,k} + (p_i^{-\alpha}\sigma_i^n)_{,k}\Delta t_k \,. \tag{158.1}$$

Now solve the following equations simultaneously:

$$\sigma_{1,k+1} + 2\sigma_{2,k+1} = \sigma \,,$$

$$e_{2,k+1} = 2e_{1,k+1} \,,$$

$$e_{1,k+1} - \sigma_{1;k+1} = p_{1,k+1} \,,$$

$$e_{2,k+1} - \sigma_{2,k+1} = p_{2,k+1} \,,$$

and find values of σ_i and e_i for the instant t_{k+1}, and so on.

This device can be modified somewhat as follows. Integrating the creep equation we obtain

$$\frac{p^{1+\alpha}}{1+\alpha} = \int \sigma^n \, \mathrm{d}t \,.$$

Hence, for the interval Δt_k we obtain

$$(p_{k+1})^{1+\alpha} = (p_k)^{1+\alpha} + (\sigma_k)^n \Delta t_k (1+\alpha) \,. \tag{158.2}$$

The calculation is commenced at the instant $t = 0$, when creep deformation is still absent. But the quantity p in the right-hand side of formula (158.1) is of negative power and is, therefore, unsuitable for the first step. The advantage of formula (158.2) is that it is free from this defect. We shall not calculate this example as it is not of particular interest, a similar method will be used later on for other problems.

159. **Transient creep in bending**

The simplest problems of transient creep of a rectangular beam in pure bending does not permit of a closed solution on the basis of strain-hardening theory. Shchetinin [157] who was apparently the first to work on this problem, used the following creep law:

$$\dot{p} = p^{\alpha}(\exp \sigma - 1) \,.$$

The solution was obtained in the form of series and it was found that σ changes sign close to the neutral axis. This is because the initial equation is unsuitable for small values of σ because as was shown in section 50, for small values of σ the right-hand side should be of the form const $\times\ \sigma^n$, where $n > 1 + \alpha$. In the work [157] the equation was corrected to overcome this difficulty. A noteworthy device, used in this work, is to linearise the initial creep equation, which makes it possible to obtain the solution in the form of series and to demonstrate their convergency.

The power creep law was used in a later work on this subject [246], namely,

$$\dot{p} = p^{-\alpha}\sigma^n . \tag{159.1}$$

A beam of arbitrary cross-section is considered, it has two axes of symmetry and the plane of action of the bending couple is assumed to coincide with one of the principal planes, for example, the plane yOz.

Put $e = \kappa(t)y$ and $\int \sigma(y, t)y \, \mathrm{d}F = M$. The bending moment is applied suddenly at the time $t = 0$ and is maintained constant, therefore, $\kappa(0) = 0$.

Denote by I_s a moment of order s of cross-sectional area relative to the x axis

$$I_s = \int_F |y|^s \mathrm{d}F .$$

Then, of course, the stress in the elastic state immediately after application of the bending couple is

$$\sigma(y, 0) = \frac{M}{I_2} y .$$

We shall seek a solution for stress in the form

$$\sigma(y, t) = \frac{M}{I_2} y \left[1 + \sum_{r=1}^{\infty} F_r(y) t^{rm} \right] . \tag{159.2}$$

Since $\sigma(-y, t) = \sigma(y, t)$ all the functions $F_r(y)$ should be even or

$$F_r(y) = F_r(|y|) .$$

From the condition of equilibrium it follows that:

$$\sum_{1}^{\infty} t^{rm} \int_F F_r(y) y^2 \, \mathrm{d}F = 0 ,$$

and because this equation is valid for any value of t,

$$\int_F F_r(y) y^2 \, \mathrm{d}F = 0 . \tag{159.3}$$

The author now substitutes expression (159.2) into the creep eq. (159.1) and writes the result as follows:

$$\frac{\mathrm{d}}{\mathrm{d}t}(p^{1+\alpha}) = (1+\alpha)\left(\frac{M}{I_2}\right)^n y^n \left[1 + \sum_{1}^{\infty} F_r(y) t^{rm}\right]$$

$$= (1+\alpha)\left(\frac{M}{I_2}\right)^n y^n \left[1 + nF_1 t^m \right.$$

$$\left. + \left(\frac{n(n-1)}{1 \cdot 2} F_1^2 + nF_2\right) t^{2m} + \ldots\right].$$

This equation is integrated with the initial condition $p(y, 0) = 0$, the series in the right-hand side is raised to the power $1/(1+\alpha)$ and the following expansion is obtained for p:

$$p = \lambda \left(\frac{M}{I_2}\right)^{\nu} y^{\nu} \left\{ t^{\frac{1}{1+\alpha}} + \frac{1}{1+\alpha} \frac{n}{m+1} F_1 t^{m+\frac{1}{1+\alpha}} \right.$$

$$+ \ldots \left[\frac{1}{1+\alpha} \frac{n}{2m+1} \left(\frac{n-1}{2} F_1^2 + F_2\right)\right.$$

$$\left.\left. - \frac{\alpha}{2(\alpha+1)^2} \frac{n^2}{(m+1)^2} F_1^2 \right] t^{2m+\frac{1}{1+\alpha}} + \ldots \right\}. \tag{159.4}$$

Here

$$\nu = \frac{n}{1+\alpha}, \qquad \lambda = (1+\alpha)^{\frac{1}{1+\alpha}} .$$

Now the curvature can be determined as

$$\kappa = \frac{e}{y} = \frac{\sigma + p}{y}.$$

Here σ is given by formula (159.2) and p by formula (159.4). But κ should not depend on y and, therefore,

$$m = \frac{1}{1+\alpha},$$

and

$$\frac{M}{I_2}F_1 + \lambda\left(\frac{M}{I_2}\right)^{\nu} y^{\nu-1} = C_1,$$

$$\frac{M}{I_2}F_2 + \lambda\left(\frac{M}{I_2}\right)^{\nu} y^{\nu-1} \frac{n}{\alpha+2}F_1 = C_2, \tag{159.5}$$

$$\frac{M}{I_2}F_3 + \lambda\left(\frac{M}{I_2}\right)^{\nu} y^{\nu-1}\left[\frac{n}{\alpha+3}\left(\frac{n-1}{2}F_1^2 + F_2\right) - \frac{\alpha n^2}{2(\alpha+2)^2}F_1^2\right] = C_3$$

where C_1, C_2 are constants.

To determine these constants use condition (159.3), multiplying each of the expressions (159.5) by y^2 and integrating over the area. Bearing in mind condition (159.3) we obtain

$$C_1 = \lambda\left(\frac{M}{I_2}\right)\frac{I_{\nu+1}}{I_2},$$

$$C_2 = -\lambda^2\left(\frac{M}{I_2}\right)^{2\nu-1}\frac{n}{\alpha+2}\left(\frac{I_{2\nu}}{I_2} - \frac{I_{\nu+1}^2}{I_\nu^2}\right), \tag{159.6}$$

$$C_3 = \lambda^3\left(\frac{M}{I_2}\right)^{3\nu-2}\frac{n}{2}\,\frac{n(3m+8)-(m+2)^2}{(m+3)(m+2)^2}$$

$$\times\left(\frac{I_{\nu+1}^3}{I_2^3} - \frac{2I_{2\nu}I_{\nu+1}}{I_2^2} + \frac{I_{3\nu-1}}{I_2}\right),$$

. . . .

Now

$$\kappa = \kappa_0 + C_1 t^m + C_2 t^{2m} + \dots . \tag{159.7}$$

Here

$$\kappa_0 = \frac{M}{I_2} .$$

When the constants C_k have been found explicit expressions for the functions F_k can be obtained from the relationships (159.5) and thus we can write a formula for the stress as function of y and of t. We do not give these cumbersome formulae but note that the author included only three terms of the expansion. It is possible of course to write formulae for $C_4, C_5, \dots$ and correspondingly for $F_4, F_5, \dots$, but with increasing numbers the expressions become more and more complicated and a general term of the series cannot be written. Convergency of the series is not proved.

The attempts to solve the problem of pure bending by the strain-hardening theory that we have mentioned are not of great practical value and a quicker result is obtained by numerical solution based on the device that was considered in section 158.

160. Numerical solution of the problem of transient creep of a beam

Let the section of the beam have two axes of symmetry, put the centre-line of the rod along the z axis, let the plane of action of the bending couple coincide with the plane yOz. Suppose that at the initial time $\sigma = \sigma_0(y)$ these stresses may be elastic or may be associated with instantaneous plastic deformation. Here we shall consider that $p = 0$. Take the creep law in the form of eq. (51.1), namely

$$\dot{\hat{p}} \hat{p}^{\alpha} = \exp \hat{\sigma} . \tag{160.1}$$

Here $\hat{p}$ and $\hat{\sigma}$ are given by the formulae of section 51. We shall use dimensionless variables, the dimensionless bending moment being

$$\hat{M} = \frac{2}{hF} \int \hat{\sigma} y \, \mathrm{d}F .$$

Here F is the cross-sectional area, h is any linear dimension (the height of the section), the integral is taken over half the cross-sectional area $y > 0$. Because of the law of plane sections $\hat{e} = ky$, here k is the curvature of the bent centre-line, when $t = 0$ and $\sigma = \sigma_0(y)$ we have $k = 0$. The elastic deformation is $e-p$, therefore,

$$\hat{\sigma} = ky + \hat{\sigma}_0 - \hat{p} \, .$$

Substituting this expression into eq. (160.1) and separating variables we obtain

$$\dot{\hat{p}} \hat{p}^{\alpha} \exp \hat{p} = \exp (ky + \hat{\sigma}_0) \, . \tag{160.2}$$

If the bending moment is constant

$$\int \hat{\sigma} y \, \mathrm{d}F = \int \hat{\sigma}_0 y \, \mathrm{d}F \, ,$$

and consequently,

$$k = \frac{2}{I} \int_0^h \hat{p} y \, \mathrm{d}F \, . \tag{160.3}$$

Here I is the moment of inertia of the cross-sectional area. Eq. (160.1) is only valid for fairly large values of $\hat{\sigma}$. As was stated in section 50, if $\dot{\hat{p}} \hat{p}^{\alpha} < 1$, then $\hat{\sigma} = 0$. Therefore, strictly speaking, the integral in eq. (160.3) should be calculated between the limits y_0 and h, where y_0 is determined from the condition $\dot{\hat{p}} \hat{p}^{\alpha} = 1$. It was shown in section 106 that making this allowance makes only an insignificant correction in bending moment calculations.

Now put

$$\chi(\hat{p}) = \int_0^{\hat{p}} \hat{p}^{\alpha} \exp \hat{p} \, \mathrm{d}\hat{p} \, .$$

Then eq. (160.2) can be integrated numerically by the formula

$$\chi(\hat{p}_m) - \chi(\hat{p}_{m-1}) = \exp (k_{m-1} y + \sigma_0)(\hat{t}_m - \hat{t}_{m-1}) \, . \tag{160.4}$$

After each step of numerical integration the distribution of $\hat{p}_m$ is found, k_m is calculated by formula (160.3) and the process is repeated.

Shorr [155] developed a method of numerical integration of the transient creep equations in bending with more general assumptions. The rod section is completely arbitrary, the rod is loaded with bending moments M_x and M_y and a tensile force N, the temperature distribution across the section can also be arbitrary. Because of the plane sections hypothesis

$$e = e_0 + k_x y - k_y x \,.$$

On the other hand

$$e = \frac{\sigma}{E} + \gamma T + e_{\mathrm{p}} + e_{\mathrm{c}} \,.$$

Here γ is the coefficient of thermal expansion, T is the temperature, e_{p} is the instantaneous plastic deformation, e_{c} the creep deformation. Combining the above expressions we find

$$\sigma = E[e_0 + k_x y - k_y x - \gamma T - e_{\mathrm{p}} - e_{\mathrm{c}}] \,. \tag{160.5}$$

In this formula E depends on the coordinates x and y. The rod equilibrium equations are given in section 107, rewriting them we obtain

$$N = \int_F \sigma \,\mathrm{d}F \,, \quad M_x = -\int_F \sigma y \,\mathrm{d}F \,, \quad M_y = \int_F \sigma x \,\mathrm{d}F \,.$$

Substitute here the expression (160.5). So far no conditions have been attached to the selection of the x and y axes, now let us require that

$$\int_F Ex \,\mathrm{d}F = \int_F Ey \,\mathrm{d}F = \int_F Exy \,\mathrm{d}F = 0 \,.$$

If E is constant, the x and y axes are principal central axes of inertia of the section. As a result we find

$$e_0 = \frac{1}{E_0 F}\left(N + \int_F E\gamma T \,\mathrm{d}F + \int_F Ee_{\mathrm{p}} \,\mathrm{d}F + \int_F Ee_{\mathrm{c}} \,\mathrm{d}F\right),$$

$$\kappa_x = \frac{1}{E_0 I_x}\left(-M_x + \int_F E\gamma Ty \,\mathrm{d}F + \int_F Ee_{\mathrm{p}} y \,\mathrm{d}F + \int_F Ee_{\mathrm{c}} y \,\mathrm{d}F\right), \tag{160.6}$$

$$\kappa_y = \frac{1}{E_0 I_y}\left(-M_y - \int_F E\gamma Tx \,\mathrm{d}F - \int_F Ee_{\mathrm{p}} x \,\mathrm{d}F - \int_F Ee_{\mathrm{c}} x \,\mathrm{d}F\right).$$

The following notation is introduced:

$$E_0 = \frac{1}{F}\int_F E\,\mathrm{d}F\,, \quad I_x = \frac{1}{E_0}\int_F Ey^2\,\mathrm{d}F\,, \quad I_y = \frac{1}{E_0}\int_F Ex^2\,\mathrm{d}F\,.$$

Now the stress formula (160.5) can be rewritten

$$\sigma = \sigma_1 + R(Ee_c)\,. \tag{160.7}$$

Here

$$\sigma_1 = \frac{E}{E_0}\left(\frac{N}{F} - \frac{M_x y}{I_x} + \frac{M_y x}{I_y}\right) + R(E\gamma T) + R(Ee_p)\,,$$

and the operator R is defined by the expression

$$R(f) = \frac{E}{E_0}\left(\frac{1}{F}\int f\,\mathrm{d}F + \frac{y}{I_x}\int fy\,\mathrm{d}F + \frac{x}{I_y}\int fx\,\mathrm{d}F\right). \tag{160.8}$$

If the external forces and temperature distribution remain unaltered the quantity σ_1 can only vary because creep is accompanied by instantaneous plastic deformation, as in principle it can be. During stress redistribution the stresses increase in one part of the section and there is additional plastic deformation. The actual situation is that in the region where stresses are great they relax; the stress increase occurs in the part of the section where they were relatively small and so caused only elastic instantaneous deformation. In any case it is recommended to consider the quantity σ_1 constant and equal to the initial stress when $t = 0$. Therefore, creep calculation reduces to determination of the change with time of the quantity e_c, considered as function of x, y and t.

The notation e_c was introduced for creep deformation so as to retain the notation p for an essentially positive quantity that characterises the amount of strain hardening. With stress of constant sign $p = |e_c|$, but during the creep in bending, the stress may change sign at some places and accordingly the creep rate also changes sign. Shorr's hypothesis for this case was given in section 60 and here we shall not consider the general case, but only the simple assumption that creep in tension has no work-hardening effect on creep in subsequent compression and vice versa (in the formulae of section 60 we take $\lambda = 0$). Therefore, p is the absolute plastic deformation accumulated during the period of tension if $\sigma > 0$, and the absolute plastic deformation during the period of compression if $\sigma < 0$. A graph of the change in e_c and p is

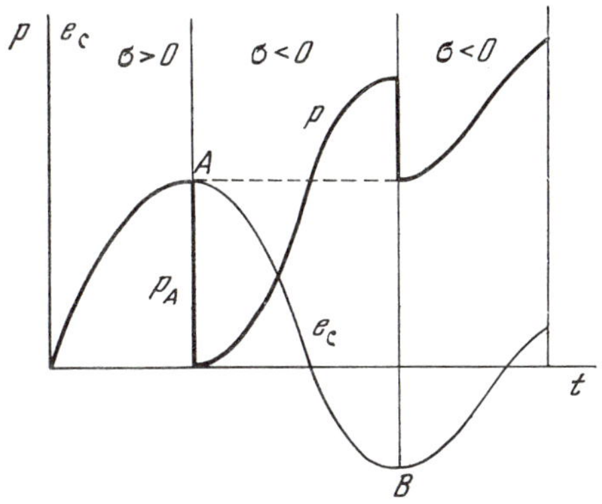

Fig. 233.

shown in fig. 233. In the first segment $\sigma > 0$ and $p = e_c$, the point A corresponds to the instant of change of sign of stress, as soon as $\sigma < 0$ work-hardening must be considered to disappear; in the second segment p is the magnitude of the accumulated compressive deformation, that is $p = p_A - e_c$. This expression is valid up to point B when again $\sigma < 0$. The work-hardening that has accumulated during the period of compression ceases to act but the tensile deformation p' during the first period contributes its share to the work-hardening with respect to tensile stresses; this work-hardening is increased by the increase in deformation, therefore, in the third segment $p = p_A + (e_c - e_{cB})$.

The creep law is now written in the form

$$\dot{p}h(p) = k \exp(-\beta)\left[\exp\frac{|\sigma|}{A} - 1\right]. \tag{160.9}$$

It will be seen that the function of $|\sigma|$ in the right-hand side is selected so as to avoid the need to exclude a low stress zone near to the neutral axis. It has already been mentioned (see section 50) that this selection of function of $|\sigma|$ is not quite correct, however, in numerical integration the region of very small stresses is practically excluded and there are no difficulties of the kind that were encountered in the work [157]. The quantity β in eq. (160.9) is a function of temperature, the form of this function is not important, it is usual to take $\beta = T^{-1} \times$ const. The function $h(p)$ is selected in the form

$$h(p) = \begin{cases} p^{\alpha}, & p < p_c, \\ p_c^{\alpha}, & p > p_c, \end{cases} \tag{160.10}$$

in order to describe both the first and second parts of the creep curve (section 50). The quantity p_c is a function of stress and temperature, it is determined directly from the creep curve. If it is found that with change in σ the quantity p_c diminishes with time, the author recommends determination of $h(p_c)$ from the greatest previous value of p_c and making separate determinations for $\sigma > 0$ and $\sigma < 0$.

The time after $t = 0$ is subdivided into intervals Δt_i from t_{i-1} to t_i, so that $\Delta t_i = t_i - t_{i-1}$. A numerical solution can now be obtained in various ways, one of which is as follows. Put

$$H(p) = \int_0^p h(p)\,\mathrm{d}p\,. \tag{160.11}$$

If the function $h(p)$ is defined by expressions (160.10) we have

$$H(p) = \begin{cases} \dfrac{1}{1+\alpha}\, p^{1+\alpha}\,, & p < p_c\,, \\ H(p_c) + p_c^{\alpha}(p - p_c)\,, & p > p_c\,. \end{cases} \tag{160.12}$$

We shall consider that the quantity $|\sigma|$ varies according to a linear law in the interval from t_{i-1} to t_i so that

$$|\sigma| = |\sigma_{i-1}| + \frac{\Delta|\sigma_i|}{\Delta t_i}(t - t_{i-1})\,,$$

$$t_{i-1} \leqslant t \leqslant t_i\,.$$

Here

$$\Delta|\sigma_i| = |\sigma(t_i)| - |\sigma(t_{i-1})|\,.$$

Integrating eq. (160.9) over the appropriate range and bearing in mind eq. (160.11) we obtain

$$H(p_i) = H(p_{i-1}) + k\exp(-\beta)\left[\exp\frac{|\sigma_{i-1}|}{A}\;\frac{\exp\dfrac{\Delta|\sigma_i|}{k} - 1}{k\Delta|\sigma_i|} - 1\right]\Delta t_i\,. \tag{160.13}$$

If the sign of the stress does not alter during the time Δt_i, then by definition of p we obtain

$$e_{ci} = e_{ci-1} + (p_i - p_{i-1}) \operatorname{sign} \sigma_{i-1} . \tag{160.14}$$

Now find the stress σ_i by formula (160.7):

$$\sigma_i = \sigma_1 + R(Ee_{ci}) .$$

If the time intervals Δt_i are small enough and the increments $\Delta\sigma_i$ are correspondingly small, the right-hand side of expression (160.13) can be simplified giving

$$H(p_i) = H(p_{i-1}) + k \exp(-\beta)\left[\exp\frac{|\sigma_{i-1}|}{A} - 1\right]\Delta t_i . \tag{160.15}$$

Calculation by formula (160.15) is extremely simple, because the right-hand side includes known quantities σ_{i-1} found in approximation number $(i-1)$, however, here the intervals Δt_i must be sufficiently small. When using the complete formula (160.13) the intervals Δt_i may be fairly large, however, $\Delta\sigma_i$ must be found by the method of successive approximations. A similar method can also be used when the load varies reasonably smoothly with time. Here the quantity σ_1 is a given function of time and the remainder of the procedure remains unaltered. However, there are fundamental difficulties in extending the method to the case of temperature varying with time.

161. Transient creep of discs

The simplest hypothesis of transient creep in a state of complex stress is that creep deformation is pure shear produced by the maximum tangential stress τ_{max}. The rate of shear $\dot{\gamma}_{max}$ depends on the magnitude of the shear γ_{max} and on the tangential stress τ_{max}. If the principal stresses are $\sigma_1 > \sigma_2 > \sigma_3$, this assumption means that:

$$\dot{p}_1 = -\dot{p}_3 , \quad \dot{p}_2 = 0 .$$

For the equivalent stress we select the quantity $s = 2\tau_{max} = \sigma_1 - \sigma_3$; and for the equivalent deformation we select the quantity $p = p_1 = -p_3$. In simple tension, s is the tensile stress, p is the elongation, therefore, s and p are related in the same way as σ and p in tension. Taking the strain-hardening law in the form of eq. (75.4) and converting to dimensionless quantities, as in section 51,

$$\hat{\sigma}_{ij} = \frac{\sigma_{ij}}{A}, \quad \hat{e}_{ij} = \frac{Ee_{ij}}{A}, \quad \hat{p}_{ij} = \frac{Ep_{ij}}{A}, \quad \hat{t} = k\left(\frac{E}{A}\right)^{1+\alpha} t\,.$$

Hereafter we shall omit the sign denoting dimensionless quantities, so that the fundamental creep equation is written

$$\dot{p}p^{\alpha} = \exp(s-\beta)\,. \tag{161.1}$$

If two of the principal stresses, for example, σ_1 and σ_2, are equal and positive, the state of stress of the element is compression in the third principal direction by a compressive stress $\sigma_1 - \sigma_3$ with a superimposed overall hydrostatic tensile stress of intensity σ_1. The rate of compression in the direction of the σ_3 axis is determined by the quantity $\sigma_1 - \sigma_3$ which is taken as $\dot{p}$; the rates of deformation in directions σ_1 and σ_2 are indeterminate and derive only from the condition of incompressibility. Therefore, the following terms must be added to eq. (161.1):

$$\begin{aligned} &p_1 = p\,, \quad p_3 = -p\,, \quad p_2 = 0 \qquad (\sigma_1 > \sigma_2 > \sigma_3)\,, \\ &p_3 = -p\,, \quad p_1 = \lambda p\,, \quad p_2 = (1-\lambda)p \quad (\sigma_1 = \sigma_2 > \sigma_3)\,. \end{aligned} \tag{161.2}$$

Eqs. (161.2) are written in integrated form and this is only possible when the inequality signs between the principal stresses remain unchanged during creep. But it can happen that for a certain time, for instance, $\sigma_1 > \sigma_2$, and at some instant these stresses become equal after which $\sigma_2 > \sigma_1$. This means that creep is a result of slip in other planes and it is not certain whether work-hardening acquired in the first stage is maintained. For simplicity we shall assume that work-hardening is entirely additive, irrespective of the systems in which shear occurs, and that it leads to an accumulated work-hardening creep deformation. Certain published results indicate that in crystals, shear in one slip system strengthens that system and all others. Therefore, in the general case, eqs. (161.2) should be written for rates and integrated in stages.

Now proceed to the problem of a rotating disc in which the dimensionless stresses are σ_r and σ_θ. Assume that initial stresses σ_r^0 and σ_θ^0 exist in the disc and there is also an initial distribution of work-hardening creep strain p_0. If σ_r^0 and σ_θ^0 are related to the instantaneous creep deformation, we must put $p_0 = 0$. Consider the following possibilities:

$$1)\ \sigma_\theta > 0 > \sigma_r\,,\quad s = \sigma_\theta - \sigma_r\,,\quad 2(p - p_0) = e_{\theta c} - e_{rc}\,,\quad e_{zc} = 0\,;$$

$$2)\ \sigma_\theta > \sigma_r > 0\,,\quad s = \sigma_\theta\,,\quad 2(p - p_0) = e_{\theta c} - e_{zc}\,,\quad e_{rc} = 0\,;$$

$$3)\ \sigma_r > \sigma_\theta > 0\,,\quad s = \sigma_r\,,\quad 2(p - p_0) = e_{rc} - e_{zc}\,,\quad e_{\theta c} = 0\,;$$

$$4)\ \sigma_\theta = \sigma_r > 0\,,\quad s = \sigma_\theta = \sigma_r\,,\quad 2(p - p_0) = e_{\theta c} - e_{rc}\,.$$

Write the equation of equilibrium for an element of the disc in the following way:

$$\frac{\mathrm{d}}{\mathrm{d}\xi}(\xi\sigma_r) - \sigma_\theta + m\xi^2 = 0\,.$$

Here $\xi = r/b$, where r is the instantaneous radius and b is the external radius of the disc, $m = \rho\omega^2 b^2/A$. Put

$$\sigma_r = z - \tfrac{1}{2} m\xi^2\,,$$
$$\sigma_\theta = \xi\frac{\mathrm{d}z}{\mathrm{d}\xi} + z - \tfrac{1}{2} m\xi^2\,, \tag{161.3}$$

and the equation of equilibrium is satisfied identically.

Denote by u the radial displacement divided by a characteristic linear dimension bA/E, the $\mathrm{d}u/\mathrm{d}\xi = \hat{e}_r$, $u/\xi = \hat{e}_\theta$ or

$$\frac{u}{\xi} = \sigma_\theta - \nu\sigma_r - (\sigma_\theta^0 - \nu\sigma_r^0) + e_{\theta c} + (\varphi - \varphi_0)\,,$$
$$\frac{\mathrm{d}u}{\mathrm{d}\xi} = \sigma_r - \nu\sigma_\theta - (\sigma_r^0 - \nu\sigma_\theta^0) + e_{rc} + (\varphi - \varphi_0)\,. \tag{161.4}$$

Here φ is the temperature strain

$$\varphi = \frac{E\gamma T}{A}\,.$$

Now consider particular cases.

Case 1. In eqs. (161.4) put $e_{\theta c} = (p - p_0)$, $e_{rc} = -(p - p_0)$, substitute expression (161.3) for σ_θ and σ_r and eliminate the displacement u. We obtain the following equation for $z(\xi)$:

$$\frac{d^2z}{d\xi^2}+\frac{3}{\xi}\frac{dz}{d\xi}=m(1-\nu)-\frac{1}{\xi}\left(\frac{dp^*}{d\xi}+\frac{2p^*}{\xi}\right)-\frac{1}{\xi}\frac{d(\varphi-\varphi_0)}{d\xi}.$$

Here $p^*=p-p_0-(\sigma_\theta^0-\sigma_r^0)$. By integration we find

$$z=\frac{m(1-\nu)}{8}\xi^2-\int\frac{p^*d\xi}{\xi}-\frac{1}{\xi^2}\int\xi(\varphi-\varphi_0)d\xi+B-\frac{C}{\xi^2}.$$

Now write the stress formulae

$$\sigma_r=\sigma_r^0-P(\xi)-\frac{3+\nu}{8}m\xi^2-\frac{1}{\xi^2}\int\xi(\varphi-\varphi_0)d\xi+B-\frac{C}{\xi^2},$$

$$\sigma_\theta=\sigma_\theta^0-p-P(\xi)-\frac{1+3\nu}{8}m\xi^2-(\varphi-\varphi_0)+\frac{1}{\xi^2}\int\xi(\varphi-\varphi_0)d\xi+B+\frac{C}{\xi^2}. \tag{161.5}$$

Here we take

$$P(\xi)=\int\frac{p-p_0}{\xi}d\xi.$$

The quantity s is expressed as

$$s=s^*-p+\frac{2C}{\xi^2}. \tag{161.6}$$

Here s^* is a known function of the coordinates and time, namely

$$s^*=\sigma_\theta^0-\sigma_r^0-p_0-(\varphi-\varphi_0)+\frac{2}{\xi^2}\int\xi(\varphi-\varphi_0)d\xi+\frac{m(1-\nu)}{4}\xi^2.$$

Now substitute expression (161.6) into the creep eq. (161.1), after separating variables it assumes the form

$$\dot{p}p^\alpha\exp(p)=\exp\left(s^*+\frac{2C}{\xi^2}-\beta\right).$$

This equation is integrated numerically, just as in the case of bending in section 160; once again as in section 51 we introduce the function $\chi(p)$, and the formulae for calculating successive steps are written

$$\chi(p_i) = \chi(p_{i-1}) + \exp\left(s_{i-1}^* + \frac{2C_{i-1}}{\xi^2} - \beta_{i-1}\right)(t_i - t_{i-1}) . \quad (161.7)$$

After each step the constants C_i must be calculated numerically by some integral expression that follows from the boundary conditions. Let us consider a few examples.

a) *Disc with internal relative radius α shrunk onto a rigid shaft with interference fit.* Take as the initial state that which occurs when the disc is shrunk onto the shaft but is not rotated and not heated. The boundary conditions are

$$\sigma_r(1) = 0 , \qquad u_r(\alpha) = 0 .$$

By eqs. (161.4) and (161.5) we find

$$C = \left\{ -P(1) - \frac{3+\nu}{8} m + \frac{1+\nu}{8} m\alpha^2 - \int_\alpha^1 \xi(\varphi - \varphi_0)\,\mathrm{d}\xi \right\} : \left\{ 1 + \frac{1+\nu}{1-\nu}\,\frac{1}{\alpha} \right\} ,$$

$$B = C + P(1) + \frac{3+\nu}{8} m + \int_\alpha^1 \xi(\varphi - \varphi_0)\,\mathrm{d}\xi .$$

b) *Disc of internal relative radius α shrunk on an elastic shaft.* The temperatures of the shaft and of parts of the disc near to the shaft are low and creep in this part of the disc is quite unimportant, stress redistribution results from creep in peripheral parts of the disc. Adopt the usual procedure of replacing the shaft by a cylinder of finite length which exceeds the disc thickness by a factor $1/\kappa$. The values of κ are given, for instance in [108]. The condition $\sigma_r(1) = 0$ is retained, but it can readily be replaced by the condition that $\sigma_r(1)$ is equal to the given blade-load stress. Denote the initial fit pressure by q_0, then the second condition is written:

$$C = -\left\{\frac{m}{8}\left[(3+\nu)\left(\frac{\kappa\alpha^2}{1-\kappa}+1\right) - \alpha^2\frac{1+\nu}{1-\kappa}\right] + P(1)\right.$$

$$\left. + \int_{\alpha}^{1} \xi(\varphi-\varphi_0)\,\mathrm{d}\xi\right\} : \left\{1 + \frac{1}{\alpha^2}\left[\frac{1+\nu}{1-\nu}\,\frac{1}{1-\kappa} + \frac{\kappa}{1-\kappa}\right]\right\}.$$

The value of B is found as in the previous example.

c) *Disc of stepped profile.* Formulae (161.5), (161.6) and (161.7) remain valid but the constants B and C are different on each section. If the thickness changes suddenly by a factor of η on the boundary between two sections, then in addition to the boundary conditions it is necessary to consider the conditions of conjunction. If the relative radius of the boundary between the sections is c, then

$$\sigma_r'(c) = \eta\sigma_r''(c)\,, \qquad u'(c) = u''(c)\,.$$

The single and double primes respectively denote stresses and displacements for the boundary regions. Introducing the boundary conditions, we obtain general formulae for the constants B and C on each section, but they are so cumbersome that in practice it is better to solve the resulting system of equations numerically each time for particular values of the parameters.

162. **Transient creep of discs. Cases 2, 3 and 4**

Case 2. The case considered in section 161 in which $\sigma_\theta > 0 > \sigma_r$ is realised in a disc with a central hole with a shrink fit on the shaft. In a free disc with uniform temperature distribution the second case is encountered in which $\sigma_\theta > \sigma_r > 0$. Here there is no creep in the radial direction and in formulae (161.4) we take $e_{\theta c} = (p - p_0), e_{rc} = 0$. Proceeding in exactly the same way as before we obtain the following differential equation for $z(\xi)$:

$$\frac{\mathrm{d}^2 z}{\mathrm{d}\xi^2} + \frac{3}{\xi}\frac{\mathrm{d}z}{\mathrm{d}\xi} = m(1-\nu) - \frac{1}{\xi^2}\frac{\mathrm{d}}{\mathrm{d}\xi}\left[\xi(p-p_0)\right]$$

$$+ \frac{1}{\xi^3}\frac{\mathrm{d}}{\mathrm{d}\xi}\left[\xi^2(\sigma_\theta^0 - \sigma_r^0)\right] - \frac{1}{\xi}\frac{\mathrm{d}(\varphi-\varphi_0)}{\mathrm{d}\xi}.$$

The integral of this equation is

$$z = \frac{1-\nu}{8} m\xi - \tfrac{1}{2}\int \frac{p-p_0}{\xi}\,\mathrm{d}\xi - \frac{1}{2\xi^2}\int \xi(p-p_0)\,\mathrm{d}\xi$$

$$+ \sigma_r^0 - \frac{1}{\xi^2}\int \xi(\varphi-\varphi_0)\,\mathrm{d}\xi + B - \frac{C}{\xi^2}.$$

The stress formulae are

$$\sigma_r = \sigma_r^0 - P_1(\xi) - \frac{3+\nu}{8} m\xi^2 - \frac{1}{\xi^2}\int \xi(\varphi-\varphi_0)\,\mathrm{d}\xi + B - \frac{C}{\xi^2},$$

$$\sigma_\theta = \sigma_\theta^0 - p - P_2(\xi) - \frac{1+3\nu}{8} m\xi^2 - (\varphi-\varphi_0) \tag{162.1}$$

$$+ \frac{1}{\xi^2}\int \xi(\varphi-\varphi_0)\,\mathrm{d}\xi + B + \frac{C}{\xi^2} = s\,.$$

Here

$$P_1(\xi) = \tfrac{1}{2}\int \frac{p-p_0}{\xi}\,\mathrm{d}\xi + \frac{1}{2\xi^2}\int (p-p_0)\xi\,\mathrm{d}\xi\,,$$

$$P_2(\xi) = \tfrac{1}{2}\int \frac{p-p_0}{\xi}\,\mathrm{d}\xi - \frac{1}{2\xi^2}\int (p-p_0)\xi\,\mathrm{d}\xi\,.$$

Put

$$s^* = \sigma_\theta^0 - \frac{1+3\nu}{8} m\xi^2 - (\varphi-\varphi_0) + \frac{1}{\xi^2}\int \xi(\varphi-\varphi_0)\,\mathrm{d}\xi - P_2(\xi)\,. \tag{162.2}$$

Then the creep equations are

$$\dot{p}p^\alpha \exp(p) = \exp\left(s^* + B + \frac{C}{\xi^2} - \beta\right), \tag{162.3}$$

and the formula used for numerical calculation is

$$\chi(p_i) = \chi(p_{i-1}) + \exp\left(s_{i-1}^* + B_{i-1} + \frac{C_{i-1}}{\xi^2} - \beta_{i-1}\right)(t_i - t_{i-1})\,.$$

In this case the calculation is much more laborious than in the previous one, because after each step we must again calculate s_i^* by formula (162.2). The constants B and C are also determined from the boundary conditions for each stage. For example, if the disc is free we have

$$\sigma_r(\alpha) = \sigma_r(1) = 0 .$$

By the first of the formulae (162.1) we find

$$C = \frac{3+\nu}{8} m\alpha^2 + \frac{\alpha^2}{1-\alpha^2}\left[P(1) + \frac{1}{\xi^2}\int_0^1 \xi(\varphi-\varphi_0)\,\mathrm{d}\xi\right],$$

$$B = \frac{C}{\alpha^2} + \frac{3+\nu}{8} m\alpha^2 .$$

Case 3. If the disc has a heavy rim it can happen that $\sigma_r > \sigma_\theta > 0$. Proceeding as before we find

$$\sigma_r = s = \sigma_r^0 - P_2(\xi) - \frac{3+\nu}{8} m\xi^2 - \frac{1}{\xi^2}\int \xi(\varphi-\varphi_0)\,\mathrm{d}\xi + B - \frac{C}{\xi^2},$$

$$\sigma_\theta = \sigma_\theta^0 + P_1(\xi) - \frac{1+3\nu}{8} m\xi^2 + \frac{1}{\xi^2}\int \xi(\varphi-\varphi_0)\,\mathrm{d}\xi \qquad (162.4)$$

$$-(\varphi-\varphi_0) + B + \frac{C}{\xi^2} .$$

Generally speaking, zones with different relationships between the principal stresses can occur in a single disc and the additional difficulty arises of establishing the boundary between the zones. A continuous disc will serve as a typical example. In the elastic state with uniform temperature distribution, $\sigma_\theta > \sigma_r > 0$ everywhere, and the condition $\sigma_\theta = \sigma_r$ is fulfilled only at the centre of the disc and is not disturbed during creep. We have seen in section 123 that the problem of creep in a continuous disc is solved by a Saint Venant type criterion, the equation $\sigma_\theta = \sigma_r$ must be valid over some finite region near the centre. Exactly the same situation arises during transient creep, the region in which the principal stresses are equal to one another gradually extends and tends to some limiting position, which can be found by solving the problem of quasi-steady-state creep, as was done in section 123.

Case 4. The case of a continuous disc was considered in the work [135]. In the central region $0 < \xi < c$, $\sigma_r = \sigma_\theta = \sigma$ and from formulae (161.3) it follows that $z = \text{const} = Z$, therefore,

$$\sigma = Z - \tfrac{1}{2} m\xi^2 . \tag{162.5}$$

As $\sigma_z = 0$, therefore, $s = \sigma$, $e_{rc} = \lambda p$, $e_{\theta c} = (1-\lambda)p$. On the other hand

$$e_r = \frac{\mathrm{d}u}{\mathrm{d}\xi} = (1-\nu)\sigma + \lambda p ,$$

$$e_\theta = \frac{u}{\xi} = (1-\nu)\sigma + (1-\lambda)p . \tag{162.6}$$

Adding we obtain

$$\frac{\mathrm{d}u}{\mathrm{d}\xi} + \frac{u}{\xi} = 2(1-\nu)\sigma + p = 2(1-\nu)Z - (1-\nu)m\xi^2 + p .$$

Integrating we find

$$u = Z(1-\nu)\xi - \frac{1-\nu}{4} m\xi^3 + \frac{1}{\xi}\int_0^\xi p\xi \,\mathrm{d}\xi . \tag{162.7}$$

The constant of integration is zero because $u(0) = 0$. When $\xi \geqslant c$ the stresses are no longer equal and we have $\sigma_\theta > \sigma_r > 0$, consequently, $e_{rc} = 0$. To ensure continuity of displacement, in the inner region when $\xi = c$ we must have $e_{rc} = 0$ and consequently, $\lambda = 0$. By the first of expressions (162.6) we find that when $\xi = c$

$$\frac{\mathrm{d}u}{\mathrm{d}\xi} = (1-\nu)\sigma ,$$

or, bearing in mind eqs. (162.5) and (162.7)

$$p(c)c^2 - \frac{1-\nu}{4} m\xi^4 = \int_0^c p\xi \,\mathrm{d}\xi . \tag{162.8}$$

In the outer part of the disc when $\xi > c$ we have everywhere $\sigma_\theta > \sigma_r$, and consequently, the stress formulae (162.1) which contain two constants B and

C, remain valid. Denote by a single prime quantities that relate to the region $0 < \xi < c$, and by two primes those that relate to the region $c < \xi < 1$, in the first region the equation

$$\dot{p}p^{\alpha} = \exp\left(Z - \tfrac{1}{2}m\xi^2\right)$$

is integrated numerically and in the second, eq. (162.3) is integrated. After each integration step the constants are determined from the boundary conditions and the conditions of compatibility, namely

$$u'(c) = u''(c)\,, \quad \sigma_r'(c) = \sigma_r''(c)\,, \quad \sigma_r''(1) = q\,.$$

A graph taken from the work [135] is given in fig. 234. The calculations were made for a disc of constant thickness 900 mm diameter of steel grade 40N at a temperature of 400°C, running at 3000 rpm. The blade load was taken to be 4 kg/mm. We give only three curves of σ_r and σ_θ; for the initial condition, for a time of 1 hour, and for the steady-state condition calculated by the formulae of section 123.

A somewhat different method of solving transient creep problems numerically on the basis of the same variant of the strain hardening theory is developed in the work [136]. Here, unlike the method discussed above, differential equations are formulated for functions $u(\xi)$. In principle, the method of doing this is quite clear; stresses are determined from eqs. (161.4) and substituted into the equilibrium equation. The thickness of the disc, the temperature and the elastic constants are considered as functions of the radius and are

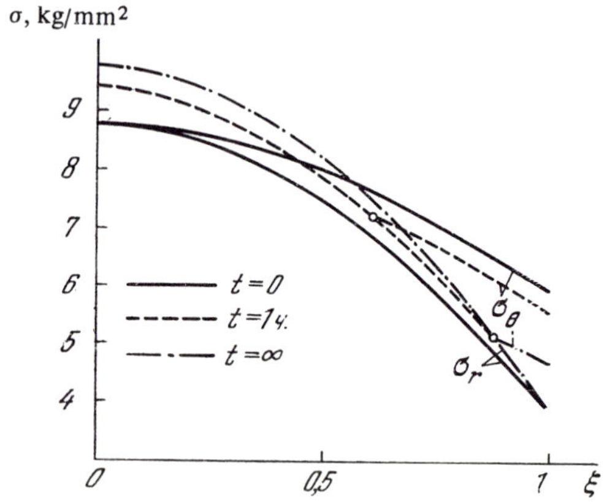

Fig. 234.

given as power series, the solution is also sought in the form of a power series, the coefficients of which are functions of time.

163. **Creep of a rotating cylinder and tube**

In a rotating cylinder or thick-walled tube all three principal stresses differ from zero. Put

$$e_{\theta c} = 2\lambda\,, \quad e_{zc} = \mu - \lambda\,, \quad e_{rc} = -\mu - \lambda\,.$$

Retaining the notation of section 161, we obtain

$$\begin{aligned} e_\theta &= \frac{u}{\xi} = \sigma_\theta - \nu(\sigma_r + \sigma_z) + (\varphi - \varphi_0) + 2\lambda\,, \\ e_r &= \frac{du}{d\xi} = \sigma_r - \nu(\sigma_\theta + \sigma_z) + (\varphi - \varphi_0) - \lambda - \mu\,, \\ e_z &= \sigma_z - \nu(\sigma_\theta + \sigma_r) + (\varphi - \varphi_0) - \lambda + \mu\,. \end{aligned} \tag{163.1}$$

Here for simplicity we have considered that initial stresses are absent. Assuming that the cylindrical section remains plane we consider that e_z depends only on the time t. Eliminate σ_z from the two first equations of (163.1) and formulate the condition of compatibility. Here σ_θ and σ_r are expressed in terms of the function $z(\xi, t)$ using formulae (161.3), which remain valid because the equation of equilibrium is the same as for a disc. As a result the following differential equation is obtained:

$$\frac{d^2z}{d\xi^2} + \frac{3}{\xi}\frac{dz}{d\xi} = \frac{1-2\nu}{1-\nu} m\xi^2 - \frac{1}{1-\nu}\frac{d(\varphi-\varphi_0)}{d\xi} - \frac{2-\nu}{1-\nu^2}\frac{1}{\xi}\frac{d\lambda}{d\xi}$$

$$- \frac{\nu}{1-\nu^2}\frac{1}{\xi}\frac{d\mu}{d\xi} - \frac{1}{1-\nu^2}\left(\frac{\mu}{\xi^2} + \frac{3\lambda}{\xi^2}\right) = 0\,.$$

Write the integral of this equation for an incompressible material when $\nu = \frac{1}{2}$ as:

$$z = -\frac{2}{\xi^2}\int (\varphi - \varphi_0)\xi\, d\xi - 2\int \frac{\lambda\, d\xi}{\xi} - \frac{2}{3}\int \frac{\mu\, d\xi}{\xi} + B - \frac{C}{\xi^2}\,.$$

Here the stress formulae are

$$\sigma_r = -\tfrac{1}{2} m\xi^2 - \frac{2}{\xi^2}\int(\varphi-\varphi_0)\xi\,\mathrm{d}\xi - \tfrac{2}{3}\int\frac{\mu\,\mathrm{d}\xi}{\xi}$$

$$-2\int\frac{\lambda\,\mathrm{d}\xi}{\xi} + B - \frac{C}{\xi^2},$$

$$\sigma_\theta = -\tfrac{1}{2} m\xi^2 - 2\left[(\varphi-\varphi_0) - \frac{1}{\xi^2}\int(\varphi-\varphi_0)\xi\,\mathrm{d}\xi\right] \tag{163.2}$$

$$-\tfrac{2}{3}\left[\mu + \int\frac{\mu\,\mathrm{d}\xi}{\xi}\right] - 2\left(\lambda + \int\frac{\lambda\,\mathrm{d}\xi}{\xi}\right) + B + \frac{C}{\xi^2},$$

$$\sigma_z = -\tfrac{1}{2} m\xi^2 - 2(\varphi-\varphi_0) - \tfrac{2}{3}\left(2\mu + \int\frac{\mu\,\mathrm{d}\xi}{\xi}\right) - 2\int\frac{\lambda\,\mathrm{d}\xi}{\xi} + B + e_z(t).$$

We can now consider various particular cases, namely

1) $\sigma_r > \sigma_\theta > \sigma_z$, $s = \sigma_r - \sigma_z$, $e_{\theta c} = 0$, consequently, $\lambda = 0$;
$e_{rc} = -e_{zc} = p$, consequently, $\mu = -p$.

2) $\sigma_\theta > \sigma_r > \sigma_z$, $s = \sigma_\theta - \sigma_z$, $e_{rc} = 0$, consequently, $\mu + \lambda = 0$;
$e_{\theta c} = -e_{zc} = p$, consequently, $\lambda = \tfrac{1}{2}p$, $\mu = -\tfrac{1}{2}p$.

3) $\sigma_\theta > \sigma_z > \sigma_r$, $s = \sigma_\theta - \sigma_r$, $e_{zc} = 0$, consequently, $\mu - \lambda = 0$;
$e_{\theta c} = -e_{rc} = p$, consequently, $\lambda = \mu = \tfrac{1}{2}p$.

Similarly formulae are found for the three cases that are obtained from those considered by changing the sign of the inequality. Here numerical integration is performed just as in the problem of creep of a disc. Formulae (163.2) can also be used to consider the problem of a thick-walled tube subject to internal pressure. Putting $m = 0$, $\varphi = \varphi_0$, $\sigma_\theta > \sigma_z > \sigma_r$, we suppose that the third case is realised in the tube and consequently, $\lambda = \mu = \tfrac{1}{2}$, $s = \sigma_\theta - \sigma_r$. By formulae (163.2) we obtain

$$\sigma_\theta = -\tfrac{4}{3}p - \tfrac{4}{3}P + B + \frac{C}{\xi^2},$$

$$\sigma_r = -\tfrac{4}{3}P + B - \frac{C}{\xi^2}, \tag{163.3}$$

$$\sigma_z = -\tfrac{2}{3}p - \tfrac{4}{3}P + B + e_z(t).$$

Here

$$P(\xi) = \int_{\alpha}^{\xi} \frac{p\,\mathrm{d}\xi}{\xi}.$$

The boundary conditions are $\sigma_r(\alpha) = -\hat{q}$, $\sigma_r(1) = 0$, consequently,

$$\hat{q} = -B + \frac{C}{\alpha^2}, \quad 0 = -\tfrac{4}{3}P(1) + B - C. \tag{163.4}$$

If the ends of the tube are closed the equation of equilibrium in the axial direction is

$$\hat{q}\alpha^2 = 2\int_{\alpha}^{1} \sigma_z \xi\,\mathrm{d}\xi.$$

Integrating by parts $\int \xi P(\xi)\,\mathrm{d}\xi$, we find that because of eq. (163.4) $e_z(t) = 0$. Note that with this condition it follows from (163.3) that

$$\sigma_z = \tfrac{1}{2}(\sigma_\theta + \sigma_r).$$

Thus the assumption that the principal stresses are not equal is always fulfilled.

Now the quantity s is

$$s = -\tfrac{4}{3}p + \frac{2C}{\xi^2}; \tag{163.5}$$

here it follows from (163.4) that

$$C = \left(\hat{q} + \tfrac{4}{3}P(1)\right)\frac{\alpha^2}{1-\alpha^2}. \tag{163.6}$$

Substitute eq. (163.5) into the creep equation and separate variables, we obtain

$$\dot{p}p^{\alpha} \exp\frac{4p}{3} = \exp\left(\frac{2C}{\xi^2} - \beta\right). \tag{163.7}$$

Therefore, we integrate (163.7) numerically with the condition (163.6).

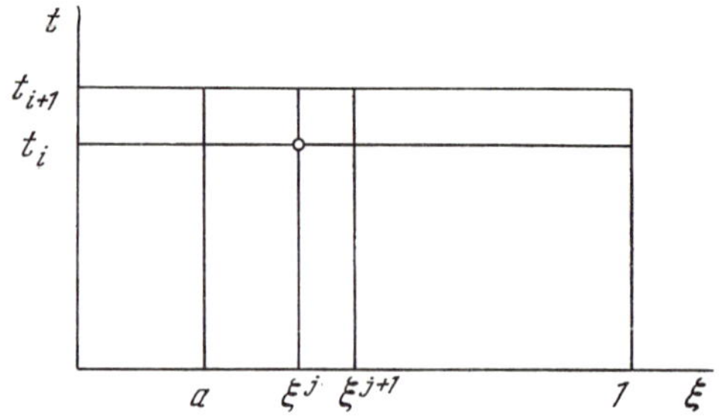

Fig. 235.

The procedure of calculation is explained with reference to the diagram of fig. 235. Divide the radius $\alpha < \xi < 1$ into a number of intervals, denote the boundaries of the intervals ξ^j, and the boundaries of time intervals t_i. Suppose that at an instant of time t_i we are given the values of the constant C_i, and of p for different values of ξ^j which are denoted p_i^j. Now determine the quantity p_{i+1}^j for all possible values of the index j by the formula

$$\chi_1(p_{i+1}^j) - \chi_1(p_i^j) = \exp\left(\frac{2C_i}{\xi^{j2}} - \beta_i\right)(t_{i+1} - t_i) .$$

Here

$$\chi_1(p) = \int \exp\left(\tfrac{4}{3}p\right)p^\alpha \, \mathrm{d}p = \left(\tfrac{3}{4}\right)^{1+\alpha}\chi\left(\tfrac{4}{3}p\right) .$$

Now by numerical integration we find the function $P(\xi)$ by the formulae

$$P_{i+1}^{j+1} = P_{i+1}^j + p_{i+1}^j \frac{1}{\xi^j}(\xi^{j+1} - \xi^j) .$$

Returning to the general case of a rotating cylinder we note that in all three cases given above it follows from formulae (163.2) that

$$s = s^* - \tfrac{4}{3}p .$$

In the first case

$$s^* = -2(\varphi - \varphi_0) + \frac{2}{\xi^2}\int(\varphi - \varphi_0)\xi \, \mathrm{d}\xi + \frac{C}{\xi^2} + e_z(t) ,$$

in the second case

$$s^* = \frac{2}{\xi^2}\int(\varphi-\varphi_0)\xi\,\mathrm{d}\xi + \frac{C}{\xi^2} - e_z(t)\,,$$

and in the third

$$s^* = 2(\varphi-\varphi_0) + \frac{4}{\xi^2}\int(\varphi-\varphi_0)\xi\,\mathrm{d}\xi + \frac{2C}{\xi^2}\,.$$

Therefore, it is always necessary to integrate the equation

$$\dot{p}p^{\alpha}\exp\frac{4p}{3} = \exp\,(s^*-\beta)\,. \qquad (163.8)$$

If $\sigma_r(\alpha) = \sigma_r(1) = 0$, then

$$B = \tfrac{1}{2}m(1+\alpha^2) + \frac{1}{1-\alpha^2}\left(2\int_{\alpha}^{1}(\varphi-\varphi_0)\xi\,\mathrm{d}\xi\right)$$

$$+\tfrac{2}{3}\int_{\alpha}^{1}\frac{\mu}{\xi}\,\mathrm{d}\xi + 2\int_{\alpha}^{1}\frac{\lambda}{\xi}\,\mathrm{d}\xi\,, \qquad (163.9)$$

$$C = -\tfrac{1}{2}m\alpha^4 + C_1\alpha^2\,.$$

The quantity $e_z(t)$ is found from the condition that the resultant axial stresses are zero, namely

$$e_z(t) = \tfrac{1}{4}m(1+\alpha^2) + \frac{1}{1-\alpha^2}\Bigg[4\int_{\alpha}^{1}(\varphi-\varphi_0)\xi\,\mathrm{d}\xi + \tfrac{8}{3}\int_{\alpha}^{1}\mu\xi\,\mathrm{d}\xi$$

$$+\tfrac{4}{3}\int_{\alpha}^{1}\xi\,\mathrm{d}\xi\int_{\alpha}^{\xi}\frac{\mu}{\xi}\,\mathrm{d}\xi + 4\int_{\alpha}^{1}\xi\,\mathrm{d}\xi\int_{\alpha}^{\xi}\frac{\lambda}{\xi}\,\mathrm{d}\xi\Bigg] - B\,. \qquad (163.10)$$

Graphs of the stress distributions σ_θ and σ_z during starting of a rotor, obtained by Danilova [23], are given in figs. 236 and 237. The starting period

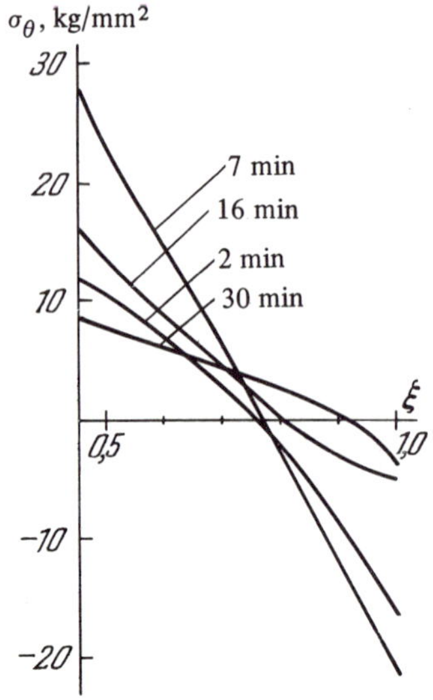

Fig. 236.

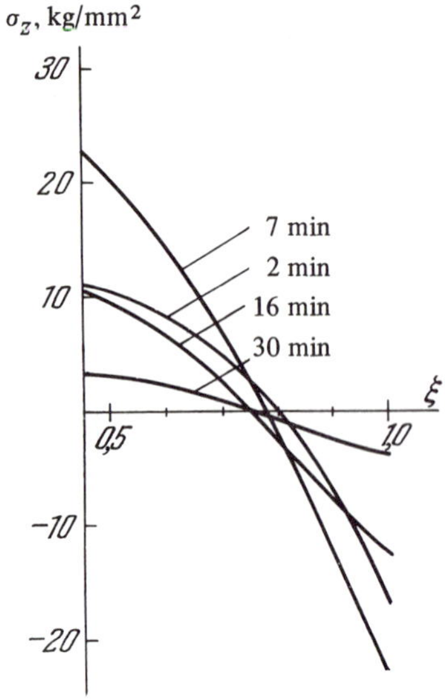

Fig. 237.

lasts for seven minutes and during this time the angular velocity increases linearly, the gas flow temperature also increases linearly from 180 to 550°C.

164. Kachanov's variational principle in the theory of transient creep

Assume that in the most general case, deformation of the body consists of instantaneous deformation, including elastic and plastic components, and creep deformation. For simplicity, we shall assume that the instantaneous deformation follows the equation of the deformation theory, consequently,

$$e_{ij}^{p} = \frac{\partial \varphi}{\partial \sigma_{ij}} \, .$$

The rate of creep deformation depends on the rate potential Φ so that

$$\dot{p}_{ij} = \frac{\partial \Phi}{\partial \sigma_{ij}} \, .$$

The rate potential may depend on any number of structural parameters, in particular on the components of the creep deformation tensor p_{ij}. The rate of total deformation is given by the expression

$$\epsilon_{ij} = \frac{\partial}{\partial \sigma_{ij}} \left(\frac{\mathrm{d}\varphi}{\mathrm{d}t} + \Phi \right) . \tag{164.1}$$

We now reproduce the derivation of Castigliano's variational principle, given in section 97. Suppose that the body is acted upon by surface forces T_i and body forces F_i and that there are set up in it systems of stresses σ_{ij} and velocity distributions v_i which define the distribution of the rate of strain tensor ϵ_{ij}. In addition to the fundamental state of stress consider an infinitesimally close, statically possible state corresponding to the same body forces and the same surface forces on the part of the surface for which these forces are given. Then the quantities $\delta\sigma_{ij}$ form a self-equilibrating system of stresses, and $\delta T_i = 0$ on part of the surface Σ_T. Formulate the equation of equilibrium in Lagrange form for this self-equilibrating state, and select as possible rates the actual rates that are realised in the body. We obtain

$$\int_{\Sigma} \delta T_i v_i \mathrm{d}\Sigma - \int_{V} \delta\sigma_{ij} \epsilon_{ij} \mathrm{d}V = 0 .$$

The surface integral becomes 0 on Σ_T where $\delta T_i = 0$, and substituting expression (164.1) for ϵ_{ij} into the volume integral we obtain

$$\int_{\Sigma_v} \delta T_i v_i \mathrm{d}\Sigma - \delta \int_{V} \left(\frac{\mathrm{d}\varphi}{\mathrm{d}t} + \Phi \right) \mathrm{d}V = 0 . \tag{164.2}$$

This is the variational equation of transient creep derived by Kachanov [52]. It should be emphasised that only stresses vary and not their rates of change; the structural parameters in the expression for Φ do not vary. Therefore, the variational eq. (164.2) remains valid for any creep theory. Shesterikov [148] derived this equation for strain hardening theory. However, in applying the variational equation the introduction of strain-hardening parameters raises serious difficulties and it is practically impossible to obtain effective results. If the potential Φ depends only on the stress, that is the creep rate at constant stress is constant, and creep is transient only because it is accompanied by elastic deformation, then eq. (164.2) can be used to obtain effective approximate methods for investigating various problems. Note that a completely analogous situation also arises if we adopt the standpoint of Kachanov's flow theory, and there are no fundamental difficulties if we consider the function Φ to depend on time as a parameter. We have seen in section 73 that when the creep curves are similar the explicit dependence on time can be eliminated by appropriate choice of scale. If in eq. (164.2) we replace t by some function $\tau(t)$ and consider that Φ depends only on the stress, we obtain a formulation of the variational principle which is applicable to the Kachanov flow theory equations. Numerous applications of this principle are contained in the book [52]. Practical applications of eq. (164.2) usually proceed as follows. A certain statically possible stress distribution is given that contains k arbitrary parameters $c_s(t)$. These parameters are usually present in the stress expressions in a linear manner. Now $\widetilde{\varphi} = \widetilde{\varphi}(c_s)$ and $\widetilde{\Phi} = \widetilde{\Phi}(c_s)$. By formula (164.2) the functional is

$$\int \left(\frac{d\varphi}{dt} + \Phi \right) dV = \frac{\partial \widetilde{\varphi}}{\partial c_s} \dot{c}_s + \widetilde{\Phi} .$$

In taking the variation of this expression we consider that stress values alter instantaneously but not their rates of change, therefore

$$\delta \int_V \left(\frac{d\varphi}{dt} + \Phi \right) dV = \left(\frac{\partial^2 \widetilde{\varphi}}{\partial c_s \partial c_q} \dot{c}_s + \frac{\partial \widetilde{\Phi}}{\partial c_q} \right) \delta c_q = 0 .$$

We assume that the boundary conditions for stress are fulfilled irrespective of the values of the parameters c_s and, therefore, the variation of the first term of eq. (164.2) is zero and the functions $c_s(t)$ are determined from the following equations:

$$\frac{\partial^2 \widetilde{\varphi}}{\partial c_s \partial c_q} \dot{c}_s + \frac{\partial \widetilde{\Phi}}{\partial c_q} = 0 . \tag{164.3}$$

In the book [52] numerous problems are considered in the following way. Let σ_{ij}^0 be the initial elastic stress distribution, σ_{ij}^∞ the stress distribution in the steady-state creep. Put

$$\sigma_{ij} = \sigma_{ij}^0 + c(t)(\sigma_{ij}^\infty - \sigma_{ij}^0) . \tag{164.4}$$

The function $c(t)$ is determined from one single equation of form (164.3). Obviously we should have $c(0) = 0$, $c(\infty) = 1$.

The variational eq. (164.2) is applicable when the creep is accompanied by instantaneous elastic deformation, then $\varphi(\sigma_{ij})$, the elastic potential, is a quadratic function. If creep is accompanied by plastic deformation this latter must be described by equations of the deformation theory type, for which there is a potential $\varphi(\sigma_{ij})$. In this case the analysis given above is incomplete because in addition to zones of active plastic deformation it is also necessary to consider zones of unloading. This does not cause fundamental difficulty, in the loading region the function φ is the plastic potential of deformation theory and in the unloading region an elastic potential. This question is considered in the book [52] so we shall not develop it here.

Finally, note that eq. (164.2) can be used to investigate transient creep by the Odquist scheme (see section 47), considering the segment of steady-state creep and allowing for the first phase of creep as instantaneous initial plastic deformation. Here the objection may be raised that in treating steady-state creep as a flow process it is illogical to consider the state that corresponds to completion of the first phase by equations of the deformation type.

165. The Wang and Prager variational principle

If the instantaneous plastic deformation follows equations of the flow theory type, it is impossible to formulate a variational principle of the kind derived in section 164. However, in transient creep theory the possibility of formulating variational equations is wider than in theories of steady-state creep or non-linear elasticity, because in addition to the tensors σ_{ij} and ϵ_{ij} the equations include the tensors $\dot{\sigma}_{ij}$ and e_{ij}. Here we consider certain new variational principles found relatively recently and based on variation of tensors of rate of change of stress and deformation. Differentiating the equation of equilibrium and the boundary conditions with respect to time we obtain

$$\dot{\sigma}_{ij,j} + \dot{F}_i = 0 , \qquad \dot{\sigma}_{ij}\nu_j = \dot{T}_i . \tag{165.1}$$

The form of these equations indicates that rates of change of stress $\dot{\sigma}_{ij}$ can be considered as stresses caused by body forces $\dot{F}_i$ and surface forces $\dot{T}_i$.

The deformation consists of an instantaneous strain ϵ^p_{ij}, which includes elastic and plastic components, and creep deformation p_{ij} so that

$$e_{ij} = e^p_{ij} + p_{ij} .$$

From the standpoint of the theory of plasticity of the flow theory type we obtain the following relationship between the increments of instantaneous deformation and of stress:

$$\mathrm{d}e^p_{ij} = a_{ijkl}\,\mathrm{d}\sigma_{kl}\,;$$

here $a_{ijkl} = a_{klij}$. Indeed if the instantaneous deformation is elastic, a_{ijkl} are elastic constants. If the instantaneous plastic deformation follows the flow theory, because of eq. (17.1) we have

$$a_{ijkl} = H\,\frac{\partial f}{\partial \sigma_{ij}}\,\frac{\partial f}{\partial \sigma_{kl}}\,.$$

The tensor a_{ijkl} depends on the stress and possibly on the previous deformation, but it is independent of the increment $\mathrm{d}\sigma_{ij}$. The rest of the investigation will be conducted in terms of rates and therefore, we write

$$\dot{e}^p_{ij} = a_{ijkl}\dot{\sigma}_{kl}\,. \tag{165.2}$$

The rate of creep deformation $\dot{p}_{ij}$ in its turn depends only on the stress and work-hardening parameters and is independent of the rate of change of stress.

In addition to the actual distribution of rates of stress $\dot{\sigma}_{ij}$ consider an infinitesimally close, statically possible distribution $\dot{\sigma}_{ij} + \delta\dot{\sigma}_{ij}$. These quantities satisfy eqs. (165.1) with changed rates of change of surface and body forces $\delta\dot{F}_i$ and $\delta\dot{T}_i$ so that

$$\delta\dot{\sigma}_{ij,j} + \delta\dot{F}_i = 0\,, \qquad \delta\dot{\sigma}_{ij}\nu_j = \delta\dot{T}_i\,.$$

Considering the quantities $\delta\dot{F}_i$ and $\delta\dot{T}_i$ as external forces and the quantities $\delta\dot{\sigma}_{ij}$ as stresses, applying to them the true rate distribution that is realised in the body as a possible rate distribution, we obtain

$$\int_V \delta\dot{\sigma}_{ij}\dot{e}_{ij}\,\mathrm{d}V = \int \delta\dot{F}_i\dot{u}_i\,\mathrm{d}V + \int \delta\dot{T}_i\dot{u}_i\,\mathrm{d}\Sigma\,. \tag{165.3}$$

In the right-hand side the variation sign can be taken outside the integral, this cannot be done in the left-hand side unless it is appropriately converted. For simplicity of presentation we represent the tensors by vectors in a nine-dimensional space; the vectors $\dot{\boldsymbol{\sigma}}$, $\dot{\mathbf{e}}_p$, $\dot{\mathbf{p}}$, $\dot{\mathbf{e}}$ have as components $\dot{\sigma}_{ij}$, $\dot{e}^p_{ij}$, $\dot{p}_{ij}$, $\dot{e}_{ij}$ respectively and the subintegral expression in the left-hand side of eq. (165.3) can be written $\delta\dot{\boldsymbol{\sigma}}\dot{\mathbf{e}}$.

Consider the following expression

$$\dot{\boldsymbol{\sigma}}(\dot{\mathbf{e}} + \dot{\mathbf{p}})\,.$$

Calculate the variation of this expression, bearing in mind that rates of change of stress $\dot{\boldsymbol{\sigma}}$ vary and consequently, so do the rates of total deformation $\dot{\mathbf{e}}$ that depend on them, but not the creep rates $\dot{\mathbf{p}}$ which depend on stresses but not on their rate of change. We obtain

$$\delta[\dot{\boldsymbol{\sigma}}(\dot{\mathbf{e}}+\dot{\mathbf{p}})] = \delta\dot{\boldsymbol{\sigma}}\dot{\mathbf{e}} + \delta\dot{\boldsymbol{\sigma}}\dot{\mathbf{p}} + \dot{\boldsymbol{\sigma}}\delta\dot{\mathbf{e}} = \delta\dot{\boldsymbol{\sigma}}\dot{\mathbf{e}} + \delta\dot{\boldsymbol{\sigma}}(\dot{\mathbf{p}} + \dot{\mathbf{e}} - \dot{\mathbf{e}}) + \dot{\boldsymbol{\sigma}}\delta\dot{\mathbf{e}}$$

$$= 2\delta\dot{\boldsymbol{\sigma}}\dot{\mathbf{e}} - \delta\dot{\boldsymbol{\sigma}}(\dot{\mathbf{e}} - \dot{\mathbf{p}}) + \dot{\boldsymbol{\sigma}}\delta\dot{\mathbf{e}}\,.$$

The quantity $\delta\dot{\boldsymbol{\sigma}}(\dot{\mathbf{e}} - \dot{\mathbf{p}})$ is $\delta\dot{\boldsymbol{\sigma}}\dot{\mathbf{e}}_p = \delta\dot{\sigma}_{ij}\dot{e}^p_{ij}$. Because of eq. (165.2) we find

$$\delta\dot{\boldsymbol{\sigma}}(\dot{\mathbf{e}} - \dot{\mathbf{p}}) = a_{ijkl}\dot{\sigma}_{kl}\delta\dot{\sigma}_{ij}\,.$$

In exactly the same way

$$\dot{\boldsymbol{\sigma}}\delta\dot{\mathbf{e}} = \dot{\boldsymbol{\sigma}}\delta\dot{\mathbf{e}}_p = \dot{\sigma}_{ij}\delta\dot{e}^p_{ij} = a_{ijkl}\dot{\sigma}_{ij}\delta\dot{\sigma}_{kl}\,.$$

These expressions are equal to one another, therefore,

$$\delta\dot{\sigma}_{ij}\dot{e}_{ij} = \tfrac{1}{2}\delta[\dot{\sigma}_{ij}(\dot{e}_{ij} + \dot{p}_{ij})]\,. \tag{165.4}$$

Now we can rewrite eq. (165.3) as follows

$$\delta\,\tfrac{1}{2}\int_V \dot{\sigma}_{ij}(\dot{e}_{ij} + \dot{p}_{ij})\mathrm{d}V = \delta\left[\int_V \dot{F}_i\dot{u}_i\,\mathrm{d}V + \int_\Sigma \dot{T}_i\dot{u}_i\,\mathrm{d}\Sigma\right]. \tag{165.5}$$

Eq. (165.5) expresses the first variational principle given in the work of Wang and Prager [380]. The rates of change of stress $\dot{\sigma}_{ij}$ must be varied and not the actual stresses, as is done in variation of the functional (164.2), because the derivation of eq. (165.5) is based on more general assumptions concerning the law of plasticity for instantaneous deformation. If as before we consider that

$$e^p_{ij} = \frac{\partial\varphi}{\partial\sigma_{ij}},$$

then using variational eqs. (164.2) and (165.5) we arrive at the identical conclusion. Indeed, put

$$\sigma_{ij} = \sigma^0_{ij} + \sigma^k_{ij}c_k\,.$$

For simplicity we shall consider that σ_{ij}^k form a self-equilibrating system of stresses, therefore, when c_k are varied the external forces are not varied. First apply variational principle (164.2), the functional being written as

$$I_1 = \int \left[\frac{\partial \varphi}{\partial \sigma_{ij}} \sigma_{ij}^k \dot{c}_k + \Phi \right] \mathrm{d}V .$$

We shall now vary c_k leaving the rates $\dot{c}_k$ unaltered, we obtain

$$\delta I_1 = \int \left[\frac{\partial^2 \varphi}{\partial \sigma_{ij} \partial \sigma_{rs}} \sigma_{ij}^k \dot{c}_k + \frac{\partial \Phi}{\partial \sigma_{rs}} \right] \sigma_{rs}^l \delta c_l \, \mathrm{d}V = 0 .$$

Since δc_l are arbitrary and independent, the condition $\delta I_1 = 0$ splits into the following system of equations:

$$\int \left[\frac{\partial^2 \varphi}{\partial \sigma_{ij} \partial \sigma_{rs}} \sigma_{ij}^k \dot{c}_k + \frac{\partial \Phi}{\partial \sigma_{rs}} \right] \sigma_{rs}^l = 0 . \tag{165.6}$$

Now use variational eq. (165.5). Rewrite the functional in the left-hand side as

$$I_2 = \tfrac{1}{2} \int \dot{\sigma}_{ij} (\dot{e}_{ij}^p + 2\dot{p}_{ij}) \, \mathrm{d}V .$$

But

$$\dot{e}_{ij}^p = \frac{\partial^2 \varphi}{\partial \sigma_{ij} \partial \sigma_{rs}} \sigma_{rs}^l \dot{c}_l , \qquad \dot{p}_{ij} = \frac{\partial \Phi}{\partial \sigma_{ij}} ,$$

therefore,

$$I_2 = \tfrac{1}{2} \int \left(\frac{\partial^2 \varphi}{\partial \sigma_{ij} \partial \sigma_{rs}} \sigma_{ij}^k \sigma_{rs}^l \dot{c}_k \dot{c}_l + \frac{\partial \Phi}{\partial \sigma_{ij}} \sigma_{ij}^k \dot{c}_k \right) \mathrm{d}V .$$

Varying the rates $\dot{c}_k$, we obtain

$$\delta I_2 = \int \left(\frac{\partial^2 \varphi}{\partial \sigma_{ij} \partial \sigma_{rs}} \sigma_{rs}^l \dot{c}_l + \frac{\partial \Phi}{\partial \sigma_{ij}} \right) \sigma_{ij}^k \delta \dot{c}_k \, \mathrm{d}V = 0 .$$

Hence it follows that:

$$\int \left[\frac{\partial^2 \varphi}{\partial \sigma_{ij} \partial \sigma_{rs}} \sigma_{rs}^l \dot{c}_l + \frac{\partial \Phi}{\partial \sigma_{ij}} \right] \sigma_{ij}^k \, \mathrm{d}V = 0 \, . \tag{165.7}$$

Eqs. (165.6) and (165.7) are identical.

166. Approximate solution of relaxation problems. Shesterikov's method

Shesterikov [151] proposed an approximate method of solving stress relaxation problems by the variational eq. (164.2). By relaxation problems we mean those for which, on the supposition of steady-state creep, there is a unique solution $\sigma_{ij}^{\infty} = 0$. Suppose that the initial elastic distribution is σ_{ij}^{*}, we shall seek an approximate solution in the form

$$\sigma_{ij} = c(t) \sigma_{ij}^{*} \, . \tag{166.1}$$

Considering the instantaneous deformation to be elastic in eq. (164.2) we take

$$\varphi = \left(\frac{1}{2G} \sigma_0^2 + \tfrac{3}{2} K \sigma^2 \right) .$$

Because of eq. (166.1) we obtain

$$\varphi = c^2 \varphi^* \, . \tag{166.2}$$

Here

$$\varphi^* = \frac{\sigma_0^{*2}}{2G} + \tfrac{3}{2} K \sigma^{*2} \, .$$

We suppose that the creep potential $\Phi = \Phi(\sigma_0, q_0)$ where q_0 is given by formula (75.3). Then, as shown in section 73,

$$\dot{q}_0 = \frac{\partial \Phi}{\partial \sigma_0} \, . \tag{166.3}$$

Formulate the functional (164.2):

$$I = \int [2c\dot{c}\varphi^* + \Phi] \, \mathrm{d}V \, .$$

The variation of this functional is

$$\delta I = \int \left[2\dot{c}\varphi^* + \frac{\partial \Phi}{\partial \sigma_0} \sigma_0^* \right] \delta c V .$$

Here substitute $\partial\Phi/\partial\sigma_0$ in terms of $\dot{q}_0$ as given by eq. (166.3) and integrate with respect to t. We obtain

$$\int [2(c-1)\varphi^* + q_0 \sigma_0^*] \, \mathrm{d}V = 0 . \tag{166.4}$$

The functional eq. (166.4) can only be solved with particular assumptions concerning the creep law. Suppose for example, that in the uniaxial case we have

$$\dot{p}p^\alpha = A\sigma^n . \tag{166.5}$$

Substituting q_0 for p and σ_0 for σ, we obtain a partial form of expression (166.3). Put $\sigma_0 = \sigma_0^* c(t)$ and integrate eq. (166.5). We find

$$q_0 = \left(\frac{A}{m}\right)^m \sigma_0^{*mn} \left[\int_0^t c^n \, \mathrm{d}t \right]^m .$$

Substitute into (166.4), we obtain

$$2(c-1) \int_V \varphi^* \, \mathrm{d}V + \left(\frac{A}{m}\right)^m \left[\int_0^t c^n \, \mathrm{d}t \right]^m \int_V \sigma_0^{*mn+1} \, \mathrm{d}V = 0 .$$

Put

$$\left(\frac{A}{m}\right)^m \int_V (\sigma_0^*)^{mn+1} \, \mathrm{d}V = I_1 ,$$

$$2 \int \varphi^* \, \mathrm{d}V = I_2 .$$

Then the previous equation is written

$$I_1 \left[\int_0^t c^n \, \mathrm{d}t \right]^m = I_2(1-c) ,$$

or

$$\int_0^t c^n\,\mathrm{d}t = \left(\frac{I_2}{I_1}\right)^{1+\alpha} (1-c)^{1+\alpha} .$$

Differentiate with respect to time, separate variables and integrate, to give,

$$\int_c^1 \frac{(1-\tau)^\alpha}{\tau^n}\,\mathrm{d}\tau = \frac{1}{1+\alpha} \times \left(\frac{I_1}{I_2}\right)^{1+\alpha} t . \qquad (166.6)$$

A graph of the function $c(t)$ is shown in fig. 238 for a disc of constant thickness with a central hole, of radius one fifth of the external radius, and with a temperature drop from 450° on the outer radius to 350° on the inner [151]. The disc is made of steel grade 40Kh. Eq. (166.5) is written in the form

$$\dot{p}p^\alpha = \left(\frac{\sigma}{s}\right)^n \exp B(T - T_0) .$$

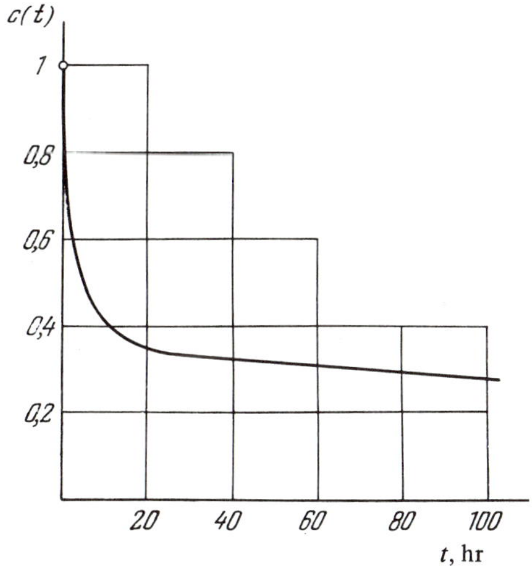

Fig. 238.

The following values were obtained on working out the experimental creep curves

$$\alpha = 0.6\,,\qquad B = 0.06\,,\qquad n = 6\,,\qquad T_0 = 840°\,.$$

167. Mixed variational principle

Let $\dot{e}_{ij} = \frac{1}{2}(\dot{u}_{i,j} + \dot{u}_{j,i})$ be the true distribution of rates of strain that corresponds to the distribution of rates $\dot{u}_i$. An infinitesimally close rate distribution is $\delta\dot{u}_i$ and an infinitesimally close rate of strain distribution $\delta\dot{e}_{ij}$. Considering $\delta\dot{u}_i$ as possible displacements, formulate the variational equation of equilibrium in the form

$$\int\limits_V \dot{\sigma}_{ij}\delta\dot{e}_{ij}\mathrm{d}V = \int\limits_V \dot{F}_i\delta\dot{u}_i\mathrm{d}V + \int\limits_\Sigma \dot{T}_i\delta\dot{u}_i\mathrm{d}\Sigma\,. \tag{167.1}$$

The left-hand side of this equation can no longer be represented as the variation of a certain functional because $\dot{\sigma}_{ij}$ are not expressed in terms of rates of total deformation. Therefore, eq. (167.1) does not contain any principle that can be used to select, from the various kinematically possible velocity distributions, the distribution that best approximates to the true one. The amount of information that can be extracted from this equation is relatively small. We shall now describe a new variational principle for creep that was formulated in the work of Sanders, McComb and Schlechte [345], which is based on simultaneous variation of the creep rate and stress rate distributions. The authors indicate that in deriving the variational equation for creep they start on the one hand from Reissner's variational principle in the theory of elasticity and on the other from the Wang and Prager theorem given in section 165. Therefore, we commence by presenting Reissner's ideas [340, 341]. Assume that on part of the surface Σ_u we are given displacements u_i^e and on part of the surface Σ_T we are given forces T_i^e. Mass forces are assumed absent. Formulate the functional

$$K = \int\limits_V [\sigma_{ij}e_{ij} - \varphi(\sigma_{ij})]\mathrm{d}V - \int\limits_V F_iu_i\mathrm{d}V$$

$$- \int\limits_{\Sigma_u} T_i(u_i - u_i^e)\mathrm{d}\Sigma - \int\limits_{\Sigma_T} T_i^e u_i\mathrm{d}\Sigma\,. \tag{167.2}$$

Find the variation of this functional on the assumption that the quantities σ_{ij} and u_i vary independently and note $e_{ij}=\frac{1}{2}(u_{i,j}+u_{j,i})$. We obtain

$$\delta K=\int_V\left[\left(e_{ij}-\frac{\partial\varphi}{\partial\sigma_{ij}}\right)\delta\sigma_{ij}+\sigma_{ij}\delta e_{ij}\right]\mathrm{d}V$$

$$-\int_{\Sigma_u}\delta T_i(u_i-u_i^e)\mathrm{d}\Sigma-\int_{\Sigma_u}T_i\delta u_i\mathrm{d}\Sigma-\int_{\Sigma_T}T_i^e\delta u_i\mathrm{d}\Sigma\,.$$

Converting as usual the volume integral of $\sigma_{ij}\delta e_{ij}$:

$$\int_V\sigma_{ij}\delta e_{ij}\mathrm{d}V=\int_V\sigma_{ij}\delta u_{i,j}\mathrm{d}V=\int_\Sigma\sigma_{ij}\nu_j\delta u_i\mathrm{d}\Sigma-\int\sigma_{ij,j}\delta u_i\mathrm{d}V\,.$$

We obtain

$$\delta K=\int_V\left[\left(e_{ij}-\frac{\partial\varphi}{\partial\sigma_{ij}}\right)\delta\sigma_{ij}-\sigma_{ij,j}\delta u_i\right]\mathrm{d}V$$

$$-\int_{\Sigma_u}\delta T_i(u_i-u_i^e)\mathrm{d}\Sigma+\int_{\Sigma_T}(\sigma_{ij}\nu_j-T_i)\delta u_i\mathrm{d}\Sigma\,.$$

Thus, the following condition is fulfilled:

$$\delta K=0\,, \tag{167.3}$$

which is equivalent to fulfilling the coupling equation $e_{ij}=\partial\varphi/\partial\sigma_{ij}$, the equation of equilibrium $\sigma_{ij,j}=0$, and also the boundary conditions $u_i=u_i^e$ on Σ_u and $\sigma_{ij}\nu_j=T_i^e$ on Σ_T.

Now turn to the case of creep. The Wang and Prager functional plays a part similar to that of the Castigliano functional, this governs the selection of the expression K for creep, namely,

$$K=\int_V\left[\dot{\sigma}_{ij}\dot{e}_{ij}-\tfrac{1}{2}(\dot{e}_{ij}^p+2\dot{p}_{ij})\dot{\sigma}_{ij}\right]\mathrm{d}V$$

$$-\int_{\Sigma_u}\dot{T}_i(\dot{u}_i-\dot{u}_i^e)\mathrm{d}\Sigma-\int_{\Sigma_T}\dot{T}_i^e\dot{u}_i\mathrm{d}\Sigma\,. \tag{167.4}$$

Here it is assumed that rates of displacements of points are given on part of the surface Σ_u and rates of change of surface loads are given on part of the surface Σ_T.

We shall consider that in expression (167.4) $\dot{\sigma}_{ij}$ and $\dot{u}_i$ are independent variables, $\dot{e}_{ij} = \frac{1}{2}(\dot{u}_{i,j} - \dot{u}_{j,i})$ whereas $\dot{e}^p_{ij}$ depends on σ_{ij} and $\dot{\sigma}_{ij}$, and $\dot{p}_{ij}$ depends on σ_{ij} and possibly on certain work-hardening parameters. Calculating the variation of the functional (167.4) we obtain

$$\delta K = \int_V \left[\delta\dot{\sigma}_{ij}\dot{e}_{ij} + \dot{\sigma}_{ij}\delta\dot{e}_{ij} - \tfrac{1}{2}(\dot{e}^p_{ij} + 2\dot{p}_{ij})\delta\dot{\sigma}_{ij} - \tfrac{1}{2}\,\delta\dot{e}^p_{ij}\dot{\sigma}_{ij} \right] \mathrm{d}V$$

$$- \int_{\Sigma_u} \delta\dot{T}_i(\dot{u}_i - \dot{u}^e_i)\mathrm{d}\Sigma - \int_{\Sigma_u} \dot{T}_i\delta\dot{u}_i\,\mathrm{d}\Sigma - \int_{\Sigma_T} \dot{T}^e_i\,\delta\dot{u}_i\mathrm{d}\Sigma \, .$$

Noting that

$$\delta\dot{e}^p_{ij}\dot{\sigma}_{ij} = \dot{e}^p_{ij}\delta\dot{\sigma}_{ij} \, ,$$

we rewrite the expression for δK as follows:

$$\delta K = \int_V (\dot{e}_{ij} - \dot{e}^p_{ij} - \dot{p}_{ij})\delta\dot{\sigma}_{ij} - \int_V \dot{\sigma}_{ij}\delta\dot{e}_{ij}\delta V$$

$$- \int_{\Sigma_u} \delta\dot{T}_i(\dot{u}_i - \dot{u}^e_i)\mathrm{d}\Sigma - \int_{\Sigma_u} \dot{T}_i\delta\dot{u}_i\mathrm{d}\Sigma - \int_{\Sigma_T} \dot{T}^e_i\,\delta\dot{u}_i\mathrm{d}\Sigma \, .$$

In varying rates of displacement we require the boundary conditions to be fulfilled, that is $\delta\dot{u}_i = 0$ on Σ_u. In varying the stress we do not infringe the boundary conditions $\dot{T}_i = \dot{T}^e_i$ on Σ_T, consequently, $\delta\dot{T}_i = 0$ on Σ_T. The first integral becomes identically zero because $\dot{e}_{ij} = \dot{e}^p_{ij} + \dot{p}_{ij}$. Converting the second integral in the usual way we obtain

$$\delta K = -\int_V \dot{\sigma}_{ij,j}\delta\dot{u}_i\mathrm{d}V + \int_\Sigma [\dot{\sigma}_{ij}\nu_j - \dot{T}_i]\delta\dot{u}_i\mathrm{d}\Sigma - \int_\Sigma \delta\dot{T}_i(\dot{u}_i - \dot{u}^e_i)\mathrm{d}\Sigma \, .$$

That the variation δK becomes zero is equivalent to fulfilment of the equation of equilibrium and the boundary conditions for $\dot{\sigma}_{ij}$, if the variation $\delta\dot{u}_i$ and $\delta\dot{\sigma}_{ij}$ are subject to the limitations formulated above.

168. Application of the mixed variation principle to the problem of bending

As the simplest example, consider the problem of bending a beam with a section that has two axes of symmetry. First consider the simple case of a material that does not work-harden for which

$$\dot{e} = g'(\sigma)\dot{\sigma} + v(\sigma) .$$

Because of the law of plane sections $e = \kappa y$. Suppose that we are given a linear stress distribution across the section of the rod, that is we put $\sigma = \beta y$. Now formulate the functional (167.2) for an element of beam of unit length under conditions of pure bending moment M:

$$K = \dot{\beta}\dot{\kappa} \int_F y^2 \,\mathrm{d}F - \tfrac{1}{2}\dot{\beta}^2 \int g'(\beta y) y^2 \,\mathrm{d}F - \dot{\beta} \int v(\beta y) y \,\mathrm{d}F - \dot{M}\dot{\kappa} . \tag{168.1}$$

Equate to zero the variation δK, calculated on the assumption that only $\dot{\beta}$ and $\dot{\kappa}$ vary independently of one another:

$$\delta K = I(\delta\dot{\beta}\dot{\kappa} + \dot{\beta}\delta\dot{\kappa}) - \dot{\beta}\delta\dot{\beta} \int g'(\beta y) y^2 \,\mathrm{d}F$$
$$- \delta\dot{\beta} \int v(\beta y) y \,\mathrm{d}F - \dot{M}\delta\dot{\kappa} = 0 .$$

Here I is the moment of inertia of the cross-sectional area. Hence it follows that:

$$\dot{\beta} = \frac{\dot{M}}{I} ,$$

$$\dot{\kappa} = \dot{\beta} \frac{1}{I} \int g'(\beta y) y^2 \,\mathrm{d}F + \frac{1}{I} \int v(\beta y) y \,\mathrm{d}F .$$

Put

$$\frac{1}{I} \int g(\beta y) y^2 \,\mathrm{d}F = \kappa_0(\beta) = \kappa_0\left(\frac{M}{I}\right) .$$

The quantity $\kappa_0(\beta)$ is the instantaneous curvature of the rod caused by application of the bending moment; if the initial deformation is elastic then $\kappa_0(\beta)$ is the accurate value of the curvature; if instantaneous plastic deformation is present $\kappa_0(\beta)$ is the approximate curvature caused by this instantaneous plastic deformation. We can now rewrite the relationship between curvature and bending moment as follows:

$$\dot{\kappa} = \dot{\kappa}_0 + \frac{1}{I}\int v(\beta y) y \, \mathrm{d}F \,. \tag{168.2}$$

It is evident that here, as in other similar cases, the functions $g(\sigma)$ and $\kappa_0(M/I)$ can only correspond to the initial plastic deformation or to the first period of creep in Odquist's scheme.

We shall now consider that the instantaneous deformation is elastic and that creep follows a power law. Then from eq. (168.2) it follows that

$$\dot{\kappa} = \frac{\dot{M}}{EI} + \kappa_n \left(\frac{M}{M_n}\right)^n . \tag{168.3}$$

Here

$$\kappa_n = \frac{\epsilon_n}{h}, \qquad M_n = \sigma_n I : \left\{\frac{h}{I}\int_F |y|^{n+1} \, \mathrm{d}F\right\}^{1/n} . \tag{168.4}$$

In assuming a linear stress distribution, we of course introduced a certain error. For steady-state creep there follows from eq. (168.3) an expression which coincides in form with the exact expression (106.2), the only difference is in the definition of the quantity M_n. For example, the following expression was derived in section 106 for a rod of rectangular cross-section of width b and height $2h$:

$$M_n = \frac{2n}{2n+1} bh^2 \sigma_n \,.$$

By formula (168.4) we obtain

$$M_n' = \tfrac{2}{3}\left(\frac{n+2}{3}\right)^{1/n} bh^2 \sigma_n \,.$$

The error of determination of the quantity M_n by formula (168.4) is given in the table on page 699 for several values of n.

Taking as unity the magnitude of the moment that is required to produce the given creep rate, the figures given in the second row are approximate values of the moment required to give the same rate on the assumption of a

n	1	2	3	4	5	6
$\dfrac{M_n'}{M_n}$	1	0.963	0.921	0.894	0.869	0.849
κ'/κ	1	1.08	1.28	1.57	2.04	2.69

linear law of stress distribution. The figures given in the third row are approximate rates of steady-state creep for the given moment if the true rate for this moment is unity. The error in determining the moment that corresponds to the given rate is relatively small, whereas the error in determining the rate that corresponds to a given moment is considerable and increases markedly with increase in n. A similar situation was encountered in analysing example 1 of section 102.

Returning to eq. (168.3) we note that application of the variational principle has led to a definite structure of this equation analogous to the initial relationship between e and σ. Indeed writing this expression in the form

$$\dot{e} = \frac{\dot{\sigma}}{E} + \epsilon_n \left(\frac{\sigma}{\sigma_n}\right)^n ,$$

we find that the generalised displacement κ and the generalised force M in eq. (168.3) are related to one another in the same way as deformation and stress. Because of the assumption that the law of stress distribution is linear the first term in the right-hand side of eq. (168.3) gives an accurate value of the instantaneous elastic deformation, but the second term that corresponds to steady-state creep is approximate, and we have explained the amount of error. If to find M_n we now use not formula (168.4), but the exact formula given in section 106, namely

$$M_n = h\sigma_n \int_F \left|\frac{y}{h}\right|^{1+\frac{1}{n}} \mathrm{d}F , \qquad (168.5)$$

then eq. (168.4) also gives an accurate result for steady-state creep and therefore it should be used.

169. **Bending of beams in creep with work-hardening**

The mixed variational principle that was formulated in section 167 is also applicable when creep is accompanied by work-hardening, the form of the work-hardening law does not matter but it is important that the rate of creep should not depend on the rate of change of stress. Assume, for example, that the work-hardening law is given by the equation

$$\dot{p}p^{\alpha} = A\sigma^{n} .$$

Put $A = \dfrac{1}{1+\alpha}\dfrac{1}{t_{*}\sigma_{*}^{n}}$, and note that the dot notation represents differentiation with respect to the dimensionless time $\hat{t} = t/t_{*}$. We also put $\hat{\sigma} = \sigma/\sigma_{*}$. Then the basic creep equation is rewritten as

$$\dot{p}p^{\alpha} = \frac{1}{1+\alpha}\hat{\sigma}^{n} . \tag{169.1}$$

From eq. (169.1) we find

$$p = \left[\int_0^t \hat{\sigma}^n \, \mathrm{d}t\right]^{\frac{1}{1+\alpha}} ,$$

and consequently

$$\dot{p} = \frac{1}{1+\alpha}\sigma^{n}\left[\int_0^t \hat{\sigma}^n \, \mathrm{d}t\right]^{-\frac{\alpha}{1+\alpha}} . \tag{169.2}$$

Now put

$$e = \frac{\kappa y}{h} , \qquad \hat{\sigma} = \frac{\beta y}{h} .$$

Note that here the notation is changed somewhat compared with section 168, the arbitrary quantity h with the dimension of length is introduced in order to make κ and β dimensionless. Substituting the expression for $\hat{\sigma}$ into eq. (169.2) we find

$$\dot{p} = y^{\frac{n}{1+\alpha}}\frac{1}{1+\alpha}\beta^{n}\left[\int_0^t \beta^n \, \mathrm{d}t\right]^{-\frac{\alpha}{1+\alpha}} .$$

Now formulate the functional K. To within a constant factor

$$K = \dot{\kappa}\dot{\beta} - \tfrac{1}{2}\dot{\beta}^2\frac{\sigma_*}{E} - \frac{\dot{\beta}}{1+\alpha}\beta^n\left[\int_0^t \beta^n\,\mathrm{d}t\right]^{-\frac{\alpha}{1+\alpha}} R - \dot{\hat{M}}\kappa\,. \qquad (169.3)$$

Here

$$R = \frac{h^2}{I}\int_F \left|\frac{y}{h}\right|^{1+\frac{n}{1+\alpha}} \mathrm{d}F\,, \qquad \hat{M} = \frac{Mh}{I\sigma_*}\,. \qquad (169.4)$$

Equating to zero the variation of the functional (169.3) we arrive at the following equations:

$$\begin{aligned} &\dot{\beta} = \dot{\hat{M}}\,,\\ &\dot{\kappa} = \dot{\beta}\frac{\sigma_*}{E} + \frac{R}{1+\alpha}\beta^n\left[\int \beta^n\,\mathrm{d}t\right]^{-\frac{\alpha}{1+\alpha}}. \end{aligned} \qquad (169.5)$$

Here $\beta(\sigma_*/E)$ is the elastic change of curvature and therefore the quantity

$$\kappa_c = \kappa - \beta\frac{\sigma_*}{E}$$

is the change of curvature that accumulates during creep. Rewrite the second expression of eqs. (169.5) as follows:

$$\dot{\kappa}_c = \frac{R}{1+\alpha}\hat{M}^n\left[\int_0^t \hat{M}^n\,\mathrm{d}t\right]^{-\frac{\alpha}{1+\alpha}}. \qquad (169.6)$$

For the time being put

$$\int_0^t \hat{M}^n\,\mathrm{d}t = q^{1+\alpha}\,.$$

Hence

$$\hat{M}^n = (1+\alpha)q^{\alpha}\dot{q}\,.$$

Substituting into eq. (169.6) we find that

$$\dot{\kappa}_c = R\dot{q} \, ,$$

and consequently,

$$q = \frac{\kappa_c}{R} \, .$$

We can now eliminate the integral factor from eq. (169.6) and obtain

$$\dot{\kappa}_c \kappa_c^{\alpha} = \frac{R^{1+\alpha}}{1+\alpha} \hat{M}^n \, . \tag{169.7}$$

The structure of eq. (169.7) is exactly the same as that of the initial work-hardening law (169.1). This equation is somewhat approximate because in deriving it we neglected the stress redistribution over the cross-section of the rod, and considered that the stress distribution law always remains linear. This is valid for the first instant after application of load and, therefore, the expression for the instantaneous deformation is accurate. However, in course of time the part played by elastic deformation becomes infinitesimally small and the stress distribution tends to the steady-state distribution that corresponds to quasi-steady-state creep. The value of the factor R in formula (169.7) is such that an accurate value of quasi-steady-state creep rate is not obtained. It is not difficult to determine the corresponding correction. Considering that $e = p$ the initial creep equation is replaced by the following:

$$\dot{e} e^{\alpha} = \frac{1}{1+\alpha} \left(\frac{\sigma}{\sigma_*} \right)^n \, .$$

As before put $e = \kappa y/h$, find σ from the above equation and calculate the moment. As a result we obtain the following relationship between curvature and moment:

$$\dot{\kappa} \kappa^{\alpha} = \frac{R_1^{1+\alpha}}{1+\alpha} \hat{M}^n \, . \tag{169.8}$$

Eq. (169.8) may be considered as a limiting case of eq. (169.7) when the elastic deformation of bending $\hat{M}(\sigma_*/E)$ becomes infinitesimally small compared with κ. However, unlike eq. (169.7), eq. (169.8) is exact and correspondingly the value of R_1 in this equation is not given by formula (169.4) but by the following formula:

$$R_1 = 1 : \left[\frac{h^2}{I} \int_F \left| \frac{y}{h} \right|^{1+\frac{1+\alpha}{n}} \mathrm{d}F \right]^{\frac{n}{1+\alpha}} . \tag{169.9}$$

Now if in using eq. (169.7) R is determined by expression (169.9) the equation gives accurate results for the extreme cases; at the initial instant of time when the deformation is purely elastic and after a sufficiently long interval of time, when the elastic deformation need not be taken into account compared with creep deformation.

170. **Transient creep of plates**

The approximate theory of transient creep of plates can be constructed in exactly the same way. Write the fundamental creep law with power dependence of work-hardening and with power dependence of rate on stress in the following way (section 74):

$$\dot{p}_{ij} = \frac{3}{2(1+\alpha)} p^{-\alpha} \left(\frac{\sigma_0}{\sigma_*} \right)^n \frac{\bar{\sigma}_{ij}}{\sigma_0} . \tag{170.1}$$

Here

$$p = \int_0^t (\tfrac{2}{3} \dot{p}_{ij} \dot{p}_{ij})^{\frac{1}{2}} \, \mathrm{d}t ,$$

although an analogous result can be obtained with another selection of work-hardening law. It follows from eq. (170.1) that

$$\dot{p} = \frac{1}{1+\alpha} \left(\frac{\sigma_0}{\sigma_*} \right)^n p^{-\alpha} ,$$

or

$$p = \left[\int \hat{\sigma}_0^n \, \mathrm{d}t \right]^{\frac{1}{1+\alpha}} . \tag{170.2}$$

Here we take $\hat{\sigma}_{ij} = \sigma_{ij}/\sigma_*$. Assume that the stress is distributed linearly across the thickness so that

$$\hat{\sigma}_{ij} = \hat{M}_{ij}\frac{z}{h}. \tag{170.3}$$

The bending and twisting moments in the plates are expressed as

$$M_{ij} = \tfrac{2}{3}\sigma_* h^2 \hat{M}_{ij}. \tag{170.4}$$

In accordance with Kirchoff's hypothesis

$$e_{ij} = \hat{\kappa}_{ij}\frac{z}{h}. \tag{170.5}$$

Here

$$\hat{\kappa}_{ij} = \hat{w}_{,ij},$$

and $\hat{w}$ is the dimensionless deflection $\hat{w} = w/h$.

Formulate the expression for the functional K. To within a constant factor

$$K = \int_F \Bigg[\dot{\hat{M}}_{ij}\dot{\hat{\kappa}}_{ij} - \frac{1+\nu}{2}\frac{\sigma_*}{E}\left(\dot{\hat{M}}_{ij}\dot{\hat{M}}_{ij} - \frac{\nu}{1+\nu}\dot{\hat{M}}_{ss}\dot{\hat{M}}_{ss}\right)$$

$$- \frac{3R}{2(1+\alpha)}\hat{M}_0^{n-1}\left(\int_0^t \hat{M}_0^n \,\mathrm{d}t\right)^{-\frac{\alpha}{1+\alpha}}\hat{\bar{M}}_{ij}\dot{\hat{M}}_{ij} - \dot{\hat{w}}\dot{\hat{q}}\Bigg]\mathrm{d}x\,\mathrm{d}y\,. \tag{170.6}$$

Here

$$\hat{\bar{M}}_{ij} = \hat{M}_{ij} - \tfrac{1}{3}\delta_{ij}\hat{M}_{ss}, \quad \hat{q} = \frac{q}{\sigma_*}, \quad R = \frac{3}{2 + \dfrac{n}{1+\alpha}}.$$

Vary this expression considering $\delta\dot{\hat{M}}_{ij}$ and $\delta\dot{\hat{\kappa}}_{ij} = \delta\dot{\hat{w}}_{,ij}$ independently to obtain

$$\int_F (\dot{\hat{M}}_{ij}\dot{\hat{w}}_{,ij} - \dot{\hat{w}}\dot{\hat{q}})\,\mathrm{d}x\,\mathrm{d}y = 0\,, \tag{170.7}$$

$$\dot{\hat{\kappa}}_{ij} = (1+\nu)\frac{\sigma_*}{E}\left[\dot{\hat{M}}_{ij} - \frac{\nu}{1+\nu}\delta_{ij}\dot{\hat{M}}_{ss}\right]$$

$$+\frac{3R}{2(1+\alpha)}\hat{M}_0^{n-1}\left[\int_0^t \hat{M}_0^n \, \mathrm{d}t\right]^{-\frac{\alpha}{1+\alpha}} \hat{\overline{M}}_{ij}\,. \tag{170.8}$$

From eq. (170.7), in the usual way, we derive the equation of equilibrium and the boundary conditions and eq. (170.8) can be converted in the same way as was done in the case of bending of a beam. Put

$$\hat{\kappa}_{ij}^p = \hat{\kappa}_{ij} + (1+\nu)\frac{\sigma_*}{E}\left[\hat{M}_{ij} - \frac{\nu}{1+\nu}\delta_{ij}\hat{M}_{ss}\right],$$

and form the invariant

$$\kappa_0^2 = \tfrac{4}{3}\,[\kappa_{11}^2 + \kappa_{22}^2 + \kappa_{11}\kappa_{22} + \kappa_{12}^2]\,.$$

By eq. (170.8) we find

$$\dot{\hat{\kappa}}_0^p = \frac{R}{1+\alpha}\hat{M}_0^n\left[\int_0^t \hat{M}_0^n \, \mathrm{d}t\right]^{-\frac{\alpha}{1+\alpha}}.$$

The structure of this equation is fully similar to that of eq. (170.6), therefore, it follows that

$$\int_0^t \hat{M}_0^n \, \mathrm{d}t = \frac{\kappa_0^p}{R}\,.$$

We can now write eq. (170.8) in the form

$$\dot{\hat{\kappa}}_{ij}^p = \frac{3R^{1+\alpha}}{2(1+\alpha)}(\hat{\kappa}_0^p)^{-\alpha}\hat{M}_0^{n-1}\hat{\overline{M}}_{ij}\,. \tag{170.9}$$

Whence it follows that

$$\dot{\hat{\kappa}}_0^p = \frac{R^{1+\alpha}}{1+\alpha}(\hat{\kappa}_0^p)^{-\alpha}\hat{M}_0^n\,.$$

In structure, eq. (170.9) is exactly similar to the initial expression for the work-hardening law (170.1).

The assumption that the stress distribution is linear across the section predetermined the structure of these equations but introduced some distortion. In the case of an elastic deformation that tends to zero, eqs. (170.9) should coincide with the equations of quasi-steady-state creep, in this case the hypothesis of normal elements leads to a completely definite, non-linear stress distribution across the thickness of the plates. Putting $\kappa_{ij}^{p} \approx \kappa_{ij}$ we readily arrive at formulae which differ from (170.9) only in the value of R, namely

$$R = \tfrac{1}{2}\left(2 + \frac{1+\alpha}{n}\right). \tag{170.10}$$

Therefore, we obtain more accurate results by using the corrected formulae (170.9) in which R is given by (170.10).

171. A general device for investigating transient creep on the basis of strain-hardening theory

We have started from a mixed variational principle and used a certain unchanged stress distribution. For the case of bending of beams and plates we derived expressions that directly relate plastic deformation, rate of plastic deformation and bending moments. These relationships were corrected so as to be valid in the limiting case of large creep deformations. An analogous procedure can be extended to arbitrary systems in a state of transient creep.

Consider a body that is subject to generalised forces Q_l $(l = 1, 2, ..., k)$. The corresponding generalised displacements are q_l. The elastic displacements are linearly expressed in terms of forces, therefore,

$$q_l^p = \beta_{li} Q_i\,. \tag{171.1}$$

Put

$$p_l = q_l - \beta_{li} Q_i\,. \tag{171.2}$$

The quantities p_l are displacements resulting from creep. If $p_l = q_l$ and $Q_l =$ const, then with a power work-hardening law quasi-steady-state creep is realised in the body. This means that the stress distribution remains unaltered and the same as in steady-state creep whereas the displacements p_l vary in proportion to $t^m[m = (1/1+\alpha)]$. Construct the corresponding relationship between p_l and Q_l. Consider first steady-state creep and to be specific select the power creep law with exponent r, so that

$$\Phi = \frac{1}{r+1} s^{r+1} .$$

Here s is a uniform first power function of stress. Assume that the corresponding problem of steady-state creep is solved. The stresses are expressed as functions of the external forces as follows:

$$\sigma_{ij} = \sigma_{ij}^l Q_l .$$

(The summation is performed with reference to the subscript l.) The rate potential for the entire body is expressed as:

$$\widetilde{\Phi} = \int \Phi(\sigma_{ij})\, \mathrm{d}V .$$

Differentiating this expression with respect to Q_l, we obtain

$$\frac{\partial \widetilde{\Phi}}{\partial Q_l} = \int \frac{\partial \Phi}{\partial \sigma_{ij}} \sigma_{ij}^l \, \mathrm{d}V .$$

Multiply by Q_l and sum with respect to the subscript l; we obtain

$$\frac{\partial \widetilde{\Phi}}{\partial Q_l} Q_l = \int \frac{\partial \Phi}{\partial \sigma_{ij}} \sigma_{ij} \, \mathrm{d}V = (1+r) \int \Phi \, \mathrm{d}V = (1+r) \widetilde{\Phi} .$$

Therefore $\widetilde{\Phi}$ is a function of Q_l of power $1+r$ which can always be represented in the form

$$\widetilde{\Phi} = \frac{1}{1+r} Q^{r+1} . \tag{171.3}$$

Here Q is a homogeneous first power function of Q_l. By definition the rate potential is

$$\dot{p}_l = Q^r \frac{\partial Q}{\partial Q_l} . \tag{171.4}$$

The quantities $\partial Q/\partial Q_l$ are homogeneous, zero-power functions of Q_l and therefore, the k quantities $\partial Q/\partial Q_l$ are expressed in terms of $(k-1)$ ratios, for example, $Q_1/Q_k, Q_2/Q_k, \ldots$. Consequently the following identical relationship exists between them:

$$f\left(\frac{\partial Q}{\partial Q_l}\right) = 0 .$$

By eq. (171.4) we find that the quantity Q^r is a function of $\dot{p}_l$:

$$Q^r = \varphi(\dot{p}_l) , \tag{171.5}$$

where $\varphi(\dot{p}_l)$ is a uniform first power function. Put

$$p = \int_0^t \varphi(\dot{p}_l)\,\mathrm{d}t .$$

Substituting the expression for $\dot{p}_l$ by formula (171.4) into eq. (171.5) we find

$$\varphi\left(\frac{\partial Q}{\partial Q_l}\right) \equiv 1 . \tag{171.6}$$

Now assume that the creep deformation, the creep rate and the forces are associated by the following relationships:

$$\dot{p}_l = \frac{1}{1+\alpha} p^{-\alpha} Q^n \frac{\partial Q}{\partial Q_l} . \tag{171.7}$$

Hence, because of expression (171.6) we obtain

$$\dot{p} = \frac{1}{1+\alpha} p^{-\alpha} Q^n . \tag{171.8}$$

If the forces that act on the body do not alter (171.8) can be integrated, giving

$$p = Q^{\frac{n}{1+\alpha}} t^{\frac{1}{1+\alpha}}$$

and

$$p_l = p \frac{\partial Q}{\partial Q_l} .$$

This solution corresponds to quasi-steady-state creep. Here the stress distribution is the same as in steady-state creep with exponent $r = n/(1+\alpha)$ and this is the value of the exponent that must be selected in constructing the function Q. If the strain hardening law is selected in the form of eq. (170.1) the constant σ_r in the creep law, which is adopted in solving the auxiliary problem of steady-state creep, should be equal to the constant σ_* in eq. (170.1).

The situation is much simplified if creep is not accompanied by work-hardening, then in place of eq. (171.7) we obtain

$$\dot{q}_l - \beta_{li}\dot{Q}_i = Q^n \frac{\partial Q}{\partial Q_l}. \qquad (171.9)$$

If we desire to reproduce the actual form of the primary creep curves, from the standpoint of the flow theory we should replace the time t by the modified time τ. The equations for creep in shells in the work [116] were formulated in this way.

If the external forces are constant, as we have already seen, eqs. (171.7) immediately give the quasi-steady-state creep which accompanies the instantaneous elastic deformation. It is here, of course, that the equations are inconsistent, the situation is what it would be if elastic deformation were followed by instantaneous stress redistribution which caused no further elastic displacement. All the same these equations can be used to find fully satisfactory approximate solutions for relaxation problems and for redistribution of reactions in statically indeterminate systems subjected to loads. However, practical solutions can only be found by numerical methods.

172. Application of ageing (time-hardening) theory to transient creep calculations

In the earlier sections of this chapter we have considered some numerical methods of solving transient creep problems and the application of variational principles to these problems. The examples were fairly simple and whilst, in principle, steady-state creep of more complicated systems can be studied in a similar way the amount of calculation becomes very great. The principle of these calculations should be clear from what has been said, the calculation is made in steps of time, the amount of creep deformation that is added at each step depends on the creep rate calculated from the values of stresses and accumulated creep deformation from the previous steps. The general procedure of these calculations is given in the work [62]. Examples of application to problems of plates and shells are contained in the articles [287, 330]. However, it is unlikely that transient creep calculations by the strain-hardening theory will soon come into fairly general use in design practice, except perhaps for relaxation problems. Indeed, the general scheme of design with transient creep under given loads is that the initial stress distribution, which depends on the elastic and instantaneous plastic properties, gradually alters and approximates to the condition of steady-state or quasi-steady-state creep. An accurate description of the stress redistribution is usually not so important. Of course, the dependence of displacements on time is to some extent governed by the stress redistribution, but this influence is not very great. Therefore, in determining total displacements of points on the system it usually suffices to use the equations of steady-state or quasi-steady-state creep. The situation is different in relaxation problems, although it is stress redistribution and the corresponding reduction in the reactions with time which are of primary interest.

Because of the scatter of experimental results there is little point in striving for accurate solutions, and so there is good reason to make extensive use of the time-hardening theory in transient creep calculations. The error that results from using this theory is greatest in relaxation problems but even then it can be used for approximate assessment. As was shown in section 72, creep calculations by the time-hardening theory are performed in just the same way as calculations by theories of the plastic deformation type; isochronous creep curves constructed in coordinates of $\sigma_0 - e_0$ correspond to the plasticity diagram, the calculation is repeated for each instant of time. We shall not give the various methods of creep calculation by the time-hardening theory, for this would be a simple paraphrase of the corresponding solutions in the plasticity theory, to which a considerable literature has been devoted. The special feature of these calculations is that, as a rule, the isochronous creep curves are used directly for solution without any analytical approximation. It is a major advantage of the time-hardening theory that this is possible, because the inevitable error that arises in fitting experimental curves to some analytical expression often annuls the advantages of the better physical justification and internal consistency of the creep theory.

The specific feature of creep calculation methods compared with methods of the theory of plasticity is essentially in the choice of objects. Most of the work in this field relates to turbine components; pipe-work, discs and diaphragms. Several variants of the method of successive approximations are widely used for discs. One of these variants is described in section 127 in application to steady-state creep, and it is transferred without change to transient creep using the strain-hardening theory if by the quantities ϵ_θ and

ϵ_r we understand not rates of deformation but the deformations themselves, and if as the functions $\nu(\sigma_0)$ we take the functions $e_0(t, \sigma_0)$, that figure in the equations of the isochronous creep curve for the time t:

$$e_0 = e_0(t, \sigma_0) .$$

If temperature is also allowed for in making the calculations a series of isochronous curves is required for different temperatures,

$$e_0 = e_0(t, \sigma_0, T) .$$

The idea of another method of calculating discs, described in [1], is that a differential equation is formulated for the radial displacements. This is a second order equation that can be written in the form of a linear equation, though in actual fact the coefficients depend in a fairly complicated nonlinear manner on the function that is to be determined. In calculating each successive approximation these coefficients are expressed in terms of a function found in the previous approximation, and therefore, each time it is necessary to integrate a linear equation with variable coefficients. This process converges fairly slowly and the method is not particularly good for manual calculation. However, the author has developed a calculation programme for the Strela computer which is fairly convenient and which is now used by a number of organisations. Boguslavskiy [7] proposed a method of designing turbine diaphragms by successive approximations. Our presentation of the application of ageing theory to transient creep calculations is confined to these few remarks. The calculated results have frequently been satisfactorily compared with experimental data. Mention may be made of one experimental investigation [11] on the possibility of modelling creep on the basis of the time-hardening theory. According to this theory, in order to find the states of stress and deformation of a body at a time t, we require only one isochronous curve that corresponds to the time t; the stress and displacement distributions are completely independent of the previous history of the specimen. Now suppose that there are two different materials and that we have isochronous creep curves at the time t_1 for the first, and at the time t_2 for the second. Then for models of these materials which are geometrically similar and similarly loaded the displacement at time t_1 for the first material and time t_2 for the second are in a certain ratio which is easily found from the conditions of similarity. It was found that the $\sigma_0 - e_0$ curve for steel grade 20 at 480° and $t_1 = 300$ hours is similar to the same curve for copper at 250°C and $t_2 = 150$ hours, although the curves are not similar at other instants of time.

Tests were made on annular plates, half-rings, models of nozzle diaphragms and springs. For plates the ratio of the deflection of a copper plate at 150°C to a steel plate at 300°C (with allowance for scale) was 0.96, for a half-ring the average value of the ratio was 0.75–0.76 for a model diaphragm 1.20 and for a spring 0.82. It will be seen that the best agreement was obtained for plates, which have the simplest shape, but even on other models the agreement must be considered satisfactory, because the scatter of properties of individual specimens is of the same order.

These results provide indirect confirmation of the applicability of the time-hardening theory to practical calculations.

Of course the conclusion that the state at the instant t is independent of all previous states, that follows from the ageing hypothesis is only valid because we are considering sufficiently smooth and monotonic processes.

173. Determination of time to failure

As was shown in Chapter 6, depending on the material, the stress and the temperature, failure due to creep may be brittle and occur at very small creep deformations, or viscous and accompanied by considerable change in the shape and size of the product. In the same way as the time to viscous failure of a rod was determined by the method of section 85 we may set the problem of determining the state of any body under creep conditions at which equilibrium becomes unstable and failure occurs. Certain problems of this kind will be considered in Chapter 12 where it is necessary to consider finite deformations and to allow for geometrical non-linearity. Calculations of this kind are of limited practical value; as a rule the object becomes unsuitable for its intended function if the deformation exceeds a certain value, which depends on the application. Modern turbine construction materials are characterised by small deformation at the instant of failure and by brittle type failure. Therefore, for example the strength of a turbine disc should be assessed in two ways. Firstly, in the case under consideration the change in external diameter should not exceed a certain rational limit; secondly, there should be a guarantee against failure in the normal meaning of the word. The simplest procedure for calculating the long-term strength of a product is to use one of the creep theories to calculate the maximum normal stress which is compared with the long-term ultimate strength for the time of operation of the disc, found by testing specimens. This method is particularly suitable for turbine discs because both the principal stresses are positive and consequently the criterion of maximum normal strength coincides with the criterion of maximum tangential stress, which is approximately valid for assessing the possibility of viscous failure.

The calculated maximum normal stress varies with time and therefore the long-term ultimate strength must be compared with the equivalent stress determined by the principle of damage summation, namely

$$\sigma_e = \left(\frac{1}{t}\int_0^t \sigma^k \, dt\right)^{1/k} . \tag{173.1}$$

It should be noted that here the long-term strength diagram plotted in coordinates of $\log \sigma - \log t$ need not be strictly linear. We compare the quantity σ_e with the true long-term strength curve whereas in formula (173.1) the quantity k is found by rather crude linearisation of the interesting part of the diagram. As we have seen, stress distribution takes place fairly rapidly in turbine discs operating under steady-state conditions and σ_e differs by only a few percent from the value σ that corresponds to the steady-state conditions. Therefore, error in the determination of the exponent k is not very important.

In order to check this method of predicting the life of discs an experimental study was made [95] on model discs of four materials widely used in turbine manufacture; pearlitic steel grade R-3 at a temperature of 575°C, austenitic steels EI-405 and EI-572 at 650°C and heat-resistant alloy EI-437B at 700°C. The test temperatures approximate to those of typical service. All the materials were given the standard heat treatment and in addition, the steel R-3 was tested in the brittle state.

The tests were made in the overspeed test equipment of TsNIITMASH (Central Scientific Research Institute of Engineering Technology), the disc models were flat with a central hole. The external diameter of the models was 280 mm, some of the models were notched on the rim to imitate the blade load.

The tests were interrupted periodically to measure the change in external diameter; most of the models were tested to failure, some were removed from the test and cut to determine the residual stresses. From the magnitude of the residual stresses the stresses under creep conditions were calculated and compared with calculated values. Altogether 25 discs were tested, the test duration ranged from several hours to 2000 hours. Creep tests were made on about 200 specimens cut from different parts of different forgings and in different directions, so as to obtain reasonably reliable mean isochronous creep curves.

These isochronous curves were used for disc calculations on the basis of time-hardening theory. The calculations showed that the stress redistribution indeed terminates very quickly and the stresses remain unchanged throughout most of the life of the discs. The value of σ_e calculated by formula (173.1) did not differ by more than 3% from the stress in the steady-state condition. The calculated values of residual stress were in good agreement with values determined experimentally, the difference between calculated and experimental values of deformation on the outer and inner radii of the disc did not exceed the limits of experimental scatter.

The times to failure were worked out in the following way. The time to failure was determined for each disc, and for this time the failure stress σ_f was determined from the long-term strength curve, found by testing standard specimens. Then the maximum normal stress in the disc σ_{max} was determined and the ratio σ_{max}/σ_f was calculated.

The value of σ_{max} was found in three different ways:

1. On the assumption that the behaviour of the material is elastic until the instant of failure.

2. On the assumption that the stress distribution is the same as that in the ideal-plastic case, i.e. $\sigma_\theta = \text{const.}$

3. With allowance for creep on the basis of time-hardening theory.

The first method is clearly incorrect, but in design office practice, until quite recently, the influence of creep on stress redistribution was not taken into account. It was of interest to assess the additional safety factor introduced by this method of calculation.

The second assumption corresponds to the limiting case of the power creep law when $n \to \infty$. This kind of approximation was frequently used in

Material	σ_{max}/σ_f			Material	σ_{max}/σ_f		
	1	2	3		1	2	3
Steel R-3				Steel EI-405	1.92	0.96	1.05
standard	1.78	0.89	0.95		1.83	0.94	1.04
heat	1.85	0.92	1.00		1.87	0.96	1.06
treatment	1.67	0.83	0.94	Steel EI-572	1.52	0.81	0.95
Steel R-3	1.32	0.72	0.96	Alloy EI-437B	1.57	0.82	1.02
brittle	1.35	0.74	0.98		1.38	0.72	0.93
	1.33	0.73	0.97		1.44	0.82	0.95
	1.33	0.73	0.97		1.52	0.86	0.97
Steel EI-405	1.89	0.93	1.03		1.58	0.89	1.01
	1.92	0.96	1.05				

Chapter 6, for large values of n it gives a reasonable result. The extreme simplicity of this method of calculation is adequate justification for it.

The results of failure tests are given in the table on page 714.

It will be seen from the table that calculation with allowance for creep gives very satisfactory agreement with experiment. For Steel EI-405 calculation by the second method gives about the same percentage error (on the other side) as calculation with allowance for creep. But discs of this steel fail with a large deformation, of the order of 12%, when on the one hand the equations of creep theory of the ageing type, which are derived on the assumption that the deformation is small, cease to be accurate and on the other the type of failure can hardly be considered brittle. For materials of this kind calculation by the mean stress gives a satisfactory assessment.

174. **More accurate determination of the time to brittle failure of a rotating disc**

The usual method of strength calculation under creep conditions takes no account of the circumstance that failure is preceded by crack formation, which in its turn influences the rate of creep and causes stress redistribution (see section 88). This is partly allowed for in calculations by the time-hardening theory because the entire creep curves, including the third section, are used in constructing the isochronous curves. On the basis of the considerations that were given in section 94, a scheme for calculating the life of a product can be proposed in which this circumstance is fully allowed for. There are certain difficulties in realising this scheme, and the example given below is more in the nature of an illustration of the formulation of the problem, which, in this case, is solved to the end. Following the presentation of section 94, we write the creep equation for a rotating disc on the assumption that $\sigma_\theta > \sigma_r > 0$:

$$\dot{e}_\theta = -\dot{e}_z = A\left(\frac{\hat{\sigma}_r}{1-\omega}\right)^n, \quad \dot{e}_r = 0, \tag{174.1}$$

and the equation of embrittlement is

$$\dot{\omega} = \left(\frac{\hat{\sigma}_\theta}{1-\omega}\right)^n. \tag{174.2}$$

Here the units of stress measurement are selected such that $c = 1$, which means that the stresses are referred to the quantity $c^{1/n}$, $\hat{\sigma} = \sigma/c^{1/n}$.

Introduce the dimensionless radius of the disc ξ so that on the periphery of the hole $\xi = \alpha$ and on the outer periphery $\xi = 1$. Denote the outer radius of the disc by b and the radial displacement by u and take

$$\hat{u} = \frac{u}{Ab}\,.$$

Then from eq. (174.1) it follows that

$$\frac{\dot{\hat{u}}}{\xi} = \left(\frac{\hat{\sigma}_\theta}{1-\omega}\right)^n, \quad \frac{\partial \dot{\hat{u}}}{\partial \xi} = 0\,. \tag{174.3}$$

The second equation shows that u is a function of time, comparing the first with eq. (174.2) we note that

$$\omega = \frac{\hat{u}}{\xi}\,,$$

because when $t = 0$, $\hat{u} = 0$, $\omega = 0$. Now from the first of eqs. (174.3) it follows that:

$$\hat{\sigma}_\theta = \left(1 - \frac{\hat{u}}{\xi}\right)\left(\frac{\dot{\hat{u}}}{\xi}\right)^{1/n}.$$

The equation of equilibrium of an element of the disc is

$$\frac{\partial}{\partial \xi}(\xi\hat{\sigma}_r) = \hat{\sigma}_\theta - m\xi^2 \qquad (m = \rho v^2 c)\,. \tag{174.4}$$

Here ρ is the density of the material, v the peripheral speed of the disc. Substitute the expression for $\hat{\sigma}_\theta$ into the equation of equilibrium (174.4) and integrate, we obtain

$$\hat{\sigma}_r = \dot{\hat{u}}^{1/n}\left[\frac{n}{n-1}\,\xi^{-1/n} + nu\xi^{-1-\frac{1}{n}}\right] - \frac{m\xi^2}{3} + \frac{c}{\xi}\,.$$

Now use the boundary conditions $\hat{\sigma}_r(1) = \hat{\sigma}_r(\alpha) = 0$. Eliminating the constant we obtain the following differential equation for the function $\hat{u}(t)$:

$$\dot{\hat{u}} = \left(\frac{mk(\alpha)}{1 - \hat{u}\rho(\alpha)}\right)^n. \tag{174.5}$$

Here

$$k(\alpha) = \frac{1}{3}\frac{n-1}{n}\,\frac{1-\alpha^3}{1-\alpha^{1-\frac{1}{n}}},$$

$$\rho(\alpha) = (n-1)\,\frac{\alpha^{-\frac{1}{n}}-1}{1-\alpha^{1-\frac{1}{n}}},$$

Integrate eq. (174.5) with the initial condition $\hat{u}(0) = 0$. The result is

$$\hat{u} = \frac{1}{\rho(\alpha)}\left[1-\left(1-\frac{t}{\tau}\right)^{\frac{1}{1+n}}\right]. \tag{174.6}$$

Here we take

$$\frac{1}{\tau(\alpha)} = \rho(\alpha)\,[mk(\alpha)]^n\,(1+n)\,.$$

At a certain instant of time t_* on the periphery of the hole when $\xi = \alpha$ we have $\omega = 1$, and consequently, $\hat{u} = \alpha$. From eq. (174.6) we find

$$t_* = \tau[1-(1-\alpha\rho)^{n+1}]\,.$$

It is readily shown directly that $\rho\alpha < 1$, if $\alpha < 1$. Bearing in mind eq. (174.5) we can now write the stress formulae

$$\sigma_\theta = \frac{mk}{1-\hat{u}\rho}\left(1-\frac{\hat{u}}{\xi}\right)\xi^{-1/n},$$

$$\sigma_r = \frac{mn}{n-1}\,\frac{1}{\xi}\left(1-\xi^{1-\frac{1}{n}}\right)\left(k(\xi)-k\,\frac{1-\hat{u}\rho(\xi)}{1-\hat{u}\rho}\right), \tag{174.7}$$

whilst the quantity $\hat{u}$ as function of t is given by formula (174.6). In formulae (174.7) we take $\rho = \rho(\alpha)$, $k = k(\alpha)$. The solution given by these formulae is suitable provided that the initial assumption $\sigma_\theta/(1-\omega) > \sigma_r$ is fulfilled. It is known that in the case of creep without embrittlement the corresponding solution obtained on the assumption that $\sigma_\theta > \sigma_r$ is valid when $\alpha > 0.25$ for any value of n [378]. In our problem such a stress distribution is obtained

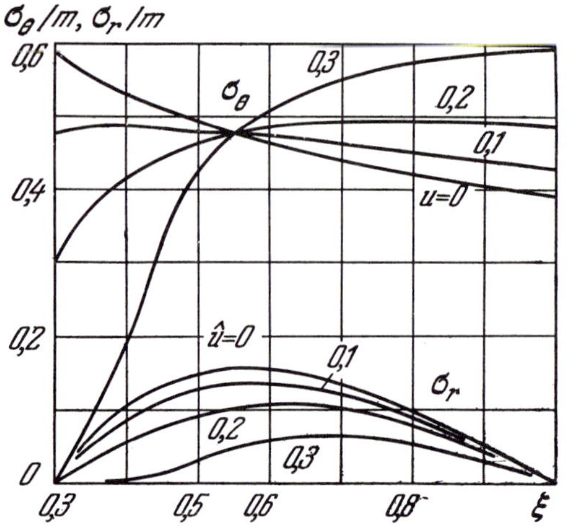

Fig. 239.

when $t = 0$ and there is still no embrittlement. There is reason to suppose that the required inequality is also fulfilled at subsequent instants of time. Stress distribution diagrams when $n = 3$ and $\alpha = 0.3$ for values of $\hat{u} = 0, 0.1, 0.2, 0.3$ are given in fig. 239. At the instant when $\hat{u} = \alpha = 0.3$ consequently $t = t_*$, the stress σ_θ on the bore falls to zero because here $\omega = 1$.

When $t > t_*$ three regions are formed in the disc. In the first region which is adjacent to the edge of the hole, when $\alpha < \xi < \gamma$ total radial cracking has occurred, $m = 1$, and $\sigma_\theta = 0$. From the equation of equilibrium (174.4) and the boundary condition $\sigma_r(\alpha) = 0$, it follows that:

$$\sigma_r = \frac{m}{3}\left(\frac{\alpha^3}{\xi} - \xi^2\right), \quad \sigma_\theta = 0, \quad \alpha < \xi < \gamma. \tag{174.8}$$

Since when $\xi = \gamma$, $\sigma_r < 0$, there should be an annular second region when $\gamma < \xi < \beta$, in which $\sigma_\theta > 0$, $\sigma_r < 0$; consequently in this region $e_\theta = -e_r$, whence

$$\frac{\partial \hat{u}}{\partial \xi} + \frac{\hat{u}}{\xi} = 0, \qquad \hat{u} = \frac{w(t)}{\xi}.$$

Here the creep equation is

$$\dot{e}_\theta = -\dot{e}_r = A\left(\frac{\sigma_\theta}{1-\omega} - \sigma_r\right)^n.$$

But in the second region the value of ω is close to unity and, therefore, the first term in brackets is predominant and we may take approximately that:

$$\dot{e}_\theta = -\dot{e}_r = A\left(\frac{\sigma_\theta}{1-\omega}\right)^n .$$

Hence bearing in mind the expression obtained for $\hat{u}$ in the second region we obtain

$$\frac{\dot{\omega}}{\xi} = \left(\frac{\sigma_\theta}{1-\omega}\right)^n = \dot{\omega} .$$

When $t = t_*$ we take $w = 0$, at this instant $\omega = \alpha/\xi$, therefore

$$\omega = \frac{w}{\xi^2} + \frac{\alpha}{\xi} .$$

Now the stress σ_θ is expressed as

$$\sigma_\theta = \left(1 - \frac{w}{\xi^2} - \frac{\alpha}{\xi}\right)\dot{w}^{1/n}\xi^{-2/n} . \qquad (174.9)$$

Substitute expression (174.9) into the equation of equilibrium (174.4) and integrate it, bearing in mind the boundary conditions

$$\sigma_r(\gamma) = \frac{m}{3}\left(\frac{\alpha^3}{\gamma} - \gamma^2\right), \qquad \sigma_r(\beta) = 0 .$$

As a result we obtain the following differential equation for the function $w(t)$:

$$\dot{w} = \left(\frac{m\varphi}{1 + w\psi}\right)^n . \qquad (174.10)$$

Here

$$\varphi = \frac{1}{3}\,\frac{\alpha^3 - \beta^3}{\dfrac{n}{n-2}\left(\gamma^{1-\frac{2}{n}} - \beta^{1-\frac{2}{n}}\right) + \dfrac{\alpha n}{2}\left(\gamma^{-2/n} - \beta^{-2/n}\right)} ,$$

$$\psi = \frac{n}{n+2}\,\frac{\gamma^{-1-\frac{2}{n}} - \beta^{-1-\frac{2}{n}}}{\frac{n}{n-2}\left(\gamma^{1-\frac{2}{n}} - \beta^{1-\frac{2}{n}}\right) + \frac{\alpha n}{2}(\gamma^{-2/n} - \beta^{-2/n})}\,.$$

In the third region when $\beta < \xi < 1$ again $\sigma_r = 0$ and therefore the solution is of the same form as in the first phase of failure, only the boundary condition $\sigma_r = 0$ is fulfilled when $\xi = \beta$. Consequently the function $\hat{u}(t)$ is determined in this region by eq. (174.5), in which the constant quantity α is replaced by the time function β, namely

$$\dot{\hat{u}} = \left(\frac{mk(\beta)}{1 - u\rho(\beta)}\right)^n . \tag{174.11}$$

Now write the conditions on the boundaries of the region. When $\xi = \gamma$ we have $\omega = 1$, consequently

$$w = \gamma^2 - \alpha\gamma\,. \tag{174.12}$$

When $\xi = \beta$ the peripheral deformation, and consequently ω, are continuous, therefore

$$\hat{u} = \alpha + \frac{w}{\beta}\,. \tag{174.13}$$

We have thus obtained a system of four eqs. (174.10)–(174.13) to determine four time functions: $\hat{u}(t)$, $w(t)$, $\gamma(t)$ and $\beta(t)$. When $\beta(t) = 1$ the third zone disappears, consequently there remain only eqs. (174.10) and (174.12); in the first of these we put $\beta = 1$. The instant at which $\gamma(t) = 1$ is the instant of failure of the disc and everywhere $\omega = 1$.

The time to failure can be assessed approximately as follows. Neglect the size of the second zone, that is put $\beta = \gamma$. Then from eqs. (174.12) and (174.13) it follows that $\hat{u} = \beta$, and only eq. (174.11) remains, from which the duration of the second phase of failure is determined in the following way:

$$t_{**} = \frac{1}{m^n}\int_{\alpha}^{1}\left(\frac{1 - u\rho(u)}{k(u)}\right)^n \mathrm{d}u\,. \tag{174.14}$$

The following values were calculated for the case $n = 3$, $\alpha = 0.3$, to which the stress diagrams in fig. 239 correspond:

$$\tau = \frac{2.32}{m^3}, \qquad t_* = \frac{2.22}{m^3}, \qquad t_{**} = \frac{0.144}{m^3}.$$

As might be expected, the second phase of failure is of short duration, when $\omega = 1$ has been reached on the bore failure takes place at catastrophic speed.

12

Geometrically Non-Linear Problems of Creep

Just as an elastic or elastic-plastic system may lose stability under certain conditions of external loading, so that slight changes in the parameters of the system cause very large displacements, similarly under creep conditions rates of displacement may become impermissibly great. The critical state, in this narrow meaning of the term, may occur after the body has been subjected to load for a certain time and accordingly the concept of critical time is introduced. In this chapter we consider certain simple problems of a fundamental nature which serve to explain the essence of the problem. Methods of solving more complicated problems are also indicated.

175. Simple geometrically non-linear creep problems

In this chapter we consider creep problems in which displacements are not small and change in geometry of the body during creep must be taken into account. Creep in metals is unlimited in the sense that however small the stress, after a sufficiently long time the deformation becomes indefinitely great and, therefore, any system is unstable under creep conditions. The change in geometrical dimensions that is associated with creep leads to a change in the effective stresses even though the external forces remain constant and so the creep rate increases and either displacements or rates of displacements tend to infinity in a finite time, which is called the critical time.

In analysing schemes of brittle failure in section 85 we encountered an example of determination of the critical time of a rod in tension. Other simple problems may be considered in a similar way, for example, determination of the critical time for thin-walled tubes. Here, as in section 85, we shall consider that the law of steady-state creep is valid for any deformation, however large. This assumption necessarily leads to the use of a logarithmic measure of deformation.

Assume that in the initial state the length of the tube is l_0, its radius r_0, the wall thickness δ_0; during deformation these quantities will be denoted l, r and δ.

From the condition of incompressibility

$$l r \delta = l_0 r_0 \delta_0 \,.$$

Logarithmic deformations in the longitudinal direction, round the periphery and across the thickness are respectively denoted:

$$e_z = \ln \frac{l}{l_0}, \qquad e_\theta = \ln \frac{r}{r_0}, \qquad e_r = \ln \frac{\delta}{\delta_0}\,.$$

Consider the following cases separately:

1. *An internal pressure of q acts on the tube.* The tube is not loaded in the axial direction, consequently $\sigma_\theta = qr/\delta$, $\sigma_z = 0$. Put $\sigma_\theta^0 = qr_0/\delta_0$, then $\sigma_\theta = \sigma_\theta^0 \exp(e_\theta - e_r)$. Since $\sigma_z = \sigma_r = 0$, the creep rates $\dot{e}_r$ and $\dot{e}_z$ are equal to one another. It follows from the condition of incompressibility and the equality of $\dot{e}_r = \dot{e}_z$ that $e_\theta - e_r = \frac{3}{2} e_\theta$; therefore, $\sigma_\theta = \sigma_\theta^0 \exp 3e_\theta/2$. Substitute this stress value into the creep law, which we shall assume to be a power law

$$\dot{e}_\theta = \epsilon_n \left(\frac{\sigma_\theta^0}{\sigma_n}\right) \exp \frac{3ne_\theta}{2}\,.$$

Separating variables and integrating from $e_\theta = 0$ to $e_\theta = \infty$, we find the time to failure

$$\frac{1}{t_k} = \frac{3}{2} n\epsilon_n \left(\frac{\sigma_\theta^0}{\epsilon_n}\right)^n . \tag{175.1}$$

The time to failure is two thirds of the time to failure of a rod with the same nominal stress.

2. *An axial force acts on the tube.* This force develops a nominal stress σ_z^0. The time to failure determined by the Hoff formula is

$$\frac{1}{t_k} = n\epsilon_n \left(\frac{\sigma_z^0}{\sigma_n}\right)^n . \tag{175.2}$$

Now the stress rises only because the wall becomes thinner, whereas in the first case increase in radius also plays a part. This must be borne in mind in interpreting failure tests on tubular specimens. With existing designs of test machine the hoop stress is developed by internal pressure and the axial stress by a combination of internal pressure and an independent axial force.

3. *The ends of the tube are sealed.* Now the load is internal pressure, considerations similar to those given for the first case show that

$$\sigma_\theta = \sigma_\theta^0 \exp(2e_\theta + e_z),$$

$$\sigma_z = \sigma_z^0 \exp(2e_\theta + e_z),$$

$$\sigma_\theta^0 = \frac{qr_0}{\delta_0}, \qquad \sigma_z^0 = \frac{1}{2}\frac{qr_0}{\delta_0}.$$

Because of the creep law

$$\dot{e}_\theta = \epsilon_n \left(\frac{\sigma_0^0}{\sigma_n}\right)^n \frac{1}{\sigma_0^0} (\sigma_\theta^0 - \tfrac{1}{2}\sigma_z^0) \exp[n(2e_\theta + e_z)],$$

$$\dot{e}_z = \epsilon_n \left(\frac{\sigma_0^0}{\sigma_n}\right)^n \frac{1}{\sigma_0^0} (\sigma_z^0 - \tfrac{1}{2}\sigma_\theta^0) \exp[n(2e_\theta + e_z)].$$

Consequently $e_z = 0$, and the first equation assumes the form

$$\dot{e}_\theta = \frac{\sqrt{3}}{2}\epsilon_n \left(\frac{qr_0\sqrt{3}}{2\delta_0\sigma_n}\right)^n \exp 2ne_\theta .$$

Integrating we find

$$\frac{1}{t_k} = n\sqrt{3}\epsilon_n \left(\frac{qr_0\sqrt{3}}{2\delta_0\sigma_n}\right)^n = \frac{3^{\frac{n+1}{2}}}{2^n} n\epsilon_n \left(\frac{\sigma_0^0}{\sigma_n}\right)^n . \tag{175.3}$$

The case of a thin-walled tube has been considered by a number of authors [52, 55] and we shall not consider it further.

176. A flexible cable

Suppose that a flexible cable, stretched between two supports is subjected to a constant normal load q. It is evident that after deformation the cable assumes the shape of a circular arc. The initial length of the cable $l_0 = 2\rho \sin \varphi$, the instantaneous length $l = 2\rho\varphi$ (fig. 240) and therefore, the logarithmic deformation is

$$e = \ln \frac{l}{l_0} = \ln \frac{\varphi}{\sin \varphi} . \tag{176.1}$$

The stress stretching the cable at any instant is

$$\sigma = \frac{q\rho}{F} .$$

But $\rho = l_0/(2 \sin \varphi)$, $Fl = F_0 l_0$, consequently, $F = F_0(l_0/l) = F_0(\sin \varphi/\varphi)$, therefore,

$$\sigma = \frac{ql_0}{2F_0} \frac{\varphi}{\sin^2 \varphi} . \tag{176.2}$$

Calculate the rate of creep from eq. (176.1),

$$\dot{e} = \frac{\sin \varphi - \varphi \cos \varphi}{\varphi \sin \varphi} \dot{\varphi} ,$$

and substitute this expression for $\dot{e}$ and also the value of σ calculated by eq.

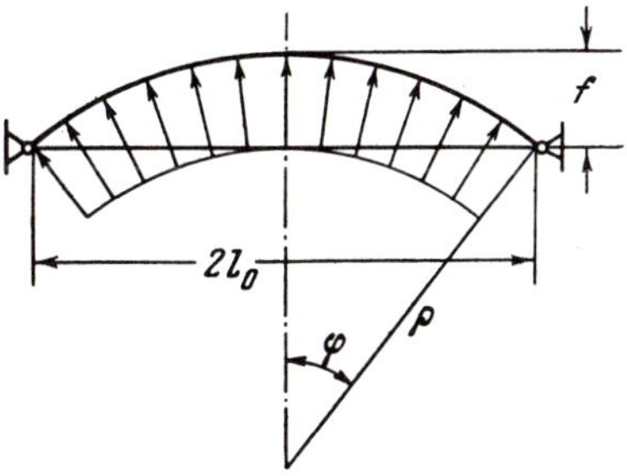

Fig. 240.

(176.2), into the formula of the power creep law, we obtain

$$\dot{\varphi}\,\frac{\sin\varphi-\varphi\cos\varphi}{\varphi\sin\varphi}\left(\frac{\sin^2\varphi}{\varphi}\right)^n=\epsilon_n\hat{q}^n\,. \tag{176.3}$$

Here

$$\hat{q}=\frac{ql_0}{2F_0\sigma_n}\,.$$

Integrating eq. (176.3) we find

$$\epsilon_n\hat{q}^n t=\int_0^{\varphi}\frac{\sin\varphi-\varphi\cos\varphi}{\varphi\sin\varphi}\left(\frac{\sin^2\varphi}{\varphi}\right)^n\mathrm{d}\varphi\,. \tag{176.4}$$

When $\varphi=\pi$, the radius of the cable becomes infinitely great, therefore, putting the upper limit of integration $\varphi=\pi$ in formula (176.4) we obtain the critical time, which is always finite.

The relationship between the dimensionless deflection $\hat{f}=2f/l_0=(1-\cos\varphi)/\varphi$ and the dimensionless time $\hat{t}=\epsilon_n\hat{q}^n t$, found by numerical integration of eq. (176.4) when $n=3$ is given in fig. 241. The critical time is $\hat{t}_{cr}=0.1973$.

It will be seen from the graph that at first the deflection increases very quickly and in assessing the real carrying capacity of a structure which in-

Fig. 241.

cludes a component of this type the critical time is not the governing factor, the life of the structure is assessed from the permissible deflection which is usually small. A simpler solution can be obtained for relatively small deflections. We shall use the usual measure of deformation instead of the logarithmic putting

$$\sin\varphi \approx \varphi - \frac{\varphi^3}{6}, \qquad \cos\varphi \approx 1 - \frac{\varphi^2}{2},$$

then from eq. (176.1) it follows that

$$e \approx \frac{\varphi^2}{6},$$

and from eq. (176.2)

$$\sigma \approx \frac{ql_0}{2F_0}\,\frac{1}{\varphi}.$$

Substituting into the creep equation we obtain

$$\varphi\dot{\varphi} = 3\epsilon_n \hat{q}^n \varphi^{-n}.$$

Hence integrating we find

$$\varphi = [3(n+2)]^{\frac{1}{n+2}} (\hat{q}^n t\epsilon_n)^{\frac{1}{n+2}}.$$

The dimensionless deflection $\hat{f} = \varphi/2$, consequently,

$$\hat{f} = \tfrac{1}{2}\,[3(n+2)\hat{t}]^{\frac{1}{n+2}}. \qquad (176.5)$$

The corresponding curve is shown dotted in fig. 241.

Note that these solutions are also suitable for a membrane in the form of a stretched rectangle loaded with uniform pressure; here the values of the constants used in determination of the dimensionless parameters are altered somewhat. The corresponding derivation is self-evident and we do not give it here.

177. Deflection of beams with hinged fixed ends

The solutions obtained in the preceding section relate to a very thin rod for which stiffness in bending may be neglected. The problem of a beam with fixed ends which cannot be displaced is much more complicated. Let the length of the beam be l and the intensity of the distributed load q. The calculation is based on an ideal I-section model of the beam and the analysis is made for the case $n = 3$. The calculations are similar for other whole-number values of n. Therefore, putting $n = 3$, from eqs. (117.3) we obtain

$$\hat{\dot{\epsilon}} = \frac{\dot{\epsilon}}{\epsilon_n'} = \hat{N}(\hat{N}^2 + 3\hat{M}^2)\,, \quad \hat{\dot{\kappa}} = \frac{h}{\epsilon_n'}\frac{\partial^2 \dot{w}}{\partial x^2} = \hat{M}(3\hat{N}^2 + \hat{M}^2)\,. \quad (177.1)$$

Here we make use of the circumstance that the power creep law leaves a certain freedom in the choice of the constants ϵ_n and σ_n, which may be replaced by two other constants ϵ_n' and σ_n', provided that

$$\frac{\epsilon_n}{\epsilon_n'} = \left(\frac{\sigma_n}{\sigma_n'}\right)^n .$$

Later we shall introduce an additional condition that defines the selection of constants. Introduce the dimensionless coordinate $\xi = x/h$, the dimensionless deflection $\hat{w} = w/h$ and the dimensionless axial displacement $\hat{u} = u/h$. Then

$$\hat{\dot{\kappa}} = \frac{1}{\epsilon_n'}\frac{\partial^2 \dot{\hat{w}}}{\partial \xi^2}\,, \qquad \hat{\dot{\epsilon}} = \frac{1}{\epsilon_n'}\left(\frac{\partial \dot{\hat{u}}}{\partial \xi} + \frac{\partial \hat{w}}{\partial \xi}\frac{\partial \dot{\hat{w}}}{\partial \xi}\right).$$

We shall seek the deflection $\hat{w}$ in the form

$$\hat{w} = -c \sin \lambda\xi\,.$$

Here c is a time function, $\lambda = \pi h/l$, ξ varies in the range from 0 to l/h.

The bending moment in the beam is

$$M = M_1 + Nw\,,$$

where M_1 is the bending moment due to the load q. Hence it follows that:

$$\hat{M} = \hat{M}_1 + \hat{N}\hat{w}\,, \quad \hat{M}_1 = \frac{M_1}{Fh\sigma_n'}\,.$$

The bending moment due to the transverse load is expressed by the formula

$$M_1 = qlx - \frac{qx^2}{2}\,.$$

Substitute for the function $M_1(x)$ the first term of its expansion into a Fourier series

$$M_1 \approx \frac{16}{\pi^3} ql^2 \sin \frac{\pi x}{l} .$$

Consequently

$$\hat{M}_1 = \frac{16}{\pi^3} \frac{ql^2}{Fh\sigma_n'} \sin \lambda\xi .$$

Now select the constant σ_n' such that the factor by which $\sin \lambda\xi$ is multiplied is unity, that is we put

$$\sigma_n' = \frac{16}{\pi^3} \frac{ql^2}{Fh} . \tag{177.2}$$

Now we can write

$$\hat{M} = a \sin \lambda\xi , \qquad a = 1 - c\hat{N} . \tag{177.3}$$

Substitute values of $\hat{\epsilon}$, $\hat{\kappa}$, $\hat{M}$ into eqs. (177.1), we obtain the following system of equations:

$$\lambda^2 \dot{c} \sin \lambda\xi = \epsilon_n' a(3\hat{N}^2 + a^2 \sin^2 \lambda\xi) \sin \lambda\xi ,$$

$$\frac{\partial \dot{\hat{u}}}{\partial \xi} + \lambda^2 c\dot{c} \cos^2 \lambda\xi = \epsilon_n' \hat{N}(\hat{N}^2 + 3a^2 \sin^2 \lambda\xi) .$$

The first equation is satisfied on average, for this purpose multiply both sides by $\sin \lambda\xi$ and integrate, we obtain

$$\lambda^2 \dot{c} = \epsilon_n' a(3\hat{N}^2 + \tfrac{3}{4} a^2) . \tag{177.4}$$

Note that because of the end fixing conditions $\hat{u}(\xi)$ becomes zero at the ends of the beam, integrating the second equation from $\xi = 0$ to $\xi = l/h$; we obtain

$$\lambda^2 c\dot{c} = \epsilon_n' \hat{N}(2\hat{N}^2 + 3a^2) . \tag{177.5}$$

It now only remains to change the time scale so that the equations contain no parameters. Put

$$\hat{t} = \frac{\epsilon_n'}{\lambda^2} t = \frac{\epsilon_n l^2}{\pi^2 h^2}\left(\frac{16}{\pi^3}\frac{ql^2}{Fh\sigma_n}\right)^n t\,.$$

Eliminating $\dot{c}$ from eqs. (177.4) and (177.5) then the resulting equation, together with eq. (177.3), can be used to express a and $\hat{N}$ in terms of c, if $\zeta(c)$ is the root of the cubic equation

$$\zeta^3 - \frac{4}{c}\zeta^2 + 4\zeta - \frac{8}{3c} = 0\,, \tag{177.6}$$

then

$$\hat{N} = \frac{1}{c + \zeta(c)}\,, \qquad a = \frac{\zeta(c)}{c + \zeta(c)}\,, \tag{177.7}$$

and it is only required to integrate the equation

$$\frac{\mathrm{d}c}{\mathrm{d}\hat{t}} = a\,[3\hat{N}^2 + \tfrac{3}{4}a^2]\,, \tag{177.8}$$

with the initial condition $c(0) = 0$. Initially $\hat{N}(0) = 0$, $a(0) = 1$. If the deflections are sufficiently great, ζ is small and it follows from eq. (177.6) that $\zeta \approx 2/3c$, and neglecting small terms we obtain

$$\frac{\mathrm{d}c}{\mathrm{d}\hat{t}} = \frac{2}{c^4}\,.$$

Whence

$$c \approx (10\hat{t})^{1/5}\,.$$

Substituting into this the expression for $\hat{t}$ and proceeding to the deflection we find that the central deflection is

$$\frac{2f}{l} \approx 0.886\left(\frac{ql^2}{2Fh\sigma_n}\right)^{3/5} t^{1/5}\,,$$

whereas when $n = 3$, formula (176.5) gives

$$\frac{2f}{l} = 0.858\left(\frac{ql^2}{2Fh\sigma_n}\right)^{3/5} t^{1/5}\,.$$

The two coefficients differ by about 3.5% because in both cases the solution is approximate but the principles by which they are derived are completely different. Curves of the functions $c(\hat{t})$, $a(\hat{t})$ and $\hat{N}(\hat{t})$ and a curve obtained by numerical integration of eq. (177.8) are given in figs. 242 and 243. Here c was found as a function of ζ by eq. (177.6), and $\hat{N}$ and a and also the

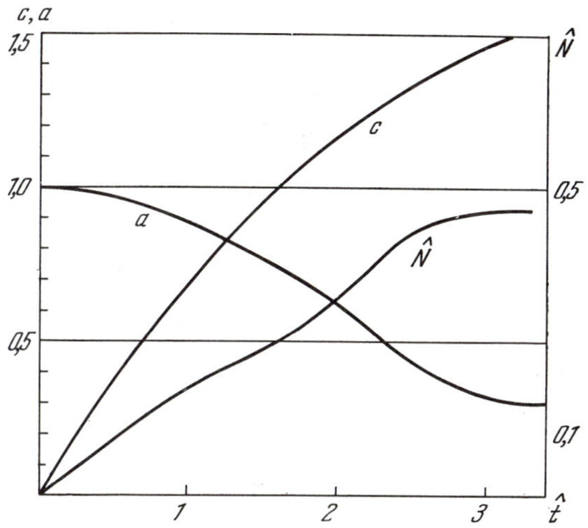

Fig. 242.

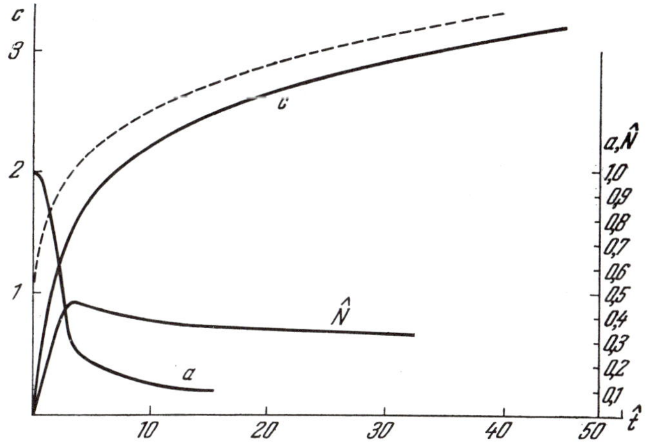

Fig. 243.

right-hand side of eq. (177.8) were calculated for different values of ζ. The dotted curve in fig. 243 is the dependence of deflection on time given by the approximate formula $c = (10\hat{t})^{1/5}$. It will be seen that commencing from about $c = 3$, the part played by deflection becomes quite insignificant, the bending moment a becomes very small and the error in determining the deflection by the asymptotic formula is about 7% when $c = 3$. Therefore, when considering problems of this kind it is not particularly important to allow for deflection. The basic behaviour of the element can be explained if the bending is neglected and can be justified since the solution of the problem of longitudinal-transverse bending is an approximation based on fairly rough assumptions. Here we have taken expressions for the rate of elongation and change of curvature as function of force and moment that apply to an ideal I-section. In section 116 we have seen that there are other possible approximations to these rèlationships and that the solutions given here can be modified in various ways. We shall return to this question in discussing variational principles in non-linear problems of creep theory, and by way of example shall consider another method of solving this problem.

Bending of a beam with fixed ends was also studied by Odquist [317]. The solution given in his work relates only to the initial stage of bending when the longitudinal force is small, the author formulates an equation of the type of eq. (177.1) for an arbitrary value of n, he introduces an expression for $\hat{M}$ which contains the term $\hat{N}\hat{w}$ and expands the right-hand sides with respect to $\hat{N}$, and considers only the terms containing up to the first power of $\hat{N}$. The deflection $\hat{w}$ is resolved into a Fourier series of cosines of odd arguments (the origin of coordinates is taken in the middle of the beam), the right-hand sides of the equations include the functions $\hat{M}_1^n$ and $\hat{M}_1^{n-1}$ which are also resolved into series of cosines; the Fourier coefficients in the deflection expressions are an infinite series of ordinary differential equations which are then truncated. Nevertheless, the system is too complicated, so that finally the author considers only a single term in the Fourier series and obtains an approximate solution which reduces to the following (in our notation):

$$c + \frac{704}{405\pi^2} c^3 = \frac{4l^2}{\pi^2 h^2} \left(\frac{ql^2}{2Fh\sigma_n} \right)^3 t c_1^{(n)} . \qquad (177.9)$$

The coefficient $c_1^{(n)}$ depends on n; values are given in the table on page 733.

An approximate solution of eq. (177.9) is sought in the form of a series of the type

$$c = c_0 t + c_1 t^3 + \dots .$$

n	0	1	2	3	4	5	6	7
$c_1^{(n)}$	1.273	1.032	0.891	0.796	0.726	0.672	0.628	0.626

If we consider only the first term of the series we obtain the case of bending by a single transverse load without longitudinal force. Odquist's solution for $n = 3$ is

$$c = 0.323 \frac{l^2}{h^2} \left(\frac{ql^2}{2Fh\sigma_n}\right)^3 t \, , \tag{177.10}$$

whereas the exact solution of this problem gives the coefficient 0.332. The case of no longitudinal force is also obtained from our equations if we put $\hat{N} = 0$, and consequently, $a = 1$. Then from eq. (177.9) it follows that:

$$c = \tfrac{3}{4} \hat{t} \, .$$

If this result is given in the form of eq. (177.10), we find that the numerical coefficient is 0.331, which is practically the same as the exact solution. We have thereby checked that the approximate solution found by this method of numerical integration of the problem of a beam subject to a transverse load with constant distance between the ends, differs very little from the exact solution at both very small and fairly large deflections.

178. Critical time of compression of a column

A compressed column of material subject to creep is absolutely unstable in the sense explained at the start of the section 175. Successive states of such a column are illustrated in fig. 244. If the deformation is measured in terms of the deflection, that is the greatest distance from the centre line of the column to the line of action of the force, it is evident that the deflection first increases and then diminishes and in a large part of the column compression is replaced by tension, and after a certain time failure occurs by thinning of the column. This is the critical time in the strict sense of the word, the problem of determining the critical time in this formulation is too complicated and the solution is hardly of practical value. In assessing the real load-carrying capacity of a column in compression it is required that the lateral deflection

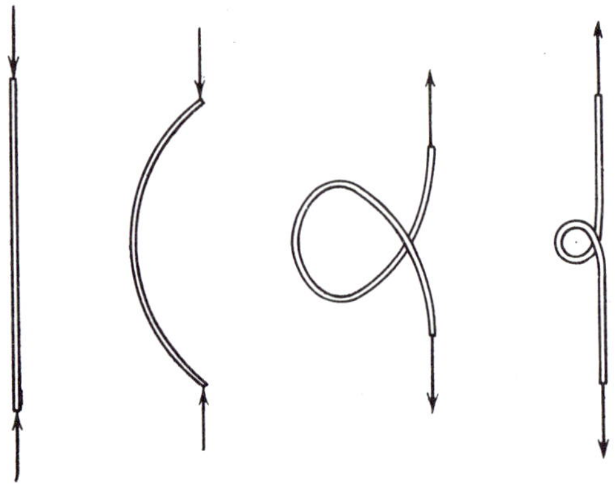

Fig. 244.

should not exceed some small value and, therefore, the practical limiting service life is the time in which the initial deflection c_0 increases to a certain prescribed value c_1. However, a special feature of the creep behaviour of a column in compression is that the initial deflection increases relatively slowly at first, but after a certain time the increase becomes catastrophically fast. If a linearised equation is used for the bent axis of the rod it is found that the deflection tends to infinity in a finite time, which is generally termed the critical time.

As we have seen, in some cases the rate of deflection becomes infinitely great, although the deflection itself remains finite. The time at which this occurs is also called the critical time. It should be borne in mind that strictly speaking the critical time defined in this way has no mechanical meaning, linearised equations are only suitable for fairly small deflections. It will be seen from fig. 244 that in practice there can be no question of infinite deflection. However, by defining the critical time in this way we do not depend on a subjective permissible deflection, but obtain an objective, though conventional assessment of the real life of the rod. Freudenthal [210] was apparently the first to discover the existence of a critical time in this meaning of the term, but he made the error of supposing that a column of linearly visco-elastic material also has a critical time, this error is corrected in the work [271]. We shall now consider some simple examples of the critical time with the assumption that the material observes the steady-state creep law and elastic deformation is neglected.

a) *Shanley's model.* Loss of stability of an inelastic system is most simply illustrated on a model proposed by Shanley [349]. A rigid block of length l is

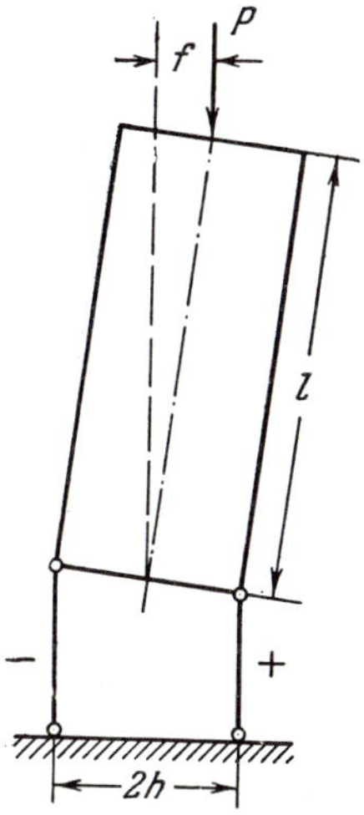

Fig. 245.

fixed to two rods of length a and area $F/2$, made of a material with creep properties (fig. 245). Denoting the relative contractions of the rods by e_+ and e_-, and the stresses in them by σ_+ and σ_-, we consider that $\dot{e} = v(\sigma)$. From the equations of equilibrium it follows that

$$M = Pf(\sigma_+ - \sigma_-)h\frac{F}{2},$$

$$P = (\sigma_+ + \sigma_-)\frac{F}{2}.$$

Whence

$$\sigma_+ = \sigma(1 + w)$$

$$\sigma_- = \sigma(1 - w),$$

$$w = \frac{f}{h}, \qquad \sigma = \frac{P}{F}.$$

On the other hand

$$\dot{w} = \frac{la}{2h^2}(\dot{e}_+ - \dot{e}_-),$$

or

$$\dot{w} = \frac{la}{2h^2} \{v[\sigma(1+w)] - v[\sigma(1-w)]\} \ . \tag{178.1}$$

Integrating eq. (178.1) we find the critical time at which the deflection becomes infinite. The critical time is readily considered with a power creep law when

$$v(\sigma) = \epsilon_n \left|\frac{\sigma}{\sigma_n}\right|^{n-1} \frac{\sigma}{\sigma_n} \ .$$

Substituting into eq. (178.1) and separating variables, we find:

$$\int_{w_0}^{\infty} \frac{dw}{(1+w)^n - |1-w|^{n-1}(1-w)} = \frac{\epsilon_n la}{2h^2} \left(\frac{\sigma}{\sigma_n}\right)^n t_{cr} \ .$$

Introduce the notation

$$\frac{2h^2}{al} = e_e \ .$$

The meaning of this notation will become clear later on. Note that

$$\epsilon_n \left(\frac{\sigma}{\sigma_n}\right)^n t = p \ ,$$

where p is the creep deformation of the fundamental state of compression of the rod that has accumulated at the time t. Therefore the above formula can be rewritten

$$p_{cr} = e_e \int_{w_0}^{\infty} \frac{dw}{(1+w)^n - |1-w|^{n-1}(1-w)} \ . \tag{178.2}$$

For any value of $n > 1$ the integral in formula (178.2) is finite. Thus, when $n = 3$ this integral is $\frac{1}{12} \ln w_0^2/(3 + w_0^2)$. Consequently, for $n = 3$ we obtain:

$$p_{cr} = \frac{e_e}{12} \ln \frac{3 + w_0^2}{w_0^2} \ .$$

In our analysis we have not allowed for elastic deformation. We shall now consider that

$$\dot{e}_\pm = \frac{1}{E}\dot{\sigma}_\pm + v(\sigma_\pm) = \pm\frac{\sigma}{E}\dot{w} + v[\sigma(1 \pm w)] \; .$$

Now in place of eq. (178.1) we have

$$\dot{w} - \frac{la}{2Eh^2}\frac{\mathrm{d}}{\mathrm{d}t}(w\sigma) = \frac{la}{2h^2}\{v[\sigma(1+w)] - v[\sigma(1-w)]\} \; . \quad (178.3)$$

Suppose that the initial deflection before application of the load is w_{00}. If the load is applied in a sufficiently short time the right-hand side of eq. (178.3) is negligibly small and we obtain

$$w_0 - w_{00} - \frac{\sigma la}{2Eh^2} w_0 = 0 \; .$$

Whence

$$w_0 = \frac{w_{00}}{1 - \dfrac{\sigma la}{2Eh^2}} \; .$$

The numerator in this formula becomes zero when

$$\sigma = \frac{2Eh^2}{la} \; .$$

But this is elastic loss of stability in the Euler sense. The quantity

$$\sigma_e = \frac{2Eh^2}{la} \, ,$$

represents the Euler critical stress for the Shanley model and the quantity $e_e = 2h^2/la$ is the elastic deformation that corresponds to the Euler critical stress. Put

$$\frac{\sigma}{\sigma_e} = \frac{P}{P_e} = \omega \; .$$

Then

$$w_0 = \frac{w_{00}}{1-\omega}. \tag{178.4}$$

Now substitute $\dot{\sigma} = 0$ into eq. (178.3) and integrate with the initial condition $w = w_0$. It will readily be seen that the result differs from that of eq. (178.2) only by a factor of $(1-\omega)$, namely

$$p_{cr} = e_e(1-\omega) \int_{w_0}^{\infty} \frac{dw}{(1+w)^n - |1-w|^{n-1}(1-w)}. \tag{178.5}$$

It should be borne in mind that the lower limit of integration w_0 is now determined by formula (178.4).

b) *A column of ideal I-section. Power law.* Assume that the column is an ideal *I*-section and consider the case when both ends are hinged and the length of the column is l [272]. Into the second of formulae (117.3) substitute $\hat{N} = P/F\sigma_n$, $\hat{M} = -Pw/2F\sigma$, $\kappa = \partial^3 w/\partial z^2 \partial t$, and we obtain the following differential equation:

$$-\frac{h}{\epsilon_n}\frac{\partial^3 w}{\partial z^2 \partial t} = \frac{1}{2}\left(\frac{P}{F\sigma_n}\right)\left[\left(1+\frac{w}{h}\right)^n - \left|1-\frac{w}{h}\right|^{n-1}\left(1-\frac{w}{h}\right)\right].$$

Introduce the dimensionless deflection $\hat{w} = w/h$ and the dimensionless coordinate $\xi = z/h$, then the equation is rewritten

$$\frac{\partial^3 w}{\partial \xi^2 \partial t} = -\tfrac{1}{2}\epsilon_n\left(\frac{\sigma}{\sigma_n}\right)^n [(1+\hat{w})^n - |1-\hat{w}|^{n-1}(1-\hat{w})]\,. \tag{178.6}$$

Here $\sigma = P/F$. Various devices are used to obtain an approximate solution of this equation, the simplest case corresponds to the value $n = 3$, and this is the only one we shall consider. When $n = 3$ it follows from (178.6) that

$$\frac{\partial^3 \hat{w}}{\partial \xi^2 \partial t} = -\epsilon_n\left(\frac{\sigma}{\sigma_n}\right)^3 [3\hat{w} + \hat{w}^3]\,. \tag{178.7}$$

We shall seek a solution in the form

$$\hat{w} = c(t)\sin\frac{\pi h \xi}{l}\,.$$

Substituting into (178.7) we obtain

$$\left(\frac{\pi h}{l}\right)^2 \dot{c} \sin\frac{\pi h\xi}{l} = \epsilon_n \left(\frac{\sigma}{\sigma_n}\right)^3 \left[3c \sin\frac{\pi h\xi}{l} + c^3 \sin^3\frac{\pi h\xi}{l}\right]. \quad (178.8)$$

We may now proceed in various ways. We may require that eq. (178.8) is fulfilled at a single point, namely the middle of the beam when $\xi = l/2h$. We then obtain

$$\dot{c} = \epsilon_n \left(\frac{l}{\pi h}\right)^2 \left(\frac{\sigma}{\sigma_n}\right)^3 (3c + c^3). \quad (178.9)$$

Separating variables and integrating we find

$$\epsilon_n \left(\frac{\sigma}{\sigma_n}\right)^3 t = \frac{1}{6}\left(\frac{\pi h}{l}\right)^2 \ln \frac{c^2(3 + c_0^2)}{c_0^2(3 + c^2)}. \quad (178.10)$$

Here c_0 is the initial dimensionless deflection. Hence we obtain the critical time by putting $c = \infty$, so that

$$\epsilon_n \left(\frac{\sigma}{\sigma_n}\right)^3 t_{cr} = \frac{1}{6}\left(\frac{\pi h}{l}\right)^2 \ln \frac{3 + c_0^2}{c_0^2}. \quad (178.11)$$

On the other hand, we may require fulfilment of eq. (178.8) on average and for this purpose we must multiply both sides by $\sin \pi h\xi/l$ and integrate over the length. As a result the following expression is obtained instead of eq. (178.5):

$$\dot{c} = \epsilon_n \left(\frac{l}{\pi h}\right)^2 \left(\frac{\sigma}{\sigma_n}\right)^3 (3c + \tfrac{3}{4} c^3), \quad (178.12)$$

and the formula for the critical time assumes the form

$$\epsilon_n \left(\frac{\sigma}{\sigma_n}\right)^3 t_{cr} = \frac{1}{6}\left(\frac{\pi h}{l}\right)^2 \ln \frac{4 + c_0^2}{c_0^2}. \quad (178.13)$$

Calculations for other values of n are given in the work [270], and analogous solutions are contained in the articles [234, 235].

c) *Column of ideal I-section. Hyperbolic sine law.* Putting in formulae (117.1) $v = 2\epsilon_e \sinh \sigma/\sigma_e$, in the second of these formulae we obtain

$$\kappa h = \epsilon_e \left[\sinh\left(\frac{N}{F\sigma_e} + \frac{M}{Fh\sigma_e}\right) - \sinh\left(\frac{N}{F\sigma_e} - \frac{M}{Fh\sigma_e}\right)\right].$$

In our case $N = \sigma F, M = -\sigma Fh\hat{w}$, consequently,

$$\kappa h = \frac{\partial^3 \hat{w}}{\partial \xi^2 \partial t} = \epsilon_e \left\{ \sinh\left[\frac{\sigma}{\sigma_e}(1-\hat{w})\right] - \sinh\left[\frac{\sigma}{\sigma_e}(1+\hat{w})\right]\right\}$$

$$= -\epsilon_e \cosh\frac{\sigma}{\sigma_e} \sinh\frac{\sigma\hat{w}}{\sigma_e}. \tag{178.14}$$

Once again put $\hat{w} = c(t) \sin \pi h\xi/l$ and require that eq. (178.14) is fulfilled at the centre of the beam. We obtain

$$\left(\frac{\pi h}{l}\right)^2 \dot{c} = \epsilon_e \cosh\frac{\sigma}{\sigma_e} \sinh\frac{\sigma c}{\sigma_e}.$$

Separating variables and integrating we obtain

$$\epsilon_e \cosh\left(\frac{\sigma}{\sigma_e}\right) t_{cr} = \left(\frac{\pi h}{l}\right)^2 \frac{\sigma_e}{\sigma} \ln\left[\frac{1}{\tanh\dfrac{\sigma c_0}{2\sigma_e}}\right]. \tag{178.15}$$

Note that in all the cases the formula for the critical time can be written in more compact form. For an ideal I-section the Euler load $P_e = \pi^2 EFh^2/l^2$ and, therefore, the factor $(\pi h/l)^2$ is the elastic deformation that corresponds to the Euler force. Denote it e_e. With a power creep law the quantity $\epsilon_n(\sigma/\sigma_n)^3\, t_{cr}$, that figures in the left-hand side of eq. (178.13), is the creep deformation with uniform compression of the column by a force P for a time t_{cr}; we shall term this quantity the critical deformation and denote it p_{cr}. Then formula (178.14) assumes the form

$$p_{cr} = \frac{e_e}{6} \ln\frac{4 + c_0^2}{c_0^2}. \tag{178.16}$$

Formula (178.15) cannot be reduced to such a simple form. However, it should be noted that for stresses at which creep is appreciable $\cosh \sigma/\sigma_e$ is little different from $\sinh \sigma/\sigma_e$ and, therefore, we may approximately put

$$p_{cr} = \frac{e_e}{\hat{\sigma}} \ln \tanh \frac{c_0}{2\hat{\sigma}} . \tag{178.17}$$

Here $\hat{\sigma} = \sigma/\sigma_e$.

As we have mentioned, solutions of the kind given above have been found by various authors, the results differ a little from one another because eqs. (178.6) or (178.14) can be integrated approximately in various ways. The method of satisfying the equation at a single point is the simplest and roughest. More accurate results are obtained by taking the deflection in the form of a Fourier series and finding the coefficients by the Galerkin method. Numerous works have been published about this matter but we shall not consider them here.

A column of ideal I-section is statically determinate and, therefore, the stress in the flanges in terms of the compressive force and the deflection is uniquely given by formula (117.3). Therefore, no difficulty arises in allowing for the initial deflection or for elastic-plastic deformation within the framework of this scheme. Indeed assume that the instantaneous deformation of the rod is elastic. If the initial deformation is c_{00}, then immediately after application of the load the deflection becomes c_0 and

$$c_0 = \frac{c_{00}}{1-\omega}, \qquad \omega = \frac{P}{P_e} .$$

Here P_e is the Euler critical load. The creep deformation is determined in terms of the effective stress, in the same way as above, but creep deformation is accompanied by an increase in elastic deformation as was explained in the example of the Shanley model. A similar investigation can readily be made in the general case, the difference is that an additional factor $(1-\omega)$ must be introduced into all the critical time formulae. Therefore, the critical time formulae of this section remain valid provided that we assume that the elasticity of the rod is unlimited. If the rod material can deform plastically the position is somewhat more complicated.

179. Allowance for instantaneous plastic deformation in determining the critical time

To explain the essence of the matter we shall first consider again the simple Shanley model. We shall first assume that the rod material is simply plastic so that $e = e_0(\sigma)$. The function $e_0(\sigma)$ is assumed to describe both active plastic deformation and unloading. Reproducing the considerations of the previous section, we find that the relationship between force and deflection is

$$w - w_0 = \frac{l}{2h^2}\{e_0[\sigma(1+w)] - e_0[\sigma(1-w)]\} . \tag{179.1}$$

This method of writing the equation is conventional because it still remains to establish whether loading or unloading is occurring in each of the columns. Either way eq. (179.1) defines the relationship between σ and w; at a certain value of w the force is maximum, consequently, $\mathrm{d}\sigma/\mathrm{d}w = 0$. Differentiating eq. (179.1) with respect to w, and putting $\mathrm{d}\sigma/\mathrm{d}w = 0$, we obtain

$$1 = \sigma \frac{l}{2h^2}\{e_0'[\sigma(1+w)] + e_0'(1-w)]\} . \tag{179.2}$$

Eq. (179.2) defines the critical state of the rod, i.e. the maximum deflection at which the column withstands the compressive load σ or the maximum force which the system can withstand for a given deflection. In practical assessment of stability or instability of systems eqs. (179.1) and (179.2) are solved simultaneously, w is eliminated from them, and a relationship is found between the force σ and the initial deflection w_0 that corresponds to the critical state. Now assume that the instantaneous elastic-plastic deformation is accompanied by creep, which causes the deflection to increase until it reaches the value that corresponds to the given force and expression (179.2). Equilibrium of the system then becomes impossible and buckling is instantaneous. The critical time of the column is the time required to reach the final critical deflection, the rate of buckling becomes infinitely great when the critical time is reached.

These intuitively obvious considerations are readily checked. Assume that the instantaneous deformation is accompanied by creep and that eq. (47.4) is valid,

$$\dot{e} = e_0'(\sigma)\dot{\sigma} + v(\sigma) .$$

The deflection $w(t)$ satisfies the following differential equation:

$$\dot{w}\ \{1 - \frac{l\sigma}{2h^2}[e_0'(\sigma(1+w)) + e_0'(\sigma(1-w))]$$

$$= \frac{l}{2h^2}\{v[\sigma(1+w)] - v[\sigma(1-w)]\} . \tag{179.3}$$

When condition (179.2) is fulfilled the factor multiplying $\dot{w}$ becomes zero and consequently, $\dot{w} = \infty$.

This simple result, which is that the critical time for a rod of ideal I-section is the time to reach a critical deflection that depends only on the elastic-plastic properties of the material and not on its creep characteristics, was formulated by De Veubeke [374] by analysing Hoff's [237] solution. In his solution Hoff considers a column of ideal I-section with hinged ends, the qualitative side of the question remains the same as in the Shanley column model considered above. The following relationship between stress and deformation (see section 47) is used as a basis:

$$\dot{e} = \frac{\dot{\sigma}}{E} + k_1 \left(\frac{\sigma}{\mu}\right)^m \frac{\dot{\sigma}}{\mu} + k_2 \epsilon_n \left(\frac{\sigma}{\sigma_n}\right)^n . \tag{179.4}$$

Here k_1 can be $-1, 0, +1$; k_2 can be $-1, +1$, as shown in section 47. If n is a whole odd number, $k_2 = +1$. When the stress σ is increasing we take $k_1 = +1$ and when unloading, $k_1 = 0$. In his calculations Hoff took $n = 3$ and $m = 1$.

As we have seen, Odquist proposed eq. (179.4) and also applied it to the problem of stability [315]. The second term in the right-hand side of eq. (179.4) represents with equal success both the instantaneous plastic deformation and the deformation of the first phase of creep. With the same assumptions as were made in section 178 we obtain the following differential equation of bending:

$$\frac{\partial^3 \hat{w}}{\partial \xi^2 \partial t} = \tfrac{1}{2}(\dot{e}_+ - \dot{e}_-) = \frac{\dot{\sigma}_+ - \dot{\sigma}_-}{2E} + \tfrac{1}{2}\left[k_1^+ \left(\frac{\sigma_+}{\mu}\right)^m \frac{\dot{\sigma}_+}{\mu} - k_1^- \left(\frac{\sigma_-}{\mu}\right)^m \frac{\dot{\sigma}_-}{\mu}\right] + \frac{\epsilon_n}{2}\left[k_2^+ \left(\frac{\sigma_+}{\sigma_n}\right)^n - k_2^- \left(\frac{\sigma_-}{\sigma_n}\right)^n\right] . \tag{179.5}$$

Here we substitute

$$\sigma_+ = \sigma(1 + \hat{w}) , \qquad \sigma_- = \sigma(1 - \hat{w}) .$$

If $\hat{w}$ is small, σ_+ and σ_- are of the same sign; as usual in longitudinal bending problems we shall consider compressive stresses positive. With increase in $\hat{w}$ the stress σ_+ increases and σ_- diminishes, therefore, $k_1^+ = 1$, $k_1^- = 0$, the quantities k_2^+ and k_2^- are both unity. Substituting the expressions for σ_+ and σ_- into eq. (179.5) we obtain

$$-\frac{\partial^3 \hat{w}}{\partial \xi^2 \partial t} = \frac{\sigma \dot{\hat{w}}}{E} + \tfrac{1}{2}\left(\frac{\sigma}{\mu}\right)^{m+1} (1+\hat{w})^m \dot{\hat{w}}$$

$$+ \tfrac{1}{2}\left(\frac{\sigma}{\sigma_n}\right)^n [(1+\hat{w})^n - (1-\hat{w})^n] \ . \tag{179.6}$$

It is assumed that the initial curvature of the beam axis after application of the load can be represented by a single half-sine wave

$$\hat{w}_0 = c_0 \sin \frac{\pi h \xi}{l} \ .$$

The deflection is also written in the form

$$\hat{w} = c \sin \frac{\pi h \xi}{l} \ ,$$

where c is a function of time.

In the subsequent calculations it is taken that $m = 1, n = 3$ which, as Hoff indicates, correspond to aluminium alloy 24-S-T4 at 314°C. The expression for $\hat{w}$ is substituted into eq. (179.6) and all non-linear terms are resolved into Fourier series of sines and the coefficient of $\sin \pi h\xi/l$ is equated to zero. As a result we obtain the following equation for the function $c(t)$:

$$\left(\frac{\pi h}{l}\right)^2 \dot{c} = \frac{\sigma}{E}\dot{c} + \tfrac{1}{2}\left(\frac{\sigma}{\mu}\right)^2 \left(1 + \frac{8c}{3\pi}\right)\dot{c} + \epsilon_n \left(\frac{\sigma}{\sigma_n}\right)^3 (3c + \tfrac{3}{4}c^3) \ .$$

As we have seen the quantity $(\pi h/l)^2$ is the elastic deformation e_e that corresponds to the Euler load. Rewrite the differential equation as follows:

$$\dot{c}\left[e_e - \frac{\sigma}{E} - \tfrac{1}{2}\left(\frac{\sigma}{\mu}\right)^2 \left(1 + \frac{8c}{3\pi}\right)\right] = \epsilon_n \left(\frac{\sigma}{\sigma_n}\right)^3 (3c + \tfrac{3}{4}c^3) \ . \tag{179.7}$$

This equation is integrated giving

$$\epsilon_n t = -\tfrac{1}{3}\left(\frac{\sigma_n}{\sigma}\right)^3 \left\{\frac{A}{2} \ln\left[\frac{c^2(4+c_0^2)}{c_0^2(4+c^2)}\right] - 2B \arctan\left[\frac{2(c-c_0)}{4+cc_0}\right]\right\} . \tag{179.8}$$

Here

$$A = e_e - \frac{\sigma}{E} - \frac{1}{2}\left(\frac{\sigma}{\mu}\right)^2, \qquad B = \frac{4}{3\pi}\left(\frac{\sigma}{\mu}\right)^2.$$

The factor by which $\dot{c}$ is multiplied in the left-hand side of eq. (179.7) becomes zero at a certain value $c = c_{cr}$, which is found by the formula

$$c_{cr} = \frac{A}{B} = \frac{3\pi}{4}\left(\frac{\mu}{\sigma}\right)^2 \left[e_e - \frac{\sigma}{E} - \frac{1}{2}\left(\frac{\sigma}{\mu}\right)^2\right]. \tag{179.9}$$

When $c = c_{cr}$, in accordance with formula (179.7), the rate of deflection becomes infinite, and, therefore, the critical time t_{cr} is given by formula (179.8) in which we put $c = c_{cr}$. However, this analysis is only valid when formula (179.9) gives $c_{cr} < 1$. Indeed in formulating eq. (179.6) we essentially assumed that both flanges are in compression. If the condition $c_{cr} < 1$ is not fulfilled we must consider two stages of the process. Denote by t_1 the time of reaching the value $c = 1$, this value t_1 is found by eq. (179.8) in which we put $c = 1$. When $t > t_1$ over a certain segment near to the centre of the span the stress σ_- becomes tensile and, therefore, for this segment we should put $k_1^- = 1$ whereas for the remainder as before, $k_1^- = 0$. In this formulation it is impossible to obtain a relatively simple approximate solution, like that which is obtained in the case $c_{cr} < 1$. The author proposes that upper and lower assessments be made for the critical time. The upper assessment is found by neglecting the change in sign of the stress in one of the flanges and considering the solution of (179.8) also valid when $c > 1$. Neglecting plastic deformation due to tension of the flange we overestimate the rigidity of the beam, and therefore increase the critical time. The lower assessment for the critical time is found by taking $k_1^- = 1$ for the entire beam and not just for its central part. Eq. (179.7) is then replaced by the equation

$$\dot{c}\left[e_e - \frac{\sigma}{E} - \frac{8}{3\pi}\left(\frac{\sigma}{\mu}\right)^2 c\right] = \frac{3}{4}\,\epsilon_n \left(\frac{\sigma}{\sigma_n}\right)^3 (4c + c^3). \tag{179.10}$$

The critical deflection is found from the condition that the factor by which $\dot{c}$ is multiplied becomes zero, namely

$$c_{cr} = \frac{3\pi}{8}\left(\frac{\mu}{\sigma}\right)^2 \left[e_e - \frac{\sigma}{E}\right]. \tag{179.11}$$

Now the critical time is $t_{cr} = t_1 + t_2$. Here t_2 is the time at which the deflection alters from $c = 1$ to $c = c_{cr}$, given by formula (179.11). Integrating (179.10) we find

$$\epsilon_n t_2 = -\tfrac{1}{3}\left(\frac{\sigma_n}{\sigma}\right)^3 \left[\frac{A'}{2}\ln\left(\frac{5c_{cr}^2}{4+c_{cr}^2}\right) - 4B\arctan\left(\frac{2(c_{cr}-1)}{4+c_{cr}}\right)\right]. \tag{179.12}$$

In his article Hoff [237] gives a calculation of the critical time for a column of aluminium alloy, values taken were $l/h = 50$, $c_0 = 0.15$. The following assessments of the critical time were obtained by the method described above:

$$15.85 < t_{cr} < 16.3 \text{ min}.$$

It was found that $t_1 = 11.95$ min.

The quantity c_0 is the deflection of the column immediately after application of the load. What is actually given is the initial curvature of the column, the deflection corresponding to the initial curvature is denoted c_{00}. The quantity c_0 is found by solving the plasticity problem. If c_{00} is small enough, and the compressive force is not too great $c_0 < 1$, and at the first instant after application of the load both flanges are under compression and the analysis given above is valid. However, it may happen that the initial eccentricity is so great that immediately after application of the load, tensile stresses are set up in one or other flange. For details of determination of c_0 and a consideration of all the possibilities that can arise in this case the reader is referred to the original article.

180. **Critical time of a column in compression. Further results**

As we have mentioned, the critical deformation at which the rate of deflection becomes infinite depends only on the instantaneous elastic-plastic properties of the column material. This conclusion was derived for a column of ideal *I*-section and does not strictly apply to real columns. The point is that in an ideal *I*-section the stress distribution in the flanges is uniquely determined for a given deflection and a given value of compressive force, and therefore, for a given deflection, instantaneous loss of stability occurs in the same way irrespective of the route by which the deflection is reached. In actual columns a given deflection may correspond to different stress distributions across the section and the critical state depends both on the magnitude of the deflection and on the stress distribution. Therefore, the critical deflection also depends on the progress of creep. This circumstance was illustrated by Shesterikov [153] on a model column consisting of three flanges (fig. 246). The thickness of each flange δ is considered small compared with h, the

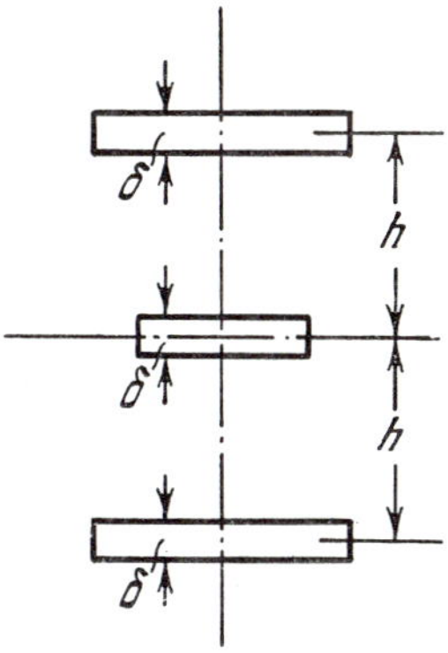

Fig. 246.

stress is assumed to be uniformly distributed in each flange. Such a rod is statically indeterminate, the stress distributions between flanges can be different for given values of compressive force and bending moment, and so its condition can become critical at various values of deflection. Analysis of even this simple model is fairly complicated and a closed solution cannot be achieved.

However, certain experimental results indicate that the range of critical deflection of real columns is quite narrow. The work [238] makes reference to these results. Thus, according to the results of [321] the ratio of the critical deflection to the radius of inertia of the cross-section ranged from 0.49 to 0.83, whereas the ratio of the initial deflection to the radius of inertia ranged from 0.035 to 0.131. A similar result was obtained in the work [182], for columns of the same flexibility and the same compressive force, a sudden increase in deflection occurred at about the same deflection.

Generally speaking, the instantaneous plastic properties of the material depend on previous creep (see section 57), therefore, it is not entirely accurate to determine the critical state from an instantaneous compression diagram.

On the basis of remarks made at the start of this section, Hoff recommends a procedure for determining the critical time of a compressed column which is based on replacing it by an ideal *I*-section [238, 241]. Selection of the dimensions of the *I*-section was explained in section 117. Retaining the notation of section 179, we write the differential equation of deflection of a column in compression as follows:

$$-2\dot{k}h = \dot{\sigma}\{(1+w)e_0'[\sigma(1+w)] - (1-w)e_0'[\sigma(1-w)]\}$$
$$+ \sigma\dot{w}\{e_0'[\sigma(1+w)] + e_0'[\sigma(1-w)]\}$$
$$+ \{v[\sigma(1+w)] - v[\sigma(1-w)]\} . \qquad (180.1)$$

Here k is the curvature of the rod axis, w the dimensionless deflection. Put

$$k = hw'' , \qquad w = cV .$$

Here, unlike section 179, the primes denote differentiation with respect to the coordinate z and V satisfies the equation

$$EIV'' + PV = 0 .$$

The constants of this equation define the Euler force P_e. For an ideal I-section $I = Fh^2$, consequently

$$\frac{P_e}{EI} = \frac{\sigma_e}{Eh^2} = \frac{e_e}{h^2} .$$

Substituting into eq. (180.1) we obtain

$$2e_e\dot{c} = \dot{\sigma}\{(1+w)e_0'[\sigma(1+w)] - (1-w)e_0'[\sigma(1-w)]\}$$
$$+ \sigma\dot{w}\{e_0'[\sigma(1+w)] + e_0'[\sigma(1-w)]\}$$
$$+ \{v[\sigma(1+w)] - v[\sigma(1-w)]\} . \qquad (180.2)$$

An approximate solution of eq. (180.2) can be obtained in various ways, the simplest is to require fulfilment of the equation at a single section $z = z_1$. Obviously the section to be selected is the one that is subject to maximum bending moment. The function $V(z)$ can always be defined in such a way that $V(z_1) = 1$. Then eq. (180.2) gives the following ordinary differential equation for the function $c(t)$:

$$2e_e\dot{c} = \dot{\sigma}\{(1+c)e_0'[\sigma(1+c)] - (1-c)e_0'[\sigma(1-c)]\}$$
$$+ \sigma\dot{c}\{e_0'[\sigma(1+c)] + e_0'[\sigma(1-c)]\}$$
$$+ \{v[\sigma(1+c)] - v[\sigma(1-c)]\} . \qquad (180.3)$$

Another possibility is to multiply eq. (180.2) by $V(z)$ and to integrate over the length of the beam. This method was explained in section 178. Naturally the calculations are considerably more complicated.

Consider first the process of loading a column. Assume that the compressive force is applied so quickly that in expression (180.3) the terms that contain derivatives with respect to time are large compared with other terms. Integrate (180.3) and ignore the part of it which depends on the rate of creep v, to give

$$2e_e(c_0 - c_{00}) = e_0[\sigma(1+c)] - e_0[\sigma(1-c)] \ . \qquad (180.4)$$

Solving eq. (180.4) we find the deflection c_0 which is taken as the initial value in analysing subsequent creep. In most cases the initial deflection can be determined by assuming the material elastic, and consequently, by the formula

$$c_0 = \frac{c_{00}}{1-\omega}, \qquad \omega = \frac{\sigma}{\sigma_e} .$$

Initial plastic deformation can only be permitted in rods of very short life. If initial plastic deformation has to be allowed for in using formula (180.4) it must be borne in mind that in the flange on the convex side the stress first increases then diminishes. Therefore, the function $e_0[\sigma(1-c)]$ first corresponds to loading and then to unloading. In place of eq. (180.4) it may be more convenient to use the differential equation which follows from (180.3) retaining in it only terms that contain derivatives, and to consider the stress σ as an independent variable. For brevity introduce the notation

$$\frac{1}{E_+} = e_0'[\sigma(1+c)] , \qquad \frac{1}{E_-} = e_0'[\sigma(1-c)] ,$$

$$\frac{1}{E_*} = \frac{1}{2}\left(\frac{1}{E_+} + \frac{1}{E_-}\right) = \frac{E_+ + E_-}{2E_+E_-} .$$

We then obtain

$$\left(e_e - \frac{\sigma}{E_*}\right)\frac{dc}{d\sigma} = \frac{c}{E_*} + \frac{1}{2}\left(\frac{1}{E_+} - \frac{1}{E_-}\right) . \qquad (180.5)$$

Then, depending on whether loading or unloading takes place in the least loaded flange, the quantity E may be the tangential modulus on the plasticity diagram or the elastic modulus. As σ increases the stress $\sigma_- = \sigma(1-c)$ first increases then starts to diminish at the time when the condition

$$\frac{d\sigma_-}{d\sigma} = (1-c) - \sigma\frac{dc}{d\sigma} = 0$$

is fulfilled and consequently

$$\frac{dc}{d\sigma} = \frac{1-c}{\sigma}.$$

Substituting into eq. (180.5) we obtain the following condition of change of sign of $y(d\sigma_-/d\sigma)$:

$$\frac{1}{\sigma}(1-c)\left(e_e - \frac{\sigma}{E_*}\right) = \frac{c}{E_*} + \frac{1}{2}\left(\frac{1}{E_+} - \frac{1}{E_-}\right). \tag{180.6}$$

The example of an elastic column can be used to show that the sign does indeed change. Putting $E_+ = E_- = E$, $\sigma/\sigma_e = \omega$, we find that eq. (180.6) is fulfilled provided that

$$c = 1 - \omega .$$

But for an elastic column

$$c = \frac{c_{00}}{1-\omega};$$

therefore, the condition of change of sign is

$$\omega = 1 - \sqrt{c_{00}} .$$

Given the compression diagram of the material we construct a graph of the function $e_0'(\sigma)$ and consequently we can determine the quantities E_+ and E_- for any values of σ and c; therefore, numerical integration of eq. (180.5) presents no difficulty.

If $c > 1$, $\sigma_- > 0$ and the quantity E_- should be calculated from the tensile diagram following unloading, with allowance for the Bauschinger effect in some form or other. For reasons that will be explained later this case is not considered in detail.

Accordingly, assume that the quantity c_0 has been determined. Further motion of the rod is described by eq. (180.3) in which we put $\dot{\sigma} = 0$. Consequently, we have

$$2\left(e_c - \frac{\sigma}{E_*}\right)\dot{c} = v[\sigma(1+c)] - v[\sigma(1-c)] \; . \tag{180.7}$$

Hence

$$t_{cr} = 2e_e \int_{c_0}^{c_k} \frac{1 - \dfrac{\sigma}{E_* e_e}}{v[\sigma(1+c)] - v[\sigma(1-c)]} \, dc \; . \tag{180.8}$$

In formula (180.8), in calculating E_* we must always put $E_- = E$. Indeed when $\sigma = \text{const}$, $\dot{\sigma}_- = -\sigma\dot{c} < 0$. The upper limit of integration in formula (180.8) is the value of c at which the numerator of the subintegral expression becomes zero, i.e. there is instantaneous loss of stability, as has already been explained and as can be seen from eq. (180.5). The integral in formula (180.8) can be calculated in closed form only with simple assumptions concerning the creep law and the shape of the instantaneous deformation curve. The example considered in section 179 relates to this particular case. In principle, formula (180.8) is applicable for any values of c_k, however, when $c > 1$ allowance must be made for the change in sign of the stress σ_-, and E_- must be appropriately selected. However, large deflections cannot be permitted in columns in structural elements and, therefore, if the deflection is great the limiting service life of the rod should not depend on its reaching a critical state, but on its reaching a certain deflection that depends on other considerations; for example, $c = c_k = 1$, which is a deflection equal to half the thickness of the rod (Rozenblyum [52]). Thus if the numerator of the subintegral expression in formula (180.8) becomes a zero when $c = c_k < 1$, c_k may be taken as the upper limit of the integral. If the critical state is achieved when $c > 1$, we take in this formula $c_k = 1$.

In his work [241] Hoff recommends a still simpler procedure for determining the critical time, namely he divides the problem into two independent parts. The critical deflection is first sought from the condition

$$\frac{\sigma}{E_* e_e} = 1 \, ,$$

and then the time to achieve the critical deflection is calculated approximately. Here no attention is paid to the change in plastic deformation that accompanies creep and deflection is found as in section 178, that is the numerator is assumed unity in the subintegral expression of formula (180.8). This method obviously gives too long a critical time. Experimental values of the critical time were shorter than calculated values, although it is difficult to say whether the difference can be entirely attributed to the assumption that we have mentioned.

The point of this very simple scheme is as follows. The complete formulation of the buckling problem is very difficult to solve and a complete solution can only be found by making very simple and rough assumptions. Given the deformation diagram, the critical deflection of an elastic-plastic rod can be determined very accurately and an extensive literature has been devoted to this subject [6, 156]. Longitudinal deflection in creep can also be determined by numerical methods and the solution is not tied to an over-idealised creep law of the real material; since the change in stress is monotonic, calculations based on time-hardening theory and actual creep curves are sufficiently accurate.

Attempts to determine the critical time more accurately on the basis of restrictive assumptions concerning the shape of the cross-section and of the initial law, have been made in a number of works. For example, in the works [282, 283] the critical time is determined from the strain-hardening theory in the form $\dot{p}p^{\alpha} = k \exp \sigma/A$. In the first of these works this is done for a rod of ideal *I*-section. It is assumed that the column is hinged at the two ends and is deflected to a sinusoidal shape, the equation of equilibrium is satisfied at the centre of the span as in the examples considered above. The second work relates to buckling of a column of rectangular cross-section, the solution is obtained by numerical integration. Other solutions are given in the works [146, 284].

181. Mixed variational principles applied to buckling problems

In section 167 we considered a recently proposed, mixed variational principle by which hypothetical distributions of stress and displacement can be given independently. However, in stability problems derivatives of displacements must be considered small and only linear terms are retained in expressions for components of deformation in terms of displacements. We shall use more accurate expressions for components of deformation (see section 5), which in Cartesian coordinates are

$$e_{ij} = \tfrac{1}{2} [u_{i,j} + u_{j,i} + u_{k,i}u_{k,j}] \, . \qquad (181.1)$$

Now, strictly speaking, the equilibrium equation cannot be written in the original metric, because by giving the distribution of displacements we have defined a certain curvilinear system of coordinates in the body. The simplest way of deriving a more accurate system of equilibrium equations is as follows. Suppose that surface forces T_i act on the body, then the variational equation of equilibrium is

$$\int_V \sigma_{ij} \delta e_{ij} \, \mathrm{d}V - \int_\Sigma T_k \delta u_k \, \mathrm{d}\Sigma = 0 \, .$$

But by formula (181.1)

$$\delta e_{ij} = \tfrac{1}{2} \left[\delta u_{i,j} + \delta u_{j,i} + u_{k,i} \delta u_{k,j} + u_{k,j} \delta u_{k,i} \right]$$
$$= \tfrac{1}{2} \left[(\delta_{kj} + u_{k,j}) \delta u_{k,i} + (\delta_{ki} + u_{k,i}) \delta u_{k,j} \right] .$$

Here δ_{ks} is unit tensor.

Introduce the asymmetric tensor

$$s_{ki} = \sigma_{ij} (\delta_{kj} + u_{k,j}) \, . \tag{181.2}$$

Now the variational equation of equilibrium is written

$$\tfrac{1}{2} \int_V s_{ki} \delta u_{k,i} \, \mathrm{d}V + \tfrac{1}{2} \int_V s_{kj} \delta u_{k,j} \, \mathrm{d}V - \int_\Sigma T_k \delta u_k \, \mathrm{d}\Sigma = 0 \, .$$

Converting volume integrals in the usual way we obtain

$$s_{ij,j} = 0 \, , \tag{181.3}$$

$$s_{ij} \nu_j = T_i \quad \text{or} \quad \Sigma \, . \tag{181.4}$$

If in the functional K, defined by formula (167.4), we understand by e_{ij} expression (181.1), then in order to derive the condition of equilibrium (181.3) and boundary conditions (181.4) from the variational equation it is necessary to add further terms. This was done in the work cited [345]; for non-linear problems the expression of the functional is

$$K = \int_V \left[\dot{\sigma}_{ij} \dot{e}_{ij} + \tfrac{1}{2} \sigma_{ij} \dot{u}_{k,i} \dot{u}_{k,j} - \tfrac{1}{2} (\dot{e}^p_{ij} + 2\dot{p}_{ij}) \dot{\sigma}_{ij} \right] \mathrm{d}V$$
$$- \int_{\Sigma_u} \dot{T}_i (\dot{u}_i - \dot{u}^e_i) \, \mathrm{d}\Sigma - \int_{\Sigma_T} \dot{T}^e_i \dot{u}_i \, \mathrm{d}\Sigma \, . \tag{181.5}$$

Here all the notation is the same as in section 167, the quantities $\dot{u}_i$ and $\dot{\sigma}_{ij}$, which are considered to be independent quantities, are varied and the derivatives of the deformation components with respect to time are calculated by means of formula (181.1), namely

$$\dot{e}_{ij} = \tfrac{1}{2}(\dot{u}_{i,j} + \dot{u}_{j,i} + u_{k,i}\dot{u}_{k,j} + u_{k,j}\dot{u}_{k,i}) .$$

The demonstration that the variation of the functional K becomes zero is equivalent to fulfilling the equation of equilibrium, the boundary conditions and the conditions that link rates of deformation, stresses and rates of change of stress, in exactly the same way as in section 167. The additional terms in formula (181.5) do not contain rates of change of stress, therefore, the difference consists only in that part of δK which corresponds to change in rate distribution. Denote this part $\delta K'$, we obtain

$$\delta K' = \int_V [\dot{\sigma}_{ij} \tfrac{1}{2}(\delta\dot{u}_{i,j} + \delta\dot{u}_{j,i} + u_{k,i}\delta\dot{u}_{k,j} + u_{k,j}\delta\dot{u}_{k,i})$$

$$+ \tfrac{1}{2}\sigma_{ij}(\dot{u}_{k,i}\delta\dot{u}_{k,j} + \dot{u}_{k,j}\delta\dot{u}_{k,i})]\, \mathrm{d}V - \int_{\Sigma_u} \dot{T}_k \delta\dot{u}_k \,\mathrm{d}\Sigma$$

$$- \int_{\Sigma_T} \dot{T}_k^e \delta\dot{u}_k \,\mathrm{d}\Sigma = \tfrac{1}{2}\int_V \dot{s}_{ki}\delta\dot{u}_{k,i}\,\mathrm{d}V + \tfrac{1}{2}\int_V \dot{s}_{kj}\delta\dot{u}_{k,j}\,\mathrm{d}V$$

$$- \int_{\Sigma_u} \dot{T}_k \delta\dot{u}_k \,\mathrm{d}\Sigma - \int_{\Sigma_T} \dot{T}_k^e \delta\dot{u}_k \,\mathrm{d}\Sigma .$$

Converting the volume integral we find from the condition $\delta K' = 0$ that

$$\dot{s}_{ij,j} = 0 , \quad \dot{s}_{ij}\nu_j = \dot{T}_i \text{ or } \Sigma .$$

These equations are obtained by differentiating formulae (181.3) and (181.4) with respect to time.

182. **Buckling of a column in compression. Application of variational method**

To illustrate the application of the variational principle described in section 181, reconsider the stability in creep of a column compression with an initial curvature. Denote by u and w the axial displacement and the lateral deflection respectively, we obtain

$$e = u_{,z} + \tfrac{1}{2} w_{,z}^2 - y w_{,zz} .$$

Suppose that a force N is applied to the end of the section, and moreover, there is a transverse load q. The expression for the functional K is written

$$K = \int_0^l \int_F [\dot{\sigma}(\dot{u}_{,z} + w_{,z}\dot{w}_{,z} - y\dot{w}_{,zz}) + \tfrac{1}{2}\,\sigma\dot{w}_{,z}^2 - \dot{p}\dot{\sigma}]\,\mathrm{d}F\,\mathrm{d}z$$

$$- \int_0^l \dot{q}\dot{w}\,\mathrm{d}z - \dot{N}[\dot{u}(l) - \dot{u}(0)] . \qquad (182.1)$$

The volume integral extends over the cross-sectional area. Note that

$$\int \dot{\sigma}\,\mathrm{d}F = \dot{N} , \qquad \int \dot{\sigma} y\,\mathrm{d}F = \dot{M} .$$

Here M is the bending moment in the section of the rod. The rate of creep is expressed in terms of the creep potential by the formula

$$\dot{p} = \frac{\partial \Phi}{\partial \sigma} .$$

The product

$$\dot{p}\dot{\sigma} = \frac{\partial \Phi}{\partial \sigma}\,\dot{\sigma} = \frac{\mathrm{d}\Phi}{\mathrm{d}t} ,$$

therefore the corresponding integral in formula (182.1) assumes the form

$$\int_F \dot{p}\dot{\sigma}\,\mathrm{d}F = \frac{\mathrm{d}}{\mathrm{d}t}\int_F \Phi\,\mathrm{d}F .$$

But the integral of Φ over the area is the creep potential per unit length of rod, that is a function of N and M

$$\int_F \Phi\,\mathrm{d}F = \Phi(M, N) .$$

Consequently,

$$\int_F \dot{p}\sigma \, dF = \frac{\partial \Phi}{\partial M} \dot{M} + \frac{\partial \Phi}{\partial N} \dot{N} .$$

Integrating in this way over the area in formula (182.1) and integrating the first term with respect to z, we find the following expression for the functional K:

$$K = \int_0^l \left[\dot{N} w_{,z} \dot{w}_{,z} + \dot{M} \dot{w}_{,zz} + \tfrac{1}{2} \dot{N} w_{,z}^2 - \frac{\partial \Phi}{\partial N} \dot{N} - \frac{\partial \Phi}{\partial M} \dot{M} - \dot{q} \dot{w} \right] dz . \tag{182.2}$$

The problem of creep with combined bending and tension or compression was considered in section 116, in particular it was shown that an elliptical approximation to the creep potential can be taken, namely

$$\Phi = \Phi(R) , \quad R^2 = M^2 + N^2 k^2 h^2 .$$

Then

$$\frac{\partial \Phi}{\partial M} = \frac{\kappa(R)}{R} M , \quad \frac{\partial \Phi}{\partial N} = \frac{\kappa(R)}{R} N k^2 h^2 . \tag{182.3}$$

The function $\kappa(R)$ is defined such that when $R = M$, $\kappa(M)$ is the rate of change of curvature.

Now put

$$w = hc(t) \sin \frac{\pi z}{l} , \quad M = hb(t) \sin \frac{\pi z}{l} .$$

Here h is a certain characteristic transverse dimension, for instance half the depth of the section. Substituting into formula (182.2) we find

$$K = \left(\frac{\pi h}{l} \right)^2 \frac{l}{2} [-\dot{N} c \dot{c} - N \dot{c}^2 + \dot{b} \dot{c}] - N \dot{N} k^2 h^2 \int \frac{\kappa(R)}{R} dz$$

$$- b \dot{b} h^2 \int \frac{\kappa(R)}{R} \sin^2 \frac{\pi z}{l} dz - \dot{c} h \int \dot{q} \sin \frac{\pi z}{l} dz . \tag{182.4}$$

In solving the buckling problem we shall consider that $N = \text{const} = -P$ and

$\dot{q} = 0$, therefore, the corresponding terms in formula (182.4) cancel out. Only the parameters $\dot{c}$ and $\dot{b}$ are now subject to variation. As a result we obtain the following system of equations:

$$P\dot{c} + \dot{b} = 0\,, \quad \dot{c} + b\,\frac{2l}{\pi^2}\int_0^l \frac{\kappa(R)}{R}\sin^2\frac{\pi z}{l}\,dz = 0\,. \tag{182.5}$$

From the first equation it follows that

$$Pc = -\,b\,,$$

and the second assumes the form

$$\dot{c} = \frac{2l}{\pi^2}\,Pa\int_0^l \frac{\kappa(R)}{R}\sin^2\frac{\pi z}{l}\,dz\,. \tag{182.6}$$

The right-hand side is a function of c and P because

$$R^2 = P^2h^2\left[k^2 + c^2\sin^2\frac{\pi z}{l}\right].$$

The critical time can be determined numerically from eq. (182.6) for any creep law; as usual the power law gives the simplest result. Put

$$f(c) = \frac{1}{l}\int_0^l \left[k^2 + c^2\sin^2\frac{\pi z}{l}\right]^{n-1}\sin^2\frac{\pi z}{l}\,dz\,.$$

After simple transformations we derive the following expression for the critical creep deformation

$$p_{cr} = \frac{k^{n+1}}{2}\left(\frac{\pi h}{l}\right)^2\int_{c_0}^{\infty}\frac{dc}{cf(c)}\,. \tag{182.7}$$

The following result is obtained when $n = 3$:

$$p_{cr} = \frac{3k^2}{16}\left(\frac{\pi h}{l}\right)^2\ln\frac{\frac{4}{3}k^2 + c_0^2}{c_0^2}\,.$$

The method of determining k was explained in section 116. For an idealised I-section $k = 1$ and the formula above can be compared with formula (178.11) or (178.13).

Now consider how variational equations can be used to solve the problems of bending beams with fixed ends, as was considered in section 177. Assume the same approximate expressions for deflection and bending moment, so that for the functional K we again obtain expression (182.4), but now $\dot{N} \neq 0$ and must be varied. By variation with respect to the three parameters we obtain the following equations:

$$-\dot{N}c - N\dot{c} + \dot{b} - \frac{2l}{h}\int \dot{q} \sin\frac{\pi z}{l}\, \mathrm{d}z = 0\,,$$

$$\dot{c} + b\,\frac{2l}{\pi^2}\int \frac{\kappa(R)}{R} \sin^2\frac{\pi z}{l}\, \mathrm{d}z = 0\,,$$

$$\left(\frac{\pi h}{l}\right)^2 \frac{l}{2} c\dot{c} + Nk^2h^2 \int \frac{\kappa(R)}{R}\, \mathrm{d}z = 0\,.$$

The first equation is readily integrated and it follows that:

$$b = Nc + Q\,, \qquad Q = \frac{2l}{\pi h}\int q \sin\frac{\pi z}{l}\, \mathrm{d}z\,.$$

The following system for c and N is obtained from the other two equations:

$$\dot{c} = -(Nc + Q)\,\frac{2l}{\pi^2}\int \frac{\kappa(R)}{R} \sin^2\frac{\pi z}{l}\, \mathrm{d}z\,,$$

$$c\dot{c} = -N\,\frac{2l}{\pi^2}\,k^2 \int \frac{\kappa(R)}{R}\, \mathrm{d}z\,. \qquad (182.8)$$

The right-hand sides of these equations are known functions of c and N and they may be expressed in explicit form with the power creep law with odd whole exponents. Dividing the one equation by the other we obtain a finite ratio between c and N, after which we can always integrate any of the equations (182.8) numerically; here the course of the calculation is very close to that of section 177, where the same problem is solved by another method.

Note that in the work [345], use of the mixed variational principle is also illustrated by an example of buckling of a column in compression. However, this work assumes a linear distribution across the depth of the column. The final formula obtained for $n = 3$ is of the same structure as that given above but the numerical values are somewhat different. This was explained in section 168 when applied to transverse bending of a beam. Its extension to the case of buckling under the action of a longitudinal force is self-evident and requires no special explanation. It should be mentioned that Rozenblyum solved the problem of buckling of a column in compression on the basis of flow theory using a somewhat different variational method, and also assumed a linear stress distribution across the section. This work is described in detail in the book [52].

183. **Stability of arches**

A more complicated stability problem that can be solved by the variational method described above is the stability of a flat arch, uniformly loaded with distributed pressure. Structures of this kind are subject to sudden loss of stability which can only be explained by taking geometrical non-linearity into account. The solution given below [327] is interesting because it illustrates some of the difficulties that are encountered in solving stability problems of thin shells although the difficulties are much more severe in shells. Accordingly, consider the arch illustrated in fig. 247, with hinged supports at the two ends having a span l and a height c_0. The arch axis is assumed to be sinusoidal.

$$w_0 = c \sin \frac{\pi z}{l} .$$

The arch carries a uniformly distributed vertical load of intensity q. Denote by y the distance of a certain point on the arch axis along a normal. Because of the plane sections law we have

$$e = u_{,z} + \tfrac{1}{2} [w_{,z}^2 - w_{0,z}^2] - y [w_{,zz} - w_{0,zz}] ,$$

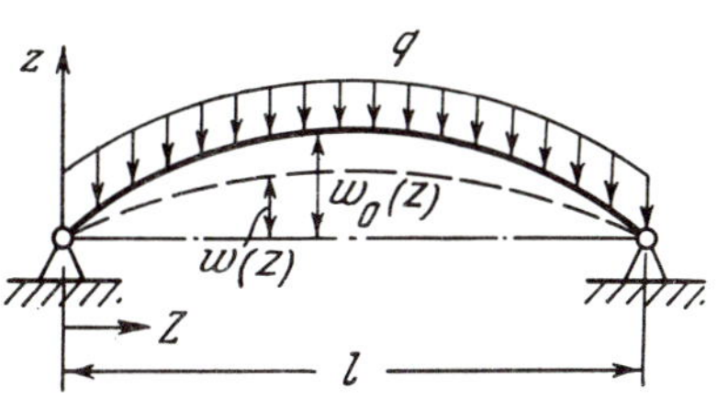

Fig. 247.

and the functional (181.5) takes the form

$$K = b \int_{-h}^{+h} \int_{0}^{l} \left[\dot{\sigma}(\dot{u}_{,z} + w_{,z}\dot{w}_{,z} - y\dot{w}_{,zz}) + \tfrac{1}{2}\,\sigma\dot{w}^2_{,z} - \frac{\dot{\sigma}^2}{2E} - \epsilon_n \left(\frac{\sigma}{\sigma_n}\right)^n \right] \mathrm{d}y\, \mathrm{d}z\,. \tag{183.1}$$

Here it is assumed that the instantaneous deformation is elastic, and creep follows a power law. The arch section is assumed rectangular of height $2h$ and width b. It is natural to assume that during deformation the axis of the arch remains sinusoidal and also to assume a sinusoidal law of distribution of moment over the length of the arch. The author assumes that the stress distribution across the thickness is linear. Therefore

$$\begin{aligned} &w = c \sin\frac{\pi z}{l} = \eta c_0 \sin\frac{\pi z}{l}\,,\\ &u = 0\,,\\ &\sigma = \sigma_e\left[\sigma_0 + \sigma_1 \frac{2y}{h} \sin\frac{\pi z}{l}\right]. \end{aligned} \tag{183.2}$$

Here $\sigma_e = \pi^2 h^2 E/3l^2$ is a critical Euler stress for a beam of length l, therefore, η, σ_0 and σ_1 are dimensionless parameters.

The procedure of calculation is that expressions (183.2) are substituted into formula (183.1) and the functional K is found as a function of η, σ_0, σ_1 and derivatives of these parameters with respect to time. Further K is varied with respect to $\dot{\eta}$, $\dot{\sigma}_0$ and $\dot{\sigma}_1$ giving a system of three equations. Omitting elementary calculations we directly give these equations for $n = 1$ and $n = 3$.

For $n = 1$

$$\begin{aligned} &6\left(\frac{c_0}{2h}\right)^2 \eta\dot{\eta} - \dot{\sigma}_0 - \epsilon_n \frac{E}{\sigma_n}\sigma_0 = 0\,,\\ &\left(\frac{c_0}{2h}\right)\dot{\eta} - \tfrac{1}{12}\dot{\sigma}_1 - \tfrac{1}{12}\epsilon_n \frac{E}{\sigma_n}\sigma_1 = 0\,,\\ &\left(\frac{c_0}{2h}\right)\dot{\sigma}_0\eta + \tfrac{1}{2}\dot{\sigma}_1 + \left(\frac{c_0}{2h}\right)\sigma_0\dot{\eta} = 0\,. \end{aligned} \tag{183.3}$$

For $n = 3$

$$6\left(\frac{c_0}{2h}\right)^2 \eta\dot{\eta} - \dot{\sigma}_0 - \epsilon_n \frac{E\sigma_e^2}{\sigma_n^3}\left(\sigma_0^3 + \frac{\sigma_0\sigma_1^2}{8}\right) = 0\,,$$

$$\left(\frac{c_0}{2h}\right)\dot{\eta} - \tfrac{1}{12}\dot{\sigma}_1 - \epsilon_n \frac{E\sigma_e^2}{\sigma_n^3}\left(\frac{\sigma_0^2\sigma_1}{4} + \tfrac{3}{320}\sigma_1^3\right) = 0\,, \tag{183.4}$$

$$\left(\frac{c_0}{2h}\right)\dot{\sigma}_0\eta + \tfrac{1}{12}\dot{\sigma}_1 + \left(\frac{c_0}{2h}\right)\sigma_0\dot{\eta} = 0\,.$$

The initial conditions for eqs. (183.3) and (183.4) are written

$$\eta(0) = \eta_i\,, \qquad \sigma_0(0) = \sigma_{0i}\,, \qquad \sigma_1(0) = \sigma_{1i}\,.$$

The quantities η_i, σ_{0i} and σ_{1i} are values of the corresponding parameters immediately after application of load. If we consider that the instantaneous deformation is elastic, it is natural to apply the Reissner variational method to determine the required parameters (see section 167). The same hypothetical distributions of stress and displacement are assumed as in the analysis of creep, that is w and σ are given by formulae (183.2). In our case the Reissner functional assumes the form

$$K = b\int_{-h}^{+h}\int_0^l \left\{\sigma\left[\tfrac{1}{2}w_{,z}^2 - \tfrac{1}{2}w_{0,z}^2 - y(w_{,zz} - w_{0,zz})\right]\right.$$

$$\left. \frac{\sigma^2}{2E}\right\} \mathrm{d}y\,\mathrm{d}z + \int_0^l qw\,\mathrm{d}z\,. \tag{183.5}$$

Calculating this functional and carrying out variation with respect to η_i, σ_{0i} and σ_{1i} we obtain the following system of equations:

$$\sigma_{1i} = 12\left(\frac{c_0}{2h}\right)(\eta_i - 1)\,, \quad \sigma_{0i} = 3\left(\frac{c_0}{2h}\right)^2(\eta_i^2 - 1)\,,$$

$$\sigma_{0i}\left(\frac{c_0}{2h}\right)\eta_i + \tfrac{1}{12}\sigma_{1i} + \frac{4}{\pi^5}\,\frac{ql^4}{EI}\,\frac{1}{2h} = 0\,.$$

Excluding σ_{0i} and σ_{1i} we arrive at the following algebraic equation for η_i:

$$3\left(\frac{c_0}{2h}\right)^2 (\eta_i^3 - \eta_i) + (\eta_i - 1) + \alpha = 0 . \tag{183.6}$$

Here

$$\alpha = \frac{4}{\pi^5} \frac{ql^4}{EI} \frac{1}{c_0} .$$

It follows from formula (183.6) that α considered as a function of η_i assumes an extremal value at

$$\eta_i = \eta_{icr} = \pm \frac{\sqrt{3\left(\frac{c_0}{2h}\right)^2 - 1}}{3\left(\frac{c_0}{2h}\right)^2} . \tag{183.7}$$

If $(c_0/2h)^2 < \frac{1}{3}$, the deflection increases steadily with change in load, if $(c_0/2h)^2 > \frac{1}{3}$ when a critical value of the load parameter α_{cr} is reached, collapse occurs and the deflection immediately changes by a finite amount as shown by arrows in fig. 248. The critical load is found from formula (183.6), by substituting into it the value of η_{icr} given by formula (183.7). There is reason to consider stability in creep when the applied load is less than the critical value. Here we determine η_i as the maximum root of eq. (183.6) after which we determine σ_{0i} and σ_{1i}; these quantities serve as initial values of the corresponding variables in integrating the systems (183.3) or (183.4).

It is evident that the case of linear creep is only of limited practical interest. Nevertheless a considerable number of works have been devoted to the stability of linear viscous-elastic bodies, since the calculations are relatively simple. Thus by eliminating σ_1 from formula (183.3) we obtain the following system:

$$\frac{d\sigma_0}{d\eta} = \beta\eta + \frac{\sigma_0}{\alpha}(1 + \sigma_0 + \beta\eta^2) , \quad \frac{d\eta}{d\tau} = -\frac{\alpha}{1 + \sigma_0 + \beta\eta^2} . \tag{183.8}$$

Here

$$\beta = 6\left(\frac{c_0}{2h}\right)^2 , \qquad \tau = \epsilon_n \frac{E}{\sigma_n} t .$$

The first system of eqs. (183.8) has no closed solution and it must be integrated numerically. When we have found $\sigma_0(\eta)$ from this equation we deter-

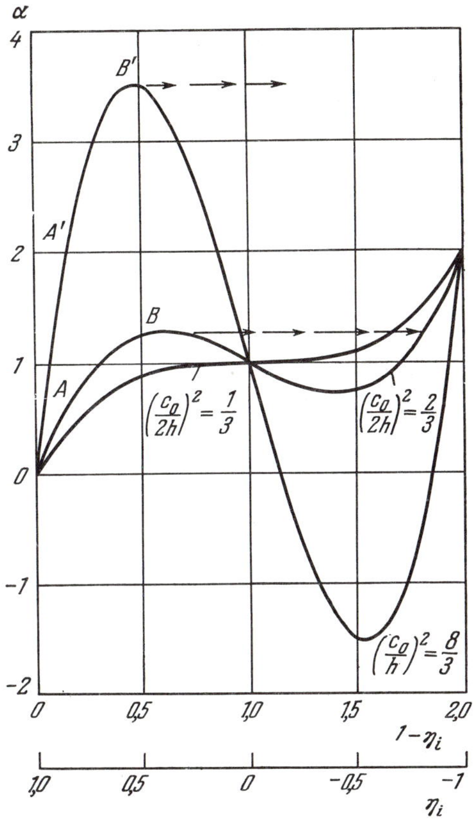

Fig. 248.

mine the dimensionless time τ required to achieve a deflection η by the formula

$$\tau = -\int_{\eta_i}^{\eta} \frac{1 + \sigma_0(\eta) + \beta\eta^2}{\alpha} \, d\eta \,. \tag{183.9}$$

Two possibilities arise in determining the functions $\tau(\eta)$; either the subintegral expression retains the same sign, or the sign changes at certain values of η. In the first case the deflection increases steadily with time, and the quantity η tends to $\eta = -1$. In the second case when the critical time is reached at which

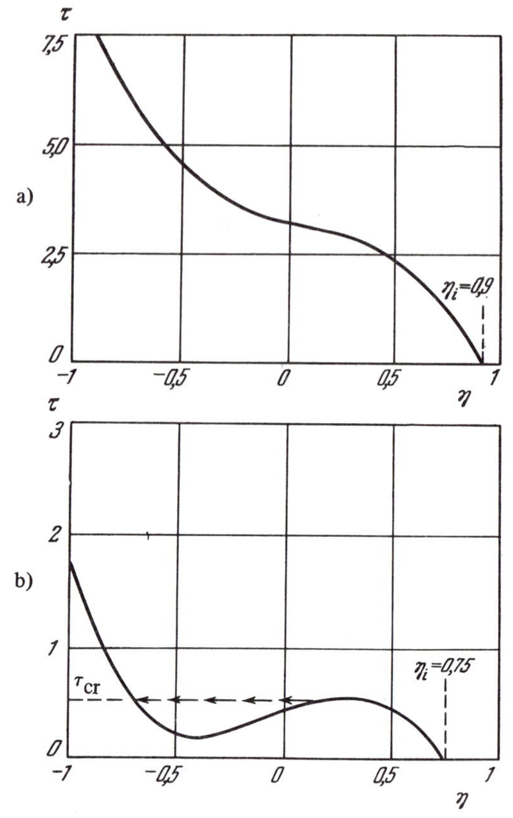

Fig. 249.

the sign of the derivative $\mathrm{d}\tau/\mathrm{d}\eta$ changes, there is a sudden jump. Results of calculations for $\beta = \frac{2}{3}$ when $\alpha = 0.442$ ($\eta_i = 0.9$), and $\alpha = 0.906$ ($\eta_i = 0.75$) are plotted in fig. 249.

The problem of non-linear creep when $n = 3$ is solved in exactly the same way, but the system (183.4) is of more complicated form. The results of calculations when $\beta = \frac{2}{3}$, $\alpha = 0.906$ are plotted in the graph of fig. 250. Here the dimensionless time is defined as

$$\tau = \epsilon_n \frac{E\sigma_e^2}{\sigma_n^3} t .$$

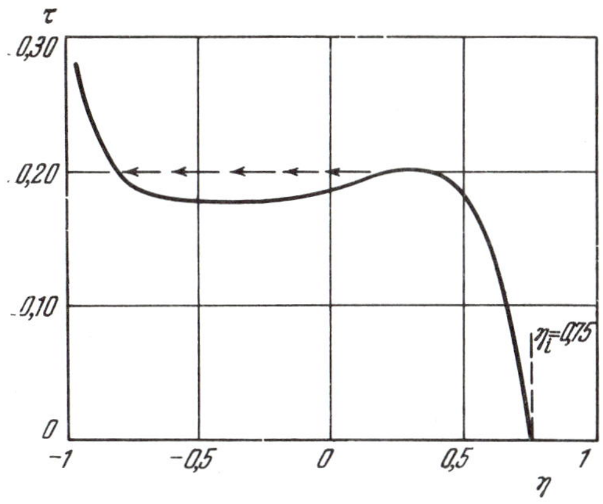

Fig. 250.

In a similar way Shepelenko [147] considered the stability of an arch with clamped ends. He gave certain calculations for the cases of linear visco-elasticity.

184. Certain investigations on the stability of shells in creep

Studies of the stability of shells in creep involve great difficulties of calculation and so the few that have been made are based on rather restrictive hypotheses. Hoff's [236, 238] work on loss of stability of a cylindrical shell in bending was one of the first in this field. He took as the basic mechanism of loss of stability the influence of flattening of the cross-section on the bending stiffness. A similar solution for an elastic shell was proposed in an earlier work of Brazier [174]. Denote the radius of the shell by a and put the generating line along the x axis. According to eq. (147.1) the condition that the peripheral deformation is zero is written as follows:

$$\frac{\partial v}{\partial \varphi} - w = 0 .$$

This condition may be satisfied by putting

$$w = c \cos 2\varphi , \qquad v = \frac{c}{2} \sin 2\varphi . \tag{184.1}$$

Here the circular section of the shell becomes an ellipse with half-axes $a \pm c$. The change of curvature in the peripheral direction is

$$k_\varphi = \frac{1}{a^2}\left(\frac{\partial^2 w}{\partial \varphi^2} + \frac{\partial v}{\partial \varphi}\right) = -\frac{3c}{a^2}\cos 2\varphi\,. \tag{184.2}$$

The important simplification that is made in considering this problem is that the stress in the direction of the generator is considered to be associated only with rates of deformation in this direction, whereas rates of change of curvature k_φ are associated only with transverse bending moments, namely

$$\dot{e} = \epsilon_* \left(\frac{\sigma}{\sigma_*}\right)^n, \qquad \dot{k}_\varphi = \kappa_* \left(\frac{M_\varphi}{M_*}\right)^p. \tag{184.3}$$

This assumption is close to the truth for reinforced shells when reinforcing members in tension participate in the longitudinal deformation and struts in compression in the transverse deformation, both working independently. Formulae (184.3) relate to a reference shell and include the rigidity of the reinforcement. In using the first of these formulae the stress σ is considered to be distributed over a certain reference thickness $2h$. Before deformation the distance of a point on the mean surface of the shell from the axis of bending is $y = a\cos\varphi$, after deformation it is

$$y = a\cos\varphi - w\cos\varphi - v\sin\varphi\,,$$

or

$$y = a\cos\varphi\left[1 - \frac{c}{a}\cos^2\varphi\right].$$

Denoting by k the change in curvature of the axis of the shell that results from the bending moment M, we find that

$$e = ky = ka\cos\varphi\left(1 - \frac{c}{a}\cos^2\varphi\right). \tag{184.4}$$

Apply the commencement of possible displacements to formulate the conditions of equilibrium of the shell. We vary only the parameter c, retaining the moment M constant. The equation of equilibrium is written

$$h\int_0^{\pi/2} \sigma\delta e\, \mathrm{d}\varphi + 2\int_0^{\pi/4} M_\varphi \delta k_\varphi\, \mathrm{d}\varphi = 0\,. \tag{184.5}$$

Here e and k_φ are determined by formulae (184.4) and (184.2) whereas σ and M_φ must be determined from formulae (184.3). The parameters c and k and their derivatives $\dot{c}$ and $\dot{k}$ appear in eq. (184.5). The second equation is obtained by equating the moment due to the stress σ in the cross-section to the constant applied moment M

$$M = 4ha \int_0^{\pi/2} \sigma y \,\mathrm{d}\varphi . \tag{184.6}$$

Self-evident calculations give the following system of two differential equations for the unknown functions $k(t)$ and $c(t)$:

$$\dot{k} = \frac{\epsilon_*}{a}\left(\frac{M}{ha^2\sigma_*}\right)^n \left[\frac{I_{1n}}{1 - \frac{n+1}{n}\frac{c}{a}\left(1 + \frac{z}{n+1}\right)\frac{I_{3n}}{I_{1n}}}\right]^n , \tag{184.7}$$

$$\dot{c} = \kappa_* a^2 \tfrac{1}{3}\left(\frac{Mk}{M_*}\right)^p \left[\frac{I_{3n}}{3I_{1n}I_{1p}} \frac{1 - \frac{c}{na}(1+z)\frac{I_{5n}}{I_{3n}}}{1 - \frac{n+1}{n}\frac{c}{a}\left(1 + \frac{z}{n+1}\right)\frac{I_{3n}}{I_{1n}}}\right]^p . \tag{184.8}$$

Here

$$z = \frac{k\dot{c}}{\dot{k}c}, \qquad I_{kn} = 4\int_0^{\pi/2} (\cos\varphi)^{\frac{kn+1}{n}} \,\mathrm{d}\varphi .$$

The system (184.7)–(184.8) can be integrated numerically. Hoff [236] did this for the case of a smooth shell when $n = p$. Determination of the critical time, that is the time at which the rate of flattening becomes infinitely great, is not associated with a given initial flattening. Indeed the system (184.7)–(184.8) has a solution that satisfies initial conditions $k(0) = c(0) = 0$ of the type

$$c = c_0 t^{p+1} + \dots , \qquad k = k_0 t + \dots .$$

Calculations of the critical time of a smooth shell of aluminium alloy were in very satisfactory agreement with experimental results (fig. 251).

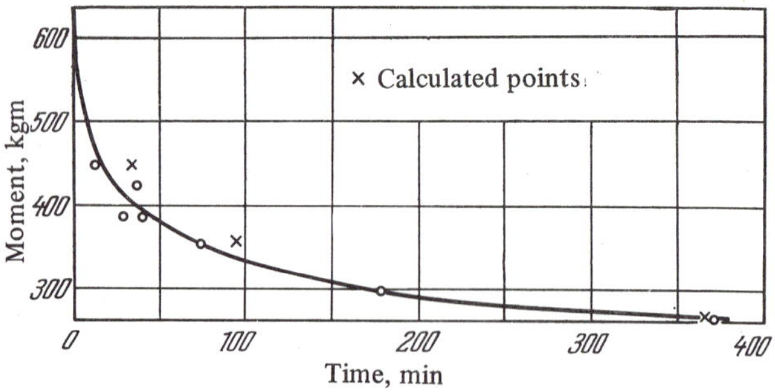

Fig. 251.

We shall mention the few other works on the buckling of shells without considering them in detail. Sundstrom [359] studied the stability of shells in compression and external pressure and also considered displacements in which the mean surface is not deformed; Hoff, Jahsman, and Nachbar [240] and also Thein Wan [371] considered the behaviour of circular cylindrical shells with external pressure. It is assumed that there is initial ellipticity. A number of simplifying assumptions are made in all these works. The mixed variational principle described in section 181 is applied in the work of Lianis [281] which is an extension to the case of creep of the well-known Wagner-Vlasov theory of loss of stability of thin-walled beam of open profile by combined bending and torsion. This variational principle was used to solve the problem of buckling of a long cylindrical tube in Serpico's work [348].

185. **Axially symmetric buckling of short cylindrical shells**

The equations of axially symmetric deflections of cylindrical shells in the presence of a longitudinal force, derived in section 154, are easily extended to axially-symmetric instability. Denote by w the deflection of the shell. Then evidently,

$$\epsilon_* u = \dot{w} \,.$$

Now the curvature of the generator must be allowed for in the equations of equilibrium, and for this purpose we add the quantity $-T_{11}(\partial^2 w/\partial x^2)$ to the applied load q. With the dimensionless notation used in sections 150 and 154 we find that an additional term is added to the quantity p in eq. (155.1), namely

$$-\frac{a}{h}w''.$$

Here the primes denote differentiation with respect to the dimensionless coordinate ξ. It is evident that the variational equation remains valid for the stability problem, and the functional (155.4) is rewritten as follows:

$$N = \int_0^l \left[u'm' + \frac{n}{n+1} u^2 \left(\frac{1}{\omega^+} + \frac{1}{\omega^-} \right) - \frac{1}{n+1} \{(m+\tau)^2 \omega^+ \right.$$

$$\left. + (m-\tau)^2 \omega^-\} + 2\tau \left(\frac{1}{\sqrt{3}} + \frac{a}{h} w'' \right) u \right] \mathrm{d}\xi \,. \qquad (185.1)$$

Volchkov's work [14] contains an application of the variational equation to a study of buckling of a short cylindrical shell with hinged end supports; we shall present his results briefly. Just as in section 152, we consider that u is greater than m and τ, and because $\omega < 1$ in the equation

$$(\omega^{\pm})^{\frac{2n}{n-1}} = u^2 + (m \pm \tau)^2 (\omega^{\pm})^2 \,, \qquad (185.2)$$

the second term in the right-hand side is small compared with the first. As in section 152 we put

$$\omega^{\pm} = u^{\frac{n-1}{n}} (1 + \delta^{\pm}) \,,$$

where $\delta^{\pm} \ll 1$. Substituting into formula (185.2) and retaining up to the first power of δ we find

$$\delta^{\pm} = \frac{n-1}{2n} u^{-2/n} (m \pm \tau)^2 \,.$$

Now substitute the approximate expression for $\omega^{\pm}$ into formula (185.1) and retain in it terms containing $(\omega \pm \tau)$ not higher than second power. We obtain

$$K = \int_0^l \left\{ u'm' + \frac{2n}{n+1} u^{1+\frac{1}{n}} - (m^2 + \tau^2) u^{1-\frac{1}{n}} \right.$$

$$\left. + 2\left(\frac{1}{\sqrt{3}} + \frac{a}{h} w''\right) \tau u \right\} \mathrm{d}\xi \,. \tag{185.3}$$

For a sufficiently short shell we may put

$$u = \dot{\alpha} \sin \frac{\pi\xi}{l}, \qquad m = \beta \sin \frac{\pi\xi}{l}, \qquad w = \epsilon_* \alpha \sin \frac{\pi\xi}{l}.$$

Substituting into expression (185.3) we obtain the following expression for the functional K:

$$K = \frac{\pi^2}{2l} \left[\dot{\alpha}\beta + A\dot{\alpha}^{\frac{n+1}{n}} - B\beta^2 \dot{\alpha}^{\frac{n-1}{n}} - C\tau^2 \dot{\alpha}^{\frac{n-1}{n}} + D\tau\dot{\alpha} - \frac{a}{h} \alpha\dot{\alpha} \right].$$

Here A, B, C and D are integrals of certain powers of sin $\pi\xi/l$. Varying the parameters $\dot{\alpha}$ and $\dot{\beta}$ we obtain the following system:

$$\beta + \frac{n+1}{n} A\dot{\alpha}^{1/n} - \frac{n-1}{n} B\beta^2 \dot{\alpha}^{-1/n} - \frac{n-1}{n} C\tau^2 \dot{\alpha}^{-1/n} - D\tau - \frac{a}{h} \dot{\alpha} = 0,$$

$$\dot{\alpha} - 2B\beta\dot{\alpha}^{\frac{n-1}{n}} = 0 \,.$$

Eliminating β from this we arrive at the following differential equation for the function $\alpha(t)$:

$$k_1 \dot{\alpha}^{1/n} - k_2 \dot{\alpha}^{-1/n} - \frac{a}{h} \alpha + D = 0 \,. \tag{185.4}$$

Here k_1 and k_2 are certain combinations of the constants A, B and C. The solution of eq. (185.4) can be written in parametric form:

$$\alpha = \frac{h}{a} \left[k_1 q^{1/n} - k_2 q^{-1/n} + D \right] ,$$

$$\hat{t} = \hat{t}_{cr} - \frac{h}{a} \left[\frac{q^{1/n}}{n-1} + \frac{q^{-1/n}}{n+1} \right] \frac{1}{q} . \tag{185.5}$$

Here the dimensionless time $\hat{t} = t\epsilon_* \tau^n$, and the constant of integration is the critical time at which $\alpha \to \infty$. Indeed when $q = \infty$ the first of eqs. (185.5) gives $\alpha = \infty$, but when $q = \infty$ from the second equation $\hat{t} = \hat{t}_{cr}$. Therefore, to determine the critical time we must use the initial conditions; when $\hat{t} = 0$, $\alpha = 0$, $q = q_0$. From the first equation putting $\alpha = 0$, we find q_0 and substitute its value into the second equation in which we take $\hat{t} = 0$. The critical time depends on the length of the shell and certain results of calculations are given in the following tables:

$n = 3$

l	1	1.5	2	2.5	3	3.5	4	5
$\hat{t}_{cr}\frac{a}{h}$	83.2	8.41	2.73	1.85	1.78	2.16	2.43	3.42

$n = 5$

l	2.8	2.9	3	3.1	3.2	3.3	3.4
$\hat{t}_{cr}\frac{a}{h}$	1.51	1.48	1.48	1.47	1.49	1.49	1.54

As would be expected the critical time decreases with increase in length of shell and reaches a minimum value. The increase in critical time on further increasing the length does not of course represent a mechanical fact but results from the inadequacy of representing the deflection curve by a sine curve. Indeed in fairly long shells there is an edge effect and the critical time does not depend on the length. The minimum values of critical time determined above, i.e. $\hat{t}_{cr} = 1.78\ h/a$ for $n = 3$, and $\hat{t}_{cr} = 1.47\ h/a$ for $n = 5$ appear to provide a realistic assessment of the critical time of long shells on condition that the type of loss of stability is axially symmetric.

186. Approximate method of solving buckling problems in non-linear formulation

It was mentioned in section 180 that in certain systems the problem of buckling devolves into two individual problems: determination of the deflection that is critical for the given load, and determination of the time to reach this critical deflection by creep. In a compressed column loss of stability was found to depend on the instantaneous characteristics of the system, and in

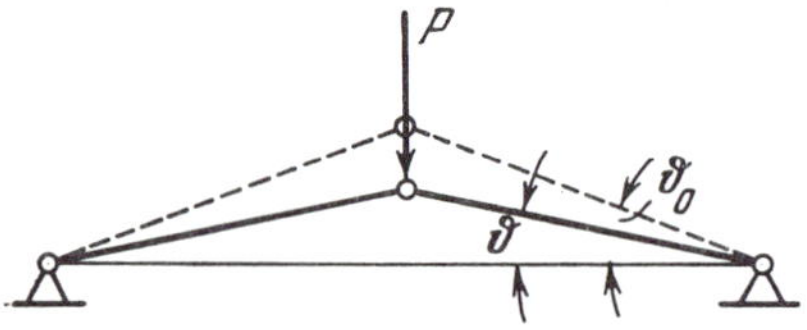

Fig. 252.

the example considered on its elastic-plastic characteristics. The effect of creep is to change the geometry, so bringing the rod into the critical condition. For this procedure to be strict the system must be statically determinate. This means that given the external forces and displacements, the stress distribution in the elements of the system is uniquely determined. The situation may be quite similar in geometrically non-linear systems; if the system is statically determinate the effect of creep is to alter the geometrical dimensions until a critical state is reached at which there is sudden loss of stability. The simplest example on which the essentials of the question can be explained is the so-called Mises girder (fig. 252). A detailed investigation of the behaviour of this system of rods under conditions of linear creep is contained in the work [245], but here we give a simpler analysis. We consider that the rods can be compressed but do not lose stability because of bending. The initial position of the rods is shown dotted, each of them is at an angle θ_0 to the horizontal. When subjected to the force P the rods contract and their angle of slope becomes θ. It is readily calculated that the relative contraction is

$$e = \tfrac{1}{2}(\theta_0^2 - \theta^2) . \qquad (186.1)$$

The compressive stress in each rod is

$$\sigma = \frac{P}{2F\theta} .$$

We consider that the angles θ and θ_0 are small so that $\sin\theta \approx \theta$, $\cos\theta \approx 1-(\theta^2/2)$.

Consider first the behaviour of the system on the assumption that the material is elastic, that is $\sigma = Ee$. Put $q = P/EF$, then

$$q = \theta(\theta_0^2 - \theta^2) . \qquad (186.2)$$

We can find the value of θ at which $\mathrm{d}q/\mathrm{d}\theta = 0$. This is when

$$\theta_0^2 - 3\theta^2 = 0\,,$$

and consequently, $\theta = \theta_{cr} = \theta_0/\sqrt{3}$. The corresponding value of the load parameter is maximum and the angle diminishes further with diminishing load. When $\theta = \theta_{cr}$ the system loses stability, the critical value of the load parameter is

$$q_{cr} = 2\theta_{cr}^3\,. \tag{186.3}$$

Now assume that the material of the rod creeps, the nature of the creep law is quite immaterial, the only important thing is that deformation is always elastic and related to the instantaneous stress by Hooke's law. Then the relationship between stress and deformation is

$$\dot{e} = \frac{\dot{\sigma}}{E} + \dot{p}\,.$$

The value of $\dot{p}$ depends on σ and any parameters except $\dot{e}$ or $\dot{\sigma}$. Substituting the expressions for e and σ we find

$$\left(\frac{q}{2\theta^2} - \theta\right)\dot{\theta} = \dot{p}\,. \tag{186.4}$$

The rate of creep $\dot{\theta}$ becomes infinitely great when the term in the brackets becomes zero, namely

$$\frac{q}{2\theta^2} - \theta = 0\,.$$

Therefore, quite irrespective of the creep law we again arrive at condition (186.3), which can now be interpreted as follows. Any load q can become critical when the angle of slope of the rod reaches a value given as function of q by formula (186.3). The critical time is that at which creep causes the deflection to reach the critical value for the given load. To determine the critical time eq. (186.4) must be integrated and, therefore, the statement that the critical state depends on the geometry and the elastic characteristics may appear useless. However, in determining the dependence of deformation on time under creep conditions for relatively complicated systems we have to use some approximate device and in the approximate equations the singularity which leads to an infinitely high rate, which is clearly evident in eq. (186.4), need not be considered. Any numerical integration process becomes difficult near to values of the independent variable for which the derivative of the required function increases indefinitely. There is, therefore, reason to formulate the stability problem as that of determining the time required to achieve a given critical deflection.

Returning to eq. (186.4) we assume that the creep law, i.e. the dependence of $\dot{p}$ on σ, t and other parameters, can be of any form. During a short interval of time Δt the stresses in the rods do not change significantly and consequently the quantity $\Delta p \approx \dot{p}\Delta t$ can be calculated. As a result eq. (186.4) is replaced by the following difference equation:

$$\Delta\theta = \frac{\Delta p}{\dfrac{q}{2\theta^2} - \theta}. \tag{186.5}$$

Formula (186.5) can be used to calculate successive stages, determining the stress each time by the formula $\sigma = Eq/2\theta$. A somewhat different interpretation can be given for formula (186.5). The influence of creep deformation Δp taking place in a time Δt is fully equivalent to the influence of change in initial angle θ_0, and it follows from eq. (186.1) that the equivalent change in θ_0 is

$$\Delta\theta_0 = -\frac{\Delta p}{\theta_0}. \tag{186.6}$$

Now assuming the system to be elastic, from eq. (186.2) when $q = \text{const}$ we find that the change in θ_0 causes a change in the angle θ which is expressed as

$$\Delta\theta = -\frac{\theta_0 \Delta\theta_0}{\dfrac{q}{2\theta} - \theta}.$$

Substituting into this the expression for $\Delta\theta_0$ from eq. (186.6) we again arrive at expression (186.5).

In this example, in which the force distribution is fully determined by giving the external forces and deflection, it is strictly correct to use the device of determining the deformation of the system by the formulae for the case of elasticity, and considering that the effect of creep is to change the initial deflection. In the general case of course, this device is not accurate, but it can be used for approximate determination of the critical time of complicated systems.

The works of Vol'mir and Zykin [16, 15] give an approximate solution of the problem of buckling a cylindrical panel supported round the edge and in compression along the generating line. The method of solution is to assume the shell elastic and to obtain a known system of two non-linear differential equations for the deflection w and the force function on the mean surface Φ. It is assumed that

$$w = ch \sin\frac{\pi x}{a} \sin\frac{\pi y}{b}, \qquad w_0 = c_0 h \sin\frac{\pi x}{a} \sin\frac{\pi y}{b}. \tag{186.7}$$

One of the equations is now integrated and an expression is found for the function Φ, and the other is satisfied approximately in the sense of the Galerkin method. As a result we obtain the dependence of the compressive force on c and c_0:

$$p = p(c, c_0). \tag{186.8}$$

The curve for the dependence of p on c with c_0 given, has a maximum corresponding to the upper critical load and a minimum corresponding to the lower critical load. For given values of p and c_0 we can now calculate the forces in the middle surface T_{ij} and the moments M_{ij}. The forces and moments that are found correspond to a particular creep rate, and consequently a certain creep deformation occurs in a time Δt. It is evident that with a non-linear creep law the resultant creep deformation inevitably distorts the shape of the panel surface given by eq. (186.7). It is supposed that the form of the function $w(x,y)$ is approximately retained. The alteration in deflection is calculated from the magnitude of the creep deformation at the centre of the panel. This change of curvature is considered equivalent to a change in the initial deflection c_0 by an amount Δc_0. The calculation is now repeated but the initial deflection is considered to be $c_0' = c_0 + \Delta c_0$. Thus successive time steps lead the calculation to the instant of failure. The condition at which failure occurs is given by eq. (186.8) because for any value of p we can find a value of c such that $dp/dc = 0$, if p is expressed in terms of c and c_0 by formula (186.8).

187. Stability of a compressed visco-elastic rod

The problems considered above are similar in character to 'stability' problems of the theories of elasticity or plasticity. However, we avoided using the term 'stability' because all the examples related to systems that are unstable in the classical sense of the term. We were not so much concerned with the stability of a compressed column as of the way in which its deflection increases with time and of establishing the instant at which the deflection becomes impermissible for one reason or another. The stability problem can be formulated in the direct and strict meaning of the word for visco-elastic rods. Suppose that the relationship between stress and deformation is given by eq. (23.5), namely

$$\dot{\sigma} + \lambda\sigma = E(\dot{e} + \mu e) . \tag{187.1}$$

The stability of a compressed rod of material of this kind was first considered by Rzhanitsyn [112]. It will be recalled that in eq. (187.1) E is the instantaneous modulus of elasticity. This means that on the instantaneous application of a force σ a deformation $e = \sigma/E$ occurs 'instantaneously' (ignoring dynamic effects). If a constant stress σ acts for a long time the deformation e tends to a value given by the formula $e = \sigma/E_f$. Here E_f is the long-term modulus of elasticity, $E_f = E(\mu/\lambda)$. It is evident, as was explained in section 23, that $\mu < \lambda$. Now assume that the column deflects in the principal plane of the load, which corresponds to a bending moment distribution $M_0(t, z)$. The initial curvature of the rod in this main plane is given by a function $v_0(z)$. Put $e = ky$, substitute into eq. (187.1), multiply both sides by y and integrate across the area of the section. We obtain

$$\dot{M} + \lambda M = EI(\dot{k} + \mu k) . \tag{187.2}$$

Here M is the bending moment, I the moment of inertia of the area of the section. If the compressive force is P and the deflection v, we have

$$M = -Pv + M_0 , \qquad k = v''(z) - v_0''(z) .$$

Now eq. (187.2) assumes the form

$$-\dot{P}v - P(\dot{v} + \lambda v) + \dot{M}_0 + \lambda M_0 = EI(\dot{v}'' + \mu v'') - EI\dot{v}_0'' - \mu EI v_0'' . \tag{187.3}$$

Let $V_k(z)$ be the fundamental function of the problem of stability of an elastic rod, that is the function which satisfies the equation

$$EIV_k''(z) + P_k V_k(z) = 0$$

and homogeneous boundary conditions of the problem. The quantities P_k are the elastic critical loads. The functions $v(z)$ and $v_0(z)$ can, of course, be expanded into series of the fundamental functions. Let us formally represent the function $M_0(z)$ in the form of such a series, namely

$$v(t,z) = \sum \tau_k(t) V_k(z) , \qquad v_0(z) = \sum c_k V_k(z) ,$$
$$M_0(t,z) = -\sum m_k(t) V_k(z) P_k . \tag{187.4}$$

Now assume that the initial curvature and the external load are such that only the first terms are retained in the series for $v_0(z)$ and $M(t,z)$; $v_0 = c_1 V_1$, $M_0 = -m_1 V_1 P_1$; it then suffices to retain the first term $v = \tau_1 V_1$ in the expression for v. Here V_1 is the first fundamental function that determines the kind of loss of stability of the elastic rod when the applied force is equal to the first Euler load. This does not limit the generality of the discussion. Following the usual method it is easy to construct a solution for the general case as well, in which the initial deflection and the moment are given by series (187.4). Hereafter we omit the subscripts of τ_1, c_1, m_1 and V_1 and instead of P_1 we write P_e. With these assumptions a differential equation of the following kind for the function τ is derived from eq. (187.3):

$$-\dot{\omega}\tau + \dot{\tau}(1-\omega) + (\mu - \omega\lambda)\tau = \mu c + (\dot{m} + \lambda m) ,$$
$$\omega = \frac{P}{P_e} . \tag{187.5}$$

Now assume that the longitudinal force increases according to an arbitrary law from zero to its final value in a short time t_1, after which it remains constant. The external bending moment changes very slightly during t_1. When $t = 0$ we have $\tau = c$. In the time interval $0 < t < t_1$ the derivatives $\dot{\omega}$ and $\dot{\tau}$ are very large and, therefore, only terms containing these derivatives need be retained in eq. (187.5), we obtain

$$\dot{\tau} - \dot{\omega}\tau - \omega\dot{\tau} = 0 . \tag{187.6}$$

Integrating eq. (187.6) between the limits $t = 0$ and $t = t_1$, bearing in mind that $P(0) = 0$, and $\tau(0) = c$, we obtain

$$\tau - c = \omega\tau .$$

Whence

$$\tau_0 = \frac{c}{1-\omega} . \tag{187.7}$$

Here τ_0 denotes the value of τ when $t = t_1$. In the limit when t_1 tends to zero we find that with a constant force P the eq. (187.5), in which $\dot{\omega} = 0$ must be integrated; the initial condition for this equation is $\tau(0) = \tau_0$ where τ_0 is given by formula (187.7).

The quantity P_e is proportional to the modulus of elasticity of the material E and therefore P_e is the critical force that corresponds to the instantaneous modulus. The critical force that corresponds to the long-term modulus can be determined

$$P_e' = P_e \frac{E_f}{E} = \frac{\mu}{\lambda} P_e .$$

Therefore, the ratio μ/λ is the critical value of the dimensionless force P/P_e calculated relative to the long-term modulus. Put

$$\frac{\mu}{\lambda} = \omega' .$$

Then when $\dot{\omega} = 0$, eq. (187.5) is rewritten as

$$\dot{\tau} + \beta\tau = \frac{1}{1-\omega} [\mu c + \dot{m} + \lambda m] . \tag{187.8}$$

Here $\beta = \lambda[(\omega' - \omega)/(1-\omega)]$ and $\tau(0) = c/(1-\omega)$.

Note that

$$\beta > 0 \quad \text{when} \quad \omega < \omega' ,$$

$$\beta < 0 \quad \text{when} \quad \omega' < \omega < 1 .$$

The terms in the right-hand side of eq. (187.8) are disturbances, and their origin is two-fold; c is the initial deflection, m the applied moment. There is reason to consider the actions of the disturbances separately, bearing in mind that they are of different nature. The initial deflection c is set up in the rod before application of the load and it cannot be altered after the force has been applied, this disturbance acts continuously. The transverse load that creates the moment m is applied externally and is given as a function of time of arbitrary form. Accordingly, assume first that $m = 0$. Then the solution of eq. (187.8) that satisfies the initial condition for τ is

$$\tau = c\left[\frac{1}{1-\dfrac{\omega}{\omega'}} + \left(\frac{1}{1-\omega} - \frac{1}{1-\dfrac{\omega}{\omega'}}\right)e^{-\beta t}\right]. \tag{187.9}$$

If $\beta > 0$, with increasing time τ tends to the limit

$$\tau(\infty) = \frac{c}{1-\dfrac{\omega}{\omega'}}.$$

This is the deflection that is obtained on the assumption that the material is elastic with a long-term modulus. If $\beta > 0$, the solution given by eq. (187.9) is stable in the sense that a small change Δc in the initial deflection c causes a small change $\Delta\tau$ in the quantity τ, and whatever the positive number η a positive number ϵ can always be found such that $|\Delta c| < \epsilon$, then $|\Delta\tau| < \eta$ for any value of t. This is the classical definition of stability in the Lyapunov sense. When $\beta < 0$ and consequently, $\omega > \omega'$, the quantity τ given by formula (187.9) increases indefinitely; the quantity $|\Delta\tau|$ also increases indefinitely, however small $|\Delta c|$. When $\omega > 1$ we again have $\beta > 0$. However, the stability of the column does not follow from this, for in this case stability is lost suddenly. Indeed formula (187.7) which is obtained by integrating formula (187.6) is valid for the process of loading the rod with a sufficiently rapidly increasing force. When $\omega \to 1$, the quantity τ_0 given by this formula, tends to infinity; therefore, when $\omega > 1$ formula (187.7) has no meaning.

Now consider the action of the disturbance due to the transverse load. In eq. (187.8) put $c = 0$, we shall consider that m is a given function of t. Without impairing generality we can take $m(0) = 0$. Indeed, if the load is applied suddenly we are always justified in assuming that this does not occur at the instant $t = 0$, but at the instant $t = \epsilon$, where ϵ is any positive quantity, however small. Then the solution of eq. (187.8) becomes

$$\tau = \frac{m}{1-\omega} + \frac{1-\omega'}{(1-\omega)^2}\, e^{-\beta t} \int_0^t m(t)e^{\beta t}\,dt. \tag{187.10}$$

To be specific we shall consider that $m(t) \neq 0$ when $0 < t < t_2$, whereas when $t > t_2$, $m(t) = 0$. Then when $t > t_2$ the integral in the right-hand side of eq. (187.10) retains a constant value and we have

$$\tau = \mathrm{e}^{-\beta t} \times \text{const}\,.$$

When $\beta > 1$, the deflection developed by the load diminishes when the load ceases to act and tends to zero as $t \to \infty$. If $\beta < 1$, the deflection increases indefinitely even though the disturbance no longer acts.

This example shows that it is very easy to assess the stability and instability of a rod of a material whose properties are described by eq. (187.1). The rod is stable when the compressive force is less than the Euler force given by the long-term modulus. If the compressive force is greater than the long-term Euler force but less than the instantaneous Euler force the rod is unstable and any disturbance causes an indefinitely great increase in the deflection with time. Finally, if the compressive force is greater than the instantaneous Euler force stability is lost instantaneously. This result is also valid for the linear visco-elastic body described by the relationships of the linear hereditary theory. If the creep kernel is unlimited the rod is always unstable. If the kernel is limited the stability or instability can be assessed, just as for an elastic rod, by substituting the long-term modulus for the instantaneous modulus [191].

If in eq. (187.1) we put $\mu = 0$, we obtain a visco-elastic body with unlimited creep; the creep rate is constant with constant stress. Compression of a column of such a material is always unstable, for $\omega' = 0$ and $\beta < 0$. Therefore, in considering the use of elements of such material in compression, the deflection must be found by formula (187.9) if the initial deflection is known, or by formula (187.10) if the transverse load is given. The time is determined beyond which the deflection starts to exceed a certain value that is selected on the basis of particular considerations.

188. Stability of a compressed column of material that observes the strain-hardening law

Assume that the material of the column creeps according to the strain-hardening law

$$\Phi(p, \dot{p}, \sigma) = 0\,, \qquad p = e - \frac{\sigma}{E}\,. \tag{188.1}$$

The fundamental state of the column is one of uniform compression, with $\sigma = \text{const} = P/F$; we suppose that eq. (188.1) is integrated for the fundamental condition which means that p is known as a function of t. We assume that deviations from the fundamental conditions are small; then quantities δe and $\delta\sigma$ that are associated with deflection satisfy the equation that is obtained by linearisation of eqs. (188.1) about the fundamental condition. This equation was given in section 61, namely

$$(E\lambda - \mu)\delta\sigma - \nu\delta\dot{\sigma} + E(\mu\,\delta e + \nu\,\delta\dot{e}) = 0\,. \tag{188.2}$$

Here $\lambda = \partial\Phi/\partial\sigma$, $\mu = \partial\Phi/\partial p$, $\nu = \partial\Phi/\partial\dot{p}$ are known functions of time. Suppose, as in section 187, that $\delta e = ky$, substitute into eq. (188.2), multiply by y and integrate over the area of the cross-section. We obtain

$$(E\lambda - \mu)M - \nu\dot{M} + EI(\mu k + \nu\dot{k}) = 0\,. \tag{188.3}$$

The difference between eqs. (188.3) and (187.3) is that the coefficients of (188.3) are variable, known functions of time. As in the preceding section, put

$$M = M_0 - Pv\,, \qquad k = v'' - v_0''\,.$$

Then eq. (188.3) assumes the form

$$(E\lambda - \mu)M_0 - \nu\dot{M}_0 - P(E\lambda - \mu)v + \nu(\dot{P}v + P\dot{v})$$
$$+ EI(\mu v'' - \mu v_0'' + \nu\dot{v}'') = 0\,.$$

Now let

$$M_0 = -P_e m(t)V(z)\,, \qquad v = \tau(t)V\,, \qquad v_0 = cV\,.$$

Here the function V is the first eigenfunction of the equation

$$EIV'' = -PV\,,$$

and P_e is the corresponding eigenvalue.

Substituting the expressions for M_0, v and v_0 into the equation of deflection we find that the function $\tau(t)$ satisfies the following differential equation:

$$\omega(E\lambda-\mu)\tau-\nu(\dot{\omega}\tau+\dot{\tau}\omega)+(E\lambda-\mu)m-\nu\dot{m}+\mu\tau-\mu c+\nu\dot{\tau}=0\,. \tag{188.4}$$

When $t=0$ we should have $\tau=c$. Assume that the longitudinal compressive force is applied for a very short time. Then in eq. (188.4) we retain only terms containing $\dot{\tau}$ and $\dot{\omega}$, namely

$$-\dot{\omega}\tau-\tau\dot{\omega}+\dot{\tau}=0\,.$$

Integrating we find that the instantaneous deflection after application of the force is

$$\tau_0=\frac{c}{1-\omega}\,.$$

Now if the compressive force is constant, we should consider eq. (188.4) with $\dot{\omega}=0$. The initial condition for integration of this equation is $\tau(0)=\tau_0$. Write the equation for the function $\tau(t)$ as follows:

$$\dot{\tau}+\frac{\tau}{\nu}\left[\frac{E\lambda\omega}{1-\omega}+\mu\right]=\frac{1}{1-\omega}\left[\dot{m}-\frac{E\lambda-\mu}{\nu}m+\frac{\mu}{\nu}c\right]. \tag{188.5}$$

We need some form of the strain-hardening law, let us put

$$\Phi(\dot{p},p,\sigma)=\dot{p}p^{\alpha}-A\sigma^{n}\,.$$

Then

$$\nu=p^{\alpha}\,,\qquad \mu=\alpha\dot{p}^{\alpha-1}\,,\qquad \lambda=-nA\sigma^{n-1}=-\frac{n}{\sigma}\dot{p}p^{\alpha}\,.$$

Substituting these expressions into eq. (188.5) and taking as the independent variable a quantity that is proportional to the creep deformation of the fundamental state of compression, namely

$$\xi=\frac{nE}{\sigma}\,\frac{\omega}{1-\omega}\,p\,,$$

then eq. (188.5) is converted to the form

$$\frac{\mathrm{d}\tau}{\mathrm{d}\xi}-\tau\left(1-\frac{\alpha}{\xi}\right)=\frac{1}{1-\omega}\left\{\frac{\mathrm{d}m}{\mathrm{d}\xi}+\left(\frac{1-\omega}{\omega}+\frac{\alpha}{\xi}\right)m+\frac{\alpha c}{\xi}\right\}. \tag{188.6}$$

The terms in the right-hand side that depend on m are disturbances due to external loads; the term containing the factor c is a disturbance which acts constantly. Let us consider these two disturbances separately.

a) *Constantly acting disturbances.* If $m = 0$, the solution of eq. (118.6) which satisfies the initial condition $\tau(0) = c/(1-\omega)$ is

$$\tau = \frac{\alpha c}{1-\omega} e^{\xi} \xi^{-\alpha} \int_0^{\xi} e^{-\xi} \xi^{\alpha} d\xi . \tag{188.7}$$

Eq. (188.7) always gives an increasing function ξ and so motion of the rod is unstable.

b) *Disturbances of the external load type.* If $c = 0$, the integral of eq. (188.6) is

$$\tau = m + \frac{1}{\omega} \xi^{-\alpha} e^{\xi} \int_0^{\xi} m(\xi) \xi^{\alpha} e^{-\xi} d\xi . \tag{188.8}$$

(In the work [338] this formula contains an error in the coefficient which does not affect the qualitative result.)

189. Nominal linearised criteria of stability

We have shown that motion of a column in compression under creep conditions is always unstable. Formulae (188.7) and (188.8) give the magnitude of the deflection which tends to infinity with time. Therefore, in practice these formulae can be used to determine the time (or compressive deformation) at which the deflection reaches a certain value. If the initial disturbance and the permitted deflection are both sufficiently small the linearised equation gives a sufficiently close approximation. However, the behaviour of the functions given by expressions (188.7) and (188.8) differ in certain singularities. There may be characteristic points on the curves of $\tau(t)$ at which the behaviour of the function alters in one way or another. The time at which these points are reached is termed the nominal critical time. Depending on the type of disturbance, the nominal critical time is defined in various ways.

1. *Quasi-static criterion.* If we do not consider the action of the disturbance, that is we consider eq. (188.5) without the right-hand side, the function $\tau(t)$ given by this equation is diminishing or increasing depending on the sign

of the factor by which τ is multiplied. In the first case let us call the condition stable and in the second unstable. The boundary of stability is given by the condition

$$\frac{E\lambda\omega}{1-\omega} + \mu = 0 \,. \tag{189.1}$$

With the power creep law used in the preceding paragraph this condition becomes

$$\xi = \xi_1 = \alpha \,. \tag{189.2}$$

This condition is immediately obtained by equating to zero the factor that multiplies τ in eq. (188.6). The criterion of stability (189.1) was introduced in the works [336, 107], but they did not give a clear explanation of the meaning of the criterion (189.1); the point is that the initial condition cannot be stated for eq. (188.5) without the right-hand side. A rod that deviates from the rectilinear condition is not in a state of equilibrium. However, if we consider the action on the rod of a certain system of disturbances, criterion (189.1) has a perfectly definite meaning, as is shown in [338]. Assume that a certain load acts for a time t^*, after which it is removed and subsequent motion of the rod is followed. The parameter that now plays the part of time is ξ, which is proportional to the creep deformation of the fundamental state; the time t^* corresponds to the value ξ^* of this parameter. When $\xi > \xi^*$, $m(\xi) = 0$, consequently, the integral in eq. (188.8) becomes a constant and formula (188.8) is written as follows:

$$\tau = \xi^{-\alpha} e^{\xi} \times \text{const} \,.$$

Whence it follows

$$\frac{d\tau}{d\xi} = \left(1 - \frac{\alpha}{\xi}\right)\tau \,.$$

If $\xi < \alpha$, $d\tau/d\xi < 0$; if $\xi > \alpha$, $d\tau/d\xi > 0$. The change in deflection as a function of ξ is illustrated schematically in fig. 253. When $\xi < \xi^*$, so long as the transverse load acts the quantity τ varies in accordance with the law of change of load, i.e. the dependence on the function $m(\xi)$ is given by eq. (188.8). If the load is removed before the critical deformation $\xi_1 = \alpha$ is reached, that is

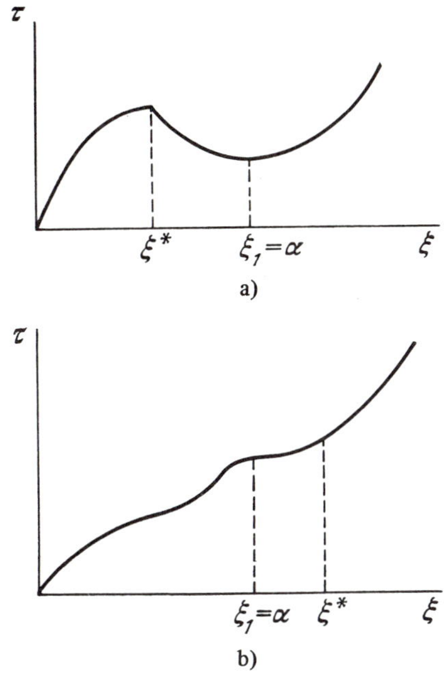

Fig. 253.

if $\xi^* < \xi_1$, as shown in fig. 253a, the deflection that is reached at the instant $\xi = \xi^*$ diminishes, it reaches a minimum value when $\xi = \xi_1$ after which it increases indefinitely. If the load is removed after the critical deformation $\xi_1 = \alpha$ has been reached, the deflection immediately commences to increase (fig. 253b).

The works mentioned above [107, 336] also give a dynamic treatment of the criterion (189.1). It should be emphasised that in this case the dynamic approach to the problem of stability does not mean that the form of the strain-hardening law which is used to describe slow processes is also physically true for describing oscillatory processes. Here consideration of the dynamic formulation of the stability problem has a purely formal meaning, all we do is to consider all the possible consequences of the particular creep law. In fact, of course, loss of stability is of quasi-static nature and consideration of dynamic effects is of interest only in special cases. The curious and rather unexpected behaviour of a compressed column of material that obeys the strain-hardening law that we have described was investigated in detail by Hoff [242].

2. *Kurshin's criterion.* Kurshin [63, 64] considered the behaviour of a

column subject to disturbance by an external load, but of a somewhat different kind. The column is assumed to remain rectilinear until the parameter ξ reaches a value $\xi = \xi_2$. At this time a load which corresponds to the bending moment m is suddenly applied. For simplicity suppose m is constant. Motion of the column is investigated during a short time interval after application of the load. From formula (188.8) we obtain

$$\tau = m + \frac{1}{\omega} \xi^{-\alpha} e^{\xi} \int_{\xi_2}^{\xi} m \xi^{\alpha} e^{-\xi} \, d\xi \, .$$

If $\xi - \xi_2$ is small, we may put

$$\tau = m + \frac{m}{\omega} \left(\frac{\xi}{\xi_2} \right)^{-\alpha} e^{\xi - \xi_2} (\xi - \xi_2) \, .$$

Calculate the derivative $d\tau/d\xi = \tau'$:

$$\tau' = \frac{m}{\omega} \left[1 + \left(1 - \frac{\alpha}{\xi} \right) (\xi - \xi_2) \right] \left(\frac{\xi}{\xi_2} \right)^{-\alpha} e^{\xi - \xi_2} \, . \tag{189.3}$$

Here put $\xi = \xi_2$, we find the rate at which the rod bends immediately after application of the load

$$\tau'(\xi_2) = \frac{m}{\omega} \, .$$

Of course the quantity τ' is not a rate because ξ is a function of time. The important point is that for any value of ξ_2, τ' and consequently $\dot{\tau}$ are both positive. However, as we shall see, the second derivative of τ with respect to time $\ddot{\tau}$, can be either positive or negative. As the parameter ξ is proportional to t^m $[m = 1/(1 + \alpha)]$ it can readily be shown that

$$\ddot{\tau} \sim \tau'' - \frac{\alpha}{\xi} \tau' \, . \tag{189.4}$$

Differentiating (189.3) once more with respect to ξ we find

$$\tau'' = \frac{m}{\omega} \left[\left(1 - \frac{\alpha}{\xi} \right) + \frac{\alpha}{\xi^2} (\xi - \xi_2) + \left(1 - \frac{\alpha}{\xi} \right)^2 (\xi - \xi_2) \right] \left(\frac{\xi}{\xi^2} \right)^{-\alpha} e^{\xi - \xi_2} .$$

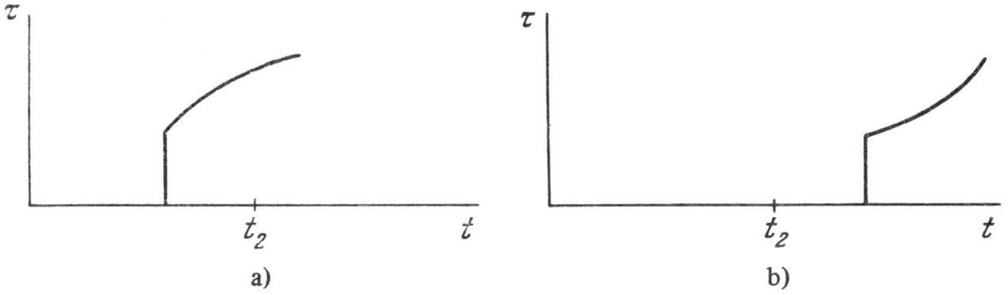

Fig. 254.

When $\xi = \xi_2$,

$$\tau''(\xi_2) = \frac{m}{\omega}\left(1 - \frac{\alpha}{\xi_2}\right).$$

Because of eq. (189.4) the condition $\ddot{\tau} = 0$ reduces to the form

$$1 - \frac{2\alpha}{\xi_2} = 0\,.$$

Whence

$$\xi_2 = 2\alpha\,. \tag{189.5}$$

The quantity ξ_2 corresponds to the time t_2, which is taken as critical. If the load is applied when $t < t_2$ motion is decelerating, as shown in fig. 254a. If the load is applied when $t > t_2$ motion is accelerating; a typical graph of the function $\tau(t)$ is shown in fig. 254b.

3. *Shesterikov's criterion.* In the most realistic case the rod has an initial deflection or initial eccentricity and its motion is described by eq. (188.6). The integral in eq. (188.7) is an incomplete gamma-function found from tables, and consequently, $\tau(\xi)$ can always be constructed. Here the rate of deflection is positive, however, at a certain value $\xi = \xi_3$ the acceleration changes sign. The point $\xi = \xi_3$ conventionally separates the region of slow rate of increase in deflection from that of rapid increase and is taken as the critical point. To calculate ξ_3 we calculate the function $\tau(\xi)$; in the work [150] this is done for several whole-number values of α, the results are

$\alpha =$	1	2	3
$\xi_3 =$	1.36	3.10	4.98

All three criteria that we have considered define the critical state of the column as one at which the parameter ξ reaches some critical value ξ_k which is a function of α. Returning to the definition of the parameter ξ we find that in the critical state

$$\frac{nE}{\sigma}\frac{\omega}{1-\omega}p = \xi_k(\alpha) . \tag{189.6}$$

But

$$p = e - \frac{\sigma}{E}, \qquad \omega = \frac{P}{P_e} = \frac{\sigma}{\sigma_e} .$$

We now use the notation $\hat{\sigma}$ for the dimensionless stress $\sigma/\sigma_e = \omega$. In the same way, we denote eE/σ_e by $\hat{e}$. Then condition (189.6) can be written in the form

$$\hat{e} = \frac{\xi_k}{n} + \left(1 - \frac{\xi_k}{n}\right)\hat{\sigma} . \tag{189.7}$$

Eq. (189.7) plotted in coordinates $\hat{\sigma}-\hat{e}$ gives a straight line which passes through the point (1.1) and which forms an intercept ξ_k/n on the abscissus (fig. 255). The bisectrix of the coordinate angle $\hat{\sigma} = \hat{e}$ is the line of instantaneous deformation. On application of a certain stress $\sigma < \sigma_e$ such that $\hat{\sigma}$ is represented by the segment OA in fig. 255, there is instantaneous elastic deformation given by the segment $\hat{e} = AB$. As a result of creep the deformation increases and when a creep deformation represented by segment BC is

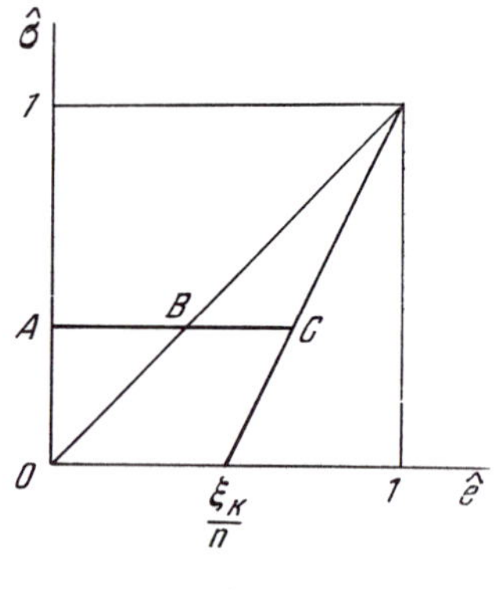

Fig. 255.

added to the elastic deformation represented by the segment AB a critical state is reached.

190. Application of linearised equations to the problem of buckling

Comparison with experiments shows that the conventional criteria of stability described above significantly underestimate the critical time. This is to be expected because the qualitative changes in the shape of the creep curve that are given by these criteria are insignificant and the actual rate of increase in deflection is much slower. Thus if we assume a type of disturbance that corresponds to the quasi-static criterion, the deflection reaches a minimum value when $\xi = \xi_1 = \alpha$. On doubling the parameter ξ and when $\alpha = 2$ (Andrade's creep law) the deflection increases by 84%, that is it approximately doubles. If $n = 3$ this increase in deflection occurs after a time which is eight times longer than the critical time. These changes of deflection are not recorded experimentally, the tests determine the actual time at which the rod suffers total loss of load-bearing capacity or the time at which very large deflections occur. On the other hand we have seen that it is also without meaning to define the critical time as that at which the deflection becomes infinite. In practice, we should take as critical the time taken to reach a certain deflection, which is either considered to be impermissible or which corresponds to instantaneous elasto-plastic loss of stability. The ratio of this deflection to half the thickness of the rod can be assessed to be of the order of $\hat{c} \approx 1$, if when $\hat{c} < 1$ there is no instantaneous loss of stability. It is of interest to assess the accuracy if a linearised equation is used to determine this deflection instead of the full creep equations. To be specific assume that there is no strain-hardening and the creep law is

$$\dot{e} = \frac{1}{E}\dot{\sigma} + \epsilon_n\left(\frac{\sigma}{\sigma_n}\right)^n . \tag{190.1}$$

The first variation of eq. (190.1) is

$$\delta\dot{e} = \frac{1}{E}\delta\dot{\sigma} + \frac{n}{\sigma}\dot{p}\,\delta\sigma . \tag{190.2}$$

Here $\dot{p} = \epsilon_n(\sigma/\sigma_n)^n$ is the rate of creep in the fundamental condition. Put $e = ky$, multiply by y and integrate, we find the following relationship between moment and curvature:

$$EI\,\frac{\partial k}{\partial p}=\frac{\partial M}{\partial p}+\frac{nE}{\sigma}\,M\,.$$

Put as above $k=v''-v_0''$, $M=-Pv$, $v=cV$, where V is the first fundamental function of the stability problem of a compressed column. As a result we obtain the following equation for $c(p)$:

$$(1-\omega)\,\frac{\mathrm{d}c}{\mathrm{d}p}=\frac{n}{e_e}\,c\,. \tag{190.3}$$

The initial condition when $t=0$ is $c(0)=c_0=c_{00}/(1-\omega)$. Integrating eq. (190.3) we obtain

$$p=(1-\omega)\,\frac{e_e}{n}\,\ln\frac{c}{c_0}\,. \tag{190.4}$$

Formula (190.4) is fully accurate for a linearised creep law irrespective of the value of the exponent n and the shape of the cross-section of the rod. In section 178, using the exact creep equation, we could only solve this problem for an ideal I-section and that by an approximate method. Even with the limitations that we have adopted, an explicit expression for the critical time can be obtained if n is a whole number. Compare formulae (190.4) and (178.10),

$$p=(1-\omega)\,\frac{e_e}{3}\,\ln\frac{c\sqrt{3+c_0^2}}{c_0\sqrt{3+c^2}}\,, \tag{190.5}$$

obtained for $n=3$. For this value of n formula (190.4) gives

$$p=(1-\omega)\,\frac{e_e}{3}\,\ln\frac{c}{c_0}\,. \tag{190.6}$$

Of course, the essential difference between formulae (190.6) and (190.5) is that when $c=\infty$ formula (190.5) gives a finite critical deformation, whereas by formula (190.6) when $c=\infty$, $p=\infty$. However, if we consider $c_0\ll 1$ and c of the order of unity the difference between the results obtained by formulae (190.5) and (190.6) is small. Indeed, considering $c_0^2\ll 3$ and $c=1$ we find that the quantity under the logarithm in formula (190.5) is:

$$\frac{c}{c_0}\sqrt{\tfrac{3}{4}}\,,$$

which corresponds to increasing the initial deflection c_0 in the approximate formula (190.6) in the ratio $2/\sqrt{3} = 1.15$. Of course, the initial deflection is never known to within 15%. Moreover, the method of deriving formula (190.5) is such that only the structure of the formula is reliable and not the numbers that appear in it. Indeed, in section 178 this formula was obtained by two methods and the numbers were different. In section 182 this problem was solved by a variational method, the result was again different. Therefore, use of a linearised creep law to determine the nominal time of buckling in the sense given above, is fully justified. The advantage of this method is that the linearised equation permits a fully accurate solution irrespective of the shape of the section, the boundary conditions and the creep law. In particular, using formula (188.7) we can also solve the problem of buckling in the presence of creep accompanied by work-hardening.

191. **Linearised equations of creep with work-hardening**

In order to extend the linearised theory to problems of buckling of plates and shells it is first necessary to write a system of equations that relate variations of stress to those of deformation for a material that observes the law of creep with hardening. The problem is simple in the unidimensional case, but the expressions are not so simple for an arbitrary state of stress. These relationships were derived in the work [107] on the assumption that the parameters that characterise the fundamental condition may be considered constant. For the unidimensional case this means that in eq. (188.2) the quantities λ, μ and $\bar{\nu}$ are considered as constants. This suffices for the study of motion during a short time after a disturbance, but not for determining the actual dependence of deflection on time. The equations given below were derived in the work [152], and they were applied to the buckling of plates. Simultaneously and independently, linearised equations of deflection of a plate were derived in the work [63].

We shall start from the power work-hardening law

$$\dot{p}_{ij} = \tfrac{3}{2} A p^{-\alpha} \sigma_0{}^{n-1} \bar{\sigma}_{ij} ,$$

$$\dot{p} = A p^{-\alpha} \sigma_0^n , \tag{191.1}$$

$$\dot{p}^2 = \tfrac{2}{3} \dot{p}_{ij} \dot{p}_{ij} = \sigma_0^2 = \tfrac{3}{2} \bar{\sigma}_{ij} \bar{\sigma}_{ij} .$$

Assume that in the fundamental state of stress the stresses are either constant or vary in proportion to a single parameter, so that

$$\sigma_{ij} = \alpha_{ij}\sigma_0 , \qquad \bar{\sigma}_{ij} = \bar{\alpha}_{ij}\sigma_0 .$$

Here α_{ij} is a constant tensor and $\bar{\alpha}_{ij}\bar{\alpha}_{ij} = \frac{2}{3}$. Let us vary the second of eqs. (191.1), we obtain

$$\delta\dot{p} = -\alpha A p^{-\alpha-1}\sigma_0^n \delta p + A p^{-\alpha} n \sigma_0^{n-1} \delta\sigma_0 ,$$

or

$$\delta\dot{p} = -\frac{\alpha}{p}\dot{p}\,\delta p + \frac{n}{\sigma_0}\dot{p}\,\delta\sigma_0 .$$

It is now convenient to consider the quantity p in the fundamental state as an independent variable. Denoting the derivative with respect to this variable by a prime we obtain

$$\delta p' = -\frac{\alpha}{p}\delta p + \frac{n}{\sigma_0}\delta\sigma_0 . \qquad (191.2)$$

On the other hand from the definition of $\dot{p}$ it follows that:

$$\dot{p}\,\delta\dot{p} = \frac{2}{3}\dot{p}_{ij}\delta\dot{p}_{ij} .$$

Divide both sides by $\dot{p}^2$ and note that

$$\frac{\dot{p}_{ij}}{\dot{p}} = \frac{3}{2}\frac{\bar{\sigma}_{ij}}{\sigma_0} = \frac{3}{2}\bar{\alpha}_{ij}$$

because of (191.1) whereas

$$\delta\dot{p}_{ij} = \delta\dot{e}_{ij} - \frac{1}{2G}\delta\dot{\bar{\sigma}}_{ij} .$$

Consequently

$$\delta p' = \bar{\alpha}_{ij}\delta e'_{ij} - \frac{1}{2G}\bar{\alpha}_{ij}\delta\sigma'_{ij} .$$

Put

$$q = E\bar{\alpha}_{ij}e_{ij} .$$

Note that from the definition of σ_0 it follows that

$$\delta\sigma_0 = \frac{3}{2\sigma_0}\,\bar{\sigma}_{ij}\delta\bar{\sigma}_{ij} = \tfrac{3}{2}\,\bar{\alpha}_{ij}\delta\bar{\sigma}_{ij}\,.$$

Now the expression for $\delta p'$ can be rewritten as

$$\delta p' = \frac{\delta q' - \delta\sigma_0'}{E}\,,$$

or, after integration

$$\delta p = \frac{1}{E}(\delta q - \delta\sigma_0)\,. \tag{191.3}$$

Eliminating δp from eqs. (191.2) and (191.3) we arrive at the following differential equation between δq and $\delta\sigma_0$:

$$\delta\sigma_0' + \left(\frac{\alpha}{p} + \beta\right)\delta\sigma_0 = \delta q' + \frac{\alpha}{p}\delta q\,. \tag{191.4}$$

Here we have introduced the notation

$$\beta = \frac{En}{\sigma_0}\,.$$

Integrating (191.4) we find

$$\delta\sigma_0 = \delta q - \beta p^{-\alpha} e^{-\beta p}\int_0^p p^{\alpha} \mathrm{e}^{\beta p}\,\delta q\,\mathrm{d}p\,. \tag{191.5}$$

Now taking the variation of the first of eqs. (191.1), we obtain

$$\delta\bar{\sigma}_{ij}' + \gamma\delta\bar{\sigma}_{ij} = \tfrac{2}{3}\,\delta e_{ij}' + \bar{\alpha}_{ij}\left[\frac{\alpha}{p}\delta q - \left(\frac{\alpha}{p} + (\beta - \gamma)\delta\sigma_0\right)\right]. \tag{191.6}$$

Here

$$\gamma = \frac{E}{\sigma_0}\,.$$

We now substitute into eq. (191.6) the expression for $\delta\sigma_0$ from formula (191.5) and integrate the resulting equation relative to $\delta\bar{\sigma}_{ij}$. Here, integrating by parts we avoid the double integral operator and the final result is written in the form

$$\delta\bar{\sigma}_{ij} = 2G\,\delta e_{ij} + \gamma e^{-\gamma p} \int_0^p e^{\gamma p}\,[\bar{\alpha}_{ij}\,\delta q - 2G\,\delta e_{ij}]\,dp$$

$$- \bar{\alpha}_{ij}\,\beta p^{-\alpha} e^{-\beta p} \int_0^p p^{\alpha} e^{\beta p}\,\delta q\,dp\,. \tag{191.7}$$

192. Buckling of plates according to the linearised theory

Apply eq. (191.7) to the problem of buckling of plates. Assume a plane state of stress and incompressible material. We put, as usual

$$e_{ij} = k_{ij} z\,, \quad M_{ij} = \int_{-h}^{+h} \sigma_{ij} z\,dz \qquad (i, j = 1, 2)\,.$$

It then follows from eq. (191.7) that

$$M_{11} = D\{(w_{,11} + \tfrac{1}{2} w_{,22}) - \gamma e^{-\gamma p} \int e^{\gamma p} (w_{,11} + \tfrac{1}{2} w_{,22})\,dp$$
$$+ \tfrac{3}{4}\alpha_{11}\,[\gamma e^{-\gamma p} \int e^{\gamma p} m\,dp - \beta p^{-\alpha} e^{-\beta p} \int p^{\alpha} e^{\beta p} m\,dp]\}\,. \tag{192.1}$$

Here D is the rigidity $D = 8Eh^3/9$,

$$m = \alpha_{11} w_{,11} + \alpha_{22} w_{,22} + 2\alpha_{12} w_{,12}\,.$$

Introduce the differential operator

$$P = \alpha_{11} \frac{\partial^2}{\partial x^2} + \alpha_{22} \frac{\partial^2}{\partial y^2} + 2\alpha_{12} \frac{\partial^2}{\partial x\,\partial y}\,,$$

then

$$m = P(w)\,.$$

The equation of equilibrium for a buckled plate subject to a force T_{ij} in the middle surface and to a transverse load q is

$$M_{ij,ij} + T_{ij} w_{,ij} + q = 0 . \tag{192.2}$$

Here $T_{ij} = 2h\sigma_{ij} = 2h\alpha_{ij}\sigma_0$.

We shall assume that the plate has an initial deflection w_0, then in the right-hand side of eq. (192.1) w must be replaced by $w - w_0$, but w remains in eq. (192.2). After simple transformations we obtain the following integral-differential equation for the deflection of the plate

$$(1-R_1^*)\Delta\Delta(w-w_0) + \tfrac{3}{4}(R_1^* - R_2^*)PP(w-w_0) + \frac{2h\sigma_0}{D} P(w) + q = 0. \tag{192.3}$$

Here the integral operators R_1^* and R_2^* are defined as follows:

$$R_1^* f = \gamma \int_0^p \exp\,[-\gamma(p-x)]\, f(x)\, \mathrm{d}x ,$$

$$R_2^* f = \beta \int_0^p \left(\frac{p}{x}\right)^{-\alpha} \exp\,[-\beta(p-x)]\, f(x)\, \mathrm{d}x , \tag{192.4}$$

$$\gamma = \frac{E}{\sigma_0}, \qquad \beta = \frac{En}{\sigma_0} .$$

The variables in eqs. (192.3) can only be separated in certain special cases. For instance, if a plate of arbitrary outline is subject to uniform compression then $P(w) = -\Delta w$. When $q = 0$ eq. (192.3) assumes the form

$$(1-R^*)\,\Delta\Delta\,(w-w_0) - \frac{2h\sigma_0}{D}\,\Delta w = 0 . \tag{192.5}$$

Here

$$R^* = \tfrac{1}{4} R_1^* + \tfrac{3}{4} R_2^* .$$

Consider the equation of elastic instability

$$\Delta\Delta w - \frac{2h\sigma_0}{D}\,\Delta w = 0 .$$

For the given boundary conditions this equation determines the magnitude of the first buckling stress σ_{0e} and the corresponding fundamental function W_1. Put $w = c(t)W_1$. We also consider that $w_0 = c_0 W_1$. Then it follows from eq. (192.5) that

$$(1 - R^*)(c - c_0) - \omega c = 0 . \tag{192.6}$$

Here, as above, $\omega = \sigma_0/\sigma_{0e}$.

The solution of the integral eq. (192.6) can be written in symbolic form

$$c = \left(1 + \frac{\omega}{1 - \omega - R^*}\right) c_0 . \tag{192.7}$$

This equation is difficult to solve; eliminating the integral operators we arrive at a second-order differential equation with variable coefficients similar to Whittaker's equations.

The behaviour of the deflection was investigated in the work [152] on the basis of the conventional criteria described in section 189, namely, the quasi-static criterion and criterion that the second derivative of the deflection with respect to time becomes zero. These investigations were only possible for cases in which ω is either small compared with unity or close to unity. In the work [63] the stability of plates was investigated by Kurshin's criteria. Bearing in mind the conventional nature of these criteria we do not give the results here. The value of eq. (192.3) is that it can be used to calculate the actual dependence of deflection on time if the initial deflection is known. The linearised theory can also be expected to give satisfactory results for plates if the deflections are small (of the order of the thickness of the plate).

Bibliography

Abbreviations:

MGU	–	Moscow State University
LGU	–	Leningrad State University
AN SSSR	–	Academy of Sciences of the USSR
OTN	–	Technical Science Division
DAN	–	Doklady A.S. USSR

1. Amel'yanchik, A.V., Turbine machine disc strength calculations on a mathematical machine. Izv. AN SSSR, OTN, Mechanics and Engineering, No. 1 (1959).
2. Andrade da K., E.N., Creep concepts, in the collection 'Creep and recovery' (Metallurgizdat, Moscow, 1961).
3. Annin, B.D., Asymptotic expansion of a fractional order exponential function, PMM, Vol. 25, Issue 4 (1961).
4. Arutyunyan, N.Kh., Some problems of creep theory (Gostekhizdat, Moscow, 1952).
5. Belyaev, N.M., Application of plastic deformation theory to creep calculations of components at high temperatures, Izv. AN SSSR, OTN, No. 7 (1943).
6. Bleich, F., The stability of metal structures (Fizmatgiz, Moscow, 1959).
7. Boguslavskiy, P.Ya., Study of creep in turbine components. Thesis of reports to All-Union Conference on the Theory of Elasticity, Structural Mechanics and Theory of Plasticity, 7–10 December 1950, Izd. vo. AN SSSR (1950).
8. Boychenko, G.A., Resistance of rolling of hereditary-elastic bodies, Izv. AN SSSR, OTN, No. 9 (1955).

9. Bronskiy, A.P., Aftereffect phenomena in solid bodies, PMM, Vol. 5, Issue 1 (1941).
10. Bryzgalin, G.I., On the description of anisotropic creep in vitreoplastics, PMTF, No. 6 (1963).
11. Varshavskiy, D.P., P.Ya.Boguslavskiy and I.G.Polumordvinova, Modelling of the deformed state of components of complex shape under creep conditions, Teploenergetika, No. 5 (1955).
12. Vilesova, N.S. and V.S.Namestnikov, Concerning a parameter of hardening, PMTF, No. 3 (1964).
13. Vlasov, V.Z., General theory of shells (Gostekhizdat, Moscow, 1949).
14. Volchkov, Yu.M., Axially symmetric problems in the creep of circular cylindrical shells, Izv. AN SSSR Mechanics, No. 5 (1965).
15. Vol'mir, A.S., Stability of elastic systems (Fizmatgiz, Moscow, 1963).
16. Vol'mir, A.S. and P.G.Zykin, Stability of large cylindrical shells during creep. In the collection 'Thermal stresses in turbine machine components', Issue 2, Kiev (1962).
17. Vul'fson, S.Z., On the question of linear creep theory, Works (Trudy) of Central Scientific Research Institute of Constructional Structures, Academy of Construction and Architecture SSSR, Issue 4 (1961).
18. Gerasimov, A.N., A generalisation of the linear laws of deformation and their application to problems of internal friction, PMM, Vol. 12, Issue 3 (1948).
19. Gol'denblat, I.I., Some problems in the mechanics of deformed media (Gostekhizdat, Moscow, 1955).
20. Gol'denveyzer, A.L., General theory of shells (Gostekhizdat, Moscow, 1949).
21. Gol'denveyzer, A.L., On the theory of thin-walled rods, PMM, Vol. 13, Issue 6 (1949).
22. Goodier, J.N. and P.G.Hodge, Elasticity and plasticity (Foreign Language Publishers, Moscow, 1960).
23. Danilova, I.N., On calculation of the stressed state in a rotor during transient creep, Izv. AN SSSR, OTN, Mechanics and Engineering, No. 5 (1959).
24. Danilovskaya, V.I., G.M.Ivanova and Yu.N.Rabotnov, Creep and relaxation of chrome-molybdenum steel, Izv. AN SSSR, OTN, No. 5 (1955).
25. Dzhrbashyan, M.M., The integral representation of functions that are continuous on several rays, Izv. AN SSSR, Mathematical series, Vol. 18, No. 5 (1954).
26. Dzhrbashyan, M.M., Concerning a new integral conversion and its application to the theory of integral functions, Izv. AN SSSR, Mathematical series, Vol. 19, No. 2 (1955).
27. Ditkin, V.A. and A.P.Prudnikov, Integral conversions and operational calculus (Fizmatgiz, Moscow, 1961).
28. Erzhanov, Zh.S., On assessment of the stressed state of a mountain massif. In the collection 'Mathematical methods in mining', Vol. 2, Izd. vo SO (Siberian Division) AN SSSR, Novosibirsk (1963).
29. Erzhanov, Zh.S., Theory of creep in rocks and its application, Izd-vo 'Nauka', Alma-Ata (1964).
30. Zhukov, A.M., Yu.N.Rabotnov and F.S.Churikov, Experimental verification of certain creep theories, Inzhenernyy sbornik, Vol. 17, Izd-vo AN SSSR, Moscow (1953).
31. Zhukov, A.M., Concerning plastic deformation of isotropic metal under complex loading, Izv. AN SSSR, OTN, No. 12 (1956).

32. Zhurkov, S.N. and B.N.Narzullayev, The time-dependence of the strength of hard substances, ZhTF, Vol. 23, No. 10 (1953).
33. Zhurkov, S.N. and T.P.Sanfirova, The temperature/time-dependence of the strength of pure metals, DAN SSSR, Vol. 101, No. 2 (1955).
34. Zhurkov, S.N., A problem in the strength of hard bodies, Vestnik AN SSSR, No. 11 (1957).
35. Zhurkov, S.N. and T.P.Sanfirova, The relationship between strength and creep in metals and alloys, ZhTF, Vol. 28, No. 8 (1958).
36. Zver'kov, B.V., The long-term strength of tubes under complex loading, Teploenergetika, No. 3 (1958).
37. Ivanova, V.S., Concerning the nature of deformation on an area of flow in metals, DAN SSSR, Vol. 94, No. 2 (1954).
38. Ivanova, G.M., Creep of alloy EI-437B at variable temperature, Izv. AN SSSR, OTN, No. 4 (1958).
39. Ivanova, R.Ya., Rolling of a viscoelastic cylinder on a base of the same material, PMTF, No. 3 (1964).
40. Ivanov, G.V. and V.N.Shepelenko, Bulging and buckling of a square cylindrical panel compressed along the generatrix under creep conditions, Inzh. zhurn. Mekhanika tverdogo tela, No. 3 (1966).
41. Ivlev, D.D., On the theory of ideal plastic anisotropy, PMM, Issue 6 (1959).
42. Ivlev, D.D., Concerning the properties of the relationships of the law of anisotropic work-hardening of plastic material, PMM, Vol. 24, Issue 1 (1960).
43. Ol'yushin, A.A., Plasticity (Gostekhizdat, Moscow, 1948).
44. Ishlinskiy, A.Yu., Longitudinal vibrations of a rod with a linear law of aftereffect and relaxation, PMM, Vol. 4, Issue 1 (1940).
45. Ishlinskiy, A.Yu., On the equations of spatial deformation of not completely elastic and of viscoplastic bodies, Izv. AN SSSR, OTN, No. 3 (1945).
46. Ishlinskiy, A.Yu., General theory of plasticity with linear work-hardening, Ukrainskiy matematicheskiy zhurnal, No. 3 (1954).
47. Kagan, V.F., Fundamentals of theory of surfaces in tensor exposition, Vol. 1 (Gostekhizdat, Moscow, 1947).
48. Kadashevich, Yu.I. and V.V.Novozhilov, Theory of plasticity with allowance for residual microstresses, PMM, Vol. 22, Issue 1 (1958).
49. Kaptelin, Yu.P., An equation of state for creep of work-hardening, copper, Sbornik trudov LIIZhT, Issue 192 (1962).
50. Karpin, E.B., Experience of autofretting of turbine discs, Energomashinostroyeniye, No. 1 (1959).
51. Kachanov, L.M., Time to failure under creep conditions, Izv. AN SSSR, OTN, No. 8 (1958).
52. Kachanov, L.M., Creep theory (Fizmatgiz, Moscow, 1960).
53. Kats, Sh.N., An investigation of the long-term strength of carbon-steel pipes, Teploenergetika, No. 11 (1955).
54. Kats, Sh.N., Failure of austenitic pipes under internal pressure under creep conditions, Energomashinostroyeniye, No. 2 (1957).
55. Kats, Sh.N., Creep and failure of pipes subject to internal pressure, Izv. AN SSSR, OTN, No. 10 (1957).

56. Kennedy, A.J., The influence of variable loads on creep and relaxation. In the collection 'Fatigue of metals' (Foreign Language Publishers, Moscow, 1961).
57. Klyushinkov, V.D., Laws of plasticity for materials subject to work-hardening, PMM, Vol. 22, Issue 1 (1958).
58. Klyushinkov, V.D., A possible method of constructing formulae for plasticity, PMM, Vol. 23, Issue 2 (1959).
59. Klyushinkov, V.D., New concepts in plasticity and deformation theory, PMM, Vol. 23, Issue 4 (1959).
60. Kostyuk, A.G., Concerning the equilibrium of an annular plate with a stepwise law of work-hardening, PMM, Vol. 14, Issue 3 (1950).
61. Kostyuk, A.G., Stresses in a rotating disc during creep, Inzhenernyy sbornik, Vol. 15, Izd. vo AN SSSR (1953).
62. Kuratov, P.S. and V.I.Rozenblyum, On the integration of the equations of transient creep in solids, PMM, Vol. 24, Issue 1 (1960).
63. Kurshin, L.M., On the stability of columns and plates under creep conditions, DAN SSSR, No. 3 (1961) 140.
64. Kurshin, L.M., The stability of columns under creep conditions, PMTF, No. 6 (1961).
65. Kurshin, L.M., A possible approach to the problem of the stability of columns under creep conditions. In the collection 'Creep and long-term strength', Izd-vo SO (Siberian Division) AN SSSR, Novosibirsk (1963).
66. Kuznetsov, A.P., Solution of the problem of the stability of a compressed surface under creep conditions by the work-hardening hypothesis, PMTF, No. 2 (1964).
67. Leybenzon, L.S., Variational methods of solving problems of the theory of elasticity, OGIZ, GTTI, Moscow (1933).
68. Lifshits, I.M., On the theory of diffusional-viscous flow of polycrystalline bodies, ZhETF, Vol. 44, Issue 4 (1963).
69. Malinin, N.N., Fundamentals of creep calculations (Mashgiz, Moscow, 1948).
70. Malinin, N.N., Certain unidimensional problems of transient creep, Inzhenernyy sbornik, Vol. 10, Izd-vo AN SSSR (1951).
71. Malinin, N.N., Steady-state creep of circular symmetrically loaded plates. In the collection 'Strength rigidity and creep calculations on components of engineering structures' (Mashgiz, Moscow, 1952).
72. Malinin, N.N., Creep calculations on rotating non-uniformly heated discs of variable thickness. In the collection 'Problems of creep of materials and structures', Izd-vo AN SSSR, Moscow (1959).
73. Malinin, N.N., Strength of turbine machines (Mashgiz, Moscow, 1962).
74. Mileyko, S.T. and V.I.Telenkov, Short-term creep in aluminium alloys, PMTF, No. 5 (1962).
75. Mileyko, S.T., Short-term creep with variable stress in the collection 'Creep and long-term strength', Izd-vo SO (Siberian Division) AN SSSR, Novosibirsk (1963).
76. Mirkin, I.L. and I.I.Trunin, Investigation of creep and failure of steel in a zone of stress concentration. In the collection 'Strength of metals', Izd-vo AN SSSR, Moscow (1956).
77. Muskhelishvili, N.I., Some fundamental problems of the mathematical theory of elasticity, Edition 4, Izd-vo AN SSSR, Moscow (1954).
78. Namestnikov, V.S., Concerning creep under constant load with complex stress conditions, Izd. AN SSSR, OTN, No. 4 (1957).

79. Namestnikov, V.S., Creep under variable load with complex stress conditions, Izd. AN SSSR, OTN, No. 10 (1957).
80. Namestnikov, V.S., Forward and reverse torsion under creep conditions, PMTF, No. 1 (1960).
81. Namestnikov, V.S., Concerning a hypothesis in the theory of tri-axial creep, Izv. SO (Siberian Division) AN SSSR, No. 2 (1960).
82. Namestnikov, V.S., Creep in an aluminium alloy under variable loads, PMTF, No. 4 (1960).
83. Namestnikov, V.S. and Yu.N.Rabotnov, On hereditary creep theories, PMTF, No. 4 (1961).
84. Namestnikov, V.S. and Yu.N.Rabotnov, Hypothesis concerning the equations of state during creep, PMTF, No. 3 (1961).
85. Namestnikov, V.S. and A.A.Khvostunkov, Creep of duralumin under constant and variable loads, PMTF, No. 4 (1960).
86. Namestnikov, V.S., Creep equations based on the criteria of maximum referred stress, Izv. AN SSSR, Mechanics and Engineering, No. 3 (1964).
87. Nemirovskiy, Yu.V. and Yu.N.Rabotnov, The ultimate load of reinforced cylindrical shells, Izv. AN SSSR, OTN, Mechanics and Engineering, No. 3 (1963).
88. Novozhilov, V.V., Theory of thin shells (Sudpromgiz, Leningrad, 1962).
89. Novozhilov, V.V., On the physical meaning of stress invariants, PMM, Vol. 15, Issue 2 (1951).
90. Novozhilov, V.V., Theory of elasticity (Sudpromgiz, Leningrad, 1958).
91. Oding, I.A., The mechanism of creep in metals, Metallurg, No. 1 (1934).
92. Oding, I.A. and G.A.Tulyakov, Creep of austenitic steel with complex stress, Izv. AN SSSR, OTN, No. 1 (1958).
93. Oding, I.A., and others, Theory of creep and long-term strength of metals (Metallurgizdat, Moscow, 1959).
94. Odquist, F.K.S., Technical theories of creep 'Mekhanika' collection. Translation and Review of Foreign Periodic Literature, No. 2 (1959).
95. Rabotnov, Yu.N. and V.P.Rabinovich, Concerning the strength of discs under creep conditions, Izv. AN SSSR, OTN, Mechanics and Engineering, No. 4 (1959).
96. Rabotnov, Yu.N., Fundamental equations of the theory of shells, DAN, SSSR, Vol. 17, No.2 (1945).
97. Rabotnov, Yu.N., Certain solutions in the momentless theory of shells, PMM, Vol. 10, Issue 5–6 (1946).
98. Rabotnov, Yu.N., Creep calculations on machine components, Izv. AN SSSR, OTN, No. 6 (1948).
99. Rabotnov, Yu.N., Certain problems of creep theory, Vestnik MGU, No. 10 (1948).
100. Rabotnov, Yu.N., Concerning a disc of uniform strength, PMM, Vol. 12, No. 4 (1948).
101. Rabotnov, Yu.N., Equilibrium of an elastic medium with aftereffects, PMM, Vol. 12, No. 1 (1948).
102. Rabotnov, Yu.N., Approximate technical theory of elastoplastic shells, PMM, Vol. 15, Issue 2 (1951).
103. Rabotnov, Yu.N., Possibilities of describing transient creep in application to creep in rotors, Izv. AN SSSR, OTN, No. 5 (1957).

104. Rabotnov, Yu.N., On the mechanism of long-term failure. In the collection 'Problems of the strength of materials and structures', Izd-vo AN SSSR, Moscow (1959).
105. Rabotnov, Yu.N., Strength of materials (Fizmatgiz, Moscow, 1962).
106. Rabotnov, Yu.N., Concerning failure by creep, PMTF, No. 2 (1963).
107. Rabotnov, Yu.N. and S.A.Shesterikov, Stability of rods and plates under creep conditions, PMM, Vol. 21, Issue 3 (1957).
108. Ponomarev, S.D. and others, Strength calculations in engineering (Mashgiz, Moscow, 1958).
109. Ratner, S.I., Strength and plasticity of metals (Oborongiz, Moscow, 1949).
110. Rashevskiy, P.K., Course of differential geometry, ONTI, Moscow (1938).
111. Reissner, E. and R.Clark, Bending of tubes with a curvilinear centre-line. In the collection 'Problems of mechanics' (Foreign Language Publishers, Moscow, 1955).
112. Rzhanitsyn, A.R., Some problems of the mechanics of systems that are deformed with time (Gostekhizdat, Moscow, 1949).
113. Rozenblyum, V.I., Time to failure of a rotating disc under creep conditions, PMM, Vol. 2, No. 3 (1957).
114. Rozenblyum, V.I., The influence of plastic deformation on time-to-failure in creep. In the collection 'Creep and long-term strength', Izd-vo SO (Siberian Division) AN SSSR, Novosibirsk (1963).
115. Rozenblyum, V.I., Concerning approximate creep equations, Izv. AN SSSR, OTN, Mechanics and Engineering, No. 5 (1959).
116. Rozenblyum, V.I., An approximate analysis of transient creep in plates and envelopes. In the collection 'Investigations on elasticity and plasticity', Collection 3, Izd-vo LGU (1964).
117. Rozovskiy, M.I., Creep and long-term failure of materials, ZhTF, Vol. 21, No. 11 (1951).
118. Rozovskiy, M.I., Some peculiarities of hereditary elastic media, Izd. AN SSSR, Mechanics and Engineering, No. 2 (1961).
119. Rozovskiy, M.I., Non-linear integral-operator equations of creep and the problem of twisting a cylinder through large angles, Izd. AN SSSR, OTN, Mechanics and Engineering, No. 5 (1959).
120. Rozovskiy, M.I., A semi-symbolic method of solving certain problems of creep theory, Izv. AN ArmSSR, Physical-Mathematical Sciences and Technical Sciences, Vol. 9, No. 5 (1956).
121. Rozovskiy, M.I., Some properties of special operators used in creep theory, PMM, Vol. 23, Issue 5 (1959).
122. Rozovskiy, M.I., Concerning a property of the power of a special operator and its application to the solution of dynamic problems. In the collection 'Creep and long-term strength', Izd-vo SO (Siberian Division) AN SSSR, Novosibirsk (1963).
123. Sazonova, N.D., Creep and long-term strength testing of heat-resistant materials, Izd-vo 'Mashinostroyeniye', Moscow (1965).
124. Sully, A., Creep of metals and heat-resistant alloys (Oborongiz, Moscow, 1953).
125. Sveshinkova, V.A., Plastic deformation of work-hardening metals, Izv. AN SSSR, OTN, No. 1 (1956).
126. Sdobyrev, V.P., Long-term strength of the alloy EI-437B under complex stress, Izv. AN SSSR, OTN, No. 4 (1958).
127. Sdobyrev, V.P., Long-term strength criteria for certain heat-resistant alloys under complex stress, Izv. AN SSSR, OTN, Mechanics and Engineering, No. 6 (1959).

128. Serensen, S.V., Strength problems in engineering, Vestnik AN SSSR, No. 4 (1951).
129. Serensen, S.V. and Ya.A.Kozlov, Load-carrying capacity and strength calculations of components subject to static and variable stress, Vestnik mashinostroyeniya, No. 12 (1953).
130. Serensen, S.V. and others, Strength under transient loading, Izd-vo AN UkrSSR, Kiev (1961).
131. Slonimskiy, G.L., On the laws of deformation of real materials, ZhTF, Vol. 9, No. 20 (1939).
132. Slonimskiy, G.L., On the law of deformation of visco-elastic polymeric bodies, DAN, SSSR, Vol. 140, No. 2 (1961).
133. Sokolovskiy, V.V., The plastic stressed state of a rotating disc, PMM, Vol. 12, Issue 1 (1948).
134. Sokolovskiy, V.V., Theory of plasticity (Gostekhizdat, Moscow, 1950).
135. Sosnin, O.V., Stress redistribution in a solid rotating disc in the primary stage of creep, PMTF, No. 2 (1960).
136. Sosnin, O.V., Transient creep of rotating discs. In the collection 'Creep and long-term strength', Izd-vo SO (Siberian Division) AN SSSR, Novosibirsk (1963).
137. Sosnin, O.V., Steady-state anisotropic creep of discs, PMTF, No. 4 (1963).
138. Timoshenko, S.P., Plates and shells (Gostekhizdat, Moscow, 1948).
139. Torshenov, N.G., Creep of aluminium alloy D-16T in compression, PMTF, No. 6 (1961).
140. Trefftz, E., Mathematical theory of elasticity, GTTI, Moscow (1934).
141. Whittaker, E.T. and G.N.Watson, Course of modern analysis, GTTI, Moscow (1934).
142. Hardy, G.G., J.Littlewood and G.Polia, Inequalities (Foreign Language Publishers, Moscow, 1948).
143. Hill, R., Mathematical theory of plasticity (Gostekhizdat, Moscow, 1956).
144. Khuan Ke-chzhi, On work-hardening of a rigid-plastic material, PMM, Vol. 22, Issue 4 (1958).
145. Churikov, F.S., On stress and deformation at high temperature, Vestnik MGU, No. 2 (1949).
146. Shanley, F.K., Analysis of the weight and strength of aircraft structures (Oborongiz, Moscow, 1957).
147. Shepelenko, V.N., The stability of an infinitely long cylindrical panel clamped at the edge, Izd. AN SSSR, Mekhanika, No. 6 (1965).
148. Shesterikov, S.A., On a variational principle in creep theory, Izv. AN SSSR, OTN, No. 2 (1957).
149. Shesterikov, S.A., Concerning a condition for creep laws, Izv. AN SSSR, OTN, No. 1 (1959).
150. Shesterikov, S.A., Concerning a criterion of stability in creep, PMM, Vol. 23, Issue 6 (1959).
151. Shesterikov, S.A., Relaxation calculations on discs, PMTF, No. 1 (1960).
152. Shesterikov, S.A., Stability of plates in creep according to the flow theory, PMTF, No. 5 (1961).
153. Shesterikov, S.A., Buckling in creep with allowance for instantaneous plastic deformation, PMTF, No. 2 (1963).
154. Shorr, B.F., The influence of non-uniform heating under creep conditions on change in the stressed state, DAN SSSR, Vol. 123, No. 5 (1958).

155. Shorr, B.F., On the calculation of transient creep of non-uniformly heated rods of arbitrary cross-section, Izv. AN SSSR, OTN, Mechanics and Engineering, No. 1 (1959).
156. Shtaerman, I.Ya. and A.A.Pikovskiy, Methods of calculating the stability of structures, Gos. izd. mestn. prom. Kiev (1938).
157. Shchetinin, N.N., Pure bending of beams under creep conditions, Izv. AN SSSR, OTN, No. 8 (1956).
158. Yagn, Yu.I. and I.N.Vinogradova, Influence of the form of stress deviator on the resistance of metals to plastic deformation, DAN SSSR, Vol. 96, No. 3 (1954).
159. Andrade, E.N., da C., On the viscous flow of metals and allied phenomena, Proc. Roy. Soc. (A) 84, No. A567, London (1910).
160. Andrade, E.N., da C., The validity of the $t^{1/3}$ law of flow of metals, Phil. Mag. 7, No. 84 (1962).
161. Bailey, R.W., Creep of steel under simple and compound stresses and the use of high initial temperature in steam power plant, Transactions Tokyo Sectional Meeting of the World Power Conference, Oct.-Nov., Tokyo (1929).
162. Bailey, R.W., The utilisation of creep test data in engineering design, Proc. Inst. Mech. Engrs. 131 (Nov.-Dec. 1935).
163. Bailey, R.W., Creep relationships and their applications to pipes, tubes and cylindrical parts under internal pressure, Proc. Inst. Mech. Engrs. 164 (1951).
164. Batdorf, S.B. and B.Budiansky, A mathematical theory of plasticity based on the concept of slip, NACA, T.N. 1871 (1949).
165. Batdorf, S.B. and B.Budiansky, Polyaxial stress-strain relations of a strain hardening metal, J. Appl. Mech. 21 (1954).
166. Bernett, E.C., Short-time, elevated temperature stress-strain behaviour of tensile, compressive and column members properties, Wright Air Dev. Center, Techn. Rep. 59-484, Ohio (1959).
167. Bernett, E.C., Tensile and short-time creep properties of N-155 alloy sheet, Trans. ASME, B-82, No. 4 (1960).
168. Besseling, J.F., A theory of elastic, plastic and creep deformations of an initially isotropic material showing anisotropic strain hardening, creep recovery and secondary creep, J. Appl. Mech. 25, No. 4 (1958).
169. Betteridge, W., The extrapolation of the stress-rupture properties of the nimonic alloys, J. Inst. of Metals (January 1958).
170. Bieniek, M.P. and A.M.Freudenthal, Creep deformation and stresses in pressurized, long cylindrical shell, JAS 27, No. 10 (1960).
171. Bishop, J.F.W. and R.Hill, A theoretical derivation of the plastic properties of a polycrystalline face centered metal, Phil. Mag. Ser. 7, XLII (1951).
172. Bont, de, R.A., A special case of uniaxial creep, including rupture, Institutionen för Hållfesthetslära Kungl. Tekniska Högskolan, Publ. No. 37, Stockholm (1962).
173. Boltzmann, L., Zur Theorie der elastischen Nachwirkung, Annalen der Physik und Chemie, Erg. Bd. 7 (1876).
174. Brazier, L.G., The flexure of thin cylindrical shells and other "thin" sections, A.R.S.R. and M., No. 1081, London (1927).
175. Breuer, S. and E.T.Onat, On uniqueness in linear viscoelasticity, Quart. Appl. Math., XIX, No. 4 (1962).
176. Budiansky, B., A reassessment of deformation theories of plasticity, J. Appl. Mech. 26, No. 2 (1959).

177. Budiansky, B. and O.L.Mangasarian, Plastic stress concentration at a circular hole in an infinite sheet subjected to equal biaxial tension, J. Appl. Mech. 27, No. 1 (1960).
178. Calladine, C.R. and D.C.Drucker, Nesting surfaces of constant rate of energy dissipation in creep, Quart. Appl. Math. XX, No. 1 (1962).
179. Calladine, C.R. and D.C.Drucker, A bound method for creep analysis of structures: direct use of solutions in elasticity and plasticity, J. Mech. Eng. Sci. 4, No. 1, (1962).
180. Calladine, C.R., Edge-load response of a thin cylindrical shell in creep, Non classical shell problems, Proc. IASS Symp., North-Holland Publ. Co.–PWN (1964).
181. Calladine, C.R., A note on the bending of beams made of nonlinear material, J. Royal Aeronaut. Soc. 67, No. 626 (February 1963).
182. Carlson, R.L. and G.K.Manning, Investigation of compressive creep properties of aluminium columns at elevated temperatures, Wright Air Dev. Center, Techn. Rep. 52-231, 2 (1954), 3 (1955), 4 (1955).
183. Carlson, R.L. and E.G.Bodine, An experimental study of compressive creep behaviour at elevated temperatures, Proc. ASTM 57 (1957).
184. Cicala, P., Sulle deformazione plastiche, Acc. dei Lincei, Rendiconti 5, 6 (1950).
185. Clapper, R.B., Isochronous stress-strain curves for some magnesium alloys showing the effect of varying exposure times on their creep resistance, Proc. ASTM 58 (1958–1959).
186. Claus, F.L., An examination of high temperature stress-rupture correlating parameters, Proc. ASTM 60 (1960).
187. Cottrell, A.H., Dislocations and plastic flow in crystals, Oxford (1953).
188. Davenport, C.C., Correlation of creep and relaxation properties of copper, J. Appl. Mec. 60, p. A-56 (June 1938).
189. Davis, E.A., Creep and relaxation of oxygen-free copper, J. Appl. Mech. 10, No. 2 (1943).
190. Davies, P.E. and B.Wilshire, An interpretation of the relationship between creep and fracture, Spec. Rep. Iron and Steel Inst., No. 70 (1961).
191. DiStefano, J.N., Sulla stabilitá in regime visco-elastico a comportamento lineare, Acc. dei Lincei, Rendiconti, Ser. VIII, XXVII, 56 (1959).
192. Dorn, J.E. and T.E.Tietz, Creep and stress-rupture investigations on some aluminium alloy sheet metals, Proc. ASTM 49 (1949).
193. Dorn, J.E., Some fundamental experiments on high-temperature creep, J. Mech. Phys. Solids 3, No. 2 (January 1955).
194. Drucker, D.C., Coulomb friction, plasticity and limit loads, J. Appl. Mech. 21, No. 1 (1954).
195. Drucker, D.C., Some implications of work hardening and ideal plasticity, Quart. Appl. Math. VII, No. 4 (January 1950).
196. Drucker, D.C., Plasticity, from "Structural Mechanics" (Pergamon Press, 1960).
197. Drucker, D.C., On uniqueness in the theory of plasticity, Quart. Appl. Math. XIV, No. 1 (1956).
198. Drucker, D.C. and E.D.Stockton, Instrumentation and fundamental experiments in plasticity, Proc. Soc. Exper. Stress Analysis X, No. 2.
199. Drucker, D.C., A more fundamental approach to plastic stress-strain relations, Proc. First USA National Congr. Appl. Mech. (1952).

200. Duffing, G., Elastizität und Reibung beim Riementrieb, Forschung auf dem Gebiete des Ingenieurwesens 2, No. 3 (1931).
201. Dushman, S., L.Dunbar and H.Huthsteiner, Creep of metals, J. Appl. Phys. 15 (1944).
202. Everett, F.L. and C.L.Clark, Report on torsion creep tests for comparison with tension creep tests on a carbonmolybdenum steel, Proc. ASTM 39 (1939).
203. Faure, G. and A.Touski, Détermination de la vitesse de fluage établi d'un acier réfractaire fluant très rapidement à haute température, La recherche aéronautique, No. 85 (1961).
204. Feltham, P. and J.D.Meakin, On the representation and extrapolation of creep data of metals and technical alloys, Rheol. Acta 1, No. 2–3 (1958).
205. Fenn, R.W. and J.A.Gusak, Effect of strain rate and temperature on the strength of magnesium alloys, Proc. ASTM, No. 58 (1958).
206. Fenn, R.W., Compression testing of sheet magnesium utilising rapid heating, Proc. ASTM 60 (1960–1961).
207. Findley, W.N. and D.B.Peterson, Prediction of long-time creep with ten years creep data on four plastic laminates, Proc. ASTM 58 (1958).
208. Findley, W.N., Creep and stress relaxation of plastics, from "High temperature effects in aircraft structures" (Pergamon Press, 1958).
209. Findley, W.N., The effect of temperature and combined stress on creep of plastics, Proc. Second Intern. Reinforced Plastics Conference, London (Dec. 1960).
210. Freudenthal, A.M., The inelastic behavior of engineering materials and structures (John Wiley, N.Y., 1950).
211. Fry, C.G. and H.K.Hughes, Asymptotic developments of certain integral functions, Duke Mathem. J. 9 (1942).
212. Garofalo, F., G.W.Smith and B.V.Royle, Validity of time compensated temperature parameters for correlating creep and creep-rupture data, Trans. ASME 78, No. 7 (1956).
213. Goldhoff, R.M., The application of Rabotnov's creep parameter, Proc. ASTM 61 (1961).
214. Goldhoff, R.M., Comparison of parameter methods for extrapolating high temperature date, Pap. Amer. Soc. Mech. Eng. A-121 (1958).
215. Gomez, M.P., Correlation and extrapolation methods of creep data, Trend. Eng. Univ. Wash. 11, No. 3 (1959).
216. Graham, A., Numerical regularities in uniaxial deformation, Proc. IX Intern. Congress Appl. Mech. 8, Brussels (1957).
217. Grant, N.J. and G.Bucklin, On the extrapolation of short time stress-rupture data, Trans. Amer. Soc. Metals 42 (1950).
218. Green, A.E. and R.S.Rivlin, The mechanics of non-linear materials with memory, Arch. Rat. Mechanics 1 (1957).
219. Green, A.E. and J.E.Adkins, Large elastic deformations and non-linear continuum mechanics (Clarendon Press, Oxford, 1960).
220. Greenwood, J.N., The influence of vibration on the creep of lead, Proc. ASTM 49 (1949).
221. Griffith, I.E. and J.Marin, Creep relaxation for combined stress, J. Mech. Phys. Solids 4, No. 4 (1956).
222. Gross, B., Mathematical structure of the theories of viscoelasticity (Hermann, Paris, 1953).

223. Guarnieri, G.I., Intermittent stressing and heating tests of aircraft structural metals, Wright Air Dev. Center, Techn. Rep. 53-24 (1954).
224. Guarnieri, C.I., The creep-rupture properties of aircraft sheet alloys subjected to intermittent load and temperature, ASTM, Spec. Techn. Publ., No. 165, Baltimore (1954).
225. Gurtin, M.E. and E.Sternberg, On the linear theory of viscoelasticity, Archive for Rational Mechanics and Analysis 11, No. 4 (1962).
226. Hanstock, R.F., The effect of vibration on a precipitation hardened aluminium alloy, J. Inst. Metals 74 (1948) 469.
227. Hencky, H., Zur Theorie plastischer Verformungen und der hierdurch im Material hervorgerufenen Nachspannungen, ZAMM 4, No. 4 (1924).
228. Hencky, H., Ueber langsame stationäre Strömungen in plastischen Massen mit Rücksicht auf die Vorgänge beim Walzen, Pressen und Ziehen von Metallen, ZAMM 5 (1925).
229. Herring, C., Diffusional viscosity of a polycrystalline solid, J. Appl. Phys. 21, No. 5 (1950).
230. Hill, R., On uniqueness and stability in the theory of finite elastic strain, J. Mech. Phys. Solids 5, No. 4 (1957).
231. Hodge, P. and B.Venkatraman, Approximate solutions of some problems in steady creep. Memorie di symposium sulla plasticitá nella scienza delle construzioni in onore di A.Danusso, Bologna (1956).
232. Hodge, P.C., Piecewise linear plasticity, Proc. IX Intern. Congress Appl. Mech. (1957).
233. Hoff, N.J., The necking and rupture of rods subjected to constant tensile loads, J. Appl. Mech. 20 (1953).
234. Hoff, N.J., Buckling and stability, Forty-first Wilbur Wright Memorial Lecture, J. Roy. Aeronaut. Soc. 58 (January 1954).
235. Hoff, N.J., Approximate analysis of structures in the presence of moderately large creep deformations, Quart. Appl. Math. 12, No. 1 (1954).
236. Hoff, N.J., Creep bending and buckling of thin circular cylindrical shells, PIBAL Report No. 555 (July 1956).
237. Hoff, N.J., Creep buckling, The Aeronautical Quarterly VII (February 1956).
238. Hoff, N.J., Buckling at high temperature, J. Roy. Aeronaut. Soc. 61, No. 563 (1957).
239. Hoff, N.J., Stress distribution in the presence of creep, from "High temperature effects in aircraft structures" (Pergamon Press, 1958).
240. Hoff, N.J., W.E.Jahsman and W.A.Nachbar, A study of creep collapse of a long circular cylindrical shell under uniform external pressure, JASS No. 10 (1952).
241. Hoff, N.J., Structure and materials for finite lifetime, Advances in Aeronaut. Sciences 2 (Pergamon Press, 1959).
242. Hoff, N.J., Reversed creep: a remark to the creep buckling theory of Rabotnov and Shesterikov, J. Mech. Phys. Solids 12, No. 2 (1964).
243. Howe, P.W.H., Mathematical techniques applying to the thermal fatigue behaviour of high temperature alloys, The Aeronautical Quarterly XIII (1962).
244. Hu, L. and J.Marin, Anisotropic loading functions for combined stresses in the plastic range, J. Appl. Mech. 22 (1955) 1.
245. Hult, J., Oil canning problems in creep, from "Creep in structures" (Springer, 1962).

246. Hult, J., Mechanics of a beam subject to primary creep, Chalmers Tekniska Högskolans handlingar No. 236 (1962).
247. Hunter, S.C., The rolling contact of a rigid cylinder with a viscoelastic half space, J. Appl. Mech. 28, No. 4 (1961).
248. Inverkan av intermittent uppvarming på hållfastheten för lättmetall, 3526-3-FFA Rapport No. HU-568 (1957).
249. Isaksson, Å., Kryptning vid variabel spänning ogh temperatur, Institutionen för Hållfasthetslära, Kungl. Tekniska Högskolans Public. No. 116 (June 1957).
250. Jenkinson, E.A., A.I.Smith and M.T.Hopkin, The long time creep properties of an 18% Cr, 12% Ni, 1% Nb steel steam pipe and superheater tube, J. of the Iron and Steel Inst. 200 (1962).
251. Johannson, A., Fatigue of steels at constant strain amplitude and elevated temperature, Proc. IUTAM Coll. on Fatigue, Stockholm, 1955; Berlin, 1956.
252. Johnson, A.E. and H.J.Tapsell, Creep under combined tension and torsion, Engineering 150, No. 3887, 3889, 3892, 3894 (1940).
253. Johnson, A.E., The creep recovery of a 0.17 per cent carbon steel, Inst. of Mech. Eng. J. Proc. 145 (1941) 5.
254. Johnson, A.E., The creep of a nominally isotropic aluminium alloy under combined stress systems at elevated temperatures, Metallurgia 40 (237), (1949).
255. Johnson, A.E., Behaviour of cast steel at elevated temperatures, Engineer, Lond. 188, No. 4879, 4880, 4881, 4882 (1949).
256. Johnson, A.E., The creep of a nominally isotropic magnesium alloy at normal and elevated temperatures under complex stress systems, Metallurgia 42, No. 252 (1950).
257. Johnson, A.E., Creep under complex stress systems at high temperatures, Proc. Inst. Mech. Engrs. 164, No. 4 (1951).
258. Johnson, A.E. and N.E.Frost, Fracture under combined stress creep conditions of a 0.5 per cent Mo steel, Engineer 191 (1955).
259. Johnson, A.E. and N.E.Frost, A tension and torsion machine for relaxation tests, Engineer 193, No. 5160 (1954).
260. Johnson, A.E. and N.E.Frost, Note on the fracture under complex stress creep conditions of a 0.5 per cent Mo steel at 550° and a commercially pure copper at 250°, Creep and fracture of metals at high temperatures, London, H.M. Stationary Office (1956).
261. Johnson, A.E., J.Henderson and V.D.Mathur, The complex stress creep fracture of a commercial copper at 250°, Engineer 202 (1956).
262. Johnson, A.E., J.Henderson and V.D.Mathur, Creep under changing complex creep systems, Engineer 206 (1958).
263. Johnson, A.E., J.Henderson and V.D.Mathur, Complex stress creep relaxation at elevated temperatures, Aircr. Eng. 31 (1959).
264. Johnson, A.E., J.Henderson and V.D.Mathur, Complex stress creep fracture of an aluminium alloy, Aircr. Eng. 32 (1960).
265. Johnson, A.E., Complex stress creep of metals, Metallurgical Reviews 5, No. 20 (1960).
266. Johnson, A.E., J.Henderson and B.Khan, Complex stress creep, relaxation, and fracture of metallic alloys, Edinburgh, H.M. Stationary Office (Department of Scientific and Industrial Research, National Engineering Laboratory) (1962).

267. Kattus, J.R., High temperature structural materials for missiles, Jet Propulsion 27, No. 6 (1957).
268. Kattus, J.R. and D.W.McDowell, Strength of structural alloys under rapid heating and loading, Proc. ASTM 60 (1960–1961).
269. Kauzman, W., Flow of solid metals from the standpoint of the chemical rate theory, Trans. AIME 143 (1941).
270. Kempner, J. and A.Patel, Creep buckling of columns, NACA T.N., 3138 (January 1954).
271. Kempner, J.E. and F.V.Pohle, On the non-existence of a finite critical time for linear viscoelastic columns, JAS 20, No. 8 (August 1953).
272. Kempner, J.E., Creep bending and buckling of nonlinearly viscoelastic columns, NACA T.N., 3137 (January 1954).
273. Kennedy, A.J., The creep of metals under interrupting stressing, Proc. Roy. Soc. A213 (1952).
274. Kennedy, C.R., W.O.Harms and D.A.Douglas, Multiaxial creep studies on inconel at 1500°F, ASME Paper No. 58-A-193 (1958).
275. Kooistra, L.F., R.U.Blazer and J.T.Tucker, High temperature stress rupture testing, Trans. ASME 74, No. 5 (1952).
276. Larson, F.R. and J.Miller, A time-temperature relationship for rupture and creep stress, Trans. ASME 74, No. 5 (1952).
277. Lazan, B.J., Fatigue of structural materials at high temperatures, from "High temperature effects in aircraft structures" (Pergamon Press, 1958).
278. Leaderman, H., Elastic and creep properties of filamentous and other high polymers, Textile Foundation, Washington (1943).
279. Leaderman, H., F.McCrackin and O.Nakada, Large longitudinal retarded elastic deformation of rubberlike network polymers, Trans. Soc. Rheol. VII (1963).
280. Lee, E.H., J.R.M.Radok and W.B.Woodward, Stress analysis for linear viscoelastic materials, Trans. Soc. Rheol. III (1959).
281. Lianis, G., Torsional creep buckling of open tubes having arbitrary cross-section, School of Aeronaut., Eng. Purdue Univ., Rep. No. S-59-1 (December 1959).
282. Libove, C., Creep buckling of columns, JAS No. 7 (1952).
283. Libove, C., Creep buckling analysis of rectangular section columns, NACA T.N. No. 2956, Washington, D.C. (1953).
284. Lin, T.H., Creep stresses and deflections of columns, J. Appl. Mech. 25, No. 2 (1958).
285. Lin, T.H., A proposed theory of plasticity based on slip, Proc. 2nd U.S. Nat. Congr. Appl. Mech., N.Y. (1954).
286. Lin, T.H., On stress-strain relations based on slips, Proc. 3rd U.S. Nat. Congr. Appl. Mech., N.Y. (1958).
287. Lin, T.H., Bending of a plate with non-linear strain hardening creep, from "Creep in structures" (Springer, 1962).
288. Ludvik, P., Element der technologischen Mechanik, Berlin (1908).
289. Ma, B.M., A further creep analysis for rotating solid discs of variable thickness, J. Franklin Inst. No. 5 (1960).
290. Machlin, E. and A.Nowick, Stress rupture of heat resisting alloys, Trans. AIME 172 (1947).
291. Manson, S.S. and A.M.Haferd, A linear time-temperature relation for extrapolation of creep and stress rupture data, NACA TN 2890, Washington, D.C. (March 1953).

292. Marin, I. and PaoJoh-Han, An analytical theory of the creep deformation of materials, J. Appl. Mech. 20, No. 2 (1953).
293. Mellgren, A., Measuring accuracy in creep test, Institutionen for Hållfasthetslära, Kungl. Tekniska Högskolans, No. 125, Stockholm (1958).
294. Michalko, R.E., Short-time, rapid heating, elevated temperature testing, J. of Metals 12, No. 8 (1960).
295. Moon, D.P. and W.F.Simmons, Methods for conducting short-time tensile, creep and creep-rupture tests under conditions of rapid heating, Battelle Memor. Inst. DMIC Report No. 121 (1959).
296. Morland, L.W., A plan problem of rolling contact in linear viscoelasticity theory, J. Appl. Mech. 29, No. 2 (1962).
297. Mott, N.F. and F.R.N.Nabarro, Dislocation theory and transient creep. Report of Conference on Strength of Solids, Phys. Soc. London (1948).
298. Nabarro, F.R.N., Deformation of crystals by the motion of single ions. Report of the Conference on Strength of Solids, Phys. Soc. London (1948).
299. Nadai, A., On the creep of solids at elevated temperatures, J. Appl. Phys. 8, No. 6 (1937).
300. Nadai, A., Plasticity (McGraw-Hill Book Co., New York, London, 1931).
301. Nakada, O., Theory of non-linear responses, J. Phys. Soc. Japan 15, No. 12 (1960).
302. Namestnikov, V.S., Combined-stress creep under changing loads, Proc. of the Joint Intern. Conference on Creep (1963).
303. Nishihara, T., K.Tanaka and Shima, Creep under tension and torsion, Proc. 5th Japan Nat. Congr. Appl. Mech. 1955, Tokyo, 1956.
304. Nishihara, T., S.Taira and K.Tanaka, Effect of unloading on the creep test of steel, Proc. 5th Japan Nat. Congr. Appl. Mech. 1955, Tokyo (1956).
305. Nishihara, T., S.Taira and K.Tanaka, Creep of mild steel, Techn. Rep. Eng. Res. Inst. Kyoto Univ. VI, No. 6 (1956).
306. Nishihara, T. and K.Tanaka, Tension and creep test on mild steel, Proc. 5th Japan Nat. Congr. Appl. Mech. 1955, Tokyo 1956.
307. Nishihara, T., S.Taira, K.Tanaka and K.Ohji, Creep of low carbon steel under interrupted stressing, Techn. Rep. Eng. Res. Inst. Kyoto Univ. VI, No. 7 (1956).
308. Nishihara, T., S.Taira, K.Tanaka and K.Ohji, On the creep of low carbon steel under fluctuating stress, Proc. 7th Japan Nat. Congr. Appl. Mech. 1957, Tokyo (1958).
309. Nishihara, T., S.Taira, K.Tanaka and M.Ohnami, Creep of low carbon steel under varying temperatures, Proc. 1st Japan Congr. Testing Mater. Kyoto (1958).
310. Nishihara, T., S.Taira, K.Tanaka and M.Ohnami, Experimental study on the relationship between creep rate determining factors of low carbon steel, Techn. Rep. Eng. Res. Inst. Kyoto Univ. 8 (1958) 2 (Rep. No. 42).
311. Norton, F.H., Creep in tubular pressure vessels, Transactions ASME 61 (1939).
312. Norton, F.H. and C.R.Soderberg, Report on tubular creep tests of 0.5% Mo, C steel, Trans. ASME 64, No. 8 (1942).
313. Odquist, F.K.G., Verfestigung von flusseisenähnlichen Körpern, ZAMM 13 (1933).
314. Odquist, F.K.G., Influence of primary creep on stresses in structural parts, Transactions of the Roy. Inst. of Technology, No. 66, Stockholm (1953).
315. Odquist, F.K.G., Influence of primary creep on column buckling, J. Appl. Mech. 21, No. 3 (1954).
316. Odquist, F.K.G., On theories of creep rupture, Institutionen för Hållfasthetslära, Kungl. Tekniska Högskolan, Public. No. 136, Stockholm (1961).

317. Odquist, F.K.G., Applicability of the elastic analogue to creep problems of plates, membranes and beams, from "Creep in structures" (Springer, 1962).
318. Onat, E.T. and T.T.Wang, The effect of incremental loading on creep behaviour of metals, from "Creep in structures" (Springer, 1962).
319. Orowan, E., Creep in metallic and non-metallic materials, Proc. 1st U.S. Nat. Congr. Appl. Mech. (1952).
320. Orr, R.L., O.D.Sherby and J.D.Dorn, Correlation of rupture data for metals at elevated temperatures, Trans. Amer. Soc. for Metals 46 (1954).
321. Patel, S.A. and J.Kempner, Correlation of creep-buckling tests with theory, PIBAL Report No. 285 (1955).
322. Patel, S.A., K.A.V.Pandalai and B.Venkatraman, Creep stress analysis of thin walled structures, J. Royal Aeronaut. Soc. 64 (November 1960).
323. Patel, S.A. and B.Venkatraman, On the creep stress analysis of some structures, from "Creep in structures" (Springer, 1962).
324. Patel, S.A., F.A.Cozzarelli and B.Venkatraman, Creep of compressible circular plates, Intern. J. of Mech. Sci. 5, No. 1 (1963).
325. Perzyna, P., The constitutive equations for rate sensitive plastic materials, Quart. Appl. Math. XX, No. 4 (1963).
326. Phillips, Aris, The shear center in creep of beams of thin-walled open cross section, from "Creep in structures" (Springer, 1962).
327. Pian, T.H.H., Creep buckling of curved beam under lateral loading, Proc. 3rd U.S. Nat. Congr. Appl. Mech., N.Y. (1958).
328. Popov, E.P., Correlation of tension tests with relaxation tests, J. Appl. Mech. 14, No. 2 (1947).
329. Popov, E.P., Stresses in turbine disks at high temperatures, J. Franklin Inst. 243, No. 5 (1947).
330. Poritsky, H., Effect of creep on stresses in cylindrical shells, from "Creep in structures" (Springer, 1962).
331. Poulignier, J., Fluage périodique rapide. Cas du nicral D-alliage réfractaire Fe–Ni–Cr, La recherche aéronautique No. 83 (1961).
332. Poulignier, J., Fluage rapide de l'alliage réfractaire Fe–Ni–Cr, nicral D, La recherche aéronautique No. 81 (1961).
333. Prager, W., Strain hardening under combined stress, J. Appl. Phys. 16, No. 12 (1945).
334. Prager, W., Recent developments in the mathematical theory of plasticity, J. Appl. Phys. 20 (1949) 235.
335. Prager, W., A new method of analyzing stress and strains in work-hardening plastic solids, J. Appl. Mech. 23 (1956).
336. Rabotnov, Yu.N. and S.A.Shesterikov, Creep stability of columns and plates, J. Mech. Phys. Solids 6 (1957).
337. Rabotnov, Yu.N., Pelzanie metali i obliczanie pelzania, Rozprawy inzynierskie, CLVIII, VIII, No. 3 (1960).
338. Rabotnov, Yu.N., The theory of creep and its applications, from "Plasticity" (Pergamon Press, 1960).
339. Rabotnov, Yu.N., On the equations of state for creep, from "Progress in Applied Mechanics" (Galley, N.Y., 1963).
340. Reissner, E., On a variational theorem in elasticity, J. Math. and Phys. 29, No. 2 (1950).

341. Reissner, E., On a variational theorem for finite elastic deformations, J. Math. and Phys. 32, No. 2–3 (1953).
342. Robinson, E.L., 100 000 hour creep test, Mech. Engineering (1943).
343. Robinson, E.L., Effect of temperature variation on the long time rupture strength of steels, Trans. ASME 74, No. 5 (1952).
344. Sanders, J.L., Plastic stress-strain relations based on linear loading functions, Proc. 2nd U.S. Nat. Congr. of Appl. Mech., N.Y. (1954).
345. Sanders, J.L., H.G.McComb and F.R.Schlechte, A variational theorem for creep with applications to plates and columns, NACA, Rep. No. 1342 (1957).
346. Sawczuk, A., Linear theory of plasticity of anisotropic bodies and its applications to problems of limit analysis, Archivum mechaniki stosowanej XI, No. 5 (1959).
347. Schwartzberg, F.R., Effects of high strain rates and rapid heating on the tensile properties of titanium alloys, Light Metal Age 17, No. 1–2 (1959).
348. Serpico, J.C., A study of creep collapse of a long cylindrical shell under various distributed stress systems, JAS 29, No. 11 (1962).
349. Shanley, F.K., Inelastic column theory, JAS 14, No. 5 (1947).
350. Shepard, D.A., C.D.Starr, C.D.Wiseman and J.E.Dorn, The creep properties of metals under intermittent stressing and heating conditions, Wright Air Dev. Center, Techn. Rep. No. 53-336 (1954).
351. Shield, R.T. and H.Ziegler, On Prager's hardening rule, Z. ang. Math. und Phys. 9a (1958).
352. Siegfried, W., Failure from creep as influenced by the state of stress, J. Appl. Mech. 10, No. 4 (1943).
353. Simmons, W.F. and H.S.Cross, Constant and cyclic-load creep tests of several materials, ASTM Spec. Techn. Publ. No. 165, Baltimore (1954).
354. Sobotka, Z., Some problems of non-linear rheology, Int. Symp. on second-order effects in elasticity, plasticity and fluid dynamics, Haifa, 21–29 April 1962, (preprint).
355. Soderberg, C.R., The interpretation of creep tests for machine design, Trans. ASME 58, No. 8 (1936).
356. Soderquist, B., An experimental study of creep under biaxial tension, Institut för Hållfesthetslära, Kungl. Tekniska Högskolan, Publ. No. 142, Stockholm (1963).
357. Sully, A.H., G.N.Cale and G.Willoughby, Creep of metals subjected to compression stress, Nature No. 162 (September 1948).
358. Sully, A.N., Creep testing in compression for simple creep assessment, Product Engineering 24, No. 4 (1953).
359. Sundstrom, B., Creep buckling of cylindrical shells, Trans. of the Roy. Inst. of Technology, Stockholm, No. 115 (1957).
360. Taira, S., K.Tanaka and K.Ohji, Creep deformation under varying stresses, Proc. 8th Japan Nat. Congr. Appl. Mech., Tokyo (1959).
361. Taira, S., K.Tanaka and R.Koterazawa, Dynamic creep characteristics of low carbon steel at elevated temperature, Proc. 2nd Japan Congr. Testing Mater. Kyoto, Japan (1959).
362. Taira, S. and M.Ohnami, Creep under rapid cycling temperatures, Proc. 3rd Japan Congr. Testing Mater. Kyoto (1960).
363. Taira, S. and R.Koterazawa, Dynamic creep and fatigue of an 18–8 Mo–Cr steel at elevated temperature, Bull. JSME 5, No. 17 (1962).

364. Taira, S., K.Tanaka and K.Ohji, A mechanism of deformation of metals at high temperatures with special reference to the creep after a sudden change in stress, Bull. JSME 4, No. 13 (1961).
365. Taira, S., Lifetime of structures subjected to varying load and temperature, from "Creep in structures" (Springer, 1962).
366. Taira, S. and R.Koterazawa, Some experiments and analysis of dynamic compression creep, Proc. 5th Japan Congr. Testing Mater. Kyoto (1962).
367. Taira, S., M.Ohnami and M.Sakato, Influence of temperature history on creep, Bull. JSME 5, No. 17 (1962).
368. Taira, S., M.Ohnami and S.Sugioka, Creep under stepwise temperature cycling, Bull. JSME 5, No. 17 (1962).
369. Taira, S. and F.Suzuki, Relaxation and stepdown tests of metals at high temperature, Proc. 4th Japan Congr. Testing Mater. Kyoto (1961).
370. Taylor, G.I., Analysis of plastic strain in a cubic crystal, Timoshenko Anniversary Volume (Macmillan Co., 1938).
371. Thein Wan, Creep collapse of cylindrical shells, J. Franklin Inst. 272, No. 1 (1961).
372. Turner, F.H. and K.E.Blomquist, A study of the applicability of Rabotnov's creep parameter for aluminium alloy, JAS 23, No. 12 (1956).
373. McVetty, P.G., Creep of metals at elevated temperatures. The hyperbolic sine relation between stress and creep rate, Trans. ASME 65 (1943).
374. De Veubeke Fraeijs, B., Creep buckling, from "High temperature effects in aircraft structures" (Pergamon Press, 1958).
375. Vitovec, F.H. and B.J.Lazan, Fatigue, creep and rupture properties of heat resisting materials, Wright Air Dev. Center, Techn. Rep. No. 36-181.
376. Volterra, V., Fonctions de lignes (Gauthier-Villard, Paris, 1913).
377. Wahl, A.M., Analysis of creep in rotating disks based on the Tresca criterion and associated flow rule, J. Appl. Mech. 23, No. 2 (1956).
378. Wahl, A.M., Stress distribution in rotating disks subjected to creep including effects of variable thickness and temperature, J. Appl. Mech. 24, No. 2 (1957).
379. Wahl, A.M., Further studies of stress distribution in rotating disks and cylinders under elevated temperature creep conditions, J. Appl. Mech. 25, No. 2 (1958).
380. Wang, A.J. and W.Prager, Thermal and creep effects in work hardening elastic-plastic solids, JAS 21, No. 5 (1954).
381. Ward, J.M. and E.T.Onat, Non-linear mechanical behaviour of oriented polypropylene, J. Mech. Phys. Solids 11 (1963).
382. Wimen, G., Ueber den Fundamentalsatz in der Theorie der Funktionen $E_\alpha(x)$, Acta Math. 29 (1905).
383. Winzer, A. and W.Prager, On the use of power laws in stress analysis beyond the elastic range, J. Appl. Mech. 14, No. 4 (1947).
384. Yerkovich, L.A., Investigation of the compressive, bearing and shear creep rupture properties of aircraft structural metals and joints at elevated temperatures, Wright Air Dev. Center, Techn. Rep. 34-279 (1958).
385. Ziegler, H., A modification of Prager's hardening rule, Quart. Appl. Math. 17 (1959).

Subject Index

Shell International E & P Inc.
Bellaire Technology Center
Houston EP Library
P.O. Box 481
Houston, TX 77001-0481